Symbols

n	An integer (1, 2, 3, . . .)	T	Tension
n	Principal quantum number	$\mathbf{v}$	Velocity
n	Turns per unit length	V	Potential difference
n	Neutron	V	Volume
N	Number of turns	w	Energy density
N	Normal (perpendicular) force	w	Weight
p	Object distance	W	Work
p	Pressure	X_C	Capacitive reactance
p	Proton	X_L	Inductive reactance
P	Power	Y	Young's Modulus
PE	Potential Energy	Z	Atomic number
q, Q	Electric charge	Z	Impedance
q	Image distance	α (alpha)	Angular acceleration
Q	Heat	γ (gamma)	Electromagnetic photon
r	Angle of reflection, refraction	Δ (delta)	"Change in"
r	Internal resistance	θ (theta)	An angle
r, R	Radius	λ (lambda)	Wavelength
R	Range	μ (mu)	Charge per unit length
R	Rate of flow	μ (mu)	Coefficient of friction
R	Electrical resistance	μ (mu)	Magnetic permeability
°R	Degrees rankine	ρ (rho)	Resistivity
s	Displacement from equilibrium position	σ (sigma)	Stefan-Boltzmann constant
s	Distance; displacement	Σ (sigma)	"Sum of"
S	Shear modulus	τ (tau)	Torque
t	Time	ϕ (phi)	Phase angle
T	Period	Φ (phi)	Magnetic flux
T	Temperature	ω (omega)	Angular velocity

Powers of Ten

10^{-10}	$= 0.000,000,000,1$	10^0	$= 1$
10^{-9}	$= 0.000,000,001$	10^1	$= 10$
10^{-8}	$= 0.000,000,01$	10^2	$= 100$
10^{-7}	$= 0.000,000,1$	10^3	$= 1000$
10^{-6}	$= 0.000,001$	10^4	$= 10,000$
10^{-5}	$= 0.000,01$	10^5	$= 100,000$
10^{-4}	$= 0.000,1$	10^6	$= 1,000,000$
10^{-3}	$= 0.001$	10^7	$= 10,000,000$
10^{-2}	$= 0.01$	10^8	$= 100,000,000$
10^{-1}	$= 0.1$	10^9	$= 1,000,000,000$
10^0	$= 1$	10^{10}	$= 10,000,000,000$

Physics
Second Edition

The Benjamin/Cummings Publishing Company, Inc.

Menlo Park, California • Reading, Massachusetts
London • Amsterdam • Don Mills, Ontario • Sydney

Physics

Second Edition

Arthur Beiser

The Benjamin/Cummings Publishing Company, Inc.
2727 Sand Hill Road
Menlo Park, California 94025

Preface

Scope of the Text

Physics is the science of matter and is concerned with its fundamental structure, properties, and behavior. Like any other scientists, physicists are in search of ideas that unify their vast subject, that permit them to understand the workings of atoms and molecules, stars and galaxies, and everything in between. Physics is thus the master science whose concepts and principles are drawn upon by such other sciences as chemistry, biology, geology, and astronomy in their explorations of narrower aspects of the universe around us. It is these same concepts and principles that constitute the foundations of engineering and have made possible the technological world of today.

This textbook is intended primarily for students who will require some competence in physics in their subsequent work. The organization is straightforward, from the laws of motion to the properties of matter in bulk, vibrations and waves, heat and thermodynamics, electromagnetism, optics, and finally modern physics. Wherever possible the origin of a significant result is described, since the essence of science is that nothing is ever pulled out of a hat but must have a firm basis in experiment or established theory. The mathematical level has been kept to the absolute minimum. Only elementary algebra and the simplest trigonometry are used, and these are reviewed in an appendix together with powers-of-ten notation.

About 600 illustrations are provided to help make the reader's task easier, together with such additional aids as marginal notes, sample problems, and lists of important terms and formulas at the end of each chapter. A comprehensive glossary is included. The more than 2000 exercises are divided into three categories: multiple-choice questions for quick review of comprehension; straightforward exercises for practice in manipulating numbers and formulas; and problems that offer a modest degree of challenge. Answers to all the multiple-choice questions and outline solutions for the odd-numbered exercises and problems are given in the text. Outline solutions for even-numbered exercises and problems are given in the Solutions Manual. Each outline solution shows how the problem can be attacked and gives the answer (usually to one more significant figure than is justified), but most of the intermediate steps are omitted. In this way the reader who is stumped by a certain type of problem can obtain guidance on how to proceed and still have something left to do on his or her own. The outline solutions thus supplement the sample problems in the body of the text.

New in the Second Edition

In revising *Physics*, I have had the benefit of guidance from a number of users of the First Edition, to whom I am grateful. A fair amount of the text was rewritten to improve the clarity of the exposition. The emphasis of the book on the fundamentals of physics remains unaltered, but a number of changes were made so that the revised Second Edition can better serve students in a variety of disciplines. In particular, the applications of physics in various aspects of the life sciences now receive more attention; this is reflected in the exercises and sample problems as well as in the text. Such topics as muscular forces, blood pressure, animal metabolism, and defects of vision should interest all readers—not just those who will later specialize in biology or medicine. New material is provided in other areas as well. Fluid flow, blackbody radiation, and impedance matching are given expanded treatments, to give a few examples. To make room for these various additions without lengthening the book, a number of subjects of peripheral interest were abbreviated, among them mechanical advantage, electrochemistry, and electrical machinery.

Contents Reorganized

In order to make *Physics* more convenient to use in the classroom, its contents were somewhat rearranged. There is now a natural division about halfway through, so that the first term can cover mechanics and heat (chapters 1–16) and the second can cover electricity and magnetism, optics, and modern physics (chapters 17–33). If modern physics is omitted, the first term can cover mechanics (chapters 1–13) and the second can cover heat, electricity and magnetism, and optics (chapters 14–26). SI units are used throughout. Because of their familiarity, British units are briefly discussed where appropriate, but they can be disregarded without loss of continuity.

The "Special Topic," an Added Feature

Most chapters now include a section titled Special Topic that examines a facet of the chapter's subject which I hope its readers find interesting but for which they are not held responsible in the exercises. Some of the Topics deal with applications in the life sciences (for instance, "The Action Potential") or in technology ("Semiconductor Devices"). Others ("More about Collisions") are extensions of the text. For those readers who know a little calculus, seven of the Topics show how helpful this branch of mathematics is to the physicist.

A. B.

Ile des Embiez, France
December 1977

Introduction

Each of us is a member of the human community, and each of us participates in some way in the continuing evolution of this community. Problems of politics and economics involve us all, and few people would wish to be ignorant of them.

Each of us is also part of the physical universe, whose evolution we can no more escape than we can escape the actions of our fellow human beings. We are made of atoms linked into molecules, liquids, and solids, and we live on a planet bound to a star which is a member of one of the galaxies that occupy the reaches of space. Atoms, galaxies, and everything in between have certain regularities of behavior in common, which are the "laws of nature." These same laws of nature underlie the technology that today dominates the lives of people and nations. An acquaintance with science is as vital to understanding our place in modern civilization as it is to understanding our place in the universe.

The scientist explores the natural world directly by experiment and observation and indirectly by abstract reasoning. Experiment and observation yield numbers, the raw material of science. Abstract reasoning yields theories—relationships which not only generalize the results of many measurements but also permit inferences to be drawn about phenomena either as yet undiscovered or impossible to examine directly. We will never visit the sun's interior or the interior of an atom, yet we know a great deal about what goes on in both places. The evidence is indirect but persuasive.

What has made science such a powerful tool for investigating nature is the constant testing of its findings, both experimental and theoretical. Nothing is accepted on anybody's personal authority, on the basis of "common sense," or because it is part of a religious or political doctrine. Because every generalization in science is subject to modification in the light of fresh evidence, science is a living body of information and not a collection of dogmas. Challengers of currently accepted ideas are not scorned—if their views are well-founded, anyway—nor is disgrace attached to a scientist whose contributions do not survive indefinitely. To rock the boat is part of the game; to overturn it is one way to win.

Broadly speaking, the science of physics is concerned with the *elementary particles* of which all matter is composed, the ways in which these particles interact with one another, and the behavior of the composite bodies (atoms, molecules, liquids, and solids) formed by elementary particles.

Consider a sample of matter of any sort—a drop of water, a grain of sand, a blade of grass—and imagine that it is somehow chopped up into finer and finer pieces. How far can this go on? Eventually each sample is reduced to its constituent atoms, of which there are 92 different varieties. If the atoms themselves are broken up, we are left with just three kinds of elementary particle which cannot be subdivided further: protons, neutrons, and electrons. (There are other kinds of elementary particles, but these are the three primarily involved in ordinary matter.)

Elementary particles interact with one another in only four different ways, which are responsible for the structure and behavior of the entire universe. These four *fundamental interactions* are the strong nuclear, the weak nuclear, the electromagnetic, and the gravitational. Protons and neutrons join together by virtue of the strong nuclear interaction to form atomic nuclei, whose exact compositions are partly determined by the weak nuclear interaction as well. Electromagnetic forces hold electrons to nuclei to form atoms. Electromagnetic interactions between nearby atoms yield stable clumps of them called molecules and larger aggregates that constitute liquids and solids. On a larger scale, gravitation pulls matter together into the planets, stars, and galaxies which populate space. Recent evidence suggests that the weak nuclear interaction is actually an aspect of the electromagnetic interaction; conceivably all are related in some as-yet unknown way.

Three particles and four, perhaps only three, interactions—these hardly seem enough to account for the span of things and events in everyday life, let alone for the evolution of the universe. Yet nothing else is needed—apparently—to explain phenomena ranging from the falling of a dropped stone to the development of plants and animals from inanimate matter.

To perceive such exquisite unity in the diversity around us took four hundred years of effort. The time scale is not arbitrary, for modern science began with Galileo (1564–1642). Before Galileo explanations of the natural world were sought in terms of self-evident principles, ideas supposed to be so clearly correct that experiments were unnecessary. Such a supposedly self-evident principle was Aristotle's assertion that a heavy object falls faster than a light one; another was the belief that all motion must be sustained by an applied force. The trouble with these principles, and many others like them, is that they are wrong—in the real world, moving bodies behave quite differently. Galileo's greatest contribution, overshadowing his specific discoveries in mechanics and astronomy, was the notion that statements about nature must be tested by observation.

It is not easy to get a firm intellectual grip on nature in the rough. Nothing is as simple as it seems. We think of the earth as being round, but in fact it is shaped more like a grapefruit with a bumpy skin than like a perfect sphere. We say that the earth moves in an elliptical orbit about the sun, but in fact the orbit has wiggles no ellipse ever had. In order to extract the essence of a phenomenon, the physicist must idealize reality with the help of *models*. By choosing a sphere as a model for the actual earth and an ellipse as a model for its actual orbit, the physicist simplifies his task of analysis by isolating the most important features of his subject. If he had to deal from the start with a squashed, corrugated earth moving in an irregular path, he would find it much more difficult to make any progress.

Another kind of model arises when the physicist tries to understand the microscopic world of atoms and molecules. A useful model of a gas, for example, pictures it as a collection of myriad tiny particles like billiard balls that fly about in all directions. This model is quite successful in accounting for many aspects of the behavior of gases. But it is still a model and not the whole story, since if we could somehow look directly at the structure of a gas, we would find that the particles of which it is composed are not at all like miniature billiard balls and in certain respects do not even act like particles in the usual sense.

There is no clear line between a model and a theory. Usually the term *theory* is reserved for a large-scale generalization which not only accounts for existing data but from which predictions about the outcome of new experiments can be made. A theory may or may not be based upon a particular model. Thus the kinetic theory of gases is a logical structure based upon the model of a gas as a collection of particles in rapid, random motion. On the other hand, there is no model directly associated with Newton's theory of gravitation, which deals with concepts that do not need to be pictured in order to be related to one another. Nevertheless, the key to formulating this theory was Newton's choice of spherical models for the sun, moon, and planets and of ellipses as models for the planetary orbits.

More than one model may be convenient or necessary in some cases. Light is a notable example. In many practical applications it is quite sufficient to imagine that light consists of "rays," thin pencils of something-or-other which are perfectly straight unless reflected or refracted, when they are bent through definite angles. A more sophisticated picture of light capable of explaining many more aspects of its behavior is the wave model. Without even specifying what kind of waves are involved, the wave model accounts for the interference, diffraction, and polarization of light, and interprets reflection and refraction in a straightforward way.

The next level of model-building brings even more rewards of understanding, but there is an unexpected (and at first glance disturbing) problem: *two* models are needed, of equal validity but different areas of application. One of them is a further development of the simple wave model, with the waves now identified as electromagnetic in character. The other model employs no waves at all but regards light as a stream of tiny particles. Certain phenomena can only be interpreted on the basis of the electromagnetic wave model, others only on the basis of the particle model.

Then what is light? The answer is that light is what it is, and no model that we can visualize in terms of our everyday experience can encompass its ultimate nature. But a purely abstract theory of light does exist whose conclusions are in superb agreement with experiment, and which makes comprehensible the need for two models. These models are not wrong, but each is limited in scope, which is true of all models in physics. Models are useful devices, but they are seldom the last word.

Nearly always in this book we shall be considering models rather than actual things. It is a good habit to try to pick out those aspects of the phenomenon under study which are incorporated in its model and those which are left out, and to ask ourselves what is gained and what is lost in each case.

Contents

Physics
Second Edition

1

Scalars and Vectors

The raw material of the physicist consists of numbers obtained by observation, and each conclusion is a quantitative statement that can be tested by measuring something and comparing the result with the predicted value. In turn, these conclusions are used by other scientists in their own investigations of nature and by engineers to design such things as bridges, aircraft, and computers. Before we actually take up the study of physics, then, we must review how physical quantities are expressed and treated mathematically. A quantitiy is called scalar when it possesses a magnitude only (time is an example), and vector when a certain direction is associated with it as well (force is an example). Many of the most important quantities in physics are vector in character.

1-1 Units

The process of measurement is essentially one of comparison. A certain standard quantity of some kind, called a *unit*, is first established, and other quantities of the same kind are compared with it. When we say a ladder is 2.6 meters long, we mean that its length is 2.6 times a certain distance called the meter whose magnitude is fixed by international agreement. The result of every measurement must therefore have two parts, a number to answer the question "How many?" and a unit to answer the question "Of what?"

Nearly all quantities in the physical world can be expressed in terms of only four fundamental measurements: length, time, mass, and electric current. Thus every unit of area (the square foot, for instance) is the product of a length unit and a length unit; every unit of speed (the mile per hour, for instance) is a length unit divided by a time unit; every unit of force (the newton, for instance) is the product of a mass unit and a length unit divided by the square of a time unit; and every unit of electric charge (the coulomb, for instance) is the product of a unit of electric current and a time unit.

Four fundamental measurements

A *system of units* is a set of specified units of length, time, mass, and electric current from which all other units are to be derived. All units are arbitrary. Thousands of different ones have been adopted at various times and places in the course of history, a number of which survive. The most widely used is the *metric system,* which was introduced in France nearly two centuries ago. The virtues of the metric system soon led to its universal adoption by scientists, more slowly to its spread throughout the world for engineering and everyday life. The Système International (SI) is the current version of the metric system and is the one emphasized in this book. The fundamental SI units are the meter for length, the kilogram for mass, the second for time, and the ampere for electric current.

Metric system

Although the *British system* is on the way out, it is still often used in commerce and industry in English-speaking countries, and British units are included here because of their familiarity to the reader. In the British system the units of length, mass, and time are respectively the foot, the slug, and the second. Electrical units are the same in both systems.

British system

Because a basic unit in a system may not always be convenient in size for a given measurement, other units have come into use within each system. Thus long distances are usually given in miles rather than in feet, or in kilometers rather than in meters. The great advantage of the SI system is that it is wholly decimalized, which facilitates calculation, whereas the British system is quite irregular in this respect. (1 kilometer = 1000 meters, for example, but 1 mile = 5280 feet) Table 1-1 contains the chief units of length and time used in each system, together with their equivalents in the other system. The conventional prefixes used with metric units of all kinds are listed in Table 1-2; power-of-ten notation is reviewed in Appendix B.

Table 1-1. Metric and British units of length and time.

LENGTH

Metric Units

1 meter (m) = The standard unit = 100 cm = 39.4 in. = 3.28 ft

1 centimeter (cm) = 0.01 m = 10 mm = 0.394 in.

1 millimeter (mm) = 0.001 m = 0.0394 in.

1 kilometer (km) = 1000 m = 0.621 mi

British Units

1 yard (yd) = The standard unit = 3 ft = 0.914 m

1 foot (ft) = 12 in. = 0.305 m

1 inch (in.) = 0.083 ft = 2.54 cm

1 statute mile (mi) = 5280 ft = 1.61 km

TIME

1 second (s) = The standard unit

1 minute (min) = 60 s

1 hour (hr) = 60 min = 3600 s

1 day = 24 hr = 86,400 s

Table 1-2. Subdivisions and multiples of metric units are widely used, and each is designated by a prefix according to the corresponding power of ten. The most common prefixes are listed here.

Prefix	Power of ten	Abbreviation	Example
pico–	10^{-12}	p	1 pf = 1 picofarad = 10^{-12} farad
nano–	10^{-9}	n	1 ns = 1 nanosecond = 10^{-9} second
micro–	10^{-6}	μ	1 μA = 1 microampere = 10^{-6} ampere
milli–	10^{-3}	m	1 mg = 1 milligram = 10^{-3} gram
centi–	10^{-2}	c	1 cm = 1 centimeter = 10^{-2} meter
kilo–	10^{3}	k	1 kw = 1 kilowatt = 10^{3} watts
mega–	10^{6}	M	1 MJ = 1 megajoule = 10^{6} joules
giga–	10^{9}	G	1 GeV = 1 gigaelectron-volt = 10^{9} electron volts

It is essential that standard units be absolutely constant in magnitude. Whenever possible in recent years they have been redefined in terms of quantities in nature that are regarded as invariant under all circumstances and not subject to change in time. The standard meter, for instance, is now defined as exactly 1,650,763.73 wavelengths of one of the spectral lines of the krypton isotope ^{86}Kr. The yard is in turn defined in terms of the standard meter—one meter represents exactly 39.37 inches. Measuring rods can be directly compared with the new "standard meter" in any well-equipped laboratory, and no longer must the entire world rely upon the measurement between scratches on a certain platinum-iridium bar kept at Sèvres, France.

Often a quantity expressed in terms of a certain unit must instead be expressed in terms of another unit of the same kind. For instance, we might find from a European map that Amsterdam is 648 km from Berlin and want to know what this distance is in miles. To carry out such a conversion we must keep in mind two rules:

Converting units

1. Units are treated in an equation in exactly the same way as any algebraic quantity, and may be multiplied and divided by one another;

2. Multiplying or dividing a quantity by 1 does not affect its value.

To convert 648 km to its equivalent in miles we note from Table 1-1:

$$1 \text{ km} = 0.621 \text{ mi}$$

Therefore

$$0.621 \, \frac{\text{mi}}{\text{km}} = 1$$

and multiplying or dividing any quantity by 0.621 mi/km does not affect its value but only changes the units in which it is given. Hence we have

$$d = 648 \, \cancel{\text{km}} \times 0.621 \, \frac{\text{mi}}{\cancel{\text{km}}} = 402 \text{ mi}$$

since km/km = 1 and so drops out.

Problem. Express the velocity of 60 mi/hr in ft/s.
Solution. Since 1 mi = 5280 ft,

$$v = 60 \, \frac{\cancel{\text{mi}}}{\text{hr}} \times 5280 \, \frac{\text{ft}}{\cancel{\text{mi}}} = 316,800 \, \frac{\text{ft}}{\text{hr}}.$$

There are 3600 seconds in an hour, and so

$$v = \frac{316,800 \; \frac{ft}{hr}}{3600 \; \frac{s}{hr}} = 88 \; \frac{ft}{s}.$$

We could have carried out the conversion in a single step:

$$v = 60 \; \frac{mi}{hr} \times 5280 \; \frac{ft}{mi} \times \frac{1}{3600 \; \frac{s}{hr}} = 88 \; \frac{ft}{s}.$$

1-2 Scalar and Vector Quantities

A scalar quantity has magnitude only

A vector quantity has magnitude and direction

Fig. 1-1.   An airplane that leaves New York and flies for a distance of 400 mi may land anywhere within the circle shown. If the airplane undergoes a displacement of 400 mi to the west, it lands in Cleveland. Distance is a scalar quantity; displacement is a vector quantity.

Some physical quantities require only a number and a unit to be completely specified. It is quite sufficient to say that the mass of a man is 85 kilograms, that the area of a farm is 160 acres, that the frequency of a sound wave is 660 cycles/second, or that a light bulb consumes electrical energy at the rate of 100 watts. These are examples of *scalar quantities*.

Quantitative statements cannot always be confined to numbers and units alone. A man's mass is the same whether he stands on his feet or on his head, but the motion of a car traveling north at 60 mi/hr is hardly the same as the motion of a car traveling in a circle at 60 mi/hr. A quantity whose direction is significant is called a *vector quantity*. Vector quantities occur frequently in physics, and their arithmetic is different from the arithmetic of scalar quantities.

The simplest example of a vector quantity is *displacement*, which is a change in position. If we are told that a certain airplane leaves New York City and flies for 400 mi (a scalar statement), we know nothing about its actual path or final destination; it could land anywhere within a circle of radius 400 mi whose center is New York. If we are told instead that the airplane flies for 400 mi to the west (a vector statement), we know where it is headed and can identify its destination as Cleveland (Fig. 1-1).

Two other vector quantities of special significance are *velocity* and *force*. The velocity of a body incorporates both its speed (a scalar quantity) and its direction: the speed of a car might be 60 mi/hr, whereas its velocity is 60 mi/hr to the north. This may seem a minor distinction, but it is not. Given its speed and direction, we can readily calculate where a moving body will be after a certain time interval; but if we do not know the direction, we cannot determine the path of the body at all.

A force is often spoken of as a "push" or a "pull." Although there is more to the concept of force than this description indicates, it is adequate for the time being, and we shall postpone a more elaborate discussion until later. Forces are important because they are responsible for all changes in motion; a force is needed to start a stationary object in motion and to deviate or stop a moving one. The direction as well as the magnitude of a force must be known to determine its effects, and vector methods must be used to treat forces as well as to treat displacements and velocities.

1-3 Vectors

The most straightforward way to represent a vector quantity is to draw a straight line with an arrowhead at one end to indicate the direction of the quantity. The length of the line is proportional to the magnitude of the quantity. A line of this kind is called a *vector*.

Figure 1-2 shows how a 400-mi westward displacement might be represented by a vector. The compass rose establishes the orientation of the displacement, and the distance scale establishes the relationship between length in the diagram and the corresponding actual length.

Vector quantities other than displacements can also be represented by vectors. The length of the vector in each case is proportional to the magnitude of the quantity it represents, and its direction is the direction of the quantity. Thus a velocity whose magnitude is 20 m/s is represented on a scale of $\frac{1}{4}$ in. = 5 m/s by an arrow 1 in. long, and a force whose magnitude is 350 lb is represented on a scale of 1 cm = 600 lb by an arrow 0.6 cm. long (Fig. 1-3). The directions of the arrows indicate the direction of the quantities.

The symbols of vector quantities are customarily printed in boldface type (**F** for force), while italics are used both for scalar quantities (*m* for

A vector is a directed line

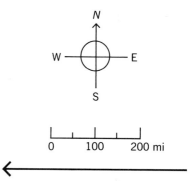
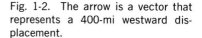

Fig. 1-2. The arrow is a vector that represents a 400-mi westward displacement.

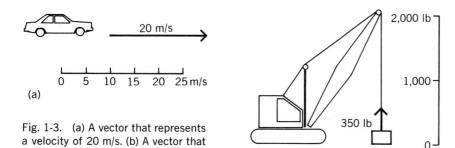

(a)

Fig. 1-3. (a) A vector that represents a velocity of 20 m/s. (b) A vector that represents a force of 350 lb.

(b)

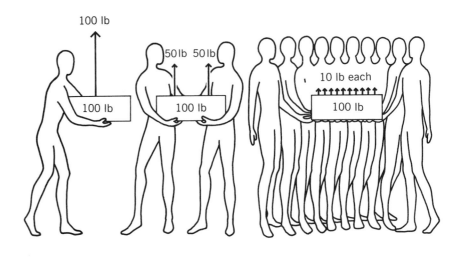

Fig. 1-4. The total force on a body rather than the individual forces themselves affects the motion of the body.

Symbols for vector and scalar quantities

mass) and for the magnitudes of vectors. Thus we might denote a 400-mi westward displacement by the symbol **A** and its magnitude of 400 mi by the symbol A. In handwriting, vector quantities are indicated by placing arrows over their symbols, for instance $\vec{A}$.

1-4 Vector Addition

Ordinary arithmetic is used to add two or more scalar quantities of the same kind together. Ten kg of potatoes plus 4 kg of potatoes equals 14 kg of potatoes. The same kind of arithmetic is used for vector quantities of the same kind when the directions are the same. Thus a total upward force of 100 lb is required to lift a 100-lb weight, regardless of whether the force is applied by one, two, or ten men (Fig. 1-4). It is the total force on a body, not the individual forces, that affects the motion of the body.

What do we do when the directions are different? A man who walks 10 mi north and then 4 mi west does *not* undergo a net displacement of 14 mi from his starting point, even though he has walked a total of 14 mi. A vector diagram provides a convenient method of determining his actual displacement. The procedure is to make a scale drawing of the successive displacements **A** and **B**, as in Fig. 1-5, and to join the starting point and terminal point with a single vector **R**. The required net displacement is **R**, whose length corresponds to 10.8 mi and whose direction corresponds to 22° west of north.

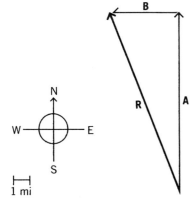

Fig. 1-5. The successive displacements of 10 mi north, **A**, and 4 mi west, **B**, result in a net displacement **R** of 10.8 mi in a direction 22° west of north.

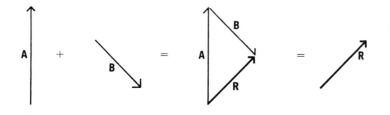

Fig. 1-6. The addition of two vectors to obtain the resultant **R**.

The general rule for adding vectors is illustrated in Fig. 1-6. To add **B** to **A**, shift **B** parallel to itself until its tail is at the head of **A**. In its new position **B** must still have its original length and direction. The vector sum **A** + **B** is a vector **R** (often called the *resultant*) drawn from the tail of **A** to the head of **B**. To find the magnitude R of the resultant **R**, measure the length of **R** on the diagram and compare this length with the scale. The direction of **R** with respect to **A** or **B** may be determined with a protractor.

The same procedure may be used with any number of vectors: place the tail of each vector at the head of the previous one, keeping their lengths and original directions unchanged, and draw a vector **R** from the tail of the first vector to the head of the last. **R** is the sum required. Figure 1-7 shows how

The sum of two or more vectors is called their resultant

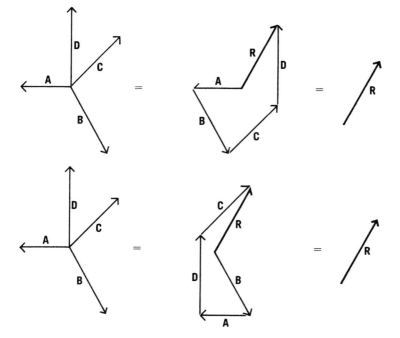

Fig. 1-7. The addition of four vectors to obtain the resultant **R**. The order in which the vectors are added does not affect the magnitude or direction of **R**.

four vectors, whose initial magnitudes and directions differ, are added together. We note that the order in which the vectors are added does not affect the result; that is,

$$\mathbf{A} + \mathbf{B} = \mathbf{B} + \mathbf{A}. \hspace{4cm} \textit{Vector addition} \hspace{0.5cm} (1\text{-}1)$$

Problem. A boat heads east at 10.0 mi/hr in a river that flows south at 3.0 mi/hr. What is the boat's velocity relative to the earth?

Solution. We add the vector that represents the river current to the vector that represents the boat's velocity relative to the river, as in Fig. 1-8. This procedure leads to a single vector which indicates that the boat's velocity relative to the earth is 10.4 mi/hr in a direction about 17° south of east.

For rough calculations it is sufficient to determine the magnitude and direction of a resultant directly from a vector diagram with ruler and protractor. More accurate results may be obtained with the help of trigonometry. In the case of this problem, the fact that $\mathbf{v}_{\text{boat}}$ is perpendicular to $\mathbf{v}_{\text{river}}$ means that we can use the Pythagorean theorem (see Appendix A for a review of trigonometry) to find the magnitude v of the resultant velocity $\mathbf{v}$:

$$v_R{}^2 = v_{\text{boat}}^2 + v_{\text{river}}^2$$
$$v_R = \sqrt{v_{\text{boat}}^2 + v_{\text{river}}^2}$$
$$= \sqrt{(10.0 \text{ mi/hr})^2 + (3.0 \text{ mi/hr})^2} = \sqrt{109 \text{ (mi/hr)}^2} = 10.4 \text{ mi/hr.}$$

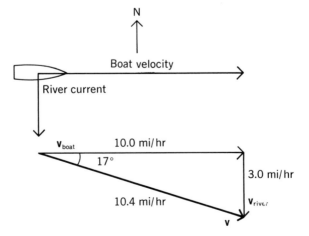

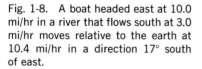

Fig. 1-8. A boat headed east at 10.0 mi/hr in a river that flows south at 3.0 mi/hr moves relative to the earth at 10.4 mi/hr in a direction 17° south of east.

The angle θ between **v** and $\mathbf{v}_{\text{boat}}$ is specified by

$$\tan \theta = \frac{v_{\text{river}}}{v_{\text{boat}}} = \frac{3.0 \text{ mi/hr}}{10.0 \text{ mi/hr}} = 0.30.$$

When we examine a table of trigonometric functions, we find that a portion of it reads as shown in Table 1-3. The angle whose tangent is closest to 0.30 is 17°, since $\tan 17° = 0.306$. Hence θ is slightly less than 17°. To the nearest degree,

$$\theta = 17°.$$

Elementary trigonometry is reviewed in Appendix A-5.

Table 1-3.

Angle	Sine	Cosine	Tangent
14°	0.242	0.970	0.249
15°	.259	.966	.268
16°	.276	.961	.287
17°	.292	.956	.306
18°	.309	.951	.325
19°	.326	.946	.344
20°	.342	.940	.364

1-5 Vector Subtraction

It is sometimes necessary to subtract one vector from another. To subtract **A** from **B**, for example, we first form the negative of **A**, denoted −**A**, a vector which has the same length as **A** but which points in the opposite direction (Fig. 1-9). We then add −**A** to **B** in the usual manner. This procedure may be summarized as

$$\mathbf{B} - \mathbf{A} = \mathbf{B} + (-\mathbf{A}). \qquad \textit{Vector subtraction} \quad (1\text{-}2)$$

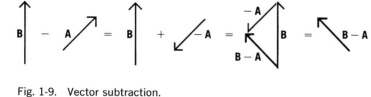

Fig. 1-9.   Vector subtraction.

1-6 Resolving a Vector

Just as we can add together two or more vectors to give a single resultant vector, so we can break up a vector into two or more others. The process of replacing one vector by two or more others is called *resolving* the vector, and the new vectors are called the *components* of the original one

Components of a vector

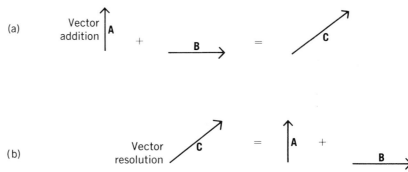

(a) Vector addition

(b) Vector resolution

Fig. 1-10. (a) In vector addition, two or more vectors are combined to form a single vector. (b) In vector resolution, a single vector is replaced by two or more others whose sum is the same as the original vector. Here **A** and **B** are the components of the vector **C**.

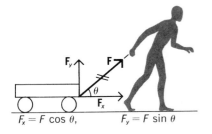

$$F_x = F \cos \theta, \qquad F_y = F \sin \theta$$

Fig. 1-11. The resolution of a force vector into horizontal and vertical components. The original vector is indicated by two short lines across it.

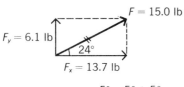

$F = 15.0$ lb

$F_y = 6.1$ lb

$24°$

$F_x = 13.7$ lb

$$F^2 = F_x^2 + F_y^2$$

Fig. 1-12.

(Fig. 1-10). Often the best way to analyze a physical problem is to resolve a vector into components. Almost invariably the components of a vector are chosen to be perpendicular to one another.

Figure 1-11 shows a boy pulling a wagon with a rope at the angle θ above the ground. Only part of the force he exerts affects the horizontal motion, since the wagon moves horizontally while the force **F** is not a horizontal one. We can resolve **F** into two components, $\mathbf{F}_x$ and $\mathbf{F}_y$, where

$\mathbf{F}_x =$ horizontal component of **F**,

$\mathbf{F}_y =$ vertical component of **F**.

(It is a good habit to distinguish between the original vector and its components when all are on the same diagram by drawing two short lines across the original vector, as in Fig. 1-11.)

The magnitudes of $\mathbf{F}_x$ and $\mathbf{F}_y$ are

$$F_x = F \cos \theta, \tag{1-3}$$

$$F_y = F \sin \theta. \tag{1-4}$$

The horizontal component $\mathbf{F}_x$ is responsible for the wagon's motion, while the vertical component $\mathbf{F}_y$ merely pulls upward on it. $\mathbf{F}_x$ is the projection of **F** in the horizontal direction, and $\mathbf{F}_y$ is the projection of **F** in the vertical direction.

Problem. If the force the boy exerts on the wagon in Fig. 1-11 is 15.0 lb and $\theta = 24°$, find F_x and F_y. Does their sum equal 15.0 lb? Should it?

Solution. We have (Fig. 1-12)

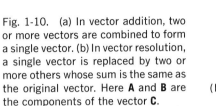

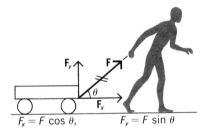

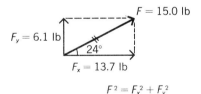

Fig. 1-12.

$F_x = F \cos \theta$

  $= 15.0 \text{ lb} \times \cos 24°$

  $= (15.0 \times 0.914) \text{ lb} = 13.7 \text{ lb},$

$F_y = F \sin \theta$

  $= 15.0 \text{ lb} \times \sin 24°$

  $= (15.0 \times 0.407) \text{ lb} = 6.1 \text{ lb}.$

The algebraic sum of the magnitudes F_x and F_y is $13.7 \text{ lb} + 6.1 \text{ lb} = 19.8 \text{ lb}$, although the boy has exerted a force of only 15 lb. Where is the mistake? The answer is that there is no mistake: $\mathbf{F}_x$ and $\mathbf{F}_y$ are vectors whose directions are different, so they can *only* be added vectorially. The algebraic sum of the magnitudes $F_x + F_y$ has no meaning. If we add $\mathbf{F}_x$ and $\mathbf{F}_y$ vectorially, we find that, from the Pythagorean theorem,

$F^2 = F_x{}^2 + F_y{}^2$

$F = \sqrt{F_x{}^2 + F_y{}^2}$

  $= \sqrt{(13.7 \text{ lb})^2 + (6.1 \text{ lb})^2} = \sqrt{225 \text{ lb}^2} = 15.0 \text{ lb}$

which is equal to the force the boy exerts.

Problem. A 2800-lb car is parked in a driveway that is at a 15° angle with the horizontal. Find the components of the car's weight parallel and perpendicular to the driveway.

Solution. The weight $\mathbf{w}$ of anything is the gravitational force the earth exerts on it, a force that acts vertically downward (Fig. 1-13). Because $\mathbf{w}$ is ver-

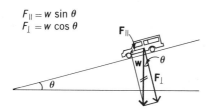

Fig. 1-13. Weight is a force that acts vertically downward.

tical and $F_\perp$ is perpendicular to the road, the angle θ between **w** and $F_\perp$ is equal to the angle $\theta = 15°$ between the road and the horizontal. Hence

$$F_\parallel = w \sin \theta = 2800 \text{ lb} \times \sin 15° = 725 \text{ lb},$$
$$F_\perp = w \cos \theta = 2800 \text{ lb} \times \cos 15° = 2705 \text{ lb}.$$

1-7 Vector Addition by Components

As we have seen, it is easy to add two or more vectors together graphically by drawing them head-to-tail and joining the tail of the first vector to the head of the last to form the resultant **R**. There are two ways to solve a problem of this kind by trigonometry. One is the direct procedure of adding two of the initial vectors together, then adding a third to the sum of the first two, a fourth to the sum of the first three, and so on. This is usually a laborious process that presents many chances for error.

A better method for adding several vectors is to work in terms of their components. The procedure is as follows for vectors that lie in the same plane:

Component method of vector addition

1. Resolve the initial vectors into their components in the x and y directions.

2. Add the components in the x direction to give R_x and add the components in the y direction to give R_y. That is,

$$R_x = x\text{-component of } R$$
$$= A_x + B_x + C_x + \cdots = \text{sum of } x\text{-components},$$

$$R_y = y\text{-component of } R$$
$$= A_y + B_y + C_y + \cdots = \text{sum of } y\text{-components}.$$

(1-5)

3. Find the magnitude and direction of the resultant **R** from the components R_x and R_y. From the Pythagorean theorem,

$$R = \sqrt{R_x^2 + R_y^2}.$$

(1-6)

The direction of **R** can be found from the values of the components by trigonometry; the best way to do this depends upon the problem at hand.

Problem. A sailboat is headed due north at a forward speed of 6.0 knots. [A *knot* (kn) is a unit of speed equal to one nautical mile per hour, which is about 1.15 statute miles per hour, since a nautical mile is 6080 ft in length.]

The pressure of the wind on its sails causes the boat to move sideways to the east at a speed of 0.5 kn. A tidal current is flowing to the southwest at a speed of 3.0 kn. What is the velocity of the sailboat relative to the earth's surface?

Solution.   For convenience we shall call north the $+y$-direction, south the $-y$-direction, east the $+x$-direction, and west the $-x$-direction. With the help of Fig. 1-14 we see that the magnitudes of the components of the three velocity vectors are

$$A_x = 0 \qquad\qquad C_x = -3.0 \cos 45° \text{ kn}$$

$$A_y = 6.0 \text{ kn} \qquad\quad = -2.1 \text{ kn}$$

$$B_x = 0.5 \text{ kn} \qquad C_y = -3.0 \sin 45° \text{ kn}$$

$$B_y = 0 \qquad\qquad\quad = -2.1 \text{ kn}$$

We then add together the components in each direction.

$$R_x = A_x + B_x + C_x \qquad\qquad R_y = A_y + B_y + C_y$$

$$= 0 + 0.5 \text{ kn} - 2.1 \text{ kn} \qquad = 6.0 \text{ kn} + 0 - 2.1 \text{ kn}$$

$$= -1.6 \text{ kn} \qquad\qquad\qquad = 3.9 \text{ kn}$$

The magnitude of the resultant velocity is

$$R = \sqrt{R_x{}^2 + R_y{}^2}$$

$$= \sqrt{1.6^2 + 3.9^2} \text{ kn} = \sqrt{17.8} \text{ kn} = 4.2 \text{ kn}.$$

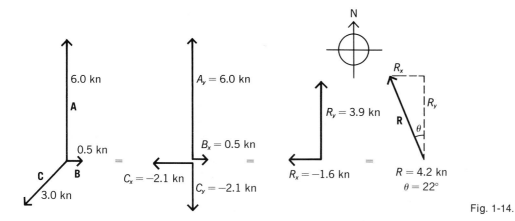

Fig. 1-14.

Thus the speed of the boat relative to the earth is 4.2 kn. Its direction, as shown in the last part of the diagram, is west of north at an angle θ with north. We can find the value of θ by first finding the value of $\tan \theta$:

$$\tan \theta = \frac{R_x}{R_y} = \frac{1.6}{3.9} = 0.410,$$

$$\theta = 22°.$$

Special Topic

Vectors in Three Dimensions

When all the vector quantities involved in a problem lie in the same plane, usually a horizontal or a vertical one, it is sufficient to consider for each one only two mutually perpendicular components, A_x and A_y for example, that lie in the plane. In general, though, three such components are required to completely describe a vector quantity A, namely A_x, A_y, and A_z as in Fig. 1-15.

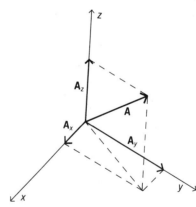

Fig. 1-15.

Sometimes a vector quantity in three dimensions will be specified by its magnitude A and the angles θ and ϕ shown in Fig. 1-16. The magnitudes of the components A_x, A_y, and A_z can be found from A, θ, and ϕ with the help of Fig. 1-17:

$$A_x = A \sin \theta \cos \phi,$$
$$A_y = A \sin \theta \sin \phi,$$
$$A_z = A \cos \theta.$$

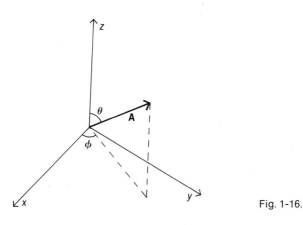

Fig. 1-16.

On the other hand, if we already know A_x, A_y, and A_z, we can calculate A, θ, and ϕ from these formulas:

$$A = \sqrt{A_x^2 + A_y^2 + A_z^2},$$

$$\theta = \cos^{-1}\frac{A_z}{A},$$

$$\phi = \tan^{-1}\frac{A_y}{A_x}.$$

The quantity "$\cos^{-1} X$" means the angle whose cosine has the value X, and similarly for $\sin^{-1}$ and $\tan^{-1}$. Sometimes $\cos^{-1} x$ is written arc cos x.

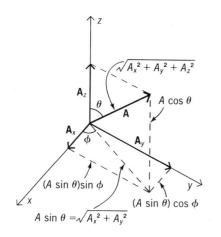

Fig. 1-17.

Important Terms

To measure a quantity means to compare it with a standard quantity of the same kind called a **unit**. Almost all physical quantities can be expressed in terms of four fundamental measurements: length, mass, time, and electric current. In a **system of units**, a set of units of length, mass, time, and electric current is specified from which all other units in the system are derived. The **SI** (metric) and **British** systems are in wide use.

A **scalar quantity** is one that has magnitude only. A **vector quantity** is one that has both magnitude and direction. Thus time is a scalar quantity, and force is a vector quantity.

A **vector** is an arrowed line whose length is proportional to the magnitude of some vector quantity and whose direction is that of the quantity. A **vector diagram** is a scale drawing of the various forces, velocities, or other vector quantities involved in the motion of a body.

In the graphical method of **vector addition,** the tail of each successive vector is placed at the head of the previous one, with their lengths and original directions kept unchanged. The **resultant** is a vector drawn from the tail of the first vector to the head of the last.

A vector can be **resolved** into two or more other vectors called the **components** of the original vector. Usually the components of a vector are chosen to be in mutually perpendicular directions.

Important Formulas

Vector addition:

$$\mathbf{A} + \mathbf{B} = \mathbf{B} + \mathbf{A}$$

Vector subtraction:

$$\mathbf{B} - \mathbf{A} = \mathbf{B} + (-\mathbf{A})$$

Components of a vector:

$$\mathbf{A} = \mathbf{A}_x + \mathbf{A}_y$$

$$A_x = \cos \theta$$
$$A_y = \sin \theta$$

Vector addition by components:

$$R = \sqrt{R_x{}^2 + R_y{}^2}$$

where $R_x = A_x + B_x + C_x + \cdots$
$R_y = A_y + B_y + C_y + \cdots$

Angle θ between $\mathbf{R}$ *and* $\mathbf{R}_x$:

$$\tan \theta = \frac{R_y}{R_x}$$

Multiple Choice

1. Of the following, the shortest is
   a. 1 mm.
   b. 0.01 in.
   c. 0.001 ft.
   d. 0.00001 km.

2. Of the following, the longest is
   a. 10^4 in.
   b. 10^4 m.
   c. 10^3 ft.
   d. 0.1 mi.

3. The number of cm³ in a ft³ is approximately
   a. 1.7×10^3.
   b. 1.7×10^4.
   c. 2.8×10^4.
   d. 1.7×10^5.

4. The number of seconds in a day is closest to
   a. 10^4.
   b. 10^5.
   c. 10^6.
   d. 10^7.

5. The number of seconds in a month is approximately
   a. 2.6×10^6.
   b. 2.6×10^7.
   c. 2.6×10^8.
   d. 2.6×10^9.

6. Of the following units, which could be associated with a vector quantity?
   a. meters/minute
   b. quarts/second
   c. hours
   d. cubic feet

7. Which of the following statements is incorrect?
   a. All vector quantities have directions.
   b. All vector quantities have magnitudes.
   c. All scalar quantities have directions.
   d. All scalar quantities have magnitudes.

8. The minimum number of unequal forces whose vector sum can equal zero is
   a. 1.
   b. 2.
   c. 3.
   d. 4.

9. Which of the following pairs of displacements *cannot* be added to give a resultant displacement of 2 m?
   - a. 1 m and 1 m
   - b. 1 m and 2 m
   - c. 1 m and 3 m
   - d. 1 m and 4 m

10. Which of the following sets of forces *cannot* have a vector sum of zero?
    - a. 10, 10, and 10 lb
    - b. 10, 10, and 20 lb
    - c. 10, 20, and 20 lb
    - d. 10, 20, and 40 lb

11. Which of the following sets of displacements might be able to return a car to its starting point?
    - a. 2, 8, 10, and 25 mi
    - b. 5, 20, 35, and 65 mi
    - c. 60, 120, 180, and 240 mi
    - d. 100, 100, 100, and 400 mi

12. A displacement of 9 ft and another 6 ft can be added to give a resultant displacement of
    - a. 0 ft.
    - b. 1.5 ft.
    - c. 4 ft.
    - d. 16 ft.

13. A 10-lb force and a 5-lb force act on a body. The resultant force on the body must be
    - a. between 5 and 10 lb.
    - b. between 5 and 15 lb.
    - c. more than 5 lb.
    - d. more than 10 lb.

14. A man walks 8 mi north and then 5 mi in a direction 60° east of north. His resultant displacement from his starting point is
    - a. 11 mi.
    - b. 12 mi.
    - c. 13 mi.
    - d. 14 mi.

15. The resultant of a 4-lb force acting upward and a 3-lb force acting horizontally is
    - a. 1 lb.
    - b. 5 lb.
    - c. 7 lb.
    - d. 12 lb.

16. The angle between the resultant of Question 15 and the vertical is approximately
    - a. 37°.
    - b. 45°.
    - c. 53°.
    - d. 60°.

17. An airplane travels 100 miles to the north and then 200 miles to the east. The displacement of the airplane from its starting point is approximately
    - a. 100 miles.
    - b. 200 miles.
    - c. 220 miles.
    - d. 300 miles.

18. At what angle east of north should the airplane of Question 17 have headed in order to reach its destination in a straight flight?
    - a. 22°
    - b. 45°
    - c. 50°
    - d. 63°

19. A vector $\mathbf{A}$ lies in a plane and has the components A_x and A_y. The magnitude A_x of $\mathbf{A}_x$ is equal to
    - a. $A - A_y$.
    - b. $\sqrt{A} - \sqrt{A_y}$.
    - c. $\sqrt{A - A_y}$.
    - d. $\sqrt{A^2 - A_y^2}$.

20. An escalator has a velocity of 10.0 ft/s at an angle of 60° above the horizontal. The vertical component of its velocity is
    - a. 1.7 ft/s.
    - b. 5.0 ft/s.
    - c. 6.0 ft/s.
    - d. 8.7 ft/s.

Exercises

1. Is it correct to say that scalar quantities are abstract, idealized quantities with no precise counterparts in the physical world, whereas vector quantities can be said to properly represent reality?

2. What kind of quantity is the magnitude of a vector quantity? What kind of quantity is the resultant of two vector quantities of the same kind?

3. The resultant of three vectors is zero. Must they all lie in a plane?

4. A bird lands on an overhead power cable, which sags slightly under its weight. Is it possible to prevent any such sagging by applying enough tension to the cable?

5. The tallest tree in the world is a Sequoia in California that is 368 ft high. How high is this in meters? In kilometers?

6. The largest sperm whale ever captured was 20.7 m long. How many ft is this?

7. An acre contains 4840 yd². How many ft² is this? How many acres are there in a mi²?

8. How many ft³ are there in a m³? How many m³ in a ft³?

9. A "board foot" is a unit of lumber measure that corresponds to the volume of a piece of wood 1 ft square and 1 in. thick. How many in³ are there in a board foot? How many ft³? How many cm³?

10. The velocity of sound in air at sea level is about 1100 ft/s. Express this velocity in m/s, mi/s, and mi/hr.

11. The speedometer of a European car is calibrated in km/hr. What is the car's velocity in mi/hr when the speedometer reads 50?

12. Geoffrey Williams won the 1968 single-handed Transatlantic race from Plymouth, England to Newport, Rhode Island in 25 days 20 hr 33 min. He sailed a total of 3784 nautical miles (abbreviated n-mi). What was his average velocity in knots? In mi/hr? (1 knot = 1 n-mi/hr; 1 ni-mi = 6080 ft.)

13. A man walks 70 m north, 60 m west, and then ascends 40 m in an elevator. What is the magnitude of his displacement from his starting place?

14. A driver loses his way and travels 10 mi north, 4 mi east, and 3 mi south. What is his displacement from his starting place?

15. A woman pushes a 50-lb lawnmower with a force of 25 lb. If the handle of the lawnmower is 45° above the horizontal, how much downward force is being exerted on the ground by the lawnmower?

16. A weight of 100 lb is at rest on an inclined plane that makes an angle of 35° with the horizontal. Find the components of this weight parallel and perpendicular to the plane.

19. A sailboat cannot sail directly to windward but must "tack" back and forth at a certain angle with respect to the direction from which the wind is blowing. Which sailboat has the greater component of velocity to windward, the *Alpha* whose velocity is 5 mi/hr at an angle of 40° off the wind, or the *Beta* whose velocity is 6 mi/hr at an angle of 50° off the wind?

20. The *Queen Elizabeth II* is heading due east at 20 mi/hr in the presence of a north wind of 15 mi/hr. What is the horizontal component of velocity of the smoke relative to the ship as it leaves the funnel? Relative to the earth?

21. A 100-lb sack of potatoes is suspended by a rope. A man pushes sideways on the sack with a force of 40 lb. What is the tension in the rope?

Problems

1. A common unit of liquid measure in the metric system is the *liter*, which is equal to 10³ cm³, and a common unit of liquid measure in the United States is the gallon, which is equal to 231 in³. How many liters are there in a gallon? How many gallons are there in a liter?

2. The resultant of two perpendicular forces has a magnitude of 20 lb. If the magnitude of one of the forces is 10 lb, what is the magnitude of the other force?

3. A man on the ground observes an airplane climbing at an angle of 37° above the horizontal. He gets in his car and by driving at 70 mi/hr is able to stay directly below the airplane. What is the airplane's velocity?

4. An airplane is heading southeast when it takes off at an angle of 25° above the horizontal. Its velocity is 200 mi/hr. (a) What is the vertical component of its velocity? (b) What is the horizontal component of its velocity? (c) What is the component of the velocity of the plane toward the south?

5. A horizontal and a vertical force combine to give a resultant force of 10 lb that acts in a direction 40° above the horizontal. Find the magnitudes of the horizontal and vertical forces.

6. A length of pipe is mounted on a truck traveling at 10 m/s. It is raining, and the raindrops fall vertically downward at 30 m/s. At what angle with the vertical should the pipe be tilted so the raindrops pass through it without touching its wall?

7. Two billiard balls are rolling on a flat table. One has the velocity components $v_x = 1$ m/s, $v_y = 2$ m/s. The other has the velocity components $v_x = 2$ m/s, $v_y = 3$ m/s. If both balls started from the same point, what is the angle between their paths?

8. Two tugboats are towing a ship. Each exerts a horizontal force of 5.0 tons, and the angle between the tow ropes is 30°. What is the resultant force exerted on the ship?

9. An airplane flies 200 mi east from city A to city B, then 200 mi south from city B to city C, and finally 100 mi northwest to city D. How far is it from city A to city D? In what direction must the airplane head to return directly to city A from city D?

10. The following forces act on an object resting on a level, frictionless surface: 10 lb to the north, 20 lb to the east, 10 lb at an angle 37° south of east, and 20 lb at an angle 53° west of south. What is the magnitude and direction of the resultant force acting on the object?

11. An airplane whose velocity is 100 mi/hr climbs from the ground at an angle of 23° with the horizontal. What is its altitude in feet 1 min after take-off? How many miles does it travel in a horizontal direction in this period of time?

12. A boat moving at 15 mi/hr is crossing a river 2 mi wide in which the current is flowing at 5 mi/hr. (a) If the boat heads directly for the other shore, how long will the trip take? (b) In what direction should the boat head if it is to reach a point on the other shore directly opposite the starting point? (c) How long will the crossing take in case (b)?

13. The ketch *Minots Light* is heading northwest at 7 knots through a tidal stream which is flowing southwest at 3 knots. Find the magnitude and direction of its velocity relative to the earth's surface.

14. The vector **A** has a magnitude of 20 cm and points 20° clockwise from the $+y$ direction. The vector **B** has a magnitude of 10 cm and points 60° clockwise from the $+y$ direction. Find the magnitude and direction of $A + B$, $A - B$, and $B - A$.

15. The vector **A** has a magnitude of 10 cm and points 37° clockwise from the $+y$ direction. The vector **B** has a magnitude of 10 cm and points 37° clockwise from the $+x$ direction. Find the magnitude and direction of $A + B$, $A - B$, and $B - A$.

Answers to Multiple Choice

1. b	8. c	15. b
2. b	9. d	16. a
3. c	10. d	17. c
4. b	11. c	18. d
5. a	12. c	19. d
6. a	13. b	20. d
7. c	14. a	

2

Describing Motion

Everything in the physical world is in motion, from the elementary particles within atoms to the largest galaxies of stars. Since the goal of physics is an understanding of the nature and behavior of the physical world, the physicist is directly concerned with things in motion in nearly all his work. For this reason we shall begin our study of this subject with an analysis of the simplest kinds of motion normally encountered. Besides obtaining useful and interesting information, we shall be introduced to the manner in which the physicist thinks, how he obtains general relationships from specific data, and how he goes about applying these relationships to other situations.

2-1 Frame of Reference

When something has changed its position with respect to its surroundings, we say it has *moved*.

There are two separate ideas here. One is that of *change*: when something has moved, the world is not exactly the same as it was before. The other idea is that of *frame of reference*: if we are to notice that something is moving, we must be able to check its position relative to something else. The choice of a frame of reference for reckoning motion depends upon the situation. For a car, the most convenient frame of reference is the earth's surface; for a sailor, it is the deck of his ship; for a planet, it is the sun; for an electron in an atom, it is the atom's nucleus.

The key to solving many problems in physics is the proper selection of a frame of reference. Newton was able to interpret the motions of the earth and planets in terms of the gravitational pull of the sun only because he recognized the sun as the center of the solar system. If instead he had considered the earth as the center, with the sun and the other planets moving around it, their paths would have appeared to be very irregular. It is unlikely that he would have been able to discover the law of gravitation by analyzing these seemingly complicated motions.

2-2 Constant Velocity

Let us imagine we have marked off 100-ft intervals with chalk on a straight road and we are now standing beside the road with a watch, pencil, and paper in hand (Fig. 2-1). A car comes along, and we note the exact time at which the front of the car passes each of the marks we made on the road. Our figures might be those given in Table 2-1.

To interpret these measurements a sheet of graph paper is helpful. Along the bottom of the sheet we establish a time scale running from 0 to 15 s, and on the left-hand side we establish a distance scale running from 0 to 500 ft. Next we plot each of our measurements as a point on the graph. In Fig. 2-2, one point is plotted, the intersection of the lines on the graph paper that pass through the time and distance figures on the two scales.

Each point on the complete graph represents the results of a measurement. The various points come very close to lying along a straight line

All motion is relative to frame of reference

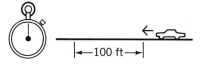

Fig. 2-1.

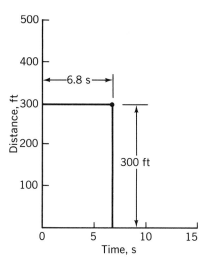

Fig. 2-2. The method by which the data of Table 2-1 are plotted as a graph. The point representing 300 ft and 6.8 s is shown.

Table 2-1. Measurements made with the arrangement in Fig. 2-1.

Total distance, ft	0	100	200	300	400	500
Elapsed time, s	0	2.3	4.5	6.8	9.1	11.4

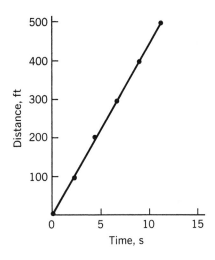

Fig. 2-3. A graph of the data of Table 2-1. Each point represents the results of one measurement.

Velocity

Speed is the magnitude of velocity

(Fig. 2-3). The line does not go through all the points, but we may reasonably attribute the deviations to experimental errors. We have no way of knowing from the graph itself whether additional intermediate measurements would also fall on the same line; but let us suppose that if we had made such additional measurements, they too would conform to this line.

When a graph of one quantity versus another results in a straight line, each quantity is said to be *directly proportional* to the other. We can verify the proportionality of distance and time in this case from the graph, where we see that doubling t means that s also doubles, tripling t means that s also triples, and so on. Thus we can write

$$s = vt \qquad (2\text{-}1)$$

Distance = velocity × time.

The *velocity* of the car, v, is the constant of proportionality: the rate at which the distance covered by the car changes with time. The greater the time t, the farther the car goes, in exact proportion to t; hence the car is said to move with *constant velocity*.

Velocity is a vector quantity and in general is denoted by the vector symbol **v**. However, since we are concerned with straight-line motion for the time being, we can consider velocity to be a scalar quantity with the provision that a given direction (here the one in which the car is headed) is to be reckoned as positive and the opposite direction is to be reckoned as negative. Thus the car's velocity is positive when it is moving ahead and is negative when it is moving in reverse. In general, the magnitude of a velocity is called *speed*.

How can we find the value of v, the velocity of the car? What we do is rewrite the equation $s = vt$ in the form

$$v = \frac{s}{t} \qquad (2\text{-}2)$$

Velocity = $\dfrac{\text{distance}}{\text{time}}$.

and then calculate s/t for each of the measurements. The results, worked out to four figures, are given in Table 2-2. Only two of these figures are significant, however, since the original measurements only had two significant figures. When the values of v are rounded off, we find that

$$v = 44 \text{ ft/s.}$$

Table 2-2. Velocities calculated from the data of Table 2-1.

Total distance, ft	0	100	200	300	400	500
Elapsed time, s	0	2.3	4.5	6.8	9.1	11.4
Velocity, ft/s		43.48	44.44	44.12	43.96	43.86

(Significant figures are discussed in Appendix B.)

Knowing that the velocity of a particular car (or other body) is constant, we can then predict exactly how far it will go in a given period of time; or, given the distance it is to travel, we can determine the time required.

Problem. A car moves at the constant speed of 44 ft/s. How far will it travel in exactly 1 hr?

Solution. First we note that

$$1 \text{ hr} = 60 \text{ min} \times 60 \; \frac{\text{s}}{\text{min}} = 3600 \text{ s.}$$

Hence we have

Distance = velocity × time

$$s = vt = 44 \; \frac{\text{ft}}{\text{s}} \times 3600 \text{ s} = 158{,}400 \text{ ft.}$$

The proper way to express this result is

$$s = 1.6 \times 10^5 \text{ ft}$$

since only two figures are significant here.

Problem. How long does it take the same car to travel exactly 1 mi?

Solution. For this problem we must rewrite the basic formula $s = vt$ as

$$t = \frac{s}{v}$$

$$\text{Time} = \frac{\text{distance}}{\text{velocity}}.$$

Since there are 5280 ft in a mile, we have

$$t = \frac{s}{v} = \frac{5280 \text{ ft}}{44 \text{ ft/s}} = 120 \text{ s}.$$

The car takes 2 min to travel 1 mi.

2-3 Instantaneous Velocity

Not all cars have constant velocities. Table 2-3 contains data that might be taken for a car that starts moving from rest on the marked stretch of road at $t = 0$. When we plot these data on a graph, as in Fig. 2-4, we find that the line joining the various points is not straight but shows a definite upward curve. In each successive interval of time (as marked off on the bottom of the graph) the car goes a greater distance than before; in other words, the

Table 2-3. Measurements made with the arrangement in Fig. 2-1 for a car which starts from rest at $t = 0$.

Total distance, ft	0	100	200	300	400	500
Elapsed time, s	0	11	16	20	23	26

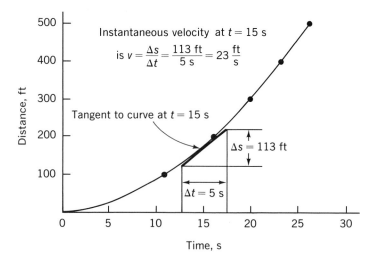

Fig. 2-4. A graph of the data of Table 2-3. The procedure for finding the instantaneous velocity of the car is shown for $t = 15$ s.

car is going faster and faster. Because s is not directly proportional to t, the car's velocity is not constant.

Even though v varies, at every instant it has a definite value. To find this *instantaneous velocity* at a particular time t, we draw a straight line tangent to the s-t curve at that value of t. (The length of the line does not matter.) Then we determine v from that straight line from the formula

Instantaneous velocity

$$v = \frac{\Delta s}{\Delta t} \qquad\qquad\qquad \textit{Instantaneous velocity} \quad (2\text{-}3)$$

where Δs is the distance interval between the ends of the line and Δt is the time interval between them. (Δ is the Greek letter *delta*.) The instantaneous velocity of the car at $t = 15$ s is, from Fig. 2-4,

$$v = \frac{\Delta s}{\Delta t} = \frac{113 \text{ ft}}{5 \text{ s}} = 23 \frac{\text{ft}}{\text{s}}.$$

Table 2-4 shows the instantaneous velocities of the car calculated for the times indicated.

Table 2-4. Instantaneous velocities at 5-s intervals as determined from Fig. 2-4.

Elapsed time, s	0	5	10	15	20	25	30
Instantaneous velocity, ft/s	0	7.5	15	23	30	38	45

When we define instantaneous velocity as the tangent to the s-t curve, we are simply applying our previous procedure to a very short segment of the curve. The shorter the segment of the curve, the more nearly straight it is. In the limit of infinite shortness, the segment is a straight line whose slope is the same as that of the tangent to the curve there. Thus the tangent is equal to the velocity v in a time interval of vanishing duration, which is what is meant by instantaneous velocity.

2-4 Average Velocity

In some situations we are interested in the instantaneous velocity of a moving body at a certain specific time, but very often our chief concern is with its *average velocity*. The average velocity, denoted $\bar{v}$, of a body during a period of time t is the net distance s it has traveled in that period of time divided by t:

$$\bar{v} = \frac{s}{t} \tag{2-4}$$

$$\text{Average velocity} = \frac{\text{net distance}}{\text{time}}.$$

We note that this is the same as the definition of v for a body moving at constant velocity, since in that case v and $\bar{v}$ are the same. For a body whose velocity varies, however, the instantaneous velocity is in general different from the average velocity.

Let us find the average velocity of the second car, whose instantaneous velocity varied from $v = 0$ at the start to $v = 45$ ft/s after 30 s had gone by. Since the car went a total distance of 500 ft in 26 s, its average velocity was

$$\bar{v} = \frac{s}{t} = \frac{500 \text{ ft}}{26 \text{ s}} = 19 \text{ ft/s}.$$

Problem.   An airplane takes off at 9:00 A.M. and flies in a straight path at 300 km/hr until 1:00 P.M. At 1:00 P.M. its velocity is increased to 400 km/hr and it maintains this velocity in the same direction as before until it lands at 3:30 P.M. What is its average velocity for the entire flight?

Solution.   To calculate the average velocity of the airplane we must know the net distance s it has covered and the total time t involved. In the first part of the flight the airplane covers

$$s_1 = v_1 t_1 = 300 \frac{\text{km}}{\text{hr}} \times 4.00 \text{ hr} = 1200 \text{ km,}$$

and in the second part it covers

$$s_2 = v_2 t_2 = 400 \frac{\text{km}}{\text{hr}} \times 2.50 \text{ hr} = 1000 \text{ km.}$$

Hence we obtain for s and t the values

$$s = s_1 + s_2 = (1200 + 1000) \text{ km} = 2200 \text{ km,}$$

$$t = t_1 + t_2 = (4.00 + 2.50) \text{ hr} = 6.50 \text{ hr,}$$

and the average velocity is

$$\bar{v} = \frac{s}{t} = \frac{2200 \text{ km}}{6.50 \text{ hr}} = 338 \frac{\text{km}}{\text{hr}}.$$

2-5 Acceleration

In the real world, few bodies move at constant velocity for very long. The velocity of a body changes only when it interacts with something else in some way, which means that such interactions occur almost all the time. For the moment our concern is with how changing velocities are described; later we shall discuss the interactions responsible for them.

A body whose velocity is not constant is said to be *accelerated*. This term is applied regardless of whether **v** is increasing, decreasing, or changing in direction. Figure 2-5 shows three examples of accelerated motion.

Just as velocity is the rate of change of displacement with time, *acceleration* is the rate of change of velocity with time. The symbol for acceleration is **a**, since it is a vector quantity. For the time being we are discussing straight-line motion only, so we shall consider acceleration as a scalar quantity *a*.

Acceleration is rate of change of velocity

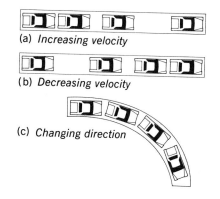

(a) *Increasing velocity*

(b) *Decreasing velocity*

(c) *Changing direction*

Fig. 2-5.   The successive positions of three accelerated cars after equal periods of time. (a) The car is going faster and faster, so it travels a longer distance in each period of time. (b) The car is going slower and slower, so it travels a shorter distance in each period of time. (c) The magnitude of the car's velocity is constant, but its direction changes.

If a body's velocity is v_0 to begin with and changes to v_f after a time interval of t, the body's acceleration is given by the formula

$$a = \frac{v_f - v_0}{t} \tag{2-5}$$

$$\text{Acceleration} = \frac{\text{change in velocity}}{\text{time interval}}.$$

When the final velocity v_f is greater than the initial velocity v_0, the acceleration is positive, which signifies that the body is going faster and faster. When the final velocity is *less* than the initial velocity, the acceleration is negative, which signifies that the body is going slower and slower.

Force and acceleration

Just what is it that causes something to be accelerated? The answer is a force, which for the moment we can think of as a push or a pull. In the absence of a force, an object at rest remains at rest and an object in motion continues in motion at constant velocity. These matters are discussed in detail in Chapter 4.

2-6 Constant Acceleration

Let us return to the data on car No. 2 and plot a graph of its instantaneous velocity v versus time t, as in Fig. 2-6. All the points lie on a straight line, which implies that v is directly proportional to t. Although the car's velocity is not constant, it varies in a uniform way with time: as time goes on, the velocity increases exactly in proportion. Therefore the car's acceleration is constant.

From its definition, acceleration is expressed in terms of

$$\frac{\text{Velocity}}{\text{time}} = \frac{\text{distance/time}}{\text{time}} = \frac{\text{distance}}{\text{time}^2}.$$

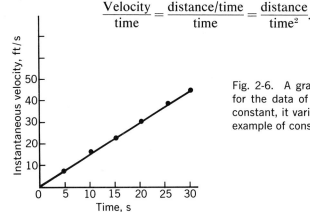

Fig. 2-6. A graph of instantaneous speed versus time for the data of Table 2-4. Although the velocity is not constant, it varies in a uniform way with time. This is an example of constant acceleration.

A body whose acceleration increases by 10 ft/s in each second would thus have its acceleration expressed as

$$10 \, \frac{\text{ft/s}}{\text{s}} = 10 \text{ ft/s}^2.$$

Let us calculate the acceleration of the second car whose motion was examined before. The initial velocity of the car is $v_0 = 0$ and after $t = 15$ s it is $v_f = 23$ ft/s. Hence we have

$$a = \frac{v_f - v_0}{t} = \frac{(23 - 0) \text{ ft/s}}{15 \text{ s}} = 1.5 \text{ ft/s}^2.$$

If we make the same calculation at the later time $t = 25$ s we have, since v is now 38 ft/s,

$$a = \frac{(38 - 0) \text{ ft/s}}{25 \text{ s}} = 1.5 \text{ ft/s}^2.$$

The value of a is the same, because the acceleration is constant. If the acceleration were not constant, different values of a would be obtained at different times.

Not all accelerations are constant, of course, but a great many real motions are best understood by idealizing them in terms of constant accelerations.

Problem. Discuss the motion of the car whose velocity-time graph is shown in Fig. 2-7.

Solution. A horizontal line in a v-t graph means that v does not change, so the acceleration is 0 at first. A line sloping upward means a positive acceleration (v increasing), and a line sloping downward means a negative acceleration (v decreasing). Hence the car at first moves with the constant velocity of 10 ft/s, then is accelerated at 2 ft/s² to the velocity of 20 ft/s, next travels at constant speed for 5 s, and finally undergoes a negative acceleration of −4 ft/s² until it comes to rest.

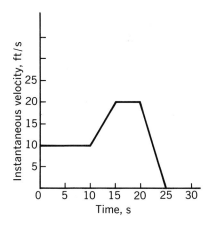

Fig. 2-7. Instantaneous velocity versus time for a certain car.

2-7 Velocity and Acceleration

In the event a body starts to accelerate from some initial velocity v_0, its change in velocity at during the time interval t in which the acceleration a (assumed constant) occurs is added to v_0 (Fig. 2-8). Hence the final velocity

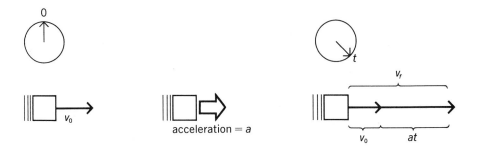

Fig. 2-8. The final velocity v_f of an accelerated body is equal to its initial velocity v_0 plus the velocity change at that occurred during the interval t.

v_f at a time t after the acceleration begins is the initial velocity v_0 plus the change in velocity at:

$$v_f = v_0 + at \qquad (2\text{-}6)$$

Final velocity = initial velocity + velocity change.

The same formula can be obtained by multiplying by t both sides of the defining equation of acceleration,

$$a = \frac{v_f - v_0}{t}$$

and then solving for v_f.

Problem.   A car accelerates at the constant rate of 5 ft/s² from an initial velocity of 20 mi/hr. Find its velocity in mi/hr after 10 s.

Solution.   After 10 s the increase at in the car's velocity is

$$at = 5\,\frac{\text{ft}}{\text{s}^2} \times 10\ \text{s} = 50\ \text{ft/s}.$$

Since 1 ft/s = 0.682 mi/hr,

$$at = 50\,\frac{\text{ft}}{\text{s}} \times 0.682\,\frac{\text{mi/hr}}{\text{ft/s}} = 34\ \text{mi/hr},$$

and the final velocity v_f of the car is

$$v_f = v_0 + at = (20 + 34)\,\frac{\text{mi}}{\text{hr}} = 54\ \text{mi/hr}.$$

Equation (2-6) can be used regardless of the algebraic signs of v_0 and a. In the case of a car, a negative velocity customarily implies motion opposite to a specified direction, which we might refer to as backward motion (though the car may actually be facing either way). A negative acceleration means that the car is slowing if v_0 is positive or increasing in backward velocity if v_0 is negative. If v_0 is negative and a is positive, the car is traveling backward at a diminishing velocity. The importance of keeping track of the algebraic signs of the various quantities is evident.

Problem.   A car has an initial velocity of 20 m/s and an acceleration of -1 m/s². Find its velocity after 10 s and after 50 s.

Solution.   The velocity after 10 s is

$$v_f = v_0 + at = 20\ \frac{\text{m}}{\text{s}} - 1\ \frac{\text{m}}{\text{s}^2} \times 10\ \text{s} = (20 - 10)\ \text{m/s} = 10\ \text{m/s},$$

which is half what it was initially. After 50 s the velocity of the car, if a stays the same, is

$$v_f = v_0 + at = 20\ \frac{\text{m}}{\text{s}} - 1\ \frac{\text{m}}{\text{s}^2} \times 50\ \text{s} = (20 - 50)\ \text{m/s} = -30\ \text{m/s},$$

which is in the opposite direction and greater in magnitude than the original velocity v_0. (This example is not very realistic, of course, since 30 m/s is 67 mi/hr, but it illustrates how negative quantities are used in calculations.)

Special Topic

Velocity and Acceleration in Calculus

In Sec. 2-3 the instantaneous velocity of an object at a certain time was defined as the value of the tangent to the s-t curve of its motion at that time. Another way to say the same thing is to define instantaneous velocity as the limit of $\Delta s/\Delta t$ as Δt approaches 0, where Δs is the object's displacement in the time interval Δt. That is,

$$v = \lim_{\Delta t \to 0} \frac{\Delta s}{\Delta t},$$

which in calculus notation is written

$$v = \frac{ds}{dt}.$$

Thus the instantaneous velocity of an object is the derivative of its displacement with respect to time. If we know how s varies with t in a particular case, we can obtain a formula for v simply by differentiating s with respect to t. In the absence of an analytical expression for s we must, of course, find the instantaneous velocity by fitting a tangent to the s-t curve at the point of interest.

In a similar way the instantaneous acceleration of an object can be defined as

$$a = \lim_{\Delta t \to 0} \frac{\Delta v}{\Delta t},$$

which in calculus notation is written

$$a = \frac{dv}{dt}.$$

The instantaneous acceleration of an object is the derivative of its velocity with respect to time. Since the instantaneous velocity is $v = ds/dt$, we can express a directly in terms of s and t as the second derivative of displacement with respect to time:

$$a = \frac{dv}{dt} = \frac{d}{dt}\left(\frac{ds}{dt}\right) = \frac{d^2s}{dt^2}.$$

Here is an example of how these expressions are used. Experiments show that the distance s an object in free fall near the earth's surface descends is proportional to t^2, where t is the time that has elapsed since it was released. We therefore can write

$$s = Kt^2$$

where K is a constant. To find how the object's downward velocity varies with time, we need only differentiate the expression for s, which gives

$$v = \frac{ds}{dt} = 2Kt.$$

The velocity increases linearly with time t. The object's acceleration is

$$a = \frac{dv}{dt} = 2K$$

so it is uniformly accelerated. As we shall find in Chap. 3, the acceleration $2K$ has the same value for all objects near the earth's surface and is called the *acceleration of gravity*, symbol g. Its value is $g = 9.8$ m/s^2 = 32 ft/s^2.

    Another example concerns an object suspended by a coil spring. Such an object oscillates up and down so that its displacement y from its equilibrium point varies with time according to the formula

$$y = A \cos \omega t$$

where A and ω are constants whose values depend on the particular situation. (ω is the Greek letter *omega*.) This is a case of simple harmonic motion, an important topic treated in Chap. 12. The object's velocity at any time t is

$$v = \frac{dy}{dt} = -\omega A \sin \omega t$$

since $d(\cos x)/dt = -\sin x \, dx/dt$ and $d(\omega t)/dt = \omega$. The object's acceleration is

$$a = \frac{dv}{dt} = -\omega^2 A \cos \omega t = -\omega^2 y$$

since $d(\sin x)/dt = \cos x \, dx/dt$. This important result tells us that the acceleration of an object in simple harmonic motion is proportional to its displacement from its equilibrium position and is always in the opposite direction to that of the displacement.

Important Terms

The **velocity** of a body is a specification of both its speed and the direction in which it is moving. Velocity is a vector quantity. The **instantaneous velocity** of a body is its velocity at a specific instant of time. The **average velocity** of a body is the total displacement through which it has moved in a time interval divided by the interval.

The **speed** of a moving body is the rate at which distance is being covered by the body. Speed is a scalar quantity, unlike velocity. When the distance is directly proportional to the elapsed time, the body moves at **constant speed**.

The **acceleration** of a body is the rate at which its velocity changes with time; the change in velocity may be a change in magnitude or a change in direction or both.

Important Formulas

Velocity:

$$v = \frac{s}{t}$$

Acceleration:

$$a = \frac{v_f - v_0}{t}$$

Final velocity under constant acceleration:

$$v_f = v_0 + at$$

Multiple Choice

1. The term motion refers to a change in position with respect to
   a. a specified reference object.
   b. the ground.
   c. the center of the earth.
   d. the sun.

2. A body whose velocity is constant
   a. has a positive acceleration.
   b. has a negative acceleration.
   c. has zero acceleration.
   d. might have any of the above accelerations.

3. An accelerated body must at all times
   a. be moving.
   b. have an increasing velocity.
   c. have a changing velocity.
   d. have a changing direction.

4. An example of a body whose motion is *not* accelerated is a car that
   a. turns a corner at the constant velocity magnitude of 10 mi/hr.
   b. descends a hill at the constant velocity magnitude of 30 mi/hr.
   c. descends a hill at a velocity magnitude that increases from 20 mi/hr to 40 mi/hr uniformly.
   d. climbs a hill, goes over its crest, and descends on the other side, all at the constant velocity magnitude of 30 mi/hr.

5. On a distance-time graph, a horizontal straight line corresponds to motion at
   a. zero velocity.
   b. constant velocity.
   c. increasing velocity.
   d. decreasing velocity.

6. On a distance-time graph, a straight line sloping upward to the right corresponds to motion at
   a. zero velocity.
   b. constant velocity.
   c. increasing velocity.
   d. decreasing velocity.

7. On a velocity-time graph, the motion of a car traveling along a straight road with the uniform acceleration of 2 m/s² would appear as a
   a. horizontal straight line.
   b. straight line sloping upward to the right.
   c. straight line sloping downward to the right.
   d. curved line whose downward slope to the right increases with time.

8. An airplane travels 250 mi in ½ hr at constant velocity. Its velocity is
   a. 125 mi/hr.
   b. 250 mi/hr.
   c. 500 mi/hr.
   d. 1000 mi/hr.

9. A ship travels 12 mi at constant velocity in 40 min. Its velocity is
   a. 0.3 mi/hr.
   b. 8 mi/hr.
   c. 18 mi/hr.
   d. 48 mi/hr.

10. A certain DC-9 airplane has a range of 910 mi at a velocity of 520 mi/hr. Its maximum time in the air at this velocity is
    a. 29 min.
    b. 34 min.
    c. 57 min.
    d. 1 hr 45 min.

11. A car that travels at 40 mi/hr for 2 hr, at 50 mi/hr for 1 hr, and at 20 mi/hr for ½ hr has an average velocity of
    a. 31.4 mi/hr.   b. 40 mi/hr.
    c. 45 mi/hr.   d. 55 mi/hr.

12. Two airplanes headed for the same destination leave an airport an hour apart. The one that leaves first travels at 300 mi/hr and the other travels at 400 mi/hr. The latter will overtake the former in
    a. 45 min.   b. 80 min.
    c. 3 hr.   d. 4 hr.

13. In question 12, how far will the airplanes have traveled when they meet?
    a. 300 mi   b. 400 mi
    c. 900 mi   d. 1200 mi

14. A police car leaves in pursuit of a holdup car ½ hr after the latter has left the scene of the crime at 60 km/hr. How fast must the police car go if it is to catch up with the holdup car in 1 hr?
    a. 90 km/hr   b. 100 km/hr
    c. 110 km/hr   d. 120 km/hr

15. In question 14, how far will the cars have traveled when they meet?
    a. 30 km   b. 60 km
    c. 90 km   d. 150 km

16. A pitcher takes 0.1 s to throw a baseball, which leaves his hand with a velocity of 30 m/s. The ball's acceleration was
    a. 3 m/s².   b. 30 m/s².
    c. 300 m/s².   d. 3000 m/s².

17. A car starts from rest and reaches a velocity of 40 ft/s in 20 s. Its acceleration is
    a. 0.5 ft/s².
    b. 2 ft/s².
    c. 8 ft/s².
    d. 80 ft/s².

18. A car accelerates uniformly from rest to 60 mi/hr in 5 s. Its acceleration is
    a. 0.083 (mi/hr)/s.   b. 6 (mi/hr)/s.
    c. 12 (mi/hr)/s.   d. 300 (mi/hr)/s.

19. A car accelerates uniformly from 20 mi/hr to 45 mi/hr in 12 min. Its acceleration is
    a. 25 mi/hr²   b. 125 mi/hr²
    c. 250 mi/hr²   d. 300 mi/hr²

20. A car has an acceleration of 15 ft/s². How long does it take the car to go from 10 ft/s to 70 ft/s?
    a. 4.0 s   b. 4.7 s
    c. 60 s   d. 900 s

Exercises

1. Can a rapidly moving body have the same acceleration as a slowly moving one?

2. The acceleration of a certain moving body is constant in magnitude and direction. Must the path of the body be a straight line? If not, give an example.

3. The velocity of light is 3×10^8 m/s. How many minutes does it take light to reach us from the sun, which is 1.5×10^8 km away?

4. How many feet does a car traveling at 35 mi/hr cover in 20 min?

5. A ship travels 400 mi in 1 day and 3 hr. What is its average velocity in mi/hr?

6. A car travels 200 mi in 4 hr and 35 min. What is its average velocity in mi/hr?

7. A boat has a crusing velocity of 9 knots (1 knot = 1 nautical mi/hr). How long will it take to go from New London, Connecticut, to Nantucket, Massachusetts, a distance of 101 nautical miles?

8. Echoes return in 4 s to a man standing in front of a cliff. How far away is the cliff? (The speed of sound in air at sea level is about 330 m/s.)

9. A man hears the sound of thunder 6 sec after seeing a lightning flash. How far away was the lightning? (The speed of light is so great, 3×10^8 m/s, that the man sees the flash a negligible time after it occurs.)

10. A woman observes lightning strike a building 1 km away from her. How long must she wait until she hears the thunder?

11. A car's velocity increases from 8 m/s to 20 m/s in 10 s. Find its acceleration.

12. A car's velocity decreases from 20 m/s to 10 m/s in 5 s. Find its acceleration.

13. A car undergoes an acceleration of 4 m/s² starting from rest. How fast does it go after $\frac{1}{2}$ s? After 5 s?

14. The tires of a certain car begin to lose their grip on the pavement at accelerations greater than about 16 ft/s². Express this acceleration in terms of the number of seconds required to reach a velocity of 60 mi/hr (88 ft/s) starting from rest.

15. The brakes are applied to a car whose initial velocity is 60 ft/s, and the car undergoes an acceleration of −5 ft/s². What is its velocity 6 s later?

16. How long does it take a car whose initial velocity is 20 m/s to slow down to a velocity of 5 m/s when its acceleration is −3 m/s²?

17. Figure 2-9 shows distance-time graphs for nine cars.
    a. Which cars are or have been moving in the forward direction?
    b. Which cars are or have been moving in the backward direction?

c. Which car has the highest constant velocity?
d. Which car has the highest constant velocity in the forward direction?
e. Which car has the highest constant velocity in the backward direction?
f. Which cars have the same velocity?
g. Which car has not moved at all?
h. Which car has been accelerated from rest to a constant velocity?
i. Which car has been brought to a stop from an initial velocity in the forward direction?
j. Which car has been brought to a stop from an initial velocity in the backward direction?

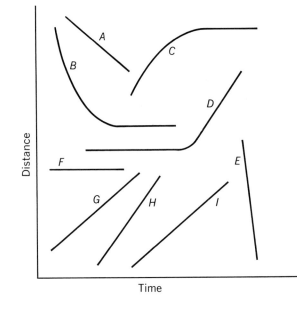

Fig. 2-9.

Problems

1. A car travels at 60 mi/hr along a straight road for 2 hr and then at 40 mi/hr for the next 3 hr. What is the car's average speed for the entire 5 hr?

2. A car travels at 40 mi/hr along a straight road for 60 mi and then at 50 mi/hr for the next 60 mi. What is the car's average speed for the entire 120 mi?

3. (a) What is the speed with respect to the ground of an airplane whose air speed is 200 mi/hr when it is bucking a head wind of 56 mi/hr? (b) What distance can it cover relative to the ground in 5 hr?

4. What is the air speed of an airplane that requires 3 hr and 30 min to go 1200 mi from one city to another when it has a 70 mi/hr tail wind?

5. A body moving along a straight line at 20 ft/s undergoes an acceleration of −4 ft/s². Find its speed 2 s and 20 s after the acceleration has started.

6. A car's speed is increased from 20 mi/hr to 50 mi/hr in 10 s. Find its acceleration in ft/s².

7. The brakes of a car moving at 50 km/hr are suddenly applied, and the car comes to a complete stop in 4 s. (a) What was its acceleration (assumed constant) in (km/hr)/s? (b) in m/s²?

8. A car starts from rest and reaches a speed of 80 km/hr in 20 s. (a) What was its acceleration (assumed constant) in (km/hr)/s? (b) in m/s²?

9. How long would the car of Problem 7 take to come to a complete stop starting from 80 km/hr?

10. How long would the car of Problem 8 take to reach 120 km/hr?

11. A passenger in an airplane flying from New York to San Francisco notes the times at which he passes over various cities and towns. With the help of a map he determines the distances between these landmarks, and compiles the table shown below. Plot the distance covered by the airplane versus time from these data, and describe the airplane's motion with the help of this graph.

Time (P.M.):	4:00	5:22	5:46	5:55
Distance (mi):	0	480	620	680

Time (P.M.):	6:30	6:48	8:32	9:04
Distance (mi):	910	1030	1550	1710

12. A train passes successive mile posts at the times indicated below. Plot the data on a graph, and determine from this whether the train's speed is constant or not. If it is not constant, plot the train's speed in each time interval versus time, determine whether its acceleration is constant, and, if so, what the value of the acceleration is.

Distance (mi):	0	1	2	3	4
Time (min):	0	1.5	2.9	4.2	5.5

Distance (mi):	5	6	7	8
Time (min):	6.7	7.8	8.9	10.0

13. The odometer of a car is checked at 1 min intervals, and the readings below are obtained. Calculate the speed of the car in mi/hr in each time interval and plot the results on a graph. Describe the motion of the car with the help of this graph.

Time (min):	0	1	2	3
Distance (mi):	42.2	42.4	42.8	43.7

Time (min):	4	5	6	7
Distance (mi):	44.6	45.8	47.0	48.2

Time (min):	8	9	10
Distance (mi):	49.4	50.4	51.2

Answers to Multiple Choice

1. a	8. c	15. c
2. c	9. c	16. c
3. c	10. d	17. b
4. b	11. b	18. c
5. a	12. c	19. b
6. b	13. d	20. a
7. c	14. a	

3

Analyzing Motion

Thus far our study of physics has largely dealt with means for organizing various kinds of quantitative information. In Chapter 1 we reviewed the process of measurement and saw how quantities that have specific directions associated with them are treated mathematically. Our concern in the previous chapter was motion; and we found that introducing the quantities speed, velocity, and acceleration to supplement the fundamental ones of distance and time permits us to describe many kinds of motion in a simple, straightforward way. These intellectual tools have a wide range of usefulness, particularly when used in combination, and we shall now consider some further examples that illustrate this range.

3-1 Distance, Time, and Acceleration

A body has some initial velocity v_0 when it begins to be uniformly accelerated. As we saw in Section 2-7, its velocity after a time t is given by

$$v_f = v_0 + at. \tag{2-6}$$

The next question to ask is, how far does the body go during the time interval t?

We know that, in general, $s = \bar{v}t$, so that if we can determine the average velocity during the time interval t we can also find s, the distance through which the body moves. Because the acceleration a is constant, v is changing at a uniform rate, and

$$\bar{v} = \frac{\text{initial velocity} + \text{final velocity}}{2}.$$

(If a is not constant, this formula does not hold.)

Here the initial velocity is v_0 and the final velocity is $v_0 + at$; hence

$$\bar{v} = \frac{v_0 + v_0 + at}{2} = v_0 + \tfrac{1}{2}at.$$

Average velocity under constant acceleration

The distance traveled is accordingly

$$s = \bar{v}t = (v_0 + \tfrac{1}{2}at) \times t,$$

which yields the very useful formula

$$s = v_0 t + \tfrac{1}{2}at^2. \qquad \textit{Distance under constant acceleration} \quad (3\text{-}1)$$

Figure 3-1 illustrates this result. When the initial velocity is $v_0 = 0$, we have simply

$$s = \tfrac{1}{2}at^2 \quad [v_0 = 0]. \tag{3-2}$$

Problem.   A car has an initial velocity of 20 m/s and an acceleration of -1 m/s². Find its displacement after the first 10 s and after the first 50 s from the moment the acceleration begins.

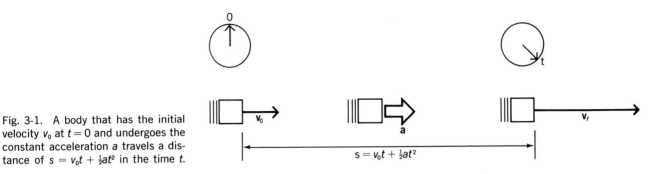

Fig. 3-1. A body that has the initial velocity v_0 at $t = 0$ and undergoes the constant acceleration a travels a distance of $s = v_0 t + \frac{1}{2}at^2$ in the time t.

Solution.   In the first 10 s the car travels

$$s = v_0 t + \tfrac{1}{2}at^2 = 20 \ \frac{\text{m}}{\text{s}} \times 10 \text{ s} - \frac{1}{2} \times 1 \ \frac{\text{m}}{\text{s}^2} \times (10 \text{ s})^2$$

$$= (200 - 50) \text{ m} = 150 \text{ m},$$

and its displacement after 50 s is

$$s = v_0 t + \tfrac{1}{2}at^2 = 20 \ \frac{\text{m}}{\text{s}} \times 50 \text{ s} - \frac{1}{2} \times 1 \ \frac{\text{m}}{\text{s}^2} \times (50 \text{ s})^2$$

$$= (1000 - 1250) \text{ m} = -250 \text{ m}.$$

This result means that the car is 250 m *behind* its starting point after 50 s have elapsed.

3-2 Distance, Velocity, and Acceleration

It is not hard to find a relationship among s, v_0, v_f, and a that does not directly involve the time t. The first step is to rewrite the defining formula for acceleration, which is

$$a = \frac{v_f - v_0}{t}, \tag{2-5}$$

in the equivalent form

$$t = \frac{v_f - v_0}{a}.$$

This formula gives the time during which the velocity of the body changed from v_0 to its final value of v_f. Now we substitute this expression for t into the formula for the distance traveled.

$$s = v_0 t + \tfrac{1}{2} a t^2$$

The result is

$$s = v_0 \frac{(v_f - v_0)}{a} + \frac{1}{2} a \frac{(v_f - v_0)^2}{a^2}$$

$$= \frac{v_0 v_f}{a} - \frac{v_0^2}{a} + \frac{v_f^2}{2a} - \frac{v_0 v_f}{a} + \frac{v_0^2}{2a}$$

$$= \frac{v_f^2}{2a} - \frac{v_0^2}{2a}.$$

Finally, by multiplying both sides of the last equation by $2a$ and rearranging the terms, we arrive at the formula

$$v_f^2 = v_0^2 + 2as. \qquad \textit{Velocity under constant acceleration} \quad (3\text{-}3)$$

Problem.   How far will the car of the previous example have gone when it comes to a stop?

Solution.   When the car is at rest, $v_f = 0$. Since $v_0 = 20$ m/s and $a = -1$ m/s², we have

$$v_f^2 = v_0^2 + 2as,$$

$$s = \frac{v_f^2 - v_0^2}{2a} = \frac{0 - (20 \text{ m/s})^2}{2 \times (-1 \text{ m/s}^2)} = \frac{400 \text{ m}^2/\text{s}^2}{2 \text{ m/s}^2} = 200 \text{ m}.$$

The car comes to a stop after it has gone 200 m. Then, since its acceleration is negative and is assumed here to continue, it begins to move in the negative (backward) direction.

3-3 The Acceleration of Gravity

Drop a stone, and it falls. Does the stone fall at a constant speed, or is it accelerated? Does the motion of the stone depend upon its weight, or its size, or its color? Over two thousand years ago questions such as these were answered by Greek philosophers, notably Aristotle, on the basis of "logical reasoning" only. To them it seemed reasonable that heavy things should fall faster than light things, for example. Almost nobody felt it necessary to perform experiments to seek information on the physical universe until Galileo (1564–1642) revolutionized science by doing just that: performing experiments. Modern science owes its success in understanding and utilizing natural phenomena to its reliance upon experiment.

What Galileo found, as the result of careful measurements made on balls rolling down inclined planes, was that *all freely-falling bodies have the same acceleration* at the same place near the earth's surface. This acceleration, which is called the acceleration of gravity (symbol g), has the value

$$g = 9.8 \text{ m/s}^2$$
$$= 32 \text{ ft/s}^2$$

to two significant figures. The value of g varies slightly around the world because of irregularities in the earth's shape and because of local deposits of materials whose densities are greater or smaller than the average. Because this variation is small, it will be ignored here, though it is important to geologists.

A body falling from rest thus has a velocity of 9.8 m/s (32 ft/s) after the first second, a velocity of 19.6 m/s (64 ft/s) after the next second, and so on. The greater the distance through which a stone falls after being dropped, the greater its velocity when it hits the ground (Fig. 3-2). But the stone's *acceleration* is always the same.

Another aspect of Galileo's work deserves comment. His conclusion that all things fall with the same constant acceleration is only an *idealization* of reality. The actual accelerations with which objects fall depend upon many factors: the location on the earth, the size and shape of the object, and the density and state of the atmosphere. For example, a bullet falls faster than a feather does in air because of the effects of buoyancy and air resistance. Galileo perceived that the essence of the phenomenon was a constant acceleration downward, with other factors acting merely to cause deviations from the constant value. In a vacuum the bullet and feather fall with exactly the same acceleration (Fig. 3-3).

The drag force due to air resistance on an object of given size and shape depends upon the velocity of the object—the faster it goes, the more the drag. In the case of a falling body, the drag force increases as the velocity increases until finally it cannot go any faster. The body then continues to fall at a constant velocity known as its *terminal velocity* (Fig. 3-4). The

The acceleration of gravity near the earth's surface is very nearly the same everywhere

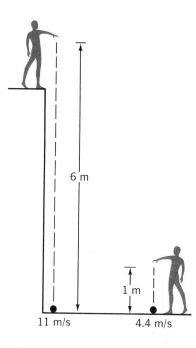

6 m

1 m

11 m/s   4.4 m/s

Fig. 3-2. All freely-falling bodies near the earth's surface have the same acceleration g. Hence, if air resistance can be neglected, the farther a body falls, the greater its final speed.

Terminal velocity

Fig. 3-3. All bodies that fall in a vacuum near the earth's surface have the same downward acceleration.

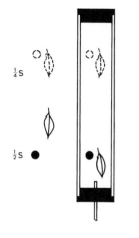

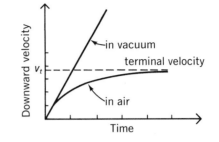

Fig. 3-4. This graph compares how the velocity of a falling body varies with time when it is in a vacuum and when it is in air. In a vacuum the velocity is given by $v = gt$ and increases without limit until the body strikes the ground. In air the terminal velocity of v_t is eventually reached and the body continues to fall at this velocity.

terminal velocity of a man in free fall is about 120 mi/hr, whereas it is only about 14 mi/hr after he has opened his parachute. In the absence of air resistance, raindrops and hailstones would reach the ground at velocities high enough to be dangerous.

3-4 Free Fall

We can apply the formulas we derived earlier for motion under constant acceleration to bodies in free fall. It must be kept in mind, of course, that the direction of **g** is always downward, no matter whether we are dealing with a dropped object or with one that is initially thrown upward.

Problem.   A stone is dropped from the top of the Empire State Building in New York City, which is 450 m high. Neglecting air resistance, how long does it take the stone to reach the ground? What is its speed when it strikes the ground? (See Fig. 3-5.)

Solution.   The general formula for the distance traveled in the time t by an accelerated object is

$$s = v_0 t + \tfrac{1}{2}at^2.$$

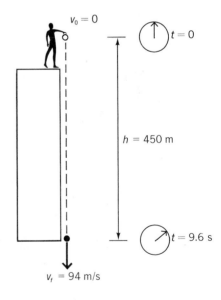

Fig. 3-5.

Here $v_0 = 0$, since the stone is simply dropped with no initial velocity, and the acceleration is $a = g$. Hence we have

$$h = \tfrac{1}{2}gt^2$$

where h represents vertical distance from the starting point. First we solve this formula for t, which yields

$$t = \sqrt{\frac{2h}{g}},$$

and then we substitute $h = 450$ m and $g = 9.8$ m/s² to obtain

$$t = \sqrt{\frac{2h}{g}} = \sqrt{\frac{2 \times 450 \text{ m}}{9.8 \text{ m/s}^2}} = \sqrt{92} \text{ s} = 9.6 \text{ s}.$$

Knowing the duration of the fall makes it simple to compute the stone's final velocity:

$$v_f = v_0 + at = 0 + gt = 9.8 \text{ m/s}^2 \times 9.6 \text{ s} = 94 \text{ m/s}.$$

The magnitude of the velocity which a dropped object has when it reaches the ground is the same as that with which it must be thrown upward from the ground to rise to the same height (Fig. 3-6). To prove this statement, we refer to the formula $v_f^2 = v_0^2 + 2as$ and replace the s with h to give

$$v_f^2 = v_0^2 + 2ah.$$

When a stone is dropped, its acceleration a equals the acceleration of gravity g, the initial velocity v_0 is zero, and so

$$v_f^2 = 0 + 2gh,$$
$$v_f = \sqrt{2gh}.$$

When the stone is thrown upward, on the other hand, $a = -g$ (since the downward acceleration is opposite in direction to the upward initial velocity), and at the top of its path $v = 0$. Hence

$$0 = v_0^2 - 2gh,$$
$$v_0 = \sqrt{2gh}.$$

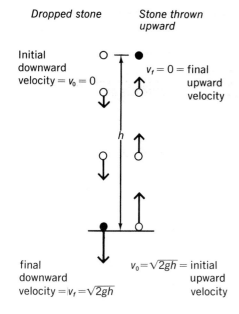

Dropped stone   *Stone thrown upward*

Initial downward velocity $= v_0 = 0$

$v_f = 0 = $ final upward velocity

h

final downward velocity $= v_f = \sqrt{2gh}$

$v_0 = \sqrt{2gh} = $ initial upward velocity

Fig. 3-6.   A stone dropped from a height h reaches the ground with the velocity $\sqrt{2gh}$. In order to reach the height h, a stone thrown upward from the ground must have the minimum velocity $\sqrt{2gh}$.

The velocity has the same magnitude in both cases.

Problem.   A stone is thrown upward with an initial velocity of 50 ft/s. What will its maximum height be? When will it strike the ground? Where will it be in $1\frac{1}{8}$ s? in 2 s?

Solution.   To find the highest point the stone will reach, we rewrite Eq. (3-3) in the form

$$s = \frac{v_f^2 - v_0^2}{2a}.$$

Here $v_0 = 50$ ft/s and $a = -32$ ft/s^2, and at the top of the path $s = h$ and $v_f = 0$. (We are reckoning up as $+$ and down as $-$ in this calculation.) Hence

$$h = \frac{0 - (50 \text{ ft/s})^2}{2(-32 \text{ ft/s}^2)} = \frac{2500 \text{ ft}^2/\text{s}^2}{64 \text{ ft/s}^2} = 39 \text{ ft.}$$

When will the stone strike the ground? It takes a body precisely as long to fall from a height h as it does to rise that high (provided that h is its maximum height, as it is here), just as an object's final speed when dropped from a height h is the same as the initial speed required for it to go that high. From Eq. (3-2),

$$h = \tfrac{1}{2}gt^2$$

$$t = \sqrt{\frac{2h}{g}} = \sqrt{\frac{2 \times 39 \text{ ft}}{32 \text{ ft/s}^2}} = \sqrt{2.44} \text{ s}^2 = 1.6 \text{ s}$$

Because the stone takes as long to rise as to fall, the total time it is in the air is twice 1.6 s, or 3.2 s.

To compute the position of the stone a given time after it has been thrown, we make use of Eq. (3-1),

$$s = v_0 t + \tfrac{1}{2}at^2,$$

rewritten as

$$h = v_0 t - \tfrac{1}{2}gt^2.$$

Substituting $1\frac{1}{8}$ s ($\frac{9}{8}$ s) for t, we have

$$h = 50 \, \frac{\text{ft}}{\text{s}} \times \frac{9}{8} \text{ s} - \frac{32}{2} \, \frac{\text{ft}}{\text{s}^2} \left(\frac{9}{8} \text{ s} \right)^2 = (56 - 20) \text{ ft} = 36 \text{ ft.}$$

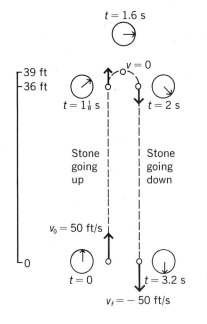

$t = 1.6$ s

$v = 0$

39 ft
36 ft

$t = 1\frac{1}{8}$ s

$t = 2$ s

Stone
going
up

Stone
going
down

$v_0 = 50$ ft/s

0

$t = 0$

$t = 3.2$ s

$v_f = -50$ ft/s

Fig. 3-7. The path of a stone thrown upward with an initial velocity of 50 ft/s. Air resistance is neglected.

When we substitute $t = 2$ s, we find that again

$$h = 50\,\frac{\text{ft}}{\text{s}} \times 2\text{ s} - \frac{32}{2}\,\frac{\text{ft}}{\text{s}^2} \times (2\text{ s})^2 = 36\text{ ft}.$$

All this apparently paradoxical result means is that at $1\frac{1}{8}$ s the stone is at a height of 36 ft on its way up, then it goes on up to its maximum height and returns to 36 ft on its way down (Fig. 3-7).

3-5 Motion in a Vertical Plane

A body that moves through space usually has a curved path rather than a perfectly straight one. Our strategy in attacking problems of this kind is to resolve the body's velocity **v**, whose vector nature we must now take into account, into two components, $\mathbf{v}_x$ in a horizontal direction and $\mathbf{v}_y$ in a vertical direction, and then to examine separately how the body behaves in each of these directions.

Suppose that we drop a ball A from the edge of a table while rolling an identical ball B off to the side (Fig. 3-8). At the moment the balls leave the table, A has zero velocity while B has the horizontal velocity v_0. The velocity components of the balls therefore have the magnitudes

$$v_{Ax} = 0, \qquad v_{Bx} = v_0, \qquad v_{Ay} = 0, \qquad v_{By} = 0.$$

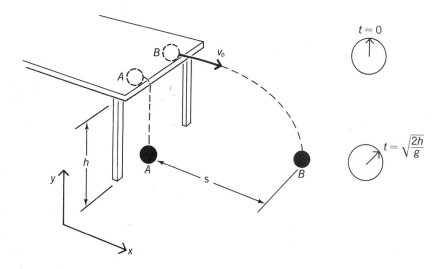

$t = 0$

$t = \sqrt{\dfrac{2h}{g}}$

B

v_0

A

h

y

A

s

B

x

Fig. 3-8. Ball A is dropped from the edge of a table while ball B is simultaneously rolled off the edge with the initial speed v_0.

Both balls reach the floor at the same time, even though B has traveled some distance s away from the table. The reason for this behavior is that the acceleration of gravity is the same for all bodies near the earth regardless of their state of motion; both A and B started out with no vertical velocity, both underwent the same downward acceleration, and so both took the same period of time to fall.

A body in free fall descends the distance

$$h = \tfrac{1}{2}gt^2$$

in the time t when it starts with no vertical component of velocity. Hence both balls require the time

$$t = \sqrt{\frac{2h}{g}}$$

to reach the floor. If the table is 1.0 m high, then

$$t = \sqrt{\frac{2 \times 1.0 \text{ m}}{9.8 \text{ m/s}^2}} = 0.45 \text{ s.}$$

While it is falling, ball B is also moving horizontally with the velocity v_0. When it strikes the floor it will have traveled the horizontal distance

$$s = v_0 t.$$

Let us say that $v_0 = 5.0$ m/s. Therefore

$$s = 5.0 \text{ m/s} \times 0.45 \text{ s} = 2.3 \text{ m.}$$

Problem.   Find the velocities with which balls A and B strike the floor.

Solution.   The final velocity of A has only the single component

$$v_{Ay} = gt,$$

and so

$$v_A = v_{Ay} = 9.8 \, \frac{\text{m}}{\text{s}^2} \times 0.45 \text{ s} = 4.4 \text{ m/s.}$$

The final velocity of B, however, has both horizontal and vertical components, namely

$$v_{Bx} = v_0, \qquad v_{By} = gt.$$

From Fig. 3-9 we see that since $\mathbf{v}_{BX}$ is perpendicular to $\mathbf{v}_{By}$, their vector sum $\mathbf{v}_B$ has the magnitude

$$v_B = \sqrt{v_{Bx}^2 + v_{By}^2} = \sqrt{v_0{}^2 + (gt)^2}.$$

Since $v_0 = 5.0$ m/s and $t = 0.45$ s,

$$v_B = \sqrt{\left(5.0\ \frac{\text{m}}{\text{s}}\right)^2 + \left(9.8\ \frac{\text{m}}{\text{s}^2} \times 0.45\ \text{s}\right)^2} = 6.7\ \text{m/s}.$$

It is worth noting that the *vector sum* of $\mathbf{v}_{Bx}$ and $\mathbf{v}_{By}$ has the magnitude 6.7 m/s, whereas the *algebraic sum* of v_{Bx} and v_{By} is 9.4 m/s. The latter figure is, of course, completely meaningless, since velocity is a vector quantity and velocity addition must obey the rules of vector addition.

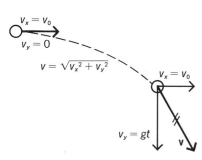

Fig. 3-9. The horizontal and vertical components of the velocity of ball B must be added vectorially to determine the magnitude of its velocity.

3-6 Projectile Flight

A more general case of motion in a vertical plane is exemplified by the flight of a projectile, for instance a rocket. Let us ignore the curvature of the earth and the frictional resistance of the atmosphere to the passage of the rocket, and assume that the rocket uses up its fuel at a distance from its launching point that is small compared with the total distance it travels. If the initial velocity $\mathbf{v}_0$ of the rocket makes an angle of θ with the ground, we can resolve $\mathbf{v}_0$ into the components

$$v_x = v_0 \cos \theta,$$

$$v_y = v_0 \sin \theta.$$

The horizontal velocity component v_x remains constant during the rocket's flight. The vertical component v_y, however, gradually drops to zero due to the downward acceleration of gravity, and then becomes more and more negative (meaning that there is faster and faster motion downward) until the rocket hits the ground (Fig. 3-10).

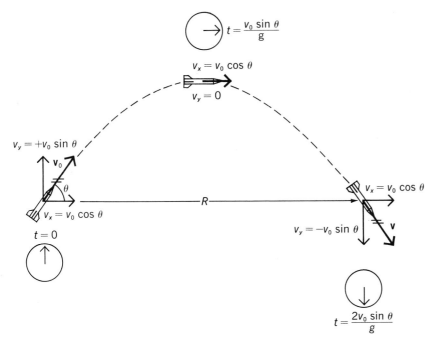

Fig. 3-10. In rocket flight, the horizontal component of a rocket's velocity is constant in the absence of air resistance.

We can once again use the formulas for straight-line motion to discuss the horizontal and vertical aspects of the rocket's motion separately, since these are independent of each other. Let us first calculate the time of flight of the rocket. The rocket will continue to rise until the vertical component of its velocity, given by

Vertical motion of projectile

$$v_y = v_0 \sin \theta - gt,$$

is zero. At this time t,

$$0 = v_0 \sin \theta - gt \qquad \text{and} \qquad t = \frac{v_0 \sin \theta}{g}.$$

The rocket requires the same period of time to return to the ground, and so its total time of flight T is

$$T = 2t = \frac{2v_0 \sin \theta}{g}. \tag{3-4}$$

Horizontal motion of projectile

We can now find the range R of the rocket; that is, we can see how far from its launching point it will strike the ground. Since the horizontal component v_x of the rocket's velocity is constant,

$$R = v_x T$$

$$= v_0 \cos \theta \times \frac{2v_0 \sin \theta}{g}$$

$$= \frac{2v_0{}^2}{g} \sin \theta \cos \theta. \tag{3-5}$$

Equation (3-5) can be simplified by making use of the trigonometric formula

$$\sin \theta \cos \theta = \tfrac{1}{2} \sin 2\theta.$$

The rocket's range may therefore be written

$$R = \frac{v_0{}^2}{g} \sin 2\theta. \tag{3-6}$$

This formula gives the range of a rocket (or any other projectile, for that matter, provided it obeys the restrictions given earlier) in terms of its initial velocity v_0 and the angle θ at which it is launched. We note that R is a maximum when $\sin 2\theta = 1$, since 1 is the highest value the sine function can have. Since $\sin 90° = 1$, the maximum range occurs when the initial angle θ is 45°. Any other angle, greater or smaller, will result in a shorter range (Fig. 3-11).

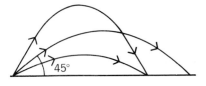

Fig. 3-11.   In the absence of air resistance, the maximum range of a projectile occurs when it is fired at an angle of 45°.

Problem.   An arrow leaves a certain bow with a velocity of 30 m/s. (a) What is its maximum range? (b) At what two angles could the archer point the arrow if it is to reach a target 70 m away?

Solution.   (a) The maximum range is

$$R_{max} = \frac{v_0{}^2}{g} = \frac{(30 \text{ m/s})^2}{9.8 \text{ m/s}^2} = 92 \text{ m.}$$

(b) From the general formula $R = (v_0{}^2/g) \sin 2\theta$ we obtain

$$\sin 2\theta = \frac{Rg}{v_0{}^2} = \frac{70 \text{ m} \times 9.8 \text{ m/s}^2}{(30 \text{ m/s})^2} = 0.762,$$

$$2\theta = 50°, \qquad \theta = 25°.$$

There is also an angle greater than 45° which will give the same range. To find this angle, we refer to Appendix A-5 and find there the trigonometric identity

$$\sin \phi = \sin (180° - \phi)$$

which holds for any angle ϕ. If we let $\phi = 2\theta$, we have

$$\sin 2\theta = \sin (180° - 2\theta)$$

from which we conclude that the two possible angles are θ and $90° - \theta$. The second angle is therefore $90° - 25° = 65°$ (Fig. 3-12).

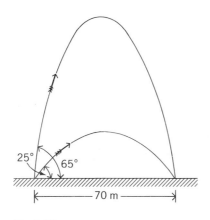

Fig. 3-12.

The examples of this chapter have been worked out not because they are in themselves especially significant, but because they illustrate the power of the mathematical approach to physical phenomena. By defining certain quantities and relating them to each other and to events that actually occur in the real world, a whole theoretical structure of equations may be built up. This structure is an instrument enabling us to solve problems which otherwise would each require a separate, perhaps difficult or impossible, experiment. We must remember that the validity of the theoretical structure depends upon its experimental basis; but once this is established, we may proceed to work out its consequences with pencil and paper.

Special Topic

Velocity and Distance under Constant Acceleration

The formula $s = v_0 t + \frac{1}{2}at^2$ for the distance an object travels in the time t under the constant acceleration a is easy to obtain by integration. We can start from the definition of acceleration as

$$a = \frac{dv}{dt},$$

which we rewrite in differential form as

$$dv = a\, dt.$$

Now we integrate the left-hand side from the initial velocity v_0 to the final velocity v and the right-hand side from the initial time $t = 0$ to the final time t. The result is

$$\int_{v_0}^{v} dv = a \int_{0}^{t} dt,$$
$$v - v_0 = at,$$
$$v = v_0 + at,$$

which is Eq. (2-6).

Now we replace v by ds/dt and multiply both sides of the above equation by dt:

$$v = \frac{ds}{dt} = v_0 + at,$$
$$ds = v_0\, dt + at\, dt.$$

Integrating from the initial position $s = 0$ to the final position s for a total displacement of s, and from the corresponding initial time $t = 0$ to the final time t, we find that

$$\int_0^s ds = v_0 \int_0^t dt + a \int_0^t t \, dt,$$

$$s = v_0 t + \tfrac{1}{2} a t^2$$

since $\int x \, dx = \tfrac{1}{2} x^2$.

Important Term

The **acceleration of gravity** is the acceleration of a freely falling body near the earth's surface. The symbol of the acceleration of gravity is g, and its value is 32 ft/s² in British units and 9.8 m/s² in metric units.

Important Formulas

Distance under constant acceleration:
$$s = v_0 t + \tfrac{1}{2} a t^2$$

Velocity under constant acceleration:
$$v_f^2 = v_0^2 + 2as$$

Free fall from rest:
$$h = \tfrac{1}{2} g t^2$$
$$v_f = \sqrt{2gh}$$

Projectile range:
$$R = \frac{v_0^2}{g} \sin 2\theta$$

[Air resistance is assumed negligible in the following exercises and problems.]

Multiple Choice

1. The acceleration of a stone thrown upward is
   a. greater than that of a stone thrown downward.
   b. the same as that of a stone thrown downward.
   c. smaller than that of a stone thrown downward.
   d. zero until it reaches the highest point in its motion.

2. Ball A is thrown horizontally and ball B is dropped from the same height at the same moment.
   a. Ball A reaches the ground first.
   b. Ball B reaches the ground first.
   c. Ball A has the greater velocity when it reaches the ground.
   d. Ball B has the greater velocity when it reaches the ground.

3. A ball is thrown horizontally from a moving car. While it is in flight (if air resistance is neglected) it is *not* true that
   a. its velocity changes.
   b. its acceleration changes.
   c. its direction of motion relative to the car changes.
   d. its direction of motion relative to the road changes.

4. A rocket is fired vertically upward. After its engine has used up its fuel, the rocket always
   a. begins to fall.
   b. ceases to be accelerated.
   c. is accelerated downward.
   d. becomes an earth satellite.

5. The acceleration of gravity is
   a. 3.2×10^{0} ft/s².
   b. 3.2×10^{1} ft/s².
   c. 3.2×10^{2} ft/s².
   d. 3.2×10^{3} ft/s².

6. A car undergoes a constant acceleration of 6 m/s² starting from rest. In the first second it travels
   a. 3 m.   b. 6 m.
   c. 18 m.   d. 36 m.

7. An airplane requires 20 s and 400 m of runway to become airborne, starting from rest. Its velocity when it leaves the ground is
   a. 20 m/s.   b. 32 m/s.
   c. 40 m/s.   d. 80 m/s.

8. A car has an initial speed of 50 ft/s and an acceleration of 4 ft/s². In the first 10 s after the acceleration begins, the car travels
   a. 200 ft.   b. 500 ft.
   c. 700 ft.   d. 900 ft.

9. A car has an initial speed of 50 ft/s and an acceleration of −4 ft/s². In the first 10 s after the acceleration begins, the car travels
   a. 100 ft.   b. 300 ft.
   c. 500 ft.   d. 700 ft.

10. A car has an initial speed of 80 ft/s when an acceleration of −5 ft/s² begins. How far will the car go before coming to a stop?
    a. 8 ft   b. 16 ft
    c. 80 ft   d. 640 ft

11. Two balls are thrown vertically upward, one with an initial velocity twice that of the other. The ball with the greater initial velocity will reach a height
    a. $\sqrt{2}$ that of the other.
    b. twice that of the other.
    c. 4 times that of the other.
    d. 8 times that of the other.

12. A wheel falls from an airplane flying horizontally at an altitude of 490 m. If there were no air resistance, the wheel would strike the ground in
    a. 10 s.   b. 50 s.
    c. 80 s.   d. 100 s.

13. The wheel of question 12 will strike the ground with a velocity of
    a. 49 m/s   b. 98 m/s
    c. 490 m/s   d. 9604 m/s

14. A stone is dropped from a cliff. After it has fallen 30 m its velocity is
    a. 17 m/s   b. 24 m/s
    c. 44 m/s   d. 588 m/s

15. A ball thrown vertically upward at 25 m/s continues to rise for approximately
    a. 2.5 s.   b. 5 s.
    c. 7.5 s.   d. 10 s.

16. In question 15, how much time will elapse before the ball strikes the ground?
    a. 2.5 s   b. 5 s
    c. 7.5 s   d. 10 s

17. A ball is rolled off the edge of a table at 3 ft/sec. After $\frac{1}{8}$ sec the ball's velocity is
    a. 3 ft/s.   b. 4 ft/s.
    c. 5 ft/s.   d. 7 ft/s.

18. A ball is thrown at a 30° angle above the horizontal with a velocity of 10 ft/s. After $\frac{1}{2}$ s the horizontal component of its velocity will be
    a. 9 ft/s.   b. 10 ft/s.
    c. 12 ft/s.   d. 19 ft/s.

19. In the absence of winds and air resistance, a projectile has its maximum range when fired at an angle with the ground
    a. of 30°.
    b. of 45°.
    c. of 60°.
    d. that depends upon the initial velocity of the projectile.

Exercises

1. A hunter aims a rifle directly at a squirrel on a branch of a tree. The squirrel sees the flash of the rifle's firing. Should it stay where it is or drop from the branch in free fall at the instant the rifle is fired?

2. A man at the masthead of a sailboat moving at constant velocity drops a wrench. The man is 60 ft above the boat's deck at the time, and the stern of the boat is 36 ft aft of the mast. The effect of air resistance on the motion of the wrench is negligible. Is there a minimum velocity the sailboat can have such that the wrench does not land on the deck? If so, what is this speed?

3. Is it true that an object dropped from rest falls three times farther in the second second after being released than it does in the first second?

4. A movie is shown that appears to be of a ball falling through the air. Is there any way to find out from what appears on the screen if the movie is actually of a ball being thrown upward but the film is being run backwards in the projector?

5. Find the acceleration of a car that slows down to a stop from a velocity of 50 ft/s in a distance of 100 ft.

6. A truck starts from rest and rolls down a hill with constant acceleration. It travels a distance of 400 m in the first 20 s. Find the acceleration and the velocity of the truck after 20 s.

7. An object undergoes an acceleration of 8 m/s² starting from rest. How far does it go in the first 0.1 s?

8. An object undergoes an acceleration of −5 m/s² from a velocity of 20 m/s. How far does it go in the first second of the acceleration?

9. In order to take off, a certain airplane must have a velocity of 50 m/s. (a) If the runway is 500 m long, what is the minimum acceleration the airplane must have? (b) How much time does the airplane spend in becoming airborne?

10. A typical DC-8 airplane has a takeoff velocity of 174 mi/hr which it reaches 35 s after starting from rest. (a) What is its average acceleration? (b) What is the minimum length the runway must have?

11. A Jaguar XJ-6 car can cover ¼ mi in 16.3 s from a standing start. What is its acceleration in ft/s²?

12. A car whose initial velocity is 40 km/hr receives an acceleration of 5(km/hr)/s. How far will it go

from the point where the acceleration is applied until its velocity is 100 km/hr?

13. The brakes of a particular car are capable of producing an acceleration of −20 ft/s². How far will the car go in the course of slowing down from 90 ft/s to 30 ft/s?

14. A body is moving at the constant acceleration of 2 ft/s². How much time is required for its velocity to increase from 10 ft/s to 16 ft/s? How far does it go while its velocity increases from 10 ft/s to 16 ft/s?

15. Divers in Acapulco, Mexico, leap from a point 36 m above the water. If there were no air resistance, what would their final speed be?

16. A stone is dropped from a cliff 490 m above the ground. How long does it take to fall?

17. A ball is dropped from the roof of a building and takes 4 s to reach the ground. Find the height of the building in m.

18. Paul Wilson set a world record with a pole vault of 17.7 ft. (a) With what velocity did he strike the ground afterward? (b) How much time did the fall take? Consider Mr. Wilson as a point object in these calculations.

19. A stone is dropped from a cliff 64 ft high. (a) How long will it take to fall to the foot of the cliff? (b) What will its speed be when it strikes the ground?

20. A stone is thrown vertically upward at 9.8 m/s. When will it reach the ground?

21. A ball is thrown vertically downward with an initial velocity of 10 m/s. What is its speed 1 s later? 2 s later?

22. A ball is thrown vertically upward with an initial velocity of 10 m/s. What is its speed and direction 1 s later? 2 s later?

23. A lead pellet is propelled vertically upward by an air rifle so that its initial velocity is 16 ft/s. Find its maximum altitude.

24. A ball is thrown upward and reaches a height of 80 ft. Find its initial velocity.

25. Does the speed of a projectile sent off at a 45° angle of elevation vary in its path? If so, where is the speed greatest and where is it least?

26. What effect does doubling the initial speed of a rocket have on its range (neglecting air resistance)?

27. The longest recorded distance a champagne cork has flown is 34 ft. What was the minimum initial velocity of this cork?

28. In April 1959, Miss Victoria Zacchini was fired 155 ft from a cannon in Madison Square Garden, New York City. What was the minimum muzzle velocity of the cannon in mi/hr?

29. A football leaves the toe of a punter at an angle of 50° above the horizontal. What was its minimum initial velocity if it travels 40 yards?

Problems

1. The engineer of a train traveling at 80 mi/hr applies the brakes when he passes an amber signal. The next signal is one mile down the track and the train reaches it 75 s later. Assuming a uniform deceleration, find the velocity of the train at the second signal.

2. An express train passes a certain station at 20 m/s. The next station is 2 km away and the train reaches it 1 min. later. (a) Did the train's velocity change? (b) If it did, what was its velocity at the second station, assuming a constant acceleration?

3. A stone is dropped in a well whose water level is 100 ft down. How much time elapses until the sound of the splash is heard? Assume the speed of sound to be 1100 ft/s.

4. A boy throws a ball to a height of 20 m. (a) How long a period elapses between the time the ball leaves the boy's hand and the moment he catches it on the way down? (b) With what velocity does he have to throw the ball?

5. A girl throws a ball vertically upward with a velocity of 10 m/s from the roof of a building 20 m high. (a) How long will it take the ball to reach the ground? (b) What will its velocity be when it strikes the ground?

6. A girl throws a ball vertically downward with a velocity of 10 m/s from the roof of a building 20 m high. (a) How long will it take the ball to reach the ground? (b) What will its velocity be when it strikes the ground?

7. A British parachutist bails out at an altitude of 500 ft and accidentally drops his monocle. If he descends at the constant velocity of 20 ft/s, how much time separates the arrival of the monocle on the ground from the arrival of the parachutist himself?

8. A Russian balloonist floating at an altitude of 500 ft accidentally drops his samovar and starts to ascend at the constant velocity of 4 ft/s. How high will the balloon be when the samovar reaches the ground?

9. A bouncing ball reaches a maximum height of 1 m. How much time does each complete up-and-down cycle take?

10. An orangutan throws a coconut vertically upward at the foot of a cliff 40 m high while his mate simultaneously drops another coconut from the top of the cliff. The two coconuts collide at an altitude of 20 m. What was the initial velocity of the coconut that was thrown upward?

11. A girl throws a ball to a height of 20 m. (a) How long a period elapses between the time the ball leaves the girl's hand and the moment she catches it on the way down? (b) With what velocity did she throw the ball?

12. An elevator has a maximum acceleration of ± 5 ft/s² and a maximum velocity of 15 ft/s. Find the shortest period of time required for it to take a passenger to the tenth floor of a building from street level, a height of 150 ft, with the elevator coming to a stop at this floor.

13. An elevator has a maximum acceleration of ±1.5 m/s² and a maximum velocity of 6 m/s. Find the shortest period of time required for it to take a passenger to the tenth floor of a building from street level, a height of 50 m.

14. A man in an elevator drops an apple 6 ft above the elevator's floor. With a stop watch he deter-

mines how long the apple takes to fall to the floor. What does he find (a) when the elevator is ascending with an acceleration of 2 ft/s²; (b) when it is descending with an acceleration of 2 ft/s²; (c) when it is ascending at the constant velocity of 8 ft/s; (d) when it is descending at the constant velocity of 8 ft/s; (e) when the cable has broken and it is descending in free fall?

15. A ball is thrown horizontally from the roof of a building 60 ft high at a velocity of 100 ft/s. What will the magnitude and direction of the ball's velocity be when it strikes the ground?

16. A ball is rolled off the edge of a table at a velocity of 1 m/s. What is the magnitude of the ball's velocity 0.1 s later?

17. A ball is thrown horizontally toward the north from a rooftop at a velocity of 8 m/s. A 10-m/s wind is blowing from the east. (a) What is the velocity of the ball relative to the ground after 2 s? (b) What angle does its velocity make relative to the vertical at this time? (c) What angle does its velocity make relative to due north at this time?

18. A bomb is dropped by an airplane in level flight whose velocity is 400 mi/hr when it is 12,000 ft above the ground. Where will the bomb strike the ground relative to the location above which it was dropped?

19. A blunderbuss can fire a slug 300 ft vertically upward. (a) What is its maximum horizontal range? (b) With what velocities will the slug strike the ground when fired upward and when fired so as to have maximum range?

20. A golf ball leaves a tee at 60 m/s and strikes the ground 200 m away. At what two angles with the horizontal could it have begun its flight? Find the time of flight in each case.

21. A shell is fired at a velocity of 300 m/s at an angle of 30° above the horizontal. (a) How far does it go? What is its time of flight? (b) At what other angle could the shell have been fired to have the same range? What would its time of flight have been in this case?

Answers to Multiple Choice

1. b	8. c	15. a
2. c	9. b	16. b
3. b	10. d	17. c
4. c	11. c	18. a
5. b	12. a	19. b
6. a	13. b	
7. c	14. b	

4

Force and Motion

Thus far we have only discussed how motion is described mathematically. But why does anything move in the first place? Why do some things move faster than others? Why are some accelerated and others not? How can a body traveling in one direction be accelerated in the opposite direction? All these are reasonable questions and ones we must be able to answer if, as scientists, we are to understand the factors at work in the world around us that produce the physical phenomena we observe, and if, as technologists, we are to harness these factors to meet our needs. Almost three centuries ago Isaac Newton (1642 – 1727) formulated three principles based upon observations he and others had made which summarize so much of the behavior of moving bodies that they have become known as the laws of motion. These laws form the subject of this chapter.

4-1 The First Law of Motion

In everyday life it is a familiar observation that stationary objects tend to remain stationary and moving objects tend to continue moving. A certain amount of effort is needed to start a cart moving on a level road, and once in motion a certain amount of effort is also required to bring it to a stop. Of course, in the case of a cart friction is a factor in its reluctance to begin to move, but friction is only part of the story. Even without friction, a cart at rest on a level road will remain at rest unless something pushes it. And without friction, a cart moving along a level road at a certain velocity will continue to move at that velocity indefinitely.

Newton's first law of motion is a statement of the above behavior:

A body at rest will remain at rest and a body in motion will continue in motion in a straight line at constant velocity in the absence of any interaction with the rest of the universe.

First law of motion

We might object that, although everyday experience does indicate that bodies in motion *tend* to remain in motion along a straight line at constant velocity, sooner or later they invariably come to a stop and often deviate from a straight path as well. Even celestial bodies such as the sun, moon, and planets are neither at rest nor pursue straight paths. But these observations do not invalidate the first law of motion; they merely emphasize how difficult it is to avoid interactions between a body and its environment. A cart rolling along a smooth, perfectly level road will not continue forever owing to friction and air resistance, but we are at liberty to imagine what would happen if the air were to disappear and the friction were to vanish.

The resistance a body offers to any change in its state of rest or uniform motion is a property of matter known as *inertia*. When a bus suddenly starts to move, standing passengers seem to be pushed backward (Fig. 4-1). What

Inertia

Fig. 4-1. (a) When a bus suddenly starts to move, the inertia of the passengers tends to keep them at rest relative to the earth. (b) When a bus comes to a sudden stop, their inertia tends to keep them moving. Both effects illustrate the first law of motion.

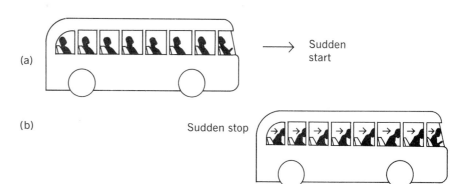

(a)

Sudden start

(b)

Sudden stop

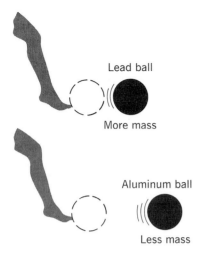

Lead ball

More mass

Aluminum ball

Less mass

Fig. 4-2. The mass of an object determines its inertia.

An operational definition of mass

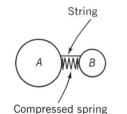

String

A [WWW] B

Compressed spring

v_A A   B v_B

Fig. 4-3. An experiment to compare the masses of two bodies. When the string is cut, the body with the greater mass moves off with the smaller velocity.

is actually happening is that inertia tends to keep their bodies in place relative to the earth while the bus carries their feet forward. When a bus suddenly stops, on the other hand, standing passengers seem to be pushed forward. What is actually happening is that inertia tends to keep their bodies moving while the bus has come to a halt.

4-2 Mass

A quantitative measure of the inertia of a body at rest is its *mass*. The greater the resistance a body offers to being set in motion, the greater its mass. The inertia of a lead ball exceeds that of an aluminum ball of the same size, as we can tell by kicking them in turn, so the mass of the lead ball exceeds that of the aluminum one (Fig. 4-2).

We can arrive at a precise definition of mass by using a simple experiment to compare the inertias of two bodies. What we do is put a small spring between them, push them together so the spring is compressed, and tie a string between them to hold the assembly in place (Fig. 4-3). Now we cut the string. The compressed spring pushes the bodies apart, and body A flies off to the left at the velocity v_A while body B flies off to the right at the velocity v_B. Body A has a lower velocity than body B, and we interpret this difference to mean that A exhibits more inertia than B.

We repeat the experiment a number of times using springs of different stiffness, so that the recoil velocities are different in each case. What we find is that each time body A moves more slowly than body B, and that, regardless of the exact values of v_A and v_B, their *ratio*

$$\frac{v_B}{v_A}$$

is always the same.

The fact that the velocity ratio v_B/v_A is constant gives us a way to specify what we mean by mass unambiguously. If we denote the masses of bodies A and B by the symbols m_A and m_B respectively, we define the ratio of these masses to be

$$\frac{m_A}{m_B} = \frac{v_B}{v_A}. \tag{4-1}$$

The body with the greater mass has the lower velocity, and vice versa. The above procedure provides a definite, experimental method for finding the ratio between the masses of any two bodies—it is an *operational definition*.

The next step is to select a certain object to serve as a standard unit of mass. By international agreement this object is a platinum cylinder at Sèvres, France, called the *standard kilogram*. The mass of any other body in the world can be determined by a recoil experiment with the standard kilogram. (There are easier ways to measure mass, needless to say, but we are interested in basic principles for the present.)

Problem. In a recoil experiment with the standard kilogram, a body of unknown mass moves off at a velocity of 2 m/s and the standard kilogram moves off at a velocity of 6 m/s (Fig. 4-4). What is the mass of the body?

Solution. If we call the body of unknown mass A and the standard kilogram B, then $m_B = 1$ kg, $v_A = 2$ m/s, and $v_B = 6$ m/s. We find that

$$m_A = 1 \text{ kg} \times \frac{6 \text{ m/s}}{2 \text{ m/s}}$$

$$= 3 \text{ kg}.$$

Standard kilogram

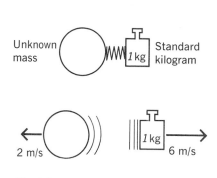

Fig. 4-4.

Why is it necessary to define mass in this seemingly roundabout way? After all, everybody knows that the mass of an object refers to the amount of matter it contains. The trouble is that "amount of matter" is a nebulous concept: it could refer to an object's volume, to the number of atoms it contains, or to yet other properties. It has proven most fruitful to choose the inertia of an object as a measure of the quantity of matter it contains, and to define the object's mass in terms of this inertia as manifested in an appropriate experiment.

In the British system of units the *slug* rather than the kilogram is the standard unit of mass. The slug and its relation to the pound (which is a unit of force, not of mass) will be discussed later in this chapter.

4-3 Force

According to the first law of motion, the velocity of a body (which may be 0) remains constant as long as the body is isolated from the rest of the universe. When the body interacts with something else, its velocity may change. We can interact with a football by kicking it, and the result is a change in the football's velocity from $v_1 = 0$ to some value v_2 (Fig. 4-5). Or the interaction can take the form of catching a moving football, in which case again the football's velocity changes. An interaction can lead to a change in the *direction* of v as well as to a change in its magnitude, as for instance when a football bounces off a tree.

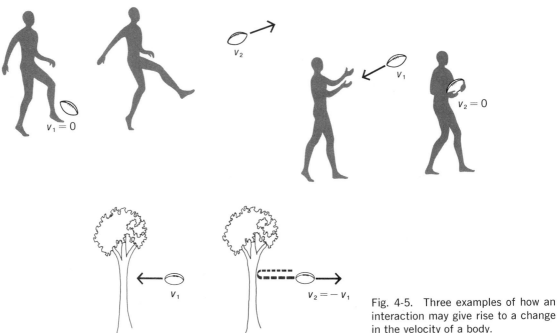

Fig. 4-5. Three examples of how an interaction may give rise to a change in the velocity of a body.

Some interactions are more effective than others in causing velocity changes. A swift kick affects the velocity of a football more than a gentle tap does. The concept of *force* can be used to put the matter on a precise basis. In general,

A force is any influence that can produce a change in the velocity of a body.

This definition is in accord with the notion of a force as a "push" or a "pull," but it goes further since no direct contact is necessarily implied. No hand reaches up from the earth to pull a dropped stone downward, yet the increasing downward velocity of the falling stone testifies to the action of a force upon it (Fig. 4-6).

It is entirely possible for two or more forces to act upon a body without affecting its state of motion; the forces may be such as to cancel one another out. What is required for a velocity change is a *net force*, often called an *unbalanced force*. When a body is acted upon by several forces whose vector sum is zero, the forces are said to be *balanced forces* and the body is then in *equilibrium* (Fig. 4-7). But each of the forces acting by itself must be capable of accelerating the body.

Force

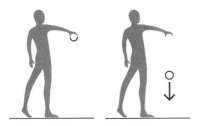

Fig. 4-6. A downward force exerted by the earth causes dropped objects to fall.

Balanced and unbalanced forces

Fig. 4-7. When several forces act on a body, they may cancel one another out to leave no net force.

We can therefore restate the first law of motion in terms of net force:

In the absence of a net force acting on it, a body at rest will remain at rest and a body in motion will continue in motion at constant velocity.

Every force, without exception, arises from one or another of the four fundamental interactions that are possible between the elementary particles of which all matter consists. Two of these interactions are effective only when the particles involved are extremely close together. These are called the "strong" and "weak" *nuclear interactions* because they are responsible for the ability of protons and neutrons to stick together to form atomic nuclei, and one is more powerful than the other. The others—the *gravitational* and *electromagnetic interactions*—are unlimited in range, but their strength decreases with distance. (The electromagnetic and weak nuclear interactions may well be different aspects of the same fundamental interaction, bringing the total number down to three.) In later sections we shall explore the properties of these various interactions.

The four fundamental forces

4-4 The Second Law of Motion

In the *second law of motion* we have a quantitative definition of force:

The net force acting upon a body is equal to the product of the mass and the acceleration of the body; the direction of the force is the same as that of the body's acceleration.

Second law of motion

In equation form,

$$\mathbf{F} = m\mathbf{a} \qquad \textit{Second law of motion} \quad (4\text{-}2)$$

Force = mass × acceleration.

The second law of motion provides us with a way to analyze and compare forces in terms of the accelerations they give rise to. Thus a force which causes a body to have twice the acceleration another force produces must be twice as great as the other one (Fig. 4-8). A body moving to the right but going slower and slower is accelerated to the left. Hence there is a force toward the left acting on it (Fig. 4-9). The first law of motion is clearly a special case of the second: when the net force on a body is zero, its acceleration is also zero.

The second law is in accord with the definition of mass given earlier since the smaller the mass of an object acted upon by a given force, the greater its acceleration and hence final velocity (Fig. 4-10).

It is convenient to have a special unit for force. In the SI system of units the appropriate unit is the *newton* (abbreviated N):

The newton

A newton is that force which, when applied to a 1-kg mass, gives it an acceleration of 1 m/s².

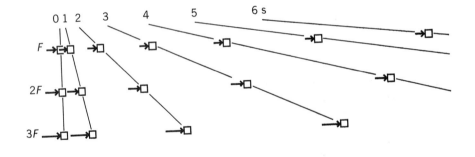

Fig. 4-8. The acceleration of a body is proportional to the net force applied to it. Successive positions of a block are shown at 1-s intervals while forces of *F*, *2F*, and *3F* are applied.

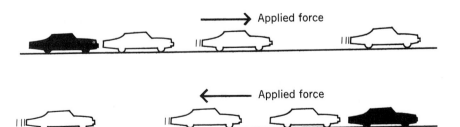

Fig. 4-9. A force and the acceleration it produces are always in the same direction.

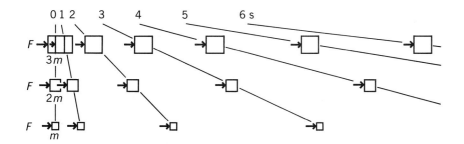

Fig. 4-10. When the same force is applied to bodies of different masses, the resulting accelerations are inversely proportional to the masses. Successive positions of blocks of mass m, $2m$, and $3m$ are shown at 1-s intervals while identical forces of F are applied.

The newton is not a fundamental unit like the meter, second, and kilogram, and in some calculations it may have to be replaced by its equivalent in terms of the latter. This equivalent may be found as follows:

$$F = ma$$

$$1 \text{ N} = 1 \text{ kg} \times 1 \frac{\text{m}}{\text{s}^2}$$

$$= 1 \frac{\text{kg-m}}{\text{s}^2}.$$

A newton is equivalent to 0.225 lb, a little less than $\frac{1}{4}$ lb.

Let us examine a few problems in order to become familiar with the application of the second law of motion. Some of these problems are, of course, artificial and oversimplified, but they do illustrate the power of the second law in situations involving force and motion.

Problem. A 60-g tennis ball approaches a racket at 30 m/s, is in contact with the racket's strings for 5 ms (1 ms = 1 millisecond = 10^{-3} s), and then rebounds at 30 m/s (Fig. 4-11). What was the average force the racket exerted on the ball?

Solution. The tennis ball experienced a velocity change of

$$\Delta v = v_f - v_0 = (-30 \text{ m/s}) - (30 \text{ m/s}) = -60 \text{ m/s}$$

so its acceleration was

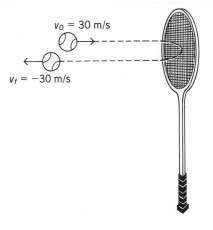

Fig. 4-11.

$$a = \frac{\Delta v}{\Delta t} = \frac{-60 \text{ m/s}}{5 \times 10^{-3} \text{ s}} = -1.2 \times 10^4 \text{ m/s}^2.$$

The corresponding force is, since $60\,g = 0.060$ kg,

$$F = ma = -0.060 \text{ kg} \times 1.2 \times 10^4 \text{ m/s}^2 = -720 \text{ N}.$$

The minus sign means that the force was in the opposite direction to that of the ball when it approached the racket. The British equivalent of 720 N is 162 lb.

Problem.   A force of 10 N is applied to a 4.0-kg block that is at rest on a perfectly smooth, level surface. Find the velocity of the block and how far it has gone after 6.0 s.

Solution.   We start from the second law of motion in scalar form, since only one direction is involved here:

$$F = ma.$$

We know what F and m are, so we find the acceleration a of the block as follows:

$$a = \frac{F}{m}$$

$$= \frac{10 \text{ N}}{4.0 \text{ kg}} = \frac{10 \text{ kg-m/s}^2}{4.0 \text{ kg}} = 2.5 \frac{\text{m}}{\text{s}^2}.$$

The direction of the acceleration is the same as that of the force.
    To find the velocity of the block after $t = 6.0$ s, we use the formula $v = at$ and obtain

$$v = at = 2.5 \frac{\text{m}}{\text{s}^2} \times 6.0 \text{ s} = 15 \frac{\text{m}}{\text{s}}.$$

    For the distance the block travels in $t = 6.0$ s at an acceleration of $a = 2.5$ m/s^2 we require the formula

$$s = v_0 t + \tfrac{1}{2}at^2. \tag{3-1}$$

The block started from rest, so $v_0 = 0$ and we have

$$s = \tfrac{1}{2}at^2$$

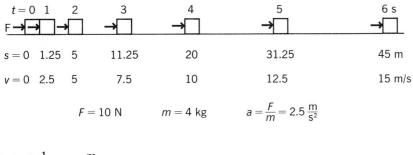

$$= \frac{1}{2} \times 2.5 \, \frac{m}{s^2} \times (6.0 \text{ s})^2$$

$$= 45 \text{ m}.$$

After 6.0 s a 4.0-kg mass acted upon by a 10-N force will have gone 45 m and have a velocity of 15 m/s (Fig. 4-12).

Problem. Next let us consider the same block sliding on a level but slightly rough surface without the 10-newton force acting. The block starts at a velocity of 15 m/s, and gradually slows down and stops at a distance of 20 m. What was the frictional resistive force that acted?

Solution. We start by determining the acceleration of the block. Since we know that its velocity has gone from 15 m/s to 0 in 20 m, we use the formula

$$v_f^2 = v_0^2 + 2as \tag{3-3}$$

from Chapter 3. Here the final velocity is $v_f = 0$, the initial velocity is $v_0 = 15$ m/s, and the distance is $s = 20$ m. Hence

$$v_f^2 = v_0^2 + 2as,$$

$$a = \frac{v_f^2 - v_0^2}{2s}$$

$$= \frac{0 - (15 \text{ m/s})^2}{2 \times 20 \text{ m}} = -5.6 \, \frac{m}{s^2}.$$

The minus sign indicates a negative acceleration, that is, an acceleration directed opposite to the velocity of the block. Now that we know the mass and the acceleration of the block, it is easy to find the resistive force:

$$F = ma = 4.0 \text{ kg} \times \left(-5.6 \, \frac{m}{s^2}\right) = -22 \text{ N}.$$

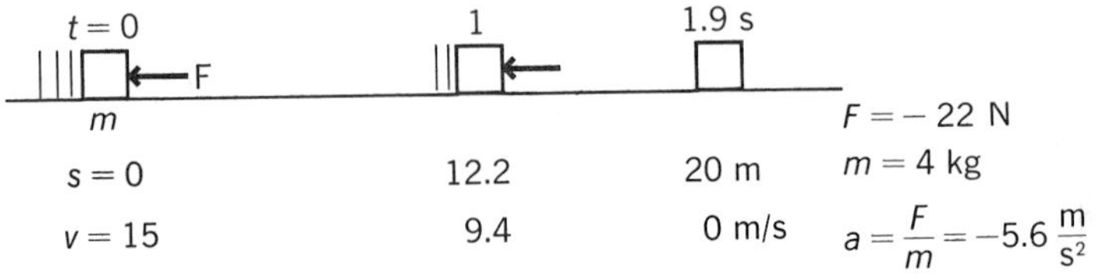

Fig. 4-13. A mass of 4 kg with an initial speed of 15 m/s comes to a stop in a distance of 20 m when a resistive force of 22 N acts on it.

The minus sign indicates that the direction of the force is opposite to that of the block's velocity (Fig. 4-13).

4-5 Weight

Weight is a force

The force with which an object is attracted to the earth is called its *weight*. Weight is different from mass, which is a measure of the inertia an object exhibits. Although they are different physical quantities, mass and weight are closely related.

The weight of a stone is the force that causes it to be accelerated when it is dropped. All objects in free fall near the earth's surface have a downward acceleration of $g = 9.8$ m/s², the acceleration of gravity. (We continue to ignore the small variation in g with geographic position.) If the stone's mass is m, then the downward force on it, which is its weight w, can be found from the second law of motion, $F = ma$, by letting $F = w$ and $a = g$. Evidently

$$w = mg \qquad\qquad\qquad \textit{Weight} \quad (4\text{-}3)$$

Weight is proportional to mass

Weight = mass × acceleration of gravity.

The weight of any object is equal to its mass multiplied by the acceleration of gravity. Since g is a constant near the earth's surface, the weight w of an object is always directly proportional to its mass m: a large mass is heavier than a small one.

The mass of an object is a more fundamental property than its weight, because its mass when at rest is the same everywhere in the universe whereas the gravitational force on it depends upon its position relative to the earth or to some other astronomical body. A 100-kg man weighs 980 N on the earth, but he would weigh 2587 N on Jupiter, 372 N on Mars, 162 N on the moon, and 0 in space far from the sun and other stars. (The mass of an object varies with its velocity with respect to an observer. This effect is significant only at velocities approaching that of light, 3×10^8 m/s, and is discussed in Chap. 27.)

Problem. How much force is needed to lift a 5.0-kg box from the ground? How much force is needed to give it an upward acceleration of 2.0 m/s²?

Solution. The weight of the box is

$$w = mg = 5.0 \text{ kg} \times 9.8 \, \frac{\text{m}}{\text{s}^2} = 49 \text{ N}.$$

Therefore an upward force of 49 N is needed to lift the box from the ground.

To give the box an upward acceleration, we must apply a force F greater than its weight w in order that there be a *net* upward force (Fig. 4-14). The net upward force required here is

$$F - w = ma = 5.0 \text{ kg} \times 2.0 \, \frac{\text{m}}{\text{s}^2} = 10 \text{ N},$$

and so the applied force must be

$$F = w + ma$$
$$= 49 \text{ N} + 10 \text{ N} = 59 \text{ N}.$$

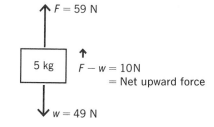

Fig. 4-14.   An upward force of 59 N must be applied to a 5-kg box in order to give it an upward acceleration of 2 m/s².

Problem. A box slides down a frictionless plane that is 20° above the horizontal. Find the acceleration of the box.

Solution. The force F that accelerates the box is the component of its weight w parallel to the plane. From Fig. 1-13 the magnitude of this force is

$$F = w \sin \theta$$

and so, since $m = w/g$, the box's acceleration is

$$a = \frac{F}{m} = \frac{F}{w/g} = \frac{w \sin \theta}{w/g} = g \sin \theta = 9.8 \text{ m/s}^2 \times \sin 20° = 3.4 \text{ m/s}^2.$$

The same answer, of course, would be obtained by finding the component of g parallel to the plane, but the above procedure is the more general one since it can be used when other forces, such as friction, are also acting.

4-6 British Units of Mass and Force

In the British system the unit of mass is the *slug* and the unit of force is the *pound* (lb). A body whose mass is 1 slug experiences an acceleration of 1 ft/s² when a net force of 1 lb acts on it. Thus

Slug and pound

$$1 \text{ lb} = 1 \, \frac{\text{slug-ft}}{\text{s}^2}.$$

Table 4-1. Units of mass and weight.

System of units	Unit of mass	Unit of weight	Acceleration of gravity g	To find mass m given weight w	To find weight w given mass m
Metric	Kilogram (kg)	Newton (N)	9.8 m/s²	$m \text{ (kg)} = \dfrac{w \text{ (N)}}{9.8 \text{ m/s}^2}$	$w \text{ (N)} = m \text{ (kg)} \times 9.8 \text{ m/s}^2$
British	Slug	Pound (lb)	32 ft/s²	$m \text{ (slugs)} = \dfrac{w \text{ (lb)}}{32 \text{ ft/s}^2}$	$w \text{ (lb)} = m \text{ (slugs)} \times 32 \text{ ft/s}^2$

Conversion of units: 1 slug = 14.6 kg 1 newton = 0.225 lb
 1 kg = 0.0685 slug 1 lb = 4.45 newtons

The slug is an unfamiliar unit because in everyday life weights rather than masses are specified in the British system: we go shopping for 10 lb of apples, not $\frac{1}{3}$ slug of apples. In the metric system, on the other hand, masses are normally specified: European grocery scales are calibrated in kilograms, not in newtons.

In order to convert a weight in pounds to a mass in slugs we make use of Eq. (4-3) to obtain

$$m \text{ (slugs)} = \frac{w \text{ (lb)}}{g \text{ (ft/s}^2)}.$$

Since g, the acceleration of gravity at the earth's surface, has the value 32 ft/s²,

$$m \text{ (slugs)} = \frac{w \text{ (lb)}}{32 \text{ ft/s}^2}. \qquad (4\text{-}4)$$

The *weight* of a 1-slug mass is 32 lb, and the *mass* of a 1-lb weight is $\frac{1}{32}$ slug (Table 4-1).

Kilograms and pounds   1 kilogram corresponds to 2.21 pounds in the sense that the weight of 1 kilogram is 2.21 pounds.

1 pound corresponds to 0.454 kilogram in the sense that the mass of 1 pound is 0.454 kilogram.

Problem.   A body that weighs 8.0 lb is acted on by a net force of 15 lb. What is its acceleration?

Solution.   First we find the mass of the body, which is

$$m = \frac{w}{g} = \frac{8 \text{ lb}}{32 \text{ ft/s}^2} = 0.25 \text{ slug}.$$

Now we employ the second law of motion to find that

$$a = \frac{F}{m} = \frac{15 \text{ lb}}{0.25 \text{ slug}} = 60 \text{ ft/s}^2.$$

Problem.   A 3000-lb car has an initial speed of 10 mi/hr. How much force is required to accelerate the car to a speed of 50 mi/hr in 9.0 s?

Solution.   The mass of the car is

$$m = \frac{w}{g} = \frac{3000 \text{ lb}}{32 \text{ ft/s}^2} = 94 \text{ slugs}.$$

Before we can compute the acceleration involved in going from 10 to 50 mi/hr in 9 s we must convert the speeds from mi/hr to ft/s. Since

$$1 \frac{\text{mi}}{\text{hr}} = 1.47 \frac{\text{ft}}{\text{s}},$$

we find that, to two significant figures,

$$10 \frac{\text{mi}}{\text{hr}} = 15 \frac{\text{ft}}{\text{s}} \quad \text{and} \quad 50 \frac{\text{mi}}{\text{hr}} = 73 \frac{\text{ft}}{\text{s}}.$$

The car's acceleration in the proper units of ft/sec² is

$$a = \frac{v_f - v_i}{t} = \frac{73 \text{ ft/s} - 15 \text{ ft/s}}{9 \text{ s}} = \frac{58 \text{ ft/s}}{9 \text{ s}} = 6.4 \frac{\text{ft}}{\text{s}^2}.$$

Hence the force that must act upon the car is

$$F = ma = 94 \text{ slugs} \times 6.4 \frac{\text{ft}}{\text{s}^2} = 600 \text{ lb}.$$

Problem.   An elevator weighing 2000 lb fully loaded is suspended by a cable whose maximum permissible tension is 5000 lb. What is the greatest upward acceleration possible for the elevator under these circumstances?

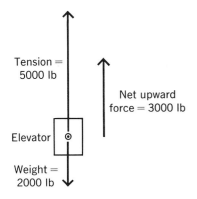

Fig. 4-15. The net upward force on an elevator weighing 2000 lb is 3000 lb when the tension in its supporting cable is 5000 lb.

What is the greatest downward acceleration?

Solution. When the elevator is at rest, or moving at constant speed, the tension in the cable is just the weight of 2000 lb. To accelerate the elevator upward, an additional tension is required in order to provide a net upward force (Fig. 4-15). Here it is stated that the maximum total tension cannot exceed 5000 lb, so the greatest accelerating force that can be applied to the elevator is 3000 lb. The mass of the elevator is

$$m = \frac{w}{g} = \frac{2000 \text{ lb}}{32 \text{ ft/s}^2} = 63 \text{ slugs},$$

and the acceleration of this mass when a force of 3000 lb is applied to it is therefore

$$a = \frac{F}{m} = \frac{3000 \text{ lb}}{63 \text{ slugs}} = 48 \text{ ft/s}^2.$$

For the elevator to exceed the downward acceleration of gravity, 32 ft/s², a downward force in addition to the weight of the elevator is required. Since this cannot be provided by a supporting cable, the maximum downward acceleration is 32 ft/s².

Problem. Figure 4-16 shows a 12-lb block, *A*, that hangs from a string which passes over a pulley and is connected at its other end to a 30-lb block, *B*, which rests on a frictionless table. Find the accelerations of the two blocks under the assumption that the string is massless and the pulley frictionless.

Solution. The blocks are joined by the string, and their accelerations are accordingly the same in magnitude even though different in direction. The force applied to the system of the two blocks is the weight of block *A* exclusively, since block *B* is resting upon the table. However, the mass being accelerated by w_A is the combined mass of both blocks, $m_A + m_B$. Hence the acceleration of the blocks is

$$a = \frac{F}{m} = \frac{w_A}{m_A + m_B} = \frac{w_A}{(w_A/g) + (w_B/g)}.$$

Here, since w_A is 12 lb and w_B is 30 lb,

$$a = \frac{12 \text{ lb}}{[(12 \text{ lb})/(32 \text{ ft/s}^2)] + [(30 \text{ lb})/(32 \text{ ft/s}^2)]} = 9.1 \text{ ft/s}^2.$$

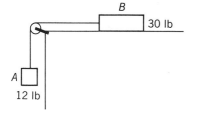

Fig. 4-16. The force on the system of the two blocks is the 12-lb weight of *A*, while the mass being accelerated is the total mass of both.

Problem. Figure 4-17 shows the same two blocks A and B suspended by a string on either side of a pulley. Find the accelerations of the two blocks.

Solution. Here the net force acting on the system of the two blocks is $w_B - w_A$, the *difference* between their weights. A single force is acting upon both blocks since they are connected by the string; the pulley merely changes the direction of the force. Hence the acceleration of each block can be found by determining the acceleration of the entire system:

$$a = \frac{F}{m} = \frac{w_B - w_A}{m_B + m_A}$$

$$= \frac{18 \text{ lb}}{(42 \text{ lb})/(32 \text{ ft/s}^2)} = 14 \text{ ft/s}^2.$$

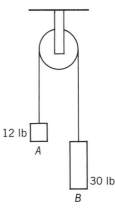

Fig. 4-17. The force on the system of the two blocks is the 18-lb difference between their weights, while the mass being accelerated is the total mass of both.

4-7 The Third Law of Motion

The third of Newton's laws of motion states that

When a body exerts a force on another body, the second body exerts a force on the first body of the same magnitude but in the opposite direction.

If we call one of the interacting bodies A and the other B, then according to the third law of motion

$$\mathbf{F}_{AB} = -\mathbf{F}_{BA} \tag{4-5}$$

whenever they interact. Here $\mathbf{F}_{AB}$ is the force body A exerts on body B and $\mathbf{F}_{BA}$ is the force body B exerts on body A (Fig. 4-18).

The third law of motion always applies to two different forces on two different bodies—the *action force* that one body exerts on another, and the equal but opposite *reaction force* that the second exerts on the first.

Action and reaction forces

There is no such thing as a single force in the universe. Four examples of action-reaction pairs of forces are shown in Fig. 4-19. (a) We push against a wall; the wall pushes back on us. (b) We throw a ball; as we are pushing it into the air, it is pushing back on our hand. (c) We fire a rifle, and the ex-

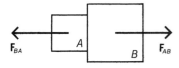

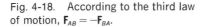

Fig. 4-18. According to the third law of motion, $\mathbf{F}_{AB} = -\mathbf{F}_{BA}$.

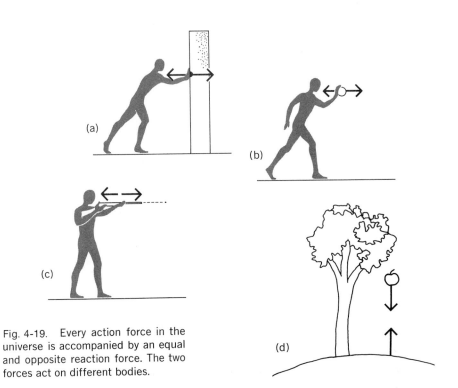

Fig. 4-19. Every action force in the universe is accompanied by an equal and opposite reaction force. The two forces act on different bodies.

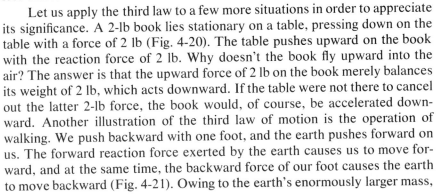

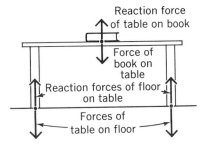

Fig. 4-20. The action-reaction forces between a book and a table and between the table and the floor.

panding gases in its barrel push the bullet out; the gases push back on the rifle, which gives rise to the recoil force we feel with our shoulder. (d) An apple falls because of the downward gravitational pull of the earth; there is an equal upward pull by the apple on the earth which we cannot detect because the earth is so much more massive than the apple, but it is nevertheless there.

Let us apply the third law to a few more situations in order to appreciate its significance. A 2-lb book lies stationary on a table, pressing down on the table with a force of 2 lb (Fig. 4-20). The table pushes upward on the book with the reaction force of 2 lb. Why doesn't the book fly upward into the air? The answer is that the upward force of 2 lb on the book merely balances its weight of 2 lb, which acts downward. If the table were not there to cancel out the latter 2-lb force, the book would, of course, be accelerated downward. Another illustration of the third law of motion is the operation of walking. We push backward with one foot, and the earth pushes forward on us. The forward reaction force exerted by the earth causes us to move forward, and at the same time, the backward force of our foot causes the earth to move backward (Fig. 4-21). Owing to the earth's enormously larger mass,

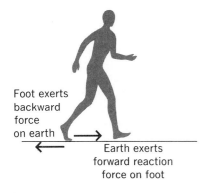

Foot exerts
backward
force
on earth

Earth exerts
forward reaction
force on foot

Fig. 4-21.   When we push backward on the earth with one foot, the opposite reaction force of the earth pushes forward on us. The latter force causes us to move forward.

its motion cannot be detected practically, but it is there. Why is it that there is no reaction force on us, responding to the earth's push on our foot, to keep us from moving? The explanation is that every action-reaction pair of forces acts on *different* bodies. We push on the earth, the earth pushes back on us. If there are no *additional* forces present to impede our motion (for instance, pressure by a wall directly in front of us), we proceed to undergo a forward acceleration.

We might conceivably find ourselves on a frozen lake with a perfectly smooth surface. Now we cannot walk because the absence of friction prevents us from exerting a backward force on the ice which would produce a forward force on us. But what we can do is exert a force on some object we may have with us, say a rock. We throw the rock forward by applying a force to it; at the same time the rock is pressing back on us with the identical force but in the opposite direction, and in consequence we find ourselves moving backward (Fig. 4-22).

We shall learn in Chapter 8 how these notions are expressed in the *principle of conservation of momentum*, one of the most useful formulations of the laws of motion.

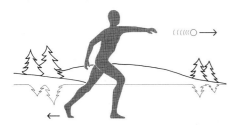

Fig. 4-22.   If we throw a rock while standing on a frozen lake, the reaction force pushes us backward.

4-8 Friction

Frictional forces are of many kinds, but they all act to impede motion. The effects of friction must be distinguished from those of inertia. The term inertia refers to the fact that bodies maintain their original states of rest or of motion in the absence of net forces on them; but even a relatively minute force is sufficient to accelerate a body despite its inertia. The term friction, on the other hand, refers to actual forces that come into being when two surfaces are in contact that act to oppose motion between them. Often friction is desirable: the fastening action of nails, screws, and bolts and the resistive action of brakes depend upon it. Walking would be impossible without friction. In many situations, however, friction merely reduces efficiency, and great efforts are made in industry to minimize it through the use of lubricants — notably grease and oil — and special devices — notably the wheel.

To clearly understand the characteristic properties of frictional forces, let us consider what happens when we attempt to move a box across a level floor (Fig. 4-23). At first the box is stationary; no horizontal forces whatever act on it. As we begin to push, the box remains in place because the floor exerts a force on the bottom of the box which opposes the force we apply. This opposition force is friction, and it arises from the nature of the contact between the floor and the box. As we push harder, the frictional force also increases to match our efforts, until finally we are able to exceed the frictional force and begin to move the box.

Fig. 4-23. As force is applied to a box on a level floor, the frictional resistive force increases to a certain maximum, decreases somewhat as the box begins to move, and then remains constant.

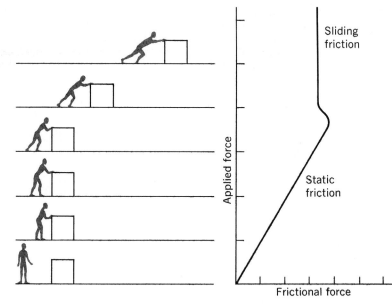

Evidently the opposing frictional force has a maximum value which it cannot go beyond; and when we apply a force greater than this maximum, the box will experience a net force. As the box moves under the influence of the net force, the frictional force usually drops to slightly less than its maximum value when the box is at rest. Because the net force on the box is the force of our push *minus* the force of friction, it is always less than (or equal to) the force we apply; it may even be zero, as we have seen.

Lubricants reduce friction by separating two contacting surfaces with an intermediate layer of a softer material. Instead of rubbing against each other, the surfaces rub against the lubricant. Depending upon the specific application, the most suitable lubricant may be a gas, a liquid, or a solid. Most lubricants are oils derived from petroleum: grease consists of oil to which a thickening agent has been added to prevent the oil from running out from between the surfaces involved. The joints of the limbs of the human body are lubricated by a substance called *synovial fluid* which resembles blood plasma.

Lubricants

There is no relative motion between the rim of a wheel and a smooth surface over which it rolls if wheel and surface are both rigid, and there is accordingly no frictional resistance to overcome. By reducing friction so drastically, the wheel makes it possible to transport loads from one place to another without the enormous forces that dragging them would require. Most aspects of our present technological civilization rely upon the wheel in one way or another.

Rolling friction

Neither wheels nor the surfaces on which they travel can ever be perfectly rigid. Figure 4-24 shows the flattening of the wheel and the indentation of the surface which both contribute to *rolling friction*. Because the wheel and surface must be constantly deformed as the wheel rolls, a force is needed to keep the wheel rolling. However, this force is usually many times smaller than that needed to overcome sliding friction. Balls and rollers are widely used to reduce the friction between a rotating shaft and the bearings that hold it in place by replacing sliding friction with rolling friction. Tire deformation accounts for most of the frictional force on a car at velocities of up to 30 to 50 mi/hr, when air resistance starts to dominate.

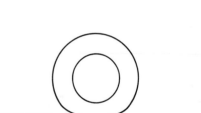

Fig. 4-24. The flattening of a wheel and the indentation of the surface it presses on both contribute to rolling friction.

4-9 Coefficient of Friction

It is a matter of experience that the frictional force exerted by one surface upon another depends upon two factors: (1) the perpendicular force holding the surfaces together, and (2) the nature of the surfaces in contact. The perpendicular force is usually called the *normal force*, symbol N.

The more tightly two objects are pressed together by a normal force, the greater the friction between them. For this reason an empty box is easier

Fig. 4-25. The greater the normal force **N** that presses two surfaces together, the greater the force of friction between them.

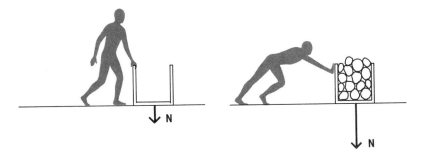

to push across a floor than a similar box loaded with something heavy (Fig. 4-25). Equally familiar is the effect of the nature of the contracting surfaces. For instance, it takes more than three times as much force to push a wooden box across a wooden floor as it does to push a steel box of the same weight across a steel floor. Interestingly enough, the area in contact between the two surfaces is not important: it is just as hard to push a small wooden box weighing 50 lb over a given floor as it is to push a large wooden box weighing 50 lb over the same floor.

To a good degree of approximation the following formula relates the normal force N holding two surfaces together with the frictional force F_f that results:

$$F_f = \mu N \qquad \textit{Maximum frictional force} \quad (4\text{-}6)$$

Frictional force = coefficient of friction × normal force.

Coefficient of friction   The quantity μ (Greek letter "mu") is called the *coefficient of friction* and is a constant for a given pair of surfaces. The value of F_f given by the above formula represents a maximum. When the applied force is less than F_f, the frictional force always equals the applied force. Otherwise, since F_f acts in the opposite direction to an applied force, things would move *backward* when pushed weakly—which, needless to say, does not happen. Table 4-2 is a list of coefficients of friction for several surfaces. Static friction is discussed later in this section.

When an object is being pushed or pulled horizontally, the normal force N holding it against the surface it is on is simply its weight mg. Thus in such cases,

$$F_f = \mu N = \mu mg.$$

Table 4-2. Approximate coefficients of static and sliding friction for various materials in contact

Materials in contact	Coefficient of static friction, μ_s	Coefficient of sliding friction, μ
Wood on wood	0.5	0.3
Wood on stone	0.5	0.4
Steel on steel (smooth)	0.15	0.09
Metal on metal (lubricated)	0.03	0.03
Leather on wood	0.5	0.4
Rubber tire on dry concrete	1.0	0.7
Rubber tire on wet concrete	0.7	0.5
Glass on glass	0.94	0.40
Steel on Teflon	0.04	0.04
Bone on bone (dry)		0.3
(lubricated with synovial fluid)		0.003

We must apply a force greater than μmg when moving a body of mass m across a level surface where the coefficient of friction is μ.

Problem.   A 100-kg wooden crate is being pushed across a wooden floor with a horizontal force of 350 N. What is its acceleration?

Solution.   The frictional force that opposes the applied force F_A here is

$$F_f = \mu mg = 0.3 \times 100 \text{ kg} \times 9.8 \, \frac{\text{m}}{\text{s}^2} = 294 \text{ N}$$

since the coefficient of friction here is 0.3. Hence the net force acting on the crate is

$$F = F_A - F_f = 350 \text{ N} - 294 \text{ N} = 56 \text{ N}$$

and the crate's acceleration is, from the second law of motion,

$$a = \frac{F}{m} = \frac{56 \text{ N}}{100 \text{ kg}} = 0.56 \text{ m/s}^2.$$

Problem.   A 500-lb sled is pulled at constant speed over level snow by a rope that makes an angle of 35° with the horizontal. If the coefficient of friction is 0.10, find the force required.

Solution.   Here the frictional force μN must be overcome by the horizontal component $F \cos \theta$ of the applied force $\mathbf{F}$ (Fig. 4-26). Since the normal force is the sled's weight w minus the upward vertical component $F \sin \theta$ of the force $\mathbf{F}$, we have

Horizontal component of $\mathbf{F}$ = frictional force,

$$F \cos \theta = \mu(w - F \sin \theta),$$

$$F = \frac{\mu w}{\mu \sin \theta + \cos \theta}$$

$$= \frac{0.10 \times 500 \text{ lb}}{0.10 \sin 35° + \cos 35°}$$

$$= 57 \text{ lb}.$$

When a body in contact with a surface is pushed, the frictional force resisting motion increases with the applied force until a limiting value is reached. If the applied force exceeds the limiting value of the frictional force, the body begins to move. The value of the coefficient of friction corresponding to the maximum frictional force between two surfaces at rest is called the *coefficient of static friction* and is denoted by the symbol μ_s.

Static friction usually exceeds sliding friction

When no lubricant is present, the coefficient of static friction μ_s is greater than that of sliding friction μ: the force needed to set a body in motion against friction is more than that needed to maintain it in motion at constant speed. (Fig. 4-23 shows the drop in frictional force when the box

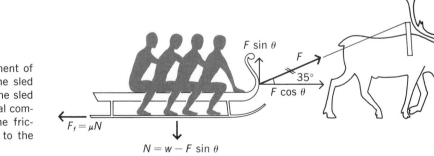

Fig. 4-26. The vertical component of $F \sin \theta$ of force applied to the sled reduces the normal force of the sled on the snow, and the horizontal component $F \cos \theta$ overcomes the frictional resistance of the snow to the sled's motion.

has begun to move.) When the surfaces in contact are smooth and properly lubricated, μ_s and μ are virtually the same. Table 4-2 contains some typical values of coefficients of static friction.

Problem.   A wooden chute is being built along which wooden crates of merchandise are to be slid down into the basement of a store. (a) What angle with the horizontal should the chute make if the crates are to slide down at constant speed? (b) With what force must a 200-lb crate be pushed in order to start it sliding down the chute if the angle of the chute is that found in (a)?

Solution.   The procedure here is first to resolve the weight of the crate, which is a force of magnitude $w = mg$ that acts downward, into a component **F** parallel to the plane and a component **N** perpendicular to the plane. With the help of Fig. 4-27 we find that

$$F = w \sin \theta, \qquad N = w \cos \theta.$$

When the crate slides down at constant speed, there is no net force acting on it, according to Newton's first law of motion. Hence the downward force along the chute must exactly balance the force of sliding friction, which means that

$$F = \mu N,$$
$$w \sin \theta = \mu w \cos \theta,$$
$$\mu = \frac{\sin \theta}{\cos \theta} = \tan \theta.$$

From Table 4-2 the value of μ for wood on wood is 0.3, and so $\theta = 17°$.
     To answer (b), we note that the coefficient of static friction here is $\mu_s = 0.5$. Hence the force of static friction to be overcome is, with the help of Fig. 4-27,

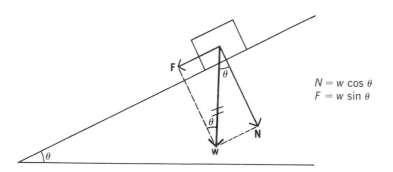

$$N = w \cos \theta$$
$$F = w \sin \theta$$

Fig. 4-27.   The weight **w** of a block on an inclined plane can be resolved into forces parallel and perpendicular to the plane. At an angle θ such that tan $\theta = \mu s$, the block slides down the plane at constant velocity.

$$F_f = \mu_s N = \mu_s w \cos \theta.$$

This is greater than the force of sliding friction, and so a crate will not begin to move without a push. The downward force along the plane due to the crate's own weight is $F = w \sin \theta$. Here F is less than F_f, and so, if we call F' the outside force parallel to the plane required to move the crate,

Outside force + forward component = backward frictional
         of weight   force,

$$F' + w \sin \theta = \mu_s \, w \cos \theta$$

$$F' = w(\mu_s \cos \theta - \sin \theta)$$

$$= 200 \text{ lb } (0.5 \times \cos 17° - \sin 17°)$$

$$= 37 \text{ lb.}$$

Special Topic

Muscular Force

The movements of an animal are produced by contractions of its skeletal muscles. A muscle is a bundle of parallel fibers that tapers at each end into a tendon, which provides the connection to a bone. In some cases a muscle end forks into two or even three tendons. The bones linked by a muscle are hinged together at a joint, and the motion of the bones relative to the joint is controlled by the muscle, usually in conjunction with another muscle on the opposite side.

A muscle fiber contracts when it is given an electrical stimulus by a nerve ending. The force of the contraction is constant for each fiber; the greater the required total force, the greater the number of fibers that are stimulated. The maximum force a muscle can exert thus depends on the number of fibers it contains, which is proportional to its cross-sectional area. Maximum forces of up to 70 N/cm² (100 lb/in²) have been reported. An athlete might have a biceps muscle in his arm 8 cm in diameter, which means it would be capable of producing forces up to 3500 N (790 lb). As will be seen in the next chapter, the geometries of animal skeletons and muscles favor range of movement over force, so the actual forces a person's hands and feet can exert are considerably smaller than those produced by the muscles themselves.

An animal of a certain type whose length (or other representative linear dimension) is L has, in general, muscles whose cross-sectional areas and hence strengths are roughly porportional to L^2. Hence another animal of the same type whose length is, say, $2L$, has muscles which are $(2L)^2/L^2 = 4$ times stronger than the corresponding ones in the first animal. To be sure, the mass of an animal depends upon its volume and so upon L^3, which means that the larger it is, the stronger its muscles have to be to carry out the same tasks. Because mass varies as L^3 whereas strength varies as L^2, large animals are weaker in relation to their masses than smaller ones. This is obvious in nature, where many insects, for instance, can carry objects several times their own weights, whereas animals the size of man are limited to loads well under their own weights. Whether the muscles of a certain kind of animal are intrinsically stronger or weaker than those of an animal of another kind is another matter; human muscles are considerably stronger than those of insects, figured on the basis of force exerted per unit cross-sectional area. Apart from the structural problems a man-sized insect would have, it would be a rather feeble creature.

Important Terms

The **inertia** of a body refers to the apparent resistance it offers to changes in its state of motion. The property of matter that manifests itself as inertia is called **mass**. The unit of mass in the SI system is the **kilogram**, in the British system the **slug**.

A **force** is any influence that can cause a body to be accelerated. The unit of force in the MKSA system is the **newton**, in the British system the **pound**.

The **weight** of a body is the gravitational force exerted on it by the earth. The weight of a body is proportional to its mass.

Newton's **first law of motion** states that, in the absence of a net force acting on it, a body at rest will remain at rest and a body in motion will continue in motion at constant velocity. The **second law of motion** states that a net force acting on a body causes it to have an acceleration proportional to the magnitude of the force and inversely proportional to the body's mass; the acceleration is in the same direction as the force. The **third law of motion** states that, when a body exerts a force on another body, the second body exerts a force on the first of the same magnitude but in the opposite direction.

The term **friction** refers to the resistive forces that arise to oppose the motion of a body past another with which it is in contact. **Sliding friction** is the frictional resistance a body in motion experiences, while **static friction** is the frictional resistance a stationary body must overcome in order to be set in motion.

The **coefficient of friction** is the constant of proportionality for a given pair of contacting surfaces that relates the frictional force between them to the normal force holding them together; usually the coefficient of static friction is greater than that of sliding friction.

Important Formulas

Second law of motion:

$$\mathbf{F} = m\mathbf{a}$$

Weight:

$$w = mg$$

Third law of motion:

$$\mathbf{F}_{AB} = -\mathbf{F}_{BA}$$

Frictional force:

$$F_f = \mu N$$

Multiple Choice

1. When a body undergoes an acceleration,
   a. its mass increases.
   b. its velocity increases.
   c. it falls toward the earth.
   d. a force acts upon it.

2. A force acts on a body that is free to move. If we know the magnitude and direction of the force and the mass of the body, Newton's second law of motion enables us to determine the body's
   a. weight.
   b. position.
   c. velocity.
   d. acceleration.

3. Which of the following is not a unit of mass?
   a. the gram
   b. the kilogram
   c. the pound
   d. the slug

4. A sheet of paper can be withdrawn from under a bottle of milk without toppling it if the paper is jerked out quickly. This is an example of
   a. inertia.
   b. weight.
   c. acceleration.
   d. the third law of motion.

5. The weight of a body
   a. is the quantity of matter it contains.
   b. refers to its inertia.
   c. is basically the same quantity as its mass but expressed in different units.
   d. is the force with which it is attracted to the earth.

6. An automobile that is towing a trailer is accelerating on a level road. The force that the automobile exerts on the trailer is

   a. equal to the force the trailer exerts on the automobile.
   b. greater than the force the trailer exerts on the automobile.
   c. equal to the force the trailer exerts on the road.
   d. equal to the force the road exerts on the trailer.

7. When a horse pulls a wagon, the force that causes the horse to move forward is the force
   a. he exerts on the wagon.
   b. the wagon exerts on him.
   c. he exerts on the ground.
   d. the ground exerts on him.

8. The action and reaction forces referred to in Newton's third law of motion
   a. act upon the same body.
   b. act upon different bodies.
   c. need not be equal in magnitude but must have the same line of action.
   d. must be equal in magnitude but need not have the same line of action.

9. When a 1-N force acts on a 1-kg body that is able to move freely, the body receives
   a. a velocity of 1 m/s.
   b. an acceleration of 0.102 m/s².
   c. an acceleration of 1 m/s².
   d. an acceleration of 9.8 m/s².

10. When a 1-N force acts on a 1-N body that is able to move freely, the body receives
    a. a velocity of 1 m/s.
    b. an acceleration of 0.102 m/s².
    c. an acceleration of 1 m/s².
    d. an acceleration of 9.8 m/s².

11. When a 1-lb force acts on a 1-slug body that is able to move freely, the body receives
    a. a velocity of 1 ft/s.
    b. an acceleration of 0.031 ft/s².
    c. an acceleration of 1 ft/s².
    d. an acceleration of 32 ft/s².

12. When a 1-lb force acts on a 1-lb body that is able to move freely, the body receives
    a. a velocity of 1 ft/s.
    b. an acceleration of 0.031 ft/s².
    c. an acceleration of 1 ft/s².
    d. an acceleration of 32 ft/s².

13. The mass of 8 lb of salami is
    a. 0.25 slug.
    b. 0.5 slug.
    c. 4 slugs.
    d. 256 slugs.

14. A certain force gives a 100-lb weight an acceleration of 20 ft/s². The same force would give a weight of 1000 lb an acceleration of
    a. 2 ft/s².
    b. 5 ft/s².
    c. 10 ft/s².
    d. 200 ft/s².

15. A certain force gives a 5-slug body an acceleration of 8 ft/s². The same force would give a 20-slug body an acceleration of
    a. 2 ft/s².
    b. 4 ft/s².
    c. 32 ft/s².
    d. 160 ft/s².

16. A force of 10 N gives a body an acceleration of 5 m/s². What force would be needed to give the body an acceleration of 1 m/s²?
    a. 1 N
    b. 2 N
    c. 5 N
    d. 50 N

17. A 4800-lb car accelerated from 20 ft/s to 50 ft/s in 6 s. The force on the car is
    a. 750 lb.
    b. 960 lb.
    c. 4500 lb.
    d. 24,000 lb.

18. The weight of a 10-kg mass is
    a. 10 N.
    b. 98 N.
    c. 10 lb.
    d. 320 lb.

19. The weight of a 10-slug mass is
    a. 10 N.
    b. 98 N.
    c. 10 lb.
    d. 320 lb.

20. To set an object in motion on a surface usually requires

    a. less force than to keep it in motion.
    b. the same force as that needed to keep it in motion.
    c. more force than to keep it in motion.
    d. only as much force as is needed to overcome inertia.

21. The frictional force between two surfaces in contact does *not* depend on
    a. the normal force holding them together.
    b. the areas of the surfaces.
    c. whether the surfaces are stationary or in relative motion.
    d. whether a lubricant is used or not.

22. The coefficient of static friction between two wooden surfaces is
    a. 0.5.
    b. 0.5 lb.
    c. 0.5 slug/lb.
    d. 0.5 lb/slug.

23. A force of 40 lb is needed to set a 100-lb steel box moving across a wooden floor. The coefficient of static friction is
    a. 0.08.   b. 0.25.
    c. 0.4.   d. 2.5.

24. The coefficient of static friction for steel on ice is 0.1. The force needed to set a 70-kg skater in motion is approximately
    a. 0.1 N.   b. 0.7 N.
    c. 7 N.   d. 70 N.

25. A horizontal force of 150 N is applied to a 51-kg carton on a level floor. The coefficient of static friction is 0.5 and that of sliding friction is 0.4. The frictional force acting on the carton is
    a. 150 N.   b. 200 N.
    c. 250 N.   d. 500 N.

26. The coefficients of static and sliding friction for wood on wood are respectively 0.5 and 0.3. If a 100-lb wooden box is pushed across a horizontal wooden floor with just enough force to overcome the force of static friction, its acceleration is
    a. 0.2 ft/s².   b. 0.5 ft/s².
    c. 6.4 ft/s².   d. 16 ft/s².

27. A toboggan reaches the foot of a hill at a speed of 4 m/s and coasts on level snow for 15 m before

coming to a stop. The coefficient of sliding friction is

a. 0.004.   b. 0.05.

c. 0.16.   d. 0.27.

Exercises

1. When a body is accelerated, a force is invariably acting upon it. Does this mean that, when a force is applied to a body, it is invariably accelerated?

2. The moon revolves around the earth in an approximately circular orbit. Is the moon accelerated in its motion? Does a force act on the moon? If so, in which direction?

3. It is less dangerous to jump from a high wall onto loose earth than onto a concrete pavement. Why?

4. Compare the tension in the coupling between the first two cars in a train with the tension in the coupling between the last two cars (a) when the train's velocity is constant; (b) when the train is accelerating.

5. Measurements are made of distance versus time for three moving objects. The distances are found to be directly proportional to t, t^2, and t^3 respectively. What can you say about the net force acting on each of the objects?

6. A lead ball has a ring at each end. The ball is suspended from the ceiling by a fine thread tied to one ring, and an identical fine thread hangs from the other ring. (a) If the lower thread is pulled steadily, which thread will break? (b) If the lower thread is given a quick jerk, which thread will break?

7. When a force equal to its weight is applied to a body free to move, what is its acceleration?

8. A force of 1 N is applied in turn to an object of mass 1 kg and an object of weight 1 N. Which receives the greater acceleration?

9. Can we conclude from the third law of motion that a single force cannot act upon a body?

10. Since the opposite forces of the third law of motion are equal in magnitude, how can anything ever be accelerated?

11. An engineer designs a propeller-driven spacecraft. Because there is no air in space, he incorporates a supply of oxygen as well as a supply of fuel for the motor. What do you think of the idea?

12. Ships are always built on ways that slope down to a nearby body of water. Normally a ship is launched before most of its interior and superstructure have been installed, and is completed when afloat. Is this done because the additional weight would cause the ship to slide down the ways prematurely?

13. An empty truck whose mass is 2000 kg has a maximum acceleration of 1 m/s². What is its maximum acceleration when it is carrying a 1000-kg load?

14. A force of 20 N gives an object an acceleration of 5 m/s². (a) What force would be needed to give the same object an acceleration of 1 m/s²? (b) What force would be needed to give an acceleration of 10 m/s²?

15. A force of 4000 N is applied to a 1400-kg car. What will the car's speed be after 10 s if it started from rest?

16. A 1200-kg car accelerates from 10 m/s to 15 m/s in 5 sec. Find the force acting on the car.

17. A 2000-kg truck is braked to a stop in 15 m from an initial speed of 12 m/s. How much force was required?

18. A force of 20 N acts upon a body whose mass is 4 kg. (a) What is the weight of the body? (b) What is its acceleration?

19. A force of 20 N acts upon a body whose weight is 8 N. (a) What is the mass of the body? (b) What is its acceleration?

20. A force of 100 lb acts upon a body whose mass is 1.5 slugs. (a) What is the weight of the body? (b) What is its acceleration?

21. A force of 100 lb acts upon a body whose weight is 96 lb. (a) What is the mass of the body? (b) What is its acceleration?

22. A 3200-lb car accelerates from 30 ft/s to 50 ft/s in 5 s. Find the force acting on the car.

23. During performances of the Bouglione Circus in 1976, John Tailor was fired from a compressed-air cannon whose barrel was 65 ft long. Mr. Tailor emerged from the cannon (twice daily on weekdays, three times on Saturdays and Sundays) at 130 ft/s. If Mr. Tailor weighed 160 lb, find the average force on him during the firing of the cannon.

24. A mass of 8 kg and another of 12 kg are suspended by a string on either side of a frictionless pulley. Find the acceleration of each mass.

25. A 240-lb wooden crate rests on a level wooden floor. What is the minimum force required to move it at constant velocity across the floor?

26. A 100-kg wooden crate rests on a level wooden floor. What is the minimum force required to move it at constant velocity across the floor?

27. An eraser is pressed against a vertical blackboard with a horizontal force of 10 N. The coefficient of friction between eraser and blackboard is approximately 0.2. Find the force parallel to the blackboard required to move the eraser.

28. A man prevents a 2-kg brick from falling by pressing it against a vertical wall. The coefficient of static friction is 0.6. What force must he use? Is this more or less than the weight of the brick?

29. A railway boxcar is set in motion along a track at the same initial velocity as a truck on a parallel road. If the only horizontal forces acting on both vehicles are due to rolling friction with the respective coefficients of 0.0045 and 0.04, which vehicle will come to a stop first? How many times farther will the other vehicle travel?

30. A tennis ball rolling along a floor is found to be slowing down with an acceleration of -1.6 ft/s^2. Find the coefficient of rolling friction.

31. A tennis ball whose initial velocity is 2 m/s rolls along a floor for 5 m before coming to a stop. Find the coefficient of rolling friction.

32. The coefficient of friction between a rubber tire and a dry concrete road is 0.7. (a) What is the maximum acceleration possible for a car on such a road? (b) What is the minimum distance in which a car on such a road can be stopped when it is moving at 60 mi/hr?

33. A wooden block whose initial velocity is 9 ft/s slides on a smooth floor for 6 ft before it comes to a stop. (a) Find the coefficient of friction. (b) If the block weighs 5 lb, how much force would be needed to keep it moving at constant velocity across the same floor?

34. A sled slides down a snow-covered hill at constant velocity. If the hillside is 10° above the horizontal, what is the coefficient of sliding friction between the runners of the sled and the snow?

35. The longest recorded skid marks were found on a road in England after an accident in 1960. The marks were 290 m long. If the coefficient of friction between tires and road was 0.7 and the car's brakes were locked, find the minimum initial speed of the car in km/hr. Why is it likely that the initial speed was actually higher than this?

Problems

1. A 12,000-kg airplane launched by a catapult from an aircraft carrier is accelerated from 0 to 200 km/hr in 3 s. (a) How many times the acceleration of gravity is the airplane's acceleration? (b) What is the average force the catapult exerts on the airplane?

2. How much force must you supply to give a 1-kg object an upward acceleration of $2g$? A downward acceleration of $2g$?

3. A sprinter presses on the ground with a force equal to three times his own weight at a 50° angle with the horizontal at the start of a race. What is his forward acceleration?

4. A cyclist finds that he is able to coast at constant speed along a road that slopes downward at an angle of 1° with the horizontal. If he and his bicycle together weigh 150 lb, find the force required to propel them at constant speed along a level road.

5. Two boxes, one weighing 10 lb and the other 100 lb, are sliding without friction down an inclined

plane that makes an angle of 60° with the horizontal. What is the acceleration of each box?

6. A 100-kg man slides down a rope at constant speed. (a) What is the minimum breaking strength the rope must have? (b) If the rope has precisely this strength, will it support the man if he tries to climb back up?

7. A 1000-kg elevator has a downward acceleration of 1 m/sec². What is the tension in its supporting cable?

8. A 1000-kg elevator has an upward acceleration of 1 m/sec². What is the tension in its supporting cable?

9. A 2400-lb elevator is supported by a cable which can safely withstand a tension of no more than 3000 lb. (a) What is the maximum upward acceleration the elevator can have? (b) The maximum downward acceleration?

10. A 200-lb man stands on a scale in an elevator. What does the scale read (a) when the elevator is ascending with an acceleration of 3 ft/s²; (b) when it is descending with an acceleration of 3 ft/s²; (c) when it is ascending at the constant velocity of 10 ft/s; (d) when it is descending at the constant velocity of 10 ft/s; (e) when the cable has broken and the elevator is descending in free fall?

11. A man stands on a scale in an elevator. When the elevator is at rest, the scale reads 160 lb. When the elevator starts to move, the scale reads 140 lb. (a) Is the elevator going up or down? (b) Does it have a constant speed? If so, what is this speed? (c) Does it have a constant acceleration? If so, what is this acceleration?

12. A parachutist whose total mass is 100 kg is falling at 50 m/s when his parachute opens. His speed drops to 7 m/s in a vertical distance of 40 m. What total force did his harness have to withstand? How many times his weight is this force?

13. A parachutist who weighs 200 lb is falling at 120 mi/hr when his parachute opens. His speed drops to 15 mi/hr in a vertical distance of 120 ft. What total force did his harness have to withstand?

14. A car moving at 10 m/s (22.4 mi/hr) strikes a stone wall. (a) The car is very rigid and the 80-kg driver comes to a stop in a distance of 0.2 m. What is his average acceleration and how does it compare with the acceleration of gravity g? How much force acted upon him? Express this force in both newtons and pounds. (b) The car is so constructed that its front end gradually collapses upon impact, and the driver comes to a stop in a distance of 1 m. Answer the same questions for this situation.

15. A 40-kg kangaroo exerts a constant force on the ground in the first 60 cm of her jump, and rises 2 m higher. When she carries a baby kangaroo in her pouch, she can rise only 1.8 m higher. What is the mass of the baby kangaroo?

16. A 10-kg block resting on a horizontal surface is attached to a 5-kg block that hangs freely by a string passing over a pulley, an arrangement similar to that shown in Fig. 4-16. (a) If there is no friction between the first block and the surface, what is the block's acceleration? What is the block's acceleration if the coefficient of friction between the first block and the surface is (b) 0.4? (c) 0.5? (d) 0.6?

17. A 5-kg block resting on a horizontal surface is attached to a 5-kg block that hangs freely by a string passing over a pulley, an arrangement similar to that shown in Fig. 4-16. (a) If there is no friction between the first block and the surface, what is the block's acceleration? (b) If the coefficient of friction between the first block and the surface is 0.2, what is the block's acceleration?

18. A 100-lb block resting on a horizontal surface is attached to a 50-lb block that hangs freely by a string passing over a pulley, an arrangement similar to that shown in Fig. 4-16. The coefficient of friction between the first block and the surface is 0.4. (a) How long does it take the blocks to move 20 ft starting from rest? (b) What is their velocity at that time?

19. A truck is carrying a 2-ton steel girder which rests on the wooden platform of the truck without any fastenings. The truck's velocity is 30 mi/hr. (a) What is the minimum distance in which the truck can come to a stop without having the girder move

forward? (b) If the girder weighed 3 tons, would the distance be any different? (Assume that $\mu_s = 0.5$.)

20. A crate weighing 500 lb is being slid down a ramp which makes an angle of 20° with the horizontal. The coefficient of friction is 0.3. How much force parallel to the plane must be applied to the crate if it is to slide down at constant velocity? In which direction must the force be applied?

21. A block slides down an inclined plane 9 m long that makes an angle of 38° with the horizontal. The coefficient of sliding friction is 0.25. If the block starts from rest, find the time required for it to reach the foot of the plane.

22. A block slides down an inclined plane 20 ft long that makes an angle of 31° with the horizontal. The coefficient of sliding friction is 0.30. If the block starts from rest, find the time required for it to reach the foot of the plane.

23. If the block of Problem 21 has a mass of 50 kg, find the minimum force required to move it upward along the plane. What should the direction of this force be?

24. If the block of Problem 22 weighs 75 lb, find the minimum force required to move it upward along the plane. What should be the direction of this force?

25. A skier starts from rest and slides 50 m down a slope that makes an angle of 40° with the horizontal; he then continues sliding on level snow. (a) If the coefficient of friction between skis and snow is 0.10 and air resistance is neglected, what is the velocity of the skier at the foot of the slope? (b) How far away from the foot of the slope does he come to a stop?

26. A block takes twice as long to slide down an inclined plane that makes an angle of 35° with the horizontal as it does to fall freely through the same vertical distance. What is the coefficient of friction?

27. A stick is used to push a block of wood toward the blade of a circular saw. (a) If the weight of the block of wood is w, the angle between the stick and the table is θ, and the coefficient of friction between block and table is μ, verify that the force that must be applied to the stick to move the block at constant speed is $\mu w/(\cos \theta - \mu \sin \theta)$. (b) Show that if the stick is held at too steep an angle, the block cannot be moved, no matter how much force is applied. (c) Find the value of the critical angle for $\mu = 0.25$.

Answers to Multiple Choice

1. d	10. d	19. d
2. d	11. c	20. c
3. c	12. d	21. b
4. a	13. a	22. a
5. d	14. a	23. c
6. a	15. a	24. d
7. d	16. b	25. a
8. b	17. a	26. c
9. c	18. b	27. b

5

Equilibrium

The next phase of our analysis of force and motion is an enquiry into the conditions under which an object acted upon by two or more forces is nevertheless not accelerated. As we shall find, it is not enough that the vector sum of the forces equal zero, since an object may be set rotating by such forces if their lines of action do not meet at a common point. While dynamics — the study of moving bodies — is naturally of primary significance to the physicist, an introduction to statics — the study of bodies at rest — is valuable both because statics has important applications in technology and because it affords further practice in the use of vector methods.

5-1 Translational Equilibrium

An object which has no net force acting on it is said to be in *translational equilibrium*. The important point is that the object has no linear acceleration. According to the first law of motion, such an object need not be at rest, but may instead be moving along a straight path at constant velocity.

Translational equilibrium

The condition for translational equilibrium may be expressed in the form

$$\Sigma \mathbf{F} = 0, \tag{5-1}$$

where the symbol Σ (Greek capital letter *sigma*) means "sum of" and $\mathbf{F}$ refers to the various forces acting on a specific object. This is simply the mathematical way of stating that, at equilibrium, the forces are such as to cancel one another out.

In many equilibrium situations all the various forces lie in the same plane. When this is the case, we can establish a set of x-y coordinate axes wherever convenient in the plane and then resolve each force $\mathbf{F}$ into the components $\mathbf{F}_x$ and $\mathbf{F}_y$. Thus we can replace Eq. (5-1), which is a vector equation, with the two scalar equations

$$\Sigma F_x = 0, \qquad \textit{Translational Equilibrium} \tag{5-2}$$

$$\Sigma F_y = 0. \tag{5-3}$$

It is usually much easier to calculate the components of each force present and to make use of Eqs. (5-2) and (5-3) than it is to work with the forces themselves in Eq. (5-1).

There are three steps to follow in working out problems concerning the equilibrium of an object:

1. Draw a sketch of the forces that act on the object. (This is called a *free-body* diagram.) Do not show the forces that the object exerts on other bodies, since such forces do not affect the equilibrium of the object itself.

Procedure for evaluating translational equilibrium

2. Choose a convenient set of coordinate axes and resolve the various forces acting on the object into components along these axes.

3. Set the sum of the force components along each axis equal to zero, as specified by Eqs. (5-2) and (5-3). Then solve the resulting equations algebraically for whatever quantities are to be found in the problem.

Several examples will make clear the above procedure. In the first, shown in Fig. 5-1, we have the simple case of a box of weight w being supported by a single rope. The box will be in equilibrium when all the forces acting on it cancel one another out, which here means that

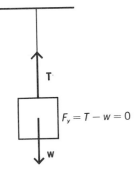

$$\Sigma F_y = 0,$$

since there are no forces in the x-direction. The tension T in the rope acts upward (the $+y$-direction) on the box and the box's weight w acts downward (the $-y$-direction) on it; hence

$$\Sigma F_y = T - w = 0$$

and

$$T = w. \tag{5-4}$$

Fig. 5-1.   A suspended object is in equilibrium when the tension in the rope is equal in magnitude to the weight of the object.

The tension in the rope must equal the weight being supported.

In Fig. 5-2 the same box is suspended from two ropes, A and B, which are at the angles θ and ϕ, respectively, with the horizontal. We begin by resolving the tension in each rope into components in the x- and y-directions, so that we have T_{Ay} and T_{By} upward, T_{Ax} to the left, and T_{Bx} to the right. From Fig. 5-2(b) we have

$$T_{Ax} = -T_A \cos \theta, \qquad T_{Bx} = T_B \cos \phi,$$
$$T_{Ay} = T_A \sin \theta, \qquad T_{By} = T_B \sin \phi.$$

We may now ignore the actual tensions T_A and T_B, and treat their components as individual forces acting at the same point as the weight w, as in

Fig. 5-2.   (a) A box of weight w is suspended by two ropes. (b) A free-body diagram of the forces acting on the box. (c) At equilibrium the sum ΣF_x of the horizontal force components and the sum ΣF_y of the vertical force components each equal zero.

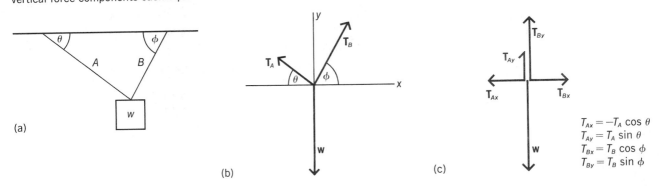

Fig. 5-2(c). For the forces in the vertical and horizontal directions to cancel separately,

$$\Sigma F_x = T_{Bx} + T_{Ax} = 0, \tag{5-5}$$

$$\Sigma F_y = T_{Ay} + T_{By} - w = 0. \tag{5-6}$$

Problem. In the above situation $w = 100$ lb, $\theta = 37°$, and $\phi = 60°$. Find the tension in each rope.

Solution. From Eq. (5-5) we find that

$$T_{Bx} + T_{Ax} = 0,$$

$$T_B \cos \phi - T_A \cos \theta = 0,$$

$$T_B = T_A \frac{\cos \theta}{\cos \phi} = T_A \frac{\cos 37°}{\cos 60°} = 1.6 T_A.$$

From Eq. (5-6) we find that

$$T_{Ay} + T_{By} - w = 0,$$

$$T_A \sin \theta + T_B \sin \phi - 100 \text{ lb} = 0.$$

Substituting $1.6 T_A$ for T_B in the last equation, we obtain

$$T_A \sin \theta + 1.6 T_A \sin \phi - 100 \text{ lb} = 0,$$

$$T_A (\sin 37° + 1.6 \sin 60°) = 100 \text{ lb},$$

$$T_A = 50 \text{ lb}.$$

The tension in rope A is 50 lb. Since $T_B = 1.6 T_A$ here,

$$T_B = 1.6 \times 50 \text{ lb} = 80 \text{ lb}.$$

The tension in rope B is 80 lb. The algebraic sum of the tensions in the two ropes is 130 lb, which is more than the weight being supported, but the vector sum of $\mathbf{T}_A$ and $\mathbf{T}_B$ is 100 lb acting vertically upward, which cancels the downward force $\mathbf{w} = 100$ lb.

Problem. A 100-lb box is suspended from the end of a horizontal strut, as in Fig. 5-3. Find the tension in the cable supporting the strut under the assumption that the strut's weight is negligible.

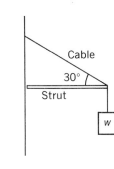

(a)

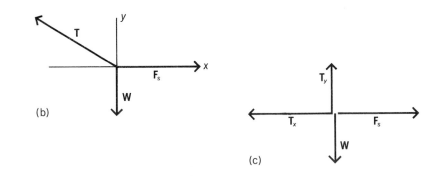

(b)

(c)

Fig. 5-3.   (a) A box of weight w is suspended from the end of a horizontal strut held in place by a cable attached to the wall. (b) A free-body diagram of the forces acting on the end of the strut. (c) At equilibrium, $\Sigma F_x = 0$ and $\Sigma F_y = 0$.

Solution. It is easiest to consider the equilibrium of the end of the strut. The three forces that act on the end of the strut are the tension **T** in the cable, the outward force **F**$_s$ exerted by the strut itself, and the weight **w** of the box. The horizontal and vertical components of the tension T are, from the diagram,

$$T_x = -T \cos 30°,$$

$$T_y = T \sin 30°.$$

The end of the strut is in equilibrium when

$$\Sigma F_x = T_x + F_s = 0,$$

$$\Sigma F_y = T_y - w = 0.$$

All we need is the second of these equations to find T:

$$T_y - w = 0,$$

$$T \sin 30° = w,$$

$$T = \frac{w}{\sin 30°} = \frac{100 \text{ lb}}{0.500} = 200 \text{ lb.}$$

5-2 Torque

When the lines of action of the various forces that act on a body intersect at a common point, they do not tend to set the body in rotation. Such forces are said to be *concurrent*.

If the lines of action of the various forces do *not* intersect, the forces are *nonconcurrent*, and the body may be set into rotation even though the vector sum of the forces may equal zero. In Fig. 5-4 force $\mathbf{F}_A$ acts to the left and force $\mathbf{F}_B$ of the same magnitude acts to the right. Their combined effect is to start the object spinning. If we want the term equilibrium to imply the absence of a rotational acceleration as well as the absence of a linear acceleration, we must supplement $\Sigma \mathbf{F} = 0$ with another condition which the forces on a body must obey if it is to be in equilibrium.

A hint as to the nature of this additional condition may be obtained by watching a seesaw in operation at a playground (Fig. 6-5). A small child can balance a large child merely by sitting farther from the pivot. Two children of the same weight will not balance unless they sit the same distance from the pivot, though the exact distance does not matter. Evidently both the magnitudes of the forces (here the weights of the children) and their lines of action determine whether or not the object is in equilibrium. If we were to try various combinations of weights and distances from the pivot, we would find that the seesaw is balanced when the product $w_1 L_1$ of the weight w_1 and distance from the pivot L_1 of one child is equal to the product $w_2 L_2$ of the other child's weight and distance from the pivot.

To make our discussion perfectly general, let us consider a force $\mathbf{F}$ acting upon a body free to rotate about some pivot point O (Fig. 5-6). The perpendicular distance L from O to the line of action of the force is called the *moment arm* of the force about O. The product of the magnitude F of the force and its moment arm L is known as the *torque* of the force about O. The symbol for torque is τ (Greek letter *tau*), so that

$$\tau = FL, \qquad \text{Torque} = \text{force} \times \text{moment arm.} \qquad (5\text{-}7)$$

In the metric system torque is expressed in newton·meters (N·m); in the British system, in lb·ft.

The greater the torque applied to an object, the greater the tendency of the object to be set into rotation.

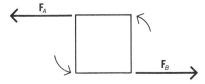

Concurrent and nonconcurrent forces

Fig. 5-4. A body will not be in equilibrium when equal and opposite forces are applied unless the forces have a common line of action.

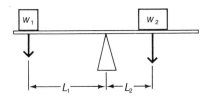

Fig. 5-5. A seesaw is balanced when $w_1 L_1 = w_2 L_2$.

Moment arm

Torque

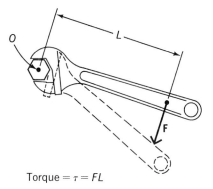

Torque $= \tau = FL$

Fig. 5-6. A measure of the turning effect of a force about a pivot point O is its *torque*, which is equal to the product FL of the magnitude F of the force and the moment arm L.

5-3 Rotational Equilibrium

Positive and negative torques

By convention a torque that tends to produce a counterclockwise rotation is considered positive and a torque that tends to produce a clockwise rotation is considered negative (Fig. 5-7). Thus the condition for a body to be in rotational equilibrium is that the sum of the torques acting upon it about any point, using the above convention for plus and minus signs, be zero:

$$\Sigma\tau = 0. \qquad\qquad\qquad\qquad \text{\textit{Rotational equilibrium}} \quad (5\text{-}8)$$

Of course, if the various forces that act do not all lie in the same plane, it is necessary that the sum of the torques in each of three mutually perpendicular planes be zero.

Calculating torques

It is possible to prove that if the sum of the torques on a body is zero about any point, it is also zero about all other points. Hence the location of the point about which torques are calculated in an equilibrium problem is completely arbitrary; *any* point will do. Let us verify this statement with an example.

Problem.   Figure 5-8 is a diagram of a rod 6.0 ft long that has weights of 10 lb at one end and 30 lb at the other. We assume that the rod is weightless. At what point should the rod be picked up if it is to have no tendency to rotate? In other words, where is the balance point of the rod?

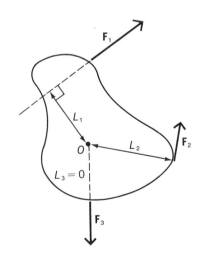

Fig. 5-7.   Forces and their moment arms about the pivot point O: Lines of action are shown as dashed lines. The line of action of $\mathbf{F}_3$ passes through O, and its moment arm is therefore zero. The torque $\tau_1 = F_1 L_1$ tends to produce clockwise rotation, and is therefore considered negative; the torque $\tau_2 = F_2 L_2$ tends to produce counterclockwise rotation, and is therefore considered positive.

Solution 1. We first compute torques about the unknown balance point. If x is the distance of the 30-lb weight from this point, the 10-lb weight is $(6.0 \text{ ft} - x)$ from it on the other side. The torques these weights exert are

$$\tau_1 = w_1 L_1 = +30x \text{ lb},$$

$$\tau_2 = w_2 L_2 = -10(6.0 \text{ ft} - x) \text{ lb},$$

and equilibrium will result when

$$\Sigma \tau = 30x \text{ lb} - 10(6.0 \text{ ft} - x) \text{ lb} = 0,$$

$$x = 1.5 \text{ ft}.$$

When the rod is picked up 1.5 ft from the 30-lb weight, the two weights exert opposite torques of the same magnitude (45 lb·ft) about this point, so the rod is in balance.

Solution 2. Let us next solve the same problem by calculating torques about the middle of the rod, as shown in Fig. 5-9. Now x represents the distance between the balance point and the center of the rod. We have

$$\tau_1 = w_1 L_1 = +30 \text{ lb} \times 3.0 \text{ ft} = +90 \text{ lb·ft},$$

$$\tau_2 = F L_2 = -40x \text{ lb},$$

$$\tau_3 = w_2 L_3 = -10 \text{ lb} \times 3.0 \text{ ft} = -30 \text{ lb·ft}.$$

Equilibrium will result when

$$\Sigma \tau = 90 \text{ lb·ft} - 40x \text{ lb} - 30 \text{ lb·ft} = 0,$$

from which we obtain

$$x = 1.5 \text{ ft}.$$

The location of the balance point is the same regardless of the particular point about which torques are calculated. It is usually wise to calculate torques about the point of application of one of the forces that act on a body, since this makes it unnecessary to consider the torque produced by that force and thereby simplifies the arithmetic.

Not all equilibrium situations are necessarily stable. For instance, a cone balanced on its apex is in equilibrium, but it will fall over when dis-

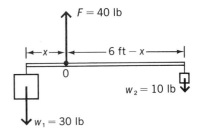

Fig. 5-8. Torques are computed about the unknown balance point of the rod in solution 1.

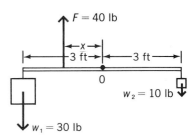

Fig. 5-9. Torques are computed about the center of the rod in solution 2.

Types of equilibrium

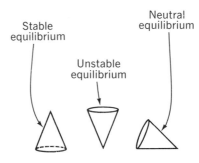

Fig. 5-10. These cones are all in equilibrium, but only one of them is in a stable position.

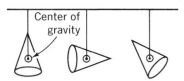

Fig. 5-11. A body suspended from its center of gravity is in equilibrium in any orientation.

For equilibrium purposes

turbed even slightly (Fig. 5-10). This is an example of an *unstable* equilibrium. The same cone on its base will return to its original position if tipped over a little; hence it is in *stable equilibrium* on its base. There is a third possibility as well, illustrated by a cone lying on its side. If such a cone is displaced, it remains in equilibrium in its new position with no tendency either to move further or to return to where it was before. A cone on its side is said to be in *neutral equilibrium*.

5-4 Center of Gravity

The *center of gravity* of a body is that point from which it can be suspended in any orientation without tending to rotate (Fig. 5-11). Each of the constituent particles of the body has a certain weight, and therefore exerts a torque about whatever point the body is suspended from. There is only a single point in a body about which all these torques cancel out no matter how the body is oriented; this is its center of gravity. For equilibrium purposes we can therefore regard the entire weight of a body as concentrated at its center of gravity.

We can now recognize the distinctions between the different kinds of equilibrium shown in Fig. 5-10. The left-hand cone is in stable equilibrium because its center of gravity has to be raised to change its orientation; the center cone is in unstable equilibrium because any change in its orientation lowers its center of gravity; and the right-hand cone is in neutral equilibrium because the height of its center of gravity does not change when it is rolled along on its side.

If the rod of Figs. 5-8 and 5-9 had the weight w instead of being weightless, we could take into account its effect on the location of the balance point by including the torque due to a force of magnitude w acting downward at

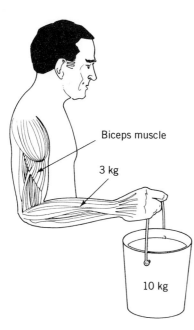

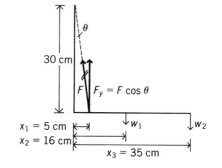

Fig. 5-12. The force exerted by the elbow on the forearm can be disregarded if torques are calculated about the elbow. The horizontal component of the tension in the biceps muscle has no moment arm about the elbow and so is also disregarded.

the center of the rod. The center of gravity of a uniform object of regular shape is located at its geometrical center. The center of gravity of an irregular object need not even be located within the object itself: the center of gravity of a seated person, for example, is a few inches in front of his abdomen.

Problem.   A person holds a 10-kg pail of water with his upper arm at his side and his forearm outstretched, as in Fig. 5-12. The palm of his hand is 35 cm from his elbow, his upper arm is 30 cm long, and his biceps muscle is attached to his forearm 5 cm from his elbow. The person's forearm (including his hand) has a mass of 3 kg and its center of gravity is 16 cm from the elbow. Find the force the biceps muscle exerts to support the forearm and pail.

Solution.   We will calculate torques about the elbow, which simplifies the calculation since we need to consider only the vertical component F_y of the muscular force, the weight $w_1 = m_1 g = 29.4$ N of the forearm, and the weight $w_2 = m_2 g = 98$ N of the pail. Since

$$\tan \theta = \frac{5 \text{ cm}}{30 \text{ cm}} = 0.167, \theta = 9.5°,$$

and we have

$$F_y = F \cos \theta = 0.986\, F.$$

The torques about the elbow, τ_1 exerted by the muscle, τ_2 by the forearm's weight, and τ_3 by the pail of water are respectively

$$\tau_1 = F_y x_1 = 0.986\, F \times 0.05\ \text{m} = 0.0493\, F\ \text{m},$$
$$\tau_2 = -w_1 x_2 = -29.4\ \text{N} \times 0.16\ \text{m} = -4.7\ \text{N·m},$$
$$\tau_3 = -w_2 x_3 = -98\ \text{N} \times 0.35\ \text{m} = -34.3\ \text{N·m}.$$

The sign of τ_1 is positive because it acts counterclockwise and those of τ_2 and τ_3 are negative because they act clockwise. The sum of the torques about the elbow must be 0 for equilibrium, hence

$$\Sigma\tau = \tau_1 + \tau_2 + \tau_3 = 0.0493\, F\ \text{m} - 4.7\ \text{N·m} - 34.3\ \text{N·m} = 0,$$

from which we find

$$F = \frac{(4.7 + 34.3)\ \text{N·m}}{0.0493\ \text{m}} = 791\ \text{N}.$$

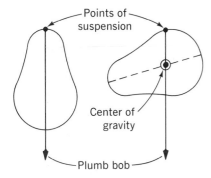

Fig. 5-13.   To find the center of gravity of a flat body, suspend it and a plumb bob successively from two different points on its edge. The center of gravity is located at the intersection of the two lines of action of the plumb bob.

This force, which is equivalent to 178 lb, is over six times the combined weights of the forearm and the pail of water. We might regard the body as being inefficiently designed, since such large musclar forces are required for ordinary tasks. However, when we reflect upon the large span this arrangement enables the hand to move through and the speed at which it can do so, it is clear that inefficiency in one sense has been traded for efficiency in another.

To obtain the center of gravity of an irregular body, we can use the experimental method shown in Fig. 5-13. An analytical procedure that follows from the definition of center of gravity can be applied to give a more accurate result. What is done is simply to break up the body into two or more separate ones whose centers of gravity are known, to consider each of these component bodies as particles joined by weightless rods, and then to calculate the balance point of the resulting object. If the body is complex, for example the hull of a ship, a great many separate strips must be considered if the result is to be accurate, but often some degree of regularity is present which facilitates the task.

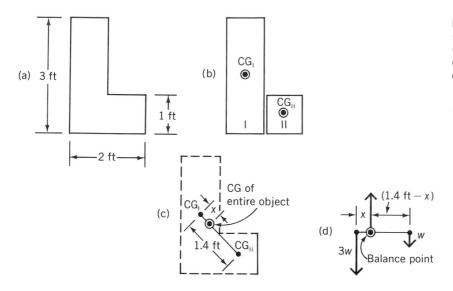

Fig. 5-14. The center of gravity of a complex object can be calculated by considering it to be composed of two or more simple objects whose centers of gravity are known.

Problem. Find the center of gravity of the L-shaped steel plate shown in Fig. 5-14.

Solution. We imagine the body to consist of two sections, one a rectangle 3 ft long by 1 ft wide and the other a square 1 ft on a side. The centers of gravity of these sections are at their geometrical centers, as shown. Now we replace each section by a particle at its center of gravity, whose weight is proportional to the area of the section. If the steel plate weighs w/ft², particle I weighs $3w$ and particle II weighs w, since their areas are respectively 3 ft² and 1 ft². The distance between CG_I and CG_{II} is, from the diagram, 1.4 ft. If the balance point of a weightless rod 1.4 ft long that has a weight of $3w$ at one end and a weight of w at the other is the distance x from the heavy end, we may compute torques about this point to find the value of x. We have

$$\Sigma\tau = 3wx - w(1.4 \text{ ft} - x) = 0,$$

$$x = 0.35 \text{ ft}.$$

The center of gravity of the entire L-shaped plate therefore lies 0.35 ft from the center of section I along a line joining this point with the center of section II. It was not necessary to know the value of w, the weight per ft² of the plate, in order to obtain the position of its center of gravity, for it is a purely geometrical quantity.

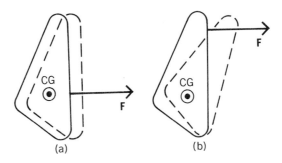

Fig. 5-15. (a) A body remains in rotational equilibrium when the line of action of an applied force passes through its center of gravity. (b) When the line of action of an applied force does not pass through its center of gravity, the body is given a rotational as well as a translational acceleration.

The response of a body to a net force acting on it depends upon whether the line of action of the force passes through the body's center of gravity or not. In the former case, the force has no effect on the body's rotational motion (Fig. 5-15); if the body is not rotating to begin with, such a force will simply accelerate it in the direction of the force. However, if the line of action of the force does not pass through the center of gravity, the body will start to spin.

5-5 Mechanical Advantage: The Lever

A machine is a device that transmits force or torque to accomplish a definite purpose. All machines, however complicated, are actually combinations of only three basic machines: the lever, the inclined plane, and the hydraulic press. Thus the train of gears that transmits power from the engine of a car to its wheels is a development of the lever, the screw jack that can raise one end of the car from the ground is a development of the inclined plane, and the brake system that permits a touch of the foot to stop the car is a development of the hydraulic press.

The *mechanical advantage* (MA) of a machine is the ratio between the output force F_{out} it exerts and the input force F_{in} that is furnished to it:

$$MA = \frac{F_{out}}{F_{in}}. \qquad \textit{Mechanical advantage} \quad (5\text{-}9)$$

A mechanical advantage greater than 1 signifies that the output force exceeds the input force, while a mechanical advantage less than 1 means that the output force is smaller. Usually the MA is greater than 1, which makes it possible for a relatively small applied force to accomplish a task ordinarily beyond its capacity, but sometimes the reverse is true, as in the case of a pair of scissors where the range of motion is increased at the expense of a reduced force.

A distinction must be made between the theoretical mechanical advantage of a machine, which is the value of its MA under ideal circumstances, and its actual mechanical advantage, which takes into account the effects of friction and any other factors present that tend to resist the transformation of F_{in} to F_{out}. The efficiency of a machine is equal to the ratio between its actual and its theoretical mechanical advantages. In some machines, such as the simple lever, the efficiency may be close to 100 percent, while in others, such as the screw, it may be less than 10 percent. In the latter case the low efficiency is actually an advantage, since it prevents the screw from backing out by itself. We shall consider only theoretical mechanical advantages here unless otherwise stated.

Machine efficiency

The *principle of equilibrium* is the basis for calcuating mechanical advantage: when it is provided with the input force F_{in}, a machine will exactly balance a load equal to F_{out}. In the case of the lever shown in Fig. 5-16 the condition for equilibrium is that the torque produced by F_{in} about the fulcrum be the same in magnitude as that produced by F_{out}. If we call the lever arms of the respective forces L_{in} and L_{out}, we have

Principle of equilibrium

$$F_{in} L_{in} = F_{out} L_{out},$$

$$\frac{F_{out}}{F_{in}} = \frac{L_{in}}{L_{out}}$$

and

$$MA = \frac{F_{out}}{F_{in}} = \frac{L_{in}}{L_{out}}. \qquad \textit{The lever} \quad (5\text{-}10)$$

The theoretical mechanical advantage of the lever is equal to the inverse ratio of the lever arms. A 4:1 ratio of lever arms, for instance, means an MA of 4 when the input force is applied to the longer arm, an MA of $\frac{1}{4}$ when it is applied to the shorter arm.

The lever of Fig. 5-16, where the fulcrum is between the load and the applied force, is called a Class I lever. In a Class II lever, of which a wheelbarrow is an example, the load is between the fulcrum and the applied force,

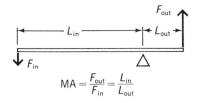

Fig. 5-16.   The lever.

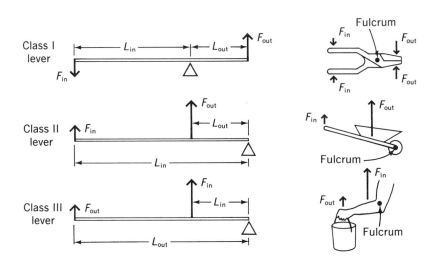

Class I lever

Class II lever

Class III lever

Fulcrum

Fig. 5-17.
The three classes of lever.

Wheel and axle

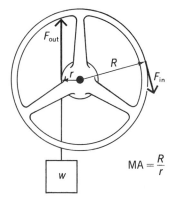

$$MA = \frac{R}{r}$$

Fig. 5-18.   The wheel and axle.

as in Fig. 5-17, while in Class III lever, of which the human forearm is an example, the load and fulcrum are at the ends with the applied force between them.

The various kinds of elementary lever are all handicapped by the limited angle through which they can operate. Certain developments of the lever, however, can readily be used on a continuous basis, and one or another of them is an important element in nearly every motor-driven machine. Perhaps the simplest is the *wheel and axle* (Fig. 5-18). A wheel of radius R is attached to an axle of smaller radius r. The input force acts tangentially on the wheel, and the output force is exerted by the rim of the axle; the center of the axle acts as the fulcrum. By the principle of equilibrium the torque of the applied force F_{in} about the center of the axle must equal that of F_{out}, which means that

$$F_{in} R = F_{out} r.$$

Hence the theoretical mechanical advantage here is

$$MA = \frac{F_{out}}{F_{in}} = \frac{R}{r} \qquad\qquad \textit{Wheel and axle} \quad (5\text{-}11)$$

The larger the wheel is relative to the axle, the greater is the mechanical advantage. For this reason trucks and buses not equipped with power

(a)   (b)

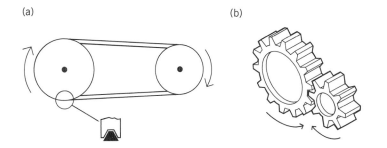

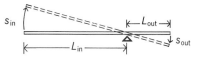

Fig. 5-19. Both V belt (a) and gear (b) drive systems utilize the principle of the lever. In each case the mechanical advantage is equal to the ratio of the diameters of the driven and driving elements.

steering have larger steering wheels than those in ordinary cars, which permits the drivers of the larger vehicles to provide the greater steering torque required. Other developments of the lever meant for torque transmission are the belt and gear drive systems (Fig. 5-19).

The ranges of motion of the ends of a lever are in porportion to their lengths: if the left-hand end of the lever of Fig. 5-20 is moved through an arc of s_{in}, the right-hand end will be moved through an arc of s_{out}, where

$$\frac{s_{in}}{s_{out}} = \frac{L_{in}}{L_{out}}.$$

Fig. 5-20. Each end of a lever moves through a different distance if its arms are not equal in length. The ratio of distances equals the ratio of lever arms.

Hence

$$\text{MA} = \frac{F_{out}}{F_{in}} = \frac{s_{in}}{s_{out}}. \qquad \text{Mechanical advantage} \quad (5\text{-}12)$$

In fact, this relationship holds not only for the lever but for all simple machines, and is often easier to apply than the principle of equivalence itself.

Problem.   Find the theoretical mechanical advantage of the block and tackle shown in Fig. 5-21.

Solution.   When the free end of the rope is pulled with the force F_{in} through a distance d, the movable block is raised through a height of $\frac{1}{4}d$, since there are four strands that must be shortened. Hence

$$\text{MA} = \frac{s_{in}}{s_{out}} = \frac{d}{\frac{1}{4}d} = 4.$$

The theoretical mechanical advantage of this block and tackle is 4.

In general, the MA of a block and tackle is equal to the number of

Fig. 5-21. The theoretical mechanical advantage of a block and tackle is equal to the number of strands of rope that support the movable block, in this case 4.

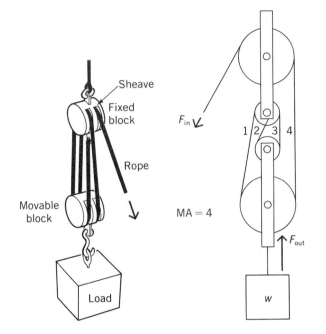

strands of rope that support the movable block and thereby the load. The strand to which the tension F_{in} is applied in Fig. 5-21 does not contribute to supporting the movable block; the upper pulley merely serves to change the direction of the applied force.

5-6 The Inclined Plane

The inclined plane is the second of the three basic machines; the third, the hydraulic press, will be considered in Chap. 11.

We are so accustomed to using the inclined plane in everyday life that it may be hard to think of it as a "machine." So instinctive an act as choosing a gradual slope of a hill to walk up instead of a steep slope is based upon the principle of the inclined plane. The most familiar adaptation of the inclined plane is a staircase, where the continuous surface of the plane is replaced by a series of steps for convenience in walking.

Figure 5-22 shows an inclined plane along which a crate of weight w is being pushed. The plane is L long and h high. From Eq. (5-12) we have

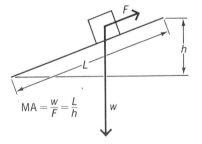

$$MA = \frac{w}{F} = \frac{L}{h}$$

Fig. 5-22. The inclined plane.

$$MA = \frac{s_{in}}{s_{out}} = \frac{L}{h}.$$ *Inclined plane*   (5-13)

The simplest development of the inclined plane is the *wedge* (Fig. 5-23). The MA of a wedge L long and h thick is L/h just as for an inclined plane, but this mechanical advantage is never even approximately realized owing to friction. Aside from its use in splitting logs, leveling objects, and holding doors open, the wedge provides the operating principle of all cutting tools. A knife or chisel is obviously a wedge, but so are the teeth of a saw and the abrasive chips of a grindstone.

Wedge

Perhaps the most important application of the inclined plane is the *screw*. In essence a screw is an inclined plane wrapped around a cylinder to form a continuous helix. The corrugations of a screw are called *threads*, and are usually (though not always) triangular in cross section. A *right-hand* screw is one that moves away from the viewer when turned clockwise, as in Fig. 5-24, while a *left-hand* screw moves away when turned counterclockwise. Standard screws are all right-handed, with left-hand ones employed only for special purposes.

Screw

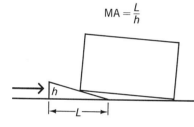

$$MA = \frac{L}{h}$$

Fig. 5-23.   The wedge.

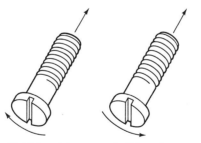

Right-hand screw   Left-hand screw

Fig. 5-24.   A right-hand screw moves into the page when turned clockwise, while a left-hand screw moves into the page when turned counterclockwise.

Important Terms

When the net force acting on a body is zero, the body is in **translational equilibrium.** When the net torque acting on it is zero, the body is in **rotational equilibrium.**

The **torque** of a force about a particular pivot point is the product of the magnitude of the force and the perpendicular distance from the line of action of the force to the pivot point. The latter distance is called the **moment arm** of the force. For a body to be in equilibrium there must be neither net force nor net torque acting upon it.

The **center of gravity** of a body is that point from which it can be suspended in any orientation without tending to rotate. The weight of a body can be considered as a downward force acting on its center of gravity.

A **machine** is a device that transmits force or torque. The three basic machines are the **lever**, the **inclined plane**, and the **hydraulic press**.

The **mechanical advantage** of a machine is the ratio between the output force (or torque) it exerts and the input force (or torque) that is furnished to it. The **theoretical mechanical advantage** is its value under ideal circumstances, while the **actual mechanical advantage** is its value when friction is taken into account.

The **efficiency** of a machine is the ratio between its actual and theoretical mechanical advantages; it is always less than 100%.

Important Formulas

Equilibrium of a particle:
$$\Sigma F_x = 0$$
$$\Sigma F_y = 0$$

Torque:
$$\tau = FL$$

Rotational equilibrium of object:
$$\Sigma \tau = 0 \text{ about any point}$$

Mechanical advantage:
$$MA = \frac{F_{out}}{F_{in}} = \frac{s_{in}}{s_{out}}$$

Multiple Choice

1. Which of the following sets of horizontal forces could leave a body in equilibrium?
   a. 25, 50, and 100 lb
   b. 5, 10, 20, and 50 lb
   c. 8, 16, and 32 lb
   d. 20, 20, and 20 lb

2. Which of the following sets of horizontal forces could not leave a body in equilibrium?
   a. 6, 8, and 10 lb
   b. 10, 10, and 10 lb
   c. 10, 20, and 30 lb
   d. 20, 40, and 80 lb

3. In general, the number of scalar equations that must be satisfied if a body free to move in a plane is to be in equilibrium is
   a. 2.   b. 3.
   c. 4.   d. 6.

4. In general, the number of scalar equations that must be satisfied if a body free to move in three dimensions is to be in equilibrium is
   a. 2.
   b. 3.
   c. 4.
   d. 6.

5. A body in equilibrium may *not* have
   a. any forces acting upon it.
   b. any torques acting upon it.
   c. velocity.
   d. acceleration.

6. Two ropes are used to support a stationary weight *W*. The tensions in the ropes must
   a. each be *W*/2.
   b. each be *W*.

c. have a vector sum of magnitude *W*.

d. have a vector sum of magnitude greater than *W*.

7. A weight is suspended from the middle of a rope whose ends are at the same level. In order for the rope to be perfectly horizontal, the forces applied to the ends of the rope

a. must be equal to the weight.

b. must be greater than the weight.

c. might be so great as to break the rope.

d. must be infinite.

8. If the sum of the torques on a body is zero about a certain point,

a. it is zero about no other point.

b. it is zero about some other points.

c. it is zero about all other points.

d. the body must be in equilibrium.

9. In an equilibrium problem, the axis about which torques are computed

a. must pass through one end of the body.

b. must pass through the center of gravity of the body.

c. must intersect the line of action of at least one force acting on the body.

d. may be located anywhere.

10. The center of gravity of a body

a. is always at its geometrical center.

b. is always in the interior of the body.

c. may be outside the body.

d. is sometimes arbitrary.

11. A 10-lb box is suspended by a string from an overhead support. If a horizontal force of 5.8 lb is applied to the box, the string will make an angle with the vertical of

a. 30°.

b. 45°.

c. 60°.

d. 75°.

12. A 5-lb picture is supported by two strings that run from its upper corners to a nail on the wall. If each string makes a 40° angle with the vertical, the tension in each is

a. 3.3 lb.   b. 3.9 lb.

c. 5 lb.   d. 10 lb.

13. A 60-lb weight is attached to one end of a steel tube 8 ft long whose weight is 40 lb. The distance from the loaded end to the balance point is

a. 0 ft.   b. 1.6 ft.

c. 2 ft.   d. 3.7 ft.

14. The output force produced by a lever does *not* depend on

a. the input force.

b. friction at the fulcrum.

c. the MA of the arrangement.

d. the class of the lever.

15. A man pries up one end of a 480-lb crate with a steel pipe 6 ft long. If he exerts a force of 80 lb, the distance of the fulcrum from the crate is

a. 6 in.   b. 9 in.

c. 18 in.   d. 24 in.

16. The minimum number of pulleys needed to achieve an MA of 6 is

a. 3.   b. 4.

c. 5.   d. 6.

17. The highest MA that can be obtained with a system of two pulleys is

a. 1.

b. 2.

c. 3.

d. 4.

18. A force of 50 lb is needed to raise a 240-lb load with a pulley system. The load ascends 1 ft for every 5 ft of rope pulled through the pulleys. The efficiency of the system is

a. 48%.

b. 50%.

c. 96%.

d. 104%.

19. A 600-rpm motor is coupled to an air compressor with a V-belt. The motor pulley is 3 in. in diameter and the compressor pulley is 9 in. in diameter. The compressor pulley rotates at

a. 200 rpm.

b. 600 rpm.

c. 1800 rpm.

d. 5400 rpm.

20. The MA of an inclined plane depends on
    a. its length.
    b. its height.
    c. the product of its length and height.
    d. the ratio between its length and height.

Exercises

1. A ladder rests against a frictionless wall. (a) In what direction must the force the ladder exerts on the wall be? (b) How is the force the ladder exerts on the ground related to the weight of the ladder? Why?

2. What are the units of torque in the metric system? In the British system?

3. What determines the moment arm of a force?

4. In an equilibrium problem, under what circumstances is it necessary to consider the torques exerted by the various forces?

5. About which point should the torques of the various forces that act on a body be calculated when this is necessary?

6. What effect does proper lubrication have on the theoretical MA of a machine? On its actual MA?

7. What must be true of the MA of a machine meant to increase force? Of a machine meant to increase velocity?

8. A *couple* consists of two forces whose magnitudes are the same that act in opposite directions along parallel lines of action a certain distance apart. If the magnitude of each force is F and the separation between their lines of action is d, find a formula for the torque exerted by a couple.

9. An airplane weighing 90,000 lb is provided with 30,000 lb of thrust by its engines as it takes off at an angle of 25° above the horizontal. If the velocity of the airplane is constant, how much upward force on it is due to the lift produced by the flow of air over its wings?

10. A 50-kg body is suspended from two ropes which each make an angle of 30° with the vertical. What is the tension in each rope?

11. A horizontal force of 18 lb acts upon an unsupported body whose weight is 10 lb. What is the direction and magnitude of the force needed to keep the body at rest?

12. A 500-lb load of bricks is lifted by a crane alongside a building under construction. A horizontal rope is used to pull the load to where it is required, and the supporting cable is then at an angle of 10° from the vertical. What is the tension in the rope? In the supporting cable?

13. A horizontal beam 6 m long projects from the wall of a building. A guy wire that makes a 40° angle with the horizontal is attached to the outer end of the beam. When a weight of 100 N is attached to the end of the beam, what is the tension in the guy wire? (Neglect the weight of the beam.)

14. A 60-lb force is applied to a flagpole 4.0 ft above its base at an angle of 60° above the horizontal. Find the torque about the base of the flagpole.

15. A 200-lb force is applied to a barber pole 3.0 ft above its base at an angle of 45° above the horizontal. Find the torque about the base of the pole.

16. A weight of 40 lb is suspended from one end of a wooden beam 8 ft long whose weight is also 40 lb. Where should the beam be picked up so that it remains horizontal?

17. Three men are carrying a horizontal ladder 12 ft long. One of them holds the front end of the ladder and the other two hold opposite sides of the ladder the same distance from its far end. What is the distance of the latter two men from the far end of the ladder if each man supports one-third of the ladder's weight?

18. A fishing rod 3 m long is attached to a pivoting holder at its lower end. The rod is 50° above the horizontal. A fisherman grasps the rod 1 m from its lower end. What horizontal force must the fisherman exert to keep the rod at this angle when reeling in a 20-lb fish from directly below the rod's upper end?

19. The wheels of a certain truck are 6 ft apart, and the truck falls over when tilted sideways at 30° from the horizontal. How high above the road is its center of gravity?

20. Find the position of the center of gravity of the flat object shown in Fig. 5-25.

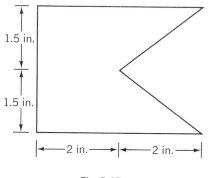

Fig. 5-25.

21. Find the position of the center of gravity of the flat object shown in Fig. 5-26.

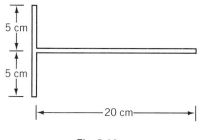

Fig. 5-26.

22. A block and tackle is used to pull a car out of the mud. A block with two pulleys is fastened to the car's bumper, and another block with three pulleys is tied to a tree. One end of a rope is made fast to the bumper and threaded through the various pulleys. Neglecting friction, how much force is applied to the car when two men exert a total force of 200 lb on the free end of the rope?

23. In a certain winch a spur gear with 10 teeth is used to drive another spur gear with 25 teeth. The efficiency of the winch is 50 per cent. (a) If the input gear is turned at 8 revolutions/min., how fast does the output gear turn? (b) If a torque of 20 lb-ft is applied to the input gear, what is the output torque?

24. A 150-kg safe on frictionless casters is to be raised 1.2 m off the ground to the bed of a truck. Planks 4 m long are available for the safe to be rolled along. How much force is required to push the safe up to the truck?

25. The pitch p of a screw is the distance between adjacent threads, and the screw travels this distance in each complete rotation. Find the theoretical mechanical advantage of a screw in terms of p and the lever arm L, which is the handle radius in the case of a screwdriver and the distance from the hand grip to the center of its jaws in the case of a wrench.

Problems

1. When a person stands on one foot with his heel raised, the entire reaction force of the floor, which is equal to his weight w, acts upward on the ball of his foot, as in Fig. 5-27. In order to raise his heel, he must apply the upward force F_1 via his Achilles tendon, so the downward force F_2 on his ankle is greater than his weight w. Find the values of F_1 and F_2 for a 180-lb man for whom $L_1 = 2$ in. and $L_2 = 6$ in. If the distance L_1 were greater, would this mean an increase or a decrease in the muscular force F_1?

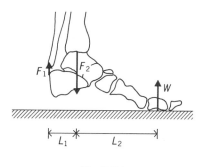

Fig. 5-27.

2. A butcher with his upper arm vertically at his side and his forearm horizontal presses down on a scale with his hand, which is 16 in. from his elbow joint. His forearm and hand together weigh 6 lb and their center of gravity is 8 in. from his elbow joint. The triceps muscle on the back of his upper arm is attached 1 in. behind the center of the elbow joint and it is exerting an upward force of 250 lb. What is the reading on the scale?

3. Find the magnitude of the force between the bones in the elbow of the person holding the pail in Fig. 5-12.

4. The arm of a certain person has a mass of 3 kg and its center of gravity is 28 cm from the shoulder joint. When the arm is outstretched so that it is horizontal, the shoulder muscle supporting it has a line of action 15° above the horizontal. (a) If this muscle is attached to the upper arm 13 cm from the shoulder joint, find the force developed by the muscle. (b) Find the force when the outstretched hand 60 cm from the shoulder joint is used to support a 1-kg load.

5. A 40-lb weight is suspended from the middle of a rope stretched between two posts 20 ft apart. The weight causes the rope to sag by 1.6 ft in the middle. Find the force exerted on each post by the rope under the assumption that the rope's weight is negligible.

6. A car is stuck in the mud. To get it out, the driver ties one end of a rope to the car and the other to a tree 100 ft away. He then pulls sideways on the rope at its midpoint. If he exerts a force of 120 lb, how much force is applied to the car when he has pulled the rope 5 ft to one side?

7. A 1-kg bird sits on a telephone wire midway between two poles 20 m apart. The wire, assumed weightless, sags by 52 cm. What is the tension in the wire?

8. The front wheels of a certain car are found to support 1600 lb and the rear wheels 1400 lb. The car's wheelbase (distance between axles) is 8 ft. (a) What is the total weight of the car? (b) How far from the forward axle is the center of gravity of the car?

9. A 30-lb child and a 50-lb child sit at at opposite ends of a 12-ft seesaw pivoted at its center. Where should a third child weighing 35 lb sit in order to balance the seesaw?

10. A 25-N bag of cement is placed on a 4-m-long plank 1.8 m from one end. The plank itself weighs 8 N. Two men pick up the plank, one at each end. How much weight must each support?

11. The front and rear axles of a 6000-lb truck are 12 ft apart. The center of gravity of the truck is located 8 ft behind its front axle. Find the weight supported by the front wheels of the truck.

12. The center of gravity of a 150-kg polar bear standing on all fours is 100 cm from her feet and 80 cm from her hands. Find the force the ground exerts on each of her hands and feet.

13. A boom 10 ft long is hinged to a vertical mast and held in position by a rope at its end that is attached to the mast 4 ft above the hinge pin. If the boom is horizontal, weighs 80 lb, and is uniform, find the tension in the rope.

14. A uniform horizontal beam 4 m long is supported at one end by a rigid post and at the other by a rope that makes an angle of 40° with the beam. A load of 1000 kg is suspended from the outer end of the beam, which itself has a mass of 200 kg. Find the tension in the rope.

15. A door 8 ft high and 3 ft wide has hinges at the top and bottom of one edge. The entire 50-lb weight of the door is supported by the upper hinge. Find the magnitude and direction of (a) the force the door exerts on the upper hinge; (b) the force the lower hinge exerts on the door.

16. A door 12 ft high and 4 ft wide has hinges on one edge that are 1 ft above the bottom and 1 ft below the top. The entire 100-lb weight of the door is supported by the lower hinge. Find the magnitude and direction of (a) the force the door exerts on the lower hinge; (b) the force the upper hinge exerts on the door.

17. A 10-ft ladder whose weight is 30 lb rests against a frictionless wall at a point 8 ft off the floor. Find the vertical and horizontal components of the force exerted by the ladder on the floor.

18. A uniform ladder that weighs 30 lb and is 10 ft long is placed against a frictionless wall with its base on the ground 2 ft from the wall. Find the magnitudes of the forces exerted on the wall and on the ground.

19. A uniform ladder that weighs 50 lb is placed against a frictionless wall at an angle of 60° from the horizontal. Find the magnitudes and directions of the forces exerted on the wall and on the ground.

20. A ladder 12 ft long rests against a vertical frictionless wall with its lower end 4 ft from the wall. The ladder weighs 40 lb and its center of gravity is at its center. If the ladder is to stay in place, what must be the minimum coefficient of friction between the bottom of the ladder and the ground?

21. A weightless ladder 20 ft long rests against a frictionless wall at an angle of 60° from the horizontal. A 150-lb man is 4 ft from the top of the ladder. What horizontal force at the bottom of the ladder is needed to keep the ladder from slipping?

Answers to Multiple Choice

1. d	8. c	15. c
2. d	9. d	16. c
3. b	10. c	17. c
4. d	11. a	18. c
5. d	12. a	19. a
6. c	13. b	20. d
7. d	14. d	

6

Circular Motion

Nearly all objects in the natural world travel in curved paths. Often these paths are either circles or are very close to being circles. For example, the orbits of the earth and the other planets about the sun are almost circular in shape, as is the orbit of the moon about the earth. On a smaller scale, a handy way to visualize an atom is to imagine it as having a central nucleus with electrons circling around. And, of course, circular motion is no novelty in our own experience; it is hard to think of any important aspect of technology in which circular motion of some kind is not involved. We shall therefore find it both interesting and essential for our later work to consider this kind of motion in some detail.

6-1 Centripetal Force

Uniform circular motion

An object traveling in a circle at a velocity whose magnitude is constant is said to undergo *uniform circular motion*.

Although the velocity of such an object has the same *magnitude* all along its path, the *direction* of the velocity changes constantly. A changing velocity means an acceleration, which in turn signifies that the object must be acted upon by a force. Since the object's path is a circle, the force on it must be directed inside the circle (Fig. 6-1). This force is called *centripetal force*, literally "force seeking the center." Without it, circular motion cannot occur. In general,

Centripetal force = inward force on an object moving in a curved path.

To verify directly the crucial role of centripetal force in circular motion, we can whirl a ball at the end of a string (Fig. 6-2). As the ball swings around, we must continually exert an inward force on it by means of the string. If we let go of the string, the ball flies off tangent to its original circular path. With no centripetal force on it, the ball then proceeds along a straight path at constant velocity as the first law of motion predicts. (Actually, of course, the ball will fall to the ground eventually because of gravity, but this is irrelevant here.)

A centripetal force is acting whenever rotational motion occurs, since such a force is required to change the direction of motion of a particle from

Centripetal force

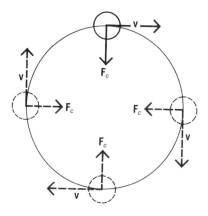

Fig. 6-1. Even though the velocity **v** of a body traveling in a circle at constant speed has the same magnitude along its path, the *direction* of **v** changes constantly. The inward force that causes this change in direction is called *centripetal force*, $\mathbf{F}_c$.

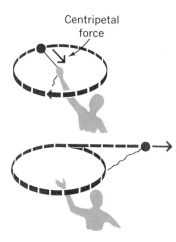

Centripetal force

Fig. 6-2. When a ball is whirled at the end of a string, the tension in the string provides the centripetal force that keeps the ball moving in a circle.

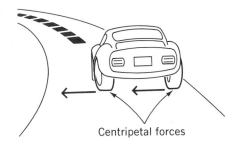

Centripetal forces

Fig. 6-3. The centripetal force exerted when a car rounds a curve on a level road is provided by friction between its tires and the road.

the straight line which it would normally follow to a curved path. Gravitation provides the centripetal forces that keep the planets moving around the sun and the moon around the earth. Friction between its tires and the road provides the centripetal force needed by a car in rounding a curve (Fig. 6-3). If the tires are worn and the road wet or icy, the frictional force is small and may not be enough to permit the car to turn.

6-2 Centripetal Acceleration

How large a centripetal force is needed to keep a given object moving in a circle with a certain speed? To find out, we must first compute the acceleration of such an object. In the following derivation we shall consider the circular motion of a particle; the same arguments and conclusions hold for the circular motion of the center of gravity of an object of definite size.

In Fig. 6-4(a) a particle is shown traveling along a circular path of radius r at the constant speed v. At $t = 0$ the particle is at the point A, where its velocity is $\mathbf{v}_A$, and at $t = \Delta t$ the particle is at the point B, where its velocity is $\mathbf{v}_B$.

The change $\Delta \mathbf{v}$ in the direction of particle's velocity in the time interval Δt is expressed in vector rotation as

$$\Delta \mathbf{v} = \mathbf{v}_B - \mathbf{v}_A,$$

and so its acceleration is

$$\mathbf{a} = \frac{\Delta \mathbf{v}}{\Delta t}$$

$$\text{Acceleration} = \frac{\text{change in velocity}}{\text{time interval}}.$$

The vector triangle whose sides are $-\mathbf{v}_A$, $\mathbf{v}_B$, and $\Delta \mathbf{v}$ is similar to the space triangle whose sides are OA, OB, and s, as we can see from Fig. 6-4(c). Since v is the speed of the particle, the magnitudes of $-\mathbf{v}_A$ and $\mathbf{v}_B$ are both v. Also, OA and OB are radii of the circle, so their lengths are both r. Corresponding sides of similar triangles are proportional, hence

$$\frac{\Delta v}{v} = \frac{s}{r} \quad \text{and} \quad \Delta v = \frac{vs}{r}.$$

The distance the particle actually covers in going from A to B is the arc joining these points, the length of which is $v\Delta t$. The distance s, however, is

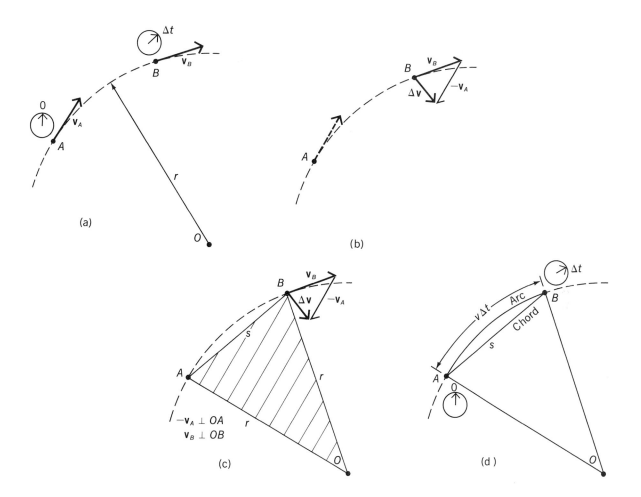

Fig. 6-4. (a) *A* and *B* are two successive positions, Δt s apart, of a particle undergoing uniform circular motion at the speed v in a circle of radius r. (b) The velocity of the particle at *A* is $\mathbf{v}_A$ and at *B* is $\mathbf{v}_B$; the change in its velocity in going from *A* to *B* is $\Delta \mathbf{v} = \mathbf{v}_B - \mathbf{v}_A$. (c) The space and vector triangles are similar because both are isosceles with the long sides of each perpendicular to the corresponding long sides of the other. (d) The chord joining *A* and *B* is s, while the actual distance the particle traverses is $v\,\Delta t$. In calculating the instantaneous acceleration of the particle we are restricted to having *A* and *B* an infinitesimal distance apart, in which case the chord and arc have the same length. (e) The magnitude of the centripetal acceleration is $a_c = \Delta v/\Delta t = v^2/r$.

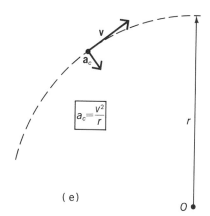

the chord joining A and B, as in Fig. 6-4(d). We are finding the *instantaneous* acceleration of the particle, and we are therefore concerned with the case where A and B are very close together, in which case the chord and arc are equal. Hence

$$s = v\Delta t$$

and we have

$$\Delta v = \frac{v^2 \, \Delta t}{r}.$$

The magnitude of the particle's acceleration is

$$a_c = \frac{\Delta v}{\Delta t}$$

and so we have

$$a_c = \frac{v^2}{r}.$$

Direction of centripetal acceleration

In Fig. 6-4(c) Δv does not quite point toward O, the center of the particle's circular path. When A and B are very close together, however, v *does* point toward O (Fig. 6-5). Because the acceleration a_c is an instantaneous acceleration, we are solely concerned with the case when Δt and hence s are extremely small, and the direction of a_c is accordingly radially inward.

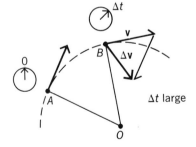

Fig. 6-5. The centripetal acceleration of a particle in uniform circular motion points toward the center of the circle.

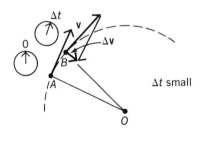

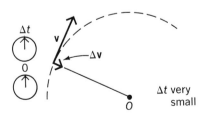

The inward, or *centripetal*, acceleration of a particle in uniform circular motion is proportional to the square of its speed and inversely proportional to the radius of its path:

$$a_c = \frac{v^2}{r}.$$ *Centripetal acceleration*   (6-1)

Problem. The moon is 3.84×10^8 m from the earth and circles the earth once every 27.3 days. Find its centripetal acceleration.

Solution. The time needed by a body in uniform circular motion to make a complete revolution and return to some starting point is called its *period*, symbol T. The distance the body travels in making a complete circle of radius r is $2\pi r$, the circumference of the circle (Fig. 6-6). The speed of the body is therefore

Period of circular motion

$$v = \frac{\text{distance}}{\text{time}} = \frac{2\pi r}{T}.$$ (6-2)

In the case of the moon, $r = 3.84 \times 10^8$ m and

$$T = 27.3 \text{ days} \times 24 \frac{\text{hr}}{\text{day}} \times 60 \frac{\text{min}}{\text{hr}} \times 60 \frac{\text{s}}{\text{min}} = 2.36 \times 10^6 \text{ s}.$$

Hence the moon's orbital speed is

$$v = \frac{2\pi r}{T} = \frac{2\pi \times 3.84 \times 10^8 \text{ m}}{2.36 \times 10^6 \text{ s}} = 1.02 \times 10^3 \frac{\text{m}}{\text{s}}.$$

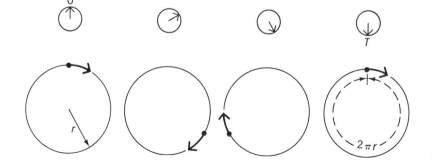

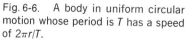

Fig. 6-6.   A body in uniform circular motion whose period is T has a speed of $2\pi r/T$.

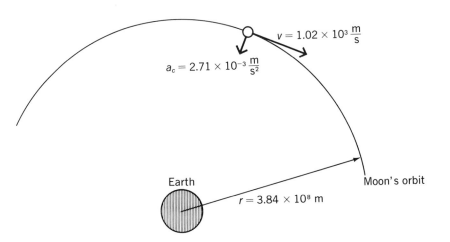

Fig. 6-7.   The moon is accelerated
toward the center of the earth.

Centripetal acceleration of moon   The centripetal acceleration of the moon is accordingly (Fig. 6-7)

$$a_c = \frac{v^2}{r} = \frac{(1.02 \times 10^3 \text{ m/s})^2}{3.84 \times 10^8 \text{ m}} = 2.71 \times 10^{-3} \frac{\text{m}}{\text{s}^2}.$$

This acceleration is directed toward the center of the earth.

6-3 Magnitude of Centripetal Force

From the second law of motion $\mathbf{F} = m\mathbf{a}$ we see that the centripetal force $\mathbf{F}_c$ which must be acting on an object of mass m that is in uniform circular motion is

$$\mathbf{F}_c = m\mathbf{a}_c.$$

Since the centripetal acceleration has the magnitude

$$a_c = \frac{v^2}{r},$$

the magnitude of the centripetal force is

$$F_c = \frac{mv^2}{r}. \qquad\qquad\qquad\qquad \textit{Centripetal Force} \quad (6\text{-}3)$$

The centripetal force that must be exerted to maintain an object in uniform circular motion increases with increasing mass and with increasing speed, with the force more sensitive to a change in speed since it is the square of the speed that is involved. An increase in the radius of the path, however, reduces the required centripetal force (Fig. 6-8).

Fig. 6-8. Centripetal force.

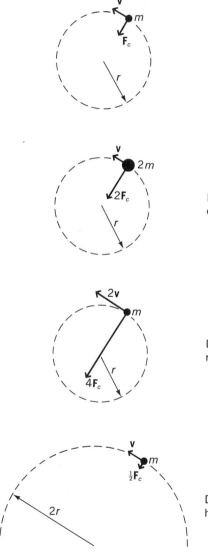

The centripetal force on an object in uniform circular motion is equal in magnitude to mv^2/r .

Doubling the mass doubles the required centripetal force.

Doubling the speed, however, quadruples the required centripetal force.

Doubling the radius of the circle halves the required centripetal force.

Problem. Find the centripetal force required by a 3000-lb car that makes a turn of radius 100 ft at a speed of 15 mi/hr (Fig. 6-9).

Solution. The mass of the car is

$$\text{Mass} = \frac{\text{weight}}{\text{acceleration of gravity}}$$

$$m = \frac{3000 \text{ lb}}{32 \text{ ft/s}^2} = 94 \text{ slugs,}$$

and its speed is

$$v = 15 \frac{\text{mi}}{\text{hr}} \times 1.47 \frac{\text{ft/sec}}{\text{mi/hr}} = 22 \text{ ft/s.}$$

The centripetal force is accordingly

$$F_c = \frac{mv^2}{r} = \frac{94 \text{ slugs} \times (22 \text{ ft/s})^2}{100 \text{ ft}} = 455 \text{ lb.}$$

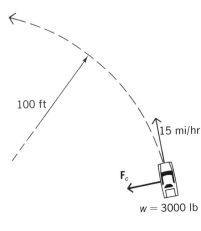

Fig. 6-9.   A centripetal force of 455 lb is required by the car to make the turn shown.

The centripetal force of 455 lb must be provided by the pavement acting on the car's tires through the agency of friction. We can readily compute the minimum coefficient of friction μ that must be present if the car is to make the turn without skidding. The frictional force is

$$F_f = \mu N$$

in general. To find μ, we substitute 455 lb for F_f and 3000 lb for the normal force N, which here is the car's weight w. Thus

$$\mu = \frac{F_f}{N} = \frac{F_c}{w} = \frac{455 \text{ lb}}{3000 \text{ lb}} = 0.15$$

to two significant figures, a coefficient of friction which is readily available under good driving conditions.

Problem. Usually the friction between its tires and the road is enough to provide a car with the centripetal force it needs to make a turn. However, if the car's speed is high or the road surface is slippery, the available frictional force may not be enough and the car will skid. To avoid the likelihood of skids, highway curves are often *banked* so that the roadbed tilts inward. The horizontal component of the reaction force of the road on the car (the action force is the car's weight pressing on the road) then furnishes the required centripetal force. Find the proper banking angle for a car making a turn of radius r at the speed v.

Solution. The reaction force $\mathbf{F}$ of the road on a car is perpendicular to the roadbed, as in Fig. 6-10. This force can be resolved into two components, $\mathbf{F}_y$ which supports the weight mg of the car, and $\mathbf{F}_x$ which is available to provide centripetal force. From the diagram,

Components of reaction force on car

$F_x = F \sin \theta = $ horizontal component of reaction force,

$F_y = F \cos \theta = $ vertical component of reaction force,

where θ is the angle between the roadbed and the horizontal. Since F_x furnishes the centripetal force F_c,

$$F_x = F_c,$$

$$F \sin \theta = \frac{mv^2}{r}. \tag{6-4}$$

The vertical component of the reaction force equals the car's weight, and so

$$F_y = mg,$$

$$F \cos \theta = mg. \tag{6-5}$$

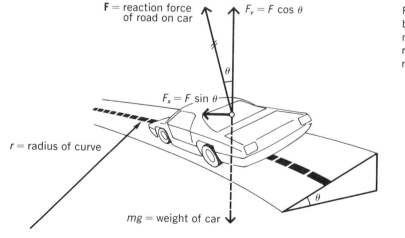

$\mathbf{F} = $ reaction force of road on car

$F_y = F \cos \theta$

$F_x = F \sin \theta$

$r = $ radius of curve

$mg = $ weight of car

Fig. 6-10.   When a car rounds a banked curve, the horizontal component $\mathbf{F}_x$ of the reaction force $\mathbf{F}$ of the road on the car provides it with the required centripetal force.

We divide Eq. (6-4) by Eq. (6-5) and obtain

$$\frac{F \sin \theta}{F \cos \theta} = \frac{mv^2}{mgr},$$

$$\tan \theta = \frac{v^2}{gr}. \qquad \qquad \textit{Banking angle} \quad (6\text{-}6)$$

The proper banking angle θ varies directly with the square of the car's speed and inversely with the radius of the curve. The mass of the car does not matter. When a car goes around a curve at precisely the design speed, the reaction force of the road provides the centripetal force. If the car goes more slowly than this, friction tends to keep it from sliding down the inclined roadway; if the car goes faster, friction tends to keep it from skidding outward.

The same considerations apply to an airplane making a turn, in which case Eq. (6-6) specifies the angle its wings should make with the horizontal.

6-4 The Centrifuge

A *centrifuge* is a device widely used to separate particles of some kind from a liquid in which they are suspended, for instance blood cells from plasma, or to separate liquids of different density from each other, for instance cream from milk. A simple type of centrifuge is shown in Fig. 6-11. As the centrifuge turns, the tubes swing upward, and the denser material migrates to the outer end of each tube.

No force pushes the denser material outward. Rather, each tube is pulled inward as the centrifuge turns, and the denser material responds less readily than the lighter material by virtue of its greater inertia and hence is left behind at the end of the tube. High speeds mean more effectiveness in separating substances having similar densities; some modern centrifuges operate at speeds exceeding 100,000 revolutions per minute.

The faster the centrifuge turns, the greater the angle between its arms and the central shaft. Why? Let us look at Fig. 6-12, which shows a particle of mass m suspended by a massless string and whirled in a horizontal circle. At a given speed, the horizontal component $\mathbf{T}_x$ of the tension $\mathbf{T}$ in the string provides the centripetal force on the particle and its vertical component $\mathbf{T}_y$ is the force that supports the particle's weight w. When the particle's speed is increased, the tension $\mathbf{T}$ increases in magnitude so that T_x equals the new, larger centripetal force. But as T increases at a given angle θ, so does T_y. Now there is a net upward force on the particle, and it rises until T_y decreases to again equal w.

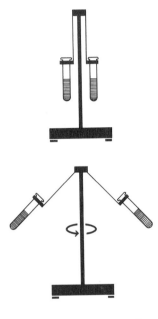

Fig. 6-11. A simple centrifuge.

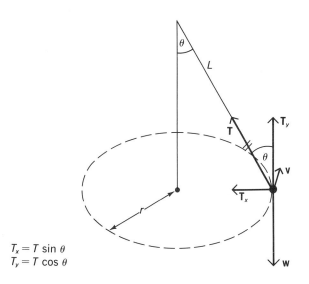

$$T_x = T \sin \theta$$
$$T_y = T \cos \theta$$

Fig. 6-12. A conical pendulum. When the particle's speed v increases, the angle θ also increases.

The arrangement of Fig. 6-12 is called a *conical pendulum* because the string traces out a cone in space as the particle moves in a circle. Reasoning similar to that in the case of a banked turn leads to the formula

$$\tan \theta = \frac{v^2}{gr}$$

for a conical pendulum. Though the string can approach close to the horizontal, it can never quite get there.

6-5 Gravitation

The earth and the other planets pursue approximately circular orbits around the sun. We conclude that the planets are being acted upon by centripetal forces that originate in the sun, since the sun is at the center of all the orbits. This much was generally understood by the middle of the 17th century, when Newton turned his mind to the question of exactly what the nature of the centripetal forces was.

Newton proposed that the inward force exerted by the sun that is responsible for the planetary orbits is merely one example of a universal interaction, called *gravitation*, that occurs between all bodies in the universe by virtue of their possession of mass. Another example of gravitation,

Gravity holds the planets in orbit around the sun and the moon in orbit around the earth

according to Newton, is the attraction of the earth for nearby bodies. Thus the centripetal acceleration of the moon and the downward acceleration g of objects dropped near the earth's surface have an identical cause, namely the gravitational pull of the earth.

Newton was able to arrive at the form of the *law of universal gravitation* from an analysis of the motions of the planets about the sun:

Law of gravitation

Every object in the universe attracts every other object with a force directly proportional to each of their masses and inversely proportional to the square of the distance separating them.

The law of gravitation is expressed in equation form as

$$F_{grav} = G \frac{m_A m_B}{r^2} \qquad \qquad \textit{Gravitational force} \quad (6\text{-}9)$$

where m_A and m_B are the masses of any two bodies and r is the distance between them. The quantity G is a universal constant whose values in the two principal systems of units are

Constant of gravitation

$$G = 6.67 \times 10^{-11} \frac{\text{N} \cdot \text{m}^2}{\text{kg}^2} \qquad \qquad \textit{(SI system)}$$

$$= 3.44 \times 10^{-8} \frac{\text{lb} \cdot \text{ft}^2}{\text{slug}^2}. \qquad \qquad \textit{(British system)}$$

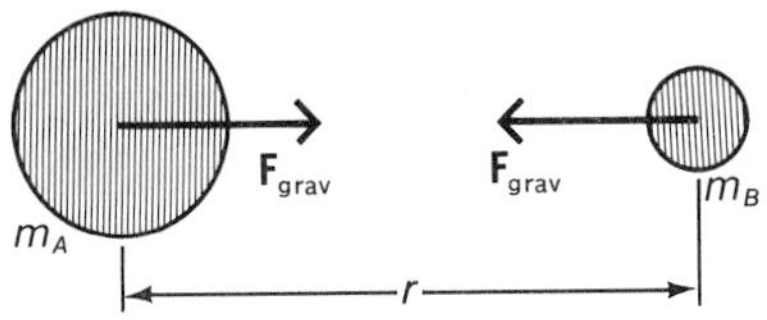

The direction of the gravitational force is always along a line joining the two objects A and B. The force on A exerted by B is equal in magnitude to that on B exerted by A, but is in the opposite direction (Fig. 6-13). A homogeneous spherical object, or one composed of homogeneous spherical shells, behaves gravitationally as if all its mass were concentrated at its center.

Fig. 6-13.   The gravitational forces between two sperical objects.

Problem.   A grocer installs a 100-kg lead block under the pan of his scale. By how much does this increase the reading of the scale when 1 kg of onions are on the pan, if the centers of mass of the lead and of the onions are 0.3 m apart?

Solution.   The gravitational force of the lead on the onions is

$$F = G \frac{m_A m_B}{r^2} = 6.67 \times 10^{-11} \frac{\text{N} \cdot \text{m}^2}{\text{kg}^2} \times \frac{100 \text{ kg} \times 1 \text{ kg}}{(0.3 \text{ m})^2} = 7.4 \times 10^{-8} \text{ N}.$$

The increase in the scale reading is therefore

$$m = \frac{F}{g} = \frac{7.4 \times 10^{-8} \text{ N}}{9.8 \text{ m/s}^2} = 7.6 \times 10^{-9} \text{ kg} = 0.0000076 \text{ g}$$

so it is hardly worth the effort. Blowing gently on the onions will increase the reading over a million times more.

Problem.   The radius of the earth is 6.4×10^6 m. Find its mass.

Solution.   Let us consider an object of mass m at the earth's surface, say an apple. The gravitational pull of the earth on the apple is the apple's weight of

$$w = mg.$$

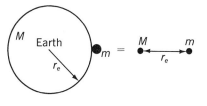

Fig. 6-14.

A spherical object behaves gravitationally as though its mass were concentrated at its center. Thus the earth-apple system can be represented by two particles of masses M and m a distance r_e apart, where M is the earth's mass and r_e is its radius (Fig. 6-14).

According to Newton's law of gravitation, the force the earth exerts on the apple is

$$F = G\,\frac{Mm}{r_e^2}.$$

This force must equal the apple's weight w, and so

$$F = w,$$

$$G\,\frac{Mm}{r_e^2} = mg.$$

When we solve this equation for M we see that the apple's mass m drops out. The mass of the earth is

$$M = \frac{gr_e^2}{G}$$

$$= \frac{9.8 \text{ m/s}^2 \times (6.4 \times 10^6 \text{ m})^2}{6.7 \times 10^{-11} \text{ N} \cdot \text{m}^2/\text{kg}^2}$$

$$= 6.0 \times 10^{24} \text{ kg}.$$

The earth's gravitational pull on an object varies inversely with the square of its distance from the center of the earth (Fig. 6-15). The moon is

Gravity is an inverse-square force

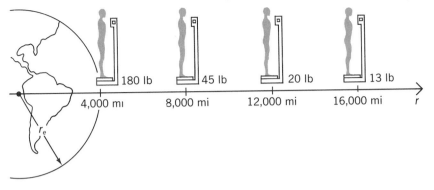

Fig. 6-15. The gravitational force of the earth on an object varies inversely with the square of the distance of the object from the center of the earth. Hence a person's weight at a distance r from the earth's center is $(r_e/r)^2$ of his weight on the earth's surface. The person's mass, of course, is the same everywhere in the universe.

3.84×10^8 m from the earth, which is 60 times the earth's radius. The gravitational force the earth exerts on the moon is therefore $(1/60)^2 = 1/3600$ as strong as the force the earth would exert on it if the moon were at the earth's surface. The acceleration a of the moon toward the earth in turn ought to be 1/3600 of the acceleration of an object at the earth's surface. Since the latter acceleration is g,

$$a = \frac{g}{3600} = \frac{9.8 \text{ m/s}^2}{3600} = 2.7 \times 10^{-3} \frac{\text{m}}{\text{s}^2},$$

which is the same as the value of the moon's centripetal acceleration that follows from its period of revolution about the earth (see Sec. 6-2). The correspondence between the observed centripetal acceleration of the moon and the acceleration inferred from the law of gravitation was used by Newton as evidence for the universal validity of the latter law, which was originally derived from data on planetary motion about the sun (see Special Topic).

What is the justification for assuming that the law of gravitation, obtained from data on the solar system, is also valid for the entire universe, describing the gravitational attraction of objects both larger and smaller than the members of the solar system?

There is no simple answer to this legitimate query; instead we must invoke a broad body of knowledge that bears upon the subject. For example, we observe that all the matter on the earth's surface experiences the same acceleration in free fall, which suggests identical gravitational behavior. Careful analysis of the light reaching us from the stars and galaxies throughout the visible universe indicates that the matter of which these bodies are composed behaves identically with matter found on the earth; and so on. Nowhere do we find reason to suspect there should be any objects in the universe that do not obey Newton's law of gravitation, and it is unreasonable to postulate the existence of such objects with no evidence whatever for the need to do so.

Strong theoretical reasons exist as well for believing in the unity of gravitational phenomena, but in physics experiment and observation are the final arbiters of the correctness of an idea, so these reasons, while welcome as corroboration, must take second place.

6-6 Earth Satellites

What keeps an artificial earth satellite from falling down? The answer, of course, is that it *is* falling down, but, like the moon, at just such a rate as to circle the earth in a stable orbit. Let us use what we know about gravitation and circular motion to investigate the orbits of earth satellites. In the following discussion the frictional resistance of the atmosphere, which ultimately brings down all artificial satellites, will be neglected.

Near the earth's surface the gravitational force on an object of mass m is its weight

$$w = mg$$

where g is the acceleration of gravity. For uniform circular motion about the earth this force must provide the object with the centripetal force

$$F_c = \frac{mv^2}{r}.$$

Hence the condition for a stable orbit is

$$w = F_c,$$

$$mg = \frac{mv^2}{r},$$

$$v = \sqrt{rg}. \qquad\qquad \text{\textit{Orbit of earth satellite}} \quad (6\text{-}10)$$

Here v is the satellite speed, r is the radius of its orbit, and g is the acceleration of gravity.

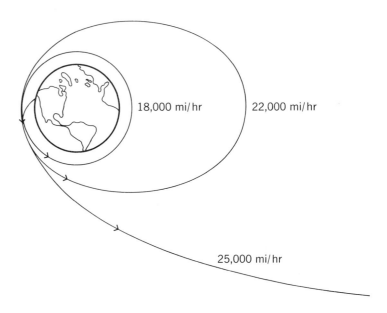

18,000 mi/hr

22,000 mi/hr

25,000 mi/hr

Fig. 6-16.   Depending upon its speed, an object projected horizontally above the earth's surface may fall back to the earth, revolve around the earth in a circular orbit, revolve around the earth in an elliptical orbit, or escape permanently from the earth into space.

Fig. 6-17.

For an orbit just above the earth's surface,

$$v = \sqrt{rg} = \sqrt{6.4 \times 10^6 \text{ m} \times 9.8 \text{ m/s}^2} = 7.9 \times 10^3 \text{ m/s}$$

which is nearly 18,000 mi/hr. Any object sent off parallel to the earth's surface with this speed will become a satellite of the earth. If it is sent off at a higher speed, its orbit will be elliptical rather than circular (Fig. 6-16). If the object's speed is 11.2×10^3 m/s, 41 percent more than the minimum orbital speed, it is able to escape from the earth permanently. This escape velocity is about 25,000 mi/hr.

Problem.   Find the minimum speed an artificial earth satellite must have if it is put in orbit 400 km above the earth's surface. What is the period of this orbit?

Solution.   Here the radius of the satellite's orbit is the earth's radius of 6.4×10^6 m plus the satellite's altitude $h = 0.4 \times 10^6$ m (Fig. 6-17). Hence

$$r = 6.4 \times 10^6 \text{ m} + 0.4 \times 10^6 \text{ m} = 6.8 \times 10^6 \text{ m}.$$

At this altitude the acceleration of gravity has decreased from its value of $g_0 = 9.8$ m/s^2 at the surface to

$$g = \left(\frac{r_{\text{earth}}}{r}\right)^2 g_0 = \left(\frac{6.4 \times 10^6 \text{ m}}{6.8 \times 10^6 \text{ m}}\right)^2 \times 9.8 \, \frac{\text{m}}{\text{s}^2} = 8.7 \, \frac{\text{m}}{\text{s}^2}.$$

Substituting these values of r and g in the basic formula

$$v = \sqrt{rg},$$

we find that

$$v = \sqrt{6.8 \times 10^6 \text{ m} \times 8.7 \text{ m/s}^2} = 7.7 \times 10^3 \frac{\text{m}}{\text{s}}$$

which is smaller than the speed needed for an orbit at the earth's surface.
To find the period T of the satellite, we note that

$$T = \frac{\text{time}}{\text{revolution}} = \frac{\text{distance/revolution}}{\text{distance/time}} = \frac{2\pi r}{v} \qquad (6\text{-}11)$$

where $2\pi r$ is the circumference of the orbit and v is the satellite speed.
Hence

$$T = \frac{2\pi \times 6.8 \times 10^6 \text{ m}}{7.7 \times 10^3 \text{ m/s}} = 5.5 \times 10^3 \text{ s} = 92 \text{ min.}$$

The satellite makes a complete circuit of the earth once every 92 min—a
little less than 16 circumnavigations per day.

6-7 Apparent Weight

Because an earth satellite is always falling toward the earth, an astronaut
inside one feels "weightless." In reality, there *is* a gravitational force acting
on him; what is missing to his senses is the upward reaction force provided
by a stationary platform underneath him—the seat of a chair, the floor of a
room, the ground itself. Instead of pushing back, the floor of the satellite
falls just as fast as he does toward the earth.

Weightlessness

It is useful to distinguish between the *actual weight* of an object, which
is the gravitational force acting on it, and its *apparent weight*, which is the
force it exerts on whatever it rests upon. Thus an astronaut in an earth
satellite has no apparent weight because he does not press down on the
floor of the satellite.

Actual and apparent weight

A person jumping off a diving board is just as "weightless" in his
descent as an astronaut, since nothing restricts his acceleration toward the
earth either. But a person standing on the ground is acted upon by *both*
the downward force of gravity and the upward reaction force of the ground:
the latter force is what prevents him from simply dropping all the way down

to the center of the earth. The human body (indeed, all living organisms on the earth) evolved in the presence of both these forces, and various body functions, such as blood circulation, do not seem to take place efficiently in a "weightless" state. Future satellites and other spacecraft designed for long journeys are likely to be set in rotation so that inertia will cause astronauts to press against the cabin sides, which will then press back (action-reaction again) and so bring about a situation corresponding to that on the earth's surface.

Problem. An airplane pulls out of a dive in a circular arc whose radius is 1000 m. If the speed of the airplane is a constant 200 m/s, find the apparent weight of the 80-kg pilot at the lowest point of the arc.

Solution. The downward force the pilot exerts on his seat has two components. The first is his weight mg, the gravitational pull of the earth on him. The second is the reaction force mv^2/r to the centripetal force that leads to the pilot's curved path through space. If we think of the situation as one of equilibrium within the airplane, we could say that the airplane pushes upward on the pilot who in turn must push down on the airplane with the same force since he is stationary in his seat (Fig. 6-18).

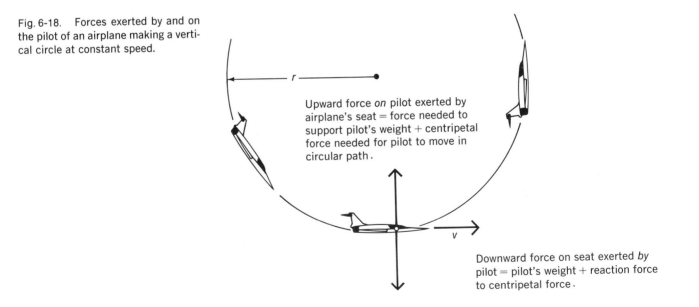

Fig. 6-18. Forces exerted by and on the pilot of an airplane making a vertical circle at constant speed.

Upward force *on* pilot exerted by airplane's seat = force needed to support pilot's weight + centripetal force needed for pilot to move in circular path.

Downward force on seat exerted *by* pilot = pilot's weight + reaction force to centripetal force.

The total downward force is therefore

$$F = mg + \frac{mv^2}{r} = 80 \text{ kg} \times 9.8 \text{ m/s}^2 + \frac{80 \text{ kg} \times (200 \text{ m/s})^2}{1000 \text{ m}} = 3984 \text{ N}.$$

The pilot presses down on his seat with a force of 3984 N. His apparent weight is therefore 3984 N, over five times his actual weight. Because there has been no compensating increase in his muscular strength, the pilot may be unable to move his arms and legs in order to control the airplane. A further complication is the tendency of the pilot's blood to leave his head because of inertia, leaving him with impaired vision ("blacked out") and perhaps unconscious. Special pressure suits have been devised that prevent disturbances in blood supply and the positions of internal organs during severe accelerations in flight.

Because the forces exerted by and on a pilot or astronaut vary with his mass, it is often convenient to speak instead of his acceleration, by custom in units of g. The above pilot's acceleration is

$$a = \frac{F}{m} = \frac{3984 \text{ N}}{80 \text{ kg}} = 49.8 \text{ m/s}^2 = 5.1 \ g.$$

Special Topic

Kepler's Laws and Gravitation

From astronomical observations which he and others (notably Tycho Brahe) had made over a long period of time, Johannes Kepler (1571–1630) discovered three laws that the planets obey as they move around the sun:

1. Each planet has an elliptical orbit with the sun at one focus.

2. Each planet moves so that a radius vector from the sun to it sweeps out equal areas in equal times.

3. The ratio between the square of a planet's period of revolution and the cube of its average distance from the sun has the same value for all the planets.

These laws were used by Newton to arrive at his law of universal gravitation. Let us see how the formula $F_{\text{grav}} = G m_A m_B / r^2$ follows from Kepler's third law.

For simplicity we will assume that the planets move in circular orbits around a stationary sun. A planet of mass m, orbital radius r, and velocity v

must be acted upon by the centripetal force

$$F = \frac{mv^2}{r}.$$

If the period of the orbit is T, then

$$v = \frac{\text{circumference of orbit}}{\text{period of orbit}} = \frac{2\pi r}{T}$$

and

$$F = \frac{4\pi^2 mr}{T^2}.$$

According to Kepler's third law,

$$\frac{T^2}{r^3} = K$$

where K has the same value for all the planets. Hence $T^2 = Kr^3$ and

$$F = \frac{4\pi^2 m}{Kr^2}.$$

According to Newton's hypothesis, this centripetal force is provided by the gravitational force exerted by the sun, from which we conclude that F_{grav} is directly proportional to the mass of a planet and inversely proportional to the square of its distance from the sun.

To complete the analysis, we refer to Newton's third law of motion, which requires that the force a planet exerts on the sun be equal in magnitude to the force the sun exerts on the same planet. If the above formula is correct, then we should be able to apply it either way for a given planet and get the same value of F_{grav}. Since r is the same in both cases, F_{grav} must be proportional to *both* the planet's mass m and the sun's mass M. Because the sun's mass is constant, we can express the quantity $4\pi^2 K$ as GM, so that

$$F_{\text{grav}} = G\,\frac{m_A m_B}{r^2}$$

where G is a universal constant. Extending this formula to *any* two bodies in the universe gives Newton's law of gravitation, Eq. (6-9).

Important Terms

A body traveling in a circle at constant speed is said to be undergoing **uniform circular motion.**

The velocity of a body in uniform circular motion continually changes in direction although its magnitude remains constant. The acceleration that causes the body's velocity to change is called **centripetal acceleration**, and it points toward the center of the body's circular path.

The inward force that provides a body in uniform circular motion with its centripetal acceleration is called **centripetal force.**

Newton's **law of universal gravitation** states that every body in the universe attracts every other body with a force directly proportional to both their masses and inversely proportional to the square of the distance separating them.

Important Formulas

Centripetal acceleration:

$$a_c = \frac{v^2}{r}$$

Centripetal force:

$$F_c = \frac{mv^2}{r}$$

Law of gravitation:

$$F_{\mathrm{grav}} = G \, \frac{m_A m_B}{r^2}$$

Satellite orbit:

$$v = \sqrt{rg}$$

Multiple Choice

1. In order to cause a moving body to pursue a circular path, it is necessary to apply
   a. inertial force.   b. gravitational force.
   c. frictional force.   d. centripetal force.

2. A body traveling in a circle at constant speed
   a. has a constant velocity.
   b. is not accelerated.
   c. has an inward radial acceleration.
   d. has an outward radial acceleration.

3. The acceleration of a body undergoing uniform circular motion is constant in
   a. magnitude only.
   b. direction only.
   c. both magnitude and direction.
   d. neither magnitude nor direction.

4. The centripetal force on a car rounding a curve on a level road is provided by
   a. gravity.
   b. friction between its tires and the road.
   c. the torque applied to its steering wheel.
   d. its brakes.

5. The centripetal force needed to keep the earth in orbit is provided by
   a. inertia.
   b. its rotation on its axis.
   c. the gravitational pull of the sun.
   d. the gravitational pull of the moon.

6. The radius of the path of a body in uniform circular motion is doubled. The centripetal force needed if its velocity remains the same is
   a. half as great as before.
   b. the same as before.
   c. twice as great as before.
   d. four times as great as before.

7. The gravitational force between two bodies does not depend upon
   a. their separation.
   b. the sum of their masses.
   c. the product of their masses.
   d. the constant of gravitation.

8. The gravitational acceleration of a body
   a. has the same value everywhere in space.
   b. has the same value everywhere on the earth's surface.
   c. varies somewhat over the earth's surface.
   d. is greater on the moon because of its smaller diameter.

9. A hole is drilled to the center of the earth and a

stone is dropped into it. When the stone is at the earth's center, compared with the values at the earth's surface
a. its mass and weight are both unchanged.
b. its mass and weight are both zero.
c. its mass is unchanged and its weight is zero.
d. its mass is zero and its weight is unchanged.

10. A $\frac{1}{2}$-kg ball moves in a circle 0.4 m in radius at a velocity of 4 m/s. Its centripetal acceleration is
a. 10 m/s².   b. 20 m/s².
c. 40 m/s².   d. 80 m/s².

11. The centripetal force on the ball of Question 10 is
a. 10 N.   b. 20 N.
c. 40 N.   d. 80 N.

12. A 3200-lb car moves in a circle 100 ft in radius at a velocity of 20 ft/s. Its centripetal acceleration is
a. 0.2 ft/s².   b. 4 ft/s².
c. 20 ft/s².   d. 400 ft/s².

13. The centripetal force on the car of Question 12 is
a. 400 lb.   b. 640 lb.
c. 3200 lb.   d. 12,800 lb.

14. A toy cart at the end of a string 0.7 m long moves in a circle on a table. The cart has a mass of 2 kg and the string has a breaking strength of 40 N. The maximum speed of the cart is approximately
a. 1.9 m/s.   b. 3.7 m/s.
c. 11.7 m/s.   d. 16.7 m/s.

15. On a rainy day the coefficient of friction between a car's tires and a certain level road surface is reduced to half its usual value. The maximum safe velocity for rounding the curve is
a. unchanged.
b. reduced to 25% of its usual value.
c. reduced to 50% of its usual value.
d. reduced to 71% of its usual value.

16. A car is traveling at 30 mi/hr on a road such that the coefficient of friction between its tires and the road is 0.5. The minimum turning radius of the car is
a. 22 ft.   b. 54 ft.
c. 108 ft.   d. 121 ft.

17. A 2-kg stone at the end of a string 1 m long is whirled in a vertical circle. The tension in the string is 52 N when the stone is at the bottom of the circle. The stone's velocity then is
a. 4 m/s.   b. 5 m/s.
c. 6 m/s.   d. 7 m/s.

18. A 3.2-lb stone at the end of a string 2 ft long is whirled in a vertical circle. The tension in the string is 5 lb when the stone is at the top of the circle. The stone's velocity then is
a. 1.8 ft/s.   b. 6 ft/s.
c. 10 ft/s.   d. 12.8 ft/s.

19. A woman has a mass of 60 kg at the earth's surface. At a height of one earth's radius above the surface her mass is
a. 15 kg.   b. 30 kg.
c. 60 kg.   d. 120 kg.

20. A man has a weight of 160 lb at the earth's surface. At a height of one earth's radius above the surface his weight is
a. 40 lb.   b. 80 lb.
c. 160 lb.   d. 320 lb.

21. The moon's mass is 1.2 percent of the earth's mass. Relative to the gravitational force the earth exerts on the moon, the gravitational force the moon exerts on the earth
a. is smaller.
b. is the same.
c. is greater.
d. depends on the phase of the moon.

Exercises

1. Under what circumstances, if any, can an object move in a circular path without being accelerated?

2. Where should you stand on the earth's surface to experience the most centripetal acceleration? The least?

3. A woman takes her bathroom scale with her on a vacation trip to a mountain resort. If her mass does not change, will her weight appear more, less, or the same as at sea level?

4. The value of *g* decreases with increasing distance from the earth's surface. What happens to the value of *G*?

5. Compare the magnitude of the gravitational force the sun exerts on the earth with the magnitude of the centripetal force involved in the earth's orbital motion.

6. Why is the earth round?

7. The earth's equatorial radius is about 21 km greater than its polar radius, a phenomenon known as the "equatorial bulge." (a) How is this fact related to the daily rotation of the earth on its axis? (b) If the earth were to rotate twice as fast as it now does, would you expect the equatorial bulge to be more than, the same as, or less than it is now?

8. Two satellites are launched from a certain station with the same initial speeds relative to the earth's surface. One is launched toward the west, the other toward the east. Will there be any difference in their orbits? If so, what will the difference be and why?

9. An artificial earth satellite is placed in an orbit whose radius is half that of the moon's orbit. Is its time of revolution longer or shorter than that of the moon?

10. For the moon to have the same orbit it has now, what would its velocity have to be if (a) the moon's mass were double its present mass, and (b) the earth's mass were double its present mass?

11. A phonograph record 12 in. in diameter rotates $33\frac{1}{3}$ times per minute. (a) What is the linear speed of a point on its rim in ft/s? (b) What is the centripetal acceleration of a point on its rim?

12. The minute hand of a large clock is 0.5 m long. (a) What is the linear speed of its tip in m/s? (b) What is the centripetal acceleration of the tip of the hand?

13. What is the centripetal force needed to keep a 3-kg mass moving in a circle of radius 0.5 m at a speed of 8 m/s?

14. What is the centripetal force needed to keep a 12-lb weight moving in a circle of radius 2 ft at a speed of 10 ft/s?

15. A certain string 4 ft long breaks when its tension is 40 lb. (a) What is the greatest speed in ft/s at which it can be used to whirl a stone weighing 3 lb? (b) Express this speed in revolutions/sec. (Neglect the gravitational pull of the earth on the stone.)

16. A 2000-kg car is rounding a curve of radius 200 m on a level road. The maximum frictional force the road can exert on the tires of the car is 4000 N. What is the highest velocity at which the car can round the curve?

17. A string 0.8 m long is used to whirl a 2-kg stone in a vertical circle. What must be the velocity of the stone at the top of the circle if the string is to be just taut? How does this velocity compare with that required for a 1-kg stone in the same situation?

18. A string 1 m long is used to whirl a $\frac{1}{2}$-kg stone in a vertical circle. What is the tension in the string when the stone is at the top of the circle moving at 5 m/s?

19. A string 2 ft long is used to whirl a 1-lb stone in a vertical circle. What is the stone's velocity at the bottom of the circle if the string tension there is 6 lb?

20. The 200-g head of a golf club moves at 45 m/s in a circular arc of 1 m radius. How much force must the player exert on the handle of the club to prevent it from flying out of his hands? Assume that the shaft of the club has negligible mass.

21. A physics instructor swings a pail of water in a vertical circle 4 ft in radius. What is the maximum time per revolution if the water is not to spill?

22. A boy swings a pail of water in a vertical circle 1 m in radius. What is the maximum time per revolution if the water is not to spill?

23. Find the gravitational force between two 1-ton lead spheres whose centers are 10 ft apart.

24. A 2-kg mass is 1 m away from a 5-kg mass. What is the gravitational force (a) that the 5-kg mass

exerts upon the 2-kg mass; (b) that the 2-kg mass exerts upon the 5-kg mass? (c) If both masses are free to move, what are their respective accelerations in the absence of other forces?

25. An object dropped near the earth's surface falls 4.9 m in the first second. How far does the moon fall toward the earth in each second? Why doesn't the moon ever reach the earth?

26. What is the acceleration of a meteor when it is one earth's radius above the surface of the earth? Two earth's radii?

27. The radius of the earth is 6.4×10^6 m. What is the acceleration of a meteor when it is 8×10^6 m from the center of the earth?

28. Show that $\cos \theta = g/4\pi^2 f^2 L$ for the conical pendulum of Fig. 6-12, where f is the number of revolutions per second the particle makes.

29. In *Alice in Wonderland* this statement appears: "Now, here, you see, it takes all the running you can do, to stay in the same place. If you want to get somewhere else, you must run at least twice as fast as that!" Check the correctness of this statement by comparing the velocity needed for a stable satellite orbit near the surface of the earth with the escape velocity. If the ratio between them is not exactly 2, what is it?

Problems

1. A dime is placed 10 cm from the center of a record. The coefficient of friction between coin and record is 0.3. Will the coin remain where it is or will it fly off when the record turns at $33\frac{1}{3}$ revolutions/min? At 78 rev/min?

2. A highway curve has a radius of 300 m. (a) At what angle should it be banked for a traffic speed of 100 km/hr? (b) If the curve is not banked, what is the minimum coefficient of friction required between tires and road?

3. An airplane traveling at 500 km/hr banks at an angle of 45° as it makes a turn. What is the radius of the turn in miles? Assume that the rudder is not used in making the turn.

4. A car whose speed is 60 mi/hr rounds a curve 600 ft in radius which is properly banked for a speed of 30 mi/hr. Find the minimum coefficient of friction between tires and road that will permit the car to make the turn.

5. A curve in a road 25 ft wide has a radius of 200 ft. How much higher than its inner edge should the outer edge of the road be if it is to be banked properly for cars traveling at 20 mi/hr?

6. A sled slides down the frictionless track shown in Fig. 6-19 and loops-the-loop without falling off. What is the minimum value of h?

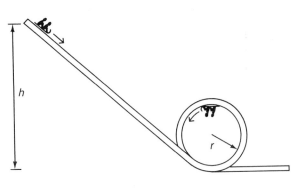

Fig. 6-19.

7. The moon's mass is 7.3×10^{22} kg and the average radius of its orbit is 3.8×10^8 m. At what point could an object be placed between the earth and the moon where it would experience no resultant force? (Neglect the gravitational attractions of the sun and the other planets.)

8. The mass of the planet Jupiter is 1.9×10^{27} kg and that of the sun is 2.0×10^{30} kg. The average distance between them is 7.8×10^{11} m. (a) What is the gravitational force the sun exerts on Jupiter? (b) Assuming that Jupiter has a circular orbit, what must its speed be for the orbit to be a stable one?

9. The moon's radius is 27% of the earth's radius and its mass is 1.2% of the earth's mass. (a) What is the acceleration of gravity (in ft/s²) on the surface of the moon? (b) How much would a boy weigh there whose weight on the earth is 100 lb?

10. The mass of the planet Jupiter is 1.9×10^{27} kg and its radius is 7.0×10^{7} m. What is the acceleration of gravity (in m/sec²) on the surface of Jupiter?

11. If a planet existed whose mass and radius were both twice those of the earth, what would the acceleration of gravity at its surface be in terms of g?

12. If a planet existed whose mass and radius were both half those of the earth, what would the acceleration of gravity at its surface be in terms of g?

13. Find the radius of a satellite orbit whose period is exactly 1 day. (Be sure to take into account the variation of g with r.) Such a satellite will remain indefinitely over a particular location on the earth; most communications relay satellites are placed in orbits of this kind.

14. A satellite is to be put into orbit around the moon just above its surface. What should its speed be? Assume that the moon's radius is half that of the earth and that the acceleration of gravity at its surface is $g/6$.

15. The gravitational force the sun exerts on the earth is nearly 180 times greater than the force the moon exerts on the earth, yet the moon is more effective in producing the tides than the sun. To see why, perform the following calculations. First find the difference between the force the sun exerts on 1 kg of water at a point on the equator nearest the sun and the force the sun

would exert on 1 kg of water at the earth's center. Then make the same calculation for the forces the moon exerts on 1 kg of water at these locations, and compare the results. The moon's mass is 7.3×10^{22} kg, the sun's mass is 2.0×10^{30} kg, the earth's radius is 6.4×10^{6} m, the earth's orbital radius is 1.5×10^{11} m, and the moon's orbital radius is 3.8×10^{8} m.

Note: There is an easy way to make these calculations. When $x \ll 1$, $1/(1-x)^2 \approx 1 + 2x$. Hence if R is the distance from the sun (or moon) to the earth's center and r is the earth's radius, then

$$\frac{1}{(R-r)^2} = \frac{1}{R^2\left(1 - \dfrac{r}{R}\right)^2} \approx \frac{1}{R^2}\left(1 + 2\frac{r}{R}\right).$$

16. Most of the stars in the galaxy of which the sun is a member (the "Milky Way") are concentrated in an assembly about 100,000 light-years across whose shape is roughly that of a fried egg. The sun is about 30,000 light-years from the center of the galaxy, and revolves around it with a period of about 2×10^{8} years. A reasonable estimate for the mass of the galaxy may be obtained by considering this mass to be concentrated at the galactic center with the sun revolving around it like a planet around the sun. On this basis, calculate the mass of the galaxy. How many stars having the mass of the sun is this equivalent to? (1 year = 3.16×10^{7} sec, 1 light-year = 9.46×10^{15} m, and $m_{\text{sun}} = 1.99 \times 10^{30}$ kg.)

Answers to Multiple Choice

1. d	8. c	15. d
2. c	9. c	16. d
3. a	10. c	17. a
4. b	11. b	18. d
5. c	12. b	19. c
6. a	13. a	20. a
7. b	14. b	21. b

7

Energy

We all use the word "energy," but how many of us know exactly what it means? We speak of the energy of a lightning bolt or of an ocean wave; we say that an active person is energetic; we hear a candy bar described as being full of energy; we read that most of the world's electricity will come from nuclear energy in the years to come. What do a lightning bolt, an ocean wave, an active person, a candy bar, and an atomic nucleus have in common?

In general terms, energy refers to an ability to accomplish change. All changes in the physical world involve energy, usually with energy being transformed from one sort into another. But "change" is not a very precise concept, and we must clarify our ideas before going further. What we shall do is first define a quantity called work, and then see how it permits us to discuss energy and its relation to change in the orderly manner of science.

7-1 Work

All changes in the physical universe are the result of forces. Forces set objects in motion, change their paths, and bring them to a stop; forces pull things together and push them apart. The quantity called work is a measure of the amount of change (in a general sense) a force gives rise to when it acts upon something.

When we push against a brick wall, nothing happens. We have applied a force, but the wall has not yielded and shows no effects. On the other hand, when we apply exactly the same force to a ball, the ball flies through the air for some distance. In the latter case something has been accomplished because of our push, whereas in the former there has been no result (Fig. 7-1).

Force and change

What is the essential difference between the two situations? In the first case, where we pushed against a wall, the wall did not move. But in the second case, where we threw the ball, the ball *did* move while the force was being applied and before it left our hand. The displacement of the body while the force acted on it was what made the difference.

If we think carefully along these lines, we will see that whenever a force acts so as to produce motion in an object, the force acts during a displacement of the object. In order to make this notion definite, a physical quantity called *work* is defined as follows:

The work done by a force acting on an object which moves in the same direction as the force is equal to the magnitude of the force multiplied by the distance through which the force acts.

Work is force times displacement

In equation form,

$$W = Fs \qquad (7\text{-}1)$$

Work = force × distance.

No work done   Work done

Fig. 7-1. Work is done by a force when the object it acts upon is displaced while the force is applied.

This definition is a great help in clarifying the effects of forces. Unless a force acts through a distance, no work is done no matter how great the force. And even if a body moves through a distance, no work is done unless a force is acting upon it or it exerts a force on something else (Fig. 7-2).

Our intuitive concept of work is in accord with the precise definition above: when something happens because a person applies a force of some kind, we say that he has done work. Here we have simply broadened the concept to include inanimate forces. (We still must be careful, though; while we may become tired after pushing against a brick wall for a long time, we still have done no work on the wall if it remains in place.)

The above definition of work has an important qualification: the force **F** must be in the same direction as the vector displacement **s**. If **F** and **s** are not parallel, we must replace F in the formula $W = Fs$ by the magnitude of its component $\mathbf{F}_s$ in the direction of the displacement s (Fig. 7-3). The magnitude of this component is

$$F_s = F \cos \theta$$

where θ is the angle between **F** and **s**. Hence the most general definition of work is

General definition of work $$W = Fs \cos \theta.$$ *Work* (7-2)

Work is a scalar quantity; there is no direction associated with work.

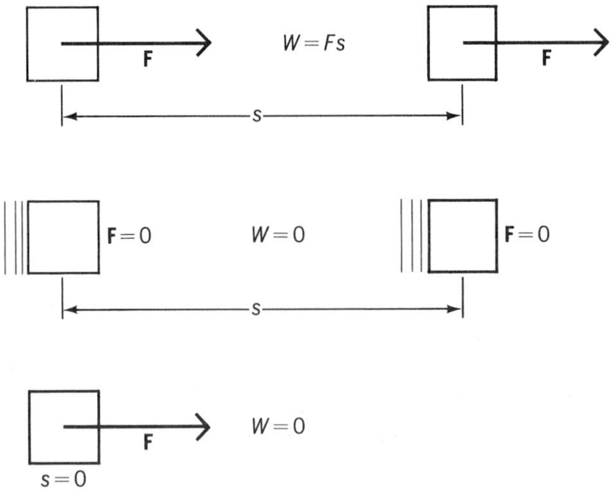

Fig. 7-2.

Work = force × distance through which force acts.

When there is no net applied force, no work is done even though the object may move.

When the object acted upon by a force remains at rest, no work is done.

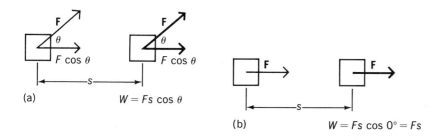

(a) $W = Fs \cos \theta$

(b) $W = Fs \cos 0° = Fs$

(c) $W = Fs \cos 90° = 0$

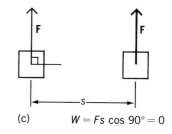

Fig. 7-3.   The work done by a force depends upon the angle θ between the force and its displacement. (a) In general, $W = Fs \cos \theta$. (b) When **F** is parallel to **s**, $W = Fs$. (c) When **F** is perpendicular to **s**, $W = 0$ since **F** then has no component in the direction of **s** and therefore does not affect its motion.

Equation (7-2) is always correct, since when the force and the displacement are parallel, $\theta = 0$ and $\cos 0 = 1$. When the force and the displacement are perpendicular, $\theta = 90°$ and $\cos 90° = 0$, so $W = 0$. In order for work to be done, the applied force must have a component in the direction of the displacement.

In the British system, the unit of work is the *foot·pound*, abbreviated ft·lb. One ft·lb is equal to the work done by a force of 1 lb acting through a distance of 1 ft.

In the SI system, work is given a special unit, the *joule*, abbreviated J. One joule is equal to the work done by a force of 1 N acting through a distance of 1 m. That is,

$$1 \text{ J} = 1 \text{ N·m}.$$

Problem.   A man pulls an 80-kg crate for 20 m across a level floor using a rope that is 30° above the horizontal. The man exerts a force of 150 N on the rope (Fig. 7-4). How much work does he perform?

Solution.   The work done by the man is

$W = Fs \cos \theta$

$= 150 \text{ N} \times 20 \text{ m} \times \cos 30°$

$= 150 \times 20 \times 0.866 \text{ J}$

$= 2.6 \times 10^3 \text{ J}.$

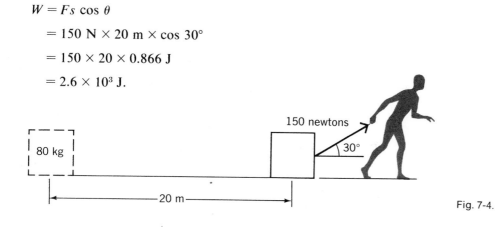

Fig. 7-4.

Evidently the 80-kg mass of the crate has no significance here — it is the force exerted by the man that determines how much work he does.

7-2 Work Done Against Gravity

It is easy to compute the work done in lifting an object against gravity. The force of gravity on an object of mass m is the same as its weight $w = mg$. Hence in order to raise the object to a height h above its original position, a force of mg must be exerted on it. Since $F = mg$ and $s = h$ here, the work done is

$$W = Fs$$
$$= mgh. \tag{7-3}$$

Thus to lift an object of mass m to a height h requires the performance of the amount of work mgh (Fig. 7-5).

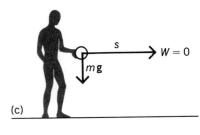

(c)

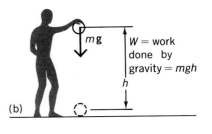

(b)

Fig. 7-5. (a) The work mgh must be done to lift an object of mass m to a height h. (b) When an object of mass m falls from a height h, the force of gravity does the work mgh on it. (c) The force of gravity does no work on objects that move parallel to the earth's surface.

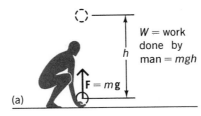

(a)

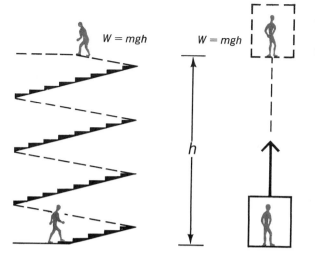

It is important to note that only the height *h* is involved in work done against the force of gravity. The particular route taken by an object being raised is not significant (Fig. 7-6); excluding any frictional effects, exactly as much work must be expended to climb a flight of stairs as to go up in an elevator to the same floor (though not by the person involved!).

Problem. How much work is done in lifting a 30-kg load of bricks to a height of 20 m on a building under construction?

Solution. Here $m = 30$ kg and $h = 20$ m, so that

$$W = mgh = 30 \text{ kg} \times 9.8 \frac{\text{m}}{\text{s}^2} \times 20 \text{ m} = 5.9 \times 10^3 \text{ joules}.$$

7-3 Power

Often the time needed to perform a task is just as significant as the actual amount of work required. Given enough time, even the feeblest motor can raise the Sphinx. However, if we want to carry out a certain operation quickly, we try to obtain a motor whose output of work is rapid in terms of the total required. The rate at which work is done is therefore an important engineering quantity. This rate is called *power*: the faster some agency can do work, the more *powerful* it is.

Power is rate of doing work

If an amount of work W is performed in a time interval t, the power involved is

$$P = \frac{W}{t} \tag{7-4}$$

$$\text{Power} = \frac{\text{work done}}{\text{time interval}}.$$

In the SI system, where work is measured in joules and time in seconds, the unit of power is the *watt*:

The watt

$$1 \text{ watt} = 1 \text{ W} = 1 \text{ J/s}.$$

The watt is rather small for most industrial purposes; even an electric clock requires several watts of power. The larger *kilowatt* (kW) is accordingly in common use, where

$$1 \text{ kW} = 1000 \text{ watts}.$$

In the British system, where energy is measured in ft·lb and time in seconds, the unit of power is the *ft·lb/s*. The ft·lb/s is also inconveniently small and has been replaced by the larger *horsepower* (hp) in engineering practice. The horsepower was introduced two centuries ago by James Watt to compare the output of the steam engine he had perfected with a more familiar source of power. Today the horsepower is defined as

The horsepower

$$1 \text{ hp} = 550 \text{ ft·lb/s}.$$

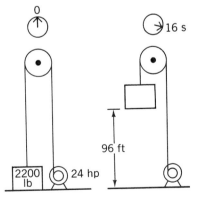

In the United States the watt and kilowatt are employed in connection with electrical power, while mechanical power is customarily specified in horsepower. To convert power figures from one system to the other, we note that

$$1 \text{ hp} = 746 \text{ W} = 0.746 \text{ kW},$$

$$1 \text{ kW} = 1.34 \text{ hp}.$$

Problem. A 24-hp motor provides power for the elevator of a 6-story building. If the total weight of the loaded elevator is 2200 lb, how long does it take to rise the 96 ft from the ground floor to the top floor (Fig. 7-7)?

Solution. The work done in raising the elevator through 96 ft is

$$W = Fs = wh = 2200 \text{ lb} \times 96 \text{ ft} = 2.11 \times 10^5 \text{ ft·lb}.$$

Fig. 7-7.

Since 1 hp = 550 ft·lb/s, the power of the lifting motor is

$$P = 24 \text{ hp} \times 550 \frac{\text{ft·lb/s}}{\text{hp}} = 1.32 \times 10^4 \text{ ft·lb/s.}$$

From the definition of power, $P = W/t$, and so, neglecting friction,

$$t = \frac{W}{P} = \frac{2.11 \times 10^5 \text{ ft·lb}}{1.32 \times 10^4 \text{ ft·lb/s}} = 16 \text{ s.}$$

When a constant force performs work on some object at a uniform rate, the power delivered is equal to the product of the force and the velocity v with which the object is moving. To verify this statement, we express the power in terms of the work done by the force in the time t:

$$P = \frac{W}{t} = \frac{Fs}{t}.$$

But the velocity of an object that travels the distance s in the time t is

$$v = \frac{s}{t}.$$

Hence

$$P = Fv \tag{7-5}$$

Power = force × velocity.

Problem.   A swimmer develops an average power of 200 W as she covers 100 m in 80 s. What is the resistive force exerted by the water on her?

Solution.   The swimmer's velocity is $v = 100$ m/80 s $= 1.25$ m/s. Since $P = Fv$, here

$$F = \frac{P}{v} = \frac{200 \text{ W}}{1.25 \text{ m/s}} = 160 \text{ N,}$$

which is about 36 lb.

An athlete who specializes in distance running, swimming, cycling, or rowing is usually capable of a sustained power output of 200–300 W, depending on his mass, perhaps more if he is of championship standard. The basic limitation on his power output is the supply of oxygen via the bloodstream from his lungs to his muscles, where it is needed for the metabolic processes that convert the energy in nutrients into work. For a momentary effort, such as that of a weightlifter, a jumper, or a golfer, the power output may be increased to over 2 kW by a process involving the reduction of pyruvic acid to lactic acid in the muscles, but the consequences of lactic acid accumulation limit such an effort to a second or two.

7-4 Energy

From the straightforward notion of work we proceed to the complicated and many-sided concept of *energy*:

Energy

Energy is that property whose possession enables something to perform work.

When we say that something has energy, we mean it is capable (directly or indirectly) of exerting a force on something else and doing work on it. On the other hand, when we do work on something, we have added to it an amount of energy equal to the work done. The units of energy are the same as those of work, the foot-pound and the joule.

Energy units

There are three broad categories of energy:

Categories of energy

1. *Kinetic energy*, which is the energy something possesses by virtue of its motion;

2. *Potential energy*, which is the energy something possesses by virtue of its position.

3. *Rest energy*, which is the energy something possesses by virtue of its mass.

In the above descriptions the word "something" was used instead of "object" because, as we shall see later, such nonmaterial entities as force fields and massless particles may also possess energy.

All modes of energy possession fit into one or another of these three categories. For instance, it is convenient for many purposes to think of heat as a separate form of energy, but what this term actually refers to is the sum of the kinetic energies of the randomly moving atoms and molecules in a body of matter.

7-5 Kinetic Energy

When we perform work on a ball by throwing it, what becomes of this work?

Let us suppose we apply the uniform force **F** to the ball for a distance s before it leaves our hand, as in Fig. 7-8(a). The work done on the ball is therefore

$$W = Fs. \tag{7-6}$$

The mass of the ball is m. As we throw it, its acceleration has the magnitude

$$a = \frac{F}{m} \tag{7-7}$$

according to the second law of motion, $\mathbf{F} = m\mathbf{a}$ (Fig. 7-8(b)).

We know from the formula

$$v_f^2 = v_0^2 + 2as \tag{3-3}$$

that when an object starting from rest ($v_0 = 0$) undergoes an acceleration of magnitude a through a distance s, its final velocity v is related to a and s by

$$v_f^2 = 2as. \tag{7-8}$$

This relationship is pictured in Fig. 7-8(c).

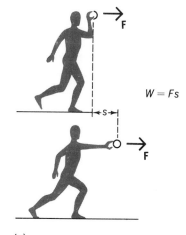

(a)

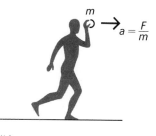

(b)

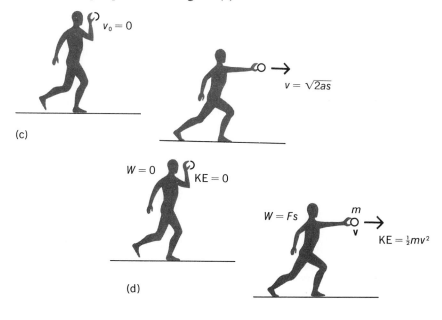

(c)

(d)

Fig. 7-8.   Successive steps in the derivation of the formula $KE = \frac{1}{2}mv^2$ for the kinetic energy of a moving body.

If we now substitute F/m for a in Eq. (7-8), we find that

$$v_f{}^2 = 2as$$

$$= 2\left(\frac{F}{m}\right)s,$$

which we can rewrite as

$$Fs = \tfrac{1}{2}mv_f{}^2. \tag{7-9}$$

The quantity on the left-hand side, Fs, is the work our hand has done in throwing the ball. The quantity on the right-hand side, $\tfrac{1}{2}mv_f{}^2$, must therefore be the energy acquired by the ball as a result of the work we did on it. This energy is *kinetic energy*, energy of motion. That is, we interpret the preceding equation as follows (Fig. 7-8(d)):

$$Fs = \tfrac{1}{2}mv_f{}^2$$

Work done on ball = kinetic energy of ball

All moving bodies possess kinetic energy

The symbol for kinetic energy is KE. The kinetic energy of an object of mass m and velocity v is therefore

$$KE = \tfrac{1}{2}mv^2. \qquad\qquad \textit{Kinetic energy} \quad (7\text{-}10)$$

A moving object is able to perform an amount of work equal to $\tfrac{1}{2}mv^2$ in the course of being stopped.

We can use the relationship $Fs = \tfrac{1}{2}mv^2$ between work done and the resulting kinetic energy to arrive at an interesting conclusion about animal velocities. According to this formula,

$$v = \sqrt{\frac{2Fs}{m}}.$$

Animal velocity and size

Let us interpret v as an animal's velocity, F as the force its leg muscles exert over the distance s, and m as its mass. As we saw in the Special Topic to Chap. 4, the mass of an animal is approximately proportional to L^3, where L is a representative linear dimension such as its length, and the forces its muscles can exert are approximately proportional to L^2. The distance through which a muscle acts is proportional to L. Hence the quantity Fs/m depends on L as $L^2 \times L/L^3 = 1$, which means that Fs/m, and hence v, should not de-

L at all! In fact, although different animals have different running abilities, there is indeed little correlation with size over a wide span. A hare can run as fast as a horse.

To help appreciate the magnitudes of the kinetic energies characteristic of various common moving bodies, let us compute a few such values. We begin with a ball of mass 1 kg which is thrown with a velocity of 5 m/s (Fig. 7-9(a)). Its kinetic energy is

$$KE = \tfrac{1}{2}mv^2 = \tfrac{1}{2} \times 1 \text{ kg} \times \left(5 \, \frac{\text{m}}{\text{s}}\right)^2 = 12.5 \text{ J.}$$

An automobile which weighs 3200 lb (Fig. 7-9(b)) is traveling at a velocity of 15 mi/hr (22 ft/s). Its kinetic energy is

$$KE = \tfrac{1}{2}mv^2 = \frac{1}{2}\frac{w}{g}v^2 = \frac{1}{2} \times \frac{3200 \text{ lb}}{32 \text{ ft/s}^2} \times \left(22\,\frac{\text{ft}}{\text{s}}\right)^2$$

$$= 24{,}200 \text{ ft·lb} = 2.4 \times 10^4 \text{ ft·lb.}$$

The same automobile at a velocity of 60 mi/hr (88 ft/s) has a kinetic energy of

$$KE = \frac{1}{2}\frac{w}{g}v^2 = \frac{1}{2} \times \frac{3200 \text{ lb}}{32 \text{ ft/s}^2} \times \left(88\,\frac{\text{ft}}{\text{s}}\right)^2$$

$$= 387{,}200 \text{ ft·lb} = 3.9 \times 10^5 \text{ ft·lb,}$$

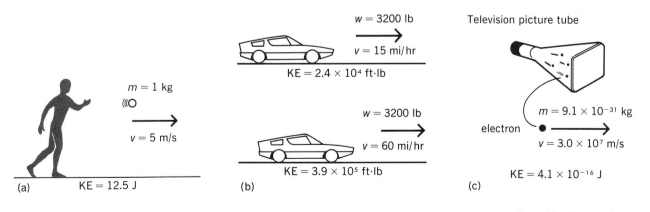

Fig. 7-9. Some kinetic energies.

which is 16 times greater than at a velocity of 15 mi/hr. The fact that kinetic energy is proportional to the *square* of the velocity is responsible for the severity of automobile accidents at high speeds.

Electrons are the smallest particles present in ordinary matter. The electron mass (provided that the electron velocity is not too close to the speed of light, a requirement we shall discuss in Chapter 27) is 9.1×10^{-31} kg. Electrons constitute the beam in a television picture tube (Fig. 7-9(c)), whose impacts on the tube screen produce tiny flashes of light. Such electrons might have speeds of 3×10^7 m/s; hence their kinetic energies are

$$KE = \tfrac{1}{2}mv^2 = \tfrac{1}{2} \times 9.1 \times 10^{-31} \text{ kg} \times \left(3 \times 10^7 \frac{\text{m}}{\text{s}}\right)^2$$

$$= 4.1 \times 10^{-16} \text{ J}.$$

Problem. What is the power output of the engine of a 1500-kg car if the car can go from rest to a velocity of 80 km/hr in 10 s?

Solution. The work needed to accelerate an object to a given velocity from rest is equal to the kinetic energy $\tfrac{1}{2}mv^2$ it has at that velocity, assuming negligible friction. Here

$$v = 80 \frac{\text{km}}{\text{hr}} \times 1000 \frac{\text{m}}{\text{km}} \times \frac{1}{3600 \text{ s/hr}} = 22.2 \text{ m/s}$$

so that

$$KE = \tfrac{1}{2}mv^2 = \tfrac{1}{2} \times 1500 \text{ kg} \times (22.2 \text{ m/s})^2 = 3.70 \times 10^5 \text{ J}.$$

To bring the car from rest to a velocity of 80 km/hr therefore requires the expenditure of 3.70×10^5 J of work. The power needed to provide this quantity of work in 10 s is

$$P = \frac{W}{t} = \frac{3.70 \times 10^5 \text{ J}}{10 \text{ s}} = 3.70 \times 10^4 \text{ W} = 37.0 \text{ kW},$$

which is equivalent to

$$P = 37.0 \text{ kW} \times 1.34 \frac{\text{hp}}{\text{kW}} = 49.6 \text{ hp}.$$

7-6 Potential Energy

When we drop a stone from a height h, it falls faster and faster and finally strikes the ground. In striking the ground the stone does work; if it is sufficiently heavy and has fallen from a great enough height, the work done by the stone is manifest as a hole (Fig. 7-10). Evidently the stone at its original location h above the ground had a capacity to do work, even though it was stationary at the time. The work the stone can perform in falling to the ground is called its *potential energy*, symbol PE.

We have already calculated that the work we must do to raise the stone to the height h is

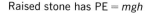

Raised stone has PE = mgh

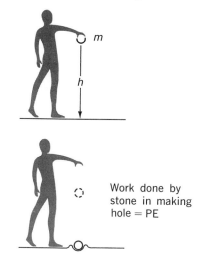

Work done by stone in making hole = PE

Fig. 7-10.

$$W = mgh, \tag{7-3}$$

where m is the stone's mass. This amount of work can also be done *by* the stone after dropping from the height h. Since the raised stone has the potentiality of doing the amount of work mgh, we define its *potential energy* as

$$PE = mgh, \qquad \text{\textit{Gravitational potential energy}} \quad (7\text{-}11)$$

the product of its mass, the acceleration of gravity, and its height. In the British system of units, weights rather than masses are usually specified. Since $w = mg$, we may also write Eq. (7-11) as

$$PE = wh, \tag{7-12}$$

which is more convenient in treating problems in this system of units.

Reference level

The gravitational potential energy of an object depends upon the reference level from which its height h is measured. For example, the potential energy of a 1.0-kg book held 10 cm above a desk is

$$PE = mgh = 1.0 \text{ kg} \times 9.8 \, \frac{\text{m}}{\text{s}^2} \times 0.10 \text{ m} = 0.98 \text{ J}$$

with respect to the desk (Fig. 7-11). However, if the book is 1.0 m above the floor of the room, its potential energy is

$$PE = mgh = 1.0 \text{ kg} \times 9.8 \, \frac{\text{m}}{\text{s}^2} \times 1.0 \text{ m} = 9.8 \text{ J}$$

Fig. 7-11. The potential energy of an object depends upon the reference level from which its height *h* is measured.

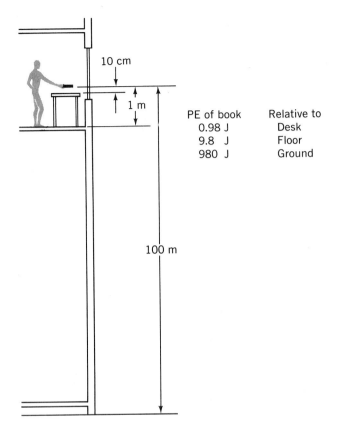

PE of book	Relative to
0.98 J	Desk
9.8 J	Floor
980 J	Ground

with respect to the floor. And the book may conceivably be 100 m above the ground, so its potential energy is

$$\text{PE} = mgh = 1.0 \text{ kg} \times 9.8 \ \frac{\text{m}}{\text{s}^2} \times 100 \text{ m} = 980 \text{ J}$$

with respect to the ground. The height *h* in the formula PE = *mgh* means nothing unless the base height *h* = 0 is specified.

In general, potential energy is a relative quantity. Just as the KE of a moving object depends upon the frame of reference in which its velocity is measured, so the PE of an object subject to a force depends upon the reference position chosen. Further, the potential energy is not a property of the object by itself but of the *system* of the object and the body that exerts the force on it. The PE of a stone held above the earth's surface is shared by both the stone and the earth. When the stone is released, both it and the earth move toward each other. However, the earth's motion is imperceptibly small because of its immense mass relative to that of the stone. It is therefore appropriate to attribute the entire PE of the system to the stone, but in the case of two objects more nearly comparable in mass, it must be kept in mind that the PE belongs to the system, not just to one of them.

We shall now compute a few potential-energy values. An apple of mass 0.5 kg is on a branch 5 m from the ground (Fig. 7-12a). Its potential energy relative to the ground is

Examples of gravitational potential energies

$$PE = mgh = 0.5 \text{ kg} \times 9.8 \frac{\text{m}}{\text{s}^2} \times 5 \text{ m} = 25 \text{ J.}$$

A 3200-lb automobile is at the top of a 100-ft hill (Fig. 7-12b). Its potential energy relative to the foot of the hill is

$$PE = mgh = wh = 3200 \text{ lb} \times 100 \text{ ft} = 320,000 \text{ ft·lb} = 3.2 \times 10^5 \text{ ft·lb,}$$

which is less than the kinetic energy of the same car when its velocity is 60 mi/hr. In other words, a crash at 60 mi/hr into a stationary obstacle will yield more work (that is, do more damage) than dropping the car 100 ft.

An electron 10 cm (0.1 m) from the bottom of a television picture tube (Fig. 7-12c) has a potential energy, relative to the bottom, of

$$PE = mgh = 9.1 \times 10^{-31} \text{ kg} \times 9.8 \frac{\text{m}}{\text{s}^2} \times 0.1 \text{ m} = 8.9 \times 10^{-31} \text{ J.}$$

This is almost 15 powers of 10 less than its kinetic energy!

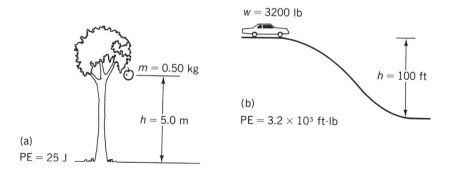

(a)
PE = 25 J

$m = 0.50$ kg

$h = 5.0$ m

$w = 3200$ lb

$h = 100$ ft

(b)
PE = 3.2×10^5 ft·lb

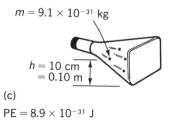

$m = 9.1 \times 10^{-31}$ kg

$h = 10$ cm
$= 0.10$ m

(c)
PE = 8.9×10^{-31} J

Fig. 7-12. Some gravitational potential energies.

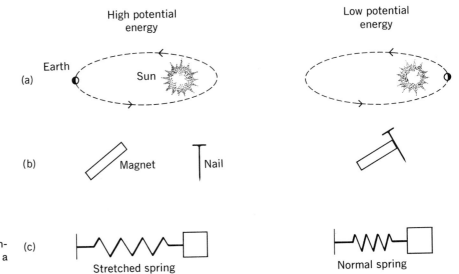

High potential
energy

Low potential
energy

Earth

(a)

Sun

(b) Magnet Nail

Fig. 7-13. Three examples of potential energy. In each case the PE is a
property of the entire system.

(c)

Stretched spring

Normal spring

Potential energy is a general concept

We have spoken of one type of potential energy only, namely that possessed by an object by virtue of being raised above some reference level in the earth's gravitational field. The concept of potential energy is a much more general one, however, for it refers to the energy something has as a consequence of its position regardless of the nature of the force acting on it. The earth itself, for instance, has potential energy with respect to the sun, since if its orbital motion were to cease it would fall toward the sun (Fig. 7-13). An iron nail has potential energy with respect to a nearby magnet, since it will fly to the magnet if released. An object at the end of a stretched spring has potential energy with respect to its position when the spring has its normal extension, since if let go the object will move as the spring contracts. In each of these cases the object in question has the potentiality of doing work in its original position.

7-7 Rest Energy

Rest energy

Every body of matter possesses a certain inherent amount of energy called *rest energy* even if it is not moving (so that KE = 0) and is not being acted upon by a force (so that PE = 0). A body whose mass when it is at rest is m_0 has a rest energy E_0 of

$$E_0 = m_0 c^2 \qquad\qquad \textit{Rest energy} \quad (7\text{-}13)$$

where c is the speed of light, 3×10^8 m/s. (As we shall learn in Chapter 27, the mass of a moving body increases with its velocity, so that m is not always equal to m_0; the difference is only significant at velocities near that of light, however.) Equation (7-13) was discovered by Albert Einstein in the early years of this century and has been verified by many experiments since then.

Why are we not aware of rest energy as we are aware of kinetic and potential energies? After all, a 1-kg object—such as this book—contains the rest energy

$$m_0c^2 = 1 \text{ kg} \times (3 \times 10^8 \text{ m/s})^2$$
$$= 9 \times 10^{16} \text{ J,}$$

which is enough energy to send a payload of perhaps a million tons to the moon. How can so much energy be bottled up without revealing itself in some manner?

In fact, all of us *are* familiar with processes in which rest energy is liberated, only we do not usually think of them in these terms. In every chemical reaction in which energy is given off, for instance a fire, a certain amount of matter is being converted into energy in the form of heat, which is molecular kinetic energy. But the amount of matter that vanishes in such reactions is so small that it escapes our notice. When 1 kg of dynamite explodes, 6×10^{-11} kg of matter is transformed into energy. The lost mass is so minute a fraction of the total mass involved as to be impossible to detect directly (hence the "law" of conservation of mass in chemistry), but it results in the evolution of

Rest energy is liberated in many familiar processes

$$m_0c^2 = 6 \times 10^{-11} \text{ kg} \times (3 \times 10^8 \text{ m/s})^2$$
$$= 5.4 \times 10^6 \text{ J}$$

of energy, which is very hard to avoid detecting.

Problem. About 4 billion kg of matter is converted into energy in the sun per second. Find the power output of the sun.

Solution. From Eq. (7-13) the energy equivalent of 4 billion kg of matter is

$$E_0 = m_0c^2 = 4 \times 10^9 \text{ kg} \times (3 \times 10^8 \text{ m/s})^2 = 3.6 \times 10^{26} \text{ J.}$$

Since 1 J/s = 1 W and 3.6×10^{26} J are evolved by the sun in each second, its power output is

$$P = 3.6 \times 10^{26} \text{ W.}$$

7-8 Conservation of Energy

One of the chief distinctions between the physical sciences and nearly all other scientific disciplines is that in the former certain very general conservation principles have been found valid. A conservation principle states that no matter what changes a system of some kind that is isolated from the rest of the universe undergoes, a certain quantity keeps the same value it had originally. For example, the law of conservation of mass revolutionized chemistry by holding that the total mass of the products of a chemical reaction is the same as the total mass of the original substances. The increase in mass of a piece of iron when it rusts therefore indicates that the iron has combined with some other material, rather than having decomposed, as the early chemists believed. In fact, the gas oxygen was discovered in the course of seeking this other material.

Conservation principles

Given one or more conservation principles that apply to a given system, we can immediately determine which classes of events can take place in the system and which cannot. Thus when iron rusts, the gain in mass means that it has combined chemically with something else. In physics it is often possible to draw some conclusions about the behavior of the particles that make up a system without a detailed investigation, basing our analysis simply upon the conservation of some particular quantities. The power of this method of approach is exemplified by the great success of physics in understanding natural phenomena, a success largely due to the variety of conservation principles that have been discovered.

The first conservation principle we shall study is that of *conservation of energy*:

Conservation of energy

The total amount of energy in a system isolated from the rest of the universe always remains constant, although energy transformations from one form to another, including rest energy, may occur within the system.

This principle is perhaps the most fundamental generalization in all of science, and no violation of it has ever been found.

In a great many physical processes the rest masses, and hence the rest energies, of the participating objects do not change. In such processes mechanical energy is conserved: the sum of the kinetic and potential energies of the objects involved is constant. An increase in potential energy means a decrease in kinetic energy, and vice versa.

A falling ball provides a simple example of conservation of mechanical energy. As it falls, its initial potential energy is converted into kinetic energy, so that the total energy of the stone remains the same. The potential energy of a 1-kg ball 50 m above the ground is $mgh = 490$ J, and its total mechanical energy is 490 J until it interacts with the ground and transfers energy to it (Fig. 7-14).

Another example is the motion of a planet about the sun. Planetary orbits are elliptical, so that at different points in its orbit the planet is at different distances from the sun. When the planet is close to the sun, it has a low potential energy, just as a stone near the ground has a low potential energy; when the planet is far from the sun, it has a high potential energy (Fig. 7-13*a*). Since the sum of the planet's PE and KE must be constant, we conclude that the kinetic energy of the planet is a maximum when it is nearest the sun and a minimum when it is farthest from the sun.

Planetary motion

Newton's laws of motion enable us — in theory — to solve all mechanical problems, that is, problems that involve forces and moving objects. However, these laws are actually useful only in the simplest cases because in order to apply them we must take into detailed account all the various forces acting on each object at every point in its path, which is usually a difficult and complicated procedure. The great advantage of the principle of conservation of mechanical energy is that it permits us to draw definite conclusions about the relationship between the initial and final states of motion of some object or system of objects without having to investigate exactly what happens in between.

Problem. A ball slides down a smooth, curved track so that it is moving horizontally when it reaches the bottom, as in Fig. 7-15. What is its final velocity?

Height	1-kg ball	PE = mgh	KE = $\frac{1}{2}mv^2$	PE + KE
50 m	○	490 J	0 J	490 J
40	○	392	98	490
30	○	294	196	490
20	○	196	294	490
10	○	98	392	490
0	○	0	490	490

Fig. 7-14.   The total energy of a falling ball remains constant as its potential energy is transformed into kinetic energy.

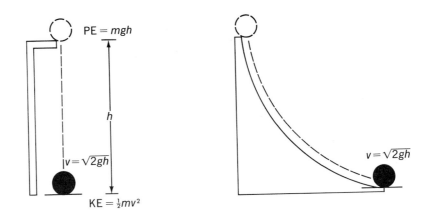

Fig. 7-15.   When a body of mass m falls from a height h, all its initial potential energy mgh has been converted to kinetic energy just as it strikes the ground. Its final velocity is therefore $\sqrt{2gh}$ (in the absence of friction) regardless of the precise path it takes.

Solution.   Because the path of the ball is curved, to apply the laws of motion directly means an involved calculation, but conservation of mechanical energy makes the problem ridiculously easy. When it is let go, the ball has a potential energy relative to the bottom of its path of

$$\mathrm{PE}_{\mathrm{top}} = mgh.$$

At the bottom the kinetic energy of the ball is

$$\mathrm{KE}_{\mathrm{bottom}} = \tfrac{1}{2}mv^2.$$

Conservation of mechanical energy requires that

$$\mathrm{KE}_{\mathrm{bottom}} = \mathrm{PE}_{\mathrm{top}}$$
$$\tfrac{1}{2}mv^2 = mgh$$
$$v = \sqrt{2gh}.$$

This is the same velocity the ball would have if it were simply dropped.

Conservative force

If the work performed in taking a body from a to b does not depend upon the path taken but only on the locations of a and b, the force acting is said to be *conservative*. Work done against a conservative force can be recovered by returning the body from b to a. Gravity is an example of a conservative force, as we have seen: if we do the amount of work $W = mgh$ to lift a body through the height h, the body can do the same amount of work when it is allowed to fall to the ground. A 5-lb brick 10 ft above the ground has a PE of 50 ft·lb relative to the ground regardless of how it got there, and all this PE can be turned into work equally well by dropping the brick or by allowing it to slide down a frictionless track.

On the other hand, when the work done *does* depend on the exact path taken, the force acting is said to be *nonconservative* (or *dissipative*), and the work cannot be recovered by reversing the path. Friction is an example of a nonconservative force: the longer the path, the more the work needed to overcome friction regardless of where the end points *a* and *b* are. Returning the body from *b* to *a* involves further work, not the recovery of the original work done.

Nonconservative force

What happens to the energy used to overcome the effects of friction? To find the answer, all we need do is rub one piece of wood against another. After a short time, it is obvious that the contacting surfaces of the pieces of wood are warmer than before. This is a quite general observation: work done against frictional forces produces a rise in the temperature of the objects involved. What is happening is that the energy that has disappeared on a macroscopic level reappears on a microscopic level as additional molecular kinetic energy, which is manifested in a rise in temperature. Temperature and heat are examined in detail in later chapters.

Special Topic

Calculating Work by Integration

Most forces are not constant but vary with position. Let us call $F(x)$ the magnitude of the force acting on a body at the point x, where by $F(x)$ we mean that the force F is a function of x. Figure 7-16 shows how $F(x)$ might vary with x in a certain case. To find the work W_{ab} done on the body in tak-

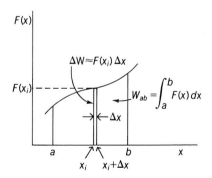

Fig. 7-16. The work done between $x = a$ and $x = b$ is equal to the area under the curve between these values.

ing it from a to b we cannot just multiply force and distance because $F(x)$ is not constant. Instead our procedure is to divide the path of the body into n short intervals each Δx long. If Δx is small enough, $F(x)$ is very nearly constant in each interval even though it changes from one interval to the next. The work ΔW done when the body moves from some position x_i to $x_i + \Delta x$ is

$$\Delta W \approx F(x_i)\,\Delta x$$

where the $\approx$ sign is needed because $F(x_i + \Delta x)$ is not quite the same as $F(x_i)$. However, in the limit of $\Delta x \to 0$, $F(x_i + \Delta x) = F(x_i)$, hence we are able to express the total work W_{ab} as the limit of the sum of all the $F(x_i)\,\Delta x$ values from $i = 1$ to $i = n$:

$$W_{ab} = \lim_{\substack{\Delta x \to 0 \\ n \to \infty}} \sum_{i=1}^{n} F(x_i)\,\Delta x.$$

Since $x_1 = a$ and $x_n = b$, we recognize this expression as the definite integral of $F(x)\,dx$ from $x = a$ to $x = b$, so that

$$W_{ab} = \int_a^b F(x)\,dx.$$

Gravitational Potential Energy

Let us use the above formula to find the work that must be done to take a particle of mass m from an initial distance of r_1 from the center of mass of a body (assumed fixed in place) of mass M to a final distance of r_2. The force acting on the particle is the gravitational force

$$F(r) = G\,\frac{mM}{r^2}.$$

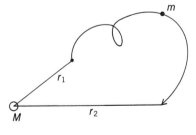

Fig. 7-17.

Since F is a function of r only, we need not specify the path of the body — only the values of r_1 and r_2 are relevant (Fig. 7-17). The work involved is found as follows:

$$W = \int_{r_1}^{r_2} F(r)\,dr = GmM \int_{r_1}^{r_2} \frac{dr}{r^2} = -GmM\left(\frac{1}{r}\right)\Big|_{r_1}^{r_2} = -GmM\left(\frac{1}{r_2} - \frac{1}{r_1}\right)$$

$$= GmM\left(\frac{1}{r_1} - \frac{1}{r_2}\right).$$

If $r_1 > r_2$, which means that m moves closer to M, the work done is negative. In this case the gravitational field of M does work on the mass m that appears as kinetic energy. On the other hand, if $r_1 < r_2$, which means that m moves farther away from M, the work done is positive. In this case work of external origin must be performed that becomes potential energy of the system relative to its original configuration. Since M is, by hypothesis, fixed in place, we can assign this increase in potential energy to the mass m, a procedure that is realistic when $M \gg m$.

We can use the above formula to determine the minimum velocity an object needs to escape permanently from the gravitational field of another body, for instance the earth. As mentioned in Chapter 6, this minimum velocity is called the *escape velocity* v_e. To escape from the earth, an object of mass m must have enough kinetic energy to reach $r_2 = \infty$ from its original distance of $r_1 = r_{\text{earth}}$ from the center of the earth, where r_{earth} is the earth's radius. The required kinetic energy $\text{KE} = \frac{1}{2}mv^2$ is equal to the work that must be done against the gravitational pull of the earth to take the object from r_1 to r_2:

$$\frac{1}{2}\,mv_e{}^2 = GmM_{\text{earth}} \left(\frac{1}{r_{\text{earth}}} - \frac{1}{\infty} \right) = \frac{GmM_{\text{earth}}}{r_{\text{earth}}}$$

$$v_e = \sqrt{\frac{2GM_{\text{earth}}}{r_{\text{earth}}}}.$$

The mass m of the object turns out to be irrelevant — *all* objects on the earth's surface have the same escape velocity, which turns out to be 1.12×10^4 m/s, or about 25,000 mi/hr. The escape velocity is $\sqrt{2}$ times greater than the velocity needed for a stable orbit just above the earth's surface, as stated in the answer to Exercise 29 of Chapter 6.

In reckoning gravitational potential energy near the earth's surface, it made sense to use the surface itself as the reference level, so that an object on the surface has $\text{PE} = 0$. The PE of an object elsewhere is equal to the work needed to bring it there from the surface. In astronomical situations, it makes sense to use an infinite separation between m and M as the reference configuration for which $\text{PE} = 0$. Accordingly the PE of mass m when it is at $r_2 = r$ is equal to the work needed to bring it there from $r_1 = \infty$, which is

$$\text{PE} = -\frac{GmM}{r}.$$

We can apply this formula to the case of an orbiting satellite, for instance the moon revolving around the earth or the earth revolving around the sun. The centripetal force on a satellite in a circular orbit of radius r is given by

$$\frac{mv^2}{r} = \frac{GmM}{r^2},$$

which means that its kinetic energy is

$$KE = \frac{1}{2}mv^2 = \frac{GmM}{2r}.$$

Hence the satellite's total energy is

$$E = KE + PE = \frac{GmM}{2r} - \frac{GmM}{r} = -\frac{GmM}{2r}.$$

The negative total energy signifies that work must be done *on* the satellite to pull it away from its orbit, hence the orbit is a stable one. This formula can be used to answer such questions as how much energy is needed to change the orbital radius of a satellite or to permanently separate it from its parent body. We shall find that a similar formula applies in the case of the electrons inside an atom, where the attractive force is electrical rather than gravitational.

Important Terms

Work is a measure of the change (in a general sense) a force gives rise to when it acts upon something. When a body undergoes a displacement while a force acts on it, the work done by the force is equal to the product of the displacement and the component of the force in the direction of the displacement. In the SI system the unit of work is the **joule** and in the British system it is the **foot-pound**.

The rate at which work is done is called **power**. The unit of power in the metric system is the **watt**, which is equal to 1 J/s, and in the British system it is the **ft·lb/s**. The **horsepower** is a unit of power equal to 550 ft·lb/s, which is 746 watts.

Energy is that which may be converted into work. When something possesses energy, it is capable of performing work or, in a general sense, of accomplishing a change in some aspect of the physical world. The units of energy are those of work.

The three broad categories of energy are **kinetic energy**, which is the energy something possesses by virtue of its motion; **potential energy**, which is the energy something possesses by virtue of its position in a force field; and **rest energy**, which is the energy something possesses by virtue of its mass.

The principle of **conservation of energy** states that the total amount of energy in a system isolated from the rest of the universe always remains constant, although energy transformations from one form to another, including rest energy, may occur within the system.

Important Formulas

Work:

$$W = Fs \cos \theta$$

Work in lifting object:

$$W = wh = mgh$$

Power:

$$P = \frac{W}{t} = Fv$$

Kinetic energy:

$$KE = \tfrac{1}{2}mv^2$$

Gravitational potential energy:

$$PE = wh = mgh$$

Rest energy:

$$E_0 = m_0c^2$$

Multiple Choice

1. According to the principle of conservation of energy (with energy interpreted as including rest energy), energy can be
   a. created but not destroyed.
   b. destroyed but not created.
   c. both created and destroyed.
   d. neither created nor destroyed.

2. A golf ball and a ping-pong ball are dropped in a vacuum chamber. When they have fallen half way down, they have the same
   a. velocity.   b. potential energy.
   c. kinetic energy.   d. rest energy.

3. In the formula $E = mc^2$, the symbol c represents
   a. the velocity of the body.
   b. the velocity of sound.
   c. the velocity of light.
   d. the rest energy of 1 kg of matter.

4. One horsepower is not equal to
   a. 550 ft-lb/s.   b. 33,000 ft-lb/min.
   c. 746 J.   d. 0.746 kW.

5. Which of the following is not a unit of energy?
   a. joule   b. foot-pound
   c. watt-hour   d. newton

6. Which of the following is not a unit of power?
   a. joule·s   b. watt
   c. ft·lb/min   d. horsepower

7. To keep a vehicle moving at the speed v requires a force F. The power required is
   a. Fv.   b. $\tfrac{1}{2}Fv^2$.
   c. F/v.   d. F/v^2.

8. A 2-lb book is held 4 ft above the floor for 50 s. The work done is
   a. 0.   b. 8 ft·lb.
   c. 12.5 ft·lb.   d. 400 ft·lb.

9. A 100-lb boy runs up a staircase to a floor 15 ft higher in 5.45 s. His power output is
   a. 1/64 hp.   b. 1/2 hp.
   c. 1 hp.   d. 275 hp.

10. A 1-kg mass has a potential energy of 1 joule relative to the ground when it is at a height of
    a. 0.102 m.   b. 1 m.
    c. 9.8 m.   d. 32 m.

11. A 1-N weight has a potential energy of 1 joule relative to the ground when it is at a height of
    a. 0.102 m.   b. 1 m.
    c. 9.8 m.   d. 32 m.

12. A 1-slug mass has a potential energy of 1 ft·lb relative to the ground when it is at a height of
    a. 0.031 ft.   b. 1 ft.
    c. 9.8 ft.   d. 32 ft.

13. A 1-lb weight has a potential energy of 1 ft·lb relative to the ground when it is at a height of
    a. 0.031 ft.   b. 1 ft.
    c. 9.8 ft.   d. 32 ft.

14. A total of 4900 joules is expended in lifting a 50-kg mass. The mass was raised to a height of
    a. 10 m.
    b. 98 m.
    c. 960 m.
    d. 245,000 m.

15. A 16-slug mass is lifted to a height of 10 ft. Its potential energy is
    a. 5 ft·lb.   b. 20 ft·lb.
    c. 160 ft·lb.   d. 5120 ft·lb.

16. A 1-kg mass has a kinetic energy of 1 joule when its speed is
    a. 0.45 m/s.   b. 1 m/s.
    c. 1.4 m/s.   d. 4.4 m/s.

17. A 1-N weight has a kinetic energy of 1 joule when its speed is
    a. 0.45 m/s.   b. 1 m/s.
    c. 1.4 m/s.   d. 4.4 m/s.

18. A 1-slug mass has a kinetic energy of 1 ft·lb when its speed is
    a. 0.25 ft/s.   b. 1 ft/s.
    c. 1.4 ft/s.   d. 8 ft/s.

19. A 1-lb weight has a kinetic energy of 1 ft·lb when its speed is
    a. 0.25 ft/s.   b. 1 ft/s.
    c. 1.4 ft/s.   d. 8 ft/s.

20. A 3200-lb automobile whose speed is 60 mi/hr has a kinetic energy of
    a. 1.8×10^5 ft·lb.
    b. 3.9×10^5 ft·lb.
    c. 7.7×10^5 ft·lb.
    d. 1.4×10^7 ft·lb.

21. The height above the ground of a child on a swing varies from 2 ft at his lowest point to 5 ft at his highest point. The maximum speed of the child is approximately
    a. 11 ft/s.
    b. 14 ft/s.
    c. 18 ft/s.
    d. dependent on the child's mass.

22. Car A has a mass of 75 slugs and a speed of 60 mi/hr and car B has a mass of 150 slugs and a speed of 30 mi/hr. The kinetic energy of car A is
    a. half that of car B.
    b. equal to that of car B.
    c. twice that of car B.
    d. four times that of car B.

23. A sedentary person requires about 6 million J of energy per day. This rate of energy consumption is equivalent to about
    a. 70 W.   b. 335 W.
    c. 600 W.   d. 250,000 W.

24. The mass equivalent of 6 million J is
    a. 6.7×10^{-11} kg.
    b. 5.4×10^{-9} kg.
    c. 6.7×10^{-3} kg.
    d. 2×10^{-2} kg.

Exercises

1. Under what circumstances (if any) is no work done on a moving body even though a net force acts upon it?

2. Does every moving body possess kinetic energy? Does every stationary body possess potential energy? Can something possess both kinetic and potential energy?

3. At what point in its motion is the kinetic energy of a pendulum bob a maximum? At what point is its potential energy a maximum?

4. Is kinetic energy a scalar or a vector quantity? Is potential energy a scalar or a vector quantity?

5. The potential energy of a golf ball in a hole is negative relative to the ground. Under what circumstances (if any) is its kinetic energy negative? Its rest energy?

6. Electrical energy is usually reckoned by utility companies in kilowatt-hours (kWh). How many joules are there in a kWh?

7. Four thousand joules are used to lift a 30-kg mass. If the mass is at rest before and after its elevation, how high does it go?

8. A man holds a 10-kg package 1.2 m above the ground for 1 min. How much work does he perform?

9. A centripetal force of 18 N is used to keep a 2-kg ball in uniform circular motion at the end of a string 1 m long. How much work does the force do in each revolution of the ball?

10. The sun exerts a force of 4×10^{28} N on the earth, and the earth travels 9.4×10^{11} m in its annual orbit of the sun. How much work is done by the sun on the earth in the course of a year?

11. (a) A force of 130 N is used to lift a 12-kg mass to a height of 8 m. How much work is done by the force? (b) A force of 130 N is used to push a 12-kg mass on a horizontal, frictionless surface for a distance of 8 m. How much work is done by the force?

12. A 20-kg wooden box is pushed a distance of 15 m on a horizontal stone floor by a force just sufficient to overcome the friction between box and floor. The coefficient of friction is 0.4. (a) What is the required force? (b) How much work does the force do?

13. An 80-kg man climbs a mountain 3000 m high in 10 hr. (a) How much work does he perform? (b) What is his average power output in watts? In hp?

14. A 120-lb woman climbs a mountain 8000 ft high in 8 hr. (a) How much work does she perform? (b) What is her average power output in ft-lb/s? In hp?

15. In 1970 approximately 2×10^{20} joules of work were performed throughout the world by inanimate devices of all kinds, perhaps 15 times as much as the muscle power provided in that year. The work was used for heat, light, transport, manufacturing, and so forth. About 98% of the work was ultimately derived from the fossil fuels coal, natural gas, and oil, the rest mainly from water power with a small (0.25% of the total) contribution from nuclear power stations. (a) Express the power consumption in 1970 in watts. (b) Find the average power consumption per person in watts and in hp on the assumption that the world's population in 1970 was 3.5×10^9.

16. A white horse has a power output of 1 hp. What is the maximum force it can exert at a velocity of 3 m/s?

17. The anchor windlass of a boat must be able to raise a total load (anchor plus chain) of 800 kg at a velocity of 0.5 m/s. What should the minimum rating of the motor be, in kW?

18. At its cruising velocity of 520 mi/hr, the two engines of a DC-9 airplane produce a thrust of 11,400 lb each. How many hp does each engine develop under these circumstances?

19. Each of the four engines of a DC-8 airplane develops 7500 hp when the cruising velocity is 240 m/s. How much thrust does each engine produce under these circumstances?

20. A motorboat requires 160 hp to move at the con-

stant velocity of 8 m/s. How much resistive force does the water exert on it at that velocity?

21. Is more work needed to bring a car's velocity from 10 mi/hr to 20 mi/hr or from 50 mi/hr to 60 mi/hr?

22. Find the kinetic energy of a 256-lb ostrich running at 50 ft/s.

23. A 3200-lb car is moving at 40 mi/hr. What is its kinetic energy?

24. A 0.02-kg bullet has a velocity of 500 m/s. What is its kinetic energy?

25. Find the average kinetic energy of a 70-kg sprinter who covers 400 m in 45 s.

26. A 90-kg pole vaulter clears the bar at a height of 5 m. Find his potential energy at this height.

27. A 160-lb diver stands on a diving board 20 ft above the surface of a lake. What is his potential energy with respect to the surface?

28. A boy slides down a sliding pond from a starting point 10 ft above the ground. His velocity at the bottom is 12 ft/s. What percentage of his initial potential energy was dissipated?

29. A man skis down a slope 100 m high. His velocity at the foot of the slope is 20 m/s. What percentage of his initial potential energy was dissipated?

30. A 3-kg stone is lifted to a height of 100 m and then dropped. What is its kinetic energy when it is 50 m from the ground?

31. A 2-kg ball is at rest when a horizontal force of 5 N is applied. In the absence of friction, what is the speed of the ball after it has gone 10 m?

32. A force of 500 newtons is used to lift a 20-kg object to a height of 10 m. There is no friction present. (a) How much work is done by the force? (b) What is the change in the potential energy of the object? (c) What is the change in the kinetic energy of the object?

33. At her highest point, a 40-kg girl on a swing is 2 m from the ground while at her lowest point she is 0.8 m from the ground. What is her maximum speed? On another swing a 50-kg boy undergoes

exactly the same motion. What is his maximum speed?

34. The source of the sun's energy (and therefore, directly or indirectly, of nearly all energy available to man) is the conversion of hydrogen to helium. As described later in the book, the nuclei of four hydrogen atoms, each of mass 1.673×10^{-27} kg, join together in a series of separate reactions to yield a helium nucleus of mass 6.646×10^{-27} kg. How much energy is liberated each time a helium nucleus is formed? How many helium nuclei are formed to produce the 10^7 J a moderately active person requires per day?

35. Sunlight falls on the earth at the rate of 1400 watts/m². Express this figure in hp/ft².

Problems

1. A horse is towing a barge with a rope that makes an angle of 20° with the canal. If the horse exerts a force of 400 N, how much work does it do in moving the barge 1 mile?

2. A man pulls a 150-lb crate for 80 ft across a level floor using a rope that is 30° above the horizontal. If the coefficient of friction between crate and floor is 0.30 and the man uses just enough force to move the crate without accelerating it, how much work does he perform? (Assume the rope is attached to the center of gravity of the crate.)

3. A boy pulls a sled with a force of 10 lb for 100 ft. The rope attached to the sled is at an angle of 30° above the horizontal. (a) How much work is done? (b) If the boy moves the sled 100 ft in 45 s, find his power output in horsepower.

4. A horizontal force of 5 N is used to push a box up a ramp 5 m long that is at an angle of 15° above the horizontal. How much work is done?

5. Two men set out to climb to the summit of a 3000-m mountain starting from sea level. One of them sets out along a slope that averages 30° above the horizontal, the other along a slope that averages 40° above the horizontal. Each man has a mass of 80 kg and carries a 10-kg knapsack. Find the work done by each of them.

6. A 160-lb man carrying a 32-lb knapsack climbs to the summit of a 16,000-ft mountain. The average slope of the mountain is 40°. If he starts from a base camp at an altitude of 8000 ft, how much work does he perform in making the ascent?

7. An escalator carries passengers from one floor of a building to another 35 ft higher. It is designed to have a capacity of 200 passengers per minute, assuming an average weight per passenger of 150 lb. Find the required horsepower of the motor if half the work it does is dissipated as heat.

8. A 15-hp motor is used to hoist a 1-ton bucket of concrete to the twentieth floor of a building under construction, a height of 300 ft. If no power is lost, how much time is required for the ascent?

9. In 1932 five members of the Polish Olympic ski team climbed from the 5th to the 102nd floor of the Empire State Building, a distance of approximately 1000 ft, in 21 min. One of these men weighed 165 lb. How many horsepower did he develop in the ascent?

10. Thirty horsepower is required to propel a 3600-lb car at 25 mi/hr on a horizontal road. (a) How much resistance must the car overcome at this speed? (b) How much power is required for the car to ascend an 8° hill at the same speed?

11. A man uses a rope and system of pulleys to lift a 160-lb object to a height of 5 ft. He exerts a force of 45 lb on the rope and pulls a total of 20 ft of rope through the pulleys in the course of lifting the object, which is at rest afterward. (a) How much work does the man do? (b) What is the change in the potential energy of the object? (c) If the answers to (a) and (b) are different, explain.

12. A man uses a rope and system of pulleys to lift an 80-kg object to a height of 2 m. He exerts a force of 220 N on the rope and pulls a total of 8 m of rope through the pulleys in the course of raising the object, which is at rest afterward. (a) How much work does the man do? (b) What is the change in the potential energy of the object? (c) If the answers to (a) and (b) are different, explain.

13. A force of 100 lb is used to lift an 80-lb weight to a height of 20 ft. There is no friction present. (a) How much work is done by the force? (b) What is the change in the potential energy of the weight? (c) What is the change in the kinetic energy of the weight?

14. (a) A force of 8 N is used to push a 0.5-kg ball over a horizontal, frictionless table a distance of 3 m. If the ball starts from rest, what is its final kinetic energy? (b) The same force is used to lift the same ball a height of 3 m. If the ball starts from rest what is its final kinetic energy?

15. A waterfall is 30 m high and 10^4 kg of water flows over it per second. (a) How much power does this flow represent? (b) If all this power could be converted to electricity, how many 100-watt light bulbs could be supplied?

16. A ball is dropped from a height of 1 m and loses 10% of its kinetic energy when it bounces on the ground. To what height does it rise?

17. In the operation of a certain pile driver, a hammer weighing 1000 lb is dropped from a height of 19 ft above the head of a pile. If the pile is driven 0.5 ft into the ground with each impact of the hammer, what is the average force on the pile when struck?

18. A sledge hammer whose head has a mass of 5 kg is used to drive a spike into a wooden beam. The workman is tired and merely allows the hammer to drop on the spike from a height 0.4 m above it. If the spike is driven 1 cm at each blow, what is the average force on it when struck?

19. Steam enters a 50,000-hp turbine at 800 m/s and emerges at 100 m/s. Assuming 100 percent mechanical efficiency, what mass of steam passes through the turbine per second?

20. Steam enters a turbine at 2000 ft/s and emerges at 300 ft/s. If 20 tons of steam pass through the turbine per hour, find its power output under the assumption of 90 percent efficiency.

Answers to Multiple Choice

1. d	9. b	17. d
2. a	10. a	18. c
3. c	11. b	19. d
4. c	12. a	20. b
5. d	13. b	21. b
6. a	14. a	22. c
7. a	15. d	23. a
8. a	16. c	24. a

8

Momentum

So complex is the physical universe that many different quantities turn out to be useful in describing its various aspects. We have already been introduced to length, time, mass, force, torque, work, and energy, and more are to come. There is nothing sacred about any of these quantities—it is entirely possible to dispense with any of them, but only at the expense of making physics a good deal more complicated than it already is. The idea behind the definition of each of the various physical quantities is to single out something that unifies a wide range of observations, so that it is then possible to boil down to a brief, clear statement a large number of separate discoveries about nature. In this chapter we shall learn how the concepts of linear momentum and impulse supplement those of work and energy to provide a particularly simple theoretical framework for analyzing the behavior of moving bodies.

8-1 Linear Momentum

We all know that a baseball struck squarely by a bat is harder to stop than the same baseball thrown gently, and that the heavy iron ball used for the shotput is harder to stop than a baseball whose velocity is the same (Fig. 8-1). These observations suggest that a measure of the tendency of a body to continue in motion at constant velocity is the product $m\mathbf{v}$ of its mass m and velocity $\mathbf{v}$.

The quantity $m\mathbf{v}$ is called the *linear momentum* of a moving body:

$$\text{Linear momentum} = m\mathbf{v}. \tag{8-1}$$

The symbol $\mathbf{p}$ is sometimes used to represent linear momentum. Linear momentum is a vector quantity whose direction is the direction of $\mathbf{v}$. The kinetic energy of a moving body, which also depends upon its mass and velocity since $KE = \frac{1}{2}mv^2$, is a scalar quantity with magnitude only. The different significances of linear momentum and kinetic energy will be discussed later in this chapter.

Because $m\mathbf{v}$ describes the tendency of a moving body to pursue a straight path at constant velocity, it is referred to as *linear* momentum. A different quantity, *angular momentum*, describes the tendency of a spinning body such as a top to continue to spin. When there is no question as to which is meant, linear momentum is usually referred to simply as momentum.

8-2 Impulse

To set something in motion from rest, a force must be applied for a period of time. We might expect that the greater the force and the longer the time, the more momentum the body will have. This expectation is correct, and the product $\mathbf{F}\Delta t$ of a constant force $\mathbf{F}$ and the time interval Δt during which it acts is accordingly given the status of a physical quantity in its own right. This quantity is called *impulse*:

$$\text{Impulse} = \mathbf{F}\Delta t. \tag{8-2}$$

Impulse, like momentum, is a vector quantity. Let us see how the momentum of a body is affected when it receives a certain impulse.

The second law of motion states that the force $\mathbf{F}$ applied to a body of constant mass m that undergoes the acceleration $\mathbf{a}$ is given by

$$\mathbf{F} = m\mathbf{a}. \tag{8-3}$$

Linear momentum is a vector quantity

High momentum

Baseball

Low momentum

High momentum

Iron shot

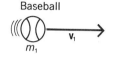

Low momentum

$m_1 = m_2$

$v_1 > v_2$

$m_1 > m_2$

$v_1 = v_2$

Fig. 8-1. The linear momentum $m\mathbf{v}$ of a moving body is a measure of its tendency to continue in motion at constant velocity.

An impulse produces a change in momentum

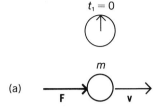

$t_1 = 0$

m

(a) F $\rightarrow$ $\bigcirc$ $\longrightarrow$ v

$t_2 = \Delta t$

m

(b) F $\rightarrow$ $\bigcirc$ $v + \Delta v$ $\longrightarrow$

Fig. 8-2.   (a) At the time $t_1 = 0$, a force **F** is applied to a body whose velocity is **v**. (b) At the later time $t_2 = \Delta t$, the velocity of the body is **v** + **Δv**.

When a force **F** is applied at the time $t_1 = 0$ to a body whose initial velocity is **v**, at the later time $t_2 = \Delta t$ its velocity will have changed to **v** + **Δv** (Fig. 8-2). The body's acceleration in this time interval is

$$a = \frac{\text{velocity change}}{\text{time interval}}$$

$$= \frac{\Delta \mathbf{v}}{\Delta t},$$

and so Eq. (8-3) becomes

$$\mathbf{F} = m\mathbf{a}$$

$$= m\frac{\Delta \mathbf{v}}{\Delta t},$$

which we can rewrite as

$$\mathbf{F}\Delta t = m\Delta \mathbf{v}.$$

Evidently the impulse provided by the force equals the momentum change of the body:

$$\mathbf{F}\Delta t = \Delta(m\mathbf{v}) \tag{8-4}$$

Impulse = momentum change.

Figure 8-3 shows the effect of applying the constant force F for the time Δt to several bodies with different momenta $m\mathbf{v}_1$. In each case, the final momentum $m\mathbf{v}_2$ is obtained by finding the vector sum $m\mathbf{v}_2 = m\mathbf{v}_1 + \mathbf{F}\,\Delta t$.

In the metric system the unit of impulse is the *newton·second* (N·s), and the unit of momentum is the kg·m/s; they are actually the same, of course, but it is often convenient to distinguish between them in this way. The corresponding British units are the lb·s and (slug·ft)/s.

Units of momentum and impulse

Problem.   The head of a golf club is in contact with a 46-g golf ball for 0.50 ms (1 ms = 1 millisecond = 10^{-3} s), and as a result the ball flies off at 70 m/s. Find the average force that was acting on the ball during the impact.

Solution.   The ball starts from rest, hence its momentum change is

$$\Delta m v = 0.046 \text{ kg} \times 70 \text{ m} = 3.22 \text{ kg·m/s}.$$

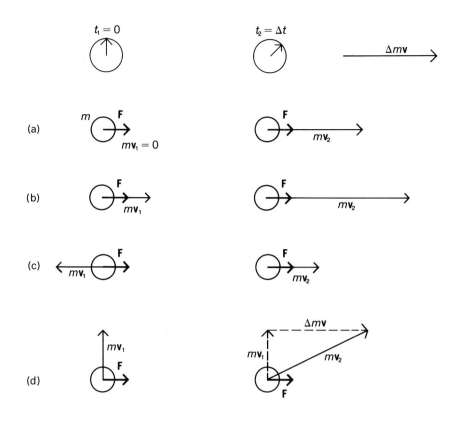

Fig. 8-3.   Applying a constant force **F** to a mass m for a time Δt changes its momentum by $\Delta m\mathbf{v} = \mathbf{F}\,\Delta t$. At (a) the mass is initially at rest, at (b) its initial momentum is in the same direction as **F**, at (c) its initial momentum is in the opposite direction to **F**, and at (d) its initial momentum is perpendicular to **F**. Since momentum is a vector quantity, the momentum change $\Delta m\mathbf{v}$ must be added to the initial momentum $m\mathbf{v}$ by the process of vector addition.

From Eq. (8-4) we therefore have

$$F = \frac{\Delta m v}{\Delta t} = \frac{3.22 \text{ kg} \cdot \text{m/s}}{5.0 \times 10^{-4} \text{ s}} = 6.44 \times 10^{3} \text{ N}$$

whose British equivalent is 1450 lb. A force of the same magnitude but acting in the opposite direction (the *recoil force*) acts on the club's head during the impact, in accordance with the third law of motion. No golf club could withstand such a static load, but the impact is so brief that its only effect on the shaft is to temporarily bend it by a few cm.

8-3 Conservation of Momentum

Energy and work are scalar quantities, having magnitude only. Despite the fundamental and all-inclusive character of the law of conservation of

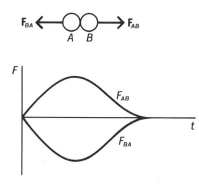

Fig. 8-4. Particles A and B collide and exert the forces F_{AB} and F_{BA} on each other. As shown in the graph, these forces are equal in magnitude and opposite in direction at all times. Their impulses are also equal and opposite, so the total momentum of A and B is left unchanged by the collision although it may be distributed differently between them.

energy, it cannot by itself provide complete solutions to most problems that involve interacting bodies. A simple example is the firing of a rifle: the requirement that energy be conserved means that the kinetic energies of the bullet and the recoiling rifle, plus the heat and sound energy that are liberated, must equal the chemical energy of the detonated explosive, but this does not tell us how the total energy is divided among the rifle, the bullet, and the atmosphere. Indeed, because energy is a scalar quantity, its conservation does not even imply that the bullet and rifle must move in opposite directions. To complete the solution of many problems in dynamics in which a detailed knowledge of the active forces is lacking, an additional principle of a vector nature is required.

Let us consider a system of two or more particles instead of a single particle. If no forces from outside the system act upon its component particles, the total linear momentum of the system, which is the sum

$$M\mathbf{V} = m_1\mathbf{v}_1 + m_2\mathbf{v}_2 + m_3\mathbf{v}_3 + \cdots \tag{8-5}$$

of the individual momenta of its particles, cannot change; with no force there is no impulse, hence no change in momentum. However, the *distribution* of the total momentum $M\mathbf{V}$ among the various particles in the system may change without $M\mathbf{V}$ changing in the absence of an external force.

Two particles A and B might collide and thereby exert forces upon each other. At every instant during their interaction these forces, $\mathbf{F}_{AB}$ acting on B and $\mathbf{F}_{BA}$ acting on A, obey Newton's third law of motion,

$$\mathbf{F}_{AB} = -\mathbf{F}_{BA}.$$

This is shown in Fig. 8-4. Hence the impulses exchanged must be equal and opposite,

$$\mathbf{F}_{AB}\,\Delta t = -\mathbf{F}_{BA}\,\Delta t,$$

and the *total* momentum of the system of A and B together is the same after the collision as it was before.

Since we almost always can include the sources of all forces relevant to a particular process within what we choose to be our "system," in such cases we have the condition that, no matter what interactions take place within the system, its total momentum never changes. Thus we have the principle of *conservation of linear momentum*:

When the vector sum of the external forces acting upon a system of particles equals zero, the total linear momentum of the system remains constant.

Let us consider a specific example readily treated with the help of conservation of momentum. Suppose that we have an isolated particle of mass m, initially at rest, that suddenly explodes into two particles of masses m_1 and m_2, which fly apart (Fig. 8-5). The forces acting on the original particle that caused it to break up were internal ones, and no external force was present. Since m has the initial momentum of zero, the final momentum of m_1 and m_2, when added together, must also be zero. Hence

$$m\mathbf{v} = 0 = m_1\mathbf{v}_1 + m_2\mathbf{v}_2,$$

and

$$\mathbf{v}_2 = -\frac{m_1}{m_2}\mathbf{v}_1, \tag{8-6}$$

where $\mathbf{v}_1$ and $\mathbf{v}_2$ are the final velocities of the two fragments. We note immediately that these velocities must be in opposite directions along the same line. This problem could *not* be solved starting from $\mathbf{F} = m\mathbf{a}$, since we do not know explicitly what forces were acting during the explosion.

Problem.   A 5-lb rifle fires a 0.03-lb bullet at a muzzle velocity of 2000 ft/s (Fig. 8-6). Find the recoil velocity of the rifle.

Solution.   Here

$$m_1 = \frac{0.03 \text{ lb}}{g}, \qquad m_2 = \frac{5 \text{ lb}}{g}, \qquad v_1 = 2000 \frac{\text{ft}}{\text{s}},$$

and, inserting these values in Eq. (8-6), we find that the recoil velocity of the rifle is

$$v_2 = -\frac{0.03 \text{ lb}/g}{5 \text{ lb}/g} \times 2000 \frac{\text{ft}}{\text{s}} = -12 \text{ ft/s}.$$

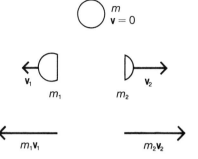

Fig. 8-5.   The total momentum of a system of two or more particles remains constant if no external forces act on the system.

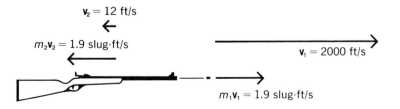

$v_2 = 12$ ft/s

$m_2v_2 = 1.9$ slug·ft/s

$v_1 = 2000$ ft/s

$m_1v_1 = 1.9$ slug·ft/s

Fig. 8-6.   When a 5-lb rifle fires a 0.03-lb bullet with a muzzle velocity of 2000 ft/s, it recoils with a velocity of 12 ft/s. The total momentum of the *system* of rifle plus bullet is zero before and after the rifle is fired, although the momenta of both rifle and bullet have changed.

Note that the g's have canceled out in the calculation, making it unnecessary for us to find the mass values numerically; the ratio of two masses is always the same as the ratio of the corresponding weights.

The conservation of linear momentum is a generalization based upon innumerable experiments and observations, and no exception to it has ever been found. With the help of advanced mathematics it is possible to show that, if the laws of nature are the same at every point in space, then the principle of conservation of linear momentum must follow as an inevitable consequence. It is also possible to show that, if the laws of nature do not change with time, so that they were always the same as they are now and always will remain the same, then energy must be conserved in all interactions. Thus these principles, as well as being useful relationships for solving practical problems, give us a hint of a profound order underlying the physical universe.

Conservation principles and the laws of nature

8-4 Rocket Propulsion

The principle underlying rocket flight is conservation of momentum. The total momentum of a rocket on its launching pad is zero. When it is fired, the exhaust gases shoot downward at high velocity, and the rocket moves upward to balance the momentum of the gases (Fig. 8-7). Rockets do not operate by "pushing" against their launching pads, the air, or anything else; in fact, they perform best in space, where there is no atmosphere to impede their motion. The energy of the rocket and its exhaust comes from chemical energy stored in the fuel. The total momentum of the system of rocket plus exhaust, which is initially zero, does not remain constant after a launch from the earth because of the impulses provided by air resistance and the earth's gravitational pull.

Rocket propulsion is a gradual rather than an instantaneous process, with the fuel burned and ejected as exhaust gases at a certain rate instead of in one lump. As a result, part of the momentum of the exhaust is "wasted" in pushing forward unburned fuel. When this factor is taken into account, the ultimate velocity of a rocket (neglecting air resistance and gravity) turns out to be directly proportional to the velocity of the exhaust gases and to the logarithm of the ratio of the rocket's initial mass to its final mass after the fuel has been consumed. A typical modern rocket might have an exhaust velocity of 3000 m/s with 75 percent of its initial mass consisting of fuel, which gives a rocket velocity of 4155 m/s.

To attain higher velocities than a single rocket is capable of, two or more rocket stages can be used. The first stage is a large rocket whose payload is another, smaller rocket. When the fuel of the first stage has been

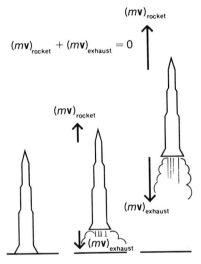

$(m\mathbf{v})_{rocket}$

$(m\mathbf{v})_{rocket} + (m\mathbf{v})_{exhaust} = 0$

$(m\mathbf{v})_{rocket}$

$(m\mathbf{v})_{exhaust}$

$(m\mathbf{v})_{exhaust}$

Fig. 8-7. Conservation of momentum in rocket flight. The downward momentum of the exhaust gases is exactly balanced by the upward momentum of the rocket itself.

Multistage rockets

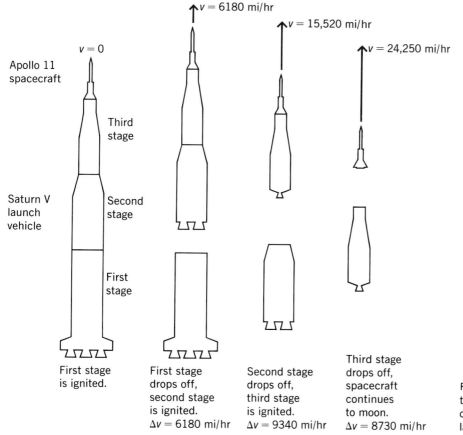

v = 0

Apollo 11
spacecraft

Third
stage

Saturn V
launch
vehicle

Second
stage

First
stage

First stage
is ignited.

v = 6180 mi/hr

First stage
drops off,
second stage
is ignited.
$\Delta v = 6180$ mi/hr

v = 15,520 mi/hr

Second stage
drops off,
third stage
is ignited.
$\Delta v = 9340$ mi/hr

v = 24,250 mi/hr

Third stage
drops off,
spacecraft
continues
to moon.
$\Delta v = 8730$ mi/hr

Fig. 8-8.   The Saturn V launch vehicle
that propelled the Apollo 11 spacecraft to the moon for the first manned
landing used three rocket stages.

consumed, its fuel tanks and engine are cast loose. Then the second stage is fired starting from a high initial velocity instead of from rest and without the burden of the fuel tanks and engine of the first stage. This process can be repeated a number of times, depending upon the final velocity required. The Saturn V launch vehicle that propelled the Apollo 11 spacecraft to the moon in July 1969 employed three stages, as shown in Fig. 8-8. At original ignition the entire assembly was 363 ft long and weighed 3240 tons.

8-5 Collisions

The law of conservation of momentum is indispensable in dealing with collisions between two or more bodies. In such cases no external forces act on the participants, and therefore their total momentum before they collide

Momentum is redistributed in
a collision

equals their total momentum afterward. *The essential effect of the collision is to redistribute the total momentum of the bodies.*

If we only know the masses and initial velocities of the bodies involved in a collision, however, momentum conservation by itself does not yield a unique result for their subsequent motion. An unknown amount of kinetic energy may be lost to heat, sound, or other forms of energy when the bodies interact, and, though the requirement that momentum be conserved does set limits to the result of the collision, it can go no further without additional information. These limiting cases, however, are worth examining.

Elastic collisions

At one extreme are *completely elastic collisions* in which kinetic energy is conserved. During the actual collision, to be sure, some kinetic energy becomes elastic potential energy as the bodies are deformed by the impact, but all of this energy is returned as the bodies move apart.

Inelastic collisions

At the other extreme are *completely inelastic collisions* in which the bodies stick together permanently upon impact. The kinetic energy loss in a completely inelastic collision is the maximum possible consistent with momentum conservation. Such a collision can be analyzed on the basis of momentum conservation only.

Problem.   A 5-kg lump of clay that is moving at 10 m/s to the left strikes a 6-kg lump of clay moving at 12 m/s to the right. The two lumps stick together after they collide. Find the final velocity of the composite body.

Solution.   This is an example of a completely inelastic collision. If we call the mass of the final body M and its velocity V, conservation of linear momentum requires that

Momentum afterward = momentum before

$$MV = m_1 v_1 + m_2 v_2.$$

Adopting the convention that motion to the right is $+$ and to the left is $-$, we have

$$m_1 = 5 \text{ kg}, \qquad m_2 = 6 \text{ kg}, \qquad M = m_1 + m_2 = 11 \text{ kg},$$

$$v_1 = -10 \, \frac{\text{m}}{\text{s}}, \qquad v_2 = +12 \, \frac{\text{m}}{\text{s}}, \qquad V = ?.$$

Solving for V yields

$$V = \frac{m_1 v_1 + m_2 v_2}{M}$$

$$= \frac{5 \text{ kg} \times (-10 \text{ m/s}) + 6 \text{ kg} \times 12 \text{ m/s}}{11 \text{ kg}} = 2 \frac{\text{m}}{\text{s}}.$$

Since V is positive, the composite body moves off to the right (Fig. 8-9).

Energy and momentum are independent concepts. In this problem the lumps of clay before the collision have the kinetic energies

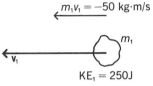

$$\text{KE}_1 = \tfrac{1}{2} m_1 v_1{}^2$$

$$= \tfrac{1}{2} \times 5 \text{ kg} \times \left(-10 \frac{\text{m}}{\text{s}}\right)^2$$

$$= 250 \text{ J},$$

$$\text{KE}_2 = \tfrac{1}{2} m_2 v_2{}^2$$

$$= \tfrac{1}{2} \times 6 \text{ kg} \times \left(12 \frac{\text{m}}{\text{s}}\right)^2$$

$$= 432 \text{ J}.$$

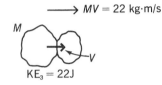

After the collision the new lump of clay has the kinetic energy

$$\text{KE}_3 = \tfrac{1}{2} M V^2$$

$$= \tfrac{1}{2} \times 11 \text{ kg} \times (2 \text{ m/s})^2$$

$$= 22 \text{ J}.$$

Fig. 8-9.   In a completely inelastic collision, the colliding bodies stick together. Kinetic energy is not conserved in such an event. Linear momentum is conserved in *all* collisions.

The total kinetic energy prior to the collision was $432 + 250$ or 682 J, while afterward it is only 22 J. The difference of 660 J was dissipated largely into heat energy in the collision, with some probably being lost to sound energy as well.

It is essential to keep in mind the directional character of linear momentum. Sometimes the problem under consideration involves bodies that move along a straight line, as in the preceding example, but in general the bodies may move in two or three dimensions and we must be sure to take this into account by a vector calculation.

In a collision, momentum is conserved in all directions

Problem.   A 60-kg man is sliding east on the frictionless surface of a frozen pond at a velocity of 0.50 m/s. He is struck by a 1.0-kg snowball whose velocity is 20 m/s toward the north. What is the man's final velocity?

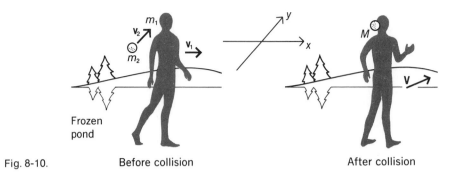

Fig. 8-10.   Before collision   After collision

Solution. Here linear momentum must be conserved separately in both the east-west and north-south directions, which we shall call the x and y axes respectively. Using the notation shown in Fig. 8-10 we have

Momentum afterward = momentum before,

x direction: $MV_x = m_1 v_{1x} + m_2 v_{2x},$

y direction: $MV_y = m_1 v_{1y} + m_2 v_{2y}.$

Since

$m_1 = 60$ kg, $m_2 = 1.0$ kg, $M = m_1 + m_2 = 61$ kg,

$v_{1x} = 0.50$ m/s, $v_{2x} = 0,$ $V_x = ?,$

$v_{1y} = 0,$ $v_{2y} = 20$ m/s, $V_y = ?,$

we have

$$V_x = \frac{m_1 v_{1x} + m_2 v_{2x}}{M}$$

$$= \frac{60 \text{ kg} \times 0.5 \text{ m/s}}{61 \text{ kg}} = 0.49 \text{ m/s},$$

$$V_y = \frac{m_1 v_{1y} + m_2 v_{2y}}{M}$$

$$= \frac{1.0 \text{ kg} \times 20 \text{ m/s}}{61 \text{ kg}} = 0.33 \text{ m/s}.$$

Hence the magnitude V of the velocity $\mathbf{V}$ of man + snowball after the collision is

$$V = \sqrt{V_x^2 + V_y^2}$$
$$= \sqrt{(0.49 \text{ m/s})^2 + (0.33 \text{ m/s})^2}$$
$$= 0.59 \text{ m/s}.$$

The direction in which the man + snowball move after the collision may be specified in terms of the angle θ between the $+y$ direction (that is, north) and $\mathbf{V}$:

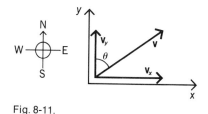

$$\tan \theta = \frac{V_x}{V_y} = \frac{0.49 \text{ m/s}}{0.33 \text{ m/s}} = 1.48,$$

$$\theta = 56°.$$

Fig. 8-11.

Thus man + snowball move in a direction $56°$ to the east of north (Fig. 8-11).

In a completely elastic collision, no kinetic energy is lost. An example is a collision between a moving billiard ball and a stationary one on a level table. The potential energies of the balls remain the same and the energy lost to heat and sound is negligible, so the sum of their kinetic energies before the collision must equal the sum of their kinetic energies afterward.

Problem. A ball rolling on a level table strikes head-on another identical ball which is stationary. What is the result of the collision?

Solution. Let us call the initial and final velocities of the balls $\mathbf{v}_1$, $\mathbf{v}_2$ and $\mathbf{v}_1$, $\mathbf{v}_2$. Conservation of momentum and of energy require that

	Before		After
Momentum:	$m_1\mathbf{v}_1 + m_2\mathbf{v}_2$	$=$	$m_1\mathbf{v}'_1 + m_2\mathbf{v}'_2,$
Energy:	$\frac{1}{2}m_1 v_1^2 + \frac{1}{2}m_2 v_2^2$	$=$	$\frac{1}{2}m_1 v_1'^2 + \frac{1}{2}m_2 v_2'^2.$

Since we have said that the balls are identical, $m_1 = m_2$, and since the second ball was originally at rest, $v_2 = 0$; hence

$$\mathbf{v}_1 = \mathbf{v}'_1 + \mathbf{v}'_2,$$
$$v_1^2 = v_1'^2 + v_2'^2.$$

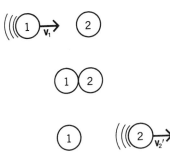

Fig. 8-12. A rolling ball makes a head-on collision with an identical stationary ball; the first ball stops and the second begins moving with the first's initial velocity.

The *only* way of solving these equations is to have either v'_1 or v'_2 equal zero. If v'_2 were zero, it would mean that the first ball traveled completely *through* the second ball. Because this is impossible, we must have as the solution

$$v'_1 = 0,$$

$$v'_2 = v_1;$$

the first ball stops, and the second begins to move with the original speed of the first ball (Fig. 8-12).

Special Topic

More About Collisions

A moving object strikes a stationary one, and as a result the second object is set in motion. What is the mass ratio between the two that will cause the struck object to have the highest possible velocity after the impact? What mass ratio will lead to the greatest transfer of energy to the struck object? These are not questions of abstract interest only but are closely connected with a variety of actual problems that range from the design of golf clubs to the design of nuclear reactors.

For simplicity we will confine ourselves to head-on collisions in which both objects move along the same straight line. We consider an object of mass m_1 and initial velocity v_1 that strikes a stationary object of mass m_2, after which their respective velocities are v_1' and v_2'. From conservation of momentum,

$$m_1 v_1 = m_1 v_1' + m_2 v_2',$$
$$m_1(v_1 - v_1') = m_2 v_2'. \tag{1}$$

If the collision is completely elastic, kinetic energy is conserved, and

$$\tfrac{1}{2}m_1 v_1^2 = \tfrac{1}{2}m_1 v_1'^2 + \tfrac{1}{2}m_2 v_2'^2,$$
$$m_1(v_1^2 - v_1'^2) = m_2 v_2'^2,$$
$$m_1(v_1 + v_1')(v_1 - v_1') = m_2 v_2'^2. \tag{2}$$

Now we divide Eq. (2) by Eq. (1) to obtain

$$v_1 + v_1' = v_2',$$
$$v_1 = v_2' - v_1'. \tag{3}$$

Since $v_2' - v_1'$ is the velocity of m_2 relative to m_1 after the collision and $-v_1$ is the same relative velocity before it, this result means that the effect of the collision is to reverse the direction of the relative velocity without changing its magnitude. Thus the relative velocity of approach is equal to the relative velocity of recession, a conclusion that holds even if m_2 had been moving before the collision (see Prob. 12). In the latter event, $-(v_2 - v_1) = v_2' - v_1'$.

Combining Eqs. (1) and (3) yields for the final velocity v_1' of m_1

$$v_1' = \frac{m_1 - m_2}{m_1 + m_2} v_1 \tag{4}$$

and for the final velocity v_2' of m_2

$$v_2' = \frac{2m_1}{m_1 + m_2} v_1. \tag{5}$$

Formulas (4) and (5) permit us to draw some general conclusions. If m_1 is less than m_2, v_1' is in the opposite direction to v_1: the lighter object rebounds from the heavier one (Fig. 8-13a). A ball striking a wall is an extreme

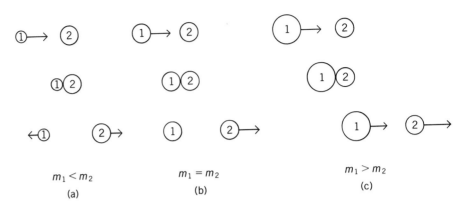

Fig. 8-13. The result of an elastic, head-on collision between a moving object of mass m_1 and a stationary object of mass m_2 depends upon the ratio of their masses.

example, where, since m_2 is virtually infinite compared with m_1, $v_1' = -v_1$. In the event that $m_1 = m_2$, $v_1' = 0$ and $v_2 = v_1$: the colliding objects stops, while the struck one moves off with the same velocity it had (Fig. 8-13b). When m_1 is greater than m_2, as in the case of a table-tennis serve, the colliding object continues on in the same direction after the impact but with reduced speed while the struck object moves ahead of it at a faster pace (Fig. 8-13c). When m_1 is much greater than m_2, the colliding object loses little speed while the struck one is given a speed nearly twice v_1.

The ratio between the kinetic energy KE_2' transferred to the initially stationary object and the kinetic energy KE_1 of the colliding object can be found with the help of Eq. (5):

$$\frac{KE_2'}{KE_1} = \frac{\frac{1}{2}m_2 v_2'^2}{\frac{1}{2}m_1 v_1^2} = \frac{4m_1 m_2}{(m_1 + m_2)^2} = \frac{4(m_2/m_1)}{(1 + m_2/m_1)^2}. \tag{6}$$

This formula is plotted in Fig. 8-14. Evidently the transfer of energy is a maximum for $m_1 = m_2$, when *all* the energy of m_1 is given to m_2. This is the situation illustrated in Fig. 8-13b.

From Eq. (6) it would seem that the best mass for the head of a golf club would be the same as that of a golf ball, in order that all the energy of the club be given to the ball (assuming an elastic collision). In this case $v_2' = v_1$. However, according to Eq. (5), the greater the mass m_2 of the clubhead, the greater the ball's velocity v_2' will exceed v_1, up to a limit of $2v_1$. The trouble with a heavy clubhead is twofold: it is hard to swing a heavy golf club as fast as a light one, and the more m_1 exceeds m_2, the smaller is the proportion of KE that is transferred to the ball—the extra effort does not provide a commensurate return. Experience has led golfers to use clubheads whose masses

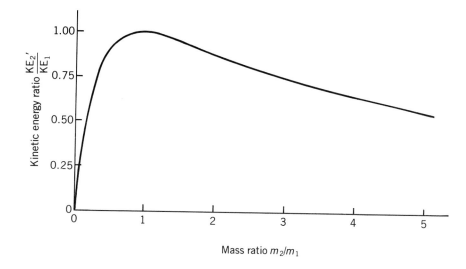

Fig. 8-14. Energy transfer in an elastic, head-on collision between a moving object and a stationary one.

are typically about four times the 46-g mass of a golf ball where maximum distance is required, though this ratio does not seem to be very critical.

If a collision is not perfectly elastic, the relative velocity of recession will be reduced by some fraction e, which is called the *coefficient of restitution*. In general, when both objects are in motion before and after the collision,

$$\text{Coefficient of restitution} = e = \frac{v_2' - v_1'}{v_1 - v_2}. \tag{7}$$

In a perfectly elastic collision, $e = 1$; and in a perfectly inelastic collision, when the objects stick together and $v_2' = v_1'$, $e = 0$. When a ball is dropped on a horizontal surface from a height h, the height of rebound is $h' = e^2 h$. (To verify this note that $v_2 = v_2' = 0$, $v_1 = -\sqrt{2gh}$, and $v_1' = \sqrt{2gh'}$, where upward is considered the positive direction.) Thus if a ball dropped from a height of 100 cm to the floor rebounds to a height of 60 cm, the coefficient of restitution must be $\sqrt{h'/h} = \sqrt{0.6} = 0.77$.

When the coefficient of restitution is taken into account in a collision between a moving object and a stationary one,

$$v_2 = \frac{(1 + e)m_1}{m_1 + m_2} v_1$$

$$\frac{\text{KE}_2'}{\text{KE}_1} = \frac{(1 + e)^2 m_1 m_2}{(m_1 + m_2)^2} = \frac{(1 + e)^2 (m_2/m_1)}{(1 + m_2/m_1)^2}. \tag{8}$$

In the case of a golf club striking a golf ball, typical values might be $m_1 = 200$ g, $m_2 = 46$ g, and $e = 0.7$, from which we find that the ball moves off with a velocity 38 percent greater than the velocity of the clubhead and carries with it 44 percent of the clubhead's original energy. If the collision were perfectly elastic, $e = 1$ and these figures would be 63 percent and 61 percent respectively.

In Chap. 32 we will see how the theory of collisions is applied to the design of nuclear reactors.

Important Terms

The **linear momentum** of a body is the product of its mass and velocity. Linear momentum is a vector quantity having the direction of the body's velocity.

The **impulse** of a force is the product of the force and the time during which it acts. Impulse is a vector quantity having the direction of the force. When a force acts on a body that is free to move, its change in momentum equals the impulse given it by the force.

The law of **conservation of momentum** states that when the vector sum of the external forces acting upon a system of particles equals zero, the total linear momentum of the system remains constant.

The **thrust** of a rocket is the force exerted by the expulsion of exhaust gases.

A **completely elastic collision** is one in which kinetic energy is conserved. A **completely inelastic collision** is one in which the bodies stick together upon impact, which results in the maximum possible kinetic energy loss. Linear momentum is conserved in all collisions.

Important Formulas

Linear momentum:

$$\mathbf{p} = m\mathbf{v}$$

Impulse and momentum change:

$$\mathbf{F}\Delta t = \Delta(m\mathbf{v})$$

Multiple Choice

1. A body at rest may possess
   a. velocity.   b. momentum.
   c. kinetic energy.   d. potential energy.

2. A body in motion need not possess
   a. velocity.   b. momentum.
   c. kinetic energy.   d. potential energy.

3. A body which has momentum must also have
   a. acceleration.   b. impulse.
   c. kinetic energy.   d. potential energy.

4. Momentum is most closely related to
   a. kinetic energy.   b. potential energy.
   c. impulse.   d. power.

5. The impulse given to a body is equal to the consequent change in its
   a. velocity.   b. momentum.
   c. kinetic energy.   d. potential energy.

6. When the velocity of a moving body is doubled,
   a. its acceleration is doubled.
   b. its momentum is doubled.
   c. its kinetic energy is doubled.
   d. its potential energy is doubled.

7. If a shell fired from a cannon explodes in midair,
   a. its total momentum increases.
   b. its total momentum decreases.
   c. its total kinetic energy increases.
   d. its total kinetic energy decreases.

8. When two or more objects collide, it is always true that
   a. the momentum of each one remains unchanged.
   b. the kinetic energy of each one remains unchanged.
   c. the total momentum of all the objects remains unchanged.
   d. the total kinetic energy of all the objects remains unchanged.

9. An elastic collision conserves
   a. kinetic energy but not momentum.
   b. momentum but not kinetic energy.
   c. neither momentum nor kinetic energy.
   d. both momentum and kinetic energy.

10. A ball whose momentum is **p** strikes a wall and bounces off. The change in the ball's momentum is
    a. 0.   b. **p**/2.
    c. **p**.   d. 2**p**.

11. An iron sphere of mass 30 kg has the same diameter as an aluminum sphere of mass 10.5 kg. The spheres are simultaneously dropped from a cliff. When they are 10 m from the ground, they have identical
    a. accelerations.   b. momenta.
    c. potential energies.   d. kinetic energies.

12. A 1280-lb white horse is cantering at 20 ft/s. Its linear momentum is
    a. 400 slug-ft/s.   b. 800 slug-ft/s.
    c. 8000 slug-ft/s.   d. 25,600 slug-ft/s.

13. A 60-lb girl and a 50-lb boy face each other on frictionless roller skates. The girl pushes the boy, who moves away at a velocity of 4 ft/s. The girl's velocity is
    a. 2.1 ft/s.   b. 3.3 ft/s.
    c. 4.0 ft/s.   d. 4.8 ft/s.

14. An astronaut whose total mass is 100 kg ejects 1 gm of gas from his propulsion pistol at a velocity of 50 m/s. His recoil velocity is
    a. 0.5 mm/s.   b. 5 mm/s.
    c. 5 cm/s.   d. 50 cm/s.

15. If a rocket of initial mass m is to rise from its launching pad, its initial thrust must exceed
    a. $\frac{1}{2}mg$.   b. mg.
    c. $2mg$.   d. $\frac{1}{2}mg^2$.

Exercises

1. Is it possible for an object to have more kinetic energy but less momentum than another object? Less kinetic energy but more momentum?

2. When the momentum of an object is doubled in magnitude, what happens to its kinetic energy?

3. When the kinetic energy of an object is doubled in magnitude, what happens to its momentum?

4. In what general way, if any, is the impulse a force imparts to an object related to the change in momentum of the object? To the change in kinetic energy of the object?

5. How is the principle of conservation of linear momentum related to the definition of mass given in Chapter 4 and to Newton's first law of motion? In what way does this principle go beyond the definition of mass and the first law of motion?

6. (a) When an object at rest breaks up into two parts which fly off, must they move in exactly opposite directions? (b) When a moving object strikes a stationary one and the two do not stick together, must they move off in exactly opposite directions?

7. A railway car is at rest on a frictionless track. A man at one end of the car walks to the other end. (a) Does the car move while he is walking? (b) If so, in which direction? (c) What happens when the man comes to a stop?

8. An empty coal car coasts at a certain speed along a level railroad track without friction. (a) It begins to rain. What happens to the speed of the car? (b) The rain stops, and the collected water gradually leaks out. What happens to the speed of the car now?

9. Find the momentum of a 256-lb ostrich running at 50 ft/s.

10. Find the momentum of a 3200-lb car moving at 30 mi/hr.

11. Find the momentum of a 0.02 kg bullet whose velocity is 500 m/s.

12. Find the average momentum of a 70-kg sprinter who covers 400 m in 45 s.

13. A certain DC-8 airplane has a mass of 160,000 kg and is flying at 870 km/hr. (a) Find its momentum. (b) If the thrust its engines can develop is

340,000 N, how much time is needed for the airplane to reach this velocity starting from rest? In this exercise and the next, ignore air resistance, changes in altitude, and the fuel consumed by the engines.

14. A certain Boeing 747 airplane weighs 768,000 lb and is flying at 550 mi/hr. (a) Find its momentum. (b) If the thrust its engines can develop is 180,000 lb, how much time is needed for the airplane to reach this velocity starting from rest?

15. A ⅓-lb baseball reaches the batter with a velocity of 75 ft/s. After it has been struck, it leaves the bat at 100 ft/s in the opposite direction. If the ball was in contact with the bat for 0.001 s, find the average force exerted on it during this period.

16. A 2000-kg truck traveling at 36 km/hr strikes a tree and comes to a stop in 0.1 sec. Find the average force on the truck during the crash.

17. A certain 8-in. howitzer has a barrel 25 ft long and fires a 1400-lb projectile at a muzzle velocity of 1400 ft/s. What is the average force on the projectile while it is in the howitzer barrel?

18. A freight car weighing 12 tons rolls at 5 ft/s along a horizontal railroad track. It collides with another freight car weighing 16 tons that is standing at rest on the track, and the two cars couple together. What is the velocity of the cars after the collision?

Problems

1. A certain cannon has a range of R. One day a shell which it fires explodes at the top of its path into two equal fragments, one of which falls vertically downward. If there is no air resistance, how far away from the cannon does the other fragment land?

2. The cannon of the previous problem fires another shell which also explodes into two equal fragments at the top of its path. One of these fragments lands next to the cannon. How far away from the cannon does the other fragment land?

3. A hunter has a rifle that can fire 0.06-kg bullets with a muzzle velocity of 900 m/s. A 40-kg leopard springs at him at a velocity of 10 m/sec. How many bullets must the hunter fire into the leopard in order to stop him in his tracks?

4. An astronaut in orbit outside an orbiting satellite throws his 1.2 lb camera away in disgust when it jams. If he and his spacesuit together weigh 225 lb and the velocity of the camera is 40 ft/s, how far away from the satellite will he be in 1 hr?

5. A driverless car weighing 4000 lb is moving along a road at 50 mi/hr. In order to stop the car, a tank weighing 16,000 lb makes a head-on collision with it. (a) What should the tank's velocity be in order that both tank and car come to a stop as a result of the collision? (b) How much kinetic energy is dissipated in the collision if the tank has the velocity of part (a)?

6. A 0.5-kg stone moving at 4 m/s overtakes a 4-kg lump of clay moving at 1 m/s. The stone becomes embedded in the clay. (a) What is the velocity of the composite body after the collision? (b) How much kinetic energy is lost?

7. A neutron of mass 1.67×10^{-27} kg and velocity 10^5 m/s collides head-on with a stationary deuteron of mass 3.34×10^{-27} kg. The particles do not stick together, and the deuteron moves off at 6.67×10^4 m/s. What is the velocity of the neutron? Is the collision elastic?

8. A neutron of mass 1.67×10^{-27} kg and velocity 10^5 km/s collides with a stationary deuteron of mass 3.34×10^{-27} kg. The two particles stick together. What is the velocity of the composite particle (called a *triton*)?

9. A 0.5-kg stone moving north at 4 m/s collides with a 4-kg lump of clay moving west at 1 m/s. The stone becomes embedded in the clay. (a) What is the velocity (magnitude and direction) of the composite body after the collision? (b) How much kinetic energy is lost?

10. A 2000-lb car traveling east at 30 mi/hr collides with a 3000-lb car traveling north at 20 mi/hr. The cars stick together after the collision. What is the

velocity (magnitude and direction) of the wreckage?

11. A ballistic pendulum consists of a wooden block of mass M suspended by long cords from the ceiling. A bullet of mass m and velocity v is fired horizontally into the block, which swings away until its height is the amount h above its original height. Find a formula that gives v in terms of g and the readily measurable quantities m, M, and h.

12. Verify that, in an elastic collision between two bodies moving along the same straight line, the relative velocity with which they move apart after the collision is equal to the relative velocity of approach before it. That is, if v_1 and v_2 are the initial velocities of the bodies and v_1' and v_2' their final velocities, show that $(v_2' - v_1') = -(v_2 - v_1)$ is in agreement with the conservation of both momentum and kinetic energy.

Answers to Multiple Choice

1. d	6. b	11. a
2. d	7. c	12. b
3. c	8. c	13. b
4. c	9. d	14. a
5. b	10. d	15. b

9

Rotational Motion

Until now we have been discussing only translational motion, motion in which the position of something changes from one moment to the next. But rotational motion is just as common as translational motion. Wheels, pulleys, propellers, drills, and phonograph records all rotate while carrying out their functions. In the atomic world protons, neutrons, and electrons all rotate; and their rotations in part govern how they interact to form nuclei and atoms and how atoms interact to form molecules, liquids, and solids. In this chapter our chief concern will be the rotational motion of a rigid body about a fixed axis. As we shall find, all of the formulas that describe such motion are exact analogs of the formulas we have already used to describe translational motion.

9-1 Angular Measure

We are accustomed to measuring angles in degrees, where 1° is defined as 1/360 of a full rotation; that is, a complete turn represents 360°. A more suitable unit for our present purposes is the *radian* (rad). The radian is defined with the help of a circle drawn with its center at the vertex of the angle in question. If the circle's radius is r and the arc cut by the angle is s as in Fig. 9-1, then the angle in radians is given by

$$\theta = \frac{s}{r} = \frac{\text{arc length}}{\text{radius}}.$$

Radian measure (9-1)

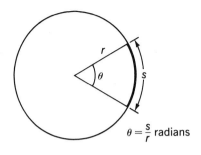

$\theta = \frac{s}{r}$ radians

Fig. 9-1.   The ratio of arc to radius gives the magnitude of an angle in radians.

That is, the angle θ between two radii of a circle, in radian measure, is the ratio of the arc s to the radius r. Evidently an angle of 1 rad has an arc that is the same as the radius.

It is easy to find the conversion factor between degrees and radians and vice versa. We observe that there are 360° in a complete circle, while the number of radians in a complete circle is

$$\theta = \frac{s}{r} = \frac{2\pi r}{r} = 2\pi,$$

because the circumference of a circle of radius r is $2\pi r$ (Fig. 9-2). Hence

$$360° = 2\pi \text{ rad,}$$

from which we find that

$$1° = 0.01745 \text{ rad,}$$
$$1 \text{ rad} = 57.30°.$$

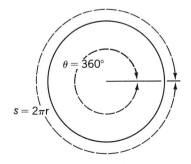

$\theta = 360°$

$s = 2\pi r$

Fig. 9-2. $360° = 2\pi$ rad, so 1 rad = 57.30°.

The radian has no dimensions   The radian is an odd kind of unit because it has no dimensions—an angle expressed in radians is specified by a ratio of lengths, so it is really a pure number. In calculations the unit "rad" is always dropped at the end unless the result is an angular quantity.

Problem.   A phonograph record 12.0 in. in diameter turns through an angle of 120°. How far does a point on its rim travel?

Solution.   First the angle is converted from degrees to radians:

$$\theta = 120° \times 0.01745 \, \frac{\text{rad}}{\text{degree}} = 2.09 \text{ rad.}$$

The radius of the record is 6.0 in., and so, from Eq. (9-1),

$$s = r\theta = 6.0 \text{ in.} \times 2.09 \text{ rad}$$
$$= 12.5 \text{ in.}$$

Problem.   A television image with 525 horizontal lines is being displayed on a picture tube whose screen is 50 cm high. If a viewer's eyes can resolve detail to 0.0003 rad—about 1′ (one minute of arc), where 60′ = 1°—how far away should he be from the screen in order to just be able to see the separate lines?

Solution.   The distance between adjacent lines is $s = 0.5$ m/525, hence

$$r = \frac{s}{\theta} = \frac{0.5 \text{ m}}{525 \times 0.0003} = 3.17 \text{ m.}$$

Sometimes it is useful to express angles in radian measure in terms of π itself. For example, an angle of 90° is $\frac{1}{4}$ of a complete circle, and so

$$90° = \tfrac{1}{4} \text{ circle} \times 2\pi \, \frac{\text{rad}}{\text{circle}} = \pi/2 \text{ rad.}$$

Of course, this has the same numerical value as

$$90° \times 0.01745 \, \frac{\text{rad}}{\text{degree}} = 1.571 \text{ rad,}$$

since $\pi/2 = 1.571$.

9-2 Angular Velocity

If a rotating body turns through the angle θ in the time t, its average angular velocity ω (Greek letter *omega*) is

Linear and angular velocities

$$\omega = \frac{\theta}{t}.$$ *Angular velocity*   (9-2)

If θ is in radians and t in seconds, which are the usual units for these quantities, the unit of ω is the rad/s.

ω is expressed in rad/s

Another common unit of angular velocity is the revolution per minute;

$$1\,\frac{rev}{min} = 1\,\frac{rev}{min} \times 2\pi\,\frac{rad}{rev} \times \frac{1}{60\ s/min} = 0.105\,\frac{rad}{s}.$$

Let us consider a particle moving with the uniform speed v in a circle of radius r, as in Fig. 9-3. This particle travels the distance

$$s = vt$$

in the time t. The angle through which it moves in that time is

$$\theta = \frac{s}{r} = \frac{vt}{r},$$

so that its angular velocity is

$$\omega = \frac{\theta}{t} = \frac{vt}{rt}$$

or

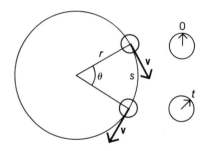

Fig. 9-3.   The angular velocity of a particle in uniform circular motion is $\omega = v/r$.

$$\omega = \frac{v}{r}$$ *Angular velocity*   (9-3)

$$\text{Angular velocity} = \frac{\text{linear velocity}}{\text{path radius}}.$$

The above relationship can be written in another way:

$$v = \omega r$$ (9-4)

Linear velocity = angular velocity $\times$ path radius.

The formulas of this section are valid only when ω is expressed in radian measure.

Axis of rotation

The *axis of rotation* of a rigid body turning in place is that line of particles which does not move (Fig. 9-4). Sometimes the axis of rotation is a line in space. All other particles of the body move in circles about the axis. Since $v = \omega r$, the farther a particle is from the axis, the greater its linear velocity, although all the particles of the body (except those on the axis) have the same angular velocity.

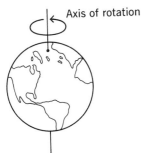

Axis of rotation

Problem. Find the linear velocities of points 1 in. and 6 in. from the axis of a phonograph record rotating at $33\frac{1}{3}$ rpm.

Solution. The angular velocity of the record is

$$33\tfrac{1}{3} \text{ rpm} \times 0.105 \frac{\text{rad/s}}{\text{rpm}} = 3.50 \frac{\text{rad}}{\text{s}}.$$

Hence a point 1 in. from the axis has a linear velocity of

$$v = \omega r = 3.5 \frac{\text{rad}}{\text{s}} \times 1 \text{ in.} = 3.5 \frac{\text{in.}}{\text{s}},$$

while a point on the record's rim, where $r = 6$ in., has a velocity of

$$v = 3.5 \frac{\text{rad}}{\text{s}} \times 6 \text{ in.} = 21 \frac{\text{in.}}{\text{s}}.$$

See Fig. 9-5.

Fig. 9-4. The axis of rotation is that line of particles which does not move in a body rotating in place. (It may be a line in space.)

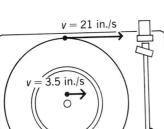
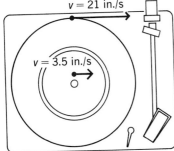

$v = 21$ in./s

$v = 3.5$ in./s

$\omega = 33\frac{1}{3}$ rpm

Fig. 9-5.

9-3 Rotational Kinetic Energy

A rotating body possesses kinetic energy because its constituent particles are in motion, even though the body as a whole remains in place. The speed of a particle that is the distance r from the axis of a rigid body rotating with the angular velocity ω is, as we know, $v = \omega r$ (Fig. 9-6). If the particle's mass is m, its kinetic energy is therefore

$$\text{KE} = \tfrac{1}{2}mv^2$$
$$= \tfrac{1}{2}m\omega^2 r^2. \tag{9-5}$$

The body consists of numerous particles which need not have the same mass or be the same distance from the axis. However, all the particles have the common angular velocity ω. Hence the total kinetic energy of all the particles may be written

$$KE = \Sigma \tfrac{1}{2}mv^2$$

$$= \tfrac{1}{2}(\Sigma mr^2)\omega^2, \qquad (9\text{-}6)$$

where the symbol Σ means, as mentioned before, "sum of." Equation (9-6) states that the kinetic energy of a rotating rigid body is equal to one-half the sum of the mr^2 values of its constituent particles multiplied by the square of its angular velocity ω.

The quantity

$$I = \Sigma mr^2 \qquad \textit{Moment of inertia} \quad (9\text{-}7)$$

is known as the *moment of inertia* of the body. It has the same value regardless of the body's state of motion. The farther a given particle is from the axis of rotation, the faster it moves and the greater is its contribution to the kinetic energy of the body. The moment of inertia of a body depends upon the way in which its mass is distributed relative to its axis of rotation; it is perfectly possible for one body to have a greater moment of inertia than another even though its mass may be much the smaller of the two.

The kinetic energy of a body of moment of inertia I rotating with the angular velocity ω is therefore

$$KE = \tfrac{1}{2}I\omega^2. \qquad \textit{Rotational kinetic energy} \quad (9\text{-}8)$$

Evidently the rotational analog of mass is moment of inertia, just as the rotational analog of velocity is angular velocity. We shall find further support for the correspondence of mass and moment of inertia later in this chapter.

9-4 Moment of Inertia

A rigid body may be considered to be made up of a large number of separate particles whose masses are m_1, m_2, m_3, and so on (Fig. 9-7). To compute the moment of inertia of such a body about a specified axis, we multiply the mass of each of these particles by the square of its distance from the axis—r_1^2, r_2^2, r_3^2, and so on—and add all the mr^2 values. That is,

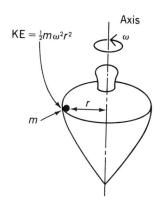

Fig. 9-6.   The kinetic energy of each particle of a rotating body depends upon the square of its distance r from the axis of rotation.

I is rotational analog of mass

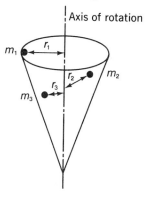

Fig. 9-7.   A rigid body consists of a large number of particles each of which has a certain mass m and distance r from the axis.

$$I = \Sigma mr^2 = m_1 r_1^2 + m_2 r_2^2 + m_3 r_3^2 + \cdots. \qquad (9\text{-}9)$$

Units of moment of inertia   The unit of I is the kg·m² in the metric system and the slug·ft² in the British system.

Problem. Determine the moment of inertia of a thin ring of mass M and average radius R about an axis passing through its center and perpendicular to the plane in which it lies.

Solution. As in Fig. 9-8, we proceed by subdividing the ring into n segments, each of which is at a distance R from the axis. Hence

$$I = m_1 R^2 + m_2 R^2 + m_3 R^2 + \cdots + m_n R^2$$
$$= (m_1 + m_2 + m_3 + \cdots + m_n)R^2.$$

But the sum of the masses of the segments is the same as the total mass M of the ring, and so

$$I = MR^2.$$

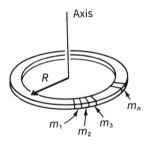

Fig. 9-8.

When a body consists of a continuous distribution of matter, the more particles we imagine it to contain, the more accurate will be our value of its moment of inertia. While I can be calculated for a few simple bodies without difficulty by Eq. (9-9), in general either considerable labor or the use of integral calculus is required. Figure 9-9 gives the moments of inertia of several regularly shaped bodies in terms of the total mass M and dimensions of each. In Section 12-6 we shall learn of a simple experimental way to determine the moment of inertia of an object whose shape is too complex to permit I to be calculated readily.

Problem. The earth's mass is 6×10^{24} kg, and its radius is 6.4×10^6 m. Find its rotational kinetic energy.

Solution. Considering the earth to be a uniform sphere, its moment of inertia is

$$I = \tfrac{2}{5}MR^2$$
$$= \tfrac{2}{5} \times 6 \times 10^{24} \text{ kg} \times (6.4 \times 10^6 \text{ m})^2$$
$$= 9.8 \times 10^{37} \text{ kg·m}^2.$$

The angular velocity of the earth is 7.3×10^{-5} rad/s, corresponding to one rotation per day. Hence its rotational kinetic energy is

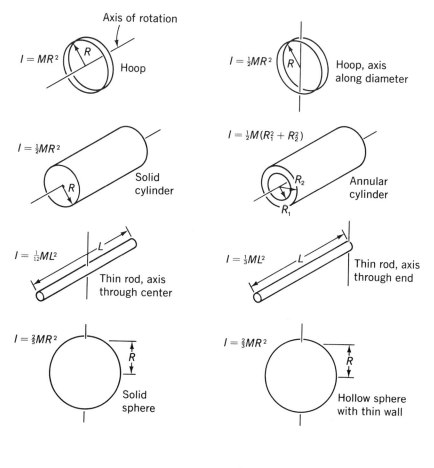

Fig. 9-9. Moments of inertia of various bodies each of mass M, about indicated axes.

$$\mathrm{KE} = \tfrac{1}{2}I\omega^2 = \tfrac{1}{2} \times 9.8 \times 10^{37}\ \mathrm{kg \cdot m^2} \times \left(7.3 \times 10^{-5}\ \frac{\mathrm{rad}}{\mathrm{s}}\right)^2$$

$$= 2.6 \times 10^{29}\ \mathrm{J}.$$

The kinetic energy of the earth's orbital motion is about 10,000 times greater than that of its rotational motion.

9-5 Combined Translation and Rotation

When a rigid body is both moving through space and undergoing rotation, its total kinetic energy is the sum of its translational and rotational kinetic energies. The translational KE is calculated on the basis that the body is a particle whose linear velocity is the same as that of the body's center of gravity; the rotational kinetic energy is calculated on the basis that the body is rotating about an axis that passes through the center of gravity. Thus

$$KE = KE_{translation} + KE_{rotation}$$

$$= \tfrac{1}{2}mv^2 + \tfrac{1}{2}I\omega^2 \tag{9-10}$$

where m is the body's mass, v is the velocity of its center of gravity, I is its moment of inertia about an axis through the center of gravity, and ω is its angular velocity about that axis.

Problem.   Consider a cylinder of radius R and mass m that is poised at the top of an inclined plane (Fig. 9-10). Will it have a greater speed at the bottom if it slides down without friction or if it rolls down?

Solution.   In the first case, we set the cylinder's initial potential energy of mgh equal to its final kinetic energy of $\tfrac{1}{2}mv^2$, and find that

$$PE = KE,$$

$$mgh = \tfrac{1}{2}mv^2,$$

$$v = \sqrt{2gh}.$$

In the second case the cylinder has both translational and rotational kinetic energy at the bottom, so that

$$PE = KE,$$

$$mgh = \tfrac{1}{2}mv^2 + \tfrac{1}{2}I\omega^2.$$

The moment of inertia of a cylinder is $I = \tfrac{1}{2}mR^2$; if it rolls without slipping, its linear and angular velocities are related by the formula $\omega = v/R$. Hence

$$mgh = \tfrac{1}{2}mv^2 + \tfrac{1}{2}(\tfrac{1}{2}mR^2)\left(\frac{v^2}{R^2}\right)$$

$$= \tfrac{1}{2}mv^2 + \tfrac{1}{4}mv^2 = \tfrac{3}{4}mv^2,$$

$$v = \sqrt{\tfrac{4}{3}gh}.$$

The cylinder moves more slowly when it rolls down the plane than when it slides without friction, because some of the available energy is absorbed by its rotation.

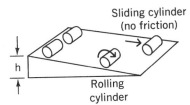

Sliding cylinder (no friction)

Rolling cylinder

Fig. 9-10.

9-6 Angular Acceleration

A rotating body need not have a uniform angular velocity ω, just as a moving particle need not have a uniform linear velocity v. If the angular velocity of a body changes by an amount $\Delta\omega$ in the time interval Δt, its average *angular acceleration* α (Greek letter "alpha") is

$$\alpha = \frac{\Delta\omega}{\Delta t}. \qquad\qquad \textit{Angular acceleration} \qquad (9\text{-}11)$$

The unit of angular acceleration is the rad/s².

A particle moving in a circle of radius r that experiences an angular acceleration α also experiences a linear acceleration a_T tangential to its path, since its orbital speed is changing (Fig. 9-11). From the definition $\omega = v/r$, if r is constant

Tangential acceleration

$$\Delta\omega = \frac{\Delta v}{r}$$

and

$$\alpha = \frac{\Delta\omega}{\Delta t} = \frac{\Delta v}{r\,\Delta t}.$$

Because the magnitude of the particle's linear acceleration along the direction of its velocity **v** is $a_T = \Delta v/\Delta t$, we see that

$$\alpha = \frac{a_T}{r} \qquad\qquad (9\text{-}12)$$

$$\text{Angular acceleration} = \frac{\text{tangential acceleration}}{\text{path radius}}.$$

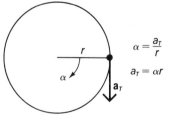

Conversely,

$$a_T = \alpha r \qquad\qquad (9\text{-}13)$$

Tangential acceleration = angular acceleration $\times$ path radius.

Fig. 9-11. The angular and tangential accelerations of a particle moving in a circle are proportional to each other.

We must be careful to distinguish between the *tangential acceleration* a_T of a particle, which represents a change in the magnitude of its linear velocity, and the *centripetal acceleration* a_C, which represents a change in its direction of motion. A particle in circular motion with the speed v has,

Tangential and centripetal acceleration

as we know, the centripetal acceleration

$$a_C = \frac{v^2}{r}$$

directed toward the center of its circular path. The accelerations $\mathbf{a}_T$ and $\mathbf{a}_C$ are therefore always perpendicular (Fig. 9-12). *All* particles in circular motion have centripetal accelerations. Only those particles whose speed changes in magnitude, however, have tangential accelerations. The centripetal acceleration of a particle moving in a circle of radius r can be expressed in terms of its angular velocity ω as

$$a_C = \omega^2 r, \tag{9-14}$$

because $v = \omega r$.

Problem.   Find the total linear acceleration of a particle moving in a circle of radius 0.4 m, at the instant when the angular velocity is 2 rad/s and the angular acceleration is 5 rad/s².

Solution.   From Eqs. (9-13) and (9-14), for the particle's tangential and centripetal accelerations, we have

$$a_T = \alpha r = 5 \, \frac{\text{rad}}{\text{s}^2} \times 0.4 \text{ m} = 2 \text{ m/s}^2,$$

$$a_C = \omega^2 r = \left(2 \, \frac{\text{rad}}{\text{s}}\right)^2 \times 0.4 \text{ m} = 1.6 \text{ m/s}^2.$$

We must add a_T and a_C vectorially, since they are vector quantities which are in different directions. As shown in Fig. 9-12, the magnitude a of the vector sum $\mathbf{a}$ of $\mathbf{a}_T$ and $\mathbf{a}_C$ is

$$a = \sqrt{a_T^2 + a_C^2}$$
$$= \sqrt{2^2 + 1.6^2} \text{ m/s}^2$$
$$= 2.6 \text{ m/s}^2.$$

Let us consider a rigid body whose angular acceleration has the constant value α. The body's angular velocity is ω_0 at $t = 0$ and at the later time $t = t$ it has changed to

$$\omega_f = \omega_0 + \alpha t. \tag{9-15}$$

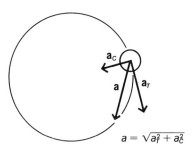

Fig. 9-12.   The tangential and centripetal accelerations on a particle in circular motion are always perpendicular.

In Chapter 3 we obtained the formula

$$s = v_0 t + \tfrac{1}{2} a t^2$$

for the displacement of an accelerated body. By means of the same reasoning, the angular displacement θ of a rigid body undergoing an angular acceleration turns out to be

Angular displacement, velocity and acceleration

$$\theta = \omega_0 t + \tfrac{1}{2} \alpha t^2. \qquad\qquad \textit{Angular displacement} \quad (9\text{-}16)$$

The angular analog of the linear formula

$$v_f^2 = v_0^2 + 2as$$

is the equally useful

$$\omega_f^2 = \omega_0^2 + 2\alpha\theta. \qquad\qquad\qquad (9\text{-}17)$$

Problem. The speed of a motor increases from 1200 rpm to 1800 rpm in 20 s. How many revolutions does it make in this period of time?

Solution. We start by noting that

$$\omega_0 = 1200 \text{ rpm} \times 0.105 \frac{\text{rad/s}}{\text{rpm}} = 126 \text{ rad/s},$$

$$\omega_f = 1800 \text{ rpm} \times 0.105 \frac{\text{rad/s}}{\text{rpm}} = 189 \text{ rad/s}.$$

The angular acceleration of the motor is therefore

$$\alpha = \frac{\Delta\omega}{\Delta t} = \frac{\omega_f - \omega_0}{\Delta t}$$

$$= \frac{(189 - 126) \text{ rad/s}}{20 \text{ s}} = 3.15 \text{ rad/s}^2.$$

We can use either Eq. (9-16) or (9-17) to find θ, since we know ω_0, ω_f, α, and t From Eq. (9-16) we have

$$\theta = \omega_0 t + \tfrac{1}{2} \alpha t^2$$

$$= 126 \frac{\text{rad}}{\text{s}} \times 20 \text{ s} + \tfrac{1}{2} \times 3.15 \frac{\text{rad}}{\text{s}^2} \times (20 \text{ s})^2 = 3150 \text{ rad}.$$

If we were to use Eq. (9-17), we would rewrite it as

$$\theta = \frac{\omega_f^2 - \omega_0^2}{2\alpha}$$

and get the same result. Since there are 2π rad in a revolution,

$$\theta = \frac{3150 \text{ rad}}{2\pi \text{ rad/revolution}} = 500 \text{ revolutions.}$$

The first and last steps of the preceding calculation would not have been necessary if we had used the revolution as the angular unit and the minute as the time unit. The advantage of always using radian measure is that then there will never be any trouble when going from linear to angular quantities or vice versa, since the formulas $\theta = s/r$, $\omega = v/r$, and $\alpha = a_T/r$ only hold when θ, ω, and α are expressed in radian measure.

9-7 Torque and Angular Acceleration

According to Newton's second law of motion, a net force applied to a body causes it to be accelerated. What can cause a body capable of rotation to experience an angular acceleration?

To fix our ideas, let us look once more at a single particle of mass m restricted to motion in a circle of radius r (Fig. 9-13). A force **F** that acts upon the particle tangent to the particle's path gives it the acceleration a_T according to the formula

$$F = ma_T. \tag{9-18}$$

The tangential acceleration a_T here is equal to αr, and so

$$F = mr\alpha.$$

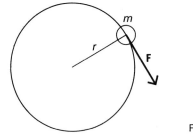

Fig. 9-13.

Multiplying both sides of this equation by r, we have

$$Fr = mr^2\alpha. \tag{9-19}$$

We recognize Fr as the torque τ of the force F about the axis of the particle's rotation (see Sec. 5-2) and mr^2 as the particle's moment of inertia I. Equation (9-19) therefore states that

$$\tau = I\alpha, \tag{9-20}$$

Torque = moment of inertia $\times$ angular acceleration.

While we have derived Eq. (9-20) for the case of a single particle, it is also valid for any rotating body, provided that the torque and moment of inertia are both calculated about the same axis.

The formula $\tau = I\alpha$ is the fundamental law of motion for rotating bodies in the same sense that $F = ma$ is the fundamental law of motion for bodies moving through space. In rotational motion, torque plays the same role that force does in translational motion. The angular acceleration experienced by a body when the torque τ acts upon it is

Torque is rotational analog of force

$$\alpha = \frac{\tau}{I};$$

α is directly proportional to the torque and inversely proportional to the body's moment of inertia.

Problem. The earth's moment of inertia is 9.8×10^{37} kg·m², and its angular velocity is 7.3×10^{-5} rad/s. How great a tangential force would have to be applied to the earth's equator in order to stop the earth from rotating in one year?

Solution. The change in the earth's angular velocity if it stopped rotating would be $\Delta\omega = -7.3 \times 10^{-5}$ rad/s, and the time interval is $\Delta t = 1$ year $= 3.2 \times 10^7$ s. Hence the angular acceleration of the earth would be

$$\alpha = \frac{\Delta\omega}{\Delta t} = -\frac{7.3 \times 10^{-5} \text{ rad/s}}{3.2 \times 10^7 \text{ s}} = -2.3 \times 10^{-12} \text{ rad/s}^2.$$

The required torque is therefore

$$\tau = I\alpha$$
$$= -9.8 \times 10^{37} \text{ kg·m}^2 \times 2.3 \times 10^{-12} \text{ rad/s}^2 = -2.3 \times 10^{26} \text{ N·m}.$$

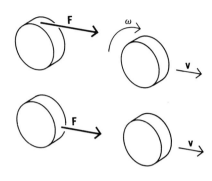

Since the earth's radius is 6.4×10^6 m, the tangential force that must be applied is

$$F = \frac{\tau}{r} = -\frac{2.3 \times 10^{26} \text{ N·m}}{6.4 \times 10^6 \text{ m}} = -3.6 \times 10^{19} \text{ N}.$$

This force is about one percent of the gravitational force that holds the earth in its orbit around the sun.

When a net force acts on a body that is able to move freely, the body will experience both linear and angular accelerations unless the line of action of the force passes through the body's center of gravity (Fig. 9-14). In the latter case the body will be in rotational equilibrium, and if it was not rotating initially, it will keep its original orientation during its motion.

Fig. 9-14. A force whose line of action passes through the center of gravity of a body cannot cause it to rotate.

9-8 Power

Mechanical energy is usually transmitted by rotary motion. The power output of almost every modern engine emerges via a rotating shaft, and more often than not this power is expended in some form of rotation as well: the tires of a car, the propeller of a ship, the bit of a drill, the vanes of a centrifugal pump, the rotor of a dynamo, all function by turning.

The relationship among power, torque, and angular velocity can be derived directly from the definitions of work and power. Let us consider a shaft of radius r on which a tangential force F acts, as in Fig. 9-15. After a time t the shaft has turned through the angle θ, the point at which the force is applied has moved through the distance s, and the force has done the amount of work

$$W = Fs = Fr\theta.$$

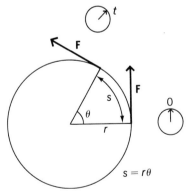

Power is the rate at which work is done, and so here

$$P = \frac{W}{t} = \frac{Fr\theta}{t} = Fr\frac{\theta}{t}.$$

Fig. 9-15. The power transmitted by a rotating shaft is the product of the torque τ on the shaft and its angular velocity ω, so that $P = \tau\omega$.

We recognize that

$$Fr = \tau = \text{torque applied to shaft},$$

$$\frac{\theta}{t} = \omega = \text{angular velocity of shaft};$$

therefore

$$W = \tau\theta,$$
Work = torque × angular displacement.   (9-21)

$$P = \tau\omega,$$ (9-22)

Power = torque × angular velocity.

When energy is to be transmitted at a certain rate P by means of a rotating shaft, the higher the angular velocity, the lower the torque needed, and vice versa. Equation (9-22) is the rotational analog of Eq. (7-5),

$$P = Fv$$

Power = force × velocity.

Problem. An automobile engine develops 230 hp at 5000 rpm. How much torque is exerted on the crankshaft?

Solution. We begin by converting 230 hp and 5000 rpm to their equivalents in lb·ft/s and rad/s:

$$230 \text{ hp} \times 550 \frac{\text{lb·ft/s}}{\text{hp}} = 1.27 \times 10^5 \text{ lb·ft/s},$$

$$5000 \text{ rpm} \times 0.105 \frac{\text{rad/s}}{\text{rpm}} = 525 \text{ rad/s}.$$

From Eq. (9-21) we obtain

$$\tau = \frac{P}{\omega} = \frac{1.27 \times 10^5 \text{ lb·ft/s}}{525 \text{ rad/s}} = 242 \text{ lb·ft}.$$

9-9 Angular Momentum

The rotational analog of linear momentum is *angular momentum*. The angular momentum L of a body depends upon its moment of inertia I and angular velocity ω in the same way that its linear momentum depends upon its mass m and linear velocity v:

$$L = I\omega$$ (9-23)

Angular momentum = moment of inertia × angular velocity.

Conservation of angular momentum

I large, ω small

I small, ω large

Fig. 9-16.   Angular momentum is conserved when a skater executes a spin.

Fig. 9-17.   Angular momentum is a vector quantity whose direction is given by the right-hand rule shown here.

When there is no net torque on a rigid body, both its angular velocity ω and its angular momentum are constant. A deeper analysis shows that the angular momentum of a body does not change in the absence of a net torque on it even if it is *not* a rigid body, but is so altered during its motion that its moment of inertia changes. In a situation of this kind the angular velocity of the body also changes so that *L* stays the same. Thus we have the useful theorem of *conservation of angular momentum*:

When the sum of the external torques acting upon a system of particles equals zero, the total angular momentum of the system remains constant.

A skater or ballet dancer doing a spin capitalizes upon conservation of angular momentum. In Fig. 9-16 a skater is shown starting his spin with his arms and one leg outstretched. By bringing his arms and extended leg inward, he reduces his moment of inertia considerably and consequently spins faster.

Because angular momentum is a vector quantity (Fig. 9-17), a torque must be applied to change the orientation of the axis of rotation of a spinning body as well as to change the magnitude of its angular velocity. The greater the magnitude of **L**, the more torque is needed for it to deviate from its original direction. This is the principle behind the spin stabilization of projectiles, such as footballs and rockets. Such projectiles are set spinning about axes in their directions of motion so that they do not tumble and thereby offer excessive air resistance. A top is another illustration of the vector nature of angular momentum. A stationary top set on its tip falls over

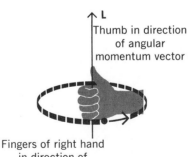

L
Thumb in direction of angular momentum vector

Fingers of right hand in direction of rotational motion

at once, but a rotating top stays upright until its angular momentum is dissipated by friction between its tip and the ground (Fig. 9-18).

Like the conservation principles of energy and of linear momentum, the conservation of angular momentum turns out to be a consequence of a symmetry property of the universe. If the laws of nature are independent of direction in space—that is, if the laws of nature are the same regardless of how an observer is oriented with respect to an event of some kind—then angular momentum must be conserved in all interactions, as observed.

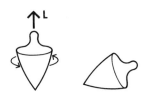

Fig. 9-18. The faster a top spins, the more stable it is. When its angular momentum has been lost through friction, the top falls over.

Special Topic

Moments of Inertia by Integration

When an object consists of a continuous distribution of matter, the more particles we imagine it to contain, the more accurate will be our value of its moment of inertia. In the limit of the finest possible subdivision the summation of Eq. (9-9) becomes the integral

$$I = \int r^2 \, dm$$

where dm is the mass of each of the infinitesimal elements of mass and r is its distance from the axis. The integral is taken over the volume of the object, and, if the object has a regular shape and a uniform density, its evaluation is usually straightforward.

A simple illustration of the process is the case of a uniform rod of length L and mass M which rotates about a perpendicular axis through one end (Fig. 9-19). We divide the rod into elements of length dx. If the rod has a cross-sectional area A and density ρ (Greek letter *rho*), each of these elements has the volume $A \, dx$ and the mass

$$dm = \rho A \, dx.$$

The moment of inertia dI of the element is

$$dI = \rho A x^2 \, dx$$

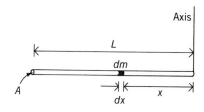

Fig. 9-19.

since the element is the distance x from the axis. Integrating dI from $x=0$ to $x=L$ yields

$$I = \int_0^L dI = \rho A \int_0^L x^2\, dx = \rho A \frac{L^3}{3}.$$

The mass of the rod is the product of its density ρ and its volume AL, so that $M = \rho AL$ and

$$I = \tfrac{1}{3}ML^2$$

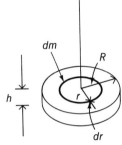

Fig. 9-20.

as given in Fig. 9-9.

The element of mass dm that is most appropriate to employ depends upon the object under consideration. For instance, the simplest way to find the moment of inertia of a solid disk or cylinder about its axis of symmetry is to choose a ring dr thick as the element of mass, as in Fig. 9-20. If the disk has the uniform density ρ and is h thick, and the ring is r in radius, the volume of the ring is $(2\pi r\, dr)h$ and its mass is

$$dm = 2\pi \rho h r\, dr.$$

The moment of inertia dI of the ring is

$$dI = r^2\, dm = 2\pi \rho h r^3\, dr.$$

Integrating dI from $r=0$ to $r=R$, the disk's radius, yields

$$I = \int_0^R dI = 2\pi \rho h \int_0^R r^3\, dr = 2\pi \rho h \frac{R^4}{4}.$$

Since the disk has the area πR^2, its mass is $M = \pi \rho R^2 h$, and its moment of inertia is

$$I = \tfrac{1}{2}MR^2.$$

Important Terms

The **radian** is a unit of angular measure equal to 57.30°. If a circle is drawn whose center is at the vertex of an angle, the angle in radian measure is equal to the ratio between the arc of the circle cut by the angle and the radius of the circle. A full circle contains 2π radians.

The **angular velocity** ω of a rotating body is the angle through which it turns per unit time. The **angular acceleration** α of a rotating body is the rate of change of its angular velocity with respect to time.

The **axis of rotation** of a rigid body turning in place is that line of particles which does not move.

All particles in circular motion experience centripetal accelerations, but only those particles whose angular velocity changes have **tangential accelerations**.

The **moment of inertia** I of a body about a given axis is the rotational analog of mass in linear motion. Its value depends upon the way in which the mass of the body is distributed about the axis.

The **angular momentum** L of a rotating body is the product $I\omega$ of its moment of inertia and angular velocity. The principle of **conservation of angular momentum** states that the total angular momentum of a system of particles remains constant when no net external torque acts upon it.

Torque:

$$\tau = I\alpha$$

Work and power:

$$W = \tau\theta$$
$$P = \tau\omega$$

Angular Momentum:

$$L = I\omega$$

Important Formulas

Radian measure:

$$\theta = \frac{s}{r}$$

Angular velocity:

$$\omega = \frac{\theta}{t}$$

Linear velocity:

$$v = \omega r$$

Moment of inertia:

$$I = \Sigma mr^2$$

Rotational kinetic energy:

$$KE = \tfrac{1}{2}I\omega^2$$

Motion under constant angular acceleration:

$$\theta = \omega_0 t + \tfrac{1}{2}\alpha t^2$$

$$\omega_f = \omega_0 + \alpha t$$

$$\omega_f^2 = \omega_0{}^2 + 2\alpha\theta$$

Angular acceleration:

$$\alpha = \frac{\Delta\omega}{\Delta t} = \frac{a_T}{r}$$

Multiple Choice

1. In a rigid body undergoing uniform circular motion, a particle that is a distance R from the axis of rotation
   a. has an angular velocity proportional to R.
   b. has an angular velocity inversely proportional to R.
   c. has a linear speed proportional to R.
   d. has a linear speed inversely proportional to R.

2. The centripetal acceleration of a particle in circular motion
   a. is less than its tangential acceleration.
   b. is equal to its tangential acceleration.
   c. is more than its tangential acceleration.
   d. may be more or less than its tangential acceleration.

3. The rotational analog of force in linear motion is
   a. moment of inertia.
   b. angular momentum.
   c. torque.
   d. weight.

4. The rotational analog of mass in linear motion is
   a. moment of inertia.
   b. angular momentum.
   c. torque.
   d. angular velocity.

5. A quantity not directly involved in the rotational motion of a body is
   a. mass.   b. moment of inertia.
   c. torque.   d. angular velocity.

6. All rotating bodies at sea level that have the same mass and angular velocity also have the same
   a. angular momentum.
   b. moment of inertia.
   c. kinetic energy.
   d. potential energy.

7. The moment of inertia of a body does not depend upon
   a. its mass.
   b. its size and shape.
   c. its angular velocity.
   d. the location of the axis of rotation.

8. Of the following properties of a Yo-Yo moving in a circle, the one that does not depend upon the radius of the circle is the Yo-Yo's
   a. angular velocity.
   b. angular momentum.
   c. linear velocity.
   d. centripetal acceleration.

9. A Yo-Yo being swung in a circle need not possess
   a. angular velocity.
   b. angular momentum.
   c. angular acceleration.
   d. centripetal acceleration.

10. The total angular momentum of a system of particles
    a. remains constant under all circumstances.
    b. changes when a net external force acts upon the system.
    c. changes when a net external torque acts upon the system.
    d. may or may not change under the influence of a net external torque, depending on the direction of the torque.

11. A full circle contains
    a. $\pi/4$ radians.   b. $\pi/2$ radians.
    c. π radians.   d. 2π radians.

12. A radian is approximately equal to
    a. $1°$.   b. $5°$.
    c. $12°$.   d. $60°$.

13. A wheel is 1 m in diameter. When it makes 30 rev/min, the linear speed of a point on its circumference is
    a. $\pi/2$ m/sec.   b. π m/sec.
    c. 30π m/sec.   d. 60π m/sec.

14. A body undergoes a uniform angular acceleration. In the time t since the body started rotating from rest, the number of turns it makes is proportional to
    a. $\sqrt{t}$.   b. t.
    c. t^2.   d. t^3.

15. A wheel that starts from rest has an angular velocity of 20 rad/s after being uniformly accelerated for 10 s. The total angle through which it has turned in these 10 s is
    a. 2π radians.
    b. 40π radians.
    c. 100 radians.
    d. 200 radians.

16. A flywheel rotating at 10 rev/s is brought to rest by a constant torque in 15 s. In coming to a stop the flywheel makes
    a. 75 rev.   b. 150 rev.
    c. 472 rev.   d. 600 rev.

17. A hoop rolls down an inclined plane. The fraction of its total kinetic energy that is associated with its rotation is
    a. $\frac{1}{4}$.   b. $\frac{1}{3}$.
    c. $\frac{1}{2}$.   d. $\frac{2}{3}$

18. A solid lead cylinder of radius R, a solid aluminum cylinder of radius R, a hollow lead cylinder of radius $R/2$, and a solid lead sphere of radius R all start rolling down an inclined plane at the same time. The one that reaches the bottom first is the
    a. solid lead cylinder.
    b. solid aluminum cylinder.
    c. hollow lead cylinder.
    d. solid lead sphere.

19. A solid iron sphere A rolls down an inclined plane, while an identical sphere B slides down the plane in a frictionless manner.
    a. Sphere A reaches the bottom first.
    b. Sphere B reaches the bottom first.
    c. They reach the bottom together.
    d. Which one reaches the bottom first depends on the angle of the plane.

20. A solid iron sphere A rolls down an inclined plane, while an identical sphere B slides down the plane in a frictionless manner. At the bottom the kinetic energy of sphere A is

a. less than that of sphere *B*.

b. equal to that of sphere *B*.

c. more than that of sphere *B*.

d. more or less than that of sphere *B*, depending on the angle of the plane.

Exercises

1. A hollow cylinder and a solid cylinder having the same mass and diameter are released from rest simultaneously at the top of an inclined plane. Which reaches the bottom first?

2. Will a car coast down a hill faster when it has heavy tires or light tires?

3. If the polar ice caps melt, how will the length of the day be affected?

4. All helicopters have two propellers; some have both propellers on vertical axes but rotating in opposite directions, and others have one on a vertical axis and one on a horizontal axis perpendicular to the helicopter body at the tail. Why is a single propeller never used?

5. Use the vector nature of angular momentum to explain the advantage of rifling gun barrels so that their bullets emerge spinning about an axis in their direction of motion.

6. The earth revolves around the sun in an elliptical orbit, and its velocity is greatest when it is closest to the sun and least when it is farthest away. What is there about the nature of the force the sun exerts on the earth to make you think that the earth's orbital motion must conserve angular momentum? How can such conservation of angular momentum be reconciled with the variation in the earth's velocity?

7. An apple pie 12 in. in diameter is cut into nine equal pieces. What angle (in radians) is included between the sides of each piece?

8. (a) Express 1.3 radians in degrees. (b) Express 198° in radians.

9. The resolution of the human eye is about 1′, where 60′ =1°. (a) Express an angle of 1′ in radians. (b) What is the length of the smallest detail that can be discerned when an object 25 cm away is being examined? (This is the distance of most distinct vision.)

10. How many radians does the second hand of a clock turn through in 30 s? In 90 s? In 105 s? Express the answers in terms of π.

11. What is the angular velocity in rad/s of the hour, minute, and second hands of a clock?

12. A car makes a U-turn in 5 s. What is its average angular velocity?

13. The shaft of a motor rotates at the constant angular velocity of 3000 rpm. How many radians will it have turned through in $\frac{1}{2}$ min?

14. The record number of consecutive turns made in skipping rope is 32,089, which required 3 hr 10 min by an Australian in 1953. What was the average angular velocity of the rope?

15. A drill bit 8 mm in diameter is rotating at 200 rad/s. What is the linear velocity of a point on its circumference?

16. A grindstone 10 cm in radius is rotating at 1725 rpm. (a) What is its angular velocity? (b) What is the linear velocity of a point on its rim?

17. A drill bit $\frac{1}{4}$ in. in diameter is rotating at 1200 rpm. (a) What is its angular velocity? (b) What is the linear velocity of a point on its circumference?

18. In 1941 a wind-driven electric power station was set up at Grandpa's Knob, Vermont. The propeller was 130 ft in diameter. When the propeller was turning at 30 rpm, what was the linear velocity in mi/hr of one of the blade tips?

19. A circular saw blade rotating at 1000 rpm is brought to a stop in 125 revolutions. How much time did this take?

20. A phonograph turntable slows down to a stop from an initial angular velocity of 45 rpm in 20 s. (a) What is its acceleration? (b) How many turns does it make while slowing down?

21. A wheel starts from rest under the influence of a constant torque and turns through 500 radians in 10 s. (a) What is its angular acceleration? (b) What is its angular velocity at the end of these 10 s?

22. The baton of a drum majorette is a 300-g uniform rod 80 cm long. What is its kinetic energy when it is twirled at an angular velocity of 10 rad/s?

23. The baton of a drum majorette is a 0.7-lb uniform rod 2 ft long. What is its kinetic energy when it is twirled at an angular velocity of 10 rad/s?

24. A 7-kg bowling ball 30 cm in diameter rolls at a speed of 5 m/sec. What is its total kinetic energy?

25. A 1-hp motor rotates at 1200 rpm. How much torque in lb·ft can it exert?

26. The first motorcycle built had an engine that developed 0.5 hp at 700 rpm. How much torque in N·m could it exert?

27. A torque of 500 N·m is applied to a turbine rotating at 200 rad/s and after 40 s its speed has doubled. What is the turbine's moment of inertia?

28. What is the angular momentum of a particle of mass m that moves in a circle of radius r at the velocity v?

Problems

1. A barrel 2 ft in diameter is rolling with an angular velocity of 5 rad/sec. What is the instantaneous linear velocity of (a) its top, (b) its center, and (c) its bottom with respect to the ground?

2. A truck whose tires have a radius of 60 cm travels at 20 m/sec. Find the angular velocity of its tires.

3. A truck whose tires have a radius of 2 ft travels at 30 mi/hr. Find the angular velocity of the tires.

4. The propeller of a boat rotates at 1000 rpm when the velocity of the boat is 20 ft/sec. The diameter of the propeller is 18 in. What is the velocity of the tip of the propeller?

5. A truck undergoes an acceleration of 0.25 m/s². If its wheels are 1 m in diameter, what is their angular acceleration?

6. A solid sphere 10 cm in radius starts from rest at the top of an inclined plane 10 m long and reaches the bottom in 7 s. What angle does the plane make with the horizontal?

7. A rope 1 m long is wound around the rim of a drum of radius 12 cm and moment of inertia 0.02 kg·m². The rope is pulled with a force of 2.5 N. (a) Assuming that the drum is free to rotate without friction, what is its final angular velocity? (b) What is its final kinetic energy? (c) How much work is done by the force?

8. A 3-kg hoop 1 m in diameter rolls down an inclined plane 10 m long that is at an angle of 20° with the horizontal. (a) What is the angular velocity of the hoop at the bottom of the plane? (b) What is its linear velocity? (c) What is its rotational kinetic energy? (d) What is its total kinetic energy?

9. The *radius of gyration* of an object is the distance from a specified axis of rotation to a point at which the object's entire mass may be considered to be concentrated from the point of view of rotational motion about that axis. Thus the moment of inertia of an object of mass M and radius of gyration k is $I = Mk^2$. (a) The radius of gyration about its center of a hollow sphere of radius R and mass M is $k = \frac{2}{3}R$. Find its moment of inertia. (b) Find the radius of gyration of a solid sphere about its center.

10. The radius of gyration of an 80-kg flywheel 35 cm in radius is 30 cm. (a) Find its moment of inertia. (b) Find the radius of a solid disk with the same mass and moment of inertia. What does this suggest about the cross-sectional form of the flywheel?

11. A uniform thin rod of length L is pivoted about a horizontal axis at one end. If it is released from a horizontal position, what will its maximum angular velocity be?

12. A 200-kg cylindrical flywheel 0.3 m in radius is acted upon by a torque of 20 N·m. (a) If it starts from rest, how much time is required to accelerate it to an angular velocity of 10 rad/s? (b) What is its kinetic energy at this angular velocity?

13. A cylinder whose axis is fixed has a string wrapped around it which is pulled with a force equal to the cylinder's weight. Show that the acceleration of the string is equal to $2g$.

14. A high-speed elevator is being planned to lift a

total load of 4 tons at 1200 ft·min. The windlass drum is to be 6 ft in diameter. Neglecting losses, how much power is required from the driving motor, and at how many rpm should the power be developed?

15. An automobile V8 engine develops 325 hp at 5600 rpm. This engine is coupled through a frictionless 5:1 reduction gear to a windlass drum 3 ft in diameter. (a) What is the heaviest load the windlass can raise? (b) At what linear velocity will it raise such a load?

16. An electric motor develops 5 kW at 2000 rpm. The motor delivers its power through a spur gear 20 cm in diameter. If two teeth of the gear transmit torque to another gear at the same instant, find the force exerted by each gear tooth.

17. An 1800-rpm motor operates a pump through a V-belt drive. The motor pulley is 8 in. in diameter and the tension in the belt is 40 lb on one side and 15 lb on the other. What is the power output of the motor in hp?

18. The alternator of a car engine produces 250 W of electric power when it turns at 4000 rpm. The alternator's pulley has a radius of 4 cm. If the alternator is 95 percent efficient, what is the difference between the tensions in the taut and slack parts of the V-belt connecting it to the engine?

19. Two 0.4-kg balls are joined by a 1-m string and set whirling through the air at 5 rev/s about a vertical axis through the center of the string. After a while the string stretches to 1.2 m. (a) What is the new angular velocity? (b) What are the initial and final kinetic energies of each ball? If these are different, account for the difference.

20. A skater has the moment of inertia 150 kg-m² when her arms are outstretched and 50 kg-m² when her arms are brought to her sides. She starts to spin at the rate of one revolution per second when her arms are outstretched, and then pulls her arms to her sides. (a) What is her final angular velocity in rev/s? (b) What are her initial and final kinetic energies? If these are different, account for the difference.

21. A disk of moment of inertia 1 slug-ft² that is rotating at 100 rad/s is pressed against a similar disk that is initially at rest but is able to rotate freely. The two disks stick together and rotate as a unit. (a) What is the final angular velocity of the combination? (b) How much kinetic energy was lost and where did it go?

Answers to Multiple Choice

1. c	8. a	15. c
2. d	9. c	16. a
3. c	10. c	17. c
4. a	11. d	18. d
5. a	12. d	19. b
6. d	13. a	20. b
7. c	14. c	

10

Mechanical Properties of Matter

The usual procedure of the physicist in approaching a complex situation is to first construct an abstract model that represents the essential features of the situation. Then, if this model proves successful in the sense that predictions obtained from it agree reasonably well with the results of observation and experiment, it is further refined until the agreement is even better. Our work in physics thus far illustrates this procedure. We began by treating objects as though they are minute particles, and went on to extend our analysis by considering them as rigid bodies. Actually, there is no such thing as a rigid body; the strongest block of steel can be stretched, compressed, or twisted by applying suitable forces. In this chapter we shall pursue reality further by examining some mechanical properties of matter which are of importance in technology.

10-1 Density

A characteristic property of every substance is its *density*, which is its mass per unit volume. When we speak of lead as a "heavy" metal and of aluminum as a "light" one, what we really mean is that lead has a higher density than aluminum: a cubic meter of lead has a mass of 11,300 kg, whereas a cubic meter of aluminum has a mass of only 2700 kg.

In the SI system the proper unit of density is the kg/m³, and in the British system it is the slug/ft³. Frequently densities are given in g/cm³. Since there are 10^3 g in a kilogram and 100 cm in a meter,

Units of density

$$1 \text{ g/cm}^3 = 10^3 \text{ kg/m}^3.$$

The densities of various common substances are given in Table 10-1. The symbol for density is *d*, so that if a volume *V* of a certain substance has the mass *m*,

Weight density

$$d = \frac{m}{V}.\tag{10-1}$$

Often *weight density*, which is weight per unit volume, is more convenient to use in a particular problem than density itself, which is mass per unit volume. Thus the density of pure water is 1.94 slugs/ft³ and its weight density is 62 lb/ft³. Since

Weight = mass × acceleration of gravity

$$w = mg,$$

weight density is related to mass density by the formula

Weight density = mass density × acceleration of gravity

$$= dg.$$

There is no special symbol for weight density, and sometimes *d* is also used to refer to this quantity. However, to avoid confusion, it is wise to represent weight density by *dg* whenever the occasion arises.

Problem. A room is 5 m long, 4 m wide, and 3 m high. What is the mass of the air it contains?

Solution. The volume of the room is

$$V = \text{length} \times \text{width} \times \text{height} = 5 \text{ m} \times 4 \text{ m} \times 3 \text{ m} = 60 \text{ m}^3.$$

Table 10-1. Densities of various substances at atmospheric pressure and room temperature.

Substance	MASS DENSITY			WEIGHT DENSITY
	kg/m^3	g/cm^3	$slugs/ft^3$	lb/ft^3
Air	1.3	1.3×10^{-3}	2.5×10^{-3}	8×10^{-2}
Alcohol (ethyl)	7.9×10^2	0.79	1.5	48
Aluminum	2.7×10^3	2.7	5.3	1.7×10^2
Balsa wood	1.3×10^2	0.13	0.25	8
Bone	1.6×10^3	1.6	3.1	9.9×10^2
Carbon dioxide	2.0	2.0×10^{-3}	3.8×10^{-3}	0.12
Concrete	2.3×10^3	2.3	4.5	1.4×10^2
Gasoline	6.8×10^2	0.68	1.3	42
Gold	1.9×10^4	19	38	1.2×10^3
Helium	0.18	1.8×10^{-4}	3.5×10^{-4}	1.1×10^{-2}
Hydrogen	0.09	9×10^{-5}	1.8×10^{-4}	5.4×10^{-2}
Ice	9.2×10^2	0.92	1.8	58
Iron	7.8×10^3	7.8	15	4.8×10^2
Lead	1.1×10^4	11	22	7×10^2
Mercury	1.4×10^4	14	26	8.3×10^2
Nickel	8.9×10^3	8.9	17	5.5×10^2
Nitrogen	1.3	1.3×10^{-3}	2.4×10^{-3}	7.7×10^{-2}
Oak	7.2×10^2	0.72	1.4	45
Oxygen	1.4	1.4×10^{-3}	2.8×10^{-3}	9×10^{-2}
Water, pure	1.00×10^3	1.00	1.94	62
Water, sea	1.03×10^3	1.03	2.00	64

The mass of the air inside is therefore

$$m = dV = 1.3 \ \frac{\text{kg}}{\text{m}^3} \times 60 \ \text{m}^3 = 78 \ \text{kg}.$$

Problem.   We want to pump water from a well 60 ft deep at the rate of 100 gal/min. Assuming 100% efficiency, how much power in horsepower is required?

Solution.   We begin by noting that

$$\text{Power} = \frac{\text{work}}{\text{time}} = \frac{\text{work}}{\text{gal}} \times \frac{\text{gal}}{\text{min}}.$$

The work that must be done to lift a gallon of water through 60 ft, since 1 gal = 0.134 ft³, is

$$\frac{\text{Work}}{\text{gal}} = \frac{\text{weight} \times \text{height}}{\text{gal}}$$

$$= 62 \frac{\text{lb}}{\text{ft}^3} \times 0.134 \frac{\text{ft}^3}{\text{gal}} \times 60 \text{ ft}$$

$$= 498 \text{ ft·lb/gal}.$$

The rate of flow is 100 gal/min, and so

$$\text{Power} = 498 \frac{\text{ft·lb}}{\text{gal}} \times 100 \frac{\text{gal}}{\text{min}} = 4.98 \times 10^4 \frac{\text{ft·lb}}{\text{min}}.$$

There are 33,000 ft·lb/min in a horsepower, and therefore

$$\text{Power} = \frac{4.98 \times 10^4 \text{ ft·lb/min}}{3.3 \times 10^4 \text{ (ft·lb/min)/hp}} = 1.5 \text{ hp}.$$

The *specific gravity* of a substance is its density relative to that of water and so is a pure number. Since the density of water is almost exactly 1 g/cm³, the specific gravity of a substance is very nearly equal to the numerical value of its density when expressed in g/cm³. Thus the density of aluminum is 2.7 g/cm³, hence its specific gravity is 2.7.

Specific gravity

10-2 Elasticity

Categories of stress

While solid bodies generally seem perfectly rigid and unyielding, it is nevertheless possible to deform them either temporarily or permanently by applying stresses. Stress forces fall into three categories: *tensions, compressions,* and *shears.* These are illustrated in Fig. 10-1.

A *tensile* stress is applied to a body when equal and opposite forces that act away from each other are exerted on its ends along the same line of action, thereby tending to elongate the body. A *compressive* stress is applied to a body when equal and opposite forces that act toward each other are exerted on its ends along the same line of action, thereby tending to decrease its length. A *shearing* stress is applied to a body when equal

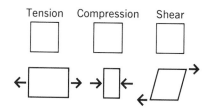

Tension   Compression   Shear

Fig. 10-1. The three types of stress.

and opposite forces are exerted on its ends along different lines of action, thereby tending to change the shape of the body without changing its volume.

In simpler terms, tensions stretch bodies on which they act, compressions shrink bodies on which they act, and shears twist bodies on which they act.

The response of a body to any of the above stresses depends upon its composition, shape, temperature, and so on, but as a general rule the amount of deformation of a crystalline solid is directly proportional to the applied stress provided that the force does not exceed a certain limit. In the case of tension, for example, we might find that supporting a 20-lb weight with a certain thin wire causes the wire to stretch 0.1 in (Fig. 10-2). Doubling the weight to 40 lb therefore will produce a total elongation of 0.2 in., tripling the weight to 60 lb will produce a total elongation of 0.3 in., and so on. When the force is removed, the wire returns to its original length.

The above proportionality is called *Hooke's law*, and may be written

$$F = ks, \hspace{3cm} Hooke's\ law \hspace{0.5cm} (10\text{-}2)$$

where F is the applied tension force, s the resulting elongation, and k a constant whose value depends upon the nature and dimensions of the object under stress. The force constant k is higher for materials such as steel than it is for materials such as lead; it is directly proportional to the cross-sectional area of the object, so that a thick wire of a given material has a higher value of k than a thin wire of the same material. Relationships similar to Eq. (10-2) are found to apply to the behavior of solids under shear stresses and to all states of matter under compressive stresses.

The term *elastic limit* refers to the maximum stress that can be applied to an object without it being permanently deformed as a result. When its elastic limit is exceeded, the object may or may not be far from rupture. Brittle substances like glass or cast iron break at or near their elastic limits. For example, a glass rod whose cross-sectional area is 1 in² obeys Hooke's law so long as the tension or compression forces on it are less than about 10,000 lb, but it will break if this figure is exceeded. Most metals may be deformed considerably beyond their elastic limits, a property known as *ductility*. Copper is a very ductile metal, and while a copper bar whose cross-sectional area is 1 in² reaches its elastic limit when a force of about 6000 lb is applied, it will not rupture until the force has reached perhaps 33,000 lb.

A graph of the elongation of an iron rod as a function of the tension applied to it is shown in Fig. 10-3. At first the graph is a straight line, which corresponds to Hooke's law. Past the elastic limit the graph flattens out, which means that each increase in tension by a given amount produces a

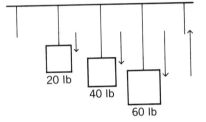

Fig.10-2.   The elongation of a wire is proportional to the stress applied to it, provided that the elastic limit is not exceeded. When the stress is removed, the wire returns to its original length.

Elastic limit

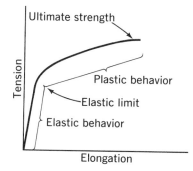

Fig. 10-3.   Graph of the elongation of an iron rod as more and more tension is applied to it. When a tension below the elastic limit is applied and then removed, the rod returns to its original length, which is elastic behavior. When a tension exceeding the elastic limit is applied and removed, the rod does not contract fully but remains permanently longer, which is plastic behavior.

proportionately greater increase in length than it did below the elastic limit: the rod stretches more readily. If the tension is removed after having exceeded the elastic limit, the rod will be permanently longer than it was originally, which is *plastic* behavior. The *ultimate strength* of the rod is the greatest tension it can withstand, and it corresponds to the highest point on the curve.

Ultimate strength

10-3 Young's Modulus

Hooke's law for each kind of stress can be expressed in such a way that only a single constant need be known for a particular material in order to relate the force applied to *any* object of this material to the resulting deformation, regardless of its size and shape. Experimentally it is found that the relative change in size of an object is proportional to the ratio between the applied force and its cross-sectional area. This ratio is called the *stress* on the object: stress is applied force per unit area. The resulting relative change in size is called *strain:* strain is change in length per unit length, change in volume per unit volume, and so on.

Thus we can summarize Hooke's law by saying that, below the elastic limit, *strain is proportional to stress*. Under a given force a thin rod will stretch more than a thick one, and a long rod will stretch more than a short one. The *modulus of elasticity* of a material subjected to a certain kind of stress is defined as the ratio between the stress and the strain that occurs because of it:

Strain is proportional
to stress below
the elastic limit

$$\text{Modulus of elasticity} = \frac{\text{stress}}{\text{strain}}.$$

Young's modulus

For tension or compression stresses, the modulus of elasticity is called *Young's modulus*. In the case of a rod of initial length L_0 and cross-sectional area A in which a tension or compression force F produces a change ΔL in its length (Fig. 10-4),

$$\text{Young's modulus} = \frac{\text{stress}}{\text{strain}},$$

$$Y = \frac{F/A}{\Delta L/L_0}. \tag{10-3}$$

Hence we have

$$\frac{\Delta L}{L_0} = \frac{1}{Y}\frac{F}{A}. \qquad\qquad \textit{Tension or compression} \quad (10\text{-}4)$$

The rod increases in length by ΔL if it is in tension and decreases by that amount if it is in compression. The value of Y depends upon the composition of the rod. Young's moduli for a number of common substances are given in Table 10-2; it is customary to express Y in units of lb/in² in the British system.

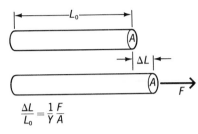

$$\frac{\Delta L}{L_0} = \frac{1}{Y}\frac{F}{A}$$

Fig. 10-4. An object under tension stretches by an amount that depends upon the value of Y for the material of which it is composed. Y is called *Young's modulus.*

Table 10-2. Typical elastic moduli for various materials.

Material	Young's modulus, Y		Shear modulus, S		Bulk modulus, B	
	$10^{10}\,\frac{N}{m^2}$	$10^6\,\frac{lb}{in^2}$	$10^{10}\,\frac{N}{m^2}$	$10^6\,\frac{lb}{in^2}$	$10^{10}\,\frac{N}{m^2}$	$10^6\,\frac{lb}{in^2}$
Aluminum	7.0	10	2.4	3.4	7.0	10
Brass	9.1	13	3.6	5.1	6.1	8.5
Copper	11	16	4.2	6.0	14	20
Glass	5.5	7.8	2.3	3.3	3.7	5.2
Iron	9.1	13	7.0	10	10	14
Lead	1.6	2.3	0.56	0.8	0.77	1.1
Nickel	21	30	7.7	11	26	34
Steel	20	29	8.4	12	16	23

According to Hooke's law, $F = k\Delta L$. We can therefore express the constant k for an object under tension in terms of Y and the dimensions of the object as $k = YA/L_0$.

The proportionality between strain and stress is valid only when the elastic limit is not exceeded. Table 10-3 is a list of the elastic limits and ultimate strengths under tension of several materials; the elastic limit is the same in both tension and compression, but the ultimate strength is greater in the latter. In general, elastic limit and ultimate strength depend upon the previous history of an object as well as upon its composition: hot-rolled and cold-rolled steel have different properties, as do annealed and tempered steel. Further, repeated cycles of stress tend to weaken an object even though superficial examination reveals no change, and "fatigue" failure is   Fatigue
accordingly a serious problem in such applications as aircraft and missile structures where vibration occurs. For these reasons the values in Table 10-3 are only approximate, though typical of each material.

Table 10-3. Typical elastic limits and ultimate strengths for materials under tension. Ultimate strengths under compression are greater.

Material	Elastic Limit		Ultimate Strength	
	$10^8 \dfrac{N}{m^2}$	$10^4 \dfrac{lb}{in^2}$	$10^8 \dfrac{N}{m^2}$	$10^4 \dfrac{lb}{in^2}$
Aluminum	1.3	1.9	1.4	2.1
Brass	3.8	5.5	4.6	6.7
Copper	1.5	2.2	3.4	4.9
Iron, wrought	1.6	2.4	3.2	4.7
Steel, annealed	2.5	3.6	5.0	7.2
Steel, spring	4.1	6.0	6.9	10

Problem. A copper wire 1 mm in diameter and 2 m long is used to support a mass of 5 kg. By how much does the wire stretch under this load? What is the minimum diameter the wire can have if its elastic limit is not to be exceeded?

Solution. From Eq. (10-4),

$$\Delta L = \frac{L_0}{Y} \frac{F}{A}.$$

Here, since the radius of a wire 1 mm in diameter is 5×10^{-4} m, we have

$$L_0 = 2 \text{ m},$$

$$F = mg = 5 \text{ kg} \times 9.8 \text{ m/s}^2 = 49 \text{ N},$$

$$Y = 1.1 \times 10^{11} \text{ N/m}^2,$$

$$A = \pi r^2 = \pi \times (5 \times 10^{-4} \text{ m})^2 = 7.85 \times 10^{-7} \text{ m}^2,$$

and so

$$\Delta L = \frac{2 \text{ m} \times 49 \text{ N}}{1.1 \times 10^{11} \text{ N/m}^2 \times 7.85 \times 10^{-7} \text{ m}^2} = 1.1 \times 10^{-3} \text{ m},$$

which is 1.1 mm.

To answer the second question, we note from Table 10-3 that the elastic limit of copper is 1.5×10^8 N/m². Hence

$$\frac{F}{A} = 1.5 \times 10^8 \text{ N/m}^2,$$

$$A = \frac{49 \text{ N}}{1.5 \times 10^8 \text{ N/m}^2} = 3.27 \times 10^{-7} \text{ m}^2.$$

Since the cross-sectional area of a wire of radius r is $A = \pi r^2$, we have

$$r = \sqrt{\frac{A}{\pi}} = \sqrt{\frac{3.27 \times 10^{-7} \text{ m}^2}{\pi}}$$

$$= \sqrt{1.04 \times 10^{-7} \text{ m}^2} = \sqrt{10.4 \times 10^{-8} \text{ m}^2} = 3.2 \times 10^{-4} \text{ m},$$

and the corresponding diameter is 6.4×10^{-4} m, which is 0.64 mm.

The compressive load on the leg bones of an animal depends upon its weight, which in turn varies as the cube L^3 of a representative linear dimension L such as its length or height. An animal three times as long as another of the same form will weigh about nine times as much. However, the strength of a bone depends upon its cross-sectional area, which for similar animals varies as L^2. Thus animals widely different in size cannot resemble one another—a large animal must have relatively thicker leg bones than a small one since L^3 increases faster than L^2. It is no accident that a hippopotamus has thicker legs for its size than a mouse does, nor that the largest animals of all, the whales, live in the oceans where their immense body weights are supported by buoyancy rather than by legs.

10-4 Shear

Shear stresses change the shape of an object upon which they act. The volume of the object is not affected. The situation is much like that of a book whose covers are pushed out of alignment, as in Fig. 10-5: the layers of atoms, which are analogous to the pages of the book, are displaced sideways, but the spacing of the layers, which corresponds to the thickness of the pages, remains the same.

Let us consider a block of thickness d whose lower face is fixed in place and upon whose upper face the force F acts (Fig. 10-6). A measure of the relative distortion of the block caused by the shear stress is the angle ϕ, called the *angle of shear*. Because this angle is always small, its value in radians is equal to the ratio s/d between the displacement s of the block's faces and the distance d between them. The greater the area A of these faces, the less they will be displaced by the shear force F. The stress-strain equation for shear stresses is therefore

$$\phi = \frac{s}{d} = \frac{1}{S}\frac{F}{A}. \qquad \textit{Shear} \quad (10\text{-}5)$$

Here the applied forces are *parallel* to the faces upon which they act and not perpendicular as in the case of tension and compression.

The quantity S is called the *shear modulus*, and values of S for various substances are given in Table 10-2. The higher the value of S, the more rigid the material; S is sometimes referred to as the *modulus of rigidity* for this reason. It is interesting to note that the shear modulus of any material is usually a good deal less than its Young's modulus. This means that it is easier to slide the atoms of a solid past one another than it is to pull them apart or squeeze them together.

The shear strength of a material is the maximum shear stress an object of that material can withstand before breaking. The greater the shear strength of a material, the more force must be applied to cut a sheet of it with a pair of scissors (or their industrial equivalent) or to punch a hole in the sheet.

Problem. Ordinary mild steel ruptures when a shear stress of about 5×10^4 lb/in^2 is applied. Find the force needed to punch a $\frac{1}{2}$ in. diameter hole in a steel sheet $\frac{1}{8}$ in. thick.

Solution. The area across which the shear stress is exerted here is the cylindrical inner surface of the hole (Fig. 10-7), so that

$$A = \pi dh = \pi \times 0.5 \text{ in.} \times 0.125 \text{ in.} = 0.196 \text{ in.}^2$$

Since we are given that

$$\left(\frac{F}{A}\right)_{max} = 5 \times 10^4 \text{ lb/in.}^2$$

Angle of shear

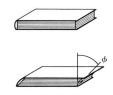

Fig. 10-5.   In shear there is a change in shape without a change in volume. The angle ϕ is the angle of shear.

Shear modulus

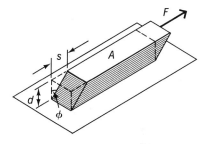

Fig. 10-6.   The angle of shear ϕ (in radians) is equal to s/d. The greater the shear modulus S, the more rigid the material.

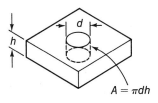

Fig. 10-7.

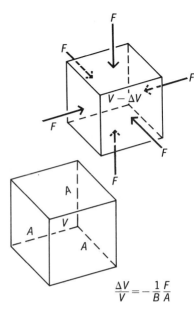

$$\frac{\Delta V}{V} = -\frac{1}{B}\frac{F}{A}$$

Fig. 10-8. Bulk compression.

Table 10-4. Bulk moduli of liquids at room temperature.

Liquid	Bulk modulus, B	
	$10^9 \frac{N}{m^2}$	$10^5 \frac{lb}{in^2}$
Alcohol, ethyl	0.90	1.3
Benzene	1.05	1.5
Kerosene	1.3	1.9
Mercury	26	38
Oil, lubricating	1.7	2.5
Sulfuric acid	3.0	4.3
Water	2.3	3.3

we have

$$F_{max} = 5 \times 10^4 \frac{lb}{in.^2} \times 0.196 \ in.^2 = 0.98 \times 10^4 \ lb$$

which is nearly 5 tons.

10-5 Bulk Modulus

When inward forces act over the entire surface of a body, its volume decreases by some amount ΔV from its original volume of V_0. Only those force components that are perpendicular to the body's surface where they act are effective in compressing it, since the parallel components lead only to shear stresses. If the compression force per unit area F/A is the same over the entire surface of the body, as in Fig. 10-8, the relationship

$$\frac{\Delta V}{V_0} = -\frac{1}{B}\frac{F}{A} \qquad \textit{Uniform compression} \quad (10\text{-}6)$$

holds. The minus sign corresponds to the fact that an increase in force leads to a decrease in volume. The quantity B is called the *bulk modulus*; typical values of B are given in Table 10-2.

The perpendicular force per unit area F/A is usually called *pressure*, symbol p, as discussed in Chap. 11. Hence we can also write

$$\frac{\Delta V}{V_0} = -\frac{p}{B}. \qquad (10\text{-}7)$$

Liquids can support neither tensions nor shears, but they do tend to resist compression. Bulk moduli for several liquids are given in Table 10-4. Interatomic forces within a liquid are smaller than within a solid, which is reflected in the considerably smaller bulk moduli of liquids. Substantial forces are nevertheless needed to compress a liquid by more than a slight amount; to compress a volume of water by one percent requires an inward force per unit area of 3300 lb/in².

Problem.   Verify the above statement.

Solution.   The bulk modulus of water is 3.3×10^5 lb/in² and, by hypothesis, $\Delta V/V_0 = -0.01$. Hence

$$\frac{F}{A} = -B\frac{\Delta V}{V_0} = -3.3 \times 10^5 \frac{lb}{in^2} \times (-0.01) = 3300 \ lb/in^2.$$

Important Terms

The **density** of a substance is its mass per unit volume. The **weight** density of a substance is its weight per unit volume.

The three categories of stress forces are **tension**, in which equal and opposite forces that act away from each other are applied to a body; **compression**, in which equal and opposite forces that act toward each other are applied to a body; and **shear**, in which equal and opposite forces that do not act along the same line of action are applied to a body. A tensile stress tends to elongate a body, a compressive stress to shorten it, and a shearing stress to change its shape without changing its volume.

Hooke's law states that the amount of deformation experienced by a body under stress is proportional to the magnitude of the stress. Thus the elongation of a wire is proportional to the tension applied to it.

The **elastic limit** is the maximum stress a solid can be subjected to without being permanently altered. Hooke's law is valid only when the elastic limit is not exceeded.

Important Formulas

Hooke's law:

$$F = ks$$

Tension or linear compression:

$$\frac{\Delta L}{L_0} = \frac{1}{Y}\frac{F}{A}$$

Shear:

$$\frac{s}{d} = \frac{1}{S}\frac{F}{A}$$

Uniform compression:

$$\frac{\Delta V}{V_0} = -\frac{1}{B}\frac{F}{A} = -\frac{p}{B}$$

Multiple Choice

1. The properties of several different materials are being compared. If the samples all have the same size and shape, the one with the greatest mass also has the greatest
   a. density.
   b. elastic limit.
   c. Young's modulus.
   d. bulk modulus.

2. When equal and opposite forces are exerted on a body along different lines of action, the body is said to be under
   a. tension.
   b. compression.
   c. shear.
   d. elasticity.

3. The ability of a material to be temporarily deformed is called
   a. elasticity.
   b. ductility.
   c. ultimate strength.
   d. viscosity.

4. The ability of a metal to be permanently deformed without rupture is called
   a. elasticity.
   b. ductility.
   c. ultimate strength.
   d. viscosity.

5. A shearing stress that acts on a body affects its
   a. length.
   b. width.
   c. volume.
   d. shape.

6. The stress on an object when a force acts on it is equal to
   a. the relative change in its dimensions.
   b. the applied force per unit area.
   c. Young's modulus.
   d. the elastic limit.

7. Another name for the shear modulus of a material is

a. Young's modulus.

b. modulus of rigidity.

c. bulk modulus.

d. ductility.

8. The only elastic modulus that applies to liquids is

a. Young's modulus.

b. shear modulus.

c. modulus of rigidity.

d. bulk modulus.

9. According to Hooke's law, the force needed to elongate an elastic body by an amount s is proportional to

a. s.

b. $1/s$.

c. s^2.

d. $1/s^2$.

10. A substance whose mass density is 8 slugs/ft^3 has a weight density of

a. 0.25 lb/ft^3.

b. 8 lb/ft^3.

c. 78 lb/ft^3.

d. 256 lb/ft^3.

11. A brass rod 5 ft long with a square cross section $\frac{1}{2}$ in. on a side is used to support a 520-lb load. Its elongation is

a. 0.0008 in.

b. 0.0048 in.

c. 0.0096 in.

d. 0.048 in.

12. Two wires are made of the same material, but wire A is half as long as and has twice the diameter of wire B. If they are to be stretched by the same amount, the required force on wire A must be

a. one-eighth that on B.

b. twice that on B.

c. four times that on B.

d. eight times that on B.

13. A wire 10 m long with a cross-sectional area of 0.1 cm^2 stretches by 13 mm when a load of 100 kg is suspended from it. The Young's modulus for this wire is

a. 0.77×10^{10} N/m^2.

b. 7.5×10^{10} N/m^2.

c. 7.7×10^{10} N/m^2.

d. 9.3×10^{10} N/m^2.

14. The ultimate strength in compression of aluminum is 5×10^4 lb/in.^{2} How much weight can a cube of aluminum 1 ft on each edge support?

a. 300 tons.

b. 600 tons.

c. 3600 tons.

d. 7200 tons.

Exercises

1. An irregular nugget of gold is dropped into a filled glass of water, and 2 in^3 of water overflow. Gold is worth $135 per ounce of weight. How much is the nugget worth?

2. Mammals have approximately the same density as fresh water. Find the volumes of a 7-lb baby, a 170-lb man, a 420-lb gorilla, and a 152-ton blue whale.

3. The radius of the earth is 6.37×10^6 m and its mass is 5.98×10^{24} kg. (a) Find the average density of the earth. (b) The average density of rocks at the earth's surface is 2.7×10^3 kg/m^3. What must be true of the matter of which the earth's interior is composed? Is it likely that the earth is hollow and peopled by another species, as the ancients believed? (Note: The volume of a sphere of radius r is $\frac{4}{3}\pi r^3$.)

4. A three-legged stool has one leg of aluminum, one of brass, and one of steel. The legs have the same dimensions. If the load on the stool is on its exact center, which leg is under the greatest stress and which under the least stress? Which leg experiences the greatest strain and which the least strain?

5. Two wires are made of the same material, but wire A is twice as long and has twice the diameter of wire B. Find the elongation of wire B relative

to that of wire *A* when both are subjected to the same load.

6. The elastic moduli of a particular material are related to one another, and each of them can be expressed in terms of the other two. For instance, the theory of elasticity shows that Young's modulus *Y* can be expressed in terms of the shear modulus *S* and the bulk modulus *B* by the formula

$$Y = \frac{9SB}{(3B + S)}.$$

Check this formula for three of the materials listed in Table 10-2. Give several reasons why you would not expect perfect agreement.

7. A nylon rope $\frac{1}{2}$ in. in diameter breaks when a force of 5000 lb is applied to it. What would you estimate for the breaking strength of a nylon rope (a) $\frac{1}{4}$ in. in diameter? (b) $\frac{3}{4}$ in. in diameter?

8. When a coil spring is used to support a 12-kg object, the spring stretches by 4 cm. What is the force constant of the spring?

9. A coil spring has a force constant of 1000 N/m. How much will it stretch when it is used to support an object whose mass is 8 kg?

10. A coil spring has a force constant of 2 lb/ft. How much will it stretch when it is used to support an object whose mass is 0.5 slug?

11. A steel column 12 ft long and of 4.3 in² cross-sectional area supports a load of 3 tons. By how much is it shortened?

12. A wire 10 ft long with a cross-sectional area of 0.12 in² stretches by 0.02 in. when a load of 600 lb is suspended from it. Find the value of Young's modulus for this wire.

13. A rectangular steel bar 12 ft long and 1 in. across on each side supports a weight of 2 tons. Find its elongation.

14. A copper wire 5 ft long and 0.05 in. in diameter supports a weight of 17 lb. By how much does it stretch?

15. A steel wire 1 m long and 1 mm square in cross section supports a mass of 6 kg. By how much does it stretch?

16. A 1-in. cube of raspberry flavored gelatin on a table is subjected to a shearing force of 0.1 lb. The upper surface is displaced by 0.2 in. What is the shear modulus of the gelatin?

17. The pressure at a depth of 1 km in the ocean exceeds sea-level atmospheric pressure by about 10^7 N/m². If an iron anchor whose volume at the surface is 400 cm³ is lowered to a depth of 1 km, by how much does its volume decrease?

18. The ultimate strength of bone in compression is typically 1.7×10^8 N/m². How safe is it for a 75-kg circus acrobat to balance on the index finger of his right hand, whose bones have a minimum cross-sectional area of 0.5 cm²?

19. The parachute of a 60-kg woman fails to open and she falls into a snowbank at a terminal velocity of 60 m/s, coming to a stop 1.5 m below the surface of the snow. (a) Find the average force on her during the impact. (b) If the area of the woman's body that strikes the snow is 0.2 m² and the stress required for serious injury to body tissues is 5×10^5 N/m², is she likely to survive the impact?

Problems

1. One gram of gold can be beaten out into a foil 1 m² in area. (Thus an ounce of gold can yield 300 ft² of foil.) How many atoms thick is such a foil? The density of gold is 1.93×10^4 kg/m³ and the mass of a gold atom is 3.27×10^{-25} kg.

2. A steel cable whose cross-sectional area is 2.5 cm² supports a 1000-kg elevator. The elastic limit of the cable is 3×10^8 N/m². What is the maximum upward acceleration that can be given the elevator if the tension in the cable is to be no more than 20% of the elastic limit?

3. An iron pipe 10 ft long is used to support a sagging floor. The inside diameter of the pipe is 4 in. and its outside diameter is 5 in. When the force on it is 3000 lb, by how much is it compressed?

4. A wall of lead bricks 1 m high is used to shield a sample of radium. Each brick was originally a cube 10 cm on an edge. What is the height of the lowest brick when the wall has been erected?

5. A vise is used to hold a 2 in. cube of iron while it is being worked on. The cube is in contact with the jaws of the vise, whose screw has a pitch of $\frac{3}{16}$ in. and whose handle is grasped 4 in. from the screw axis. If there is no friction and a force of 15 lb is applied to the handle, find the amount by which the cube is compressed. (See Exercise 25 of Chap. 5.)

6. A brass wire and a steel wire are being used side by side to support a load. (a) If they have the same diameter, what proportion of the load does each one support? (b) What should the ratio of their diameters be if they are each to support half the load?

7. A horizontal, 5-kg bar is supported by a 1.5 m wire at each end. One wire is aluminum and has a diameter of 2 mm while the other wire is steel and has a diameter of 1 mm. Find the stress and elongation for each wire.

8. A certain material has a density of d and a tensile strength of U. How long a rod of this material can be suspended from one end without breaking under its own weight? What is this length in the case of aluminum? Annealed steel? (Assume the density of steel to be the same as that of iron.)

9. A typical sample of compact bone has a tensile strength of 1.2×10^8 N/m² and a compressive strength of 1.7×10^8 N/m². (Bone is a heterogeneous substance in which fibers of a protein called collagen provide most of the tensile strength and inorganic salt crystals provide most of the compressive strength. The different properties of these materials lead to different values of Young's modulus for bone in tension and in compression as well as to different ultimate strengths.) Suppose you have cylinders of the same length and diameter of bone, aluminum, and steel. For each cylinder find the ratio between the maximum tensile

and compressive forces it can withstand and its mass. How does bone compare with aluminum and steel as a structural material?

10. Two steel plates are riveted together with ten rivets each 5 mm in diameter. If the maximum shear stress the rivets can withstand is 3.5×10^8 N/m², how much force applied parallel to the plates is needed to shear off the rivets?

11. Ordinary mild steel ruptures when a shear stress of about 5×10^4 lb/in² is applied. Find the force in tons needed to punch a 1 in. diameter hole in a steel sheet 0.060 in. thick.

12. A slot 8 in. long and $\frac{1}{2}$ in. wide is to be punched in the steel sheet of problem 11. Find the force in tons needed.

13. When a torque τ is applied to a cylinder, the angle θ (in radians) through which it twists depends upon the cylinder's length L, radius r, and shear modulus S according to the formula

$$\theta = \frac{2\tau L}{\pi S r^4}.$$

The geometry of the situation is shown in Fig. 10-9. The above formula makes it possible to determine S experimentally by applying a torque to a wire of a given material and measuring the twist that results, a much simpler procedure than one

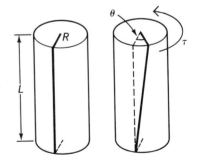

Fig. 10-9.

based directly on Eq. (10-5). Consider a rod 30 cm long that is suspended horizontally at its center by a wire 70 cm long and 1 mm in diameter. When a horizontal force of 0.02 N is applied to one end of the rod, it turns through 25°. Find the shear modulus of the wire.

14. A solid steel drive shaft is 2 in. in diameter and 10 ft long. If the angle of twist is not to exceed 2°, find (a) the maximum torque that can be applied to the shaft, and (b) the maximum power the shaft can transmit at 900 rpm.

Answers to Multiple Choice

1. a	6. b	11. c
2. c	7. b	12. d
3. a	8. d	13. b
4. b	9. a	14. c
5. d	10. d	

11

Fluids

As the name implies, a fluid is a substance that flows readily. Gases and liquids are fluids, although the dividing line between solids and liquids is not always a sharp one. Because of its ability to flow, a fluid can exert a buoyant force on an immersed body, multiply an applied force, and provide "lift" to a properly-shaped object moving through it—properties that have made possible such diverse applications as the ship, the hydraulic press, and the airplane. In what follows we shall learn that there is nothing mysterious about these properties, which are no more than natural consequences of the laws of physics.

11-1 Pressure

When a force **F** acts perpendicular to a surface whose area is A, the pressure p exerted on the surface is defined as the ratio between the magnitude F of the force and the area:

$$p = \frac{F}{A} \qquad\qquad (11\text{-}1)$$

$$\text{Pressure} = \frac{\text{force}}{\text{area}}.$$

Pressure is a scalar quantity. Often a perpendicular force is described as *normal*, which allows us to say that *pressure is the magnitude of normal force per unit area.*

Pressures may be measured in a number of ways, three of which are illustrated in Fig. 11-1. Usually what is directly determined is the difference between the unknown pressure and atmospheric pressure. This difference is

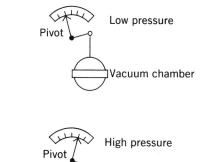

(a) Aneroid

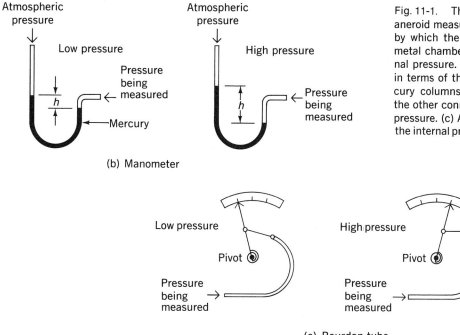

(b) Manometer

(c) Bourdon tube

Fig. 11-1.   Three types of pressure gauge. (a) An aneroid measures pressure in terms of the amount by which the thin, flexible ends of an evacuated metal chamber are pushed in or out by the external pressure. (b) A manometer measures pressure in terms of the difference in height h of two mercury columns, one open to the atmosphere and the other connected to the source of the unknown pressure. (c) A Bourdon tube straightens out when the internal pressure exceeds the external pressure.

Gauge pressure and absolute
pressure
gauge pressure, whereas the true pressure is called the *absolute pressure*.
That is,

$$p = p_{\text{gauge}} + p_{\text{atm}} \tag{11-2}$$

Absolute pressure = gauge pressure + atmospheric pressure.

Thus a tire inflated to a gauge pressure of 28 lb/in² contains air at an absolute
pressure of 43 lb/in², since sea-level atmospheric pressure is 14.7 lb/in².

The SI unit of pressure is the N/m², sometimes called the *pascal,* and
the correct British unit of pressure is the lb/ft². Because both of these units
are rather small for most purposes, a number of other pressure units are
(unfortunately) in common use. The chief ones are listed below.

Pressure units
The *atmosphere* (atm), which represents the average pressure exerted
by the earth's atmosphere at sea level and is equal to 1.013×10^5 N/m² or
14.7 lb/in².

The *bar,* equal to 10^5 N/m². The *millibar* (mb), which is widely used in
meteorology, is equal to 10^{-3} bar or 100 N/m².

The *lb/in²* (or *psi,* for "pound per square inch"), equal to 144 lb/ft² or
6.89×10^3 N/m².

The *torr,* which represents the pressure exerted by a column of mercury
1 mm high and is equal to 133 N/m². The torr was formerly referred to as the
"millimeter of mercury," abbreviated mm Hg.

Problem.   The weight of a 1000-kg car is supported equally by its four tires.
The gauge pressure of the air in the tires is 1.8 bars. Find the area of each tire
that is in contact with the ground.

Solution.   The load on each tire consists of the portion of the car's weight it
supports plus the weight of a column of the atmosphere whose cross-sectional area equals the area of the tire in contact with the ground, since this
part of the tire has no air at atmospheric pressure under it to provide an
equal upward force. Hence only the gauge pressure of the air in the tires,
which is the excess over atmospheric pressure, is effective in supporting the
car's weight. Since the car's weight is

$$w = mg = 1000 \text{ kg} \times 9.8 \text{ m/s}^2 = 9800 \text{ N},$$

each tire must support 2450 N. The area of each tire in contact with the
ground is therefore, since 1 bar $= 10^5$ N/m²,

$$A = \frac{F}{p} = \frac{2450 \text{ N}}{1.8 \times 10^5 \text{ N/m}^2} = 0.0136 \text{ m}^2 = 136 \text{ cm}^2$$

Pressure is a useful quantity because fluids flow under stress instead of being deformed elastically as solids are. The characteristic lack of rigidity exhibited by fluids has three significant consequences:

(1) The forces a fluid exerts on the walls of its container, and vice versa, always act perpendicular to the walls.

If this were not so, any sideways force by a fluid on a wall would be met, according to the third law of motion, by a sideways force back on the fluid, which would cause the fluid to move constantly parallel to the wall. But fluids may be at rest in containers of any shape, and so the sole forces they can exert on their containers must be perpendicular to the walls of the latter.

(2) An external pressure exerted on a fluid is transmitted uniformly throughout the volume of the fluid.

If this were not so, the fluid would flow from a region of high pressure to one of low pressure, thereby equalizing the pressure. We must keep in mind, however, that the above statement refers to a pressure imposed from outside the fluid. The fluid at the bottom of a container is always under greater pressure than that at the top owing to the weight of the overlying fluid. A notable example is the earth's atmosphere, although such pressure differences are ordinarily significant only for liquids.

(3) The pressure on a small surface in a fluid is the same regardless of the orientation of the surface.

If this were not so, again, the fluid would flow in such a way as to equalize the pressure (Fig. 11-2).

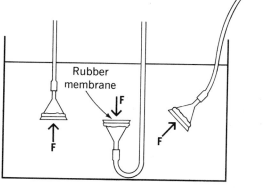

Fig. 11-2.   The force exerted by the pressure in a fluid is the same in all directions at any depth.

11-2 Hydraulic Press

We have already considered two of the three basic machines, the lever and the inclined plane. The third, the *hydraulic press*, is based upon property (2) of the previous section: An external pressure exerted on a fluid is transmitted uniformly throughout the volume of the fluid. This statement is known as *Pascal's principle*.

Pascal's principle

Figure 11-3 is a schematic diagram of a hydraulic press. A force F_{in} acts upon a piston of area A_{in} to produce the pressure

$$p = \frac{F_{in}}{A_{in}}$$

on the confined fluid. This pressure is transmitted by the fluid to the output piston upon which it acts upward. The force F_{out} on the latter piston is, if its area is A_{out},

$$F_{out} = \text{pressure} \times \text{area}$$

$$= pA_{out} = \frac{F_{in}}{A_{in}} A_{out}$$

$$= F_{in} \frac{A_{out}}{A_{in}}. \tag{11-3}$$

$$MA = \frac{F_{out}}{F_{in}} = \frac{A_{out}}{A_{in}} = \frac{L_{in}}{L_{out}}$$

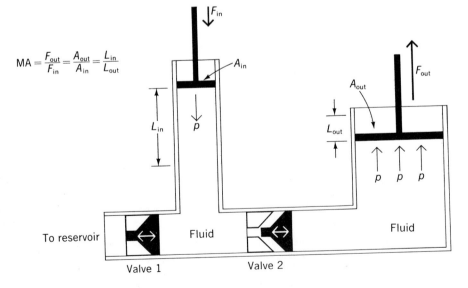

Fig. 11-3. The hydraulic press makes use of the fact that pressure exerted on a fluid is transmitted throughout the fluid. Valve 1 is closed and valve 2 open on the downstroke of the input piston; valve 1 is open and valve 2 closed on the upstroke.

The output force is equal to the input force multiplied by the ratio A_{out}/A_{in} between the piston areas. The mechanical advantage of the hydraulic press is therefore, from the principle of equilibrium,

$$\text{MA} = \frac{F_{out}}{F_{in}} = \frac{A_{out}}{A_{in}}.$$

Hydraulic press (11-4)

A small input force can be considerably increased merely by having the output piston much larger in area than the input piston.

Since the fluid is assumed incompressible, the volume of it transferred from one cylinder to the other as the pistons move must be the same. Hence The piston displacements are inversely proportional to the forces on them: the input piston must move through a large range in order to move the output piston through a small one. If the piston areas are in the ratio 100:1, a force of 1 lb can lift a weight of 100 lb, but for every inch the weight is raised the input piston must move down 100 in.

The purpose of the valves indicated in Fig. 11-3 is to permit the output piston to be raised by a series of short strokes of the input piston. When the latter moves downward, valve 1 closes and valve 2 opens, which allows fluid under pressure to move into the large cylinder. When the input piston is pulled upward, valve 1 opens and valve 2 closes, and additional fluid is drawn into the input cylinder to enable it to make another stroke.

The principle of the hydraulic press can be applied in a variety of ways. Common industrial uses include presses of various kinds, garage lifts, control systems in airplanes, and vehicle brakes.

11-3 Pressure and Depth

The pressure inside a volume of fluid depends upon the depth below the surface, since the deeper we descend, the greater the weight of the overlying fluid.

Pressure increases with depth

Suppose we have a tank of height h and cross-sectional area A which is filled with a fluid of density d (Fig. 11-4). The volume of the tank is

$V = Ah$

and the mass of fluid it contains is

$m = dV = dAh.$

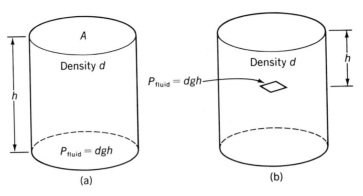

Fig. 11-4. (a) The pressure a fluid in a tank exerts on the bottom is equal to dgh. (b) The same formula holds for the fluid pressure at any depth h below the surface.

The weight of the fluid in the tank is therefore

$$w = mg = dgAh.$$

The pressure P_{fluid} the fluid exerts on the bottom of the tank is its weight divided by the area of the bottom, with the result that

$$P_{fluid} = \frac{F}{A} = \frac{w}{A} = dgh. \qquad (11\text{-}5)$$

The pressure difference between the top and the bottom of the tank is directly proportional to the height of the fluid column and to the fluid density.

The above result also applies to *any* depth h in a fluid, whether at the bottom or not, since the fluid beneath that depth does not contribute to the weight pressing down there.

The *total* pressure within a fluid, of course, also depends upon the pressure $P_{external}$ exerted on its surface by the atmosphere or, perhaps, by a piston (Fig. 11-5). Thus, in general, the total pressure at a depth h in a fluid of density d is $P = P_{external} + P_{fluid}$:

$$P = P_{external} + dgh. \qquad \textit{Pressure at depth h in a fluid} \quad (11\text{-}6)$$

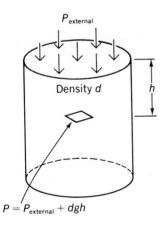

Fig. 11-5. The total pressure at a depth h in a fluid is the sum of the fluid pressure and the external pressure exerted on the fluid surface.

Problem. Find the pressure on a skin diver when he is 15 m below the surface of the sea.

Solution. The density of sea water is 1.03×10^3 kg/m³. Hence the pressure on the skin diver due to the sea alone is

$$dgh = 1.03 \times 10^3 \, \frac{kg}{m^3} \times 9.8 \, \frac{m}{s^2} \times 15 \, m = 1.51 \times 10^5 \, \frac{N}{m^2}.$$

Since atmospheric pressure is 1.01×10^5 N/m², the total pressure on the skin diver is

$$P = P_{external} + dgh$$

$$= (1.01 \times 10^5 + 1.51 \times 10^5) \, \frac{N}{m^2}$$

$$= 2.52 \times 10^5 \, \frac{N}{m^2}$$

which is about $2\frac{1}{2}$ times atmospheric pressure. The man is not crushed because the pressure within his body increases as he descends to match the pressure exerted on it.

A person's arterial blood pressures are usually measured with the help of an inflatable cuff wrapped around the upper arm at the level of the heart (Fig. 11-6). A stethoscope is used to monitor the sound of the blood flowing through an artery below the cuff. The cuff is first inflated until the flow of blood stops. Then the pressure of the cuff is gradually reduced until the blood just begins to flow, which is recognized by a gurgling sound in the stethoscope. This pressure, called *systolic*, represents the maximum pressure the heart produces in the artery. The pressure in the cuff is then further reduced until the gurgling stops, which corresponds to the restoration of normal blood flow. The pressure at this time, called *diastolic*, represents the pressure in the artery between the strokes of the heart. In a healthy person the systolic and diastolic pressures are respectively about 120 and 80 torr.

Problem. The average pressure of the blood in a person's arteries is 100 torr at the same elevation as the heart. Find the average pressure in an artery in the head of a standing person, say 40 cm above his heart, and in an artery in his foot, say 120 cm below his heart. The density of blood is 1.05×10^3 kg/m³.

Solution. The difference in pressure in each case is $\Delta p = dgh$. Here

$$\Delta p_1 = dgh_1 = 1.05 \times 10^3 \, kg/m^3 \times 9.8 \, m/s^2 \times 0.4 \, m = 4.12 \times 10^3 \, N/m^2$$
$$\Delta p_2 = dgh_2 = 1.05 \times 10^3 \, kg/m^3 \times 9.8 \, m/s^2 \times 1.2 \, m = 12.3 \times 10^3 \, N/m^2.$$

Since 1 torr = 133 N/m², $\Delta p_1 = 31$ torr and $\Delta p_2 = 93$ torr. Hence the pressure in the artery in the head is $(100 - 31)$ torr = 69 torr and that in the artery in

Systolic and diastolic pressures

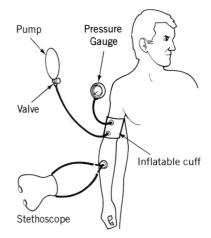

Fig. 11-6. Arterial blood pressures are measured with the help of an inflatable cuff that is wrapped around the upper arm. A pump is used to inflate the cuff until the flow of blood stops, and air is then let out by means of the valve until the flow begins again. The stethoscope is used to monitor the blood flow.

the foot is $(100 + 93)$ torr $= 193$ torr. The arteries that lead to the head expand and contract as needed to keep the flow of blood to the brain constant despite changes in the elevation of the head relative to the heart. Such expansions and contractions require a few seconds to be completed, which explains why sitting up suddenly from a horizontal position may lead to a momentary dizzy sensation.

11-4 Archimedes' Principle

Buoyancy
An object immersed in a fluid seems to weigh less than it does outside. This effect, known as *buoyancy*, permits people to swim, ships to float, and helium-filled balloons to rise through the air.

A very simple argument permits us to determine the buoyant force on an object. Suppose we have a solid object of volume V submerged in a fluid of density d. We begin by considering instead a body of fluid of the same size and shape as the object and located at the same depth, as in Fig. 11-7(a). This body of fluid is in equilibrium, which means that its weight of

$$w = Vdg$$

is supported by a buoyant force of this magnitude exerted by the rest of the fluid. The buoyant force is the vector sum of all the forces the rest of the

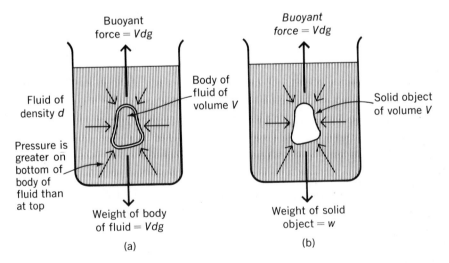

Fig. 11-7. The buoyant force on a submerged object is equal to the weight of the fluid it displaces.

fluid exerts on the body, and this force must be upward because the pressure (and hence the upward force) on the bottom of the body is greater than the pressure (and hence the downward force) on its top. (The forces on the sides of the body as a rule cancel out.) Therefore the buoyant force is Vdg, the weight of the body of fluid.

Now we replace the body of fluid by the solid object, as in Fig. 11-7(b). The various pressures remain the same, so the buoyant force of Vdg also remains the same. We conclude that

$$F_{\text{buoyant}} = Vdg \qquad\qquad \textit{Archimedes' principle} \quad (11\text{-}7)$$

Buoyant force = weight of displaced fluid.

This is Archimedes' principle:

The buoyant force on a submerged object is equal to the weight of fluid displaced by the object.

Archimedes' principle enables us to determine only the buoyant force on a submerged object, not the net force on it. If the weight of the object is greater than the buoyant force on it, it will sink; if the weight is less than the buoyant force, it will rise; if the weight is equal to the buoyant force, it will float in equilibrium.

Net force on object is difference between buoyant force and its weight

Problem. An iceberg is a chunk of freshwater ice that has broken off an icecap (such as those that cover Greenland and Antarctica) or a glacier at the edge of the sea. Find the proportion of the volume of an iceberg that is submerged.

Solution. If V_{ice} is the iceberg's total volume and V_{sub} is its submerged volume, then

Weight of iceberg = weight of displaced water,

$$V_{\text{ice}} d_{\text{ice}} g = V_{\text{sub}} d_{\text{seawater}} g,$$

$$\frac{V_{\text{sub}}}{V_{\text{ice}}} = \frac{d_{\text{ice}}}{d_{\text{seawater}}} = \frac{9.2 \times 10^2 \text{ kg/m}^3}{1.03 \times 10^3 \text{ kg/m}^3} = 0.89.$$

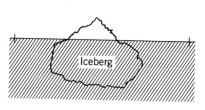

Fig. 11-8. Only 11 percent of the volume of an iceberg is above sea level.

Eighty-nine percent of the volume of an iceberg is below sea level (Fig. 11-8).

11-5 Fluid Flow

The study of fluids in motion is one of the more difficult branches of mechanics because of the diversity of phenomena that may occur. However, the fundamental aspects of fluid flow can be understood on the basis of a simple model that is reasonably realistic in many cases. This model involves liquids that are incompressible and exhibit no viscosity. (Viscosity is the term used to describe internal friction in a fluid.) In the absence of viscosity, layers of fluid slide freely past one another and past other surfaces, so that our model applies to such liquids as water but not to such liquids as molasses.

Laminar flow

Another approximation we shall make is that the fluid undergoes *laminar* (or *streamline*) *flow* exclusively. In streamline flow, which is illustrated in Fig. 11-9, every particle of liquid passing a particular point follows the same path (called a *streamline*) as the particles that passed that point previously. Furthermore, the direction in which the individual fluid particles move is always the same as the direction in which the fluid as a whole moves.

Turbulent flow

At the other extreme is *turbulent flow*, which is characterized by the presence of whirls and eddies, such as those in a cloud of cigarette smoke or at the foot of a waterfall. Turbulence generally occurs at high speeds and when there are obstructions or sharp bends in the path of the fluid.

The volume of liquid that flows through a pipe per unit time is easy to compute. If the average speed of the liquid in the pipe of Fig. 11-10 is v, each part of the stream travels the distance vt in the time interval t. The

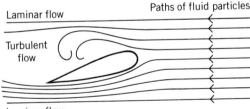

Fig. 11-9. Laminar and turbulent flows around an obstacle.

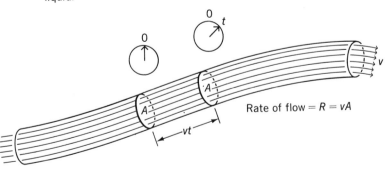

Fig. 11-10. The rate of flow of liquid through a pipe is equal to the product of the cross-sectional area of the pipe and the speed of the liquid.

Rate of flow = $R = vA$

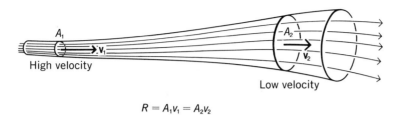

High velocity

Low velocity

$$R = A_1v_1 = A_2v_2$$

Fig. 11-11. In laminar flow, liquid speed is inversely proportional to the cross-sectional area of the pipe.

volume of liquid transported the distance vt in the time t is vt multiplied by the pipe's cross-sectional area A, or vtA. Therefore the rate of flow R of liquid through the pipe is

$$R = \frac{vtA}{t} = vA, \qquad\qquad\qquad Rate\ of\ flow \quad (11\text{-}8)$$

the product of the liquid speed and the cross-sectional area of the pipe. R is often expressed in units such as gal/min instead of ft³/s or m³/s.

If the pipe size varies, the speed of the liquid also varies so as to keep R constant, so that

$$v_1A_1 = v_2A_2. \qquad\qquad Equation\ of\ continuity \quad (11\text{-}9)$$

Hence a liquid flows faster through a constriction in a pipe and slower through a dilation. As in Fig. 11-11, streamlines drawn close together signify rapid motion while streamlines far apart signify slow motion.

Problem.   Oxygenated blood from the lungs is pumped by the left ventricle of the heart into a large artery called the aorta, whose diameter is typically 2 cm. When a person is resting, the rate of flow of blood might be 6 liters/min. (a) Find the average velocity of blood in the aorta under these circumstances. (b) Find the power output of the left ventricle of a person at rest assuming an average blood pressure of 100 torr at the aorta. (The right ventricle, which pumps blood at the same rate to the lungs where it gives up carbon dioxide and absorbs oxygen, has a smaller power output since there is less resistance to the flow of blood through the lungs than through the rest of the body.)

Solution.   (a) Since 1 liter $= 10^{-3}$ m³ and 1 min $= 60$ s, 1 liter/min $= 1.667 \times 10^{-5}$ m³/s, and $R = 1.00 \times 10^{-4}$ m³/s here. The velocity is therefore

$$v = \frac{R}{A} = \frac{1.00 \times 10^{-4}\ \text{m}^3/\text{s}}{\pi \times 10^{-4}\ \text{m}^2} = 0.318\ \text{m/s}.$$

(b) The rate at which the left ventricle does work on the blood passing through it is $P = Fv$, where the force applied is pA. Hence

$$P = pAv = 100 \text{ torr} \times 133 \, \frac{\text{N/m}^2}{\text{torr}} \times \pi \times 10^{-4} \text{ m}^2 \times 0.318 \text{ m/s} = 1.33 \text{ W}.$$

Both the pressure and the rate of flow increase during physical activity, with a corresponding increase in the power output.

11-6 Bernoulli's Equation

Pressure and velocity

When a liquid flowing through a pipe enters a region where the pipe diameter is reduced, its velocity increases. A change in velocity involves an acceleration, which means that a net force must be acting upon the liquid. This force can only arise from a difference in pressure between the different parts of the pipe. Evidently the pressure in the part of the pipe having a large diameter is the greater, since the liquid increases in velocity on its way to the constriction. Thus we expect a relationship between the pressure in a moving liquid and its velocity, which turns out to be

$$p_2 + dgh_1 + \tfrac{1}{2}dv_1{}^2 = p_2 + dgh_2 + \tfrac{1}{2}dv_2{}^2. \qquad \textit{Bernoulli's equation} \quad (11\text{-}10)$$

Here p_1, h_1, and v_1 are respectively the pressure, height above some reference level, and velocity of a liquid of density d at the point 1 in a body of the liquid, and p_2, h_2, and v_2 are the values of these quantities at another point 2.

Equation 11-10 is known as *Bernoulli's equation* after Daniel Bernoulli (1700–1782), who first derived it. According to Bernoulli's equation the quantity $(p + dgh + \tfrac{1}{2}dv^2)$ has the same value at all points in an incompressible liquid with negligible viscosity that undergoes laminar flow.

To derive Bernoulli's equation we consider a curved pipe of non-uniform cross section through which a liquid flows, as in Fig. 11-12. Let us apply the principle of conservation of energy to a parcel of the liquid of volume v_1tA_1 as it enters at the left in the time t and to the same parcel as it leaves at the right. The mass of the parcel is

Mass of liquid parcel

$$m = dV = dv_1tA_1 = dv_2tA_2,$$

since vA has the same value at 1 and 2. The net amount of work ΔW done on the liquid parcel as it passes from 1 to 2 must be equal to the net change in its potential energy ΔPE as its height goes from h_1 to h_2, plus the net change in its kinetic energy ΔKE as its velocity goes from v_1 to v_2. That is,

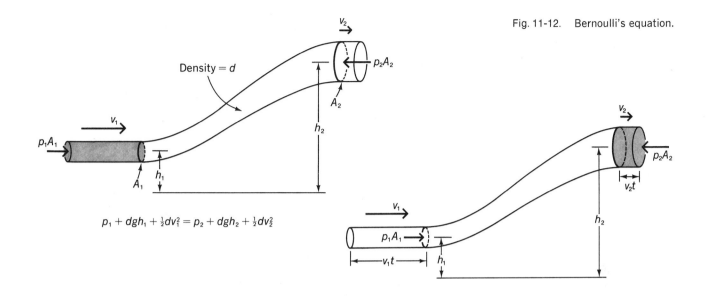

Fig. 11-12.   Bernoulli's equation.

$$\Delta W = \Delta \text{PE} + \Delta \text{KE}.$$

The work done *on* the parcel at 1 is the force $p_1 A_1$ on it multiplied by the distance $v_1 t$ through which the force acts. The work done *by* the parcel at **2** is the force $p_2 A_2$ multiplied by the distance $v_2 t$ through which the force acts. The *net* work done on the parcel is therefore

$$\Delta W = p_1 A_1 v_1 t - p_2 A_2 v_2 t, = \frac{p_1 m}{d} - \frac{p_2 m}{d}. \qquad \text{Work done on parcel}$$

The change in the potential energy of the parcel in going from 1 to 2 is

$$\Delta \text{PE} = mgh_2 - mgh_1, \qquad \text{Change of PE of parcel}$$

and the change in its kinetic energy is

$$\Delta \text{KE} = \tfrac{1}{2}mv_2^{\,2} - \tfrac{1}{2}mv_1^{\,2}. \qquad \text{Change of KE of parcel}$$

Therefore

$$\Delta W = \Delta \text{PE} + \Delta \text{KE}$$

$$\frac{p_1 m}{d} - \frac{p_2 m}{d} = mgh_2 - mgh_1 + \tfrac{1}{2}mv_2^{\,2} - \tfrac{1}{2}mv_1^{\,2}.$$

When we divide through by the common factor m, multiply by d, and re-arrange terms, the result is Bernoulli's equation,

$$p_1 + dgh_1 + \tfrac{1}{2}dv_1{}^2 = p_2 + dgh_2 + \tfrac{1}{2}dv_2{}^2.$$

The effect of viscosity is to dissipate mechanical energy into heat. If the viscosity of the liquid is not negligible, the quantity $p + dgh + \tfrac{1}{2}dv^2$ decreases in the direction of flow.

11-7 Applications of Bernoulli's Equation

In many situations the velocity, pressure, or height of a liquid is constant, and simplified forms of Bernoulli's equation hold. Thus when a liquid column is stationary, we see that the pressure difference between two depths in it is

$$p_2 - p_1 = dg(h_1 - h_2), \tag{11-11}$$

which is just what Eq. (11-5) states. Evidently the latter formula is included in Bernoulli's equation.

Another straightforward result occurs in the event $p_1 = p_2$. As an example, Fig. 11-13 illustrates a liquid emerging from an orifice at the bottom of a tank. The liquid pressure equals atmospheric pressure both at the top of the tank and at the orifice. If the orifice is small compared with the cross section of the tank, the liquid level in the tank will fall slowly enough for the liquid velocity at the top of the tank to be assumed zero. If the velocity of the liquid as it leaves the orifice is v and the difference in height between the top of the liquid and the orifice is h, Bernoulli's equation reduces to

$$\tfrac{1}{2}dv^2 = dgh,$$
$$v = \sqrt{2gh}. \tag{11-12}$$

Torricelli's theorem

The velocity with which the liquid is discharged is the same as the velocity of a body falling from rest from the height h. This result is called *Torricelli's theorem*, and, like the relationship between pressure and depth, it is a special case of Bernoulli's equation. The rate at which liquid flows through the orifice may be found from Eq. (11-8) if the orifice area A is known. The volume of liquid being discharged per unit time is

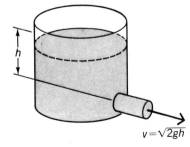

$v = \sqrt{2gh}$

Fig. 11-13.   Torricelli's theorem.

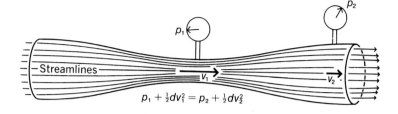

Fig. 11-14. In a horizontal pipe, the pressure is greatest when the velocity is least and vice versa.

$$R = vA = A\sqrt{2gh}. \tag{11-13}$$

Problem. How fast will water leak through a hole 1 cm² in area at the bottom of a tank in which the water level is 3 m high?

Solution. From the above formula,

$$R = 1 \text{ cm}^2 \times \frac{1}{10^4 \text{ m}^2/\text{cm}^2} \times \sqrt{2 \times 9.8 \text{ m/s}^2 \times 3 \text{ m}} = 7.7 \times 10^{-4} \text{ m}^3/\text{s},$$

which is almost a liter per second, an appreciable amount.

The most interesting special case of Bernoulli's equation occurs when there is no change in height during the motion of the liquid (Fig. 11-14). Here

$$p_1 + \tfrac{1}{2}dv_1{}^2 = p_2 + \tfrac{1}{2}dv_2{}^2, \tag{11-14}$$

which means that the pressure in the liquid is least where the speed is greatest, and vice versa.

A familiar application of Eq. (11-14) is the lifting force produced by the flow of air past the wing of an airplane, as in Fig. 11-15. (Air is, of

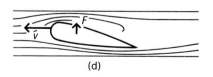

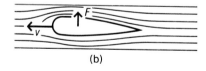

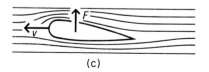

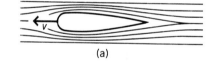

Fig. 11-15. Air flow past a wing. At (a) the flow is the same on both surfaces, so that no lift results. The lift at (c) is greater than that at (b) because of the greater pressure difference between upper and lower surfaces (the pressure is least where the streamlines are closest together). At (d) turbulence reduces the available lift.

How an airplane's wing develops
lift
course, a compressible fluid and so does not fit our model exactly, but the behavior predicted by Bernoulli's equation is not a bad approximation for gases at moderate speeds.) Air moving past the upper surface of the wing must travel faster than air moving past the lower surface; this is indicated by the closeness of the streamlines near the former. The difference in speed leads to a decreased pressure over the top of the wing, a pressure which is equivalent to a suction force that lifts the wing. The greater the difference in air speeds around the upper and lower surfaces, the greater the lift that is produced, provided the wing shape is not so extreme that turbulence results (Fig. 11-15d).

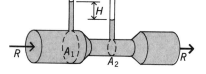

Fig. 11-16.   The Venturi meter.

Problem.   The device shown in Fig. 11-16, called a *Venturi meter*, provides a convenient method for determining the rate of flow R of a liquid through a pipe. Derive an equation that gives R in terms of the difference in height H between the manometer levels and the cross-sectional areas A_1 and A_2.

Solution.   The first step is to apply Bernoulli's equation to the liquid flowing through the pipe. Since the pipe is horizontal, $h_1 = h_2$, and

$$p_1 + \tfrac{1}{2}dv_1{}^2 = p_2 + \tfrac{1}{2}dv_2{}^2$$
$$p_1 - p_2 = \tfrac{1}{2}d(v_2{}^2 - v_1{}^2).$$

The difference between the heights of the liquid in the vertical manometer tubes reflects the greater pressure at A_1 than at A_2. From Eq. 11-11 we have

$$p_1 = p_2 = dgH.$$

Setting equal the above two expressions for $p_1 - p_2$ gives

$$dgH = \tfrac{1}{2}d(v_2{}^2 - v_1{}^2)$$
$$2gH = v_2{}^2 - v_1{}^2.$$

The velocities v_1 and v_2 are related to the cross-sectional areas A_1 and A_2 by $v_1 A_1 = v_2 A_2$, and so

$$v_2 = v_1 \frac{A_1}{A_2}, \; v_2{}^2 = v_1{}^2 \left(\frac{A_1}{A_2}\right)^2.$$

Therefore

$$2gH = v_2{}^2 - v_1{}^2 = v_1{}^2 \left[\left(\frac{A_1}{A_2}\right)^2 - 1\right].$$

Solving for v_1 yields

$$v_1 = \sqrt{\frac{2gH}{(A_1/A_2)^2 - 1}}.$$

The rate of flow R is given by $R = v_1 A_1$ and so

$$R = A_1 \sqrt{\frac{2gH}{(A_1/A_2)^2 - 1}},$$

which is what is required. Since A_1 and A_2 are fixed quantities, a measurement of the difference H between the manometer levels is all that is needed to find R.

11-8 Viscosity

The viscosity of a fluid is a kind of internal friction that prevents neighboring layers of the fluid from sliding freely past one another. Fig. 11-17 shows how the velocity of a fluid in a pipe varies with distance from the axis. The fluid in contact with the pipe's wall is stationary and, as we would expect, the velocity increases to a maximum along the pipe's axis. The smaller the viscosity of the fluid, the greater the various velocities will be, but the characteristic parabolic shape of the velocity profile will be maintained as long as the flow remains laminar.

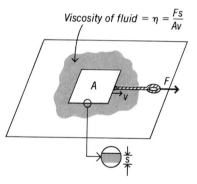

The viscosity of a fluid is given a quantitative definition in terms of the experiment shown in Fig. 11-18, in which a plate of area A is being pulled across a layer of fluid s thick. For most fluids it is found that the force F required to pull the plate at the constant velocity v is proportional to A and v and inversely proportional to s: the faster the motion and the thinner the layer of fluid, the more force is needed for a given plate area. Different fluids offer different degrees of resistance to the motion, but the force needed varies as Av/s for most of them provide the velocity is not so great that turbulence occurs. Thus we can write

$$F = \frac{\eta A v}{s} \tag{11-15}$$

where η, the constant of proportionality, is called the *viscosity* of the fluid (η is the Greek letter *eta*). The SI unit of viscosity is evidently the N·s/m². An older unit, the *poise*, remains in common use, where 10 poise = 1 N·s/m²; the *centipoise*, 0.01 poise, is equal to 10^{-3} N·s/m².

Fig. 11-17. Because of viscosity, the velocity of fluid in a pipe varies from 0 at the pipe wall to a maximum along the axis.

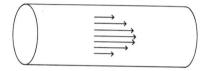

$$\text{Viscosity of fluid} = \eta = \frac{Fs}{Av}$$

Fig. 11-18. The viscosity of a fluid is defined in terms of the force needed to pull a flat plate at constant velocity across a layer of the fluid.

The viscosities of some common fluids are listed in Table 11-1. The viscosity of a liquid decreases with temperature as its molecules become less and less tightly bound to one another (see Chap. 15), a phenomenon familiar to anyone who has heated honey or molasses. One of the reasons a drop in body temperature is so dangerous is that the viscosity of the blood is thereby increased, impeding the flow. The viscosity of a gas, in contrast to that of a liquid, increases with temperature because the higher the temperature, the faster the gas molecules move and the more often they collide with one another (again, see Chap. 15).

Table 11–1. Viscosities of various fluids at atmospheric pressure and the indicated temperature.

Substance	Temperature (°C)	Viscosity (N·s/m²)
Gases		
Air	0	1.7×10^{-5}
	100	2.2×10^{-5}
Water vapor	100	1.3×10^{-5}
Liquids		
Alcohol (ethyl)	20	1.2×10^{-3}
Blood plasma	37	1.5×10^{-3}
Blood, whole (varies with velocity)	37	$\sim 4 \times 10^{-3}$
Glycerin	20	0.83
Water	0	1.8×10^{-3}
	20	1.0×10^{-3}
	40	0.66×10^{-3}
	60	0.47×10^{-3}
	80	0.36×10^{-3}
	100	0.28×10^{-3}

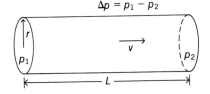

$\Delta p = p_1 - p_2$

Fig. 11-19.

Let us return to a fluid moving through a pipe. If the pipe is cylindrical with the length L and inside radius r, and a fluid of viscosity η is flowing through it under the influence of a pressure difference $\Delta p = p_1 - p_2$ as in Fig. 11-19, the rate of flow is

$$R = \frac{\pi r^4 \, \Delta p}{8 \eta L}.$$

(11-16)

Equation (11-16) is known as *Poiseuille's law*. The dependence of the rate of flow on the viscosity η and on the pressure gradient $\Delta p/L$ are both about what we might expect, but the variation with r^4 is remarkable: halving the radius of a pipe reduces R by a factor of 16 if Δp stays the same. The radius of the pipe plays a far more important role than its length does with respect to viscous resistance.

Special Topic

Surface Tension

The surface of a liquid behaves remarkably like a membrane under tension. Thus a steel sewing needle placed horizontally on the surface of some water in a dish does not sink even though its density is nearly eight times that of water (Fig. 11-20). The needle rests in a depression in the water surface just as if that surface were a sheet of rubber stretched across the dish. The term *surface tension* for this effect is quite appropriate. Another manifestation of surface tension is the tendency of a liquid drop to assume a spherical shape, just as an inflated balloon does.

The origin of surface tension lies in the fact that molecules on the surface of a body of liquid are acted upon by a net inward force whereas those in the interior are acted upon by forces in all directions (Fig. 11-21). Since the surface molecules are all being pulled inward, the surface tends to contract to the minimum possible area. A sphere has the least area relative to its volume of any object, and so a liquid sample not acted upon by any external forces (such as gravitation or air resistance if it is moving) will take on a spherical form. A liquid surface can support a small object such as a needle because the weight of the object is insufficient to rupture the surface, which would involve first stretching it to a greater degree than that shown in Fig. 11-20.

A liquid surface has a certain potential energy due to surface tension, just as anything else under tension does. This energy is proportional to the surface area A, so that for a particular liquid

$$PE = \gamma A.$$

The constant of proportionality γ (Greek letter *gamma*) is defined as the surface tension of the liquid. In the case of water, $\gamma = 0.073$ J/m^2 at 20°C. Sur-

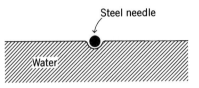

Fig. 11-20. The surface tension of a water surface supplements buoyancy in supporting a steel needle. The elastic character of a liquid surface resembles that of a membrane under tension.

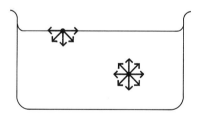

Fig. 11-21. Because a molecule on the surface of a body of liquid has no molecules above it to interact with, the net force on it is inwards.

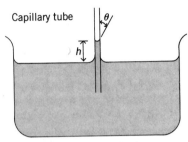

Capillary tube

θ

h

Adhesion > cohesion

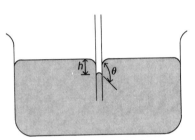

h θ

Cohesion > adhesion

Fig. 11-22(a) and (b).

face tensions generally decrease with temperature, so that $\gamma = 0.059$ J/m² for water at 100°C.

In order to expand a liquid surface by an area ΔA, we might imagine applying a force **F** parallel to the surface along a line of length L perpendicular to **F**. When we have moved the line through the distance Δs, $\Delta A = L\Delta s$ and the work done $F\Delta s$ will have increased the potential energy of the surface by ΔPE. Since ΔPE $= \gamma \Delta A$,

$$F \Delta s = \gamma L \Delta s$$

$$\gamma = \frac{F}{L}.$$

Thus we can interpret γ as the force of contraction per unit length. In this case the appropriate unit for γ is the N/m, which has the same dimensions as the J/m².

A familiar phenomenon is the rise of most liquids in a capillary tube. Capillarity is responsible for many familiar effects, such as the ability of paper and cloth fibers to absorb water. Two factors are involved: the *cohesion* of the liquid, which refers to the attractive forces its molecules exert on one another; and the *adhesion* of the liquid to the surface of a solid, which refers to the attractive forces the solid exerts on the liquid molecules. If the adhesive forces exceed the cohesive ones, as they do in the case of water and glass, then the liquid tends to stick to the solid and will rise in a capillary tube of that material since the attraction of a liquid molecule to the wall of the tube is greater than the attraction of this molecule to its brethren. In such an event the liquid surface is concave, as in Fig. 11-22(a), with an angle of contact θ that is characteristic of the liquid–solid combination. For water and clean glass, $\theta \approx 0$, as it is for any liquid that "wets" a particular solid, because adhesion must be much greater than cohesion for this to occur. For kerosene and glass, $\theta = 26°$, which reflects the smaller amount by which adhesion exceeds cohesion for these substances. If cohesion is greater than adhesion, as it is for mercury and glass, the liquid level in a capillary tube is lower than the level of the surrounding liquid, and the liquid surface in the tube is convex with an angle of contact greater than 90°; $\theta = 140°$ for mercury and glass.

It is straightforward to calculate the height h to which a liquid rises (or falls) in a capillary tube. Let us consider the situation shown in Fig. 11-22(a). The upward force on the liquid column is exerted by the vertical component $F \cos \theta$ of the surface tension of the ring of liquid that is adhering to the tube at the top of the column. If the tube has an inner radius of r, the ring is $2\pi r$ in length, and the surface tension force is $F = \gamma L = 2\pi \gamma r$. The downward force is the weight of the elevated liquid, which is $w = dgV = \pi dgr^2h$ where d is the density of the liquid and $\pi r^2 h$ its volume. Hence

$$F \cos \theta = w,$$

$$2\pi\gamma r \cos \theta = \pi dg r^2 h,$$

$$h = \frac{2\gamma \cos \theta}{rdg}.$$

This formula agrees with our intuitive expectations: the greater the surface tension of the liquid, the greater its adhesion to the tube (which means a small θ and a large $\cos \theta$), the narrower the tube, and the less the liquid density, the higher the liquid column. If cohesion exceeds adhesion, as in Fig. 11-22(b), the same formula applies. Now $\theta > 90°$ so $\cos \theta$ is negative, and h is negative also.

Important Terms

The **pressure** on a surface is the perpendicular force per unit area that acts upon it. **Gauge pressure** is the difference between true pressure and atmospheric pressure.

Pascal's principle states that an external pressure exerted on a fluid is transmitted uniformly throughout its volume.

The **hydraulic press** is a machine consisting of two fluid-filled cylinders of different diameters connected by a tube. The input force is applied to a piston in one of the cylinders, and the output force is exerted by a piston in the other cylinder. The MA of a hydraulic press is equal to the inverse ratio of the cylinder diameters.

Archimedes' principle states that the buoyant force on a submerged object is equal to the weight of fluid it displaces.

In **laminar** (or **streamline**) **flow** every particle of fluid passing a particular point follows the same path, whereas in **turbulent flow** irregular whirls and eddies occur. The greater the velocity of a fluid in streamline flow, the lower its pressure.

The **viscosity** of a fluid is a measure of its internal friction.

Important Formulas

Pressure:

$$p = \frac{F}{A} = p_{\text{gauge}} + p_{\text{atm}}$$

Pressure at depth h in a fluid:

$$p = p_{\text{external}} + dgh$$

Archimedes' principle:

$$F_{\text{buoyant}} = Vdg$$

Equation of continuity:

$$v_1 A_1 = v_2 A_2$$

Bernoulli's equation:

$$p + dgh + \tfrac{1}{2}dv^2 = \text{constant}$$

Poiseuille's law:

$$R = \frac{\pi r^4 \Delta p}{8\eta L}$$

Multiple Choice

1. Which of the following is not a pressure unit?
   a. millibar   b. atmosphere
   c. lb/ft^2 d. N·m^2

2. The fluid at the bottom of a container is
   a. under less pressure than the fluid at the top.
   b. under the same pressure as the fluid at the top.
   c. under more pressure than the fluid at the top.
   d. any of the above, depending upon the circumstances.

3. The pressure at the bottom of a vessel filled with liquid does *not* depend on the
   a. acceleration of gravity.
   b. liquid density.
   c. height of the liquid.
   d. area of the liquid surface.

4. A man stands on a very sensitive scale and inhales deeply. The reading on the scale
   a. does not change.
   b. increases.
   c. decreases.
   d. depends on the expansion of his chest relative to the volume of air inhaled.

5. Bernoulli's equation is based upon
   a. the second law of motion.
   b. the third law of motion.
   c. conservation of momentum.
   d. conservation of energy.

6. Bernoulli's equation includes as a special case
   a. Newton's third law of motion.
   b. Hooke's law.
   c. Torricelli's theorm.
   d. Archimedes' principle.

7. An express train goes past a station platform at high speed. A person standing at the edge of the platform tends to be
   a. attracted to the train.
   b. repelled from the train.
   c. attracted or repelled, depending on the ratio between the speed of the train and the speed of sound.
   d. unaffected by the train's passage.

8. The volume of liquid flowing per second out of an orifice at the bottom of a tank does *not* depend on
   a. the area of the orifice.
   b. the height of liquid above the orifice.
   c. the density of the liquid.
   d. the value of the acceleration due to gravity.

9. The hydraulic press is able to produce a mechanical advantage because
   a. the force a fluid exerts on a piston is always parallel to its surface.
   b. an external pressure exerted on a fluid is transmitted uniformly throughout its volume.
   c. at any depth in a fluid the pressure is the same in all directions.
   d. the pressure in a fluid varies with its speed.

10. In the operation of a hydraulic press, it is impossible for the output piston to exceed the input piston's
    a. displacement.   b. speed.
    c. force.   d. work.

11. The input piston of a hydraulic press is 2 in. in diameter and the output piston is 1 in. in diameter. An input force of 1 lb will produce an output force of
    a. 0.25 lb.   b. 0.50 lb.
    c. 2 lb.   d. 4 lb.

12. A manometer whose upper end is evacuated and sealed can be used to measure atmospheric pressure. Such an instrument is called a mercury barometer, and the average height of the mercury column in it is 760 mm. If water were used in a barometer instead of mercury, the height of the column of water would be about
    a. 56 mm.   b. 760 mm.
    c. 10 m.   d. 20 m.

13. Atmospheric pressure does not correspond to approximately
    a. 14.7 lb/in.^{2} b. 98 N/m^2.
    c. 1013 mb.   d. 2120 lb/ft^2.

14. A 4000-lb car is supported equally by its four tires, each of which is inflated to a gauge pressure of 30 lb/in.^{2} The area of each tire that is in contact with the ground is
    a. 1.04 in.^{2} b. 22 in.^{2}
    c. 33 in.^{2} d. 133 in.^{2}

15. A viewing window 1 ft in diameter is installed 10 ft below the surface of an aquarium tank filled with sea water. The force the window must withstand is approximately
    a. 500 lb.   b. 640 lb.
    c. 810 lb.   d. 2000 lb.

16. A force of 1000 lb is required to raise a concrete block to the surface of a fresh-water lake. The force required to lift it out of the water is approximately
    a. 700 lb.   b. 1062 lb.
    c. 1140 lb.   d. 1800 lb.

17. The depth in fresh water at which the water density is 1% greater than its value at the surface is approximately
    a. 2.3×10^2 m.   b. 2.3×10^3 m.
    c. 2.3×10^4 m.   d. 2.3×10^5 m.

18. The total cross-sectional area of all the capillaries of a certain person's circulatory system is 0.25 m². If blood flows through the system at the rate of 100 cm²/s, the average velocity of blood in the capillaries is
    a. 0.4 mm/s.   b. 4 mm/s.
    c. 25 mm/s.   d. 400 mm/s.

19. Water leaves the safety valve of a boiler at a velocity of 30 m/s. The gauge pressure inside the boiler is
    a. 4.5 millibars.   b. 1.5 bars.
    c. 4.5 bars.   d. 450 bars.

Exercises

(Assume laminar flow and negligible viscosity unless otherwise noted.)

1. A little water is boiled for a few minutes in a tin can, and the can is sealed while it is still hot. Why does the can collapse as it cools?

2. A helium-filled balloon rises to a certain altitude in the atmosphere and floats there instead of rising indefinitely. Why?

3. An ice cube floats in a glass of water filled to the brim. What will happen when the ice melts?

4. A wooden block is in such perfect contact with the bottom of a water tank that there is no water beneath it. Is there a buoyant force on the block?

5. Two spheres of the same diameter but of different mass are dropped from a tower. If air resistance is the same for both, which will reach the ground first? Why?

6. Does a fluid whose density is greater than that of another fluid necessarily have a greater viscosity as well?

7. The height of water at two identical dams is the same, but dam A holds back a lake containing 2 mi³ of water, while dam B holds back a lake containing 1 mi³ of water. What is the ratio of the total force exerted on dam A to that exerted on dam B?

8. A 120-lb woman balances on the heel of her left shoe, which is 0.4 in. in diameter. What pressure (in atm) does she exert on the ground?

9. The force on a phonograph needle whose point is 0.1 mm in radius is 0.2 N. What is the pressure it exerts on the record (in atm)?

10. A piston weighing 12 N rests on a sample of gas in a cylinder 5 cm in diameter. (a) What is the gauge pressure in the gas? (b) What is the absolute pressure in the gas?

11. A cork 2 cm in radius is used to close one end of a tube whose other end is connected to a vacuum pump. The pump removes virtually all the air from the tube. How much force would be needed to pull the cork out? [*Note:* The area of a circle of radius r is πr^2.]

12. A DC-9 airplane weighing 90,000 lb is in level flight. The area of its wings is 990 ft². What is the average difference in pressure between the upper and lower surfaces of its wings? Express the answer in lb/ft².

13. A Super Constellation airplane whose mass is 50,000 kg is in level flight. The area of its wings is 153 m². What is the average difference in pressure between the upper and lower surfaces of its wings? Express the answer in atm.

14. Find the pressure at a depth of 8 ft in a swimming pool filled with fresh water.

15. In 1960 the U.S. Navy bathyscaphe *Trieste* descended to a depth of 10,920 m in the Pacific Ocean near Guam. Neglecting the increase in water density with depth, find the pressure on the *Trieste* at the bottom of its dive.

16. A rectangular swimming raft 3 m long and 2 m wide is floating in a fresh water lake. When a man climbs aboard it, the raft sinks 1 cm further into the water. Find the man's mass.

17. The densities of people are slightly less than that of water. Assuming that these densities are the same, compute the buoyant force of the atmosphere on a 160-lb man.

18. What is the minimum area of an ice floe 3 in. thick that can support a 120-lb girl without getting her feet wet? The floe is in a fresh-water lake.

19. A barge 120 ft long and 20 ft wide weighs 200 tons. What is the depth of sea water required to float it?

20. The input and output pistons of a hydraulic jack are respectively 2 cm and 8 cm in diameter. A lever with a mechanical advantage of 6 is used to apply force to the input piston. How much mass can the jack lift if a force of 150 N is applied to the lever and friction is negligible in the system?

21. A lever with a mechanical advantage of 5 is attached to the pump piston of a hydraulic press. The area of the pump piston is 1.6 in². and that of the output piston is 20 in². (a) If the press is perfectly efficient, find the force the output piston exerts when a force of 30 lb is applied to the pump lever. (b) If each stroke of the lever moves the pump piston 1 in., how many strokes are needed to move the output piston 1 ft?

22. Water flows through a hose whose internal diameter is 1 cm at a velocity of 1 m/s. What should the diameter of the nozzle be if the water is to emerge at 5 m/s?

23. During the pumping phase of the heart's action the pressure of blood in the major arteries of a normal person at the level of the heart is about 120 torr. If one of these arteries is cut and blood spurts out vertically, how high will it go?

Problems

1. A submarine is at a depth of 100 ft in sea water. The interior of the submarine is maintained at normal atmospheric pressure. Find the force that must be withstood by a square hatch 2 ft on a side.

2. Calculate the density of sea water at a depth of 3 mi. Use the bulk modulus for water given in Table 10-4.

3. Calculate the density of mercury at the bottom of a mercury column 1.3 m high.

4. A sailboat has a ton of lead ballast attached to the bottom of its keel, the center of gravity of which is 6 ft below the water surface when the boat is level. What is the torque exerted by this ballast when the boat is heeled by 20° from the vertical? [*Hint:* Why is this problem here instead of in Chapter 5?]

5. A 30-kg balloon is filled with 100 m³ of hydrogen. How much force is needed to hold it down?

6. The largest rigid airship ever built was the German *Hindenburg*, which was 245 m long and had a capacity of 2×10^5 m³ of hydrogen. It was destroyed in a fire at Lakehurst, N.J. in 1937. What was the total payload of the *Hindenburg* including its structure but not including the hydrogen it contained?

7. A design has been proposed for a modern helium-filled airship which is to have a useful lift at sea level of 500 tons. (a) How many ft³ of helium must the airship contain? (b) The length of the airship would be 1300 ft. If it were a cylinder, what would its diameter be?

8. A tank of height H is filled with water; it is open at the top. A hole is made a distance y from the top. How far from the tank does the water strike the ground?

9. A horizontal stream of water leaves an orifice 1 m above the ground and strikes the ground 2 m away. (a) What is the velocity of the water when it leaves the orifice? (b) What is the gauge pressure behind it?

10. A barrel filled with kerosene is 4 ft high. A crack 1 in. long and 0.07 in. wide appears at its base.

How many lb of kerosene per minute flow out? The density of kerosene is 1.55 slugs/ft³.

11. A man's brain is approximately $\frac{1}{3}$ m above his heart, while this distance is approximately 2 m in a giraffe. What is the minimum pressure required to circulate blood between the heart and brain of (a) a man? (b) a giraffe?

12. A 1360-hp pump throws a jet of water 130 m into the air in Geneva, Switzerland. (a) With what velocity does the water leave the mouth of the fountain? (b) If the over-all efficiency is 60%, how many kg of water per minute are thrown into the air?

13. Water emerges from a fire hose at a speed of 64 ft/sec. If the rate of flow is 180 gal/min, find (a) the force with which the nozzle must be held, and (b) the required horsepower of the pump motor assuming 50% overall efficiency. (1 gal = 0.134 ft³)

14. The bilge pump on a boat is able to raise 100 liters of seawater per minute through a height of 1.5 m. If the overall efficiency of the system is 30%, what is the power output of the pump motor?

15. The left ventricle of a certain running man pumps 20 liters of blood per minute into his aorta at an average pressure of 140 torr. Find the total power output of his heart under the assumption that his right ventricle has an output 20 percent as great as that of his left ventricle.

16. A horizontal pipe 2 cm in radius at one end gradually increases in size so that it is 5 cm in radius at the other end. Water is pumped into the small end of the pipe at a velocity of 8 m/s and a pressure of 2 bars. Find the velocity and pressure of the water at the large end of the pipe.

17. The pipe of Problem 16 is turned so as to be vertical with the small end underneath, so the water flows upward. Find the velocity and pressure of the water at the large end of the pipe now.

18. (a) Milk flows through a Venturi meter whose cross-sectional areas are 6 cm² and 2 cm². Find the rate of flow in liters/s when the difference between the manometer heights is 10 cm. (b) Find the rate of flow when the difference between the

manometer heights is the same but the liquid is water.

19. Find the pressure gradient needed to pump 16 liters of water at 20° C per minute through a pipe 6 mm in radius. Take the viscosity of the water into account.

20. Atherosclerosis is the medical term for a common condition in which arteries are narrowed by deposits of tissue called plaque. If the flow of blood is to continue at its usual rate, a higher pressure is required. A consequence of atherosclerosis is therefore an elevated blood pressure, which has many undesirable effects on the body, one of them being that the heart must work harder to circulate the blood. (a) Find the increase in pressure needed to maintain R constant when the radius of an artery is decreased by 10%. (b) What is the corresponding increase in the power needed to maintain the flow of blood through that artery?

21. The drag force on an object moving through a fluid at a velocity v low enough for turbulence not to occur is proportional to both v and the viscosity η of the fluid, so that $F_{drag} = K\eta v$. The constant K depends on the size and shape of the object, but not upon its composition since a thin layer of fluid sticks to its surface so that the friction that occurs is between successive layers of fluid. In the case of a spherical object of radius r, $K = 6\pi r$. (a) Verify that a sphere of density d and radius r falling freely in a fluid of viscosity η and density d' has a terminal velocity of $2r^2 g(d - d')/9\eta$. (b) An iron ball is dropped into a tank of water at 20°C and another identical ball is dropped into a tank of ethyl alcohol at the same temperature. Which ball will have the greater terminal velocity?

Answers to Multiple Choice

1. d	6. c	11. a	16. d
2. c	7. a	12. c	17. b
3. d	8. c	13. b	18. a
4. d	9. b	14. c	19. c
5. d	10. d	15. a	

12

Harmonic Motion

Many events in both nature and technology are periodic, with a certain motion repeating itself over and over again. The term simple harmonic motion describes the most fundamental kind of oscillatory behavior, and all periodic events are either examples of simple harmonic motion or else the result of several such motions superimposed upon one another. In harmonic motion, the energy of the vibrations is continually transformed from kinetic to potential and back again. The potential energy may be elastic rather than gravitational in character; we shall find later that elasticity and gravitation are not the only phenomena in which potential energy is a useful concept. The electrical equivalent of potential energy, in particular, makes possible electrical oscillations that closely resemble harmonic motion.

12-1 Elastic Potential Energy

When we stretch a spring, it resists being lengthened, and if we then let it go, the spring returns to its original length. As we know, this is an example of *elastic* behavior. On the other hand, when we stretch a piece of taffy, it also resists being lengthened, but if we then let it go, nothing happens: the deformation is permanent. This is an example of *plastic* behavior.

Elastic and plastic behavior

In the case of the stretched spring, a *restoring force* comes into being that tries to return the spring to its normal length. The farther we stretch the spring, the greater the restoring force we must overcome (Fig. 12-1). Exactly the same phenomenon occurs when we compress the spring: it resists being shortened, and if we let it go, the spring returns to its normal length. Again a restoring force arises, and again the more the compression, the stronger the restoring force to be overcome.

Restoring force

The amount s by which an elastic solid is stretched or compressed by a force is directly proportional to the magnitude F of the force, provided the elastic limit is not exceeded. This proportionality is called Hooke's law, as mentioned in Chap. 10. Thus we can write

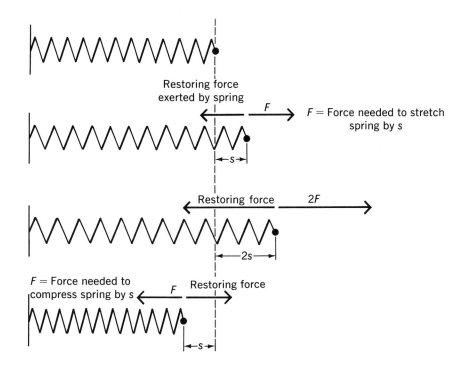

Restoring force exerted by spring

F

F = Force needed to stretch spring by s

s

Restoring force $2F$

$2s$

F = Force needed to compress spring by s F Restoring force

s

Fig. 12-1. When a spring (or other elastic body) is stretched or compressed, a restoring force comes into being that tries to return the spring to its normal length.

$$F = ks \qquad\qquad\qquad \textit{Hooke's law} \quad (12\text{-}1)$$

where k is a constant whose value depends upon the nature and dimensions of the object. A stiff spring has a higher value of k than a weak one.

The work done in stretching (or compressing) an object that obeys Hooke's law is easy to calculate. The work done by a force is the product of the magnitude of the force and the distance through which it acts. Here the force used in stretching the object is not constant, but is proportional to the elongation s at each point in the stretching process. Because F is proportional to s, the *average* force $\bar{F}$ applied while the body is stretched from its normal length by an amount s to its final length is

$$\bar{F} = \frac{F_{\text{initial}} + F_{\text{final}}}{2}$$

$$= \frac{0 + ks}{2}$$

$$= \tfrac{1}{2}ks$$

since the initial force is 0 and final force is ks (Fig. 12-2). The work done in stretching the spring is the product of the average force $\bar{F} = \tfrac{1}{2}ks$ and the total elongation s, so that

$$W = \tfrac{1}{2}ks^2. \qquad\qquad \textit{Elastic potential energy} \quad (12\text{-}2)$$

This formula is most often used in connection with springs: to stretch (or compress) a spring whose force constant is k by an amount s from its

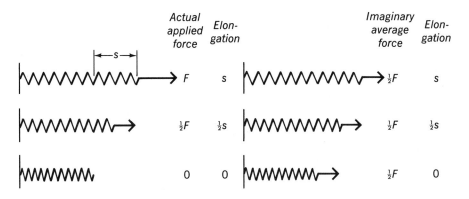

Fig. 12-2. To compute the work done in stretching a body that obeys Hooke's law, the varying force that actually acts during the expansion may be replaced by the average force.

normal length requires $\frac{1}{2}ks^2$ of work to be done. This work goes into *elastic potential energy*. When the spring is released, its potential energy of $\frac{1}{2}ks^2$ is transformed into kinetic energy or into work done on something else (Fig. 12-3); work done against frictional forces within the spring itself always absorbs some fraction of the available potential energy.

Problem. The horizontal spring shown in Fig. 12-4 has a force constant k of 90 N/m. Attached to the free end of the spring is a 1.4 kg block. If the spring is pulled out 50 cm from its equilibrium position and then released, what will the block's speed be when it returns to the equilibrium position?

Solution. When the spring is released, its elastic potential energy starts to be converted into kinetic energy of the block. We shall assume that the spring's mass is small compared with that of the block and that its internal friction may be neglected. At the equilibrium position of the spring, $s = 0$, and all the initial potential energy of $\frac{1}{2}ks^2$ is now kinetic energy $\frac{1}{2}mv^2$. Hence

$$\tfrac{1}{2}mv^2 = \tfrac{1}{2}ks^2,$$

$$v = \sqrt{\frac{k}{m}}\, s = \sqrt{\frac{90 \text{ N/m}}{1.4 \text{ kg}}} \times 0.50 \text{ m} = 4.0 \text{ m/s}.$$

12-2 Simple Harmonic Motion

When a spring with an object attached to it is stretched and then released, it does not simply return to its equilibrium position and come to a stop there. What happens is that the elastic potential energy $\frac{1}{2}ks^2$ of the spring is converted into kinetic energy $\frac{1}{2}mv^2$ of the moving object, and as the latter's momentum compresses the spring on the other side of the equilibrium position, this kinetic energy is converted back into elastic potential energy (Fig. 12-4). The amount of compression $-s$ will have the same magnitude as the original extension s, since

$$\tfrac{1}{2}ks^2 = \tfrac{1}{2}k(-s)^2.$$

Fig. 12-3.   Some devices that make use of elastic potential energy in their operation.

Left to itself in the absence of friction, the spring-object combination will continue oscillating back and forth indefinitely. The behavior of a system oscillating in this way is called *simple harmonic motion*.

Simple harmonic motion occurs whenever a force acts on a body in the opposite direction to its displacement from its normal position, with the magnitude of the force proportional to the magnitude of the displacement.

Condition for simple harmonic motion

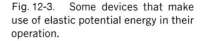

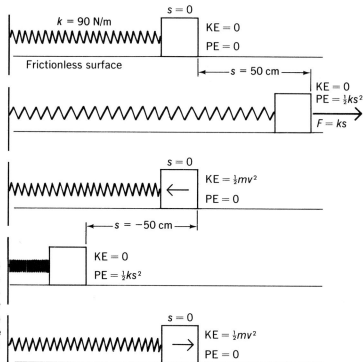

Fig. 12-4. A 1.4 kg block attached to a spring whose force constant is 90 N/m is pulled 50 cm from its equilibrium position. When the spring is released, its elastic potential energy of $\frac{1}{2}ks^2$ is converted into kinetic energy $\frac{1}{2}mv^2$, and as the block's momentum compresses the spring on the other side of the equilibrium position, the kinetic energy is converted back into elastic potential energy.

The elastic restoring force of a stretched or compressed spring always tends to return the spring to its normal length, but the momentum associated with the moving mass compels it to overshoot and thus to oscillate.

Period   The *period* of a body undergoing simple harmonic motion is the time required for it to make one complete oscillation. (A complete oscillation is often called a *cycle*.) In the case of a spring, the period is the time the spring spends in going from its maximum extension, say, through its maximum compression and back to its maximum extension once more (Fig. 12-5). For all types of simple harmonic motion, the period T is given by

$$T = 2\pi \sqrt{-\frac{s}{a}},\qquad(12\text{-}3)$$

$$\text{Period} = 2\pi \sqrt{-\frac{\text{displacement}}{\text{acceleration}}},$$

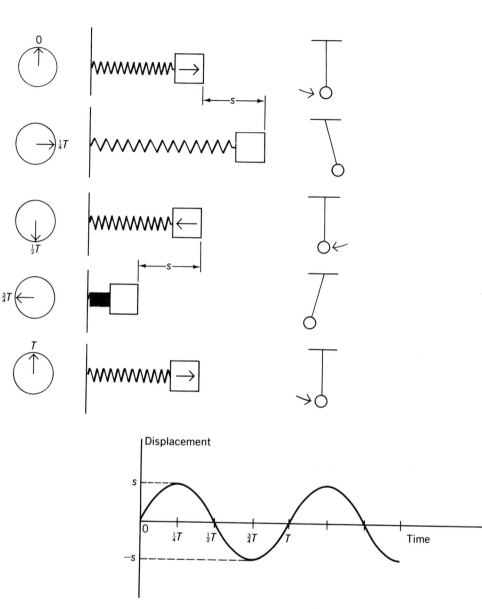

Fig. 12-5.   The period T of a body undergoing simple harmonic motions is the time required for it to make one complete oscillation.

where the acceleration is that experienced by the body when it is at the specified displacement from its equilibrium position. This formula is derived in Sec. 12-3.

To calculate the acceleration of a stretched spring we start with the second law of motion,

$$F = ma,$$

and substitute the restoring force $F_r = -ks$ since it is the restoring force that causes the body to be accelerated. This procedure yields

$$F_r = ma,$$

$$a = \frac{F_r}{m} = \frac{-ks}{m}. \qquad (12\text{-}4)$$

With this result we find that the period of a body of mass m attached to a spring of force constant k is

$$T = 2\pi \sqrt{-\frac{s}{a}} = 2\pi \sqrt{-\frac{s}{-ks/m}}$$

$$T = 2\pi \sqrt{\frac{m}{k}}. \qquad \textit{Oscillating spring} \quad (12\text{-}5)$$

Period is independent of amplitude

It is worth noting that the period T does not depend upon the maximum displacement s; no matter how much or how little the spring is initially pulled out, precisely the same amount of time is required for each cycle. If s is small, the maximum acceleration is also small and the body moves back and forth very slowly through its range, while if s is large, the acceleration is also large and the body moves correspondingly rapidly through the larger range. (The maximum displacement A of a body undergoing harmonic motion on either side of its equilibrium position is called the *amplitude* of the motion.) This peculiarity of simple harmonic motion is capitalized upon in the design of mechanical clocks and watches, which use the rotational oscillations of a coil spring or the swings of a pendulum—both examples of simple harmonic motion—to maintain a constant rate independent of any fluctuations in amplitude.

Frequency

A quantity often used in describing harmonic motion is *frequency*. The frequency is the number of cycles executed per unit time. Hence frequency, whose symbol is f, is the reciprocal of period T,

$$f = \frac{1}{T}. \qquad \textit{Frequency} \quad (12\text{-}6)$$

The unit of frequency is the *hertz* (Hz), where 1 Hz = 1 cycle/s.
    The position of a particle undergoing simple harmonic motion varies with time as shown on the graph in Fig. 12-5. The curve has exactly the same shape as a curve of sin θ plotted versus θ.

Problem. When a 1.0-kg ball is suspended from a spring, the spring stretches by 7.0 cm (Fig. 12-6). If the ball oscillates up and down, what is its period? What is its frequency?

Solution. The force exerted on the spring is the ball's weight of

$$mg = 1.0 \text{ kg} \times 9.8 \text{ m/s}^2 = 9.8 \text{ N}.$$

Since $F = ks$, the force constant of the spring is

$$k = \frac{F}{s} = \frac{mg}{s} = \frac{9.8 \text{ N}}{0.070 \text{ m}} = 140 \text{ N/m}.$$

The period of the oscillations is therefore

$$T = 2\pi\sqrt{\frac{m}{k}} = 2\pi\sqrt{\frac{1.0 \text{ kg}}{140 \text{ N/m}}} = 0.53 \text{ s},$$

and their frequency is

$$f = \frac{1}{T} = \frac{1}{0.53 \text{ s}} = 1.9 \text{ cycles/s} = 1.9 \text{ Hz}.$$

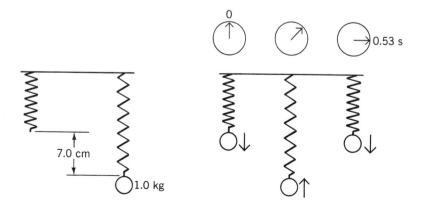

Fig. 12-6.

12-3 A Model of Simple Harmonic Motion

Figure 12-7 shows a particle moving in a vertical circle at constant speed. The particle is illuminated from above, and it casts a shadow on a horizontal screen below its orbit. As the particle travels around the circle, its shadow oscillates back and forth. The shadow moves fastest at the center, slows down as it approaches each end of the path, comes to a stop, and then reverses its direction. We might suspect that the shadow is executing simple harmonic motion. To verify this suspicion, we must show that the acceleration a of the shadow at any time is proportional to its displacement s from the center of its path and opposite in direction.

The acceleration of the shadow at any point is simply the horizontal component of the particle's acceleration a_c. As we know, a particle in

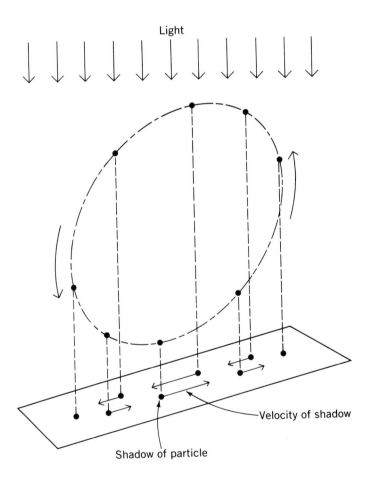

Fig. 12-7. The shadow of a particle undergoing uniform circular motion executes simple harmonic motion.

uniform circular motion at the speed V in a circle of radius R experiences the centripetal acceleration

$$a_c = \frac{V^2}{R}$$

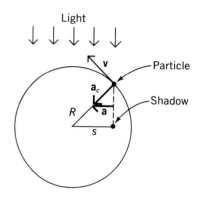

where the minus sign indicates that the acceleration is inward toward the center of the circle. The horizontal component of this acceleration is the acceleration a of the shadow. Since corresponding sides of similar triangles are proportional, we see from Fig. 12-8 that

$$\frac{a}{a_c} = \frac{s}{R}$$

$$a = \frac{s}{R} a_c$$

Because $a_c = -V^2/R$, the acceleration of the shadow is

$$a = -\frac{s}{R} \frac{V^2}{R} = -\frac{V^2}{R^2} s. \tag{12-7}$$

Fig. 12-8. The acceleration of the shadow is the horizontal component of the particle's centripetal acceleration. The shadow's acceleration is proportional to its displacement s and in the opposite direction.

The shadow's acceleration is proportional to its displacement s and in the opposite direction, which means that the shadow is indeed executing simple harmonic motion.

The value of the above analysis is that it gives us an easy way to find the period of an object in simple harmonic motion. The circumference of a circle of radius R is $2\pi R$, and a particle moving around the circle with the constant speed V covers this distance in the time

$$T = \frac{2\pi R}{V}$$

$$\text{Period} = \frac{\text{distance}}{\text{speed}}.$$

From Eq. (12-7) we find that

$$\frac{R}{V} = \sqrt{-\frac{s}{a}},$$

and so

$$T = 2\pi \sqrt{-\frac{s}{a}}. \tag{12-3}$$

This is the general formula for the period of simple harmonic motion given in the preceding section.

12-4 Position, Velocity, and Acceleration

The principle of conservation of energy permits us to express the velocity of a body in simple harmonic motion in terms of its frequency f, amplitude A, and displacement s. The total energy of the oscillator (body plus spring) is the sum of its kinetic and potential energies at any time, which are respectively $\frac{1}{2}mv^2$ and $\frac{1}{2}ks^2$. At either extreme of the motion, when $s = +A$ or $s = -A$, the body is stationary and has only the potential energy $\frac{1}{2}kA^2$. Hence

$$\text{Total energy} = \text{KE} + \text{PE},$$

$$\tfrac{1}{2}kA^2 = \tfrac{1}{2}mv^2 + \tfrac{1}{2}ks^2,$$

$$mv^2 = k(A^2 - s^2),$$

and

$$v = \sqrt{k/m}\ \sqrt{A^2 - s^2}.$$

From Eqs. (12-5) and (12-6) we know that

$$f = 1/T = 1/2\pi \sqrt{m/k},$$

which can be rewritten as

$$\sqrt{k/m} = 2\pi f. \tag{12-8}$$

The velocity of the body when it has the displacement s is accordingly

$$v = 2\pi f \sqrt{A^2 - s^2}. \tag{12-9}$$

This formula gives only the absolute value of the velocity v; whether the sign of v is $+$ or $-$ depends upon whether the body is at $+s$ or $-s$ and upon whether it is on its way toward or away from the equilibrium position from there.

Maximum velocity of oscillating body   From Eq. (12-9) we see that the maximum velocity $v_{\max}$ of the body, which occurs at the equilibrium position when $s = 0$, is

$$v_{\max} = 2\pi f A. \tag{12-10}$$

The maximum velocity is proportional to both the frequency and the amplitude of the motion.

The energy of an oscillating body shifts back and forth between kinetic and potential forms. To find the total energy, we can calculate either KE_{max} or PE_{max}, with the help respectively of Eq. (12-10) or (12-8):

$$KE_{max} = \tfrac{1}{2}mv_{max} = 2\pi^2mf^2A^2 \tag{12-11}$$

$$PE_{max} = \tfrac{1}{2}kA^2 = 2\pi^2mf^2A^2. \tag{12-12}$$

Total energy

To determine the acceleration of a body in simple harmonic motion, we refer back to Eq. (12-4), which states that

$$a = -ks/m.$$

In view of Eq. (12-8) this formula becomes

$$a = -4\pi^2f^2s. \tag{12-13}$$

The acceleration is always opposite in direction to the displacement, which, of course, is one of the conditions for simple harmonic motion to occur. The maximum acceleration occurs at either extreme, when $s = \pm A$, and has the magnitude

Maximum acceleration of oscillating body

$$a_{max} = 4\pi^2f^2A. \tag{12-14}$$

The maximum acceleration is proportional to the square of the frequency and to the amplitude.

Although the above formulas were derived for the case of a vibrating spring, their validity is perfectly general and they apply to any type of harmonic oscillator, from the bob of a pendulum to an atom in a molecule.

Problem. Atoms in a crystalline solid are in constant vibration at room temperature, where their motion has amplitudes in the neighborhood of 10^{-11} m. If the frequency of oscillation of one of the atoms in an iron bar is 2.5×10^{12} Hz, find its maximum speed and acceleration.

Solution. From Eqs. (12-10) and (12-14),

$$v_{max} = 2\pi fA = 2\pi \times 2.5 \times 10^{12} \text{ s}^{-1} \times 10^{-11} \text{ m}$$

$$= 157 \text{ m/s},$$

$$a_{max} = 4\pi^2f^2A = 4\pi^2 \times (2.5 \times 10^{12} \text{ s}^{-1})^2 \times 10^{-11} \text{ m}$$

$$= 2.5 \times 10^{15} \text{ m/s}^2.$$

Problem. A piston undergoes simple harmonic motion in a vertical direction with an amplitude of 3 in. A small mass is placed on top of the piston. What is the lowest frequency at which the mass will be left behind by the piston on its downstroke?

Solution. The mass will leave the piston when the downward acceleration of the latter exceeds the acceleration of gravity g. The maximum downward acceleration of the piston occurs at the highest point of its motion, when

$$a_{max} = 4\pi^2 f^2 A$$

according to Eq. (12-14). Hence we set a_{max} equal to g and obtain

$$a_{max} = g = 4\pi^2 f^2 A,$$

$$f = \frac{1}{2\pi} \sqrt{\frac{g}{A}} = \frac{1}{2\pi} \sqrt{\frac{32 \text{ ft/s}^2}{0.25 \text{ ft}}} = 1.8 \text{ Hz}.$$

The variations of the displacement, velocity, and acceleration of a particle undergoing simple harmonic motion are plotted versus time in Fig. 12-9. The graphs are plotted on the assumption that the particle is at $s = +A$ when $t = 0$. This corresponds to pulling out the particle and letting it go at $t = 0$. At this instant the particle's acceleration is a maximum and is

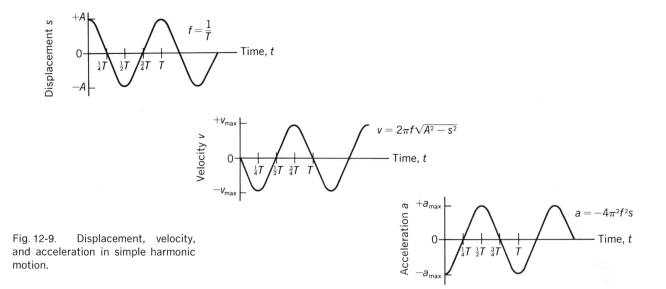

Fig. 12-9.   Displacement, velocity, and acceleration in simple harmonic motion.

opposite in direction to s, while the velocity is zero since the particle has not yet started to move.

When the particle is at the origin, $s = 0$ and the spring is at its normal length. Because the spring exerts no force on the particle at this time, its acceleration is zero. The velocity of the particle is now a maximum, which follows from the lack of any of the potential energy associated with a deformed spring.

When $s = -A$, the particle is at the other extreme of its range, and its acceleration, again a maximum, is positive, which means that it is once more in the direction of the origin, though now from the other side. All the energy of oscillation is potential, and the body is accordingly stationary at this instant.

12-5 The Simple Pendulum

A pendulum executes simple harmonic motion as it swings back and forth, provided that the arc through which the pendulum bob moves is a fairly small one. We shall see why this limitation arises if we use Eq. (12-3) to calculate the period of a pendulum.

Figure 12-10 shows a pendulum of length L whose bob has a mass m, together with a diagram of the forces acting on the bob. (It is assumed that the entire mass of the pendulum is concentrated in the bob.) The weight of the bob, $m\mathbf{g}$, which acts vertically downward, may be resolved into two forces, $\mathbf{T}$ and $\mathbf{F}$, which act respectively parallel to and perpendicular to the supporting string L. That is,

$$\mathbf{T} + \mathbf{F} = m\mathbf{g}.$$

A pendulum undergoes simple harmonic motion when it swings through a small arc

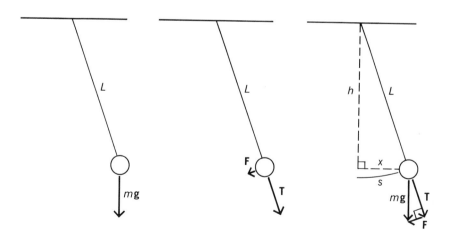

Fig. 12-10. A pendulum executes simple harmonic motion when its oscillations are so small in amplitude that the chord x is very nearly equal in length to the arc s.

The force **F** is the restoring force that acts to return the bob to the midpoint of its motion. The space triangle hLx and the vector triangle **T**m**gF** are similar, since each contains a right angle and two sides of one are parallel to the two corresponding sides of the other, and so

$$\frac{F}{x} = \frac{mg}{L}.$$

The restoring force acting on the bob is therefore

$$F = -\frac{mgx}{L},$$

where the minus sign indicates that F points in the direction of decreasing x.

If the bob is not far from the midpoint of its motion, the horizontal distance x is almost exactly equal to the actual path length s, and F then is given by

$$F = -\frac{mgs}{L}.$$

The acceleration of the bob that results from this force is

$$a = \frac{F}{m} = -\frac{gs}{L}.$$

Substituting in Eq. (12-3) we find that

$$T = 2\pi \sqrt{-\frac{s}{a}}$$

$$T = 2\pi \sqrt{\frac{L}{g}}. \qquad\qquad \textit{Simple pendulum} \quad (12\text{-}15)$$

Provided that s is small enough so that it is close to x, the motion of a pendulum is simple harmonic in character with a period proportional to the square root of the pendulum's length and independent of the mass of the bob. If the arc through which the pendulum swings on either side of the vertical is 5°, the actual period will exceed that predicted by Eq. (12-15) by only 0.05 percent; if the arc is 10° the discrepancy will be 0.2 percent; and even if the arc is as much as 20° the discrepancy is only 0.8 percent. Only when the arc on either side of the vertical is about 50° does the discrepancy reach 5 percent.

Problem.   How long should a pendulum be for it to have a period of exactly 1 s?

Solution.   We first solve Eq. (12-15) for L:

$$T = 2\pi\sqrt{\frac{L}{g}}, \quad T^2 = \frac{4\pi^2 L}{g}, \quad L = \frac{gT^2}{4\pi^2}.$$

Inserting the values $g = 9.8$ m/s^2 and $T = 1$ s, we find that

$$L = \frac{9.8 \text{ m/s}^2 \times 1 \text{ s}^2}{4\pi^2} = 0.25 \text{ m}.$$

12-6 The Torsion Pendulum

A *torsion pendulum* consists of an object suspended by a wire or thin rod, as in Fig. 12-11. When the object is turned through an angle and released, it will oscillate back and forth. The analogies between linear and angular quantities make it easy to find a formula for the period of these oscillations.

From Hooke's law the restoring torque τ that comes into being when the wire is twisted through an angle θ is

$$\tau = -K\theta$$

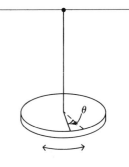

Fig. 12-11. A torsion pendulum.

where the value of the torsion constant K depends on the material and dimensions of the wire. If the moment of inertia of the object about its axis of suspension is I (the moment of inertia of the suspending wire is usually negligible), then the angular acceleration α of the object when the torque τ acts on it is, from Eq. (9-20),

$$\alpha = \frac{\tau}{I} = -\frac{K}{I}\,\theta. \tag{12-16}$$

Comparing this formula with the equivalent result for a harmonic oscillator,

$$a = -\frac{k}{m}\,s \tag{12-4}$$

suggests that the period of a torsion pendulum can be given by the general formula for the period of a harmonic oscillator,

$$T = 2\pi \sqrt{-\frac{s}{a}} \qquad (12\text{-}5)$$

with θ/α replacing s/a. This idea turns out to be correct, and we have

$$T = 2\pi \sqrt{-\frac{\theta}{\alpha}} = 2\pi \sqrt{\frac{I}{K}}. \qquad Torsion\ pendulum \quad (12\text{-}17)$$

The simplest procedure for finding the moment of inertia of an irregular object is often to suspend it by a wire, measure the torsion constant K and the period of oscillation T, and then use Eq. (12-17) to find I.

Problem. A grindstone is suspended by a wire from its center. When a torque of 0.12 N·m is applied to the grindstone, it turns through 8°, and when it is released, it oscillates with a period of 1.0 s. Find the moment of inertia of the grindstone.

Solution. Since $1° = 0.01745$ rad, $\theta = 0.14$ rad, and the torsion constant of the wire is

$$K = \frac{\tau}{\theta} = \frac{0.12\ \text{N·m}}{0.14\ \text{rad}} = 0.86\ \text{N·m/rad}.$$

From Eq. (12-17)

$$I = \frac{KT^2}{4\pi^2} = \frac{0.86\ \text{N·m/rad} \times (1.0\ \text{s})^2}{4\pi^2} = 0.022\ \text{kg·m}^2.$$

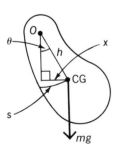

$$\tau = -(mg)(h\sin\theta)$$

Fig. 12-12. A physical pendulum pivoted at O. The center of gravity is marked CG.

12-7 The Physical Pendulum

An object of any shape will oscillate back and forth when it is pivoted at some point other than its center of gravity and given an initial displacement to one side. Such an object is called a *physical pendulum*. The formula for the period of a simple pendulum also applies to a physical pendulum provided that the length L is properly interpreted.

Figure 12-12 shows a physical pendulum pivoted about a horizontal axis at O. The pendulum is displaced so that the line from O to its center of gravity is at the angle θ from the vertical. The pendulum's weight mg acts from the center of gravity and produces the restoring torque

$$\tau = -mgx$$

where x is the horizontal distance between O and the center of gravity. The minus sign reflects the fact that the restoring torque is always opposite in direction to the angular displacement θ of the pendulum.

When θ is small, the chord x is very nearly equal to the arc s, where

$$s = h\theta$$

(with θ expressed in radians, of course). This is the same approximation made in analyzing the simple pendulum (see Fig. 12-10). Hence

$$\tau = -mgh\theta.$$

If I is the moment of inertia of the pendulum about O and α is the angular acceleration produced by the restoring torque, then

$$\alpha = \frac{\tau}{I} = -\frac{mhg}{I}\,\theta.$$

By the same reasoning used in the preceding section, the period of the physical pendulum is

$$T = 2\pi \sqrt{-\frac{\theta}{\alpha}} = 2\pi \sqrt{\frac{I}{mgh}}. \qquad \textit{Physical pendulum} \quad (12\text{-}18)$$

The simple pendulum whose period is the same as that of a given physical pendulum may be found by setting equal the formulas for their respective periods of oscillation:

$$2\pi \sqrt{\frac{L}{g}} = 2\pi \sqrt{\frac{I}{mgh}},$$

$$L = \frac{I}{mh}. \qquad\qquad (12\text{-}19)$$

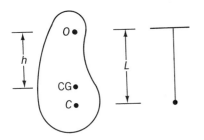

Fig. 12-13.   A simple pendulum of length $L = I/mh$ has the same period as that of a physical pendulum.

Fig. 12-14.   The pivot of a physical pendulum struck at its center of percussion experiences no reaction force.

Thus the mass of a physical pendulum can be regarded as concentrated at a point C the distance $L = I/mh$ from the point O. This point is called the center of oscillation (Fig. 12-13) and has two interesting properties:

1. If the pendulum is pivoted at C instead of at O, it will oscillate with the same period as before and O will be the new center of oscillation;

2. If the pendulum is struck along a line of action through C, there will be no reaction force on the pivot at O. A baseball that strikes a bat at the latter's center of oscillation does not produce a sting in the batter's hands, for example (Fig. 12-14). The center of oscillation is often called the *center of percussion* for this reason. The concept of center of percussion plays an important part in the design of many mechanical devices.

Problem.   Analyze the process of walking by considering the leg as a physical pendulum.

Solution.   A leg L long may be approximated by a thin rod hinged at one end. From Fig. 9-9 the moment of inertia of such a rod is

$$I = \tfrac{1}{3}mL^2$$

where m is its mass. If the center of gravity of the leg is at its middle, $h = L/2$, and the natural period of oscillation of the leg is

$$T = 2\pi \sqrt{\frac{I}{mgh}} = 2\pi \sqrt{\frac{mL^2/3}{mgL/2}} = 2\pi \sqrt{\frac{2L}{3g}}.$$

The period of a leg 1.0 m long is

$$T = 2\pi \sqrt{\frac{2 \times 1.0 \text{ m}}{3 \times 9.8 \text{ m/s}^2}} = 1.6 \text{ s}.$$

Each step represents only half a complete cycle, hence if the leg is swinging freely it takes 0.8 s and the rate of walking is 75 steps/min. Walking at a slower or a faster rate than this involves more effort. With a stride 80 cm long, 75 steps/min means a velocity of 60 m/min, which is 2.24 mi/hr.

The longer the legs of an animal, the longer its stride, but T is increased as well. The natural velocity of walking varies as L/T, and since T is proportional to $\sqrt{L}$, this velocity is proportional to $L/\sqrt{L} = \sqrt{L}$. The larger an animal, then, the faster it walks, although it takes an increase by a factor of 4 in leg length to double the natural velocity. Running is a quite different matter because the muscular strength and the mass of an animal are involved as well as its size. As we saw in Section 7-5, all animals have roughly similar running velocities.

Special Topic

Trigonometric Notation
for Simple Harmonic Motion

Figure 12-15 shows how the displacement, velocity, and acceleration of an oscillating object can be expressed as functions of the time t since it was

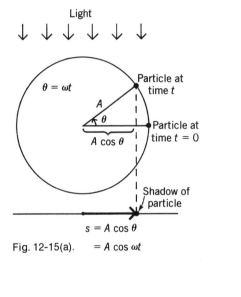

Fig. 12-15(a).

$$s = A \cos \theta$$
$$= A \cos \omega t$$

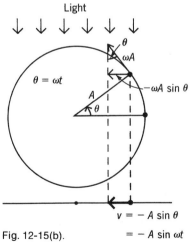

Fig. 12-15(b).

$$v = -A \sin \theta$$
$$= -A \sin \omega t$$

Fig. 12-15(a)–(c). The displacement, velocity, and acceleration of the shadow of the particle of Fig. 12-7 can be expressed in terms of the time t as shown here. The angular velocity of the particle is ω, the magnitude of its linear velocity is ωA, and the magnitude of its centripetal acceleration is $\omega^2 A$.

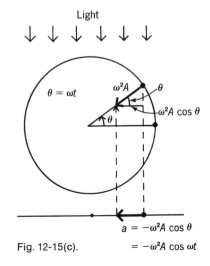

Fig. 12-15(c).

$$a = -\omega^2 A \cos \theta$$
$$= -\omega^2 A \cos \omega t$$

set in motion. The model is the same as the one in Fig. 12-7, where we saw that the shadow of a particle moving in a circle undergoes simple harmonic motion. The particle has an angular velocity of $\omega = 2\pi f$, and ω is called the *angular frequency* of the harmonic motion of its shadow. The unit of ω is, as usual, the radian/s. The angular displacement of the particle at any time t is $\theta = \omega t$, its linear velocity has the constant magnitude ωA, and its centripetal acceleration is $\omega^2 A$, where A is the radius of the circle. From Fig. 12-15 it is clear that

$$s = A \cos \theta = A \cos \omega t,$$
$$v = -\omega A \sin \theta = -\omega A \sin \omega t,$$
$$a = -\omega^2 A \cos \theta = -\omega^2 A \cos \omega t.$$

The sign convention used is that a quantity directed to the right is considered positive and one directed to the left is considered negative. Graphs of s, v, and a versus time were given in Fig. 12-9.

Once we had found that $s = A \cos \omega t$ we could have obtained the formulas for v and a by differentiation, making use of the rules

$$\frac{d}{dx} (\cos u) = -\sin u \frac{du}{dx}$$

$$\frac{d}{dx} (\sin u) = \cos u \frac{du}{dx}.$$

Thus

$$v = \frac{ds}{dt} = \frac{d}{dt} (A \cos \omega t) = -A \sin \omega t \frac{d}{dt} (\omega t) = -\omega A \sin \omega t,$$

$$a = \frac{dv}{dt} = \frac{d}{dt} (-\omega A \sin \omega t) = -\omega A \cos \omega t \frac{d}{dt} (\omega t) = -v^2 A \cos \omega t.$$

It is not necessary for us to be restricted to situations in which the motion begins (or we start our clock) at the instant when the particle is located at $s = +A$, corresponding to $\theta = 0$. If the position of the particle corresponds to $\theta = \phi$ when $t = 0$, then at any future time t

$$\theta = \omega t + \phi.$$

The initial angle ϕ is called the *phase angle* of the motion (Fig. 12-16). The

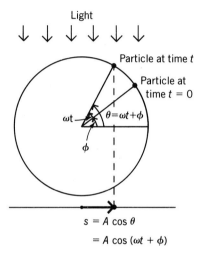

Light

$s = A \cos \theta$

$= A \cos (\omega t + \phi)$

Fig. 12-16. In general, $\theta = \omega t + \phi$, where $\theta = \phi$ at $t = 0$. The initial angle ϕ is called the *phase angle*.

general formulas that describe simple harmonic motion are therefore as follows:

$$s = A \cos (\omega t + \phi)$$
$$v = -\omega A \sin (\omega t + \phi)$$
$$a = -\omega^2 A \cos (\omega t + \phi).$$

Important Terms

A body under stress possesses **elastic potential energy,** which is equal to the work done in deforming it.

Simple harmonic motion is an oscillatory motion that occurs whenever a force acts on a body in the opposite direction to its displacement from its equilibrium position, with the magnitude of the force proportional to the magnitude of the displacement. Thus the force on a body in simple harmonic motion always tends to return it to its equilibrium position.

The **period** T of a body undergoing simple harmonic motion is the time required for it to make one complete oscillation. The **frequency** f of such a body is the number of complete oscillations it makes per unit time.

The **amplitude** of a body undergoing simple harmonic motion is its maximum displacement on either side of its equilibrium position. The period of the motion is independent of the amplitude.

The **center of percussion** of a pivoted object is that point at which it can be struck without producing a reaction force on its pivot.

Important Formulas

Elastic potential energy:

$$PE = \tfrac{1}{2}ks^2$$

Harmonic oscillator:

$$T = \frac{1}{f} = 2\pi \sqrt{\frac{m}{k}}$$

Simple pendulum:

$$T = 2\pi \sqrt{\frac{L}{g}}$$

Torsion pendulum:

$$T = 2\pi \sqrt{\frac{I}{K}}$$

Physical pendulum:

$$T = 2\pi \sqrt{\frac{I}{mgh}}$$

Multiple Choice

1. The period of a simple harmonic oscillator is independent of its
   a. frequency.   b. amplitude.
   c. force constant.   d. mass.

2. An object undergoes simple harmonic motion. Its maximum speed occurs when its displacement from its equilibrium position is
   a. zero.
   b. a maximum.
   c. half its maximum value.
   d. none of the above.

3. In simple harmonic motion, there is always a constant ratio between the displacement of the mass and its
   a. velocity.
   b. acceleration.
   c. period.
   d. mass.

4. An object attached to a horizontal spring executes simple harmonic motion on a frictionless surface. The ratio between its kinetic energy when it passes through the equilibrium position and its potential energy when the spring is fully extended is
   a. less than 1.
   b. equal to 1.
   c. more than 1.
   d. equal to the ratio between its mass and the spring constant.

5. The period of a simple pendulum depends on its
   a. mass.
   b. length.
   c. total energy.
   d. maximum speed.

6. The amplitude of a body undergoing harmonic motion is
   a. its total range of motion.
   b. its maximum displacement on either side of the equilibrium position.
   c. its minimum displacement on either side of the equilibrium position.
   d. the number of cycles per second it describes.

7. The amplitude of a simple harmonic oscillator is doubled. Which of the following is also doubled?
   a. its frequency
   b. its period
   c. its maximum velocity
   d. its total energy

8. A pendulum executes simple harmonic motion provided that
   a. its bob is not too heavy.
   b. the supporting string is not too long.
   c. the arc through which it swings is not too small.
   d. the arc through which it swings is not too large.

9. The product of the period and the frequency of a harmonic oscillator is always equal to
   a. 1.
   b. π.
   c. 2π.
   d. the amplitude of the motion.

10. A spring whose force constant is k is cut in half. Each of the new springs has a force constant of
    a. $\frac{1}{2}k$.
    b. k.
    c. $2k$.
    d. $4k$.

11. A force of 0.2 N is needed to compress a certain spring by 2 cm. Its potential energy when compressed is
    a. 2×10^{-3} J.
    b. 2×10^{-5} J.
    c. 4×10^{-5} J.
    d. 8×10^{-5} J.

12. The period of a harmonic oscillator of mass m is proportional to
    a. $\sqrt{m}$.
    b. $1/\sqrt{m}$.
    c. m^2.
    d. $1/m^2$.

13. In a refrigeration compressor, a 2-lb piston undergoes 20 cycles/s in which its total travel is 6 in. The maximum force on the piston
    a. is 247 lb.
    b. is 494 lb.
    c. is 3948 lb.
    d. cannot be calculated from the given data.

14. A boy swings from a rope 4.9 m long. His approximate period of oscillation is
    a. 0.5 s.
    b. 3.1 s.
    c. 4.4 s.
    d. 12 s.

15. When a 1-kg mass is suspended from a spring, the spring stretches by 5 cm. The force constant of the spring is
    a. 0.2 N/m.   c. 49 N/m.
    b. 1.96 N/m.   d. 196 N/m.

16. If the suspended mass of Question 15 oscillates up and down, its period will be approximately
    a. 0.032 s.   b. 0.071 s.
    c. 0.45 s.   d. 4.5 s.

17. The maximum speed of a particle that undergoes simple harmonic motion with a period of 0.5 s and an amplitude of 2 cm is
    a. π cm/s.   b. 2π cm/s.
    c. 4π cm/s.   d. 8π cm/s.

18. A lead sphere is suspended by a wire and set into rotational oscillation. If the sphere were flattened into a disk, the period of the oscillations would be
    a. shorter.
    b. the same.
    c. longer.
    d. any of the above, depending on the radius of the disk.

Exercises

1. Must a spring obey Hooke's law in order to oscillate?

2. At what point or points in its motion is the energy of a harmonic oscillator entirely potential? At what point or points is its energy entirely kinetic?

3. Upon what, if anything, does the ratio between the maximum kinetic energy and maximum potential energy of a harmonic oscillator depend?

4. A body pivoted at some point is given an initial displacement and then released. Under what circumstances will it oscillate back and forth? Under what circumstances will the oscillations be simple harmonic in character? Under what circumstances will it behave like a simple pendulum?

5. A wooden object is floating in a bathtub. It is pressed down and then released. Under what circumstances will its oscillations be simple harmonic in nature?

6. At what displacement relative to the amplitude is the kinetic energy of a harmonic oscillator three times the potential energy?

7. A toy rifle employs a spring whose force constant is 200 N/m. In use, the spring is compressed 5 cm, and when released, it propels a 5-g rubber ball. What is the ball's velocity when it leaves the rifle?

8. A force of 0.5 lb is required to push a Jack-in-the-box into its box, an operation in which the spring is compressed 4 in. If the Jack-in-the-box weighs 0.2 lb, what will its maximum velocity be when it pops out?

9. A 5-kg object is dropped on a vertical spring from a height of 2 m. If the maximum compression of the spring is 40 cm, what is its force constant?

10. The periods of the earth's rotation on its axis and revolution about the sun are 24 hr and 365 days respectively. What is the frequency of each of these motions?

11. When a 2-lb weight is suspended from a spring, the spring stretches by 3 in. If the weight oscillates up and down, what is its period? What is its frequency?

12. A body whose mass is 0.4 kg is suspended from a spring and oscillates with a period of 2 s. By how much will the spring contract when the body is removed?

13. A 160-lb gymnast jumps on a trampoline from a height of 3 ft. The trampoline sags 9 in. when the gymnast strikes it, and he then bounces up and down. If the motion is simple harmonic, find its period.

14. A harmonic oscillator has a period of 0.2 s and an amplitude of 10 cm. Find the velocity of the body when it passes through the equilibrium position.

15. A particle undergoes simple harmonic motion with a period of 2 s and an amplitude of 1 ft. Find its maximum velocity.

16. A chandelier is suspended from a high ceiling with a cable 20 ft long. What is its period of oscillation?

17. A pendulum whose length is 1.53 m oscillates 24 times per minute in a particular location. What is the acceleration of gravity there?

18. A pendulum 0.82 ft long has a period of 1 s in a certain place. What is the value of g there?

19. What is the frequency of a pendulum whose normal period is T when it is in an elevator in free fall? When it is in an elevator descending at constant velocity? When it is in an elevator ascending at constant velocity?

20. Find the period of the rotational oscillations of an aluminum sphere 8 cm in diameter whose mass is 725 g which is suspended by a wire whose torsion constant is 0.1 N·m/rad.

Problems

1. A spring has a 1-s period of oscillation when a 20-lb weight is suspended from it. Find the elongation of the spring when a 50-lb weight is suspended from it.

2. In an automobile engine, each piston undergoes simple harmonic motion. If a piston weighs 1 lb, has a total travel of 4 in., and completes 60 cycles/s, find the maximum force on it.

3. A body whose mass is 0.2 slug is suspended from a spring and oscillates with a frequency of 5 Hz and an amplitude of 0.5 ft. (a) What is the total energy of the motion? (b) What is the maximum acceleration of the body?

4. A body whose mass is 1 kg hangs from a spring. When the body is pulled down 5 cm from its equilibrium position and released, it oscillates once per second. (a) What is the force constant of the spring? (b) What is the body's velocity when it passes through its equilibrium position? (c) What is the maximum acceleration of the body?

5. A body whose mass is 0.005 kg is in simple harmonic motion with a period of 0.04 s and an amplitude of 0.01 m. (a) What is its maximum

acceleration? (b) What is the maximum force on the body? (c) What is its acceleration when it is 0.005 m from its equilibrium position? (d) What is the force on it at that point?

6. The prongs of a tuning fork vibrate in simple harmonic motion at a frequency of 660 Hz and with an amplitude of 1 mm at their tips. Find the maximum speed and acceleration of the prong tips. Express the acceleration in terms of g.

7. A pendulum has a length of 50 cm. Find its period when it is suspended in (a) a stationary elevator; (b) an elevator falling at the constant velocity of 5 m/s; (c) an elevator falling at the constant acceleration of 2 m/s²; (d) an elevator rising at the constant velocity of 5 m/s; (e) an elevator rising at the constant acceleration of 2 m/s².

8. A pendulum has a length of 1 ft. Find its period when it is suspended in (a) a stationary elevator; (b) an elevator falling at the constant velocity of 20 ft/s; (c) an elevator falling at the constant acceleration of 7 ft/s²; (d) an elevator rising at the constant velocity of 20 ft/s; (e) an elevator rising at the constant acceleration of 7 ft/s².

9. A 1/2-lb broomstick 4 ft long is suspended from one end. What is the period of its oscillations? What would be the length of a simple pendulum with the same period?

10. A 200-g brass hoop 40 cm in diameter is suspended on a knife edge on which it rocks back and forth. If the period of the oscillations is 1.27 s, find the moment of inertia of the hoop about an axis through its circumference perpendicular to its plane.

11. A 1-kg iron bar 80 cm long is suspended from one end. What is the period of its oscillations? What would be the length of a simple pendulum with the same period?

12. A wooden cube of density d that is L long on each edge floats in a liquid of density d' so that its upper and lower faces are horizontal. The cube is pushed down and released. Verify that the cube then oscillates up and down in simple harmonic motion with a period of $2\pi\sqrt{Ld/gd'}$.

13. A hole is bored through the earth along a diameter. Inside the hole the acceleration of gravity varies as rg/R, where r is the distance from the center of the earth, R is the earth's radius of 6.4×10^6 m, and g is the acceleration of gravity at the earth's surface. A stone is dropped into the hole and executes simple harmonic motion about the center of the earth. Why? Find the period of this motion.

14. A ball bearing is dropped on a steel plate from a height h. Verify that the ball bounces up and down with a frequency of $\frac{1}{2}\sqrt{g/2h}$.

Answers to Multiple Choice

1. b	10. c
2. a	11. c
3. b	12. a
4. b	13. a
5. b	14. c
6. b	15. d
7. c	16. c
8. d	17. d
9. a	18. c

13

Waves

The properties of a vibrating system and those of a wave traveling through a medium are similar in a number of essential respects. In both, a certain characteristic motion recurs at regular intervals; in both, the motions that occur represent the continuous conversion of potential energy into kinetic energy and back; in both, the properties of matter play important roles in providing restoring forces. The chief difference is that the energy in a vibrating system remains localized in space, whereas waves carry energy from one place to another without any actual transport of matter.

13-1 Wave Motion

Energy can be transmitted from one place to another in a variety of ways. Suppose we wish to supply energy to a boat in the center of a lake from a position on the shore, with the provision that the precise form in which the energy arrives does not matter. The most obvious thing to do is to throw a stone at the boat, thereby providing it with kinetic energy. Another method is to pour hot water into the lake, thereby providing the boat with thermal energy. Or we can simply drop a stone in the water near the shore; the waves that are produced transfer energy to the boat by causing it to move up and down (Fig. 13-1). When the stone strikes the water, a deformation of the water surface begins to spread. The energy that reaches the boat arrives as a periodic deformation which contains both kinetic and potential energy. Energy propagation by means of the motion of a change in a medium is called *wave motion*, and it occurs in many forms in nature.

Waves transport energy

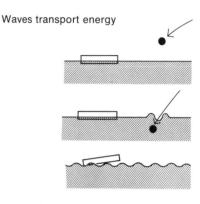

Fig. 13-1.   Waves transmit energy from one place to another through the motion of a change in a medium.

13-2 Pulses in a String

If we give one end of a stretched string a quick shake, a kink or *pulse* travels down the string at some velocity v (Fig. 13-2). If the string is uniform and completely flexible, the pulse keeps the same shape as it moves. It is worth examining the behavior of pulses in a string both because this is the simplest kind of wave phenomenon and because it is easy to visualize what is going on.

The velocity v of a pulse depends upon the properties of the string—how heavy it is and how tightly it is stretched—rather than upon the shape of the pulse or upon exactly how it is produced. Pulses move slowly down a slack, heavy rope; they move rapidly down a taut, light string. By "heavy" and "light" are meant the mass per unit length of a string, not its total weight; a pulse has the same velocity in a long string under a given tension as in a short string of the same kind under the same tension. When the mass per unit length of a string is high, the pulse velocity is low because the inertia of each segment of the string is large and it therefore responds slowly to the forces acting on it. When the string is tightly stretched, the pulse velocity is high because the tendency of the string to straighten out is greater.

If T is the tension in a certain stretched string of mass m and length L, the velocity of pulses in the string is given by the formula

$$v = \sqrt{\frac{T}{m/L}}.$$ (13-1)

Pulse velocity

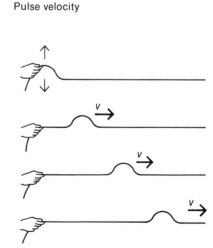

Fig. 13-2.   A pulse moves along a stretched string with a constant velocity v.

The energy content of a moving pulse is partly kinetic and partly potential. As the pulse travels, its forward part is moving upward and its rear part is moving downward; because the string has mass, there is a certain amount of kinetic energy associated with these up-and-down motions (Fig. 13-3). The potential energy is due to the tension in the string. Work had to be done in order to produce the pulse by pulling against the tension, and the deformed string accordingly possesses potential energy.

When a pulse reaches the end of a string, it may be reflected and travel back toward its starting point. Depending upon how the end of the string is held in place, the reflected pulse may be inverted (upside-down) or erect (right-side-up). Under just the right conditions, of course, the energy of the pulse may all be absorbed by the support and the pulse will then disappear.

Suppose the end of the string is held firmly in place. When the pulse arrives there, the string exerts an upward force on the support (assuming the pulse is upward, as in Fig. 13-4). By the third law of motion, the support

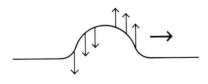

Fig. 13-3.   The forward part of a traveling pulse is moving upward and the rear part is moving downward.

Fig. 13-5.   A pulse reaching a free end of the string is not inverted on reflection.

Fig. 13-4.   A pulse reaching a fixed end of the string is inverted upon reflection.

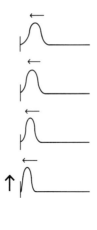

String exerts upward force on support when pulse arrives there.

Support exerts downward reaction force on string to produce inverted pulse moving in opposite direction.

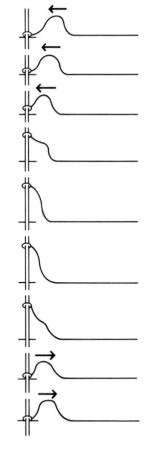

then exerts an equal and opposite reaction force on the string. The effect of this reaction force is to produce a pulse whose displacement is opposite to that of the original pulse but otherwise with the same shape. The new inverted pulse proceeds back along the string in the reverse direction of that of the original pulse. Thus an upward pulse becomes a downward one upon reflection, and vice versa.

If the end of the string is not held firmly in place, however, the reflected pulse is not inverted. Figure 13-5 shows the end of a string attached to a ring free to move up and down a frictionless rod. When the pulse arrives at this end, the string moves upward until its kinetic energy is completely converted into elastic potential energy, whereupon the end of the string moves downward again to send out a pulse that is reversed in direction but otherwise the same as the original one. If the end of the string is held in a manner exactly in between complete rigidity and complete freedom, then, the pulse will not be reflected at all but will disappear when it reaches the end.

A little experimentation with pulses in an actual string will show that, whereas it is very easy to hold the far end of the string so that the reflected pulse is smaller than the original one, it is not easy at all to keep some reflection from taking place. Similar difficulty is experienced in trying to construct surfaces that completely absorb sound, light, or water waves.

So far we have been considering pulses in a uniform string. Now let us connect two different strings together, one of them light (that is, with a low mass per unit length) and the other heavy (high mass per unit length). One end of the combination is fastened to something, and the other is given a shake to produce a pulse. Not surprisingly, the pulse passes from the first string to the second at the junction between them: the pulse is *transmitted*. But the transmission is not complete, since a reflected pulse also appears at the junction that proceeds in the opposite direction.

If the first string is the lighter one, the reflected pulse is inverted (Fig. 13-6). The greater inertia of the heavy string does not permit it to respond to the pulse as rapidly as the light string does, and an opposite reaction force occurs that causes the reflected pulse to be inverted, as though the junction were a rigid support. The energy of the original pulse is then split between the reflected and transmitted pulses. Since both strings have the same tension, the pulse travels slower in the heavy string. The length of

Wave absorption without any reflection is difficult

Pulse transmission

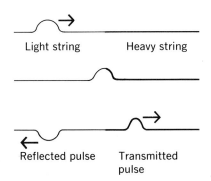

Light string   Heavy string

Reflected pulse   Transmitted pulse

Fig. 13-6. When a pulse passes from a light to a heavy string, reflection occurs with the reflected pulse being inverted. The transmitted pulse is right-side-up. Since both strings have the same tension, the pulse travels slower in the heavy string.

Fig. 13-7. When a pulse passes from a heavy to a light string, the reflected pulse stays right-side-up. The pulse velocity is again less in the heavy string.

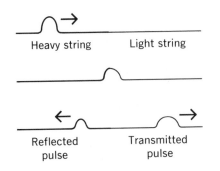

Heavy string   Light string

Reflected   Transmitted
pulse   pulse

the reflected pulse is the same as that of the original one, though its height is smaller since it has lost energy. The transmitted pulse, however, is shorter, because its velocity is less than that of the original pulse while the time interval in which it comes into being is the same as that of the original pulse.

On the other hand, if the first string is the heavy one, the reflected pulse is erect (Fig. 13-7). The smaller inertia of the light string permits it to follow the movements of the heavy one readily, and the situation is like that of a string whose end is able to move up and down freely. However, the light string does have some inertia, and so a reflected pulse again comes into being as well as a transmitted one. The pulse velocity is higher in the light string, and in consequence the pulse length is longer there than in the heavy one.

All types of waves, not just pulses in a stretched string, exhibit reflection and transmission at junctions between different media. For example, light waves are partially reflected and partially transmitted when they pass from air to glass, which is why we can see our images in a clear pane of glass (such as a shop window) even though the glass is transparent to light. The inversion of a pulse when it is reflected at a junction with a medium in which its velocity is smaller also has a counterpart in the behavior of light waves.

13-3 Principle of Superposition

Each end of a stretched string is given an upward shake, and the pulses thus produced move along the string toward each other. What happens when they meet? The result is a larger pulse at the moment the pulses come together, and then the separate pulses reappear and continue unchanged in their original directions of motion. Each pulse proceeds as though the other does not exist (Fig. 13-8).

The *principle of superposition* is a statement of the above behavior. This principle can be phrased as follows:

When two pulses travel past a point in a string at the same time, the displacement of the string at that point is the sum of the displacements each pulse would produce there by itself.

What if one of the pulses is inverted relative to the other? According to the superposition principle, if the pulses have the same sizes and shapes, their displacements ought to cancel out when they meet, only to reappear later on after they have passed the crossing point. Such behavior is indeed observed in practice. At the instant of complete cancellation, the total energy of both pulses resides in the kinetic energy of the string segment where the cancellation occurs (Fig. 13-9).

Principle of superposition for pulses

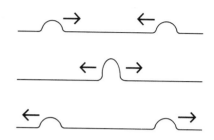

Fig. 13-8. Two pulses moving in opposite directions along a stretched string. The pulses are unaffected by their crossing.

13-4 Periodic Waves

In a periodic wave, one pulse follows another in regular succession. Sound waves, water waves, and light waves are almost always periodic, although in each case a different quantity varies as the wave passes.

In periodic waves, a certain waveform—the shape of the individual waves—is repeated at regular intervals. Periodic waves of all kinds usually have *sinusoidal* waveforms; a stretched string down which such waves move presents exactly the same appearance as a graph of sin x (or cos x) versus x that is moved along the x axis with the wave velocity v (Fig. 13-10).

Sinusoidal waves are common because the particles of matter in a medium that waves can travel through undergo simple harmonic motion

Sinusoidal waves

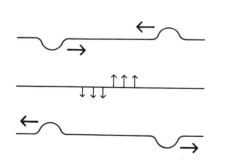

Fig. 13-9.   Complete cancellation occurs when two identical pulses with opposite displacements meet. At the instant of complete cancellation, the total energy of both pulses resides in the kinetic energy of the string segment where the cancellation occurs.

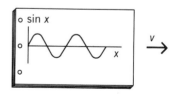

Fig. 13-10.   Most periodic waves have sinusoidal waveforms.

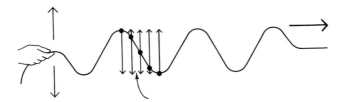

Fig. 13-11. Each particle in the path of a sinusoidal wave executes simple harmonic motion perpendicular to the wave direction.

when momentarily displaced from their equilibrium positions. The passage of a wave sets up coupled harmonic oscillations in the medium, with each particle behaving like a harmonic oscillator that has begun its cycle just a trifle later than the particle behind it (Fig. 13-11). The result in the case of waves in a stretched string is a waveform that is in essence a graph of how the position of a harmonic oscillator varies with time, which is a sine curve. (Light waves are also sinusoidal in character even though their existence does not involve the motion of material particles and they can travel through empty space.)

Three related quantities are useful in describing periodic waves:

Velocity, wavelength, and frequency

1. The *wave velocity v*, which is the distance through which each wave moves per second;

2. The *wavelength* λ (Greek letter *lambda*), which is the distance between adjacent crests or troughs;

3. The *frequency f*, which is the number of waves that pass a given point per second.

The wave velocity, wavelength, and frequency of a train of waves are not independent of one another. In every second, *f* waves (by definition) go past a particular point, with each wave occupying a distance of λ (Fig. 13-12). Therefore a wave travels a total distance of *f*λ per second, which is the wave velocity *v*. Thus

$$v = f\lambda \hspace{3cm} \textit{Wave velocity} \quad (13\text{-}2)$$

Wave velocity = frequency × wavelength,

which is a basic formula that applies to all periodic waves.

Wave period

Sometimes it is more useful to consider the *period T* of a wave, which is the time required for one complete wave to pass a given point (Fig. 13-13).

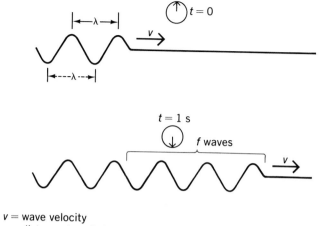

v = wave velocity
  = distance traveled per s
  = (number of waves passing a point per s) × (length of each wave)
  = $f\lambda$

Fig. 13-12. The velocity of a wave is equal to the product of its frequency and wavelength.

Since f waves pass by per second, the period of each wave is

$$T = \frac{1}{f}.$$

(13-3)

If there are five waves per second passing by, for example, each wave has a period of $\frac{1}{5}$ s. In terms of period T, the formula for wave velocity is

$$v = \frac{\lambda}{T}.$$

(13-4)

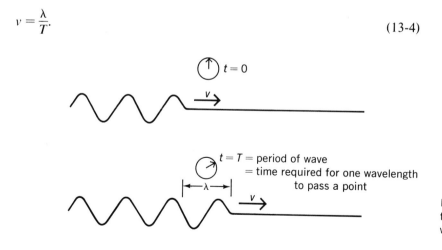

$t = T$ = period of wave
      = time required for one wavelength
         to pass a point

Fig. 13-13. The period of a wave is the time required for one complete wave to pass by a given point.

The hertz   The unit of frequency is the *cycle/s*, or the *hertz* (Hz) after Heinrich Hertz, one of the pioneers in the study of electromagnetic waves. Multiples of the cycle/s and of the Hz are used for high frequencies:

$$1 \text{ kilocycle/s (kc/s)} = 1 \text{ kilohertz (kHz)} = 10^3 \text{ cycles/s}$$

$$1 \text{ megacycle/s (mc/s)} = 1 \text{ megahertz (MHz)} = 10^6 \text{ cycles/s}$$

Thus a frequency of 50 MHz is equal to

$$50 \text{ MHz} \times 10^6 \, \frac{\text{Hz}}{\text{MHz}} = 5 \times 10^7 \text{ Hz} = 5 \times 10^7 \, \frac{\text{cycles}}{\text{s}}.$$

Problem.   One day at a certain place on the ocean the distance between adjacent wave crests is 160 ft and a crest passes by every 4.5 s. Find the frequency and velocity of the waves.

Solution.   The frequency of the waves is

$$f = \frac{1}{T} = \frac{1}{4.5 \text{ s}} = 0.22 \text{ Hz},$$

and their velocity is

$$v = \frac{\lambda}{T} = \frac{160 \text{ ft}}{4.5 \text{ s}} = 36 \text{ ft/s}.$$

Amplitude   The *amplitude A* of a wave refers to the maximum displacement from their normal positions of the particles which oscillate back and forth as the wave travels by (Fig. 13-14). The amplitude of a wave in a stretched string is the height of the crests above the original line of the string (or the depth of the troughs below the original line).

Energy   The correspondence between harmonic and wave motion permits us to obtain an interesting result. As we found in Section 12-4, a particle of mass m that undergoes simple harmonic motion of frequency f and amplitude A

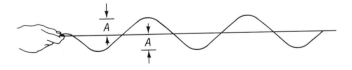

Fig. 13-14. The quantity A is the amplitude of the wave.

has a total energy of

$$E = 2\pi^2 m f^2 A^2.$$

This dependence of energy on f^2 and on A^2 is also true for mechanical waves of all kinds. (A mechanical wave is one that involves moving matter, in contrast to, say, an electromagnetic wave.) Waves in a string are an example: the energy per unit length in a string due to waves of frequency f and amplitude A is $2\pi^2\mu f^2 A^2$, where μ is the mass per unit length of the string. Energy considerations in sound and light waves will be treated later when these types of wave phenomena are examined.

13-5 Types of Waves

Waves in a stretched string are *transverse waves* since the individual segments of the string vibrate perpendicular to the direction in which the waves travel, that is, from side to side. *Longitudinal waves* occur when the individual particles of a medium vibrate back and forth in the direction in which the waves travel (Fig. 13-15). Longitudinal waves are easy to produce in a long coil spring; each portion of the spring is alternately compressed and extended as the waves pass by. Longitudinal waves, then, are essentially density fluctuations.

Transverse and longitudinal waves

Waves on the surface of a body of water (or other liquid) are a combination of longitudinal and transverse waves. If we were somehow to tag individual water molecules and follow them when a train of waves pass by, we would find that their paths are like those shown in Fig. 13-16. Each molecule describes a circular orbit with a period equal to the period of the wave, and does not undergo a permanent displacement. Because successive molecules reach the tops of their orbits at slightly different times, the water surface takes the form of a series of crests and troughs. At the crest of a wave the molecules move in the direction the wave is traveling, while in a trough the molecules are moving in the opposite direction. The passage of a wave

Water waves

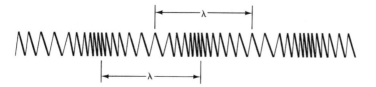

Fig. 13-15. Longitudinal waves in a coil spring.

Fig. 13-16. Water molecules move in circular orbits about their original positions when a typical deep-water wave passes by. At the crest of a wave the molecules are moving in the direction the wave is traveling, while in the trough the molecules are moving in the opposite direction. There is no net motion of water involved in the motion of such a wave.

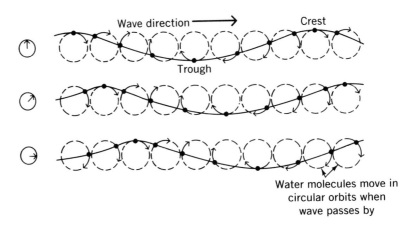

Water molecules move in circular orbits when wave passes by

across the surface of a body of water, like the passage of a wave through any medium, involves the motion of a pattern: energy is transported by virtue of the changing pattern, but there is no transport of matter.

Problem. The water waves in the sample problem of the previous section have an amplitude of 4 ft. Find the velocity of an individual molecule of water on the surface.

Solution. The water molecules on the surface are moving in circles of radius 4 ft, so that as each wave passes by, the molecules travel a distance s equal to the circumference $2\pi r$ of the circle (Fig. 13-17). Hence

$$s = 2\pi r = 2\pi \times 4\text{ ft} = 25\text{ ft}.$$

Each wave takes $T = 4.5$ s to go past a given point, which means that the molecules must cover the 25-ft circumference of their orbits in 4.5 s. The velocity V of each molecule is therefore

$$V = \frac{s}{T} = \frac{25\text{ ft}}{4.5\text{ s}} = 5.6\ \frac{\text{ft}}{\text{s}}.$$

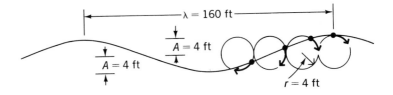

Fig. 13-17.

The velocity of the *wave*, however, is 36 ft/s, over six times greater. Thus the motion of the pattern that constitutes a wave in a medium can be much more rapid than the motions of the individual particles of the medium, and energy can be transported by wave motion faster than might be possible through the net transport of matter.

13-6 Standing Waves

When we pluck a string whose ends are fixed in place, the string starts to vibrate in one or more loops (Fig. 13-18). These *standing waves* may be thought of as the result of waves that travel down the string in both directions, are reflected at the ends, proceed across to the opposite ends and are again reflected, and so on.

In order to understand how standing waves come into being, we must transfer to the case of waves our knowledge of how pulses in a string are reflected and of what happens when two pulses traveling in opposite directions meet. When a pulse in a string is reflected at a rigid support, the reflected pulse is inverted; a similar inversion occurs for waves which means that, although their waveform and wavelength stay the same, the wave train is in effect shifted by $\frac{1}{2}\lambda$ so that a crest arriving at the end of the string is reflected as a trough and vice versa.

Now let us look into the manner in which two waves in the same string interact. Applied to waves, the principle of superposition states that

When two or more waves of the same nature travel past a point at the same time, the displacement at that point is the sum of the instantaneous displacements of the individual waves.

The principle of superposition holds for all types of waves, including waves in a stretched string, sound waves, water waves, and light waves.

What the principle of superposition signifies is that every wave train proceeds independently of any others that may also be present. Should two waves with the same wavelength come together in such a way that crest meets crest and trough meets trough, the resulting composite wave will have an amplitude greater than that of either of the original waves. When this occurs the waves are said to *interfere constructively* with each other. Should the waves come together in such a way that crest meets trough and trough meets crest, the composite wave will have an amplitude less than that of the larger of the original waves. When this occurs the waves are said to *interfere destructively* with each other (Fig. 13-19).

Let us apply these ideas to a stretched string whose ends are fixed in

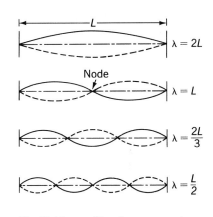

Fig. 13-18.   Standing waves in a stretched string.

Principle of superposition for waves

Constructive interference

Destructive interference

place. When the string is plucked, waves move back and forth between its ends, undergoing an inversion each time they are reflected, and they interfere with each other in such a way that destructive interference always occurs at the ends. If the string is L long, destructive interference will occur at the ends when the waves have wavelengths of $\lambda = 2L, L, 2L/3, L/2$, and so on (Fig. 13-18). In the case of the shorter wavelengths, other points, called *nodes*, are present at intermediate positions where no motion of the string takes place.

Figure 13-20 shows how a standing wave pattern comes into being: the dotted curve represents a wave moving to the right, and the dashed curve a wave of the same wavelength and amplitude moving to the left. The sum of these waves is found by adding together their displacements at each point on the string, and is shown as a solid line. The addition is performed in Fig.

Origin of standing wave

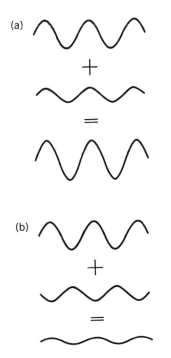

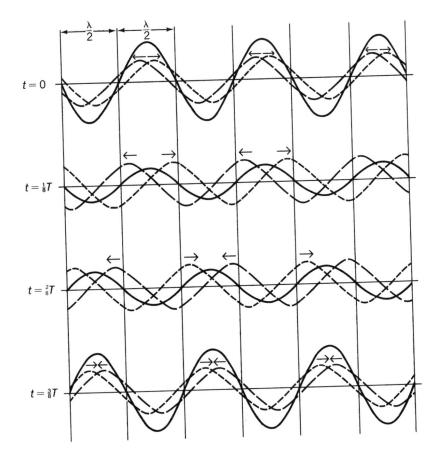

Fig. 13-20. (a) Constructive interference. (b) Destructive interference.

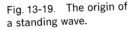

Fig. 13-19. The origin of a standing wave.

13-20 for four successive instants one-eighth of a period apart. We note that the dotted and dashed curves do not always have displacements equal to zero at the ends of the string. There is no contradiction here since it is their resultant, the solid curve, that corresponds to reality, and this curve exhibits the proper behavior.

Any type of wave can occur as a standing wave between suitable reflectors. The vibrating air columns in wind instruments and organ pipes—and in the throat, mouth, and nose of a person speaking—are standing sound waves, for instance. Standing light waves play an important role in the operation of lasers, as we shall find later. And standing waves of a rather remarkable kind are the key to understanding the structure of the atom.

13-7 Resonance

The condition that nodes occur at each end of the string restricts the possible wavelengths of standing waves to

$$\lambda = \frac{2L}{n}, \qquad n = 1, 2, 3, \ldots \tag{13-5}$$

The lowest possible frequency of oscillation f_1 of a stretched string corresponds to the longest wavelength, $\lambda = 2L$. Thus

$$f_1 = \frac{v}{\lambda} = \frac{v}{2L}. \tag{13-6}$$

Higher frequencies correspond to shorter wavelengths. From Eq. (13-5) we see that these higher frequencies can be represented in terms of f_1 by

$$f_n = nf_1, \qquad n = 2, 3, 4, \ldots \tag{13-7}$$

The frequency f_1 is called the *fundamental frequency* of the string, and the higher frequencies f_2, f_3, and so on, are called *overtones*.

Fundamental frequency and overtones

Since the wave velocity v is $\sqrt{T/(m/L)}$, we can express the fundamental frequency of a stretched string in terms of its length, linear density, and tension by the formula

$$f_1 = \frac{1}{2L} \sqrt{\frac{T}{m/L}}. \tag{13-8}$$

This formula is the basis for the design of stringed musical instruments such as pianos and violins. A short, light, taut string means a high fundamental frequency of vibration, while a long, heavy, slack string means a low fundamental frequency. The "tuning" of a stringed instrument involves changing the tensions in the various strings until their fundamental frequencies are correct.

Natural frequencies

The fundamental frequency and overtones of a stretched string are its natural frequencies of oscillation: if the string is plucked arbitrarily, one or more of these frequencies will be simultaneously excited. Eventually internal friction in the string will cause the various vibrations to die out. However, we can cause vibrations to persist by applying a periodic force to the string whose frequency is exactly the same as that of one of its natural frequencies. When this is done, the standing waves continue so long as the periodic force supplies energy to the string. If the energy provided exceeds that dissipated by internal friction, the amplitude of the standing wave will increase until the string may rupture. (A column of soldiers can destroy a flimsy bridge by marching across it in step with one of its natural frequencies, although the bridge may be capable of safely holding the static load of the soldiers.) This phenomenon is called *resonance*.

Resonance

When periodic impulses are given to a string at frequencies other than those of its fundamental frequency and overtones, hardly any response occurs: the situation then is like pushing a child's swing at a frequency different from its natural one, which produces oscillations of negligible amplitude. All rigid structures possess characteristic natural frequencies of oscillation even though their vibrations may be more complex in character than those of a stretched string, and these vibrations can be excited by a stimulus of the proper frequency. The traditional example of a goblet shattering when a violin is played with just the right frequency is an illustration of resonance.

13-8 Sound

Sound waves are longitudinal and consist of pressure fluctuations. The air (or other medium) in the path of a sound wave becomes alternately denser and rarer; the resulting changes in pressure cause our eardrums to vibrate with the same frequency, which produces the physiological sensation of sound.

Most sounds are produced by vibrating objects. An example is the diaphragm of a loudspeaker (Fig. 13-21). When it moves outward, it pushes the air molecules directly in front of it closer together to form a region of high pressure that spreads out in front of the loudspeaker. The diaphragm then moves backward, thereby expanding the volume available to nearby

Fig. 13-21. Sound consists of longitudinal waves, representing condensations and rarefactions in the air in its path.

air molecules. Air molecules now flow toward the diaphragm, and consequently a region of low pressure spreads out directly behind the high-pressure region. The continued vibrations of the diaphragm thus send out successive layers of condensation and rarefaction.

The velocity of sound in air at sea level and 0°C is 331 m/s, which is 1086 ft/s. This velocity increases with temperature because the random velocities of air molecules increase with temperature and so make the passage of pressure fluctuations more rapid.

The musical note A, whose frequency is 440 Hz, represents a wavelength λ in air of

$$\lambda = \frac{v}{f} = \frac{331 \text{ m/s}}{440 \text{ Hz}} = 0.75 \text{ m},$$

which is about 30 in. A normal ear responds to sound waves with frequencies from about 20 Hz to about 20,000 Hz, which correspond to wavelengths from 17 m (54 ft) to 17 mm (0.65 in.). Sound waves whose frequencies are above 20,000 Hz are called *ultrasonic* and can be detected by appropriate electromechanical devices (such as microphones). Ultrasonic waves are audible to many animals whose hearing organs are smaller than those of human beings and so are better suited to high frequencies.

Ultrasonic waves

Sound waves are transmitted by solids, liquids, and gases. In general, the stiffer the material, the faster the waves travel, which is reasonable when we reflect that stiffness implies particles tightly coupled together and therefore more immediately responsive to one another's motions. This notion is borne out by a detailed analysis, which shows that the velocity of sound in a fluid medium is given by

$$v = \sqrt{\frac{B}{d}},$$

where B is the bulk modulus of the fluid and d is its density. In the case of a solid rod, the velocity of sound is

$$v = \sqrt{\frac{Y}{d}},$$

where Y is the Young's modulus of the material. Table 13-1 lists the velocity of sound in various materials.

The wavelength of the musical note A in sea water is

$$\lambda = \frac{v}{f} = \frac{1531 \text{ m/s}}{440 \text{ Hz}} = 3.5 \text{ m},$$

Table 13-1. The velocity of sound.

MEDIUM	VELOCITY OF SOUND	
	m/s	ft/s
Gases (0°C):		
Air	331	1086
Chlorine	206	676
Helium	965	3165
Liquids (25°C):		
Ethyl alcohol	1207	3960
Water, pure	1498	4913
Water, sea	1531	5022
Solid rods		
Copper	3800	12,500
Glass, pyrex	5170	16,960
Lead	1200	4000
Steel	5200	17,060

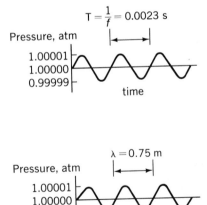

$$T = \frac{1}{f} = 0.0023 \text{ s}$$

$\lambda = 0.75$ m

Fig. 13-22. The pressure variations that constitute the musical note A in air. The amplitude is approximately that which a singer would produce.

which is nearly five times the wavelength of the same note in air. When a sound wave produced in one medium enters another in which its velocity is different, the frequency of the wave remains the same while the wavelength changes. This is in accord with the behavior of pulses that pass from one stretched string to another, as discussed earlier.

Because sound waves in air consist of pressure variations, a graph of pressure versus time at a certain place (or of pressure versus distance at a certain time) for a sound wave that contains a single frequency is sinusoidal in shape. A person singing the musical note A produces 440 Hz waves whose amplitude might be 1 N/m², which is 0.001% of sea-level atmospheric pressure. The pressure in the path of such waves increases to 1 N/m² above normal and decreases to 1 N/m² below normal 440 times per second. Figure 13-22 contains graphs showing the pressure variations of such waves.

13-9 Beats

Interference can occur in longitudinal as well as in transverse waves. A striking example of interference can be demonstrated in sound waves. If two tuning forks (or other sources of single-frequency sound waves) whose frequencies are slightly different are struck at the same time, the sound that we hear fluctuates in intensity. At one instant we hear a loud tone, then virtual silence, then the loud tone again, then virtual silence, and so on. The origin of this behavior is shown schematically in Fig. 13-23; the loud tones occur when the waves from the two forks interfere constructively, thus reinforcing one another, and the quiet periods occur when the waves interfere destructively, thus partially or wholly canceling one another out. These regular pulsations are called *beats*.

The sound we hear when beats occur has a frequency that is the average of the two original frequencies, and the number of beats per second equals the difference between the two original frequencies. If tuning forks are used whose frequencies are, say, 440 and 444 Hz, there will be a 442 Hz tone that rises and falls four times per second.

Beats are hard to distinguish when the two frequencies differ by more than perhaps 10 Hz. However, when the frequencies are very far apart, the beats may form a "difference tone." Thus two sound waves whose fre-

The interference of sound waves causes beats

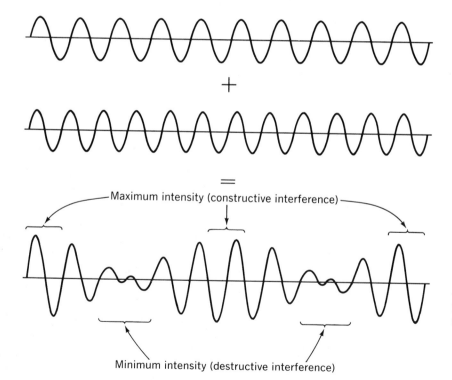

Maximum intensity (constructive interference)

Minimum intensity (destructive interference)

Fig. 13-23.   The origin of beats in sound waves.

quencies are 9000 and 10,000 Hz can combine to yield a sound of frequency 1000 Hz in addition to one of frequency 9500 Hz.

13-10 Doppler Effect

If we stand beside a road when a police car goes by with its siren blowing, we cannot help but notice that the pitch of the siren drops suddenly as the car passes by. A careful study would reveal that the pitch of the siren as the car approaches is *higher* than the pitch when the car is stationary, and the pitch when the car recedes is *lower* than its normal one. We also observe a change of pitch if the siren is at rest and we go past it in a rapidly moving car: we find a higher pitch than usual as we approach the siren, and a lower one as we recede. The change in frequency of a sound brought about by relative motion between source and listener is called the *Doppler effect*.

Origin of Doppler effect

The origin of the Doppler effect is straightforward. As a moving source emits sound waves, it is tending to overtake those traveling in the same direction (Fig. 13-24a). Hence the distance between successive waves is

(a)

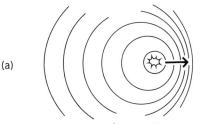

Moving source   Stationary listener

(b)

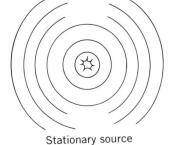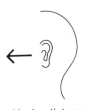

Stationary source   Moving listener

Fig. 13-24.   The Doppler effect occurs when there is relative motion between a source of sound and a listener.

smaller than usual; since this distance is the wavelength of the sound, the corresponding frequency is higher than usual. At the same time the source is moving away from those of its waves that travel in the opposite direction, increasing the distance between successive waves behind it and thereby reducing their frequency.

If the source is stationary, a listener moving toward it intercepts more sound waves per unit time than if he were at rest, and accordingly he hears a higher frequency (Fig. 13-24b). When the observer moves away from the source, fewer of the waves catch up with him per unit time, and he hears a lower frequency.

The relationship between the frequency f_L the listener hears and the frequency f_S produced by the source is

$$f_L = f_S \left(\frac{v + v_L}{v - v_S} \right). \qquad\qquad \textit{Doppler effect in sound} \quad (13\text{-}9)$$

Here v is the velocity of sound in air, v_L is the velocity of the listener (reckoned as $+$ if he moves toward the source, as $-$ if he moves away from the source), and v_S is the velocity of the source (reckoned as $+$ if it moves toward the listener, as $-$ if it moves away from the listener). If the listener is stationary, $v_L = 0$, while if the source is stationary, $v_S = 0$.

Problem.   The frequency of a train's whistle is 1000 Hz. (a) The train is approaching a stationary man at 40 m/s. What frequency does the man hear? (b) The train is stationary and the man is driving toward it in a car whose velocity is 40 m/s. What frequency does the man hear now?

Solution.   (a) Here $f_S = 1000$ Hz, $v = 331$ m/s, $v_S = 40$ m/s, and $v_L = 0$. Hence the apparent frequency of the whistle is

$$f_L = f_S \left(\frac{v + v_L}{v - v_S} \right) = 1000 \text{ Hz} \times \frac{331 \text{ m/s}}{(331 - 40) \text{ m/s}}$$

$$= 1137 \text{ Hz.}$$

(b) Again $f_S = 1000$ Hz and $v = 331$ m/s, but now $v_S = 0$ and $v_L = 40$ m/s. The apparent frequency of the whistle is therefore

$$f_L = f_S \left(\frac{v + v_L}{v - v_S} \right) = 1000 \text{ Hz} \times \frac{(331 + 40) \text{ m/s}}{331 \text{ m/s}}$$

$$= 1121 \text{ Hz.}$$

The Doppler effect is an important tool of the astonomer, who is able to determine the velocity of approach or recession of stars and galaxies from

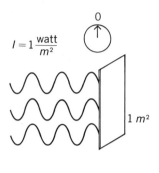

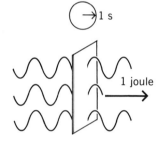

$$\text{Wave intensity} = \frac{\text{Energy/s}}{\text{area}}$$

Fig. 13-25.   The intensity of a wave is a measure of the rate at which it transports energy.

shifts in the characteristic frequencies of the light they emit. (These characteristic frequencies are discussed in Chapter 29.) This is the method by which the apparent expansion of the universe was detected. Throughout the sky, characteristic frequencies in the light from distant galaxies are lower than normal, a phenomenon called the "red shift" because red light has the lowest frequencies. The magnitude of the frequency change increases with distance from the earth, which suggests that the entire universe is expanding so that all the objects in it recede from one another.

The Doppler effect in light differs from that in sound because light, which does not depend upon a material medium for its transmission, has the same relative velocity c to all observers regardless of their state of motion (Chapter 27). The Doppler effect in light obeys the formula

$$f = f_S \sqrt{\frac{1 + v/c}{1 - v/c}} \qquad \textit{Doppler effect in light} \quad (13\text{-}10)$$

where f is the observed frequency, f_S is the frequency of the source, and v is the relative velocity between source and observer. If source and observer are approaching each other, v is reckoned as $+$, and if they are receding from each other, v is reckoned as $-$. In vacuum, the velocity of light is $c = 3.00 \times 10^8$ m/s; its value in air is very close to this. Unlike the case of sound, the Doppler effect in light cannot be used to distinguish between motion of a source and motion of an observer.

13-11 Sound Intensity and the Ear

The rate at which a wave of any kind transports energy per unit cross-sectional area is called its *intensity I*. The unit of intensity is the watt/m². One joule of energy per second flows through a 1-m² surface perpendicular to the path of a wave whose intensity is 1 W/m² (Fig. 13-25). The minimum intensity a sound wave must have in order to be audible is about 10^{-12} W/m²; at the other extreme, sound waves whose intensities exceed about 1 W/m² damage the ear.

Because sound waves spread out as they move away from their source, their intensity decreases with distance. Let us consider the sound waves from a source whose power output is P; that is, P joules of energy flow from the source per second. At the distance r from the source, the total power P is distributed over the $4\pi r^2$ area of a sphere of radius r. Hence the intensity I of the sound at this distance is

$$I = \frac{P}{4\pi r^2}.$$

The sound intensity is inversely proportional to the square of the distance from the source. If we go from 10 ft away from a source to 20 ft away, the intensity drops to $\frac{1}{4}$ its former value. This inverse-square law holds for the intensity of all waves that spread out freely in three dimensions—the intensity of the light from a lamp also varies as $1/r^2$, for example. The purpose of the concave mirror in a searchlight is to avoid the $1/r^2$ decrease in intensity by concentrating the light waves from a lamp into as nearly parallel a beam as possible.

The human ear does not respond linearly to sound intensity; doubling the intensity of a particular sound produces the sensation of a somewhat louder sound, but one that is far less than twice as loud. For this reason the scale customarily used to measure the intensity level of a sound is logarithmic, which is a reasonable approximation of the actual response of the human ear. The unit is the *decibel* (dB), with a sound that is barely audible ($I_0 = 10^{-12}$ W/m^2) being assigned a value of 0 dB. A 10-dB sound is, by definition, 10 times more intense than a 0-dB sound; a 20-dB sound is 10^2 (or 100) times more intense; a 30-dB sound is 10^3 (or 1000) times more intense; and so on. Formally, the intensity level I(dB) of a sound wave whose intensity in W/m^2 is I is given by

$$I \text{ (dB)} = 10 \log \frac{I}{I_0}. \tag{13-11}$$

(Logarithms are discussed in Appendix B-5.)

Ordinary conversation is usually about 60 dB, which is a sound intensity a million times greater than the faintest sound that can be heard. City traffic noise is about 80 dB, a rock band using amplifiers produces an intensity of as much as 125 dB, and the noise of a jet airplane is about 140 dB at a distance of 100 ft (Fig. 13-26). An extended exposure to sound intensities of over 85 dB usually leads to permanent hearing damage.

Problem. The minimum sound intensity level needed for reasonable audibility is 20 dB. If a certain person speaking normally produces an intensity level of 40 dB at a distance of 1 m, what is the maximum distance at which he can be heard?

Solution. A difference of 20 dB is equivalent to an intensity ratio of $10^2 = 100$. Since $I_2/I_1 = r_1^2/r_2^2$,

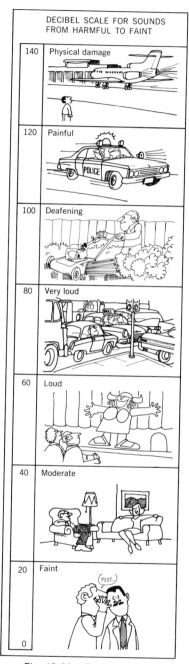

DECIBEL SCALE FOR SOUNDS FROM HARMFUL TO FAINT

140 Physical damage

120 Painful

100 Deafening

80 Very loud

60 Loud

40 Moderate

20 Faint

Fig. 13-26. Decibel scale.

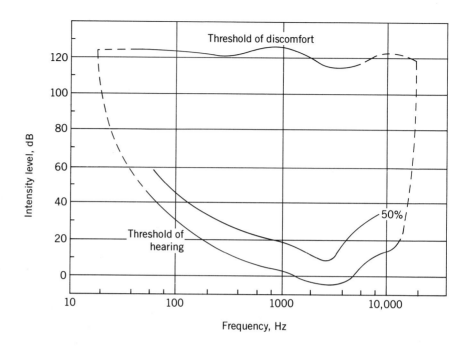

Figure 13-27. The response of the ear to sound varies with frequency. Only 1 percent of the U.S. population can hear sounds with intensity levels that fall below the lower curve; 50 percent can hear sounds with intensity levels that fall below the curve above it. Hearing acuity decreases with age.

$$r_2 = r_1 \sqrt{\frac{I_1}{I_2}} = 1 \text{ m} \times \sqrt{100} = 10 \text{ m}.$$

The ear is not equally sensitive to sounds of different frequencies. Maximum response occurs for sounds between 3000 and 4000 Hz, when the threshold for hearing is somewhat less than 0 dB. A 100-Hz sound, however, must have an intensity level of at least 40 dB to be heard. Sounds whose frequencies are below about 20 Hz *(infrasound)* and above about 20,000 Hz *(ultrasound)* are inaudible to almost everybody regardless of intensity. The intensity level at which a sound produces a feeling of discomfort in the ear is relatively constant at approximately 120 dB for all frequencies. Figure 13-27 shows how the thresholds of hearing and of discomfort vary with frequency. Hearing deteriorates with age, most notably at the high-frequency end of the spectrum. A typical person 60 years old has a threshold of hearing about 10 dB higher than the lowest curve of Fig. 13-27 for 2000-Hz tones, nearly 30 dB higher for 8000-Hz tones, and nearly 70 dB higher for 12,000-Hz tones.

Structure of the ear The structure of the human ear is shown in Fig. 13-28. The *eardrum* vibrates when sound waves reach it from the ear canal. Inside the eardrum is the *middle ear*, a small air-filled cavity that contains three linked bones called the *hammer*, the *anvil*, and the *stirrup* because of their respective shapes. The *Eustachian tube*, which opens when a person swallows, permits the middle ear to stay at atmospheric pressure. The vibrations of the eardrum pass in succession from the hammer to the anvil to the stirrup and finally to the membrane over the *oval window* between the middle and inner ears. The liquid-filled *inner ear* consists of the *semicircular canals*, whose function is to act as a reference system in providing a sense of balance, and the *cochlea*, a snail-shaped organ in which pressure waves in the inner-ear liquid excite nerve impulses that pass along the *auditory nerve* to the brain.

The ear amplifies sound waves in three ways. The ear canal itself acts as a resonant cavity—that is, standing waves are set up in it—for sound frequencies in the neighborhood of 3000-4000 Hz, and consequently the pressure fluctuations at the ear drum may be twice as great as those outside the ear. This is the reason why the ear is most sensitive to such frequencies. Second, the linked bones in the middle ear have a mechanical advantage of two to three, which correspondingly increases the forces exerted on the oval-window membrane over those exerted on the eardrum. Third, the area of the eardrum is typically 20 times greater than that of the oval window, so that (by analogy with the hydraulic press) there is a further mechanical advantage of this amount. The total amplification of pressure changes may thus be over 100, and the amplification of changes in sound intensity may be over 10,000.

The actual mechanism by which the ear discriminates among sound frequencies depends upon the physical properties of the *basilar membrane* in the cochlea. This membrane varies in thickness and tension along its length, with a different portion resonating at each frequency. Thousands of tiny fibers distributed along the basilar membrane transmit disturbances in the membrane to nerve endings, and the brain interprets the impulses it receives as sound of a certain pitch according to the position of the stimulated nerve ending. The amplitude of the disturbance of the basilar membrane is registered as the loudness of the sound that produced it.

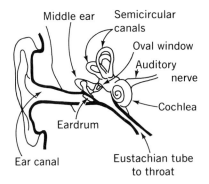

Fig. 13-27. The ear.

Special Topic

Fourier Series

In this chapter only sinusoidal waves were considered. Nevertheless, in a medium whose properties do not vary with wave frequency, everything that is true for a sinusoidal wave is also true for all other periodic waves. An interesting and important theorem by Fourier shows why this should be so.

The displacement y of a particle in a medium when a sinusoidal wave passes by varies with time in the same way as does an object undergoing simple harmonic motion. If the amplitude of the wave is A, in general

$$y = A \sin (\omega t + \phi)$$

where $\omega = 2\pi f$ is the *angular frequency* of the wave. (We could equally well have used the cosine function.) The phase angle ϕ depends on the value of y when $t = 0$; for simplicity we will let $\phi = 0$, which means that $y = 0$ at $t = 0$ since $\sin 0 = 0$. What Fourier proved is that *any* periodic wave, regardless of its waveform, can be represented as a superposition of sinusoidal waves whose angular frequencies are ω, 2ω, 3ω, and so on. The amplitudes of the various component waves depend on the precise character of the composite wave, and there exists a mathematical procedure for finding these amplitudes.

To give an example, the *Fourier series* that is equivalent to a square wave of amplitude A and frequency ω is as follows:

$$y = A \sin \omega t + \frac{A}{3} \sin 3\omega t + \frac{A}{5} \sin 5\omega t + \ldots$$

Figure 13-29 shows how well just the first three terms of this series reproduce the square wave; the more terms are included, the more exact the result. Even nonperiodic disturbances that propagate in a medium ("pulses") can be represented as a superposition of sinusoidal waves, but in such a case

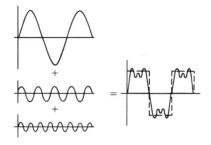

Fig. 13-29. Fourier synthesis of a square wave.

the successive frequencies must be infinitesimally close to each other instead of being multiples of ω, and the integral

$$y = \int_0^\infty A(\omega) \sin \omega t \, d\omega$$

is needed instead of a summation. In the latter formula $A(\omega)$ signifies that the amplitude of each component wave depends upon its frequency ω.

Important Terms

Wave motion is characterized by the propagation of a change in a medium. Waves transport energy from one place to another.

The **frequency** of a series of periodic waves is the number of waves that pass a particular point per unit time, while their **wavelength** is the distance between adjacent crests or troughs. The **period** is the time required for one complete wave to pass a particular point. The **amplitude** of a wave is the maximum displacement of a particle of the medium on either side of its normal position when the wave passes.

Longitudinal waves occur when the individual particles of a medium vibrate back and forth in the direction in which the waves travel. **Transverse waves** occur when the individual particles of a medium vibrate from side to side perpendicular to the direction in which the waves travel. The vibrations of a stretched string are transverse waves.

The **principle of superposition** states that when two or more waves of the same nature travel past a given point at the same time, the amplitude at the point is the sum of the amplitudes of the individual waves. The interaction of different wave trains is called **interference**: **constructive interference** occurs when the resulting composite wave has an amplitude greater than that of either of the original waves, and **destructive interference** occurs when the resulting composite wave has an amplitude less than that of either of the original waves.

Resonance occurs when periodic impulses are given to an object at a frequency equal to one of its natural frequencies of oscillation.

Sound is a longitudinal wave phenomenon which results in periodic pressure variations. Sound intensity in air is measured in *decibels*.

The **Doppler effect** refers to the change in frequency of a wave when there is relative motion between its source and an observer.

Important Formulas

Waves in a string:

$$v = \sqrt{\frac{T}{m/L}}$$

Wave motion:

$$v = f\lambda = \frac{\lambda}{T}$$

Sound intensity level:

$$I(\text{dB}) = 10 \log \frac{I}{I_0}$$

Multiple Choice

1. The velocity of waves in a stretched string depends upon
   a. the tension in the string.
   b. the amplitude of the waves.
   c. the wavelength of the waves.
   d. the acceleration of gravity.

2. The higher the frequency of a wave,
   a. the smaller its velocity.
   b. the shorter its wavelength.
   c. the greater its amplitude.
   d. the longer its period.

3. In a transverse wave, the individual particles of the medium
   a. move in circles.
   b. move in ellipses.
   c. move parallel to the direction of travel.
   d. move perpendicular to the direction of travel.

4. Of the following properties of a wave, the one that is independent of the others is its
   a. amplitude.
   b. velocity.
   c. wavelength.
   d. frequency.

5. Waves transmit from one place to another
   a. mass.
   b. amplitude.
   c. wavelength.
   d. energy.

6. Standing waves in a stretched string occur because of
   a. interference.
   b. Doppler effect.
   c. beats.
   d. friction.

7. Two waves meet at a time when one has the instantaneous amplitude A and the other has the instantaneous amplitude B. Their combined amplitude at this time is
   a. $A + B$.
   b. $A - B$.
   c. between $A + B$ and $A - B$.
   d. indeterminate.

8. Sound waves do not travel through
   a. solids.
   b. liquids.
   c. gases.
   d. a vacuum.

9. The pitch of a sound wave is related to its
   a. frequency.
   b. amplitude.
   c. velocity.
   d. beats.

10. A pure musical tone causes a thin wooden panel to vibrate. This is an example of
    a. an overtone.
    b. harmonics.
    c. resonance.
    d. interference.

11. The amplitude of a sound wave determines its
    a. pitch.
    b. loudness.
    c. overtones.
    d. resonance.

12. Beats are the result of
    a. diffraction.
    b. constructive interference.
    c. destructive interference.
    d. both constructive and destructive interference.

13. When a sound wave goes from air into water, the quantity that remains unchanged is its
    a. velocity.
    b. amplitude.
    c. frequency.
    d. wavelength.

14. Sound waves are
    a. longitudinal.
    b. transverse.
    c. partly longitudinal and partly transverse.
    d. sometimes longitudinal and sometimes transverse.

15. Sound waves whose frequency is 300 Hz have a velocity relative to sound waves in the same medium whose frequency is 600 Hz that is
    a. half as great.
    b. the same.
    c. twice as great.
    d. four times as great.

16. Sound waves whose frequency is 300 Hz have a wavelength relative to sound waves in the same medium whose frequency is 600 Hz that is
    a. half as great.
    b. the same.
    c. twice as great.
    d. four times as great.

17. Radio amateurs are permitted to communicate on the "10-meter band." What frequency of radio waves corresponds to a wavelength of 10 meters? The speed of radio waves is 3×10^8 m/s.
    a. 3.3×10^{-8} Hz
    b. 3.0×10^7 Hz
    c. 3.3×10^7 Hz
    d. 3.0×10^9 Hz

18. A certain radio station broadcasts at a frequency of 660 kHz. The wavelength of the above waves is
    a. 2.2×10^{-3} m.
    b. 4.55×10^2 m.
    c. 4.55×10^3 m.
    d. 1.98×10^{14} m.

19. A boat at anchor is rocked by waves whose crests are 100 ft apart and whose velocity is 25 ft/s. These waves reach the boat once every
    a. 2500 s.
    b. 75 s.
    c. 4 s.
    d. 0.25 s.

20. Two tuning forks of frequencies 310 and 316 Hz vibrate simultaneously. The number of times the resulting sound pulsates per second is
    a. 0.
    b. 6.
    c. 313.
    d. 626.

21. The pulsating sound of Question 20 is an example of
    a. resonance.
    b. Doppler effect.
    c. pitch.
    d. beats.

22. How many times more intense is a 90-dB sound than a 40-dB sound?
    a. 5
    b. 50
    c. 500
    d. 10^5

Exercises

1. A pulse sent down a long string eventually dies away and disappears. What happens to its energy?

2. Show that the formula $\sqrt{T/(m/L)}$ is the only combination of tension T and linear density m/L that has the dimensions of velocity.

3. The amplitude of a wave is doubled. If nothing else is changed, how is the flow of energy affected?

4. A wave of frequency f and wavelength λ passes from a medium in which its velocity is v to another medium in which its velocity is $2v$. What are the frequency and wavelength of the wave in the second medium? [*Hint:* Consider what happens at the interface between the two media when the wave passes through to decide whether the frequency or the wavelength or both change.]

5. How can constructive and destructive interference be reconciled with the principle of energy conservation?

6. In general, in what state of matter does sound travel fastest? Why?

7. What is the wavelength in ft of sound waves of frequency 8,000 Hz in (a) air, (b) sea water, (c) steel?

8. A violin string is set in vibration at a frequency of 440 Hz. How many vibrations does it make while its sound travels 200 m in air?

9. A certain groove in a phonograph record moves past the needle at a speed of 0.3 m/s. If the wiggles in the groove are 0.1 mm apart, what is the frequency of the sound that is produced?

10. A certain groove in a phonograph record moves past the needle at a speed of 0.4 m/s. The sound produced has a frequency of 3000 Hz. What is the wavelength of the wiggles in the groove?

11. Water waves are observed approaching a lighthouse at a speed of 18 ft/s. There is a distance of 20 ft between adjacent crests. (a) What is the frequency of the waves? (b) What is their period?

12. The stainless steel forestay of a racing sailboat is 20 m long and 1 cm in diameter, and its mass is 10 kg. In order to determine its tension, it is struck with a hammer at the lower end and the return of the pulse is timed. If the time interval is 0.2 sec, what is the tension in the stay?

13. A workman strikes a steel rail with a hammer, and the sound reaches an observer 0.5 km away both through the air and through the rail. How much time separates the two sounds?

14. A mine explodes at sea, and there is an interval of 5 s between the arrival of the sound through the water and its arrival through the air at a nearby ship. How many feet away is the ship?

15. A tuning fork vibrating at 440 Hz is placed in distilled water. (a) What are the frequency and wavelength in feet of the waves produced within the water? (b) What are the frequency and wavelength in feet of the waves which are produced in the surrounding air when the water waves reach the surface?

16. A man has two tuning forks, one marked "256 Hz" and the other of unknown frequency. He strikes them simultaneously, and hears 10 beats per second. "Aha," he says, "the other tuning fork has a frequency of 266 Hz." What is wrong with his conclusion?

17. The faintest audible sound has an amplitude of about 2×10^{-5} N/m^2. What percentage of sea level atmospheric pressure is this?

18. A fire engine has a siren whose frequency is 500 Hz. What frequency is heard by a stationary observer when the engine moves toward him at 40 ft/s? When it moves away from him at 40 ft/s?

19. A latecomer to a concert hurries down the aisle toward his seat so fast that the note middle C (256 Hz) appears 1 Hz higher in frequency. How fast is he going in ft/s?

20. The amplitude and wavelength of the wiggles in the grooves of a phonograph record correspond to the same quantities in the sound waves they represent. Inspection of a certain record reveals grooves whose wiggles have the same amplitude but one has a wavelength three times shorter than the other. (a) If the audio system is linear, what will be the difference in the sound intensity the two sets of wiggles produce? (b) How will the difference in apparent loudness correspond to the difference in intensity?

Problems

1. How many times more intense is a 60-db sound than a 50-db sound? Than a 40-db sound? Than a 20-db sound?

2. What is the equivalent in dB of a sound whose intensity is 5×10^{-6} W/m^2?

3. A siren produces a 120-dB sound. What is its intensity in W/m^2?

4. One of the strings of a violin has an effective length of 1 ft and weighs 0.05 oz. What tension should the string be under if its fundamental frequency is to be 440 Hz, the musical note A?

5. A stretched wire 1 m long has a fundamental frequency of 300 Hz. (a) What is the speed of the waves in the wire? (b) What are the frequencies of the first three overtones?

6. A steel wire 1 m long whose mass is 10 gm is under a tension of 400 N. (a) What is the wavelength of its fundamental mode of vibration? (b) What is the frequency of this mode? (c) What is the wavelength of the sound waves produced at 0°C when the string vibrates in this mode?

7. Two identical steel wires have fundamental frequencies of vibration of 400 Hz. The tension in one of the wires is increased by 2%, and both wires are plucked. How many beats per second occur?

8. Train A is heading east at the speed v_A and train B is heading west at the speed v_B on an adjacent track. The locomotive on train A is blowing its whistle continuously. A passenger on train B

observes the frequency of the sound from the whistle to be 307 Hz when the trains approach each other, 256 Hz when they are abreast, and 213 Hz when they recede from each other. From these figures find the speed of each train relative to the ground, taking the speed of sound as 1100 ft/s.

9. A motorist goes through a red light and, when he is arrested, claims that the color he actually saw was green ($\lambda = 5.4 \times 10^{-7}$ m) and not red ($\lambda = 6.2 \times 10^{-7}$ m) because of the Doppler effect. The judge accepts this explanation and instead fines him for speeding at the rate of $1.00 for each mi/hr he exceeded the speed limit of 50 mi/hr. What was his fine?

10. A moving reflector approaches a stationary source of sound of frequency f with the speed u.

A listener nearby hears both the original sound waves and the reflected sound waves. Obtain a formula for the number of beats per second the listener hears.

Answers to Multiple Choice

1. a	9. a	17. b
2. b	10. c	18. b
3. d	11. b	19. c
4. a	12. d	20. b
5. d	13. c	21. d
6. a	14. a	22. d
7. a	15. b	
8. d	16. c	

14

Temperature and Heat

Every body of matter contains internal energy in the form of the kinetic energies of its atoms or molecules. The higher the temperature of the body, the faster its constituent particles move, and the more internal energy it contains. Heat is, in a sense, internal energy in transit: when heat is added to a body, its temperature increases, and when heat is removed, its temperature decreases. In this chapter various aspects of the thermal behavior of matter are discussed from a purely empirical point of view, leaving their ultimate explanation in terms of the kinetic-molecular theory of matter for the next chapter.

14-1 Temperature

Temperature, like force, is a key concept in physics which requires an elaborate definition in order to be specified precisely although we have a clear idea of its meaning in terms of our sense impressions. We do not need such precision yet, and so we shall dodge the issue for the moment and consider temperature merely as that which is responsible for sensations of hot and cold.

There are a number of properties of matter that vary with temperature, and these can be used to construct *thermometers*, devices for measuring temperature. For example, when an object is heated sufficiently, it glows, at first a dull red, then bright red, and finally, at a high enough temperature, it becomes "white hot." By measuring the color of the light it gives off, we can accurately determine the temperature of an object. This method can only be used at rather high temperatures, however.

Of wider application is the fact that matter usually expands when its temperature is increased and contracts when its temperature is decreased. Concrete roads must be laid with regular gaps to allow for expansion in the summer; heated air above a radiator rises as it expands and becomes lighter than the surrounding air; a column of mercury in a glass tube changes length with a change in temperature. All three of these observations have resulted in practical thermometers (Fig. 14-1).

Fig. 14-1.   Three types of thermometer.

Thermometer

(a) Two strips of different metals that are joined together bend to one side with a change in temperature owing to different rates of expansion in the two metals, a fact employed in constructing household oven thermometers. The higher the temperature, the greater the deflection. When cooled, such a bimetallic strip bends in the opposite direction.

(c) Mercury (or colored alcohol) expands more when heated than glass does, and so the length of the liquid column in a liquid-in-glass thermometer is a measure of the temperature of the thermometer bulb.

(b) In a constant-volume gas thermometer, which is a very sensitive laboratory instrument, the height of the mercury column at the left is adjusted until the mercury column at the right just touches the gas bulb. The difference in heights of the two mercury columns is a measure of the pressure needed to maintain the gas in a fixed volume, and hence a measure of the temperature.

Before we can use any of these or other thermal properties of matter to construct a practical thermometer, we must begin by specifying a temperature scale and the method by which we shall calibrate the thermometer in terms of this scale. Water is a readily available liquid which freezes into a solid, ice, and vaporizes into a gas, steam, at definite temperatures at sea level atmospheric pressure. We can establish a temperature scale by defining the freezing point of water at 1 atm pressure (or, more exactly, the point at which a mixture of ice and water is in equilibrium, with exactly as much ice melting as water freezing) as 0° and the boiling point of water at 1 atm pressure (or, more exactly, the point at which a mixture of steam and

Celsius (centigrade) scale

water is in equilibrium) as 100°. This scale is called the *celsius* scale, and temperatures measured in it are written, for example, "40°C." In the United States, the celsius scale is sometimes called the *centigrade* scale.

To calibrate a thermometer, say an ordinary mercury thermometer, we first plunge it into a mixture of ice and water. When the mercury column has come to rest we mark the position of its top 0°C on the glass (Fig. 14-2). Then we plunge it into a mixture of steam and water, and when the mercury column has again come to rest, we mark the new position of the top of the mercury column 100°C. Finally we divide the interval between the 0°C and 100°C markings into 100 equal parts, each representing a change in temperature of 1°C, and extend the scale with divisions of the same length beyond

Fig. 14-2. Calibrating a thermometer on the celsius scale. A mixture of ice and water is, by definition, at 0°C, and a mixture of steam and water is, again by definition, at 100°C.

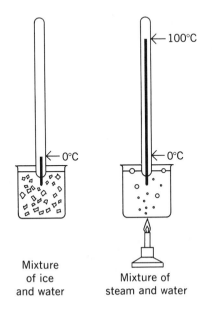

Mixture
of ice
and water

Mixture of
steam and water

0°C and 100°C as far as is convenient. In doing this we have, of course, assumed that changes in the length of the mercury column are always directly proportional to the changes in temperature that brought them about.

Although the celsius scale is used in most of the world, a different temperature scale called the *fahrenheit* scale is commonly used for non-scientific purposes in English-speaking countries. In the fahrenheit scale the freezing point of water is 32°F and the boiling point of water is 212°F (Fig. 14-3). This means that 180°F separates the freezing and boiling points of water, whereas 100°C separates them in the celsius scale. Therefore fahrenheit degrees are 100/180 or 5/9 as large as celsius degrees. We can convert temperatures from one scale to the other with the help of the formulas

$$T_F = \frac{9}{5} T_C + 32°,$$

$$T_C = \frac{5}{9} (T_F - 32°).$$

(14-1)

For instance, a fahrenheit temperature of 70°F, which is normal room temperature, on the celsius scale is

$$\tfrac{5}{9}(70° - 32°) = 21°C.$$

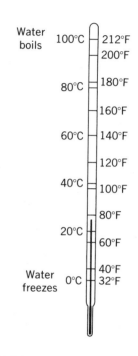

Fig. 14-3. The fahrenheit and celsius temperature scales.

14-2 Internal Energy and Heat

Every body of matter contains a certain amount of *internal energy* in addition to any kinetic or potential energy it may possess by virtue of its motion or position. This internal energy resides in the random translational motions of the atoms or molecules of which the body is composed, and in their rotations and vibrations as well. The total amount of internal energy a body contains depends upon its temperature, upon its composition, upon its mass, and upon its physical state (solid, liquid, or gas). However, the temperature of the body alone is what determines whether internal energy will br transferred from it to another body with which it is in contact or vice versa. A large block of ice at 0°C has far more internal energy than a cup of hot water, yet when the water is poured on the ice some of the ice melts and the water becomes cooler, which signifies that energy has passed from the water to the ice (Fig. 14-4).

Nature of internal energy

When the temperature of a body increases, it is customary to say that *heat* has been added to it; when the temperature of a body decreases, it is customary to say that heat has been removed from it. Thus we can think

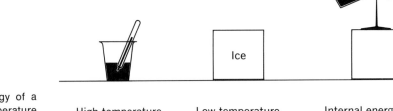

Internal energy

Ice

High temperature
but small mass,
hence little
internal energy

Low temperature
but large mass,
hence much
internal energy

Internal energy
always flows from
hot body to cold
body, regardless of
internal energy content

Fig. 14-4. The internal energy of a body depends upon its temperature and mass. However, the direction of internal energy flow depends only on the temperatures of the bodies involved.

of heat as internal energy in transit. In fact, we need not even know that heat is a form of energy in order to make an adequate working definition of heat:

Definition of heat

Heat is a quantity that causes an increase in the temperature of a body of matter to which it is added and a decrease in the temperature of a body of matter from which it is removed, provided that the matter does not change state during the process.

The latter part of the definition is required because changes of state (for instance, from ice to water or water to steam) involve the transfer of heat to or from a body without any change in temperature. We should keep in mind that heat transfer is not the only way to change the temperature of a body of matter. A body that has work done on it may become hotter as a result, and a body that does work on something else may become cooler.

The term heat remains in the vocabulary of physics partly because of convenience and partly because of tradition. Temperature, on the other hand, is a unique concept both in a macroscopic sense as an indicator of the direction of internal energy flow and (as we shall see later) in a microscopic sense as a measure of average molecular kinetic energy.

The kilocalorie

In the metric system the unit of heat, called the *kilocalorie* (abbreviated kcal), is that amount of heat required to raise the temperature of 1 kg of water through 1°C. Similarly, 1 kcal of heat must be removed from 1 kg of water to reduce its temperature by 1°C. Because this amount of heat actually varies slightly with temperature, the kilocalorie is formally defined as the amount of heat involved in changing the temperature of 1 kg of water from 14.5°C to 15.5°C; the difference is insignificant for most purposes, however.

The unit of heat in the British system is the *British thermal unit*,   The Btu
abbreviated Btu. One Btu is the amount of heat required to raise the temperature of 1 lb of water by 1°F; when 1 Btu of heat is removed from 1 lb of water, its temperature falls by 1°F. Thus

  1 kcal = heat transferred when the temperature
       of 1 kg of water changes by 1°C,

  1 Btu = heat transferred when the temperature
       of 1 lb of water changes by 1°F.

The kcal and Btu are related as follows:

  1 Btu = 0.252 kcal,

  1 kcal = 3.97 Btu.

We note that weight rather than mass is specified in the definition of the Btu. In practice this is an unimportant difference since what is being measured is the quantity of matter involved without regard to its dynamical properties. For simplicity we shall follow the custom of interpreting m in the equations of heat as mass in kilograms when metric units are employed and as weight in pounds when British units are employed.

Problem. How much heat is required to raise the temperature of 3 lb of water from 70°F to 120°F?

Solution. The temperature of the water must be raised by 50°F. Since 1 Btu of heat raises the temperature of 1 lb of water by 1°F, 50 Btu is required for each lb of water here. There are 3 lb of water in all, and so 3×50 Btu = 150 Btu of heat is required (Fig. 14-5).

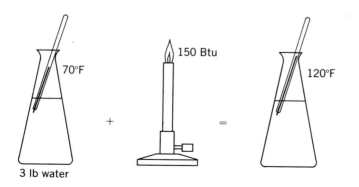

70°F   150 Btu   120°F

+   =

3 lb water   Fig. 14-5.

Problem.   A bathtub contains 70 kg of water at 26°C. Ten kg of water at 90°C is poured in (Fig. 14-6). What is the final temperature of the mixture?

Solution.   The final temperature T of the mixture will be more than 26°C and less than 90°C. The 70 kg of cold water will have gained heat, and the 10 kg of hot water will have lost heat. The heat gained by the cold water is

Heat gained = mass of water × change in temperature

$$= 70 \text{ kg} \times (T - 26°C)$$

$$= (70\,T - 1820) \text{ kcal.}$$

The heat lost by the hot water is

Heat lost = mass of water × change in temperature

$$= 10 \text{ kg} \times (90°C - T)$$

$$= (900 - 10\,T) \text{ kcal.}$$

If we neglect heat losses to the air and to the bathtub itself, the heat gained by the cold water must equal the heat lost by the hot water in order that energy be conserved. Hence

Heat gained = heat lost,

$$70\,T - 1820 = 900 - 10\,T$$

$$80\,T = 2720$$

$$T = 34°C.$$

Fig. 14-6.

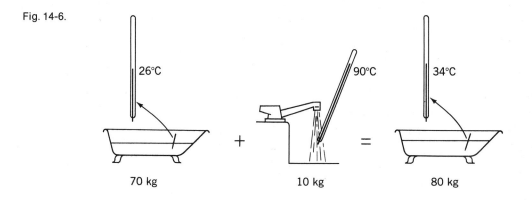

70 kg   10 kg   80 kg

14-3 Mechanical Equivalent of Heat

It is entirely possible to use the joule and ft-lb as units of heat instead of the kcal and Btu, and in fact the joule is the correct SI unit for the purpose. However, despite their inconvenience, the kcal and Btu are widely employed, and for the time being at least it is necessary to be familiar with them. The *mechanical equivalent of heat*, symbol J, is the ratio between the energy and heat units in each system:

$$J = 4185 \frac{\text{joule}}{\text{kcal}},$$

$$J = 778 \frac{\text{ft-lb}}{\text{Btu}}.$$

Mechanical equivalent of heat

Problem.   Radiant energy from the sun arrives at the earth's upper atmosphere at the rate of about 1400 W per m² of surface perpendicular to the sun's rays. On a clear day a reflector 1 m² in area is used to concentrate sunlight on a vessel that contains 1 kg of water at 20°C. How long will it take for the water to reach 100°C under the assumption that the water absorbs energy at the rate of 700 W?

Solution.   The energy absorbed by the water in the time t is Pt. The corresponding internal energy gain of the water is $Jm\,\Delta T$, where the mechanical equivalent of heat J is needed in order that this gain be expressed in energy rather than heat units. Hence

$$Pt = Jm\,\Delta T,$$

$$t = \frac{Jm\,\Delta T}{P} = 4185 \frac{\text{J}}{\text{kcal}} \times \frac{(1 \text{ kg} \times 80°\text{C}) \text{ kcal/kg}\cdot°\text{C}}{700 \text{ J/s}} = 478 \text{ s} = 8.0 \text{ min.}$$

14-4 Specific Heat Capacity

Samples of other substances respond to the addition or removal of a given amount of heat with temperature changes greater than that of an equal mass of water. One kg of water increases in temperature by 1°C when 1 kcal of heat is added to it, but 1 kcal of heat increases the temperature of 1 kg of helium (its volume held constant) by 1.3°C, of 1 kg of ice by 2°C, and of 1 kg of gold by 33°C (Fig. 14-7).

The *specific heat capacity* (symbol c) of a substance refers to what we might think of as its thermal inertia:

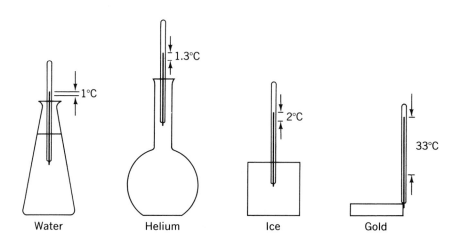

Fig. 14-7.   When 1 kcal of heat is added to 1 kg of each of the substances shown, their respective rises in temperature differ considerably.

Specific heat capacity

The specific heat capacity of a substance is the amount of heat that must be added or removed from a unit mass of it to change its temperature by 1°.

A high specific heat capacity means a relatively small change in temperature for a given change in internal energy content, just as a large inertial mass means a relatively small acceleration when a given force is applied.

In the metric system, c is the heat in kcal involved in the change in temperature of 1 kg of something by 1°C, and its units are kcal/kg·°C. In the British system, c is the heat in Btu involved in the change in temperature of 1 lb of something by 1°F, and its units are Btu/lb·°F. The numerical value for the specific heat of a substance is the same in both systems of units. Table 14-1 is a list of specific heats for some common substances. The actual values vary somewhat with temperature, and the ones given in the table represent averages.

With the help of specific heat capacity we can write a formula for the quantity of heat Q involved when a quantity m of a substance undergoes a change in temperature of ΔT. This formula is simply

Quantity of heat

$$Q = mc \ \Delta T, \tag{14-2}$$

Heat transferred = mass × specific heat capacity × temperature change.

Problem.   How much heat must be removed from 14 lb of aluminum in order to cool it from 80°F to 15°F?

Table 14-1. Specific heat capacities of various substances.

Substance	Specific heat capacity kcal/kg·°C Btu/lb·°F	Substance	Specific heat capacity kcal/kg·°C Btu/lb·°F
Alcohol (ethyl)	0.58	Marble	0.21
Aluminum	0.22	Mercury	0.033
Copper	0.093	Silver	0.056
Glass	0.20	Steam	0.48
Gold	0.030	Sulfuric acid	0.27
Granite	0.19	Turpentine	0.42
Ice	0.50	Water	1.00
Iron	0.11	Wood	0.42
Lead	0.030	Zinc	0.092

Solution. From Table 14-1 the specific heat capacity of aluminum is 0.22 Btu/lb°F, and so

$$Q = mc \, \Delta T = 14 \text{ lb} \times 0.22 \, \frac{\text{Btu}}{\text{lb} \cdot {}^\circ\text{F}} \times (-65^\circ\text{F})$$

$$= -200 \text{ Btu}.$$

The minus sign means that this quantity of heat is to be removed to achieve the temperature change of $-65°F$.

Problem. 0.20 kg of coffee at 90°C is poured into a 0.30-kg cup at 20°C. Assuming that no heat is transferred to or from the outside, what is the final temperature of the coffee?

Solution. We shall take the specific heat of coffee to be that of water and the specific heat of the cup to be that of glass. To solve the problem, we begin by noting that

Heat gained by cup = heat lost by coffee.

If the final temperature of both coffee and cup is T, then

Fig. 14-8.

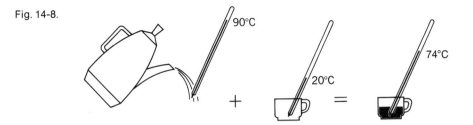

$$\text{Heat gained by cup} = m_{\text{cup}}c_{\text{cup}}\,(T - 20°C)$$

$$= 0.30 \text{ kg} \times 0.20 \,\frac{\text{kcal}}{\text{kg-°C}} \times (T - 20°C)$$

$$= (0.06\,T - 1.2) \text{ kcal,}$$

and

$$\text{Heat lost by coffee} = m_{\text{coffee}}c_{\text{coffee}}\,(90°C - T)$$

$$= 0.20 \text{ kg} \times 1.0 \,\frac{\text{kcal}}{\text{kg-°C}} \times (90°C - T)$$

$$= (18 - 0.20\,T) \text{ kcal.}$$

Now we set the heat gained by the cup equal to the heat lost by the coffee and solve for T:

$$0.06\,T - 1.2 = 18 - 0.20\,T,$$

$$0.26\,T = 19.2,$$

$$T = 74°C.$$

The temperature of the coffee drops by 16°C as it warms the cup (Fig. 14-8). Evidently it is necessary to preheat the cup if one wants really hot coffee.

14-5 Change of State

Not always does the addition or removal of heat from a sample of matter lead to a change in its temperature. Instead the sample may change

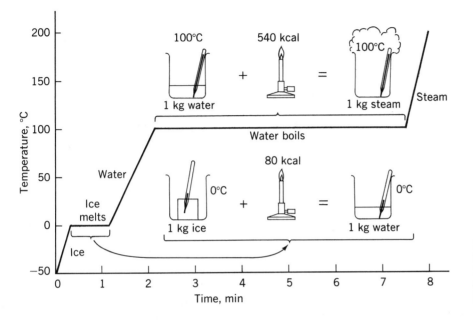

Fig. 14-9. A graph of temperature versus time for 1 kg of water, initially ice at −50°C, to which heat is being added at the constant rate of 100 kcal/min.

its state from solid to liquid or from liquid to gas when heat is added, or it may change from gas to liquid or from liquid to solid when heat is taken away. Such changes of state take place at definite temperatures for most substances, but for a few (glass or wax, for instance) there is only a gradual softening or hardening over a range of temperatures. Substances of the latter kind are not true solids, however; their structures are really those of liquids, and their hardness at room temperature is really a kind of exaggerated viscosity.

Figure 14-9 shows what happens when we add heat at the constant rate of 100 kcal/min to 1 kg of ice that is initially at −50°C. The specific heat capacity of ice is 0.5 kcal/kg-°C, and so 25 kcal is needed to bring the ice to 0°C.

At 0°C the ice begins to melt. The temperature remains constant until all the ice has melted, which requires a total of 80 kcal. Thus 80 kcal/kg is the *heat of fusion* of water: the amount of heat needed to convert 1 kg of ice into 1 kg of water at its melting point of 0°C.

When all the ice has turned to water, the temperature goes up once more as further heat is supplied. Since the specific heat of water is 1 kcal/kg-°C, there is now a rise of 1°C per kcal of heat. This rate of change is less than that of ice, since the specific heat of water is greater than that of ice, and so the slope of the graph is less steep.

When 100°C is reached, the water begins to turn into steam. The temperature stays constant until a total of 540 kcal is added, at which time all the water has become steam. Thus 540 kcal/kg is the *heat of vaporization* of water: the amount of heat needed to convert 1 kg of water into 1 kg of steam at its boiling point of 100°C (at atmospheric pressure).

After the water has become steam, its temperature rises again. The specific heat of steam is 0.48 kcal/kg-°C, so the temperature increase is 2.1°C per kcal of heat, and the slope of the graph is therefore steeper than it was for ice or water.

Heat of fusion

In general, the *heat of fusion* of a substance is the amount of heat that must be supplied to change a unit amount (1 kg or 1 lb) of the substance at its melting point from the solid to the liquid state; the same amount of heat must be removed from a unit amount of the substance in the liquid state at its melting point to change it to a solid. The usual symbol for heat of fusion is L_f.

Heat of vaporization

The *heat of vaporization* of a substance is the amount of heat that must be supplied to change a unit amount of the substance at its boiling point from the liquid to the gaseous (or vapor) state; the same amount of heat must be removed from a unit amount of the substance in the gaseous state at its boiling point to change it into a liquid. The usual symbol for heat of vaporization is L_v. The heats of fusion and vaporization for a number of substances are listed in Table 14-2 together with their melting and boiling points in both systems of units.

Sublimation

Under certain circumstances most substances can change directly from the solid to the vapor state, or vice versa. Both processes are called *sublimation*. For example, "dry ice" (solid carbon dioxide) evaporates directly to gaseous carbon dioxide at temperatures above −78.5°C, and does not pass through the liquid state. With the exception of carbon dioxide and a few other substances, however, sublimation does not occur except at pressures well below that of the atmosphere.

Problem. What is the minimum amount of ice at −10°C that must be added to 0.50 kg of water at 20°C in order to bring the temperature of the water down to 0°C?

Solution. We begin, as before, with the statement of energy conservation,

Heat absorbed by ice = heat lost by water.

The heat Q_1 absorbed by the unknown mass of ice in going from −10°C to its melting point of 0°C is

$$Q_1 = m_{ice} c_{ice} \Delta T_{ice}$$

Table 14-2. Heats of fusion and vaporization and melting and boiling points of various substances at atmospheric pressure.

Substance	Melting point, °C	Melting point, °F	L_f, kcal/kg	L_f, Btu/lb	Boiling point, °C	Boiling point, °F	L_v, kcal/kg	L_v, Btu/lb
Alcohol (ethyl)	−114	−173	25	45	78	172	204	367
Bismuth	271	520	12.5	22.5	920	1688	190	342
Bromine	−7	19	16	28	60	140	43	77
Lead	330	626	5.9	10.6	1170	2138	175	315
Lithium	186	367	160	288	1336	2437	511	920
Mercury	−39	−38	2.8	5.0	358	676	71	128
Nitrogen	−210	−346	6.1	11	−196	−320	48	86
Oxygen	−219	−362	3.3	5.9	−183	−306	51	92
Sulfuric acid	8.6	47	39	70	326	618	122	220
Water	0	32	80	144	100	212	540	972
Zinc	420	787	24	43	918	1684	475	855

$$= m_{ice} \times 0.50 \, \frac{kcal}{kg\text{-}°C} \times 10°C,$$

and the heat Q_2 absorbed by the ice in melting at 0°C is

$$Q_2 = m_{ice} L_{f \, ice}$$

$$= m_{ice} \times 80 \, \frac{kcal}{kg}.$$

Hence

$$\text{Heat absorbed by ice} = Q_1 + Q_2$$

$$= (5 + 80) \, m_{ice} \, \frac{kcal}{kg}.$$

The heat lost by the water in cooling to 0°C from 20°C is

$$\text{Heat lost by water} = m_{water} c_{water} \Delta T_{water}$$

$$= 0.50 \, kg \times 1.0 \, \frac{kcal}{kg\text{-}°C} \times 20°C$$

$$= 10 \, kcal.$$

Equating the heat absorbed with the heat lost and then solving for m_{ice} yields (Fig. 14-10)

$$85 \; m_{ice} \; \frac{kcal}{kg} = 10 \; kcal$$

$$m_{ice} = 0.12 \; kg.$$

Fig. 14-10.

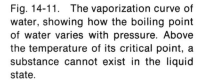

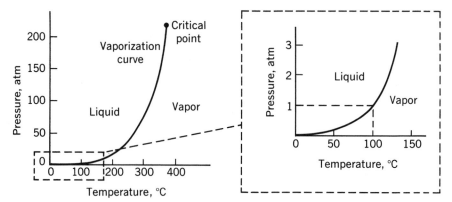

0.50 kg
water

+

0.12 kg
ice

=

0.62 kg
water

14-6 The Triple Point

Let us examine the effect of pressure on changes of state. Figure 14-11 shows how the boiling point of water varies with pressure, behavior typical of other liquids as well. The upper limit of this *vaporization curve*, which occurs at a temperature of 374°C and a pressure of 218 atm, is known as the

Fig. 14-11.   The vaporization curve of water, showing how the boiling point of water varies with pressure. Above the temperature of its critical point, a substance cannot exist in the liquid state.

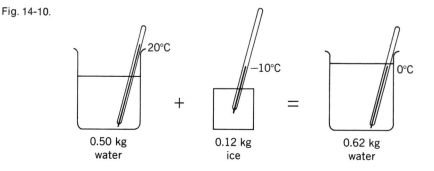

critical point. A substance cannot exist in the liquid state at a temperature above that of its critical point, regardless of how great the pressure may be. Helium has the lowest critical temperature, −268°C, and is therefore a gas at all temperatures above that.

The melting points of solids also depend upon pressure, although to a smaller extent than the boiling points. The variation of the melting point of ice with pressure is shown in the *fusion curve* of Fig. 14-12. Ice, together with gallium and bismuth, is unusual in that its melting point *decreases* with increasing pressure; the melting points of all other substances increase with increasing pressure. Hence it is possible to melt ice by applying pressure to it as well as by heating it. An ice skater makes use of this fact in an interesting way. His entire weight is supported by skate blades of very small area, and the resulting pressure on the ice may exceed 1000 atm. The ice under the blades melts because of the great pressure, which creates a thin film of water that acts as an efficient lubricant. On unusually cold days even such pressures may not be sufficient to melt the ice, and skating then becomes impossible.

The fusion and vaporization curves of water intersect at a temperature of 0.01°C and a pressure of 4.6 torr, as shown on the combined plot of Fig. 14-13. Along the fusion curve both ice and water can simultaneously exist, and along the vaporization curve both water and water vapor can simultaneously exist; hence under conditions corresponding to those of the intersection of the two curves, the solid, liquid, and vapor states of water can all exist together. This intersection is accordingly called the *triple point* of water.

Critical point

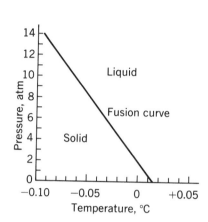

Fig. 14-12. The fusion curve of water, showing how the melting point of ice varies with pressure.

Triple point

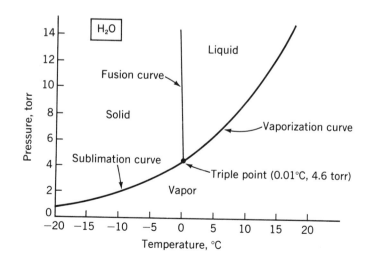

Fig. 14-13. Triple-point diagram of water. The solid, liquid, and vapor phases of water can exist simultaneously at the temperature and pressure of the triple point. (1 torr = 1 mm Hg)

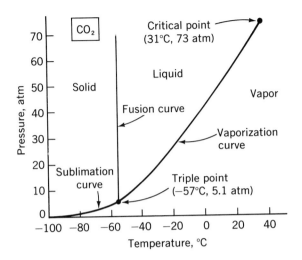

Fig. 14-14. Triple-point diagram of carbon dioxide.

At pressures below that of its triple point, no substance can exist as a liquid. The dividing line on a pressure-temperature graph between the solid and vapor states is called the *sublimation curve*, since it represents the conditions required for a solid to vaporize directly or a vapor to solidify directly. At atmospheric pressure the addition of heat causes ordinary ice to melt, since the triple point of water lies well below 1 atm, but the addition of heat causes solid carbon dioxide to sublime, since its triple point lies above 1 atm (Fig. 14-14).

At a temperature below that of its triple point, a substance passes directly from the solid state to the vapor state if the pressure is sufficiently low. This phenomenon is the basis of the *freeze-drying* process widely used for the preservation of foods, blood plasma, and biological samples. The usual procedure is to cool the material below the triple point of water in a gas-tight chamber, and then to evacuate the chamber with a vacuum pump. Freeze drying affects the structure of a material of biological origin less than other methods of dehydration do.

Special Topic

Animal Metabolism

The conversion of the metabolic energy of an animal into mechanical work varies in efficiency with the type of muscle and how fast it contracts —

the efficiency is least at high and low speeds. An efficiency of 10 to 20 percent is usual. Most of the energy liberated by an animal's metabolic processes thus ends up as heat, the greater part of which escapes through the animal's skin. The ability of an animal to dissipate the heat its body produces accordingly is proportional to its surface area, which suggests that the maximum metabolic rate of an animal, and hence its power output, depends upon this area.

If a representative linear dimension (such as its length) of an animal is L, its surface area varies as L^2. Because the animal's mass is proportional to its volume and hence varies as L^3, the metabolic rate per kg ought to depend upon $L^2/L^3 = 1/L$ by the above reasoning. Thus small animals should have higher metabolic rates per kg than large ones, which is indeed the case in nature. Typical basal metabolic rates (which correspond to an animal resting) are 5.2 W/kg for a pigeon, 1.3 W/kg for a dog, 1.2 W/kg for a man, 1.1 W/kg for a woman, and 0.67 W/kg for a cow. It is clear why birds are small: past a certain size, a bird's weight outstrips its ability to perform the work needed for it to fly. It is also clear why natural selection has led to large sizes for whales and porpoises, which are air-breathing mammals that live in the sea. About 0.3 g of oxygen is needed for each kcal of energy released by the metabolism of food in an animal's body. The amount of oxygen an animal can store varies with its volume and hence with L^3. The larger an animal, the more reserve oxygen it has relative to its metabolic rate (which varies as L^2), so a large aquatic mammal has the advantage over a small one of being able to remain submerged for a longer period.

Important Terms

The **temperature** of a body of matter is a measure of the average kinetic energy of random motion of its constituent particles. When two bodies are in contact, heat flows from the one at the higher temperature to the one at the lower temperature.

A **thermometer** is a device for measuring temperature. The two temperature scales in common use are the **celsius** (centigrade) scale, in which the freezing point of water is assigned the value 0°C and its boiling point the value 100°C, and the **fahrenheit** scale, in which these points are assigned the values 32°F and 212°F respectively.

Heat is a quantity whose addition to a body of matter causes its internal energy to increase and whose removal causes its internal energy to decrease,

provided the body neither does work nor has work done on it. If the matter does not change state during the process, the change in internal energy results in a corresponding change in temperature. The unit of heat in the metric system is the **kilocalorie** (kcal), which is that amount of heat required to change the temperature of 1 kg of water by 1°C; the unit of heat in the British system is the **British thermal unit** (Btu), which is that amount of heat required to change the temperature of 1 lb of water by 1°F.

The **specific heat capacity** of a substance is the amount of heat required to change the temperature of a unit quantity of it by 1°. The unit of specific heat in the metric system is the kcal/kg·°C; in the British system it is the Btu/lb·°F.

The **heat of fusion** of a substance is the amount of heat that must be supplied to change a unit quantity of it at its melting point from the solid to the liquid state;

the same amount of heat must be removed from a unit quantity of the substance in the liquid state at its melting point to change it to a solid.

The **heat of vaporization** of a substance is the amount of heat that must be supplied to change a unit quantity of it at its boiling point from the liquid to the gaseous (or vapor) state; the same amount of heat must be removed from a unit quantity of the substance at its boiling point to change it into a liquid.

Sublimation is the direct conversion of a substance from the solid to the vapor state, or vice versa, without it first becoming a liquid.

The **critical point** is the upper limit of the vaporization curve of a substance; a substance cannot exist in the liquid state at a temperature above that of its critical point.

The **triple point** is the intersection of the vaporization, fusion, and sublimation curves of a substance. All three states of a substance may exist in equilibrium at the temperature and pressure of its triple point.

Important Formulas

Celsius and fahrenheit scales:

$$T_F = \frac{9}{5} T_C + 32°$$

$$T_C = \frac{5}{9} (T_F - 32°)$$

Heat and temperature change:

$$Q = mc\Delta T$$

Multiple Choice

1. Heat is a physical quantity most closely related to
   a. temperature.
   b. friction.
   c. energy.
   d. momentum.
2. Two thermometers, one calibrated in the celsius scale and the other in the fahrenheit scale, are used to measure the same temperature. The numerical reading on the fahrenheit thermometer
   a. is proportional to that on the celsius thermometer.
   b. is greater than that on the celsius thermometer.
   c. is less than that on the celsius thermometer.
   d. may be greater or less than that on the celsius thermometer.

3. Oxygen boils at $-183°C$. This temperature is
   a. $-215°F$.   b. $-297°F$.
   c. $-329°F$.   d. $-361°F$.

4. A temperature of $100°F$ is almost exactly
   a. $38°C$.   b. $56°C$.
   c. $122°C$.   d. $212°C$.

5. Two blocks of lead, one twice as heavy as the other, are both at $50°C$. The ratio of the internal energy content of the heavier block to that of the lighter block is
   a. $\frac{1}{2}$.   b. 1.
   c. 2.   d. 4.

6. The quantity of heat required to change the temperature of a unit amount of a substance by $1°$ is called its
   a. specific heat capacity.
   b. heat of fusion.
   c. heat of vaporization.
   d. mechanical equivalent of heat.

7. Of the following substances, the one which requires the greatest amount of heat per kg for a given increase in temperature is
   a. ice.
   b. water.
   c. steam.
   d. copper.

8. A cup of hot coffee can be cooled by placing a cold spoon in it. A spoon of which of the following materials would be most effective for this purpose, assuming the spoons all have the same mass?
   a. aluminum.
   b. copper.
   c. iron.
   d. silver.

9. Body *A* is at a higher temperature than body *B*.

When they are placed in contact, heat will flow from A to B
a. only if A has the greater internal energy content.
b. only if both are fluids.
c. only if A is on top of B.
d. until both have the same temperature.

10. When a vapor condenses into a liquid,
   a. it absorbs heat.
   b. it evolves heat.
   c. its temperature rises.
   d. its temperature drops.

11. Sublimation refers to
   a. the vaporization of a solid without first becoming a liquid.
   b. the melting of a solid.
   c. the vaporization of a liquid.
   d. the condensation of a gas into a liquid.

12. The heat of vaporization of a substance is
   a. less than its heat of fusion.
   b. equal to its heat of fusion.
   c. greater than its heat of fusion.
   d. any of the above, depending on the nature of the substance.

13. The freezing point of a substance is always lower than its
   a. melting point.
   b. boiling point.
   c. heat of fusion.
   d. heat of vaporization.

14. The ratio between the energy dissipated in some process and the heat that appears as a result is called the
   a. joule.
   b. kilocalorie.
   c. specific heat capacity.
   d. mechanical equivalent of heat.

15. 10 lb of ice at 0°F are added to 100 lb of water at 50°F. The temperature of the resulting mixture is
   a. 19°F.   b. 31°F.
   c. 32°F.   d. 33°F.

16. 10 kg of ice at 0°C are added to 2 kg of steam at 100°C. The temperature of the resulting mixture is

a. 0°C.   b. 23°C.
c. 28°C.   d. 40°C.

17. A 1-kg lead bar at 80°C is placed in 2 kg of water at 20°C. The final temperature of the lead bar is
   a. 22°C.   b. 28°C.
   c. 40°C.   d. 50°C.

18. 20 lb of punch of specific heat capacity 0.7 at a temperature of 40°F is placed in a 10 lb silver punch bowl at 70°F. The final temperature of the punch is
   a. 41°F.   b. 52°F.
   c. 63°F.   d. 69°F.

19. A 61-kg woman eats a banana whose energy content is 100 kcal. If this energy were used to raise her from the ground, her approximate height would be
   a. 0.7 m.   b. 7 m.
   c. 70 m.   d. 700 m.

20. Under conditions corresponding to its triple point, a substance
   a. is in the solid state.
   b. is in the liquid state.
   c. is in the gaseous state.
   d. may be in any or all of the above states.

Exercises

1. A jar of water is shaken vigorously. What becomes of the work that is done?

2. Why will the engine of a car whose cooling system is filled with an alcohol antifreeze be more likely to overheat in summer than one whose cooling system is filled with water?

3. Which is more effective in cooling a drink, 10 g of water at 0°C or 10 g of ice at 0°C?

4. How does perspiration give the body a means of cooling itself?

5. Why does turning the flame higher under a pan of boiling water not reduce the time needed to cook an egg in the water?

6. When a certain quantity of a vapor condenses into a liquid, what happens to its internal energy content and to its temperature?

7. The melting point of lead is 330°C and its boiling point is 1170°C. Express these temperatures on the fahrenheit scale.

8. The normal temperature of the human body is 98.6°F. What is this temperature on the celsius scale?

9. At what temperature would celsius and fahrenheit thermometers give the same reading?

10. Mercury freezes at −40°C. What is this temperature on the fahrenheit scale?

11. Dry ice (solid carbon dioxide) vaporizes at −112°F. What is this temperature on the celsius scale?

12. Two lb of water is to be heated from 70°F to 212°F to make a pot of coffee. How much heat is needed?

13. Two hundred grams of water is to be heated from 15°C to 100°C to make a cup of tea. How much heat is needed?

14. A 60-kg woman is on a diet that provides her with 2500 kcal daily. If a corresponding amount of heat were added to 60 kg of water at 37°C, what would its final temperature be?

15. How much heat must be added to 400 lb of iron to raise its temperature from 50°F to 350°F?

16. How much heat must be removed from 60 lb of ice at 20°F to lower its temperature to −20°F?

17. How much heat must be added to 1 g of silver to raise its temperature from −5°C to 65°C?

18. How much heat must be removed from 20 kg of marble at 20°C to lower the temperature of the marble to 8°C?

19. One hundred Btu of heat is removed from a 50-lb block of ice at 30°F. Find its final temperature.

20. Eight Btu of heat is added to a 10-lb copper bar at 60°F. Find its final temperature.

21. Six kcal of heat is added to a 3 kg lead ball at 10°C. Find its final temperature.

22. 10^4 kcal of heat is removed from a metric ton (10^3 kg) of iron at 300°C. What is its final temperature?

23. A 0.6-kg copper container holds 1.5 kg of water at 20°C. A 0.1-kg iron ball at 120°C is dropped into the water. What is the final temperature of the water?

24. A 0.1-kg piece of silver is taken from a bath of hot oil and placed in a 0.08-kg glass jar containing 0.2 kg of water at 15°C. The temperature of the water increases by 8°C. What was the temperature of the oil?

25. How many kcal are evolved per hour by a 60-watt light bulb?

26. A 60-lb storage battery has an average specific heat of 0.2 Btu/lb·°F. When fully charged the battery contains 10^6 ft-lb of electrical energy. If all this energy were dissipated within the battery, find the increase in its temperature.

27. A man is sitting in the shade in an ambient air temperature of 37°C, which is the same as his body temperature. Under these circumstances the chief way his body gets rid of the 120 W his metabolic processes liberate is through the evaporation of sweat. How much sweat per hour is required? At 37°C the heat of vaporization of water is 580 kcal/kg.

28. The metabolic processes of a 55-kg sleeping woman consume 60 kcal/hr, essentially all of which ends up as heat. (a) How many watts is this? (b) If none of this heat is lost to the outside world, what would her temperature be after 8 hr of sleep? Normal body temperature is 37°C and the specific heat capacity of the human body is about 0.83 kcal/kg·°C.

29. Carbon dioxide is usually shipped in tanks under a pressure of approximately 70 atm. At 20°C, is the carbon dioxide a solid, a liquid, or a gas?

30. The pressure and temperature of the atmosphere at 115,000 ft are 4 torr and −23°C respectively. What is the state of water under those conditions? Of carbon dioxide?

Problems

1. A sedentary person requires about 30 kcal of energy in his diet per day per kg of body mass. If this energy were used to raise a 1-kg mass above the ground, find its height. [The "calorie" that dieticians use is the same as the kilocalorie. Thus the energy content of a 130-calorie cupcake is really 130 kcal. A heat unit once widely used is also called the calorie. This is equal to the heat needed to raise the temperature of 1 g of water by 1°C, so that 1,000 cal = 1 kcal. The dietician's calorie is sometimes written Calorie to distinguish it from the ordinary calorie.]

2. A 50-kg woman is on a diet that provides her with 2500 kcal daily. (a) If this amount of energy were used to raise her above the ground, how high would she go? (b) If this amount of energy were used to provide her with kinetic energy, what would her velocity be?

3. A person decides to lose weight by eating only cold food. A 100-gram piece of apple pie yields about 350 kcal of energy when eaten. If its specific heat capacity is 0.4 kcal/kg·°C, how much greater is its energy content at 50°C than at 20°C? What percentage difference is this?

4. A 1-kg block of ice at 0°C falls into a lake whose water is also at 0°C, and 0.01 kg of ice melts. What was the minimum altitude from which the ice fell?

5. If all the heat lost by 1 kg of water at 0°C when it turns into ice at 0°C could be turned into kinetic energy, what would the velocity of the ice be?

6. The minimum velocity an artificial earth satellite can have is 7.9×10^3 m/sec. Aluminum melts at 660°C. If an aluminum satellite re-enters the earth's atmosphere when its temperature is 0°C, can it be brought to rest rapidly by air resistance without melting? If not, find out how actual spacecraft avoid this dilemna.

7. A lead bullet at 100°C strikes a steel plate and melts. What was its minimum velocity?

8. A 100-kg wooden beam is pushed across a stone floor by a force just sufficient to overcome friction. The coefficient of friction is 0.4. Assuming that half the work done against friction goes into heating the beam, what is its rise in temperature for each foot that it is pushed?

9. A 50-kg block of ice at 0°C is pushed across a wooden floor also at 0°C for a distance of 20 m. A total of 25 g of ice melts as a result of the friction of the block on the floor. What is the minimum coefficient of friction in this case?

10. What is the difference in temperature between the water at the top and at the bottom of a waterfall of height h?

11. How much more heat must be added to 1 kg of ice at 0°C to convert it to steam at 100°C than is required to raise the temperature of 1 kg of water from 0°C to 100°C?

12. Six kilograms of ice at −10°C are added to 6 kg of water at +10°C. Find the temperature of the resulting mixture.

13. How much ice at −10°C is required to cool a mixture of 0.1 kg ethyl alcohol and 0.1 kg water from 20°C to 5°C?

14. A 5-kg iron bar is taken from a forge at a temperature of 1000°C and plunged into a pail containing 10 kg of water at 60°C. How much steam is produced?

15. By mistake, 0.2 kg of water at 0°C is poured into a vessel containing liquid nitrogen at −196°C. How much nitrogen vaporizes?

16. How much steam at 250°F is required to melt 5 lb of ice at 32°F?

17. Ten lb of steam at 220°F is passed through 10 lb water at 200°F. What is the temperature and physical state of the mixture afterward?

Answers to Multiple Choice

1. c	6. a	11. a	16. d
2. d	7. b	12. c	17. a
3. b	8. a	13. b	18. a
4. a	9. d	14. d	19. d
5. c	10. b	15. c	20. d

15

Thermal Properties of Matter

Nearly all substances expand when they are heated. Why does this happen? The fact that a phenomenon is familiar does not mean that its explanation need be obvious, and, as we shall see in this chapter, it is necessary to inquire deeply into the structure of matter in order to understand its thermal behavior. Interestingly enough, the details of why a gas tends to expand when heated turn out to be rather different from the reasons a solid or liquid tends to expand, although in both cases the effect is ultimately due to the close relationship between molecular energy and temperature.

15-1 Thermal Expansion

Nearly all substances expand when they are heated and contract when they are cooled. This is a familiar effect: sidewalks buckle on a hot summer day; the mercury column rises in a heated thermometer; warm air rises because, owing to its expansion, it is less dense than the surrounding cool air and so is buoyant.

Experiments show that, to a good approximation, a change in temperature ΔT causes most solids to change in length by an amount proportional to both their original lengths and to ΔT. If the original length of a rod of a certain material is L_0, its change in length ΔL after its temperature changes by ΔT is

$$\Delta L = aL_0\Delta T \qquad \textit{Thermal expansion} \quad (15\text{-}1)$$

Change in length $= a \times$ original length $\times$ temperature change.

The quantity a, called the *coefficient of linear expansion*, is a constant whose value depends upon the nature of the material. Different substances expand (and contract) to different extents; a lead rod, for example, changes in length by 60 times as much as a quartz rod of the same initial length when both are heated or cooled through the same temperature interval. Table 15-1 lists coefficients of linear expansion for various substances.

Coefficient of linear expansion

Table 15-1. Coefficients of linear expansion.

Substance	Coefficient, $\times 10^{-5}/°C$	Coefficient, $\times 10^{-5}/°F$
Aluminum	2.4	1.3
Brass	1.8	1.0
Concrete	0.7–1.2	0.4–0.7
Copper	1.7	0.94
Iron	1.2	0.67
Lead	3.0	1.7
Quartz	0.05	0.028
Silver	2.0	1.1
Steel	1.2	0.67

Problem. What is the increase in length of a steel girder that is 50 ft long at 40°F when its temperature rises to 75°F?

Solution. The coefficient of linear expansion of steel is $0.67 \times 10^{-5}/°F$, and so, from Eq. (15-1),

$$\Delta L = aL_0 \, \Delta T = 0.67 \times \frac{10^{-5}}{°F} \times 50 \text{ ft} \times 35°F$$

$$= 0.012 \text{ ft},$$

which is a little over $\frac{1}{8}$ in. (Fig. 15-1).

Problem. How much force is associated with the expansion of the girder if its cross-sectional area is 40 in²?

Solution. Since the girder increases in length by 0.012 ft, the force is the same as that required to stretch it by 0.012 ft. Equation (10-2) gives the change in length ΔL of a rod that is subjected to a tension or compression force F as

$$\Delta L = \frac{L_0}{Y} \frac{F}{A},$$

where A is the cross-sectional area of the rod and Y is Young's modulus for the material of the rod. Since the girder has a cross-sectional area of 40 in²

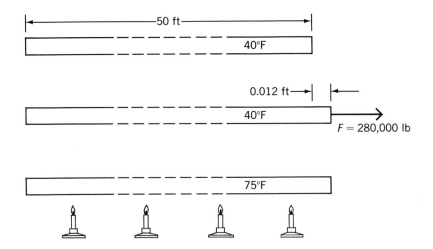

Fig. 15-1. A 50-ft steel girder expands 0.012 in. when its temperature is increased by 35°F. At constant temperature, a force of 280,000 lb would be required to produce the same increase in length if the cross-sectional area of the girder were 40 in².

and, from Table 10-2, Young's modulus for steel is 2.9×10^7 lb/in^2,

$$F = \frac{YA\,\Delta L}{L_0} = \frac{2.9 \times 10^7 \text{ lb/in}^2 \times 40 \text{ in}^2 \times 0.012 \text{ ft}}{50 \text{ ft}}$$

$$= 2.8 \times 10^5 \text{ lb};$$

a force of 280,000 lb, which is 140 tons, is associated with the expansion of the girder. Clearly thermal expansion can involve very considerable forces.

A formula similar to Eq. (15-1) holds for the changes in volume, ΔV, of a solid or liquid whose temperature changes by an amount ΔT. Here we have

$$\Delta V = b V_0 \, \Delta T, \tag{15-2}$$

where V_0 is the original volume and b is the *coefficient of volume expansion*. Table 15-2 is a list of coefficients of volume expansion for various substances. In general, the coefficients of linear and volume expansion are related by

Coefficient of volume expansion

$$b = 3a,$$

so that we can readily determine the values of b for the materials of Table 15-1.

Table 15-2. Coefficients of volume expansion.

Substance	Coefficient, $\times 10^{-4}/°C$	Coefficient, $\times 10^{-4}/°F$
Ethyl alcohol	11	6.1
Glass (average)	0.2	0.1
Glycerin	5.1	2.8
Ice	0.5	0.3
Mercury	1.8	1.0
Pyrex glass	0.09	0.05
Water	2.1	1.2

Problem. Calculate the volume of water that overflows when a Pyrex beaker filled to the brim with 250 cm^3 of water at 20°C is heated to 60°C.

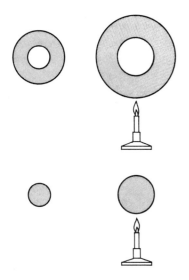

Fig. 15-2. A cavity in a body expands or contracts with a change in temperature precisely as a solid object of the same size, shape, and composition would.

Solution. First, we note that a cavity in a body expands or contracts by precisely as much as a solid object having the same volume as the cavity and having the composition of the body (Fig. 15-2). This means that for the increase in capacity of the beaker we can write

$$\Delta V_b = b_P V_b \, \Delta T$$

$$= 0.09 \times \frac{10^{-4}}{°C} \times 250 \text{ cm}^3 \times 40°C$$

$$= 0.09 \text{ cm}^3.$$

The increase in the volume of the water is

$$\Delta V_w = b_w V_w \, \Delta T$$

$$= 2.1 \times \frac{10^{-4}}{°C} \times 250 \text{ cm}^3 \times 40°C$$

$$= 2.1 \text{ cm}^3,$$

and so the volume of water that overflows is

$$\Delta V_w - \Delta V_b = 2.0 \text{ cm}^3.$$

15-2 Boyle's Law

A peculiar difficulty arises when we attempt to measure the coefficient of volume expansion of a gas. Unlike solids and liquids, gases do not have specific volumes at a particular temperature, but expand to fill their containers. The only way to change the volume of a gas is to change the capacity of its container. However, even though its volume may remain the same, another property of a confined gas varies with its temperature, namely the pressure it exerts on the container walls. The air pressure in an automobile tire drops in cold weather and increases in warm, an illustration of this property.

When the temperature of a sample of gas is held constant, the absolute pressure it exerts on its container is very nearly inversely proportional to the volume of the container. Expanding the container lowers the pressure; shrinking the container raises the pressure. Conversely, increasing the pressure on a gas sample reduces its volume; decreasing the pressure increases its volume (Fig. 15-3). This relationship is called *Boyle's law* after its discoverer, Robert Boyle (1627-1691). Though not exact, Boyle's law is an excellent approximation over a wide range of temperatures and pressures.

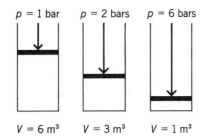

$p = 1$ bar $p = 2$ bars $p = 6$ bars

$V = 6$ m³ $V = 3$ m³ $V = 1$ m³

Fig. 15-3.   Boyle's law states that the volume of a gas sample is inversely proportional to its pressure at constant temperature. Thus $p_1 V_1 = p_2 V_2 = p_3 V_3$ as shown.

Boyle's law can be expressed in the form

$$pV = \text{constant}, \qquad [T = \text{constant}] \tag{15-3}$$

or alternatively as

$$p_1 V_1 = p_2 V_2. \qquad [T = \text{constant}] \qquad \textit{Boyle's law} \quad (15\text{-}4)$$

In the latter case p_1 is the absolute gas pressure when the volume of the gas is V_1, and p_2 is the gas pressure when the volume is V_2.

Problem. How much air at the sea-level atmospheric pressure of 15 lb/in² can be stored in the 12-ft³ tank of an air compressor which can withstand an absolute pressure of 100 lb/in²?

Solution. Letting state 1 represent the air in the compressor tank and state 2 the air at atmospheric pressure, we have

$$V_2 = \frac{p_1}{p_2} V_1 = \frac{100 \text{ lb/in}^2 \times 12 \text{ ft}^3}{15 \text{ lb/in}^2} = 80 \text{ ft}^3.$$

A total of 80 ft³ of air at atmospheric pressure can be stored in the 12-ft³ compressor tank at the high pressure.

15-3 Charles's Law

Now let us see what happens to a gas when its temperature is changed. As mentioned earlier, if the volume of a gas is held constant, the pressure it exerts on its container depends upon its temperature. According to Boyle's law, then, if we hold the gas pressure constant, its volume should vary with temperature. When this prediction is experimentally tested, which was first done over 150 years ago by Charles and Gay-Lussac, it is found that the change in volume ΔV of a gas sample is in fact related to a change ΔT in its temperature by the same formula, Eq. (15-2),

$$\Delta V = bV_0\,\Delta T,$$

that holds for solids and liquids.

The significant thing about gases at constant pressure is that they *all* have very nearly the same coefficient of volume expansion b; by contrast, as Tables 15-1 and 15-2 indicate, the thermal coefficients for solids and liquids may have markedly different values for different substances. At 0°C the coefficient of volume expansion b_0 of all gases is

$$b_0 = \frac{0.0037}{°C} = \frac{1/273}{°C}.$$

If we vary the temperature of a gas sample while holding its pressure constant, its volume changes by 1/273 of its volume at 0°C for each 1°C temperature change. A child's large balloon filled with air whose volume at 0°C is 1.000 m³ has a volume of 1.037 m³ at 10°C and 0.963 m³ at −10°C (Fig. 15-4).

What happens when the balloon is cooled to −273°C? At that temperature the air in the balloon should have lost 273/273 of its volume at 0°C, and therefore should have vanished entirely! Actually, all gases condense into liquids at temperatures above −273°C, so the question has no physical meaning. But −273°C is still a significant temperature. Let us set up a new temperature scale, the *absolute temperature scale*, and designate −273°C as

Absolute temperature scale

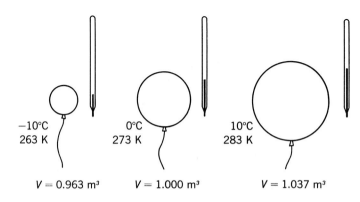

Fig. 15-4. Charles's law states that the volume of a gas sample is directly proportional to its absolute temperature at constant pressure. Thus $V_1/T_1 = V_2/T_2 = V_3/T_3$ as shown.

−10°C
263 K
$V = 0.963$ m³

0°C
273 K
$V = 1.000$ m³

10°C
283 K
$V = 1.037$ m³

the zero point (Fig. 15-5). Temperatures in the absolute scale are expressed in *kelvins*, denoted K, after Lord Kelvin (1824-1907), a noted British physicist. To convert temperatures from one scale to the other we note that

$$T_K = T_C + 273° \qquad T_C = T_K - 273°. \tag{15-5}$$

The reason for setting up the absolute temperature scale is that, provided the pressure is constant, *the volume of a gas sample is directly proportional to its absolute temperature* (Fig. 15-6). This relationship is called *Charles's law*. Like Boyle's law, Charles's law is not a basic physical principle but deviations from it are usually quite small.

We can express Charles's law in the form

$$\frac{V}{T} = \text{constant} \qquad [p = \text{constant}] \tag{15-6}$$

or alternatively as

$$\frac{V_1}{T_1} = \frac{V_2}{T_2}. \qquad [p = \text{constant}] \qquad \textit{Charles's law} \quad (15\text{-}7)$$

In the latter equation, V_1 is the volume of a gas sample at the absolute temperature T_1 and V_2 is its volume at the absolute temperature T_2.

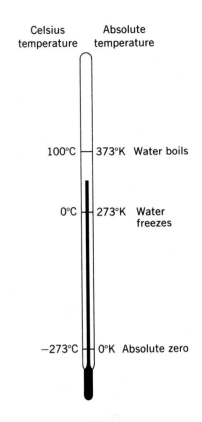

Fig. 15-5.   The absolute temperature scale.

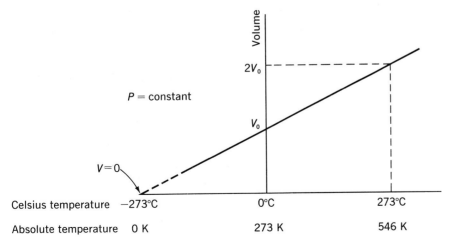

Fig. 15-6.   The volume of a gas at constant pressure is directly proportional to its absolute temperature.

Absolute zero

If there were a gas that did not liquify before reaching 0 K, then at 0 K its volume would shrink to zero. Since a negative volume has no meaning, it is natural to think of 0 K as *absolute zero*. Actually, 0 K is indeed the lower limit to temperatures capable of being attained, but on the basis of a stronger argument than one based on imaginary gases. This argument is discussed later in this chapter.

Rankine scale

An absolute temperature scale, called the *rankine scale*, may be constructed based upon the fahrenheit scale. Temperatures in this scale are designated °R. Absolute zero on the rankine scale is

$$0°R = \tfrac{9}{5}(-273°C) + 32°$$

$$= -460°F.$$

Hence the conversion formulas for °F to °R and for °R to °F are

$$T_R = T_F + 460°, \qquad T_F = T_R - 460°. \tag{15-8}$$

Charles's law holds when temperatures are expressed in the rankine scale as well as when they are expressed in the kelvin scale.

15-4 Ideal Gas Law

Boyle's law and Charles's law can be combined in a single formula called the *ideal gas law*:

$$\frac{p_1 V_1}{T_1} = \frac{p_2 V_2}{T_2}. \qquad \text{\textit{Ideal gas law}} \quad (15\text{-}9)$$

When $T_1 = T_2$, the ideal gas law becomes Boyle's law,

$$p_1 V_1 = p_2 V_2 \qquad [T = \text{constant}]$$

and when $p_1 = p_2$ it becomes Charles's law,

$$\frac{V_1}{T_1} = \frac{V_2}{T_2}. \qquad [p = \text{constant}]$$

Another way to express the ideal gas law is

$$\frac{pV}{T} = \text{constant}. \tag{15-10}$$

The ideal gas law is obeyed approximately by all gases. The significant thing is not that the agreement with experiment is never perfect, but that *all* gases, no matter what kind, behave almost identically. An *ideal gas* is defined as one that obeys pV/T = constant exactly. While no ideal gases actually exist, they do provide a target for theories of the gaseous state to aim at. It is reasonable to suppose that the ideal gas law is a consequence of the essential nature of gases. Hence the next step is to account for this law and only afterward to seek reasons for its failure to be completely correct.

Ideal gas

Problem.   (a) A tank with a capacity of 1 m³ contains helium gas at 27°C under a pressure of 20 atm. The helium is used to fill a balloon. When the balloon is filled, the gas pressure inside it is 1 atm, and its temperature has dropped to −33°C. (The gas has done work in expanding at the expense of its internal energy, and the cooling reflects this loss of internal energy.) What is the volume of the balloon at this time? (b) After a while the helium in the balloon absorbs heat from the atmosphere and returns to its original temperature of 27°C, and it expands further to maintain its pressure at 1 atm. What is the final volume of the balloon? (The gas pressure in the balloon is actually slightly greater than 1 atm to balance its tendency to contract, but this is ignored here for convenience.)

Solution.   (a) The equivalents of 27°C and −33°C on the absolute scale are 300 K and 240 K respectively. Applying the ideal gas law to the initial expansion, we obtain

$$V_2 = \frac{T_2}{T_1}\frac{p_1}{p_2}V_1 = \frac{240\ \text{K}}{300\ \text{K}} \times \frac{20\ \text{atm}}{1\ \text{atm}} \times 1\ \text{m}^3 = 16\ \text{m}^3.$$

Because the tank's capacity is 1 m², the balloon's volume after the initial expansion is 15 m³.
(b) When the helium has reached the outside air temperature of 27°C, which we shall call state 3, then $T_1 = T_3$. Hence we need only apply Boyle's law to states 1 and 3 to obtain the eventual volume of the helium:

$$V_3 = \frac{p_1}{p_3}V_1 = \frac{20\ \text{atm}}{1\ \text{atm}} \times 1\ \text{m}^3 = 20\ \text{m}^3.$$

Again we subtract the 1 m³ volume of the tank to find the volume of the balloon itself, which is 19 m³ (Fig. 15-7).

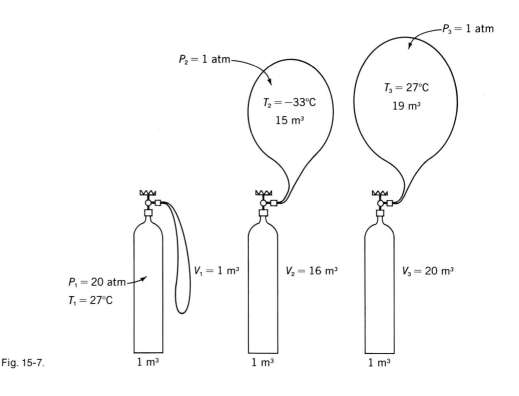

$P_2 = 1$ atm

$P_3 = 1$ atm

$T_2 = -33°C$

$T_3 = 27°C$

15 m³

19 m³

$P_1 = 20$ atm

$T_1 = 27°C$

$V_1 = 1$ m³

$V_2 = 16$ m³

$V_3 = 20$ m³

Fig. 15-7.

1 m³   1 m³   1 m³

15-5 Structure of Matter

Before we go on to see how the ideal gas law is accounted for, it is appropriate to review the notions of element, compound, and solution, which apply to bulk matter, and those of atom and molecule, which apply to matter on the microscopic level.

Liquids and gases are almost always *homogeneous*, which means that every portion of a particular sample is exactly like every other portion. Solids may be either homogeneous or *heterogeneous*; if the latter, some portions of a particular sample may be different from others. A bar of gold, for example, is a homogeneous solid, while a piece of wood is a heterogeneous one. A heterogeneous solid is not always easy to recognize as such, and instruments such as the microscope (or even more sophisticated devices) may be required for definite identification.

Homogeneous substances may be further classified into *elements*, *compounds*, and *solutions*. *Elements* are the simplest substances we encounter in bulk; they cannot be decomposed or transformed into one

Elements

another by ordinary chemical or physical means. There are 103 known elements, listed in Appendix C together with their symbols and certain of their properties, of which 92 have been found in nature and 13 artificially prepared. At room temperature and sea-level atmospheric pressure, 10 elements are in the gaseous state, namely argon, chlorine, fluorine, helium, hydrogen, krypton, nitrogen, oxygen, radon, and xenon, and two are in the liquid state, namely bromine and mercury. The rest are in the solid state, the majority being metals.

Two or more elements may combine chemically to form a *compound*, a new substance whose properties are different from those of the elements that compose it. Each constituent of a *solution*, in contrast, retains its characteristic properties (except, of course, for the mechanical properties of solids and gases dissolved in liquids), and may be separated from the other constituents by relatively simple procedures. Boiling and freezing are examples of such procedures, since the temperatures at which these changes of state occur have specific values for each element or compound. Air, for example, is a solution of several gases, chiefly nitrogen and oxygen. Oxygen boils at $-183°C$ while nitrogen boils at $-196°C$, $13°$ lower; hence if we heat a sample of liquid air to a temperature over $-196°C$ but under $-183°C$, the nitrogen will vaporize and we will be left ideally with oxygen alone. Under certain circumstances, nitrogen and oxygen unite to form the compound nitric oxide; the boiling point of nitric oxide is $-152°C$, and heating a sample of liquid nitric oxide to this temperature will result in the vaporization of the entire sample. The constituents of a solution may be elements or compounds or both.

Compounds and solutions

Another distinction between compounds and solutions is that the elements in a compound are present in certain definite proportions, always the same for a particular compound, while the constituents of a solution may be present in a wide range of proportions. At sea level the ratio by weight of the nitrogen and oxygen in the atmosphere varies slightly about an average of $3.2:1$, and is several percent greater at high elevations; the ratio by weight of the nitrogen and oxygen in nitric oxide is invariably $0.88:1$. If there is an excess of either nitrogen or oxygen when nitric oxide is being prepared, the additional amount will not combine but will be left over and can be separated out at an appropriate temperature (Fig. 15-8).

The idea that matter is not infinitely divisible, that all substances are composed of characteristic individual particles, is an ancient one. The ultimate particles of many compounds are called *molecules*. (Later chapters discuss the structure of compounds more completely.) Although molecules may be further broken down, when this happens they no longer are representative of the original substance. The molecules of a compound consist of the *atoms* of its constituent elements joined together in a definite ratio. Thus

Molecules and atoms

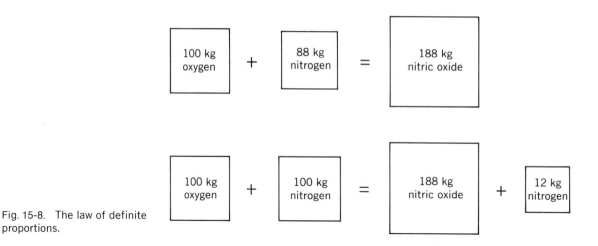

Fig. 15-8. The law of definite proportions.

each molecule of water contains two hydrogen atoms and one oxygen atom. While the ultimate particles of elements are atoms, many elemental gases consist of molecules rather than atoms. Oxygen molecules, for instance, contain two oxygen atoms each. The molecules of other gases, such as helium and argon, are single atoms. Figure 15-9 shows schematically the composition of some common molecules.

Atomic mass unit

The masses of atoms and molecules are usually expressed in *atomic mass units* (u) whose magnitude is such that the most abundant type of carbon atom has a mass of precisely 12 u. Appendix C contains a list of the atomic masses of the elements; if we know the composition of a compound, we can determine the corresponding molecular mass. The carbon atom has an actual mass of 1.992×10^{-26} kg; hence

$$1 \text{ u} = \frac{1.992 \times 10^{-26} \text{ kg}}{12.000} = 1.660 \times 10^{-27} \text{ kg.}$$

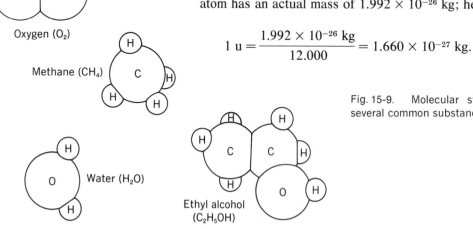

Fig. 15-9. Molecular structures of several common substances.

Problem.   How many H_2O molecules are present in 1 g of water?

Solution.   The mass of the hydrogen atom is 1.008 u and that of the oxygen atom is 16.00 u. Hence

$$2H = 2 \times 1.008\ u = \underline{\ \ 2.02\ u,}$$
$$O = 1 \times 16.00\ u = \underline{16.00\ u}$$
$$18.02\ u,$$

and the mass of the H_2O molecule in kg is

$$m = 18.02\ u \times 1.66 \times 10^{-27}\ kg/u = 2.99 \times 10^{-26}\ kg.$$

The number of H_2O molecules in 1 g $= 10^{-3}$ kg of water is

$$\text{Molecules of } H_2O = \frac{\text{mass of } H_2O}{\text{mass of } H_2O \text{ molecule}} = \frac{10^{-3}\ kg}{2.99 \times 10^{-26}\ kg}$$

$$= 3.34 \times 10^{22}\ \text{molecules.}$$

A considerable amount of experimentation and ingenious reasoning had to be carried out before the reality of atoms and molecules became definitely established. Although we will not go into the full story of the kinetic-molecular theory of matter, a large part of which involves chemistry, we shall show qualitatively that it can account for the ideal gas law. We shall also discuss briefly how the physical properties of solids and liquids, and the deviations of actual gases from the ideal gas, fit into the kinetic-molecular theory.

15-6 Kinetic Theory of Gases

According to the assumptions of the *kinetic theory of gases*, a gas consists of a great many tiny individual molecules that do not interact with one another except when collisions occur. The molecules are supposed to be far apart compared with their dimensions and to be in constant motion, incessantly hurtling to and fro as in Fig. 15-10, being kept from escaping into space only by the solid walls of a container (or, in the case of the earth's atmosphere, by gravity). A natural consequence of the random motion and large molecular separation is the tendency of a gas to completely fill its container and to be readily compressed or expanded.

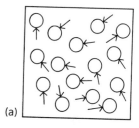
(a)

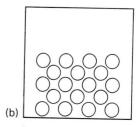
(b)

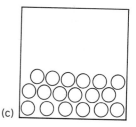
(c)

Fig. 15-10.   (a) The molecules of a gas are in constant, random motion. (b) The constituent particles of a solid are also in motion, but oscillate about definite equilibrium positions. (c) The molecules of a liquid keep a more or less constant distance apart, but move about freely.

Fig. 15-11. A simplified model of a gas.

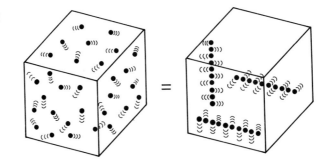

In a solid, on the other hand, the particles of which it is composed are close together, and mutual attractive and repulsive forces hold them in place to provide the solid with its characteristic rigidity. In a liquid the intermolecular forces are sufficient to keep the volume of the liquid constant; however, they are not strong enough to prevent adjacent molecules from sliding past one another, which results in the ability of liquids to flow.

Origin of Boyle's law

Boyle's law follows directly from the picture of a gas as a group of randomly moving molecules. The pressure the gas exerts originates in the impacts of its molecules; the vast number of molecules in even a tiny gas sample means that their separate blows appear as a continuous force to our senses and measuring instruments. Figure 15-11 shows a simplified model of a gas confined to a box. Although the molecules are actually traveling about in all directions, the effects of their collisions with the walls of the box are the same as if one-third of them were moving back and forth between each pair of opposite walls.

When a cylinder containing a gas is doubled in volume, as in Fig. 15-12, those molecules moving up and down have twice as far to go between impacts. Since their velocity is unchanged, the time between impacts is also doubled, and the pressure they exert on the top and bottom of the cylinder falls to half its original value. The expansion of the cylinder also means that the molecules moving horizontally are now spread over twice their former area, and the pressure on the sides of the cylinder accordingly falls to half its original value as well. Thus doubling the volume means halving the pressure, which is Boyle's law. Similar reasoning accounts for a rise in pressure when the volume is reduced.

Charles's law follows from the kinetic theory of gases when a further assumption is made:

Temperature is a measure of molecular energy

The average kinetic energy of the molecules of a gas is proportional to the absolute temperature of the gas.

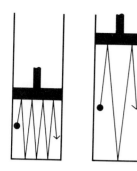
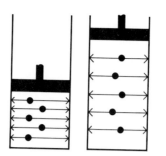

Fig. 15-12. The origin of Boyle's law according to the kinetic theory of gases.

Pressure falls on top and bottom of expanded cylinder because molecules spend more time in transit between collisions

Pressure falls on sides of expanded cylinder because molecules spread their impacts over a larger area

This assumption is reasonable, since we observe that compressing a gas quickly (so no heat can enter or leave the container) raises its temperature, and such a compression must increase the average energy of the molecules because they bounce off the inward-moving piston more rapidly than they approach it (Fig. 15-13). A familiar example of the latter effect is a baseball rebounding with greater speed when struck by a bat. On the other hand, expanding a gas lowers its temperature, and such an expansion reduces

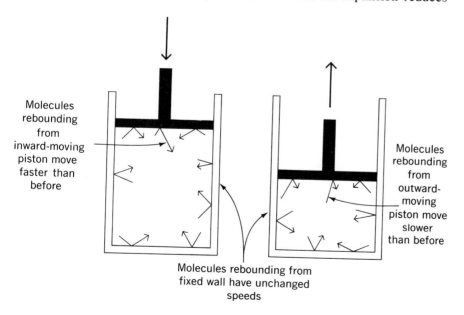

Molecules rebounding from inward-moving piston move faster than before

Molecules rebounding from outward-moving piston move slower than before

Molecules rebounding from fixed wall have unchanged speeds

Fig. 15-13. The temperature of a gas increases when it is compressed because the average energy of its molecules increases; the temperature of a gas decreases when it is expanded because the average energy of its molecules decreases.

molecular energies since molecules lose speed in bouncing off an outward-moving piston. The association between molecular energy and temperature is thus in accord with experience.

Absolute zero

The interpretation of absolute zero in terms of the elementary kinetic theory of gases is a simple one: it is that temperature at which all molecular translational movement in a gas ceases (Fig. 15-14). A more advanced analysis shows that complete cessation of movement is impossible, but the difference is not important for the discussion here. See Section 28-5 for the origin of the *zero-point energy* that particles have at 0 K.

The precise relationship between the average molecular kinetic energy KE_{av} and absolute temperature T is found to be

$$KE_{av} = \tfrac{3}{2}kT, \qquad \textit{Molecular energy} \quad (15\text{-}11)$$

where k, known as Boltzmann's constant, has the value

Boltzmann's constant

$$k = 1.38 \times 10^{-23} \text{ J/K}.$$

Equation (15-11) holds for the molecules of all gases regardless of the masses of their molecules and has been verified by direct measurements of molecular velocities.

Thus we have an interpretation of temperature in terms of molecular motion that is much more precise and definite than simply describing temperature as that which is responsible for sensations of hot and cold.

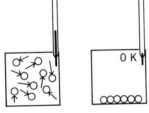

Fig. 15-14. At absolute zero, the kinetic theory of gases predicts that molecular translational motion in a gas will cease. In reality, at absolute zero the molecules would retain a small minimum amount of kinetic energy.

15-7 Molecular Speeds

We can use Eq. (15-11) to compute the average speed of gas molecules whose mass is m:

$$\tfrac{1}{2}m\overline{v^2} = \tfrac{3}{2}kT,$$

$$v_{rms} = \sqrt{\overline{v^2}} = \sqrt{\frac{3kT}{m}}. \qquad (15\text{-}12)$$

The above speed is denoted v_{rms} because it is the square root of the mean of the squared molecular speeds—the "root-mean-square" speed—and therefore different from the simple arithmetic average speed $\bar{v}$. To emphasize their difference with a simple example, we can evaluate both kinds of average for an assembly of two molecules, one with a speed of 1 m/s and the other with a speed of 3 m/s. We find that

$$\bar{v} = \frac{v_1 + v_2}{2} = \frac{(1 + 3)\ \text{m/s}}{2} = 2\ \text{m/s},$$

whereas

$$v_{\text{rms}} = \sqrt{\frac{v_1{}^2 + v_2{}^2}{2}} = \sqrt{\frac{1^2 + 3^2}{2}}\ \text{m/s} = \sqrt{5}\ \text{m/s} = 2.24\ \text{m/s}$$

so that v_{rms} and $\bar{v}$ are not at all the same. The relationship between v_{rms} and $\bar{v}$ depends upon the specific variation in molecular speeds being considered. For the distribution of molecular speeds found in a gas,

$$v_{\text{rms}} \approx 1.09\bar{v}$$

so the root-mean-square speed of Eq. (15-11) is about 9 percent greater than the arithmetic average $\bar{v}$.

Problem. Find the rms speed of oxygen molecules at 0°C.

Solution. Oxygen molecules are composed of two oxygen atoms each. Since the atomic mass of oxygen is 16.00 u, the molecular mass is 32.00 u, and an O_2 molecule has a mass in kg of

$$m = 32.00\ \text{u} \times 1.66 \times 10^{-27}\ \text{kg/u} = 5.31 \times 10^{-26}\ \text{kg}.$$

At an absolute temperature of 273 K (which corresponds to 0°C), the rms speed of oxygen molecules is therefore

$$v_{\text{rms}} = \sqrt{\frac{3kT}{m}} = \sqrt{\frac{3 \times 1.38 \times 10^{-23}\ \text{J/K} \times 273\ \text{K}}{5.31 \times 10^{-26}\ \text{kg}}} = 461\ \text{m/s},$$

which is a little over 1000 mi/hr! Evidently molecular speeds are very large compared with those of the macroscopic bodies familiar to us.

It is important to keep in mind that actual molecular velocities vary considerably on either side of v_{rms}. The graph in Fig. 15-15 shows the distribution of molecular speeds in oxygen at 273 K and in hydrogen at 273 K. The mass of an O_2 molecule is 16 times that of an H_2 molecule. Rms molecular speed decreases with molecular mass, hence at the same temperature molecular speeds in hydrogen are on the average greater than in oxygen. At the same temperature the average molecular *energy* is the same for all gases, however.

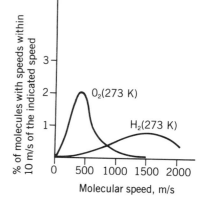

Fig. 15-15. Molecular speeds in oxygen and hydrogen at 273 K (0°C). The smaller masses of H_2 molecules means that they have higher average speeds than O_2 molecules at the same temperature, since the average kinetic energy depends only on temperature.

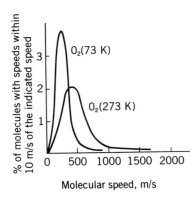

Fig. 15-16.   Molecular speeds in oxygen at 73 K (−200°C) and 273 K (0°C). The higher the temperature, the greater the average kinetic energy.

In Fig. 15-16 we see the distributions of molecular speeds in oxygen at 73 K and at 273 K. The average molecular speed increases with temperature, as predicted. The curves of Figs. 15-15 and 15-16 are not symmetrical because the lower limit to v is fixed at $v = 0$ whereas there is, in principle, no upper limit; actually, as the curves show, the likelihood of speeds many times greater than v_{rms} is small.

The distribution of molecular speeds in a gas has an interesting astronomical consequence. The higher the surface temperature of a planet, the faster the molecules of its atmosphere move, and the greater the chance they may exceed the escape velocity and disappear into space. The smaller a planet, the lower its escape velocity; and the closer a planet is to the sun, the warmer it is. Thus it is not surprising that Mercury, small and hot, has no atmosphere, while the giant outer planets of Jupiter, Saturn, Uranus, and Neptune have extremely dense atmospheres.

The kinetic theory of gases leads directly to the ideal gas law; but, as mentioned earlier, the ideal gas law is only a good approximation of reality. If we examine the initial assumptions that are made, it is easy to see why we should expect discrepancies between theory and experiment. For instance, it is assumed that gas molecules have volumes so small as to be entirely negligible; that they exert no forces upon one another except in actual collisions; and that these collisions conserve kinetic energy of translational motion. (The last assumption means that the molecules are supposed to have no internal energy of their own, such as energy of rotation or vibration, whereas in fact they often do.) When the kinetic theory is worked out starting from more realistic assumptions, the results are in excellent agreement with observational data.

15-8 Molecular Motion in Liquids

The elementary kinetic theory of matter is not as successful when applied to the liquid and solid states as it is when applied to gases; the Newtonian laws of mechanics that gas molecules obey in their translational motion are not adequate to describe the behavior of the molecules in liquids and solids. However, the concept that the internal energy of a substance resides at least in part in the kinetic energies of its molecules helps in understanding a variety of phenomena characteristic of liquids and solids.

The random motion of water molecules led to an important event in the history of science. In 1827 the British botanist Robert Brown noticed that pollen grains in water are in continual, agitated movement. Similar *Brownian motion* is apparent whenever very small particles are suspended in a fluid medium, for example smoke particles in air (Fig. 15-17). According to kinetic theory, Brownian motion originates in the bombardment of the particles by molecules of the fluid. This bombardment is completely random, with successive molecular impacts coming from different directions and contributing different impulses to the particles. Albert Einstein, in 1905, found that he could account for Brownian motion quantitatively by assuming that, as a result of continual collisions with fluid molecules, the particles themselves have the same average kinetic energy as the molecules. Surprising as it may seem, this was the first direct verification of the reality of molecules, and it convinced many distinguished scientists who had previously been reluctant to believe that such things actually exist.

Another kinetic-molecular phenomenon characteristic of the liquid state is evaporation. A dish of water well below its boiling point of 100°C will nevertheless gradually turn into vapor, growing colder as it does so. The faster the evaporation, the more pronounced the cooling effect; alcohol and ether chill the skin upon contact because of their extreme volatility. This behavior follows from the distribution of molecular velocities in a liquid. While not identical with that found in a gas, this distribution resembles those shown in Figs. 15-15 and 15-16 in that a certain fraction of the molecules in any sample have much greater and much smaller velocities than the average. The fastest molecules have enough energy to escape through the liquid surface despite the attractive forces of the other molecules. The molecules left behind redistribute the available energy in collisions among themselves, but, because the most energetic ones escape, the average energy that remains is less than before and the liquid is now at a lower temperature (Fig. 15-18). Boiling occurs when the average molecular energy in a liquid is equal to the work needed to pull the molecules apart against the forces that hold them together.

When molecules from the vapor above a liquid surface impinge on the surface, they may be trapped there, so that a constant two-way traffic of molecules to and from the liquid occurs. If the density of the vapor above the liquid is sufficiently great, as many molecules return as leave it at any time, a situation that is described by saying that the region is *saturated* with the substance. The higher the temperature, the greater the maximum vapor density: at 0°C the density of water vapor at saturation is 5 g/m³, at 20°C it is 17 g/m³, at 100°C it is 598 g/m³, and at 300°C it is all the way up to 45.6 kg/m³. If for any reason (such as a sudden drop in temperature) the vapor density exceeds the saturation value, condensation will be more rapid than evaporation until equilibrium is reestablished.

The *relative humidity* of a volume of air describes its degree of saturation with water vapor. Relative humidities of 0, 50 percent, and 100 percent mean respectively that no water vapor is present, that the air contains half as much moisture as the maximum possible, and that the air is saturated. On a

Fig. 15-17. Brownian motion.

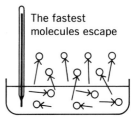

The fastest molecules escape

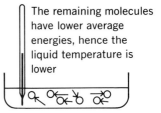

The remaining molecules have lower average energies, hence the liquid temperature is lower

Fig. 15-18. After evaporation, the remaining liquid is cooler than before.

Relative humidity

hot day the evaporation of sweat from the skin is the chief means by which the human body dissipates heat, and a high relative humidity is uncomfortable because it impedes the process. A low relative humidity is also undesirable because it leads to the drying of the skin and mucous membranes. The regulation of relative humidity is as important a function of an air-conditioning system as the regulation of temperature.

15-9 Thermal Expansion in Solids

Thermal expansion in a solid also has a straightforward explanation in terms of kinetic theory. Most solids are crystalline in nature, which means that the various atoms that compose them form a regular arrangement in space. (In some crystalline solids the basic constituents are whole molecules, rather than individual atoms, but we shall refer to them as atoms here for convenience.) The atoms behave as though they are joined together by tiny springs, thereby accounting for Hooke's law, and constantly oscillate about their equilibrium positions.

Figure 15-19 shows how the atomic potential energy of a solid varies with interatomic spacing. The spacing at room temperature a corresponds to the lower portion of the curve, where the energy per atom is least. The amplitude of the vibrations is determined by the width of the curve: when the atomic separation is a minimum or a maximum, the energy of a pair of adjacent atoms is wholly potential, as in the case of an harmonic oscillator at each end of its path, while in the middle their energy is wholly kinetic. The average interatomic spacing is what determines a and hence the dimensions of the solid.

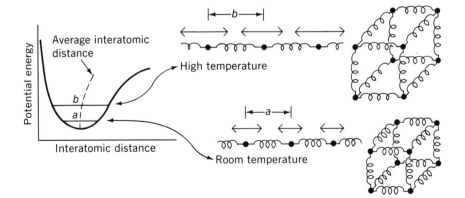

Fig. 15-19. The atomic potential energy of a crystal as a function of the spacing of its constituent atoms. The increase in the average interatomic distance with energy is the cause of thermal expansion in a solid.

When the internal energy of a solid increases, the atomic spacing alternates through a wider range than before. If the potential energy curve were symmetrical about the distance a, no change in the dimensions of the solid would occur, since a is always halfway between the two parts of the curve. However, the attractive and repulsive forces between atoms vary with distance in different ways, with the repulsive force increasing more rapidly as the atoms move closer together than the attractive force increases as the atoms move farther apart. Consequently, as shown in Fig. 15-19, the average interatomic spacing increases to b as the internal energy of the solid increases and the amplitudes of the atomic vibrations increase. As it happens, changes in the average interatomic spacing are very nearly proportional to changes in temperature, leading to the linear thermal expansion formula of Eq. (15-1).

Special Topic

The Mole

A *mole* of any substance is that quantity of it whose mass equals its molecular mass (or atomic mass if it is an elemental substance that consists of individual atoms) expressed in grams instead of in atomic mass units. A mole of water has a mass of 18.02 g because a water molecule has a mass of 18.02 u. The utility of the mole arises from the fact that a mole of any substance contains exactly as many molecules as a mole of any other substance. The mole is widely used in chemistry in place of mass as a measure of quantity because the primary interest of the chemist is usually in the relative numbers of atoms and molecules that react together.

The number of molecules in a mole is a universal constant known as *Avogadro's number*, whose value is

$$N_0 = 6.023 \times 10^{23} \text{ molecules/mole.}$$

The number of molecules in a sample of a substance is equal to the number of moles n it contains multiplied by N_0.

Under identical conditions of pressure and temperature, equal volumes of gases contain equal numbers of molecules. The reason for this is that the molecules in a sample of a gas have negligible volumes compared with the volume of the sample itself. Under the standard conditions of 0°C and atmospheric pressure, usually referred to by the abbreviation STP, one mole of any gas is found to occupy a volume of 22.4 liters. This observation makes it easy to deal with gas volumes in chemical processes. If a certain reaction produces 5 moles of a gas, for example, at STP the volume of the gas will be 5 moles × 22.4 liters/mole = 112 liters.

According to the ideal gas law, the pressure, temperature, and volume of a particular gas sample obey the relationship

$$\frac{pV}{T} = \text{constant}$$

at all times. We can find the value of this constant for a sample that contains n moles by noting that, at STP, its volume must be $V = n \times 22.4$ liters/mole, its temperature must be $T = 273$ K, and its pressure must be $p = 1$ atm. Hence

$$\frac{pV}{T} = \frac{n \times 1 \text{ atm} \times 22.4 \text{ liters/mole}}{273 \text{ K}} = nR$$

where R, the *universal gas constant*, has the value 0.0821 atm-liter/mole-K. In SI units, in which p is expressed in N/m² and V in m³, the value of R is 8.31 J/mole-K. The ideal gas law is often written

$$pV = nRT.$$

Although at first glance it is just another fact to add to our collection, the presence of 6.023×10^{23} molecules in 22.4 liters of any gas at STP is surely remarkable. If a cubic centimeter of air at STP — a thimbleful — were to be divided equally among all the four billion people on the earth, each would receive nearly 7 billion molecules! There are about as many molecules in an average breath of air as there are breaths in the entire atmosphere, so that, as James Jeans has said, "if we assume that the last breath of, say, Julius Caesar, has by now become thoroughly scattered through the atmosphere, then the chances are that each of us inhales one molecule of it with every breath we take."

Important Terms

The **coefficient of linear expansion** is the ratio between the change in length of a solid rod of a particular material and its original length per 1° change in temperature. The **coefficient of volume expansion** is the ratio between the change in volume of a sample of a particular solid or liquid and its original volume per 1° change in temperature.

Boyle's law states that, at constant temperature, the absolute pressure of a sample of a gas is inversely proportional to its volume, so that $pV =$ constant at that temperature regardless of changes in either p or V individually.

Charles's law states that, at constant pressure, the volume of a sample of a gas is directly proportional to its absolute temperature, so that $V/T =$ constant at that pressure regardless of changes in either V or T individually.

The **Kelvin absolute temperature scale** has its zero point at $-273°C$; temperatures in this scale are designated K. The **Rankine absolute temperature scale** has its zero point at $-460°F$; temperatures in this scale are designated °R. **Absolute zero** is 0 K $= 0°R = -273°C = -460°F$.

The equation $pV/T =$ constant, a combination of Boyle's and Charles's laws, is called the **ideal gas law** and is obeyed approximately by all gases.

According to the **kinetic theory of gases**, a gas consists of a great many tiny individual molecules that do not interact with one another except when collisions occur. The molecules are far apart compared with their dimensions and are in constant random motion. The ideal gas law may be derived from the kinetic theory of gases. The average kinetic energy of gas molecules is proportional to the absolute temperature of the gas. At absolute zero, gas molecules would have virtually no kinetic energy of translational motion.

The *relative humidity* of a volume of air is the ratio between the amount of water vapor it contains and the amount that would be present at saturation.

Important Formulas

Thermal expansion:

$$\Delta L = aL_0\Delta T$$

$$\Delta V = bV_0\Delta T$$

Boyle's law:

$$pV = \text{constant} \quad [T = \text{constant}]$$

Absolute temperature scale:

$$T_K = T_C + 273°$$

Rankine scale:

$$T_R = T_F + 460°$$

Charles's law:

$$\frac{V}{T} = \text{constant} \quad [p = \text{constant}]$$

Ideal gas law:

$$\frac{pV}{T} = \text{constant}$$

Molecular kinetic energy:

$$\text{KE}_{av} = \tfrac{3}{2}kT$$

Multiple Choice

1. Two elements *cannot* be combined chemically to make
   a. a compound.   b. another element.
   c. a gas.   d. a liquid.

2. The relative proportions of the elements in a compound
   a. may vary considerably.
   b. may vary only slightly.
   c. do not vary.
   d. may or may not vary, depending on the compound.

3. A pinch of salt is added to a glass of water. The result is
   a. an element.
   b. a compound.
   c. a solution.
   d. a heterogeneous substance.

4. Which of the following statements is not correct?
   a. Matter is composed of tiny particles called molecules.
   b. These molecules are in constant motion.

c. All molecules have the same size and mass.

d. The differences between the solid, liquid, and gaseous states can be attributed to the relative freedom of motion of their respective molecules.

5. The smallest subdivision of a compound that exhibits its characteristic properties is called
   a. an elementary particle.
   b. an atom.
   c. a molecule.
   d. an element.

6. The energy of molecular motion appears in the form of
   a. friction.   b. internal energy.
   c. temperature.   d. potential energy.

7. The volume of a gas sample is proportional to its
   a. fahrenheit temperature.
   b. celsius temperature.
   c. absolute temperature.
   d. pressure.

8. Absolute zero may be regarded as that temperature at which
   a. water freezes.
   b. all gases become liquids.
   c. all substances are solid.
   d. molecular motion in a gas would be the minimum possible.

9. The kinetic-molecular theory of gases predicts that, at a given temperature,
   a. all of the molecules in a gas have the same average speed.
   b. all of the molecules in a gas have the same average energy.
   c. light gas molecules have lower average energies than heavy gas molecules.
   d. light gas molecules have higher average energies than heavy gas molecules.

10. The volume of a gas is held constant while its temperature is raised. The pressure the gas exerts on the walls of its container increases because
    a. the masses of the molecules increase.
    b. each molecule loses more kinetic energy when it strikes the wall.
    c. the molecules are in contact with the wall for a shorter time.

d. the molecules have higher average speeds and strike the wall more often.

11. The temperature of a gas is held constant while its volume is reduced. The pressure the gas exerts on the walls of its container increases because its molecules
    a. strike the container walls more often.
    b. strike the container walls with higher velocities.
    c. strike the container walls with greater force.
    d. have more energy.

12. In the formula $KE = \frac{3}{2}kT$ for the average energy of a gas molecule at the absolute temperature T, the constant k is known as
    a. the atomic mass constant.
    b. Boyle's constant.
    c. Charles's constant.
    d. Boltzmann's constant.

13. A copper bar is 1 m long at 20°C. At what temperature will it be shorter by 1 mm?
    a. −17°C   b. −39°C
    c. −59°C   d. −79°C

14. An absolute temperature of 100 K is the same as a celsius temperature of
    a. −173°C.   b. 32°C.
    c. 212°C.   d. 373°C.

15. The boiling point of water on the rankine scale is
    a. −248°R.   b. −61°R.
    c. 485°R.   d. 672°R.

16. A certain container holds 1 kg of air at atmospheric pressure. When an additional kg of air is pumped into the container, the new pressure is
    a. $\frac{1}{2}$ atm. b. 1 atm.
    c. 2 atm.   d. 4 atm.

17. If the absolute pressure on 10 ft³ of air is increased from 30 lb/in² to 120 lb/in², the new volume of the air will be
    a. 2.5 ft³.   b. 5 ft³.
    c. 40 ft³.   d. 900 ft³.

18. A sample of hydrogen gas is compressed to half its original volume while its temperature is held constant. If the average velocity of the hydrogen molecules was originally v, their new average velocity is

a. $4v$.   b. $2v$.

c. v.   d. $\frac{1}{2}v$.

19. At which of the following temperatures would the molecules of a gas have twice the average kinetic energy they have at room temperature, 20°C?
    a. 40°C.   b. 80°C.
    c. 313°C.   d. 586°C.

20. The mass of a nitrogen molecule is fourteen times greater than that of a hydrogen molecule. The temperature of a sample of hydrogen whose average molecular energy is equal to that in a sample of nitrogen at 300 K is
    a. 6.5 K.   b. 21 K.
    c. 1122 K.   d. 4200 K.

Exercises

1. Classify the following substances as homogeneous or heterogeneous: salt, leather, stone, diamond, iron, blood, solid carbon dioxide, gaseous carbon dioxide, helium, rust.

2. Classify the following homogeneous liquids as elements, compounds, or solutions: mercury, alcohol, gin, pure water, sea water, bromine, tea, glycerin, liquid oxygen, liquid air.

3. In the construction of a light bulb, wires are led through the glass at the base by means of airtight seals. If the wires were made of copper, what would happen when the light is turned on and the bulb heats up? What must be true for a wire to be successfully used for this purpose?

4. Starting from the ideal gas law, obtain an equation relating the pressure and temperature of a gas at constant volume.

5. Verify that the force associated with the thermal expansion or contraction of a solid object depends upon its cross-sectional area but not upon its length.

6. At absolute zero, an ideal gas sample would occupy zero volume. Why would an actual gas not occupy zero volume at absolute zero?

7. According to the kinetic theory of gases, molecular motion virtually ceases only at absolute zero. How can this be reconciled with the definite shape and volume of a solid at temperatures well above absolute zero?

8. Molecular speeds are comparable with those of rifle bullets, yet a gas with a strong odor, such as ammonia, takes a few minutes to diffuse through a room. Why?

9. Actual molecules attract one another slightly. Does this tend to increase or decrease gas pressures from values computed from the ideal gas law? Why?

10. The air in a closed container is saturated with water vapor at 20°C. (a) What is the relative humidity? (b) What happens to the relative humidity if the temperature is reduced to 10°C? (c) If the temperature is increased to 30°C?

11. Why does the air in a heated room tend to be dry?

12. A steel bridge is 300 ft long at 80°F. What is its change in length when the temperature falls to 20°F?

13. A steel tape measure is calibrated at 70°F. A reading of 120 ft 0 in. is obtained when it is used to determine the width of a building at 0°F. What is the true width of the building at 0°F?

14. How large a gap should be left between steel rails that are 10 m long when laid at 20°C if they are to just barely touch at 30°C?

15. A rod 2 m long expands by 1 mm when heated from 8°C to 70°C. What is the coefficient of linear expansion of the material from which the rod is made?

16. The outside diameter of a wheel is 1.000 m. An iron tire for this wheel has an inside diameter of 0.992 m at 20°C. To what temperature must the tire be heated in order for it to fit over the wheel?

17. A pyrex beaker is filled to the brim with 250 cm³ of glycerin at 15°C. How much glycerin overflows at 25°C?

18. A pyrex flask holds 500 cm³ of mercury at 0°C. How much mercury will run out when it is heated to 80°C?

19. Vodka that is "100 proof" is a mixture of half ethyl alcohol and half water. How much profit per quart will a merchant make if he buys it at $5.00 per quart at 30°F and sells it at $5.00 per quart at 80°F?

20. A concrete swimming pool 40 ft × 20 ft × 8 ft is filled with water to within $\frac{1}{4}$ in. of the top when the temperature is 50°F. The coefficient of linear expansion of the concrete used is 0.5×10^{-5}/°F. What will happen to the water level as the temperature increases? If it rises, at what temperature will the water begin to overflow?

21. A sample of gas occupies 2m³ at 300 K and an absolute pressure of 2×10^5 N/m². (a) What is its pressure at the same temperature when it has been compressed to a volume of 1 m³? (b) What is its volume at the same temperature when its pressure has been decreased to 1.5×10^5 N/m²? (c) What is its volume at a temperature of 400 K and a pressure of 2×10^5 N/m²?

22. A sample of gas occupies 100 cm³ at 0°C and 1 atm pressure. What is its volume (a) at 50°C and 1 atm pressure; (b) at 0°C and 2.2 atm pressure; (c) at 50°C and 2.2 atm pressure?

23. A sample of gas occupies a volume of 8 ft³ at a temperature of 400°R and 1 atm pressure. What is its volume at 500°R and 1 atm pressure?

24. An automobile tire contains air at a gauge pressure of 24 lb/in² at 40°F. If the volume of the tire is unchanged, what will the pressure be when the temperature has increased to 80°F?

25. To what temperature must a gas sample initially at 27°C be raised in order for the average energy of its molecules to double? For their average speed to double?

26. What is the average kinetic energy of the molecules of a gas (a) at 0°C? (b) at 100°C?

27. The average speed of a hydrogen molecule at room temperature is about 1 mi/s. What is the average speed of an oxygen molecule, whose mass is 16 times greater, at this temperature?

28. Consider the following gases: CO_2, UF_6, H_2, He, Xe, NH_3. (a) Which has the highest average molecular speed at a given temperature? (b) Which has the lowest average molecular speed?

Problems

1. The density of air is 1.293 kg/m³ at 0°C and 1 atm pressure. Find its density at 100°C and 2 atm pressure.

2. The density of lead is 11.0 g/cm³ at 20°C. Find its density at 200°C.

3. An aluminum mast whose cross-sectional area is 0.05 ft² is 60 ft long at 60°F. (a) By how much does it increase in length when its temperature rises to 100°F? (b) How much force is associated with the expansion?

4. A load of 4000 kg is placed on a vertical steel column 3 m long and cross-sectional area 50 cm² when the temperature is 20°C. To what should the temperature be increased if the length of the column is to be the same after the load is applied as it was originally?

5. A diver blows an air bubble 1 cm in diameter at a depth of 10 m in a fresh-water lake where the temperature is 5°C. What is the diameter of the bubble when it reaches the surface of the lake where the temperature is 20°C? [*Note:* the volume of a sphere of radius r is $\frac{4}{3}\pi r^3$, but for this problem it is not necessary to actually calculate the volume of the bubble.]

6. Two vessels of the same size are at the same temperature. One of them holds 1 kg of H_2 gas and the other holds 1 kg of N_2 gas. (a) Which vessel contains more molecules? How many times more? (b) Which vessel is under the greater pressure? How many times greater? (c) In which vessel are the average molecular speeds greater? How many times greater?

7. (a) Find the average speed of carbon dioxide (CO_2) molecules at 0°C. (b) At what temperature would this speed be doubled?

8. Silver is a vapor at 1500 K. What is the average speed of silver atoms in a vapor at this temperature?

9. The average speed of air molecules is roughly 4×10^2 m/s, and the average distance an air molecule goes between collisions with other molecules is about 10^{-7} m. What is the average number of collisions an air molecule makes per second?

10. One of the assumptions of the kinetic theory is that the average distance between molecules is much greater than the dimensions of the molecules themselves. Oxygen and nitrogen molecules are roughly 2×10^{-10} m in diameter, and there are 2.7×10^{25} molecules in a cubic meter of air at room temperature and atmospheric pressure. (a) On the average, how far apart are the molecules in air? (b) How many molecular diameters is their average separation?

Answers to Multiple Choice

1. b	8. d	15. d
2. c	9. b	16. c
3. c	10. d	17. a
4. c	11. a	18. c
5. c	12. d	19. c
6. b	13. b	20. c
7. c	14. a	

16

Thermodynamics

Thermodynamics has as its basic concern the transformation of heat into mechanical energy. Thermodynamics thus plays a central role in technology, since almost all the "raw" energy available for our use is liberated in the form of heat. A device or system that converts heat into mechanical energy is called a heat engine, and the principles that govern its operation are the same whether it is an automobile engine whose heat source is the burning of gasoline, a steam turbine whose heat source is a nuclear reactor, or the earth's atmosphere whose heat source is the sun. Before we look into these principles, though, it is appropriate to examine how heat is transferred from one place to another.

16-1 Conduction and Convection

The three mechanisms of heat transfer are illustrated in Fig. 16-1. When we place one end of an iron rod in a fire, the other end becomes warm as a result of the conduction of heat through the iron. Conduction is a very slow process in air; a stove warms a room chiefly through the actual movement of heated air, a process called convection. Neither conduction nor convection can take place appreciably in the virtual void of interplanetary space. Instead, the heat the earth receives from the sun arrives in the form of radiation. These mechanisms all embody a fundamental fact: the natural direction of heat flow is from hot bodies to cold ones.

In most materials, conduction is a simple consequence of the kinetic behavior of matter. Molecules (or atoms, depending upon the nature of the rod) at the hot end of a rod vibrate faster and faster as the temperature there increases. When these molecules collide with their less energetic neighbors, some of their kinetic energy is transferred to the latter (Fig. 16-2). Through successive molecular collisions energy travels down the rod, and, since we perceive random molecular motion as heat, we equally well describe the situation by saying that heat travels down the rod. The average positions of the molecules themselves do not change in conduction.

For heat to be conducted through a body, its ends must be at different temperatures. If the entire body is at the same temperature, all its molecules have the same average energy, and any molecule has as much chance of losing energy in a collision with a nearby molecule as it has of gaining energy. Thus here there is no flow of energy from a region of rapidly moving molecules to an adjacent one of slowly moving molecules. In fact, we drew upon the necessity of a temperature difference for heat transport in defining temperature: one body has a higher temperature than another if, when they are placed in contact, heat flows from the former to the latter.

There are wide differences in the ability of various substances to conduct heat. Gases are poor conductors, because their molecules are relatively

Conduction

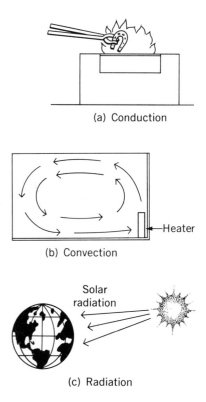

(a) Conduction

(b) Convection

(c) Radiation

Fig. 16-1. Mechanisms of heat transfer.

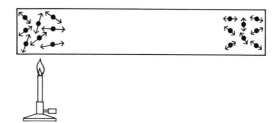

Fig. 16-2. Molecules at the hot end of a rod vibrate faster than those at the cold end.

far apart and collisions between them correspondingly infrequent. The molecules of liquids and nonmetallic solids are closer together, leading to somewhat higher thermal conductivities. Metals exhibit by far the greatest ability to conduct heat; this is why saucepans are made of metal but have handles of wood or plastic.

Conductivity of metals

The reason for the exceptional thermal conductivity of metals is the same as for their exceptional electrical conductivity: a significant number of electrons are able to move about freely instead of being bound permanently to particular atoms. Acquiring kinetic energy at the hot end of a metal object, the free electrons can travel past many atoms before giving up their energy in collisions, and thereby can speed up the rate of energy transport toward the cold end of the object. Heat conduction by free electrons in a metal compares with heat conduction by molecular interactions as travel by express train compares with travel by local train. We shall further explore the important subject of free electrons in metals in a later chapter.

Convection

Convection is a much simpler physical process than conduction, since it consists of the actual motion of a volume of hot fluid from one place to another. The hot fluid displaces cold fluid in its path, thereby setting up a *convection current*. Convection is the chief mechanism of heat transfer in fluids under most circumstances.

Convection may be either *natural* or *forced*. In natural convection, the buoyancy of a heated fluid leads to its motion; when a portion of a fluid (either gas or liquid) is heated, it expands to become of lower density than the surrounding, cooler fluid, and hence rises upward (Fig. 16-3). A steam or hot-water heating system employs radiators in each room which heat the rooms with the help of the convection currents they set up.

In forced convection, a blower or pump directs the heated fluid to its destination. In the cooling system of most cars, water is circulated between the engine block and the radiator by a pump; in the radiator, heat is transferred to the atmosphere by conduction through the thin-walled metal tubes of which it is constructed. A hot-air household heating system employs a fan to blow the air from the furnace through ducts to outlets in each room.

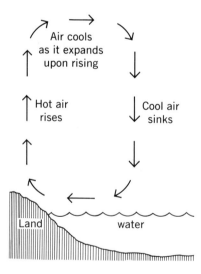

Fig. 16-3. The temperature of a land surface rises more rapidly than that of a water surface in sunlight. The resulting convection is responsible for the sea breeze experienced on sunny days near the shore of a body of water.

16-2 Radiation

In the process of radiation, energy is transported by means of *electromagnetic waves*. These waves travel at the speed of light (3×10^8 m/s = 186,000 mi/s), and require no material medium for their passage. Radio and radar waves, light waves, and x- and gamma-rays are all electromagnetic waves; they differ only in their wavelength. Figure 16-4 shows the classification of electromagnetic waves according to wavelength.

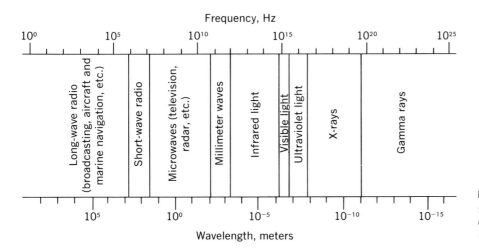

Frequency, Hz

Wavelength, meters

Fig. 16-4. The electromagnetic wave spectrum. The boundaries of the various categories are not sharp except in the case of visible light.

Electromagnetic waves

Every object radiates electromagnetic waves of all wavelengths, though the intensities of the different wavelengths vary considerably. We are all familiar with the glow of a hot piece of metal, where enough radiation is emitted as visible light for our eyes to respond, but other wavelengths are also given off. An object need not be so hot that it gives off visible light for it to be radiating electromagnetic energy — the radiation from an object at room temperature, for instance, is mainly in the infrared part of the spectrum to which the eye is not sensitive.

The ability of an object to emit radiation is proportional to its ability to absorb radiation: a good absorber is a good emitter, and vice versa. This conclusion follows from the fact that something at the same temperature as its environment must be absorbing and emitting radiation at exactly the same rates. When an object is warmer than its environment, it emits more radiation than it absorbs, and it is this difference we perceive. A perfect absorber is called a *blackbody*, and it is accordingly the best possible radiator as well.

Blackbody

The rate R (in W/m²) at which an object of surface area A and absolute temperature T emits radiation is given by the *Stefan-Boltzmann law* as

$$R = \frac{P}{A} = e\sigma T^4.$$

Stefan-Boltzmann law (16-1)

The value of the constant σ (Greek letter *sigma*) is

$$\sigma = 5.67 \times 10^{-8} \text{ W/m}^2 \cdot \text{K}^4.$$

The *emissivity e* depends on the nature of the radiating surface and ranges

Emissivity

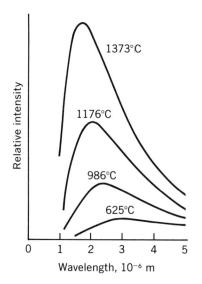

Fig. 16-5.   The intensity of the electro-magnetic radiation emitted by a heated object at various temperatures, as a function of wavelength.

from 0, for a perfect reflector which does not radiate at all, to 1, for a black-body.

Figure 16-5 shows how the intensity of the radiation emitted by an object varies with wavelength at various temperatures. The total emitted radiation (which is proportional to the area under each curve) increases with increasing temperature, while the wavelength corresponding to the peak of each curve decreases. An object that glows red is not as hot as one that glows bluish-white, since red light has the longer wavelength (see Chapter 24).

Problem.   A *thermograph* is a device that measures the amount of infrared radiation each small portion of a person's skin emits and presents this information in pictorial form by different shades of gray or different colors in a *thermogram*. The skin over a tumor is warmer than elsewhere (perhaps because of increased blood flow or a higher rate of metabolism), and thus a thermogram is a valuable diagnostic aid for detecting such maladies as breast and thyroid cancer. To verify that a small difference in skin temperature leads to a significant difference in radiation rate, calculate the percentage difference between the radiation from skin at 34°C and at 35°C.

Solution.   The emissivity of the skin is the same at both temperatures, whose values on the absolute scale are $T_1 = 34°C + 273 = 307$ K and $T_2 = 35°C + 273 = 308$ K. Since R_1 is proportional to T_1^4 and R_2 is proportional to T_2^4,

$$\frac{R_2 - R_1}{R_1} = \frac{T_2{}^4 - T_1{}^4}{T_1{}^4} = \frac{(308\ \text{K})^4 - (307\ \text{K})^4}{(307\ \text{K})^4} = 0.013,$$

which is 1.3 percent.

16-3 First Law of Thermodynamics

As mentioned earlier, a heat engine is any device or system that converts heat into work. Three characteristic processes take place in all heat engines:

1. Heat is absorbed from a source at a high temperature;
2. Mechanical work is done;
3. Heat is given off at a lower temperature.

Different heat engines carry out these processes in different ways, but the general pattern of operation is always the same (Fig. 16-6).

Two general principles have been found to apply to all heat engines. The *first law of thermodynamics* expresses the conservation of energy: Energy cannot be created or destroyed, but may be converted from one form to another. In terms of a heat engine, this law states that

Net heat input = work output + change in internal energy of engine.

If the engine operates in a cycle, energy may be alternately stored and released from storage, but the engine does not experience a net change in its internal energy. In this case

Net heat input = work output.

The net heat input equals the amount of heat the engine takes in from a reservoir at high temperature minus the amount of heat it exhausts to a reservoir at low temperature. In a steam engine the high-temperature reservoir is the boiler, and the low-temperature reservoir is the escaping steam; in a gasoline engine the high-temperature reservoir is the exploding mixture of air and gasoline vapor in each cylinder, and the low-temperature reservoir is the exhaust gas; in the earth's atmosphere the ultimate high-temperature reservoir is the sun, and the ultimate low-temperature reservoir is the rest of the universe.

A *refrigerator* is a heat engine operating in reverse. Ordinarily a heat engine absorbs heat from a high-temperature reservoir and exhausts it to a

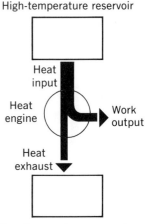

High-temperature reservoir

Low-temperature reservoir

Fig. 16-6. The work output of a heat engine is the difference between the amount of heat it takes in from a high-temperature reservoir and the heat it exhausts to a low-temperature reservoir.

First law of thermodynamics

The refrigerator

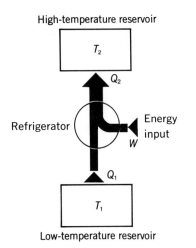

High-temperature reservoir

Refrigerator

Low-temperature reservoir

Fig. 16-7. A refrigerator takes heat from a low-temperature reservoir and transfers it to a high-temperature one. Energy must be supplied to permit this reverse flow of heat to occur.

Heat is disordered energy

low-temperature one, with an output of mechanical work being provided in the process. In a refrigerator, on the other hand, heat is transferred from a low-temperature reservoir (typically a storage chamber) to a high-temperature one (the outside world), and mechanical energy must be supplied in order to do this. In effect, heat must be "pushed uphill" if it is to go from a cold region to a warm one (Fig. 16-7).

It is important to keep in mind that a refrigerator does not "produce cold," since cold is a relative deficiency of internal energy and not something in its own right, as heat is. What a refrigerator does is to remove internal energy from a specific region and transport it elsewhere.

16-4 Second Law of Thermodynamics

Heat is the easiest and cheapest form of energy to obtain, since all we need do to liberate it is to burn a fuel such as wood, coal, or oil. The real problem is to turn heat into mechanical energy so it can power cars, ships, airplanes, electric generators, and machines of all kinds. To appreciate the problem, we recall that heat consists of the kinetic energies of moving atoms and molecules. In order to change heat into a more usable form, we must extract some of the energy of the random motions of atoms and molecules and convert it into regular motions of a piston or a wheel. Such conversions cannot take place efficiently, for the same reason that it is easier to shatter a wineglass than to reassemble the fragments: the natural tendency of all physical systems is toward increasing disorder. The *second law of thermodynamics* is an expression of this tendency, whose role in the evolution of the universe is quite as central as are those of the various conservation principles.

The first law of thermodynamics prohibits an engine from operating without a source of energy, but it does not tell us anything about the character of possible sources of energy. For instance, there is an immense amount of internal energy in the atmosphere and the oceans, yet we know that more work must be done to extract this energy than can be performed with its help. Or, to give an extreme case, it is energetically possible for a puddle of water to rise spontaneously into the air, cooling and freezing into ice as its internal energy changes into potential energy (Fig. 16-8). After all, a block of ice dropped from a sufficient height melts when it strikes the ground, its initial potential energy first being converted to kinetic energy and then into heat. Needless to say, water does not rise upward of its own accord, and we must find an appropriate way of expressing this conclusion.

The second law of thermodynamics is the physical principle, independent of the first law and not derivable from it, that supplements the

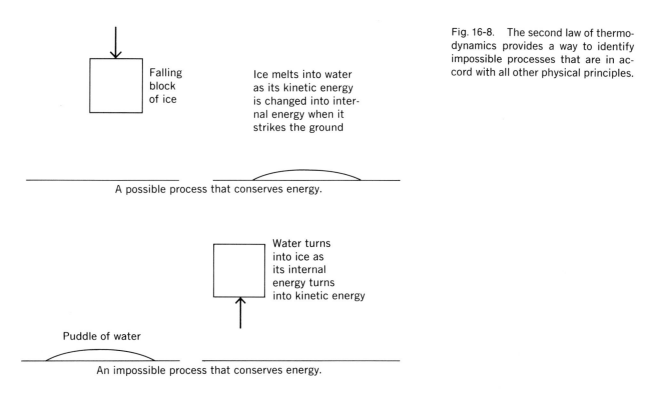

Fig. 16-8.   The second law of thermodynamics provides a way to identify
impossible processes that are in accord with all other physical principles.

first law in limiting our choice of heat sources for our engines. It can be
stated in a number of equivalent ways, a common one being as follows:

**It is impossible to construct an engine, operating in a cycle (that is,
continuously), which does nothing other than take heat from a source
and perform an equivalent amount of work.**

Second law of thermodynamics

According to the second law of thermodynamics, then, no engine can be
completely efficient—some of its heat input *must* be ejected. As we shall
see, the greatest efficiency any heat engine is capable of depends upon the
temperatures of its heat source and of the reservoir to which it exhausts
heat. The greater the difference between these temperatures, the more
efficient the engine. The second law is a consequence of the empirical fact
we have already noted:

*A heat engine extracts energy from
the flow of heat through it*

The natural direction of heat flow is from a reservoir of internal

Fig. 16-9.   The second law of thermodynamics.

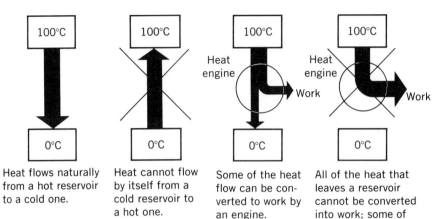

Heat flows naturally from a hot reservoir to a cold one.

Heat cannot flow by itself from a cold reservoir to a hot one.

Some of the heat flow can be converted to work by an engine.

All of the heat that leaves a reservoir cannot be converted into work; some of the heat must flow into a cold reservoir.

Engine does work

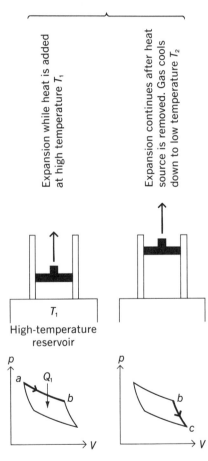

Fig. 16-10.   The Carnot cycle.

energy at a high temperature to a reservoir of internal energy at a low temperature, regardless of the total energy content of each reservoir.

The latter statement, in fact, may be regarded as an alternative expression of the second law (Fig. 16-9).

If we are to utilize the internal energy content of the atmosphere or the oceans, we must first provide a reservoir at a lower temperature than theirs in order to extract heat from them. There is no reservoir in nature suitable for this purpose, for if there were, heat would flow into it until its temperature reached that of its surroundings. To establish a low-temperature reservoir, we must employ a refrigerator (which is a heat engine running in reverse by using up energy to extract heat), and in so doing we will perform more work than we can successfully obtain from the heat of the atmosphere or oceans.

The laws of thermodynamics can be summarized by saying that the first law prohibits us from getting something for nothing, while the second law prohibits us from doing as well as breaking even.

16-5 The Carnot Engine

All heat engines behave in the same way: they absorb heat at a high temperature, convert some of it into work, and exhaust the rest at a low temperature. We may therefore simplify the task of analyzing their principles of operation by referring to an idealized heat engine free of the complications involved in actual engines. A suitably simple and straightforward model was suggested for this purpose by Sadi Carnot in 1824. The *Carnot engine* is not subject to such practical difficulties as friction and the loss of stored heat by conduction or radiation, but naturally must obey all physical laws.

Our chief concern is the efficiency of the Carnot engine: What proportion of the heat supplied to it can it transform into mechanical energy? Of course, all the effects we are neglecting reduce the performance of actual engines below that of the Carnot engine, but it is of great interest to find out the ultimate limits of engine efficiency.

A Carnot engine turns heat into work without itself undergoing a permanent change. It is distinct from, say, a dynamite blast, which is a one-shot process rather than a cyclic one that can continue indefinitely. A Carnot engine consists of a cylinder that is filled with an ideal gas and has a movable piston at one end. The four stages in its operating cycle are shown in Fig. 16-10, together with a graph of each stage on a pressure-volume diagram. These stages are as follows:

1. An amount of heat Q_1 is added to the gas, which expands at its initial temperature T_1. The heat added equals exactly the work done by the gas, which is why its temperature does not change.

2. The heat source is removed, and the expansion is allowed to continue. The second expansion takes place at the expense of the energy stored in the gas, and so the gas temperature falls from T_1 to T_2. During expansions 1 and 2 the piston exerts a force on whatever it is attached to, and thereby performs work.

3. Having done work in pushing the piston outward, the engine must now be returned to its initial state in order for it to be able to do further work. The third stage involves a compression of the gas at the constant temperature T_2 during which an amount of heat Q_2 is given off. The heat given off exactly equals the work done on the gas by the piston, which is why its temperature does not change.

4. The gas is returned to its intial temperature, pressure, and volume by a compression in which heat is neither added to it nor removed from it. Work is done on the gas in this compression, which is why its temperature rises.

In its cycle the Carnot engine performs some net amount of work W, which is the difference between the work it does during the two expansions and the work done on it during the two compressions. It has taken in the heat Q_1 and ejected the heat Q_2; we observe that the heat Q_2 *must* be ejected if the engine is to return to its initial state to await another cycle. The efficiency of any engine is

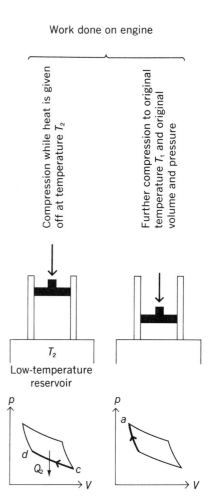

Fig. 16-10.   (continued).

$$\text{Eff} = \frac{\text{work output}}{\text{energy input}},$$

so that for a Carnot engine

$$\text{Eff} = \frac{W}{Q_1}. \tag{16-2}$$

According to the first law of thermodynamics, we have

$$W = Q_1 - Q_2, \tag{16-3}$$

since at the end of the cycle the gas has the same properties it began with (Fig. 16-11). Hence

$$\text{Eff} = \frac{Q_1 - Q_2}{Q_1} = 1 - \frac{Q_2}{Q_1}. \tag{16-4}$$

The smaller the ratio of the ejected heat Q_2 to the absorbed heat Q_1, the more efficient the engine.

The heat Q transferred to or from a Carnot engine is directly proportional to the absolute temperature T of the reservoir with which it is in contact. That is,

$$\frac{T}{Q} = \text{constant}. \tag{16-5}$$

[The derivation of Eq. (16-5) is not given here because, although it does not involve any physical principles we do not already know, a knowledge of advanced mathematics is necessary.] As a consequence of Eq. (16-5), the ratio Q_2/Q_1 between the amounts of heat ejected and absorbed per cycle by a Carnot engine is equal to the ratio T_2/T_1 between the temperatures of the respective reservoirs:

$$\frac{Q_2}{Q_1} = \frac{T_2}{T_1}. \tag{16-6}$$

The efficiency of a Carnot engine may therefore be written

$$\text{Eff} = 1 - \frac{T_2}{T_1}. \qquad \textit{Carnot efficiency} \quad (16\text{-}7)$$

The smaller the ratio between the absolute temperatures T_2 and T_1, the more efficient the engine (Fig. 16-12). No engine can be 100 percent efficient, because no reservoir can have an absolute temperature of 0 K. (Even if such a reservoir could somehow be created, the exhaust of heat to it by the engine would raise its temperature above 0 K at once.)

It is possible to prove that a Carnot engine operating between two internal-energy reservoirs has the highest efficiency permitted by the laws of thermodynamics, and also that all reversible engines have the same efficiency when operated between the same two reservoirs. Hence a Carnot engine that employs an ideal gas as its working substance is just one representative of a whole class of hypothetical reversible engines. For example, analogs of the simple Carnot cycle can be devised based upon the electrochemical changes that occur in a storage battery or upon the magnetic changes that occur in a paramagnetic substance. A real engine is never exactly reversible because of such irreversible transformations as those involved in friction and in heat losses through the engine walls, and its efficiency is less than that of a Carnot engine.

Problem. Steam enters a certain steam turbine (Fig 16-13) at a temperature of 570°C and emerges into a partial vacuum at a temperature of 95°C. What is the upper limit to the efficiency of this engine?

Solution. The absolute temperatures equivalent to 570°C and 95°C are respectively 843 K and 368 K. The efficiency of a Carnot engine operating between these two absolute temperatures is

$$\text{Eff} = 1 - \frac{T_2}{T_1} = 1 - \frac{368 \text{ K}}{843 \text{ K}} = 0.56,$$

which is 56 percent. The efficiency of a Carnot engine is the maximum possible for an engine operating between a given pair of temperatures. An actual steam turbine operating between 570°C and 95°C would have an efficiency of no more than 40 percent because of the inevitable presence of friction and heat losses to the atmosphere.

Fig. 16-13. A primitive steam turbine. In a modern turbine, steam flows past a dozen or more sets of blades on the same shaft in order to extract as much power as possible. Sets of stationary blades are interleaved between the sets of moving blades to direct the flow of steam in the most advantageous way.

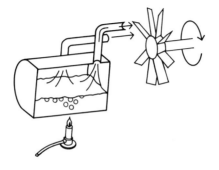

Engine efficiency depends upon temperatures of heat intake and exhaust

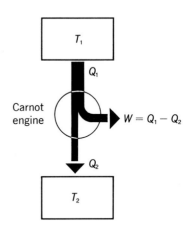

Fig. 16-11. The efficiency of a Carnot engine depends upon the ratio between Q_2 and Q_1.

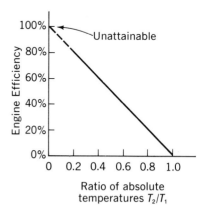

Fig. 16-12. The efficiency of a Carnot engine is equal to $1 - T_2/T_1$.

16-6 Statistical Mechanics

According to the second law of thermodynamics, it is impossible to convert heat into any other form of energy efficiently. Some of the heat input to an engine *must* be lost. Why? The reason lies in the nature of heat, which is molecular kinetic energy; the temperature of a body is a measure of the average kinetic energy of each of its constituent molecules. Let us see how the microscopic picture of matter as consisting of molecules in motion leads to the second law of thermodynamics. Although the argument will be based on molecules in a gas, the essential ideas hold for matter in any state.

The molecules of a gas are in constant random motion and undergo frequent collisions with one another. While we cannot hope to follow an individual gas molecule in its wanderings, it is possible to predict on the basis of statistical arguments what fraction of the time it will have any specified amount of kinetic energy. Hence we can calculate the distribution of molecular energies in a gas sample at a particular temperature. This distribution, which has been confirmed by experiment, has the form shown in Fig. 16-14, and holds for all equilibrium conditions in which each molecule has the same average energy over a period of time. A molecule that moves more swiftly than usual at one instant will move less swiftly at a later instant after a number of collisions have taken place.

Equilibrium is the most probable condition according to *statistical mechanics*, a branch of physics which mathematically deduces the behavior of assemblies of so many particles that deviations from statistically probable behavior are not significant. If we toss a coin a dozen times, it is unlikely that heads and tails will come up equally often; but if we toss it a million times, the percentage deviation from an equal number of heads and tails will be minute. We can appreciate why departures from the equilibrium distribution of molecular energies for more than the briefest instant are so unlikely if we look at a cubic centimeter—a thimbleful—of air at atmospheric pressure and room temperature. There are 2.7×10^{19} molecules in the cubic centimeter, and each molecule undergoes an average of 4×10^9 collisions per second (equivalent to every person in the world colliding with every other person, one at a time, in each second).

Let us consider a heat reservoir at a high temperature and a heat reservoir at a low temperature. The molecules of each are in equilibrium and have the molecular energy distributions shown in Fig. 16-15 (a) and (b). If we consider the two reservoirs as a single system, the molecular energy distribution in the system is like that of (c). This distribution is, in a statistical sense, very improbable; if they were mixed together, the molecules of the two reservoirs would soon blend their energies in collisions to attain the

Molecular energy distribution

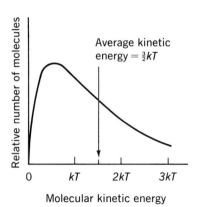

Fig. 16-14.   The distribution of molecular energies in a gas at a particular temperature.

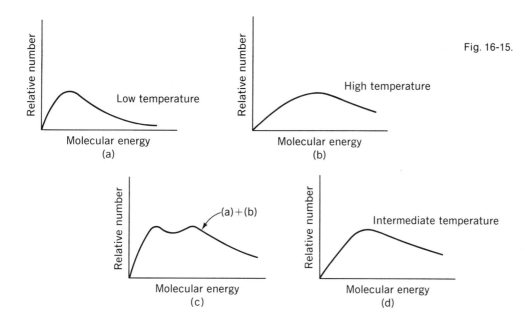

Fig. 16-15.

equilibrium distribution of (d), which corresponds to a temperature intermediate between the initial ones of the two reservoirs.

We note the important fact that the total energy contents of the distributions of Fig. 16-15 (c) and (d) are identical; the only distinction between them is the manner in which the energy is allotted to the molecules on the average. However, it is just this distinction with which the second law of thermodynamics is concerned, because this law states that a system of two heat reservoirs at different temperatures can be made to yield a net work output, while a single heat reservoir, no matter how much energy it contains, cannot be made to perform any net work. A system of molecules whose energies are distributed in the most probable way is "dead" thermodynamically, while a system having a different distribution of molecular energies is capable of doing mechanical work as it progresses to an equilibrium state (Fig. 16-16).

The universe may be thought of as a single system of molecules, and its evolution is powered by the flow of energy from high-temperature reservoirs (the stars) to low-temperature reservoirs (everything else). Ultimately the entire universe will be at the same temperature and all its constituent particles will have the same average energy, a condition sometimes called the "heat death" of the universe.

The second law of thermodynamics is evidently an unusual kind of

Probability and the second law of thermodynamics

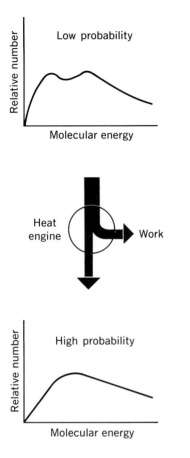

Fig. 16-16. Work can be done by a system whose distribution of molecular energies is statistically improbable.

physical principle. It does not apply to the interactions of individual particles, only to trends in the evolution of assemblies of many particles. In fact, the second law is hardly a basic principle in the usual sense, since it is the result of combining the laws of mechanics with the theory of probability, and cannot be used to predict anything specific except in the sense that it establishes upper limits to the efficiencies of various processes, shows that certain events have negligible likelihoods of occurence, and so forth.

However, the second law has the unique property of being correlated with the direction of time. Events that involve individual particles are always reversible—the same laws of motion apply to the billiard balls seen in the film of a game whether the film is run forward or backward. But events that involve systems of large numbers of particles are not always reversible—the film of an egg being dropped makes no sense at all when run backward. The transformation of a broken egg into a whole egg is not totally impossible, it is simply exceedingly unlikely.

The second law of thermodynamics is thus a statement of probability: as time goes on, order becomes disorder in an isolated system. The sequence can be reversed in parts of the universe now and then—after all, heat engines do turn heat into work, simple forms of life evolve into highly complex ones—but in the universe as a whole, which is an isolated system, increasing disorder is inevitable.

16-7 Energy and Man

The development of modern civilization has been paralleled by a steady increase in the world's use of energy. This is no coincidence: all of man's activities require energy, and the more energy that is readily available in convenient form, the more effectively he can satisfy his desires for food, clothing, shelter, warmth, illumination, transport, communication, and manufactured goods. Primitive man had only food as an energy source, and utilized a total of perhaps 8 million J/day; the mastery of fire and the harnessing of domestic animals yielded an approximately sixfold increase in this figure; the flowering of the industrial revolution a century ago meant a rise in energy consumption to about 300 million J/day in the more advanced countries; and today in the United States each person uses an average of nearly 1,000 million J/day. Energy is a more meaningful currency than money as an index of prosperity, and the fact that 35 percent of the world's energy consumption takes place in the United States, which has only six percent of its population, is more significant than any financial statistics.

Figure 16-17 shows how the world's use of energy has increased in recent times and also indicates the relative importance of different energy

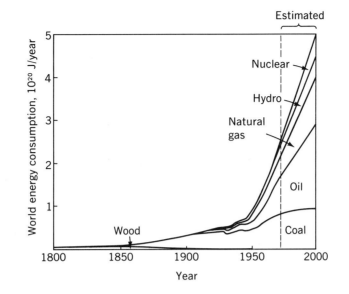

Fig. 16-17.   World energy sources and consumption.

sources. Existing trends have been projected to the year 2000, but it is not possible to go much beyond that with any confidence. There are several limiting factors here. An obvious one is the exhaustion in the foreseeable future of the fossil fuels coal, oil, and natural gas, which today furnish nearly 98 percent of the energy under man's control and which consist essentially of stored solar energy. Oil and natural gas will be the first of these to run out, in a matter of decades. Coal reserves are more abundant and may well last another century. (It is sobering to reflect that the coal currently consumed each year took about two million years to accumulate.) Nuclear fuels ought to last even longer than coal, especially in view of the practicality of "breed-er reactors" in which the normally unusable kind of uranium $^{238}_{92}$U can be converted into plutonium, which is suitable for fuel purposes—though the equal suitability of plutonium for weapons does not make the proliferation of such reactors very attractive. And if thermonuclear power becomes a practi-cal reality, or if the energy of the sun can be drawn upon economically, there will be essentially no limit to the energy potentially available. A large-scale phasing out of fossil fuels would be expensive, and inconvenient in some applications, but by no means out of the question.

The real limit to world energy production is set by the intrinsic ineffi-ciency of the means by which thermal energy is converted into mechanical energy and thence into electrical energy—in other words, by the second law of thermodynamics. The best of today's power stations have efficiencies of only about 35 percent, and this figure is unlikely to improve by much in the forseeable future. The waste heat must go somewhere, and even today heavily-industrialized countries find its disposal difficult if environmental damage is to be avoided. In the United States about 10 percent of the flow of all rivers and streams is already being used to provide cooling water for generating plants. The biological consequences of large-scale heating of inland waters are considerable. The oceans can absorb vast amounts of waste heat with minimal side effects, but if most power plants were located on their shores the transmission of electricity inland would then be a major problem. More and more power plants are discharging waste heat into the atmosphere through cooling towers, but here too there are snags in the long run, since local heating of the atmosphere alters the weather and climate of a region, not necessarily for the better.

The situation is far from hopeless with respect to both resources and environmental damage provided the world's total energy requirements do not continue to increase much longer at the present rate. It is precisely here that social rather than technical considerations enter the picture, because to taper off the growth of energy consumption without at the same time limiting population growth (which is by far the most important problem of the modern world) means a decrease in average living standards, which are too low already in most of the world. The ultimate problem is thus not a technological one at all.

Special Topic

Internal Combustion Engines

An internal combustion engine is able to achieve a relatively high operating efficiency by generating the input heat within the engine itself. In a gasoline engine a mixture of air and gasoline vapor is ignited in each cylinder by a spark plug, and the evolved heat is converted into mechanical energy by the pressure of the hot gases on a piston. The greater the ratio between the initial and final volumes of the expanding gases, the greater the engine efficiency. In a gasoline engine this ratio is limited to about 8 to 1, since the gasoline-air mixture in the cylinder will otherwise spontaneously ignite during its compression before the end of the stroke is reached. The more efficient Diesel engine circumvents this difficulty by compressing only air and injecting fuel oil into the hot, compressed air at the instant the piston has reached the end of its travel. No spark plug is required. The compression ratio in a Diesel engine might be as much as 20 to 1. The effective values of T_1 and T_2 in a modern Diesel engine are perhaps 1800 K and 800 K, respectively for a Carnot efficiency of 56%.

Figure 16-18 shows the operating cycle of a typical four-stroke gasoline engine. In the intake stroke a mixture of gasoline vapor and air from the carburetor is drawn into the cylinder through the intake valve by the suction of the downward-moving piston. In the compression stroke both valves are closed and the upward-moving piston compresses the fuel-air mixture. At the top of the stroke the spark plug is fired, which ignites the fuel-air mixture. The burning fuel expands and forces the piston down in the power stroke. At the end of the power stroke the exhaust valve opens and the upward-moving piston expels the waste gases.

Each cylinder in a four-stroke engine has one power stroke in every two shaft revolutions. When a lightweight engine is necessary a two-stroke cycle can be used, which increases the power output by permitting a power stroke in each cylinder in every shaft revolution. Usually a two-stroke cycle means reduced efficiency in the case of a gasoline engine, which is unimportant in such applications as outboard motors for boats, and greater complexity in the case of a diesel engine, which is a fair price to pay for decreased weight in an engine that must be heavily built to withstand 20-to-1 compressions.

Gasoline engine

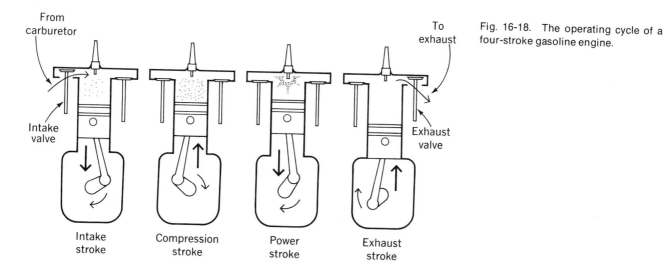

Fig. 16-18. The operating cycle of a four-stroke gasoline engine.

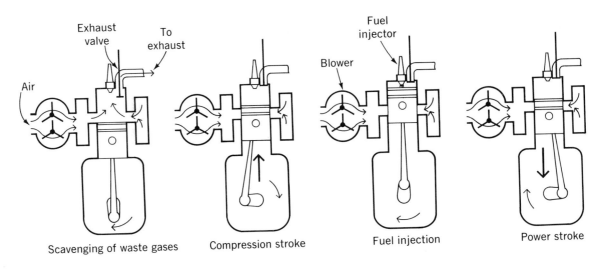

Scavenging of waste gases   Compression stroke   Fuel injection   Power stroke

Fig. 16-19.   The operating cycle of a two-stroke diesel engine.

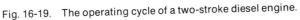

Diesel engine   Figure 16-19 illustrates the operation of a two-stroke diesel engine. When the piston is at the bottom of its path, air from a blower enters to flush waste gases from the previous power stroke out through the exhaust valve. As the piston moves upward it compresses the fresh air to a fraction of its initial volume. At the top of the compression stroke the air temperature is perhaps 550°C, and fuel oil sprayed in by the injector is ignited at once. The burning fuel presses down on the piston during the ensuing power stroke.

Gas turbine   The outstanding efficiency of reciprocating gasoline and diesel engines (nearly as high as those of steam turbines) has to some extent retarded the development of the still more efficient gas turbine. A gas turbine is similar to a steam turbine except that hot gases from the burning fuel pass through its sets of blades instead of steam. A gas turbine is lighter in weight and has fewer moving parts than a reciprocating internal combustion engine, but the high temperature and high rotational speeds at which it operates present difficulties in manufacture. Gas turbines are nevertheless coming into wider and wider use: "turboprop" aircraft engines are gas turbines, for instance, and a number of ships are already powered by gas turbines.

Jet engine   The rapidly rotating shaft of a turboprop engine is coupled to a propeller through a reduction gear. In a jet engine the propeller is eliminated and the hot gases from the burning fuel are ejected at high speed from the rear of the engine to furnish a reaction force that pushes the aircraft forward. The energy liberated by the burning fuel is thus converted directly into propulsion with no moving parts intervening (except for a turbine that powers the necessary air compressor). Rocket motors are jet engines in which the required oxygen or other oxidizing agent for fuel combustion comes from an internal reservoir instead of from the atmosphere. In a solid-fuel rocket, the ultimate in simplicity, both the components required for combustion are combined in a stable mixture whose reaction rate when ignited is relatively slow and steady rather than explosive.

The Refrigerator

In nearly all refrigerators the working substance (or *refrigerant*) is a gas that is readily liquified. Common refrigerants are ammonia, Freon 12, methyl chloride, and sulfur dioxide. Figure 16-20 shows the vaporization curve of Freon 12. Under conditions of temperature and pressure corresponding to points above the curve only liquid Freon is present, while under conditions corresponding to points below the curve only Freon vapor is present. Other refrigerants have different vaporization curves. At atmospheric pressure Freon 12 boils at $-22°F$, while at this pressure ammonia boils at $-28°F$, methyl chloride at $-11°F$, and sulfur dioxide at $14°F$. The choice of a refrigerant depends upon the precise kind of refrigerator involved and the temperatures between which it is to operate.

Let us examine a typical refrigeration system that uses Freon 12. As in Fig. 16-21, this system consists of a *compressor*, a *condenser*, an *expansion valve*, and an *evaporator*. When the piston of the compressor moves downward, Freon vapor at 20 lb/in² (gauge pressure) and approximately room temperature is sucked into the cylinder. As the piston reaches the bottom

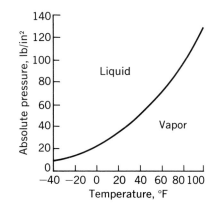

Fig. 16-20. Vaporization curve of Freon 12, a common refrigerant.

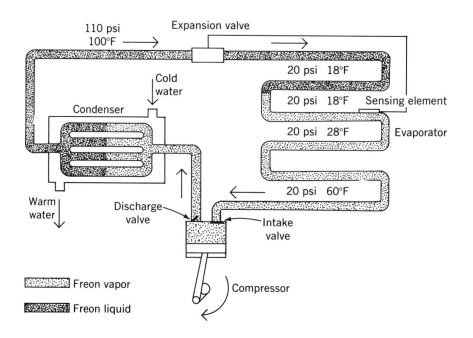

110 psi
100°F ⟶

Expansion valve

20 psi 18°F

Cold
water

20 psi 18°F   Sensing element

Condenser

20 psi 28°F   Evaporator

20 psi 60°F

Warm
water

Discharge
valve

Intake
valve

Freon vapor

Compressor

Fig. 16-21. A refrigeration system
using Freon 12. 1 psi = 1 lb/in².

Freon liquid

of its stroke and begins to move upward, the intake valve closes and the discharge valve opens. The compressed Freon, which emerges at a pressure of 110 lb/in² and a high temperature, passes into the condenser in which it is cooled until it liquifies. The condenser may be water cooled, as shown in the sketch, or air cooled. It is in this stage that the heat extracted from the refrigerated space is dissipated.

The liquid Freon then goes into the expansion valve from which it emerges at a lower pressure (20 lb/in²) and temperature (18°F). The amount of Freon supplied by the valve is regulated by a sensing element placed in the refrigerated space; this may be simply a gas-filled bulb which responds to temperature changes by pressure changes that actuate a bellows in the valve. As the cold liquid Freon flows through the evaporator tubes it absorbs heat from the region being cooled. From Fig. 16-20 we see that, at an absolute pressure of 35 lb/in², the boiling point of Freon 12 is 18°F, so that the heat absorbed by the liquid Freon in the evaporator causes it to vaporize. Farther along in the evaporator the Freon vapor itself absorbs heat, rising in temperature to perhaps 28°F. The amount of this temperature rise (called *super heat* by refrigeration engineers) is critical for the efficiency of the system, and is usually 10°F or less. Finally the Freon vapor leaves the evaporator and enters the compressor to begin another cycle.

Important Terms

Heat can be transferred from one place to another by means of **conduction**, **convection**, or **radiation**. In conduction heat is transported by successive molecular collisions, and in convection by the motion of a volume of hot fluid from one place to another. Heat transfer by radiation takes place by means of **electromagnetic waves** which require no material medium for their passage.

A **heat engine** is any device that converts heat into mechanical energy or work.

The **first law of thermodynamics** states that the work output of a heat engine is equal to its net heat input.

The **second law of thermodynamics** states that it is impossible to construct an engine, operating in a repeatable cycle, which does nothing other than take energy from a source and perform an equivalent amount of work.

A **Carnot engine** is an idealized engine which is not subject to such practical difficulties as friction or heat losses by conduction or radiation but which obeys all physical laws. No engine operating between the same two temperatures can be more efficient than a Carnot engine operating between them.

A **refrigerator** is a device that transfers heat from a cold reservoir to a hot one, and it must expend energy in order to do this. In essence, it is a heat engine operating in reverse.

Important Formulas

Stefan-Boltzmann law:

$$R = e\sigma T^4$$

Carnot efficiency:

$$\text{Eff} = 1 - \frac{T_2}{T_1}$$

Multiple Choice

1. The natural direction of the heat flow between two reservoirs depends on
   a. their temperatures.
   b. their internal energy contents.
   c. their pressures.
   d. whether they are in the solid, liquid, or gaseous state.

2. Metals are good conductors of heat because
   a. they contain free electrons.
   b. their atoms are relatively far apart.
   c. their atoms collide infrequently.
   d. they have reflecting surfaces.

3. The materials with the highest heat conductivities are the
   a. gases.
   b. liquids.
   c. woods.
   d. metals.

4. In natural convection, a heated portion of a fluid moves because
   a. its molecular motions become aligned.
   b. of molecular collisions within it.
   c. its density is less than that of the surrounding fluid.
   d. of currents in the surrounding fluid.

5. Four pieces of iron are heated in a furnace to different temperatures. The one at the highest temperature appears
   a. white.
   b. yellow.
   c. orange.
   d. red.

6. Electromagnetic radiation is emitted
   a. only by radio and television antennas.
   b. only by bodies at higher temperatures than their surroundings.
   c. only by bodies at lower temperatures than their surroundings.
   d. by all bodies.

7. A heat engine operates by taking in heat at a particular temperature and
   a. converting it all into work.
   b. converting some of it into work and exhausting the rest at a lower temperature.
   c. converting some of it into work and exhausting the rest at the same temperature.
   d. converting some of it into work and exhausting the rest at a higher temperature.

8. The first law of thermodynamics is the same as the
   a. second law of thermodynamics.
   b. law of conservation of energy.
   c. law of conservation of momentum.
   d. first law of motion.

9. The natural direction of heat flow is from a high-temperature reservoir to a low-temperature reservoir, regardless of their respective heat contents. This fact is incorporated in the
   a. first law of thermodynamics.
   b. second law of thermodynamics.
   c. law of conservation of energy.
   d. principle of superposition.

10. The work output of every heat engine
    a. equals the difference between its heat intake and heat exhaust.
    b. equals that of a Carnot engine with the same intake and exhaust temperatures.
    c. depends only upon its intake temperature.
    d. depends only upon its exhaust temperature.

11. The physics underlying the operation of a refrigerator most closely resembles the physics underlying
    a. reciprocating steam engine.
    b. diesel engine.
    c. gas turbine.
    d. Carnot engine.

12. A refrigerator
    a. produces cold.
    b. causes heat to vanish.
    c. removes heat from a region and transports it elsewhere.
    d. changes heat to cold.

13. A refrigerator exhausts
    a. less heat than it absorbs from its contents.

b. the same amount of heat it absorbs from its contents.
c. more heat than it absorbs from its contents.
d. any of the above, depending on the circumstances.

14. A Carnot engine turns heat into work
    a. with 100% efficiency.
    b. with 0% efficiency.
    c. without itself undergoing a permanent change.
    d. with the help of expanding steam.

15. In any process, the maximum amount of heat that can be converted to mechanical energy
    a. depends on the amount of friction present.
    b. depends on the intake and exhaust temperatures.
    c. depends on whether kinetic or potential energy is involved.
    d. is 100%.

16. In any process, the maximum amount of mechanical energy that can be converted to heat
    a. depends on the amount of friction present.
    b. depends on the intake and exhaust temperatures.
    c. depends on whether kinetic or potential energy is involved.
    d. is 100%.

17. A frictionless heat engine can be 100% efficient only if its exhaust temperature is
    a. equal to its input temperature.
    b. less than its input temperature.
    c. 0°C.
    d. 0°K.

18. An ideal engine absorbs heat at a temperature of 127°C and exhausts heat at a temperature of 77°C. Its efficiency is
    a. 13%.   b. 39%.
    c. 61%.   d. 88%.

19. If a heat engine exhausting heat at 100°C is to have an efficiency of 33%, it must take in heat at
    a. 149°C.   b. 284°C.
    c. 422°C.   d. 557°C.

20. When a gas is in equilibrium, its molecules
    a. all have the same energy.

b. have different energies which remain constant.

c. have a certain constant average energy.

d. do not collide with one another.

21. A system of molecules whose energies are distributed in the most probable way

a. can perform an amount of mechanical work equal to its total energy content.

b. can perform an amount of mechanical work that depends on its absolute temperature.

c. cannot perform any mechanical work.

d. is a Carnot engine.

Exercises

1. What condition is necessary for heat to flow through an object?

2. In the winter, why does the steel blade of a shovel seem colder than its wooden handle?

3. A Thermos bottle consists of two glass vessels, one inside the other, with the space between them evacuated. The vessels are both coated with thin films of silver. Why is this device so effective in keeping the contents of the bottle at a constant temperature?

4. By what mechanism or mechanisms does a man seated in front of a fire receive heat from it? A man seated in front of a radiator through which hot water is circulated?

5. Under what circumstances does an object radiate electromagnetic waves? How is the predominant wavelength in the radiation related to the temperature of the object?

6. In an attempt to cool a room in the summer, a man turns on an electric fan and leaves the room. Will the room be cooler when he returns?

7. An attempt is made to cool a kitchen during the summertime by leaving the refrigerator door open and closing the kitchen door and windows. What will happen and why?

8. The sun's corona is a very dilute gas at a temperature of about 10^6 K that is believed to extend into interplanetary space at least as far as the earth's orbit. Why can we not use the corona as the high-temperature reservoir of a heat engine in an earth satellite?

9. Two identical watches, one wound and the other unwound, are dropped into beakers of acid and completely dissolved. Is there any difference between the two reactions? Justify your answer using physical principles.

10. A copper sphere 5 cm in diameter whose emissivity is 0.3 is heated in a furnace to 400°C. At what rate does it radiate energy?

11. A small hole in a cavity behaves like a blackbody because any radiation that falls on it is trapped inside by multiple reflections until it is absorbed. At what rate does radiation escape from a hole 10 cm² in area in the wall of a furnace whose interior is at a temperature of 700°C?

12. What is the maximum possible efficiency of an engine that obtains heat at 400°F and exhausts heat at 150°F?

13. An engine operating between 300°C and 50°C is 15% efficient. What would its efficiency be if it were a Carnot engine?

14. A Carnot engine takes in 10^3 kcal of heat from a reservoir at 327°C and exhausts heat to a reservoir at 127°C. How much work does it do?

15. One of the most efficient engines ever developed operates between about 2000 K and 700 K. Its actual efficiency is 40%. What percentage of its maximum possible efficiency is this?

16. An engine is proposed which is to operate between 400°F and 100°F with an efficiency of 40%. Will the engine perform as predicted? If not, what would its maximum efficiency be?

17. A Carnot engine absorbs 200 kcal of heat at 500 K and exhausts 150 kcal. What is the exhaust temperature?

18. In a certain power station coal is consumed at the rate of 1 lb/hr for each kilowatt of electrical output. Find the overall efficiency of the power

station. The heat of combustion of coal is 7800 kcal/kg.

19. The conventional unit of refrigeration capacity is the *ton*, defined as that rate of heat removal which can freeze 1 ton of water at 32°F to ice at 32°F per day. Find the equivalent of the refrigeration ton in Btu/hr.

Problems

1. An object is at a temperature of 400°C. At what temperature would it radiate energy twice as fast?

2. Radiant energy from the sun arrives at the earth's atmosphere at a rate of about 1.4 kW per m² of area perpendicular to the sun's rays. The average radius of the earth's orbit is 1.5×10^{11} m and the radius of the sun is 7.0×10^8 m. From these figures find the surface temperature of the sun under the assumption that it radiates like a blackbody (which is approximately true).

3. A certain 50 kg woman has a surface area of 1.5 m² and a skin temperature of 34°C. When she is sitting still in a room at 20°C her metabolic activity leads to the liberation of 80 W. To see whether it is plausible that radiation is the chief mechanism by which this power is dissipated, calculate what her effective emissivity must be for her to lose energy at 80 W under the above conditions.

4. An adiabatic process is one in which heat neither enters nor leaves the system in which the process occurs. A gas sample expands from V_1 to V_2. Does it perform the most work when the expansion takes place at constant pressure, at constant temperature, or adiabatically? In which process does the gas perform the least work?

5. A Carnot engine whose efficiency is 35% takes in heat at 500°C. What must the intake temperature be if the efficiency is to be 50% with the same exhaust temperature?

6. Three designs for a heat engine to operate between 450 K and 300 K are proposed. Design *A*

is claimed to require a heat input of 0.2 kcal for each 1000 J of work output, design *B* a heat input of 0.6 kcal, and design *C* a heat input of 0.8 kcal. Which design would you choose and why?

7. The total drop of the Wollomombi Falls in Australia is 1580 ft. What would be the Carnot efficiency of an engine operating between the top and bottom of the falls if the water temperature at the top were 50°F and all the potential energy of the water at the top were converted to heat at the bottom?

8. A certain 70-kg man requires energy at the rate of 70 W when he is resting (this is his "basal metabolism"). When he is walking up a 10° hill at 2 m/s, his power requirement increases to 300 W, so that the net power input attributable to his motion is 230 W. What is the efficiency with which he converts food energy into gravitational potential energy?

9. Starting from the definition of work, show that the amount of work done by a gas that expands by ΔV at the constant pressure p is $W = p\,\Delta V$.

10. Use the result of the previous problem to find the percentage of the heat of vaporization of water that represents the work involved in expanding water into steam against the pressure of the atmosphere. At 100°C and atmospheric pressure the density of steam is 0.6 kg/m³.

11. An ideal refrigerator is a Carnot engine operating backwards. If an ideal refrigerator extracts heat from a storage chamber at the absolute temperature T_1 and ejects heat to the outside world at the absolute temperature T_2, show that the ratio between the work done on the refrigerator and heat extracted is $T_2/T_1 - 1$.

12. A Carnot refrigerator extracts heat from a freezer at −5°C and exhausts it at 25°C. How much work per kcal of heat extracted is required?

13. Three designs for a refrigerator to operate between −20°C and 40°C are proposed. Design *A* is claimed to require 300 J of work for each kcal of

heat extracted, design *B* to require 950 J, and design *C* to require 2000 J. Which design would you choose and why?

14. A Carnot refrigerator is used to make 1 kg of ice at −10°C from 1 kg of water at 20°C, which is also the temperature of the kitchen. How many joules of work must be done?

Answers to Multiple Choice

1. a	8. b	15. b
2. a	9. b	16. d
3. d	10. a	17. d
4. c	11. a	18. a
5. a	12. c	19. b
6. d	13. c	20. c
7. b	14. c	21. c

17

Electricity

The success of the laws of motion, of the law of gravitation, and of the kinetic-molecular theory of matter might tempt us into thinking that we now have, at least in outline, a complete picture of the workings of the physical universe. To dispel this notion all we need do is perform a simple experiment: on a dry day, we run a hard rubber comb through our hair, and find that the comb is now able to pick up small bits of paper and lint. The attraction is surely not due to gravity, because the gravitational force between comb and paper is far too small and should not, in any event, depend upon whether the comb is run through our hair or not. What has been revealed by this experiment is an electrical phenomenon, so called after elektron, the Greek word for amber, a substance used in the earliest studies of electricity.

17-1 Electric Charge

Electricity is familiar to all of us as the name for that which causes our light bulbs to glow, many of our motors to turn, our telephones and radios to communicate sounds, our television screens to communicate pictures. But there is more to electricity than its technological uses. Electrical forces bind electrons to nuclei to form atoms, and they hold atoms together to form molecules, solids, and liquids. All of the chief properties of matter in bulk—with the notable exception of mass—can be traced to the electrical nature of its constituent particles.

Let us begin our study of electricity by examining three basic experiments. The first experiment is shown in Fig. 17-1. By convention, we call whatever it is that a rubber rod possesses by virtue of having been stroked with a piece of fur *negative electric charge*. Part of the negative charge on the rubber rod of Fig. 17-1 flowed to the pith ball when it was touched, and the fact that the ball then flew away from the rod suggests that negative electric charges repel each other.

Negative charge

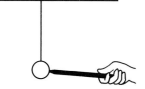

A pith ball suspended by a fine string is touched by a hard rubber rod. Nothing happens.

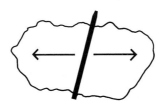

The rubber rod is stroked with a piece of fur.

Fig. 17-1. A rubber rod stroked with fur becomes negatively charged; two negatively-charged objects repel each other.

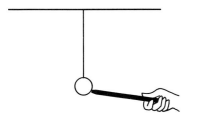

The pith ball is again touched by the rubber rod.

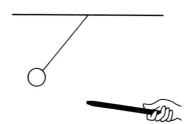

After the touch, the ball flies away from the rod.

Positive charge

The next experiment, shown in Fig. 17-2, is very similar. By convention, we call whatever it is that a glass rod possesses by virtue of having been stroked with a silk cloth *positive electric charge.* Part of the positive charge on the glass rod of Fig. 17-2 flowed to the pith ball when it was touched, and the fact that the ball then flew away from the rod suggests that positive electric charges repel each other.

Why is it assumed that the electric charge on the glass rod is different from that on the rubber rod? The reason lies in the result of the third

A second pith ball is touched by a glass rod. Nothing happens.

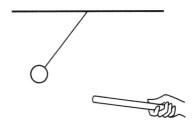

The glass rod is stroked with a silk cloth.

The pith ball is again touched by the glass rod.

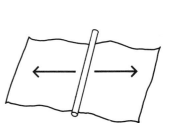

After the touch the ball flies away from the rod.

Fig. 17-2.   A glass rod stroked with silk becomes positively charged; two positively-charged objects repel each other.

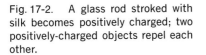

experiment, which is shown in Fig. 17-3. The attraction of the two pith balls means (1) that the charges they carry are different, since like charges have already been observed to repel, and (2) that unlike charges attract each other.

The preceding results can be summarized very simply:

Like charges repel; unlike charges attract.

Behavior of charges

Where do the charges come from when one substance is stroked with another? When we charge one pith ball with a rubber rod and another with the fur the rod was stroked with, we find (Fig. 17-4) that the two balls attract; since the rubber rod is negatively charged, this experiment indicates that the fur is positively charged. A similar experiment with a glass rod and a silk cloth indicates that the cloth acquires a negative charge during the stroking. Evidently the process of stroking serves to *separate* charges. We might infer that rubber has an affinity of some kind for negative charges and

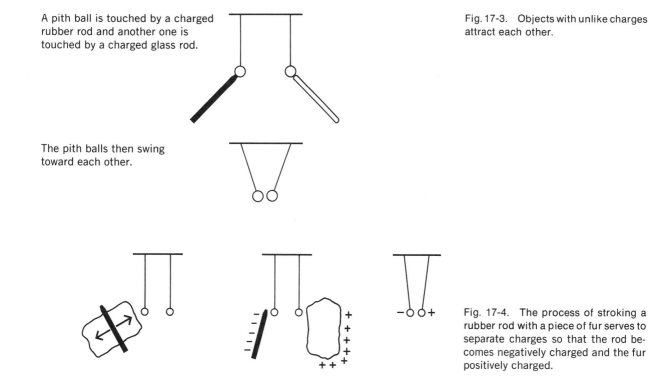

A pith ball is touched by a charged rubber rod and another one is touched by a charged glass rod.

Fig. 17-3.   Objects with unlike charges attract each other.

The pith balls then swing toward each other.

Fig. 17-4.   The process of stroking a rubber rod with a piece of fur serves to separate charges so that the rod becomes negatively charged and the fur positively charged.

fur an affinity for positive charges, so that, when rubbed together, each tends to acquire a different kind of charge.

A great many experiments with a variety of substances have shown that there are only the two kinds of electric charge, positive and negative, that we have spoken of. All electrical phenomena involve either or both kinds of charge. An "uncharged" body of matter actually possesses equal amounts of positive and negative charge, so that appropriate treatment—mere rubbing is sufficient for some substances—can leave an excess of either kind on the body and thereby cause it to exhibit electrical effects.

What is electric charge? All that can be said is that charge, like rest mass, is a fundamental property of certain of the elementary particles of which all matter is composed. Three types of particle are found in atoms, the positively-charged *proton*, the negatively-charged *electron*, and the neutral (that is, uncharged) *neutron*. The proton and electron have exactly equal amounts of charge, though of opposite sign. An atom normally contains equal numbers of protons and electrons, so it is electrically neutral unless disrupted in some way.

The *principle of conservation of charge* states that

Conservation of charge   **The net electric charge in an isolated system remains constant.**

By "net charge" is meant the algebraic sum of the charges present—the total positive charge minus the total negative charge. Net charge can be positive, negative, or zero.

Every known physical process in the universe conserves electric charge. Separating or bringing together charges does not affect their magnitudes, so such rearrangements leave the net charge unaffected. Under certain circumstances matter can be created from energy, but whenever this happens, the number of positively-charged particles created is always exactly the same as the number of negatively-charged particles created. Under other circumstances matter can be completely converted into energy, and in such events the number of positively-charged particles that disappear is again always exactly the same as the number of negatively-charged particles. Unlike rest mass, charge is invariably conserved.

Electric charge is the fourth quantity we have studied which is conserved in every physical process. The others are energy, linear momentum, and angular momentum, and their conservation can be traced to fundamental symmetry properties of nature. These properties are, respectively, the independence of physical laws to shifts in time, in space, and in orientation. Conservation of charge is also associated with a symmetry property of nature, although this property is too abstract to be described here.

17-2 Coulomb's Law

In order to arrive at the law of gravitation

$$F = G\,\frac{m_A m_B}{r^2},$$

Newton had to make use of astronomical data and an indirect argument, because gravitational forces are appreciable only when the masses involved are very large. However, the law that electrical forces obey can be readily determined in the laboratory, because these forces are so much greater in magnitude than gravitational ones.

The law of force between charges was first published by the eighteenth century French scientist Charles Coulomb, and is called Coulomb's law in his honor. If we use the symbol q for electric charge, Coulomb's law for the magnitude F of the force $\mathbf{F}$ between two charges q_A and q_B the distance r apart states that

$$F = k\,\frac{q_A q_B}{r^2}. \qquad\qquad \textit{Coulomb's Law} \quad (17\text{-}1)$$

The force between two charges is proportional to both of the charges and is inversely proportional to the square of the distance between them. The quantity k is a constant whose value depends upon the units employed and upon the medium (air, vacuum, oil, and so forth) in which the charges are located.

Electric force is, of course, a vector quantity, and the formula above gives only its magnitude. The direction of $\mathbf{F}$ is always along the line joining q_A and q_B, and the force is attractive if q_A and q_B have opposite signs and repulsive if they have the same signs (Fig. 17-5).

The unit of electric charge is the *coulomb* (abbreviated C). The formal definition of the coulomb, given in Chapter 20, is expressed in terms of magnetic forces. A more realistic way to think of the coulomb is in terms of the number of individual elementary charges that add up to this amount of charge. All charges, both positive and negative, occur only in multiples of 1.60×10^{-19} C. No elementary particle has ever been found with a charge of other than $\pm 1.60 \times 10^{-19}$ C or 0. The electron has a charge of -1.60×10^{-19} C; the proton has a charge of $+1.60 \times 10^{-19}$ C. If 6.25×10^{18} electrons were assembled, the total charge would be -1 C; if 6.25×10^{18} protons were assembled, the total charge would be $+1$ C (Fig. 17-6).

Because electric charge always occurs in multiples of $\pm 1.60 \times 10^{-19}$ C,

Gravitational force

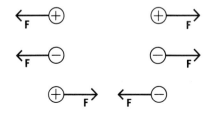

Fig. 17-5. The electric force one charge exerts on another is always along the line joining the two charges. The force is attractive if the charges have opposite signs, repulsive if they have the same sign.

Electric force

The coulomb

Electron charge

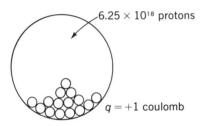

6.25 × 10¹⁸ protons

$q = +1$ coulomb

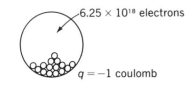

6.25 × 10¹⁸ electrons

$q = -1$ coulomb

Fig. 17-6.

Permittivity of free space

this amount of charge has been given a special name, the electron (or electronic) charge, and a special symbol, e:

$$e = 1.60 \times 10^{-19} \text{ C}. \qquad \textit{Electron charge}$$

Thus a charge of $+1.60 \times 10^{-19}$ C is abbreviated $+e$, and one of -3.20×10^{-19} C is abbreviated $-2e$.

In most processes that lead to a net charge on some object, electrons are either added to it or removed from it. Hence we can think of an object whose charge is -1 C as having 6.25×10^{18} electrons more than its normal number, and of an object whose charge is $+1$ C as having 6.25×10^{18} electrons less than its normal number. (By "normal number" is meant a number of electrons equal to the number of protons present, so that the object has no net charge.)

When q_A and q_B in Coulomb's law are expressed in coulombs and r in meters, the constant k has the value in vacuum of

$$k = 9.0 \times 10^9 \, \frac{\text{N} \cdot \text{m}^2}{\text{C}^2}.$$

The value of k in air is very slightly greater.

The constant k is often written

$$k = \frac{1}{4\pi\epsilon_0}$$

where ϵ_0, called the permittivity of free space, is equal to

$$\epsilon_0 = 8.85 \times 10^{-12} \, \frac{\text{C}^2}{\text{N} \cdot \text{m}^2}.$$

Problem. Find the force between two charges of 1 C each that are 1 m apart (Fig. 17-7).

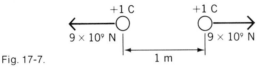

+1 C   +1 C

9 × 10⁹ N   9 × 10⁹ N

1 m

Fig. 17-7.

Solution.   From Coulomb's law,

$$F = k\,\frac{q_A q_B}{r^2}$$

$$= 9 \times 10^9 \,\frac{\text{N}\cdot\text{m}^2}{\text{C}^2} \times \frac{1\,\text{C} \times 1\,\text{C}}{1\,\text{m}^2}$$

$$= 9 \times 10^9 \,\text{N}.$$

This force is equal to about 2 billion lb. Evidently even the most highly charged objects that can be produced seldom contain more than a minute fraction of a coulomb of net charge of either sign.

At the beginning of this chapter a familiar observation was noted: a hard rubber comb that has been charged by being passed through someone's hair on a dry day is able to attract small bits of paper. Since the paper bits were originally uncharged, how could the comb exert a force on them?

The explanation depends upon Coulomb's law (Fig. 17-8). When the negatively charged comb is brought near the paper, some of the negative charges in the paper which are not tightly bound in place move as far away as they can from the comb, while some of the positive charges which are not tightly bound move toward the comb. Because electrical forces vary inversely with distance, the attraction between the comb and the closer positive charges is greater than the repulsion between the comb and the farther negative charges, and so the paper moves toward the comb. Only a small amount of charge separation actually occurs, and so, with little force available, only very light objects can be attracted in this way.

How a charge attracts an uncharged object

Fig. 17-8.   How a charged body attracts an uncharged one.

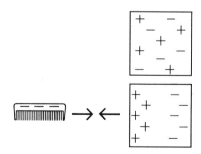

17-3 Multiple Charges

When more than two charges are in the same region, the force on any one of them may be calculated by adding vectorially the forces exerted on it by each of the others. Usually the component method of vector addition provides the most straightforward means of carrying out the calculation.

Problem. Three charges, $q_1 = +2 \times 10^{-9}$ C, $q_2 = +4 \times 10^{-9}$ C, and $q_3 = -5 \times 10^{-9}$ C, are located as shown in Fig. 17-9. Find the magnitude and direction of the net force on q_1.

Solution. If we call $\mathbf{F}_{21}$ the repulsive force exerted by q_2 on q_1 and $\mathbf{F}_{31}$ the attractive force exerted by q_3 on q_1, Coulomb's law yields, for the magnitudes of these forces,

$$F_{21} = \frac{kq_2q_1}{r_{21}^2} = \frac{9 \times 10^9 \text{ N·m}^2/\text{C}^2 \times 4 \times 10^{-9} \text{ C} \times 2 \times 10^{-9} \text{ C}}{(4 \times 10^{-2} \text{ m})^2}$$

$$= 4.5 \times 10^{-5} \text{ N},$$

$$F_{31} = \frac{kq_3q_1}{r_{31}^2} = \frac{9 \times 10^9 \text{ N·m}^2/\text{C}^2 \times 5 \times 10^{-9} \text{ C} \times 2 \times 10^{-9} \text{ C}}{(5 \times 10^{-2} \text{ m})^2}$$

$$= 3.6 \times 10^{-5} \text{ N}.$$

The directions of $\mathbf{F}_{21}$ and $\mathbf{F}_{31}$ are parallel to the 4-cm and 5-cm sides of the triangle.

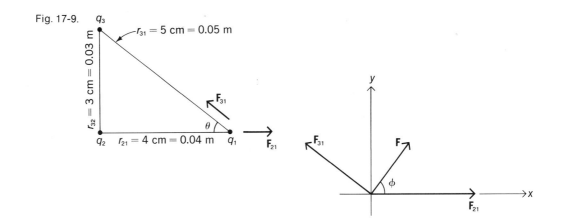

Fig. 17-9.

In order to determine **F**, the net force on q_1, we first resolve $\mathbf{F}_{21}$ and $\mathbf{F}_{31}$ into components. We have, since $\sin\theta = \frac{3}{5}$ and $\cos\theta = \frac{4}{5}$,

$$F_{21x} = F_{21} = 4.5 \times 10^{-5} \text{ N},$$

$$F_{21y} = 0,$$

$$F_{31x} = F_{31}\cos\theta = \tfrac{4}{5}F_{31} = -2.9 \times 10^{-5} \text{ N},$$

$$F_{31y} = F_{31}\sin\theta = \tfrac{3}{5}F_{31} = 2.1 \times 10^{-5} \text{ N}.$$

Hence the components of **F** are

$$F_x = F_{21x} + F_{31x} = 1.6 \times 10^{-5} \text{ N},$$

$$F_y = F_{21y} + F_{31y} = 2.1 \times 10^{-5} \text{ N},$$

and the magnitude of **F** is accordingly

$$F = \sqrt{F_x^{\,2} + F_y^{\,2}} = 2.6 \times 10^{-5} \text{ N}.$$

The direction of **F** can be specified in various ways. If ϕ is the angle between **F** and the $+x$-axis, then

$$\phi = \tan^{-1}\frac{F_y}{F_x} = \tan^{-1} 1.31 = 53°.$$

17-4 Electricity and Matter

Coulomb's law for the electric force between charges is very similar to Newton's law for the gravitational force between masses. The most striking difference is that electric forces may be either attractive or repulsive, whereas gravitational forces are always attractive. The latter fact means that matter in the universe tends to come together to form large bodies, such as stars and planets, and these bodies are always found in groups, such as galaxies of stars and families of planets.

There is no comparable tendency for electric charges of either sign to come together; quite the contrary. Unlike charges attract strongly, which makes it hard to separate neutral matter into portions of opposite signs. Furthermore, like charges repel, so it becomes harder and harder to add further charge to an already charged object. Hence the large-scale structure of the universe is largely governed by gravitational forces.

Gravitational forces dominate on a large scale

Electric forces dominate on a small scale

On an atomic scale, though, the relative importance of gravity and electricity is reversed. Elementary particles are so tiny that the gravitational forces between them are insignificant, whereas their electric charges are sufficiently great for electric forces to govern the structures of atoms, molecules, liquids, and solids.

Problem. The hydrogen atom has the simplest structure of all atoms. It consists of a proton (mass 1.7×10^{-27} kg, charge $+1.6 \times 10^{-19}$ C) and an electron (mass 9.1×10^{-31} kg, charge -1.6×10^{-19} C) whose average separation is 5.3×10^{-11} m. (For the time being we can think of the electron as circling the proton much as the moon circles the earth, as in Fig. 17-10. A more realistic model of the hydrogen atom—but one that is harder to visualize—will be given later.) Compare the electrical and gravitational forces between the proton and the electron in a hydrogen atom.

Solution. The electrical force between the electron and proton is

$$F_e = k\, \frac{q_e q_p}{r^2}$$

$$= \frac{9.0 \times 10^9 \text{ N·m}^2/\text{C}^2 \times (1.6 \times 10^{-19} \text{ C})^2}{(5.3 \times 10^{-11} \text{ m})^2}$$

$$= 8.2 \times 10^{-8} \text{ N,}$$

while the gravitational force between them is

$$F_g = G\, \frac{m_e m_p}{r^2}$$

$$= \frac{6.7 \times 10^{-11} \text{ N·m}^2/\text{kg}^2 \times 9.1 \times 10^{-31} \text{ kg} \times 1.7 \times 10^{-27} \text{ kg}}{(5.3 \times 10^{-11} \text{ m})^2}$$

$$= 3.7 \times 10^{-47} \text{ N.}$$

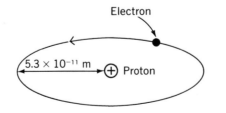

Fig. 17-10. A simple model of the hydrogen atom.

The electrical force is over 10^{39} times greater than the gravitational force. Clearly the electrical forces which subatomic particles exert upon one another are so much stronger than their mutual gravitational ones that the latter can be neglected completely.

17-5 Atomic Structure

By the beginning of this century a substantial body of evidence had been accumulated in support of the idea that the chemical elements consist of atoms. The nature of the atoms themselves, however, was still a mystery, although a significant clue had been discovered. This clue was the fact that electrons are constituents of atoms, which suggests that electrical forces are involved in atomic phenomena. J. J. Thomson, whose work had led to the identification of the electron, proposed in 1898 that atoms are spheres of positively charged matter that contain embedded electrons, much as a fruit cake is studded with raisins.

The most direct way to find out what is inside a fruit cake is simply to plunge a finger into it. In essence this is the classic experiment performed in 1911 by Geiger and Marsden at the suggestion of Ernest Rutherford. The probes they used were fast *alpha particles* spontaneously emitted by certain radioactive elements. For the present all we need to know about alpha particles is that they consist of two neutrons and two protons held tightly together, so that each one has a charge of $+2e$.

The Rutherford experiment established the structure of the atom

Geiger and Marsden placed a sample of an alpha-emitting substance behind a lead screen with a small hole in it, so that a narrow beam of alpha particles was produced. On the other side of a thin metal foil in the path of the beam they placed a zinc sulfide screen which gave off a flash of light when struck by an alpha particle, thus indicating the extent to which the alpha particles were scattered from their original direction of motion (Fig. 17-11).

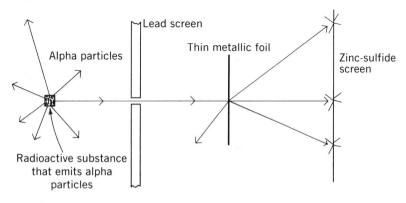

Fig. 17-11. Diagram of the Rutherford experiment.

Geiger and Marsden expected to find that most of the alpha particles go through the foil without being affected by it, with the remainder receiving only slight deflections. If the positive and negative charges within an atom are spread more or less evenly throughout its volume, only weak electric forces would be exerted on alpha particles passing through a thin foil, and their momenta would be enough to carry them through with only minor departures—at most 1° or so—from their original paths.

What Geiger and Marsden actually found was that, while most of the alpha particles indeed emerged unaffected from the foil, the others underwent deflections through very large angles, in some cases even being scattered in the backward direction. Since alpha particles are relatively heavy (almost 8000 times more massive than electrons) and have fairly high initial velocities (typically 2×10^7 m/s), it was clear that strong forces had to be exerted upon them to cause such marked deflections. To explain the results, Rutherford adopted the hypothesis that an atom is composed of a tiny *nucleus* in which its positive charge and nearly all its mass are concentrated, with the electrons some distance away (Fig. 17-12).

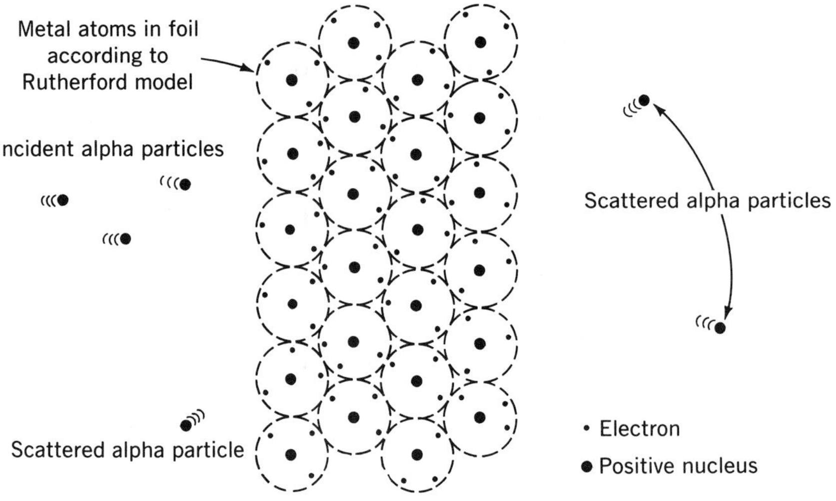

The Rutherford model of the atom

Fig. 17-12. According to the Rutherford model of the atom, positive charge is concentrated in a tiny nucleus at its center with electrons some distance away. Strong electric forces can occur within atoms on the basis of this model, and it accordingly predicts considerable deflection of alpha particles striking a thin foil. This prediction agrees with experiment.

Metal atoms in foil according to Rutherford model

incident alpha particles

Scattered alpha particles

Scattered alpha particle

• Electron

● Positive nucleus

With the atom largely empty space, it is easy to see why most alpha particles proceed right through a thin foil. On the other hand, an alpha particle that happens to come near a nucleus experiences a strong electric force, and is likely to be scattered through a large angle. (The atomic electrons, being very light, are readily knocked out of the way by alpha particles, while the situation is reversed for the nuclei, which are heavier than alpha particles.) Rutherford was able to obtain a formula for the scattering of alpha particles by thin foils on the basis of his hypothesis that agreed with the experimental results. He is therefore credited with the discovery of the nucleus.

17-6 Electrical Conduction

An electric current is a flow of charge from one place to another. Nearly all substances fall into two categories: *conductors*, through which charge can flow easily; and *insulators*, through which charge can flow only with great difficulty. Metals, many liquids, and plasmas (gases whose molecules are charged) are conductors, whereas nonmetallic solids, certain liquids, and gases whose molecules are electrically neutral are insulators. Several substances, called *semiconductors*, are intermediate in their ability to conduct charge.

In a solid metal, each atom gives up one or more electrons to a common "gas" of freely-moving electrons that pervades the entire metal. These electrons can migrate quite readily through the crystal structure of the metal, so if one end of a metal wire is given a positive charge and the other end a negative charge, electrons will flow through the wire from the negative to the positive end. This flow, of course, constitutes an electric current. By supplying new electrons to the negative end of the wire and removing electrons from the positive end as they arrive there—which can be done by connecting the wire to a battery or to a generator—a constant current can be maintained in the wire.

Conduction in a metal

In nonmetallic solids, such as salt, glass, rubber, minerals, wood, and plastics, all the atomic electrons are.bound to particular atoms or groups of atoms and cannot move from place to place. Such solids are accordingly classed as insulators. Actually, nonmetallic solids do conduct very small amounts of current, but their abilities to do this are vastly inferior to those of metals. For instance, when identical bars of copper and sulfur are connected to the same battery, about 10^{23} times more current flows in the copper bar.

Insulators

As mentioned earlier, there are a few substances called semiconductors through which current flows more readily than through insulators but still

Semiconductors

with distinctly more difficulty than through conductors. Thus about 10^7 times more current flows in a germanium bar connected to a battery than in a sulfur bar of the same size, but this is still about 10^{16} times less current than in a copper bar. The electrical conductivity of solids is discussed in more detail in a later chapter.

Superconductivity

At temperatures near absolute zero (0 K, which is $-273°C$) certain metals, alloys, and chemical compounds lose all of their resistance to the flow of electric current. This phenomenon, called *superconductivity*, was discovered by Kamerlingh Onnes in Holland in 1911. For example, aluminum is superconducting at temperatures under 1.20 K, lead at temperatures under 7.22 K, and CuS (copper sulfide) at temperatures under 1.6 K. If a current is set up in a closed wire loop at room temperature, it will die out in less than a second even if the wire is made of a good conductor such as copper or silver, whereas if the wire is made of a superconducting material and is kept cold enough, the current will continue indefinitely. Currents have persisted in superconducting loops with no apparent diminution for over two years.

Superconductivity is of immense potential importance for the transmission of electric energy and in applications where strong magnetic fields are required. Already laboratory electromagnets with superconducting coils are in use, and an experimental electric motor has been built in England whose windings are superconducting. There is no fundamental reason why superconducting magnets cannot be used to levitate trains and thereby both increase their speeds and reduce their power requirements. The immediate problem is that the best materials for the purpose thus far discovered exhibit superconductivity only at temperatures under about 20 K, though it seems likely that in time superconductors will be developed that function at more practical temperatures.

The mechanism of electrical conduction in liquids and gases is different from that in metals. The current in a metal consists of a flow of electrons past the stationary atoms in its structure. The current in a fluid medium other than a liquid metal, however, consists of a flow of entire atoms or molecules that are electrically charged. An atom or molecule that carries a charge is called an *ion*, and both positive and negative ions participate in the conduction process in liquids and gases.

Ionization

An atom or molecule becomes a positive ion when it loses one or more of its electrons; if it gains one or more electrons in addition to its usual complement, it becomes a negative ion. The fundamental positive charges in matter are protons, which are so tightly bound in the nucleus of every atom that they can be dislodged only under exceptional circumstances. Atomic electrons, however, are held more loosely, and one or two of them can be detached from an atom with relative ease. Thus the oxygen and nitrogen

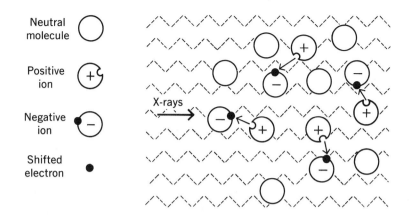

Fig. 17-13. Schematic representation of the ionization of air by x-rays. A molecule losing an electron becomes a positive ion; a molecule gaining an electron becomes a negative ion.

gases in ordinary air become ionized when a spark occurs, in the presence of a flame, and by the passage of x-rays or even ultraviolet light. These processes so disturb the air molecules that some electrons are dislodged, leaving behind positive ions. The liberated electrons almost at once become attached to other nearby molecules to create negative ions (Fig. 17-13).

The electrical attraction between positive and negative charges in time brings the ions together, and the extra electrons on the negative ions become reattached to the positive ions. The gas molecules are then neutral, as they were originally. This *recombination* is rapid at normal atmospheric pressure and temperature.

In the upper atmosphere, where air molecules are so far apart that the recombination of ions is a slow process, the continual bombardment of

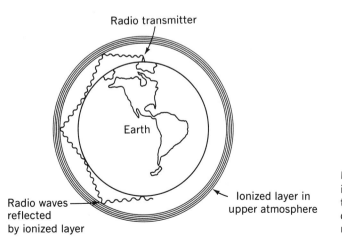

Fig. 17-14. The ionosphere is an ionized region in the upper atmosphere that makes possible long-range radio communication by its ability to reflect radio waves.

x-rays and ultraviolet light from the sun maintains a perceptible proportion of ions at all times. The layer of ions in the upper atmosphere is called the *ionosphere*, and it makes possible long-range radio communication by its ability to reflect radio waves (Fig. 17-14). The ionosphere is an example of a plasma, which, as mentioned earlier, is a gas whose constituent particles are electrically charged. The behavior of a plasma, unlike that of an ordinary gas, is strongly influenced by electric and magnetic forces. Most of the universe is in the plasma state.

Ionosphere

17-7 Ions in Solution

Many liquids contain positive and negative ions at all times and hence are able to conduct electricity. Let us look into how the ions in a liquid come into being and how they are able to resist the recombination that occurs so readily in a gas.

When atoms join together to form a molecule, their electrons are shifted in such a manner that electric forces hold the atoms together. We shall consider the details of the binding process in Chapter 31, but for the moment it is sufficient for us to note that certain molecules have asymmetrical (nonsymmetrical) distributions of charge and behave as though negatively charged at one end and positively charged at the other (Fig. 17-15). A molecule of this kind is called a *polar molecule*. A *nonpolar molecule*, on the other hand, has a uniform distribution of charge. All molecules are normally electrically neutral, and the distinction between the polar and nonpolar varieties lies solely in the way their electrons are arranged.

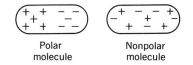

Polar
molecule

Nonpolar
molecule

Fig. 17-15. A polar molecule is one that behaves as though negatively charged at one end and positively charged at the other. The molecule as a whole is electrically neutral.

Polar and nonpolar molecules

The fact that polar molecules exist helps to explain a number of familiar phenomena. The behavior of compounds in solution is a good example. Water readily dissolves such compounds as salt and sugar, but cannot dissolve fats or oils. Gasoline readily dissolves fats and oils, but cannot dissolve salt or sugar. The key to these differences lies in the strongly polar nature of water molecules and the nonpolar nature of gasoline molecules. Water molecules tend to form aggregates under the influence of the electric forces between the ends of adjacent molecules, as shown in Fig. 17-16.

Polar molecules of other substances, such as sugar, can join in the aggregates of water molecules, and are therefore easily dissolved by water (Fig. 17-17). The nonpolar molecules of fats and oils, however, do not interact with water molecules. If samples of oil and water are mixed together, the attraction of water molecules for one another acts to squeeze out the oil molecules, and the mixture soon separates into layers of each substance. Fat and oil molecules dissolve only in liquids whose molecules are similar to theirs, which is why gasoline is a solvent for these compounds (Fig. 17-18). In general, then, "like dissolves like."

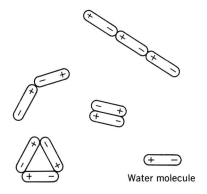

Water molecule

Fig. 17-16. Water molecules are polar and tend to clump together under the influence of electrical forces.

Like dissolves like

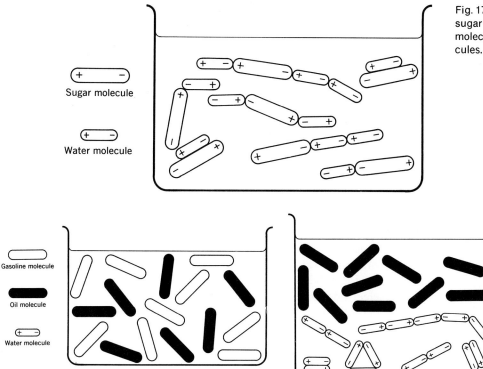

Fig. 17-17. Polar compounds such as sugar dissolve in water because their molecules can link up with water molecules.

Fig. 17-18. Nonpolar compounds dissolve only in nonpolar liquids. Thus oil dissolves in gasoline but not in water.

Many solid compounds have structures that consist of ions rather than of neutral atoms. Thus the sodium chloride (NaCl) of ordinary salt consists of Na^+ and Cl^- ions in the regular geometrical array shown in Fig. 17-19. (The symbol Na^+ refers to a sodium atom that has lost an electron to leave it with a net charge of $+e$, and the symbol Cl^- refers to a chlorine atom that has gained an electron to give it a net charge of $-e$.)

When a crystal of an ionic compound such as NaCl is placed in water, the water molecules cluster around the crystal's ions with their positive ends toward negative ions and their negative ends toward positive ions. The attraction of several water molecules is usually sufficient to pull an ion from the rest of the crystal, and it moves away surrounded by water molecules (Fig. 17-20). The resulting solution contains ions rather than molecules of the dissolved compound.

Substances that separate into free ions when dissolved in water are called *electrolytes* since they are able to conduct electric current by the

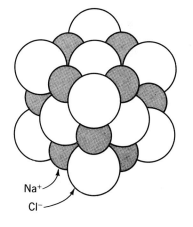
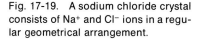

Fig. 17-19. A sodium chloride crystal consists of Na^+ and Cl^- ions in a regular geometrical arrangement.

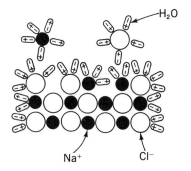

Fig. 17-20. The solution of solid NaCl.

migration of positive and negative ions. All ionic compounds soluble in water and certain other soluble compounds, such as HCl, are electrolytes. Still other compounds, such as sugar, are nonelectrolytes even though they are soluble in water.

Since the outer electron structure of an ion may be very different from that of a neutral atom of the same species, it is not surprising that the ions of an element may behave very differently from its atoms or molecules. Thus gaseous chlorine is greenish in color, has a strong, irritating taste, and is very active chemically, while a solution of chlorine ions is colorless, has a mild, pleasant taste, and is only feebly active.

17-8 Electrolysis

Let us inquire into the effect of passing an electric current through a liquid containing ions. Because water itself may participate in the events that occur in a solution, for simplicity we shall first consider molten NaCl rather than a NaCl solution. When a current flows through a bath of molten NaCl, as in Fig. 17-21, metallic sodium is observed to deposit out at the *Cathode and anode* negative electrode (or *cathode*) and gaseous chlorine to bubble up from the positive electrode (or *anode*). (At the temperature of molten NaCl, metallic sodium is also in the liquid state.) These results are not hard to understand in view of the presence of free Na^+ and Cl^- ions in the bath. The negative electrode attracts Na^+ ions and, when they arrive, neutralizes them by transferring an electron to each one:

$$Na^+ + e^- \rightarrow Na.$$

The resulting sodium atoms, unlike sodium ions, are not soluble and appear at this electrode as ordinary metallic sodium. The positive electrode at the same time attracts Cl^- ions and neutralizes them by absorbing an electron from each one:

$$Cl^- \rightarrow Cl + e^-.$$

Electrolysis The insoluble chlorine atoms are evolved as chlorine gas. The entire phenomenon is an example of *electrolysis*, the process by which free elements are liberated from a liquid by the passage of an electric current.

When a solution of NaCl in water instead of molten NaCl undergoes electrolysis, chlorine appears at the anode and hydrogen at the cathode. Hydrogen rather than sodium is liberated because hydrogen ions have the greater affinity for electrons. In effect, hydrogen and sodium in the solution compete for the available electrons, and the hydrogen wins. The leftover

Na^+ ions remain in the solution together with the OH^- (*hydroxide*) ions formed when H_2O molecules lose an H^+ ion each to produce the neutral hydrogen at the cathode. The bath acquires higher and higher concentrations of Na^+ and OH^- ions which can be recovered as solid NaOH by evaporating the solution. Thus the electrolysis of a water solution of inexpensive NaCl leads to the production of gaseous hydrogen and chlorine and the salt sodium hydroxide, all valuable industrial chemicals.

A widely-used application of electrolysis is the depositing, or *plating*, of a thin layer of an expensive metal (such as gold, silver, or chromium) on an object made of a cheaper metal. Nonmetallic objects can be plated by first coating them with a conducting substance such as graphite.

In electroplating, the object to be coated is used as the cathode and a bar of the plating metal is used as the anode in an electrolytic cell. A salt of the plating metal is dissolved in water to make the conducting bath in which the process takes place. When the cathode is connected to the negative terminal of a battery (or other source of direct current) and the anode to the positive terminal, two reactions occur. At the cathode, positive ions of the plating metal arrive and become metal atoms by absorbing one or more electrons; at the anode, atoms of the plating metal lose electrons and go into solution as ions. Thus metal ions from the bath plate out at the cathode and are replaced by new ions from the anode. The net result is the transfer of metal atoms from the anode to the cathode.

Electroplating

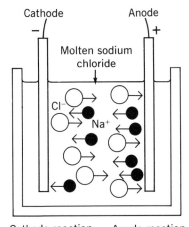

Cathode reaction: Anode reaction:
$Na^+ + e^- \rightarrow Na$ $Cl^- \rightarrow Cl + e^-$

Fig. 17-21. The electrolysis of molten sodium chloride.

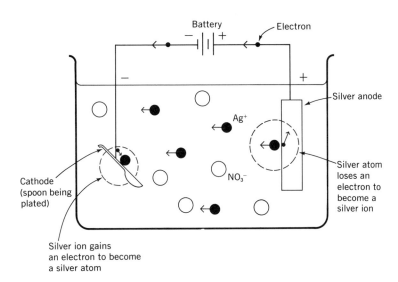

Fig. 17-22. Silver plating.

Silver plating

As an example, a silver plating bath might be prepared by dissolving silver nitrate, $AgNO_3$, to give Ag^+ and NO_3^- ions. Figure 17-22 shows what happens during the plating process. The key to the anode reaction,

$$Ag \rightarrow Ag^+ + e^-$$

is that an Ag atom loses an electron to become Ag^+ more readily than an NO_3^- ion loses its odd electron. Hence the NO_3^- ions stay in solution and do not participate in the plating process. At the cathode, the Ag^+ ions absorb electrons to become Ag atoms once more:

$$Ag^+ + e^- \rightarrow Ag.$$

17-9 Batteries and Fuel Cells

Batteries

A battery is, in a sense, the opposite of an electrolytic cell. When the terminals of a battery are connected together through an external circuit, the material of the negative electrode goes into solution and positive ions from the solution are deposited at the positive electrode in the course of producing a current. A *storage battery* is "charged" by the electrolytic reactions that occur when a current is passed through it from an outside source; when it acts as a battery, the same reactions occur backwards to produce a current in the reverse direction as the ingredients of the cell restore themselves to their initial state. In principle, all batteries can be reversed to charge them, but in practice the construction of certain types of cells prevents this, and such cells must be discarded when they are exhausted.

Fuel cells

A *fuel cell* is a type of battery in which a continuous flow of the initial reactant chemicals is possible, which means that the cell need never be exhausted (like a dry battery) or need recharging (like a storage battery); its advantage over mechanically driven generators lies in its lack of moving parts. Fuel cells are already used to supply electricity in space vehicles, where their high power/weight ratio is important, and are being developed as power sources for electric cars and as self-contained units to furnish electricity to homes.

Hydrogen-oxygen cell

The hydrogen-oxygen cell (Fig. 17-23) is convenient to illustrate how a fuel cell works. The hollow electrodes are made of inert conducting materials with microscopic pores that permit the gases to come in contact with the electrolyte at a gradual rate. The electrolyte of the cell is a potassium hydroxide (KOH) solution that contains K^+ and OH^- ions. At the negative electrode, hydrogen molecules combine with hydroxide ions to form water,

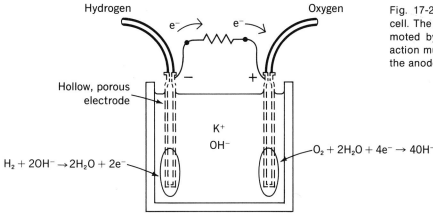

Fig. 17-23. A hydrogen-oxygen fuel cell. The electrode reactions are promoted by catalysts. The cathode reaction must occur twice for each time the anode reaction occurs.

a process that releases electrons to the external circuit:

$$H_2 + 2OH^- \rightarrow 2H_2O + 2e^-.$$

At the positive electrode, oxygen molecules combine with water molecules and incoming electrons that originated at the other electrode to produce hydroxide ions:

$$O_2 + 2H_2O + 4e^- \rightarrow 4OH^-.$$

The first of these reactions must occur twice for each time the second occurs. The net effect of both reactions may be summarized as

$$2H_2 + O_2 \rightarrow 2H_2O + \text{flow of 4 electrons.}$$

If two volumes of hydrogen gas are added to one of oxygen and the mixture ignited, a violent explosion occurs with water as the product. In the hydrogen-oxygen fuel cell the same chemical process occurs, but the liberated energy appears in the form of electric current.

Actual hydrogen-oxygen fuel cells use catalysts at each electrode to promote the reactions there; a catalyst is a substance that affects the rate of a chemical process without directly participating in it. The combination of 1 lb of hydrogen and 8 lb of oxygen in a fuel cell produces 9 lb of water and approximately 10^8 J of electrical energy, enough to power a 100-W light bulb for nearly two weeks. (Oxygen molecules weigh about 16 times as much as hydrogen molecules, so a ratio of 2 H_2 molecules per O_2 molecule is maintained by a weight ratio of 1 lb of hydrogen per 8 lb of oxygen.)

Important Terms

Electric charge, like rest mass, is a fundamental property of certain of the elementary particles of which all matter is composed. There are two kinds of electric charge, **positive charge** and **negative charge**; charges of like sign repel, unlike charges attract. The unit of charge is the **coulomb**. All charges, of either sign, occur in multiples of the fundamental **electron charge** of 1.6×10^{-19} coulomb.

The principle of **conservation of charge** states that the net electric charge in an isolated system remains constant.

Coulomb's law states that the force one charge exerts upon another is directly proportional to the magnitudes of the charges and inversely proportional to the square of the distance between them.

An **atom** consists of a tiny, positively-charged nucleus surrounded at some distance by electrons. The nucleus consists of protons and neutrons held tightly together by nuclear forces, and the number of electrons equals the number of protons so the atom as a whole is electrically neutral.

An **ion** is an atom or group of atoms that carries an electric charge. An atom or group of atoms becomes a negative ion when it picks up one or more electrons in addition to its normal number, and becomes a positive ion when it loses one or more of its usual number.

A **polar molecule** is one whose charge distribution is not uniform, so that one end is positive and the other negative even though the molecule as a whole is electrically neutral.

A substance that separates into free ions when dissolved in water is called an **electrolyte** since the resulting solution is able to conduct electric current.

Electrolysis is the process by which free elements are liberated from a liquid by the passage of an electric current.

A **battery** is a device in which chemical reactions produce an electric currnet. A **fuel cell** is a type of battery in which a continuous supply of the initial reactant chemicals is possible.

Important Formula

Coulomb's law:

$$F = k \frac{q_A q_B}{r^2}$$

Multiple Choice

1. Electric charge
   a. is a continuous quantity that can be subdivided indefinitely.
   b. is a continuous quantity but it cannot be subdivided into smaller parcels than $\pm 1.6 \times 10^{-19}$ C.
   c. occurs only in separate parcels, each of $\pm 1.6 \times 10^{-19}$ C.
   d. occurs only in separate parcels, each of ± 1 C.

2. An object has a positive electric charge whenever
   a. it contains an excess of electrons.
   b. it contains a deficiency of electrons.
   c. the nuclei of its atoms are positively charged.
   d. the electrons of its atoms are positively charged.

3. The nucleus of an atom cannot be said to
   a. contain most of the atom's mass.
   b. be small in size.
   c. be electrically neutral.
   d. deflect incident alpha particles.

4. Coulomb's law belongs in the same general category as
   a. the law of gravitation.
   b. the laws of motion.
   c. the laws of thermodynamics.
   d. the conservation principles of mechanics.

5. A negative electric charge
   a. interacts only with positive charges.
   b. interacts only with negative charges.
   c. interacts with both positive and negative charges.
   d. may interact with either positive or negative charges, depending on circumstances.

6. According to the Rutherford model of the atom, the positive charge in an atom is

a. concentrated at its center.

b. spread uniformly throughout its volume.

c. in the form of positive electrons at some distance from its center.

d. readily deflected by an incident alpha particle.

7. It is correct to say that
   a. charges carry electric current.
   b. the motion of charges produces electric current.
   c. moving charges constitute electric current.
   d. only electrons are involved in the flow of electric current.

8. All molecules are normally
   a. neutral.
   b. charged.
   c. polar.
   d. nonpolar.

9. A molecule to which an electron is added becomes
   a. a negative ion.
   b. a positive ion.
   c. a polar molecule.
   d. an electrolyte.

10. A molecule whose charge distribution is not perfectly symmetrical is called
    a. a polar molecule.
    b. a nonpolar molecule.
    c. an electrolyte.
    d. an organic molecule.

11. Water is an excellent solvent because its molecules are
    a. neutral.
    b. highly polar.
    c. nonpolar.
    d. anodes.

12. Substances that separate into free ions when dissolved in water are called
    a. polar.
    b. nonpolar.
    c. anodes.
    d. electrolytes.

13. Oil does not mix with water because
    a. their respective molecules are different in mass.

b. their respective molecules are different in size.

c. water molecules are polar whereas oil molecules are nonpolar.

d. oil molecules are polar whereas water molecules are nonpolar.

14. Crystalline solids such as NaCl that consist of ions dissolve only in liquids that are
    a. polar.   b. nonpolar.
    c. ionized.   d. oily.

15. The atoms of an element and its ions always have the same
    a. nuclei.
    b. electron structures.
    c. color.
    d. chemical behavior.

16. A fuel cell does not require
    a. an electrolyte.
    b. an anode.
    c. a cathode.
    d. recharging.

17. In the formula $F = kq_A q_B / r^2$, the value of the constant k
    a. depends upon the system of units being used.
    b. is the same no matter what medium the charges are located in.
    c. is different for positive and negative charges.
    d. has the numerical value 1.6×10^{-19}.

18. Two charges of $+Q$ are 1 cm apart. If one of the charges is replaced by a charge of $-Q$, the magnitude of the force between them is
    a. zero.   b. smaller.
    c. the same.   d. larger.

19. A charge of $+q$ is placed 2 cm from a charge of $-Q$. A second charge of $+q$ is then placed next to the first. The force on the charge of $-Q$
    a. decreases to half its former magnitude.
    b. remains the same.
    c. increases to twice its former magnitude.
    d. increases to four times its former magnitude.

20. Two charges repel each other with a force of 10^{-6} N when they are 10 cm apart. When they are brought closer together until they are 2 cm apart, the force between them becomes

   a. 4×10^{-8} N.   b. 5×10^{-6} N.
   c. 8×10^{-6} N.   d. 2.5×10^{-5} N.

21. Two charges, one positive and the other negative, are initially 2 cm apart and are then pulled away from each other until they are 6 cm apart. The force between them is now smaller by a factor of
   a. $\sqrt{3}$.   b. 3.
   c. 9.   d. 27.

22. Ten thousand electrons are removed from a neutral pith ball. Its charge is now
   a. $+1.6 \times 10^{-15}$ C.   b. $+1.6 \times 10^{-23}$ C.
   c. -1.6×10^{-15} C.   d. -1.6×10^{-23} C.

23. The force between two charges of -3×10^{-9} C that are 5 cm apart is
   a. 1.8×10^{-16} N.   b. 3.6×10^{-15} N.
   c. 1.6×10^{-6} N.   d. 3.2×10^{-5} N.

Exercises

1. Electricity was once regarded as a weightless fluid, an excess of which was "positive" and a deficiency of which was "negative." What phenomena can this hypothesis still explain? What phenomena can it not explain?

2. (a) When two bodies attract each other electrically, must both of them be charged? (b) When two bodies repel each other electrically, must both of them be charged?

3. How do we know that the inverse square force holding the earth in its orbit around the sun is not an electrical force?

4. An insulating rod has a positive charge at one end and a negative charge of the same magnitude at the other. How will the rod behave when it is placed near a fixed positive charge that is initially equidistant from the ends of the rod?

5. How can the principle of charge conservation be reconciled with the fact that a rubber rod can be charged by stroking it with a piece of fur?

6. What reasons might there be for the universal belief among scientists that there are only two kinds of electric charge?

7. Nearly all the mass of an atom is concentrated in its nucleus. Where is its charge located?

8. In what ways do the Thomson and Rutherford models of the atom agree? In what ways do they disagree?

9. Most alpha particles pass through gases and thin metal foils with no deflection. To what conclusion regarding atomic structure does this observation lead?

10. What property of the electrons in a metal enables it to conduct electric current readily? What property of the electrons in an insulator prevents it from conducting electric current readily?

11. How does electrical conduction in a metal differ from that in an ionized gas?

12. List several good conductors of electricity and several good insulators. How well do these substances conduct heat? What general relationship between an ability to conduct electricity and an ability to conduct heat can you infer?

13. What aspect of superconductivity has prevented its large-scale application thus far?

14. How could you experimentally distinguish between a solution of an electrolyte and one of a nonelectrolyte?

15. Distinguish between a molecular ion and a polar molecule.

16. In what fundamental way or ways is a fuel cell different from a battery?

17. Give an example of a polar liquid and one of a nonpolar liquid, and state several substances soluble in each but not the other.

18. A billion electrons are added to a pith ball. What is its charge?

19. Two electric charges originally 8 cm apart are brought closer together until the force between them is greater by a factor of 16. How far apart are they now?

20. Two charges attract each other with a force of 4×10^{-6} N when they are 0.4 m apart. Find the force between them when their separation is increased to 0.8 m.

21. Two charges of unknown magnitude and sign are observed to repel one another with a force of 0.1 N when they are 5 cm apart. What will the force be when they are (a) 10 cm apart? (b) 50 cm apart? (c) 1 cm apart?

22. Two charges repel each other with a force of 10^{-6} N when they are 10 cm apart. What is the force between them when the charges are 2 cm apart?

23. A charge of $+10^{-9}$ C is 5 cm from a charge of $+3 \times 10^{-9}$ C. Find the force on each charge.

24. A charge of -10^{-5} C is 10 cm from a charge of $+10^{-6}$ C. Find the force on each charge.

25. A charge of -10^{-6} C is 20 cm from a charge of -5×10^{-6} C. Find the force on each charge.

26. Two metal spheres, one with a charge of $+2 \times 10^{-5}$ C and the other with a charge of -1×10^{-5} C, are 10 cm apart. (a) What is the force between them? (b) The two spheres are brought into contact, and then separated again by 10 cm. What is the force between them now?

Problems

1. How far apart should two electrons be if the force each exerts on the other is to equal the weight of an electron?

2. At what distance apart (if any) are the electric and gravitational forces between two electrons equal in magnitude? Between two protons? Between an electron and a proton?

3. If equal amounts of positive charge were to be placed on the earth and moon until their repulsion exactly balanced the gravitational attraction between them, how much net charge would each body have?

4. According to one model of the hydrogen atom, it consists of a proton circled by an electron whose orbit has a radius of 5.3×10^{-11} m. How fast must the electron be moving if the required centripetal force is provided by the electric force exerted by the proton?

5. A particle carrying a charge of $+6 \times 10^{-7}$ C is located halfway between two other charges, one of $+1 \times 10^{-6}$ C and the other of -1×10^{-6} C, that are 40 cm apart. All three charges lie on the same straight line. What is the magnitude and direction of the force on the $+6 \times 10^{-7}$ C charge?

6. A test charge of -5×10^{-7} C is placed between two other charges so that it is 5 cm from a charge of -3×10^{-7} C and 10 cm from a charge of -6×10^{-7} C. The three charges lie along a straight line. What is the magnitude and direction of the force on the test charge?

7. Two charges, one of -1×10^{-6} C and the other of -3×10^{-6} C are 0.4 m apart. (a) Where should a charge of -1×10^{-7} C be placed on the line between the other charges in order that there be no resultant force on it? (b) Where should a charge of $+1 \times 10^{-7}$ C be placed in order that there be no resultant force on it?

8. Two charges, one of $+2 \times 10^{-8}$ C and the other of $+1 \times 10^{-8}$ C are 0.2 m apart. (a) Where should a charge of -1×10^{-7} C be placed in order that there be no resultant force on it? (b) Where should a charge of $+1 \times 10^{-7}$ C be placed in order that there be no resultant force on it?

9. Three charges, $+q, +q$, and $-q$, are at the vertexes of an equilateral triangle a long on each side. Find the magnitude and direction of the force on one of the positive charges.

10. Four charges of $+1 \times 10^{-8}$ C are at the corners of a square 0.2 m on each side. Find the magnitude and direction of the force on one of them.

Answers to Multiple Choice

1. a	9. a	17. a
2. b	10. a	18. c
3. c	11. b	19. c
4. a	12. d	20. d
5. c	13. c	21. c
6. a	14. a	22. a
7. c	15. a	23. d
8. a	16. d	

18

Electric Field

Electric forces, like gravitational forces, act between objects that may be widely separated. An appropriate way to regard such forces involves the concept of a force field. When a charge is present somewhere, the properties of space in its vicinity can be considered to be so altered that another charge brought to this region will experience a force there. The "alteration in space" caused by a charge at rest is called its electric field, and any other charge is thought of as interacting with the field and not directly with the charge responsible for it. All forces, not just electric ones, can be interpreted as arising through the intermediacy of a force field of one kind or another. Thus the sun is regarded as being surrounded by a gravitational field, and it is the forces exerted by this field on the planets which are the centripetal forces that hold them in their orbits. In the case of the "direct contact" forces between solid objects in everyday life, the force fields involved are electric fields.

18-1 What is a Field?

A force field is a model devised to provide a framework for understanding how forces are transmitted from one object to another across empty space. A successful model does more than just organize all the information we have on a certain phenomenon into a unified picture; it also enables us to predict hitherto unsuspected effects and relationships, which of course must then be verified by experiment. The creative role of a scientific model is beautifully illustrated by the electromagnetic field model proposed by James Clerk Maxwell a century ago. With the help of this model, Maxwell predicted in 1864 that electromagnetic waves should exist; he calculated that their velocity should be $c = 3 \times 10^8$ m/s, and he suggested that light consists of such waves. In 1887, after Maxwell's death, Heinrich Hertz confirmed Maxwell's hypothesis in the laboratory.

A field—in the sense that physicists use the word—is a region of space in which a certain quantity has a definite value at every point. Thus it is appropriate to speak of the temperature field in a room, of the velocity field of the water in a river, and of the gravitational field around the earth. But it is not appropriate to speak of the "chair field" in a room, or of the "rowboat field" in a river, or of the "airplane field" around the earth because chairs, rowboats, and airplanes are separate objects whose presence somewhere is not a property of a region of space that varies throughout that region in the way that temperature, water velocity, and gravitational force vary.

To determine the temperature field in a room, we must measure the temperature at a great many points with a thermometer. The results might

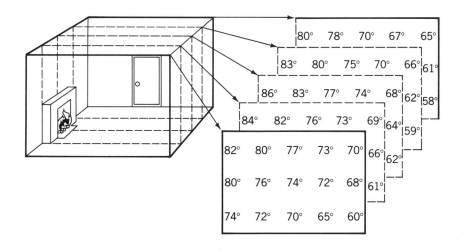

Fig. 18-1. The temperature field in a room is a scalar field, since the temperature at a point has no direction associated with it.

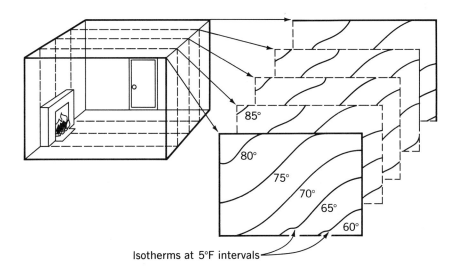

85°

80°

75°

70°

65°

60°

Isotherms at 5°F intervals

Fig. 18-2. An isotherm is a line that joins points whose temperature is the same. Here the temperature field in the room of Fig. 18-1 is displayed by means of isotherms.

be displayed by a series of cross-sectional maps of the room with temperature values written in at each point of measurement, as in Fig. 18-1.

A better way to picture the temperature field is to draw a series of lines on each map that connect points having the same temperature (Fig. 18-2). These lines are called *isotherms*, and might be drawn, for example, for temperatures that are 5°F apart. Naturally the actual measurements are not necessarily 60°, 65°, 70°, and so on; we must interpolate between the actual measurements to find the contours of the various isotherms.

Scalar field

Temperature is a scalar quantity because it involves a magnitude only. To say that the temperature somewhere is 70°F describes it completely. The field of a scalar quantity is a *scalar field*, and it can always be pictured by a plot of isolines (lines joining sets of identical values) as in the case of a temperature field.

Vector field

A vector quantity has direction as well as magnitude, so picturing a *vector field* is not a simple matter. One method is to draw lines on a map of the region occupied by the field so that the lines always point in the direction of the field quantity at each point. For example, the velocity field in a river can be shown with the help of streamlines that represent the paths of successive particles of water (Fig. 18-3). Several rubber balls thrown into a river and photographed with a movie camera would permit us to make a plot of the velocity field at the surface of the river, since each ball would follow a streamline as it moves with the current.

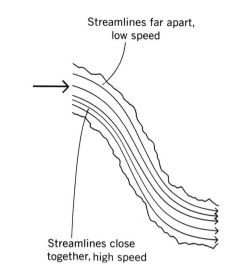

Streamlines far apart,
low speed

Streamlines close
together, high speed

Fig. 18-3.   The velocity field in a river can be visualized to some extent with the help of streamlines that represent the paths of successive particles of water.

18-2 Electric Field

An *electric field* is a force field that exists wherever an electric force acts on a charge. The field may be due to the presence of a single charge or to the presence of many charges. We would like to specify electric field in such a way that it will be possible to determine the force acting on any arbitrary charge at any point in the field. Accordingly the electric field **E** at a point in space is defined as the ratio between the force **F** on a charge q (assumed positive) at that point and the magnitude of q:

$$\mathbf{E} = \frac{\mathbf{F}}{q}. \qquad\qquad \textit{Electric field} \quad (18\text{-}1)$$

The units of **E** are newtons per coulomb. Electric field is a vector quantity that possesses both magnitude and direction.

Electric field is a vector quantity

Once we know what the electric field **E** is at some point, from the definition we see that the force which the field exerts on a charge q at that point is

$$\mathbf{F} = q\mathbf{E} \qquad\qquad\qquad (18\text{-}2)$$

Force = charge × electric field.

Problem.   The electric field between the electrodes of the gas discharge tube used in a certain neon sign has the magnitude 5.0×10^4 N/C. Find the

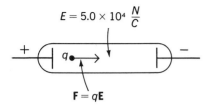

$E = 5.0 \times 10^4 \ \frac{N}{C}$

$F = qE$

Fig. 18-4.

acceleration of a neon ion of mass 3.3×10^{-26} kg in the tube if it carries a charge of $+e$ (Fig. 18-4).

Solution.   Since $e = 1.6 \times 10^{-19}$ C, the force on the ion has the magnitude

$$F = qE$$
$$= 1.6 \times 10^{-19} \ C \times 5.0 \times 10^4 \ N/C$$
$$= 8.0 \times 10^{-15} \ N.$$

Hence the acceleration of the ion is

$$a = \frac{F}{m} = \frac{8.0 \times 10^{-15} \ N}{3.3 \times 10^{-26} \ kg} = 2.4 \times 10^{11} \ m/s^2$$

which is 25 billion times greater than the acceleration of gravity. The velocity such an ion actually attains is limited by frequent collisions with the neon molecules in its path.

We can use Coulomb's law to determine the magnitude of the electric field around a single charge q. First we find the force F that q exerts upon a test charge q_0 at the distance r away, which is

$$F = k \frac{q q_0}{r^2}.$$

Since $E = F/q_0$ by definition, we have

Electric field of a charge

$$E = \frac{F}{q_0} = k \frac{q}{r^2}. \tag{18-3}$$

This formula tells us that the electric field magnitude the distance r from a point charge is directly proportional to the magnitude q of the charge and is inversely proportional to the square of r.

When more than one charge contributes to the electric field at a point P, the net field **E** is the *vector sum* of the fields of the individual charges. That is,

$$\mathbf{E} = \mathbf{E}_1 + \mathbf{E}_2 + \mathbf{E}_3 + \cdots = \Sigma \ \mathbf{E}_n. \tag{18-4}$$

Problem.   Find the electric field intensity at the position of the charge q_1 due to the other charges in the situation shown in Fig. 17-9.

Solution. Making use of our previous results, we obtain

$$E = \frac{F}{q_1} = \frac{2.6 \times 10^{-5} \text{ N}}{2 \times 10^{-9} \text{ C}} = 1.3 \times 10^4 \text{ N/C}$$

The direction of **E** is the same as that of **F** since q_1 is a positive charge; if q_1 were negative, **E** would be in the opposite direction. The usefulness of knowing **E** here is that we can determine from it what the force on *any* charge at the position of q_1 is by simply multiplying **E** and q.

18-3 Electric Lines of Force

An electric field is a region of space in which an object by virtue of its electric charge is acted upon by a force. To visualize an electric field, we may use *lines of force*. These are constructed as follows. At several points in space we draw arrows in the direction a positive charge there would experience a force (Fig. 18-5a). Then these arrows are connected to form smooth curves called lines of force (Fig. 18-5b). Two rules should be kept in mind:

1. Lines of force leave positive charges and enter negative ones.   Rules for lines of force
2. The spacing of lines of force is such that they are close together where the field is strong and far apart where the field is weak.

If we place a positive charge in an electric field, then, it will experience a force in the direction of the line of force it is on, and the magnitude of the

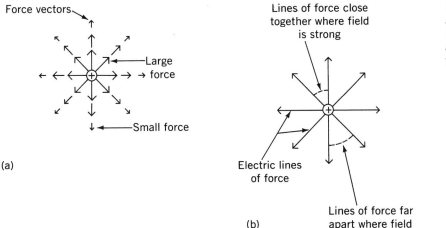

(a)

(b)

Fig. 18-5. How electric lines of force around a positive charge are drawn. In (a) the forces whose vectors are shown are those exerted by the charge on a test positive charge.

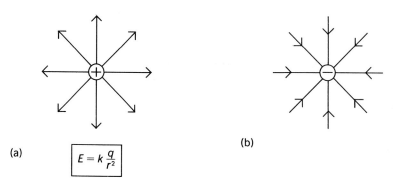

(b)

$$E = k\frac{q}{r^2}$$

Fig. 18-6. Lines of force around a positive and a negative charge.

force will be proportional to the concentration of lines of force in its vicinity. Figure 18-6 shows the patterns of lines of force around a positive and a negative charge and Fig. 18-7 shows the patterns around pairs of charges.

It is important to keep in mind that lines of force do not actually exist as threads in space; they are simply a device for giving intuitive form to our thinking about force fields. The use of lines of force is not limited to electric fields. Gravitational and magnetic fields, for instance, are often described in this manner.

Electric field around a sphere

The notion of lines of force is often helpful as a guide to our thinking when electric fields of charge distributions are concerned. For example, let us inquire into the electric field around a spherical distribution of charge whose total amount is q. Such a distribution is quite easy to arrange—all we need do is to add the charge q to a metal sphere. Since metals are good conductors of electricity, the mutual repulsion of the individual charges that make up q causes them to spread out uniformly over the surface of the

Fig. 18-7. Lines of force near two equal charges (a) of the same sign and (b) of opposite sign. At large distances relative to the separation of the charges, the field in (a) becomes that of a single charge.

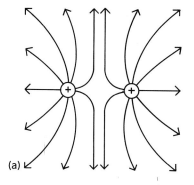

(a)

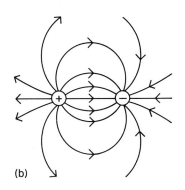

(b)

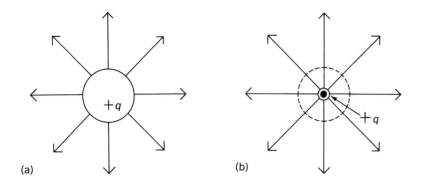

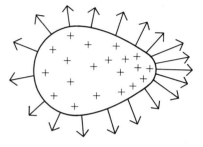

Fig. 18-8 . (a) Electric lines of force around a uniformly charged sphere. (b) Electric lines of force around a point charge.

sphere; in this way the charges are as far apart as possible. Evidently it doesn't matter whether the sphere is hollow or solid.

Whatever the number of lines of force we choose to represent the electric field around a point charge q, the same number must emerge from a body that carries the same net charge q. In the case of a charged sphere, the lines of force are symmetrically arranged, which means that the pattern of lines of force outside the sphere is identical with that around a point charge of the same magnitude (Fig. 18-8).

When an asymmetric (not symmetric) metal body is charged, the individual charges do not distribute themselves uniformly on its surface. As a general rule, the more highly curved parts have greater concentrations of charge than the less curved parts, and the electric field is accordingly most intense near the highly curved parts (Fig. 18-9).

At a point, the electric field intensity may become so great that it causes a separation of charge in nearby air molecules. The resulting electrical "discharge" is visible as a luminous glow. If two oppositely-charged pointed rods are brought close together, charge flows between them via disrupted air molecules and a "spark" occurs. If more gently-curved charged bodies, such as large spheres, are similarly brought near each other, their separation must be much smaller before a spark can occur.

Inside a charged conducting object of any shape the electric field is 0 everywhere; the electric fields due to the individual charges on the surface all cancel out in the interior, although they don't outside the object. We can see why this must be true even without a formal calculation. Suppose there *were* an electric field **E** in the interior of the object. Then the charges inside the object that are free to move (electrons in the case of a metal) would do so under the influence of the field **E**. But no currents are observed in a charged conducting object except for a moment after the charge is placed on it; since energy is needed to maintain an electric current, a supply of ener-

Fig. 18-9. The electric field near a charged asymmetric metal object is strongest near the parts of the object that have the smallest radii of curvature.

Field inside a conductor

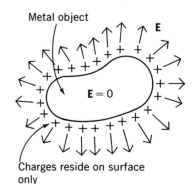

Fig. 18-10. There is no electric field inside a charged conducting object.

gy would be needed for currents to persist in such an object unless it were superconducting. The only conclusion is that the interior of an isolated conducting object is always free of electric field (Fig. 18-10).

18-4 Electric Potential Energy

In our study of mechanics we found the related concepts of work and potential energy to be useful in analyzing a wide variety of situations. These concepts are equally useful in the study of electrical phenomena, in particular electric current. Instead of potential energy itself, a related quantity called potential difference turns out to be especially appropriate in electrical problems.

Let us examine the potential energy of a charge in a uniform electric field together with a gravitational analogy. At the left in Fig. 18-11a is a

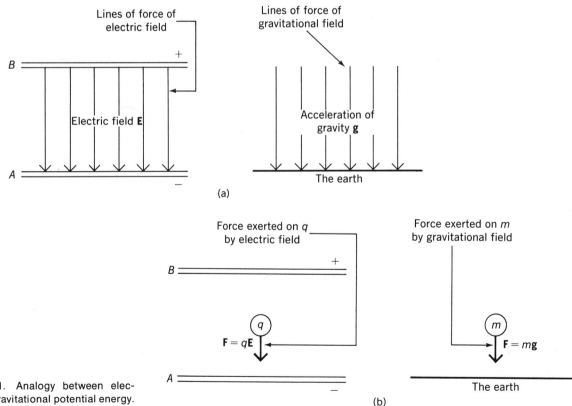

Fig. 18-11. Analogy between electric and gravitational potential energy.

uniform electric field **E** between two parallel, uniformly-charged plates *A* and *B*, and at the right is a region near the earth's surface in which the gravitational field is also uniform. Now we place a particle of charge *q* in the electric field and a particle of mass *m* in the gravitational field. As shown in Fig. 18-11b the charge is acted upon by the electric force

$$\mathbf{F}_{\text{elec}} = q\mathbf{E},$$

and the mass is acted upon by the gravitational force

$$\mathbf{F}_{\text{grav}} = m\mathbf{g}.$$

If the charge is on plate *A* and we want to move it to plate *B*, we must apply a force of magnitude *qE* to it because we have to push against a force of this magnitude exerted by the electric field. When the charge is at *B*, we will have performed the amount of work

Work = force × distance

$$W = qEs \qquad\qquad \text{Work done on charge}$$

on it, where *s* is the distance the charge has moved (Fig. 18-11c). Similarly,

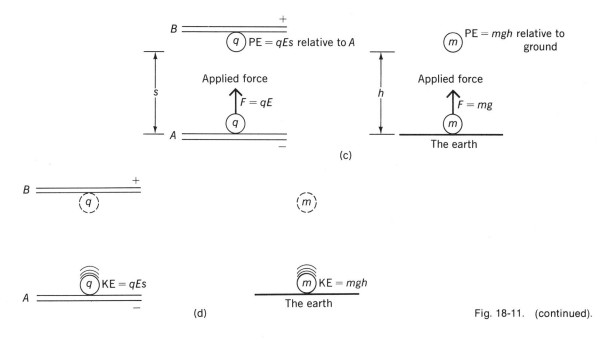

Fig. 18-11. (continued).

to raise the mass from the ground to a height h, we must apply a force of magnitude mg to it, and the work we do is

Work done on mass $\qquad W = mgh.$

At plate B the charge has the potential energy

$$PE = qEs \qquad\qquad (18\text{-}5)$$

with respect to A. If we let it go, the potential energy will become kinetic energy as the electric field $\mathbf{E}$ accelerates the charge, and when the charge is back at A it will have a kinetic energy equal to qEs (Fig. 18-11d). In the same way, the mass has potential energy *with respect to the ground* in its new location. This potential energy is equal to the work done in raising it through the height h, and is

$$PE = mgh.$$

If we let the mass go, it will fall to the ground with a final kinetic energy of mgh.

PE of charge in uniform electric field   To summarize: The amount of work which must be performed to move a charge q from A to B, the distance s apart, in a uniform electric field $\mathbf{E}$ is qEs. At B the charge accordingly has the potential energy qEs. If the charge is released at B, the force $q\mathbf{E}$ acting on it produces an acceleration such that the charge has the kinetic energy qEs when it is back at A. Thus the work done in moving the charge in the field becomes potential energy which in turn becomes kinetic energy when it is released.

There is no change in the energy of a charge moved perpendicular to an electric field, just as there is no change in the energy of a mass moved perpendicular to a gravitational field (for instance, along the earth's surface).

18-5 Potential Energy of Two Charges

A particle of charge q_A that is located the distance r from another particle of charge q_B has electrical potential energy because there is a force exerted on it by the electric field of q_B. When q_A and q_B have the same sign, the force is repulsive; when the charges have opposite signs, the force is attractive. In either case, when q_A is released, it will begin to move and acquire kinetic energy at the expense of its original potential energy (Fig. 18-12).

We have been assuming that q_B is somehow fixed in place. If instead q_A is fixed in place, then we can speak of the potential energy of q_B in the electric field of q_A. This potential energy is exactly the same as before, since by Newton's third law of motion the force one object exerts on another is equal in magnitude to the force the second object exerts on the first.

In reality, the potential energy belongs to the *system* of the two particles. If the particles are both released, *both* of them begin to move, and they share the original potential energy between them. The relative velocities of the two particles will be such as to conserve linear momentum, so the lighter particle will move faster than the heavier one. (The total amount of energy available depends upon the charges of the particles, and the division of the energy depends upon their masses.) If one of the particles has a much greater mass than the other, its velocity will be very small, and it may then be legitimate to regard it as being fixed in place. For example, the mass of a proton is 1836 times that of an electron, so in a hydrogen atom, which consists of a proton and an electron, we may think of the proton as being stationary and of the electron as possessing the potential energy of the system.

A potential energy of any kind must be specified relative to a reference location. Near the earth, for example, the gravitational potential energy of an object is usually given with the earth's surface as the reference location. In the case of individual charges interacting with one another, the reference location is chosen to be infinity, since the electric field of a charge falls to zero an infinite distance away.

Because the electric field **E** of the charge q_B (which we assume to be fixed in place) is not uniform, it is not easy to calculate the potential energy of q_A when it is the distance r away. What we must do is start with q_A at infinity and imagine we move it a short distance toward q_B. In this short distance we can consider **E** as having a constant magnitude, so we can find the work needed for this step from the formula $W = qEs$. Then we move q_A through another short interval closer to q_B; again we can consider **E** as having a constant magnitude, though a greater one than before, and we find the work needed for the second step. By continuing this process until q_A is r away from q_B and adding up all the amounts of work needed, we can determine the potential energy of q_A. With the help of calculus (see Special Topic to Chapter 7) this addition yields the result

$$PE = k\,\frac{q_A q_B}{r}.\qquad \textit{Potential energy of system of two charges}\qquad (18\text{-}6)$$

If the charges have the same sign, their potential energy is positive; thus a positive potential energy corresponds to a repulsive force. If the charges have opposite signs, their potential energy is negative; thus a negative potential energy corresponds to an attractive force. The potential energy of a

PE is always property of system

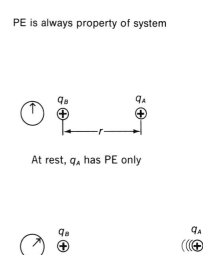

At rest, q_A has PE only

When released, q_A gains KE at the expense of its original PE

Fig. 18-12.   A charge has potential energy when it is in the electric field of another charge. Actually, the potential energy belongs to the system of the two charges. Here q_B is assumed to be fixed in place, so the PE of the system becomes kinetic energy of q_A when q_A is released.

PE relative to infinite separation

charge decreases as it moves away from another charge of the same sign and increases as it moves away from another charge of opposite sign.

18-6 Potential Difference

The quantity *potential difference* is introduced to describe the situation of a charge in an electric field in an especially convenient way. The potential difference V_{AB} between two points A and B is defined as the ratio between the work that must be done to take a charge q from A to B and the value of q:

$$V_{AB} = \frac{W_{AB}}{q} \qquad \textit{Potential difference} \quad (18\text{-}7)$$

Potential difference = work per unit charge.

The volt

The unit of potential difference is the joule per coulomb. Because this quantity is so frequently used, its unit has been given a name of its own, the *volt* (V). Thus

$$1 \text{ volt} = 1 \; \frac{\text{joule}}{\text{coulomb}}.$$

In a uniform electric field, $W_{AB} = qEs$, with the result that the potential difference between A and B is qEs/q or

$$V_{AB} = Es. \qquad \textit{Potential difference in uniform electric field} \quad (18\text{-}8)$$

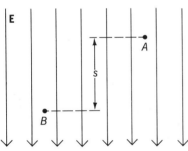

Fig. 18-13. The potential difference between two points in a uniform electric field **E** is equal to *Es*, where *s* is the component parallel to **E** of the distance between the points.

In a uniform electric field, the potential difference between two points is the product of the field magnitude E and the separation s of the two points in a direction parallel to that of **E** (Fig. 18-13).

A positive potential difference means that the energy of the charge is *greater* at B than at A; a negative potential difference means that its energy is *less* at B than at A. If V_{AB} is positive, then a charge at B tends to return to A, whereas if V_{AB} is negative, the charge tends to move further away from A.

Usefulness of potential difference

One advantage of specifying the potential difference between two points in an electric field, rather than the magnitude of the field between them, is that an electric field is normally created by imposing a difference of potential between two points in space. A *battery* is a device that uses chemical means to produce a potential difference between two terminals (Fig. 18-14). A "six-volt" battery is one that has a potential difference of six volts between its terminals. A *generator* is another device for producing a

potential difference. Batteries and generators are to electric charges what elevators are to masses: all of them increase the potential energy of what they act upon.

When a charge q goes from one terminal of a battery whose potential difference is V to the other, the work

$$W = qV$$

Work done on charge

is done on it regardless of the path taken by the charge and regardless of whether the actual electric field that caused the motion of the charge is strong or weak. Given V we can find W at once, no matter what the details of the process are. Hence the notion of potential difference permits us to simplify our analyses of electrical phenomena, just as the notion of potential energy permitted us to simplify our analysis of mechanical phenomena.

Problem. Figure 18-15 shows a tube that has a source of electrons at one end and a metal plate at the other. A 100-V battery is connected between the electron source and the plate, so that there is a potential difference of 100 V between them. The negative terminal of the battery is connected to the electron source. What is the velocity of the electrons when they arrive at the metal plate? (The tube is evacuated to prevent collisions between the electrons and air molecules.)

Solution. To find the kinetic energy, we note that the work done by the electric field within the tube on an electron is

$$W = qV$$
$$= 1.6 \times 10^{-19} \text{ C} \times 100 \text{ V}$$
$$= 1.6 \times 10^{-17} \text{ J}.$$

Since the KE of the electron is equal to the work done on it, its KE after

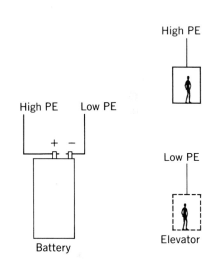

Fig. 18-14.

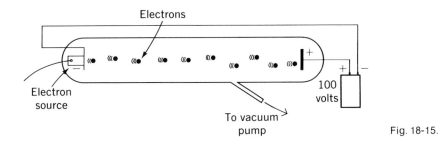

Fig. 18-15.

passing through the entire field is 1.6×10^{-17} J. Hence

$$KE = W = \tfrac{1}{2}mv^2.$$

$$v = \sqrt{\frac{2W}{m}}$$

$$= \sqrt{\frac{2 \times 1.6 \times 10^{-17} \text{ J}}{9.1 \times 10^{-31} \text{ kg}}}$$

$$= 5.9 \times 10^6 \, \frac{\text{m}}{\text{s}}.$$

Problem. The electron source and the positive electrode are 40 cm apart in the tube of the preceding problem. What is the magnitude of the electric field (assumed uniform) between them? If this distance is reduced to 20 cm, what is the new magnitude of the electric field and what effect does this change have on the final speed of the electrons?

Solution. The electric field magnitude in a uniform field is given by

$$E = \frac{V}{s}$$

from the definition of potential difference in an electric field. Hence

$$E = \frac{100 \text{ V}}{0.4 \text{ m}} = 250 \text{ V/m}.$$

If s is reduced to 20 cm, the field magnitude increases to

$$E = \frac{100 \text{ V}}{0.2 \text{ m}} = 500 \text{ V/m}.$$

The energy given to an electron by this field when it travels through the entire 20 cm is the same as the energy given to an electron that travels through the 40-cm length of the previous, weaker field, since in both cases qEs is the same. We reach the same conclusion by noting that $KE = qV$ and the potential difference V in this problem is independent of the spacing of the electrodes.

18-7 The Electron Volt

The electron volt is an energy unit

The electron volt (abbreviated eV) is a widely used energy unit in atomic physics. By definition, 1 eV is the energy acquired by an electron that has been accelerated through a potential difference of 1 volt. Hence

$$W = qV$$

$$1 \text{ eV} = 1.60 \times 10^{-19} \text{ C} \times 1.00 \text{ V},$$

and so

$$1 \text{ eV} = 1.60 \times 10^{-19} \text{ J}. \qquad \qquad \textit{Electron volt} \quad (18\text{-}9)$$

Typical quantities expressed in eV are the ionization energy of an atom (which is the work needed to remove one of its electrons) and the binding energy of a molecule (which is the work needed to break it apart into separate atoms). Thus the ionization energy of nitrogen is usually stated to be 14.5 eV and the binding energy of the hydrogen molecule, which consists of two hydrogen atoms, is usually stated to be 4.5 eV.

The eV is too small a unit for nuclear physics, where its multiples the MeV (10^6 eV) and the GeV (10^9 eV) are commonly used. The M and G respectively signify *mega* ($=10^6$) and *giga* ($=10^9$) and are used in connection with other units as well, for instance the megabuck ($\$10^6$) and the gigawatt ($10^9$ watts). The GeV was formerly called the BeV, where the B stood for "billion," but this proved confusing since in Europe a billion is 10^{12} whereas in the United States a billion is 10^9.

MeV and GeV

A typical quantity expressed in MeV is the energy liberated when the nucleus of a uranium atom splits into two parts. Such *fission* of a uranium nucleus releases an average of 200 MeV; this is the process that powers nuclear reactors and atomic bombs.

Problem. A hydrogen atom consists of a proton and an electron an average of 5.3×10^{-11} m apart. Find the potential energy of the electron in eV.

Solution. Here the two charges are $-e$ and $+e$, where $e = 1.60 \times 10^{-19}$ C. Hence

$$PE = k \frac{q_A q_B}{r} = k \frac{(-e)(e)}{r}$$

$$= 9.0 \times 10^9 \, \frac{\text{N-m}^2}{\text{C}^2} \times \frac{-(1.60 \times 10^{-19} \text{ C})^2}{5.3 \times 10^{-11} \text{ m}}$$

$$= -4.35 \times 10^{-18} \text{ J}.$$

Since $1 \text{ eV} = 1.60 \times 10^{-19}$ J,

$$PE = \frac{-4.35 \times 10^{-18} \text{ J}}{1.60 \times 10^{-19} \text{ J/eV}} = -27.2 \text{ eV}.$$

The minus sign signifies that the force on the electron is directed toward the proton. (The KE of the electron in a hydrogen atom is 13.6 eV, so the total energy of the electron is

$$PE + KE = -27.2 \text{ eV} + 13.6 \text{ eV} = -13.6 \text{ eV}.$$

The negative total energy means that the hydrogen atom is a stable system, since work must be done by an outside agency in order to liberate the electron from the proton.)

$\bullet = K^+$ $O = Cl^-$

Membrane permeable to K^+ only

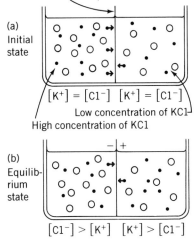

(a) Initial state

$[K^+] = [Cl^-]$ $[K^+] = [Cl^-]$

Low concentration of KCl
High concentration of KCl

(b) Equilibrium state

$[Cl^-] > [K^+]$ $[K^+] > [Cl^-]$

Fig. 18-16. (a) Initially the KCl concentration is higher in the solution to the left of the semipermeable membrane, but in each solution the numbers of K^+ and Cl^- ions are equal. Because the concentration of K^+ ions is greater in the solution at left, more K^+ ions diffuse through the membrane to the right than to the left. (b) At equilibrium, the solution at left has an excess of Cl^- ions, and the one at right has an excess of K^+ ions. The rate of diffusion of K^+ ions is now the same in both directions, and there is a potential difference across the membrane.

Special Topic

The Action Potential

Animal cells have an excess of negative charge inside their membranes and an excess of positive charge on the outside. The potential difference across the membrane of a nerve or muscle cell is normally 90 mV (0.090 V). This potential difference is a consequence of the different permeability of the membrane to different ions. To see how the process works, let us consider two solutions of potassium chloride, KCl, separated by a membrane that permits K^+ ions to pass through it but hinders Cl^- ions from doing so (Fig. 18-16). The solution at the left has the higher concentration of KCl. K^+ ions from both solutions diffuse through the membrane, but because the solution on the left has more of them per unit volume, the number going to the right at first exceeds the number going to the left. As a result the solution on the left soon has a deficiency of K^+ ions relative to its Cl^- ions and hence a net negative charge, while the solution on the right has an excess of K^+ ions and hence a net positive charge. Now an electric field exists across the membrane which favors the diffusion of K^+ ions from right to left. When the influence of the electric field exactly balances the opposite influence of the difference in concentration, as many K^+ ions pass through the membrane in one direction as in the other.

Actual cells do have higher concentrations of K^+ ions inside their membranes compared with the concentration in the fluid around them, and the observed difference is consistent with a potential difference of 90 mV. So-

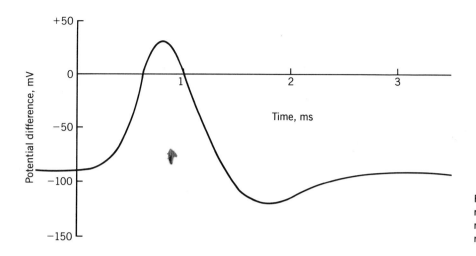

Fig. 18-17. Action potential in a nerve cell. The interior of the cell is normally at a potential of −90 mV with respect to the outside.

dium ions, Na⁺, are also present in living tissue, but their concentration is higher in the extracellular fluid, and cell walls are ordinarily impermeable to them. When a nerve or muscle cell is stimulated, its membrane momentarily becomes permeable to Na⁺ ions, which flow into the cell to neutralize the excess negative charge there. (Cl⁻ is not the only negative ion involved.) The membrane, in effect, is short-circuited by the Na⁺ ions. Because the concentration of Na⁺ ions is greater outside the cell (the opposite of the case of K⁺ ions), in a short time the potential difference across the cell membrane becomes reversed with the inside at a small positive potential relative to the outside (Fig. 18-17). Then the cell wall again becomes impermeable to Na⁺ ions, and K⁺ diffusion restores the potential difference to its normal value. The entire process involved in producing such an *action potential* takes about a millisecond (0.001 s). A mechanism which is not completely understood then "pumps" the excess Na⁺ ions out of the cell into the fluid around it.

Action potentials of the above kind are propagated from nerve cell to nerve cell at speeds of about 30 m/s. When a nerve impulse arrives at a muscle cell, an action potential is triggered in it which travels along the muscle fiber and causes it to contract. The action potentials in heart muscle are especially large and are easily detected by electrodes placed on the chest; an *electrocardiogram* is a record of these potentials.

In some fish, notably the giant electric eel, specialized cells called *electroplaques* occur whose chief purpose is to develop action potentials when stimulated. Large numbers of such cells are stacked in the same manner as the cells of a high-voltage battery, and although each cell is limited to a pulse of not much more than 0.1 V, the total can amount to several hundred volts. Such fish use these potentials to stun their prey. Over two-thirds the mass of an electric eel consists of 4,000 rows of electroplaques that can produce a 600-V pulse.

Important Terms

A **force field** is a region of space at every point of which an appropriate test object would experience a force.

An **electric field** exists wherever an electric force acts on a charged particle. The magnitude of an electric field at a point is equal to the force that would act on a charge of +1 C placed there; the direction of the field is the direction of the force on the charge. The unit of electric field is the V/m, which is equal to 1 N/C.

The electrical **potential difference** between two points is the work that must be done to take a charge of 1 C from one of the points to the other. The unit of potential difference is the **volt**, which is equal to 1 J/C.

The **electron volt** is the energy acquired by an electron which has been accelerated by a potential difference of 1 V. It is equal to 1.6×10^{-19} J.

Important Formulas

Electric field:

$$\mathbf{E} = \frac{\mathbf{F}}{q}$$

Electric field of a charge:

$$E = k \frac{q}{r^2}$$

Potential energy of a charge in uniform field:

$$PE = qEs$$

Potential energy of two charges:

$$PE = k \frac{q_A q_B}{r}$$

Potential difference:

$$V_{AB} = \frac{W_{AB}}{q}$$

$$= Es \qquad \text{[in uniform electric field]}$$

Multiple Choice

1. Of the following quantities, the one that is vector in character is electric
   a. charge.
   b. field.
   c. energy.
   d. potential difference.

2. Electric lines of force
   a. exist everywhere.
   b. exist only in the immediate vicinity of electric charges.
   c. exist only when both positive and negative charges are near one another.
   d. are imaginary.

3. The electric field at a point in space is equal in magnitude to
   a. the potential difference there.
   b. the electric charge there.
   c. the force a charge of one coulomb would experience there.
   d. the force an electron would experience there.

4. The magnitude of the electric field in the region between two parallel oppositely-charged metal plates is
   a. zero.
   b. uniform throughout the region.
   c. greatest near the positive plate.
   d. greatest near the negative plate.

5. Ten million electrons are placed on a solid copper sphere. The electrons become
   a. uniformly distributed on the sphere's surface.
   b. uniformly distributed in the sphere's interior.
   c. concentrated at the center of the sphere.
   d. concentrated at the bottom of the sphere.

6. A system of two charges has a positive potential energy. This signifies that
   a. both charges are positive.
   b. both charges are negative.
   c. both charges are positive or both are negative.
   d. one charge is positive and the other is negative.

7. From its definition, the unit of electric field E is the N/C. An equivalent unit of E is the
   a. volt·m.   b. volt·m².
   c. volt/m.   d. volt/m².

8. The electron volt is a unit of
   a. charge.
   b. potential difference.
   c. energy.
   d. momentum.

9. The force on an electron in an electric field of 200 N/C is
   a. 8×10^{-22} N.   b. 3.2×10^{-21} N.
   c. 3.2×10^{-17} N.   d. 6.4×10^{-15} N.

10. The electric field 2 cm from a certain charge has a magnitude of 10^5 N/C. The value of E 1 cm from the charge is
    a. 2.5×10^4 N/C.   b. 5×10^4 N/C.
    c. 2×10^5 N/C.   d. 4×10^5 N/C.

11. An electric field of magnitude 200 N/C can be produced by applying a potential difference of 10 V to a pair of parallel metal plates separated by
    a. 2 cm.   b. 5 cm.
    c. 20 m.   d. 2000 m.

12. A charge of 10^{-10} C between two parallel metal plates 1 cm apart experiences a force of 10^{-5} N. The potential difference between the plates is
    a. 10^{-5} V.   b. 10 V.
    c. 10^3 V.   d. 10^5 V.

13. In charging a certain storage battery, a total of 2×10^5 C is transferred from one set of electrodes to another. The potential difference between the electrodes is 12 V. The energy stored in the bat-

tery is
    a. 1.7×10^4 J.   b. 2.4×10^6 J.
    c. 2.4×10^7 J.   d. 2.9×10^7 J.

14. The rest energy of the proton is 938 MeV, which is equal to
    a. 1.67×10^{-27} J.   b. 1.50×10^{-10} J.
    c. 1.04×10^{-8} J.   d. 1.50×10^{-8} J.

15. An electron whose velocity is 10^7 m/s has an energy of
    a. 4.6×10^{-17} eV.   b. 160 eV.
    c. 284 eV.   d. 568 eV.

Exercises

1. How could you distinguish between an electric field and a gravitational field?

2. What can you tell about the force a charged object would experience at a given point in an electric field by looking at a sketch of the lines of force of the field?

3. Sketch the pattern of electric lines of force in the neighborhood of two charges, $+Q$ and $+2Q$, that are a short distance apart.

4. Sketch the pattern of electric lines of force in the neighborhood of two charges, $-Q$ and $+2Q$, that are a short distance apart.

5. Can lines of force ever intersect in space? Explain.

6. A charge of $+Q$ is placed on a cubical copper box 20 cm on an edge. What is the magnitude and direction of the electric field at the center of the box?

7. An insulating rod has a positive charge at one end and a negative charge of the same magnitude at the other. How will the rod behave when it is placed in a uniform electric field whose direction is (a) parallel to the rod; (b) perpendicular to the rod?

8. What is the electric field 0.4 m from a charge of $+7 \times 10^{-5}$C?

9. Four charges of $+1 \times 10^{-8}$ C are at the corners of a square which measures 1 m on each side. Find the electric field at the center of the square.

10. The potential difference between two parallel metal plates which are 0.5 cm apart is 10^4 V. Find the force on an electron located between the plates.

11. Two parallel metal plates are 4 cm apart. If the force on an electron between the plates is to be 10^{-14} N, what should the potential difference between them be?

12. A certain 32-V storage battery can deliver a total of 10^6 J of energy. How much charge must pass from one of its terminals to the other for this energy to be liberated?

13. A 12-V storage battery is being charged at the rate of 10 C/s. (a) How much power is needed to charge the battery at this rate? (b) How much energy is stored in the battery if it is charged for 1 hr?

14. How much energy is expended when a charge of 30 C is sent through an electric motor if the potential difference is 220 V?

15. A cloud is at a potential of 8×10^6 V relative to the ground. A charge of 40 C is transferred in a lightning stroke between the cloud and the ground. Find the energy dissipated.

16. It is necessary to do 13.6 eV of work to remove the electron from a hydrogen atom. (a) How much energy in joules would be required to remove the electrons from the 6×10^{26} hydrogen atoms in a kilogram of hydrogen? (b) How many kilocalories is this?

17. What is the velocity of an electron whose energy is 50 eV?

18. What is the energy in electron volts of an electron whose velocity is 10^6 m/s?

19. What is the energy in electron volts of a potassium atom of mass 6.5×10^{-26} kg whose velocity is 10^6 m/s?

20. What is the velocity of a neutron whose kinetic energy is 50 eV? The neutron mass is 1.67×10^{-27} kg.

Problems

1. Two charges, one of 1.5×10^{-6} C and the other of 3×10^{-6} C, are 0.2 m apart. Where is the electric field in their vicinity equal to zero?

2. Two charges of $+4 \times 10^{-6}$ C and $+8 \times 10^{-6}$ C are 2 m apart. What is the electric field intensity halfway between them?

3. A body whose mass is 10^{-6} kg carries a charge of $+10^{-6}$ C. What is the magnitude of an electric field that can hold the body suspended in equilibrium?

4. A particle carrying a charge of 10^{-5} C starts moving from rest in a uniform electric field whose intensity is 50 V/m. (a) What is the force on the particle? (b) How much kinetic energy will the particle have after it has moved 1 m?

5. A potential difference of 10 V is applied across two parallel metal plates 2 cm apart. An electron is projected at a velocity of 10^7 m/s halfway between the plates and parallel to them. How far will the electron travel before striking the positive plate?

6. What is the electric field at one vertex of an equilateral triangle whose sides are 1 m long if there are charges of $+2 \times 10^{-5}$ C at the other vertexes?

7. An electron with the initial velocity of 10 m/s is directed at another electron, whose position is fixed, from a distance of 2 m. How close to the stationary electron will the other one approach before it comes to a stop and reverses its path?

8. In charging a certain 20-kg storage battery, a total of 2×10^5 C is transferred from one set of electrodes to another. The potential difference between the electrodes is 12 V. (a) How much energy is stored in the battery? (b) If this energy were used to raise the battery above the ground, how high would it go? (c) If this energy were used to provide the battery with kinetic energy, what would its velocity be?

9. Typical chemical reactions absorb or release energy at the rate of several eV per molecular change. What change in mass is associated with the absorption or release of 1 eV?

10. Find the rest energy of the electron in MeV.

Answers to Multiple Choice

1. b	7. c	13. b
2. d	8. c	14. b
3. c	9. c	15. c
4. b	10. d	
5. a	11. b	
6. c	12. c	

19

Electric Current

An electric current consists of a flow of charge from one place to another. Currents and not static charges are involved in nearly all the practical applications of electricity. In this chapter we shall consider the chief factors that govern direct currents in conductors, and in later chapters we shall examine the magnetic effects of currents and how they are exploited in modern technology. The important topic of alternating current is discussed in Chapter 23.

19-1 Electric Current

The magnitude of an electric current, denoted i, is the rate at which charge passes a given point. If the net charge q goes past in the time interval t, then the average current is

$$i = \frac{q}{t} \qquad\qquad (19\text{-}1)$$

$$\text{Current} = \frac{\text{charge}}{\text{time interval}}.$$

The unit of electric current is the *ampere* (A), where

The ampere

$$1 \text{ ampere} = 1 \frac{\text{coulomb}}{\text{second}}.$$

The direction of a current is, by convention, taken as that in which *positive* charges would have to move in order to produce the same effects as the observed current. Thus a current is always assumed to proceed from the positive terminal of a battery or generator to its negative terminal (Fig. 19-1).

Direction of current

Despite the above convention, actual electric currents in metals consist of flows of electrons, which carry negative charges. However, a current that consists of negative particles moving in one direction is electrically the same as a current that consists of positive particles moving the other way. Since there is no overwhelming reason to prefer one way of designating current to the other, we shall follow the usual practice of considering current as a flow of positive electric charge.

Problem. How many electrons per second pass any point in a wire that carries a current of 1 A?

Solution. Since the charge on an electron is 1.6×10^{-19} C, a 1 A current corresponds to a flow of

$$\frac{1 \text{ C/s}}{1.6 \times 10^{-19} \text{ C/electron}} = 6.3 \times 10^{18} \frac{\text{electrons}}{\text{s}}.$$

This figure does not mean that 6.3×10^{18} electrons flow from one end of a copper wire to the other each second if there is a current of 1 amp in the wire. It does signify that this many electrons enter one end of the wire, and the same number leave the other end each second—they do not have to be the same electrons. A legitimate analogy is with the flow of water in a pipe.

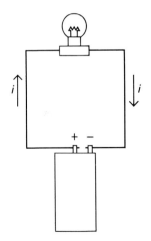

Fig. 19-1. By convention, an electric current is assumed to flow from the positive terminal of a battery or generator to its negative terminal. Actual currents in metals consist of electrons that move in the opposite direction.

Two conditions must be met in order for an electric current to exist between two points. These are:

Conditions for a current

1. There must be a path between the two points along which charge can flow. As was discussed earlier, metals, many liquids, and plasmas (gases whose molecules are charged) allow charge to pass through them readily and are classed as conductors. Nonmetallic solids, certain liquids, and gases whose molecules are electrically neutral allow charge to pass through them only with great difficulty and are classed as insulators. A few substances have an intermediate ability to permit the flow of charge and are classed as semiconductors.

2. There must be a difference of potential between the two points. (A superconductor is an exception to this requirement.) Just as the rate of flow of water between the ends of a pipe depends upon the difference of pressure between them, so the rate of flow of charge between two points depends upon the difference of potential between them. A large potential difference means a large "push" given to each charge.

The analogy between electric current and water flow is a close one. The rate of flow of water in a pipe may be increased by having the water fall through a greater height, which increases the pressure in the pipe and thereby leads to a greater force on each parcel of water. Similarly the current in a wire may be increased by increasing the potential difference across it, which means a stronger electric field in the wire and thus more force on the moving charges that constitute the current (Fig. 19-2).

Fig. 19-2. The role of potential difference in producing an electric current is analogous to the role of height in producing a flow of water.

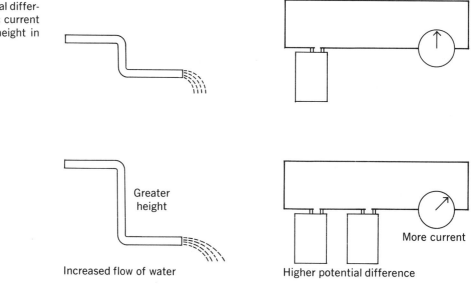

Increased flow of water   Higher potential difference

A particular conducting path—for instance, a copper wire, a light bulb, an electric heater, a transistor—is usually called a conductor, even though this is also the name of the class of substances through which current flows readily. The *resistance* of a conductor is the ratio between the potential difference V across it and the resulting current i that flows:

Resistance

$$R = \frac{V}{i} \tag{19-2}$$

$$\text{Resistance} = \frac{\text{potential difference}}{\text{current}}.$$

The unit of resistance is the *ohm* (Ω), where

The ohm

$$1 \text{ ohm} = 1 \frac{\text{volt}}{\text{ampere}}.$$

A conductor in which there is a current of 1 A when a potential difference of 1 V exists across it has a resistance of 1 Ω.

19-2 Ohm's Law

The resistance of a conductor depends in general both upon its properties—its nature and its dimensions—and upon the potential difference applied across it. In some conductors R increases when V increases, in others R decreases when V increases, and in still others R depends upon which direction the current has (Fig. 19-3).

Metallic conductors usually have constant resistances (at constant temperature), so that i is directly proportional to V in them. This relationship is called *Ohm's law,* since it was first verified experimentally by the German physicist Georg Ohm (1787-1854). Ohm's law states that

$$i = \frac{V}{R} \quad [R = \text{constant}]. \qquad \text{\textit{Ohm's law}} \quad (19\text{-}3)$$

Despite its name, Ohm's law is not a fundamental physical principle, but rather an empirical relationship obeyed by most metals under a wide range of circumstances.

Ohm's law must be distinguished from the definition of resistance,

$$R = \frac{V}{i}.$$

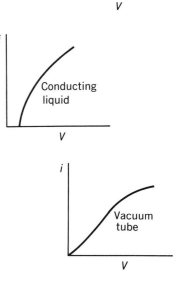

Fig. 19-3. The relationship between current and potential difference in three types of conductor. Only in a metal is i directly proportional to V, a relationship known as Ohm's law.

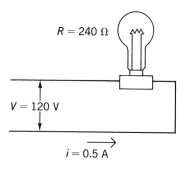

$R = 240\ \Omega$

$V = 120\ V$

$i = 0.5\ A$

Fig. 19-4.

Ohm's law only holds for conductors in which the ratio V/i is constant regardless of the values of V and i. The resistance of a certain conductor is given by $R = V/i$ whether or not it is constant as the voltage is changed.

Problem.   A light bulb has a resistance of 240 Ω. Find the current in it when it is placed in a 120-V circuit (Fig. 19-4).

Solution.   From Ohm's law

$$i = \frac{V}{R} = \frac{120\ V}{240\ \Omega} = 0.5\ A.$$

An electric current in body tissue affects it both by stimulating nerves and muscles and through the heat produced. Tissue is a fairly good conductor because of the ions in solution it contains. Dry skin has a higher resistance and so is able to protect to some extent the rest of the body in the event of accidental exposure to a high potential difference, protection that disappears when the skin is wet. A current of as little as 0.5 mA is perceptible to most people, one of 5 mA is painful, and one of 10 mA or more causes muscle contractions. Such contractions may prevent the person involved from letting go of the source of the current.

Since a closed conducting path is necessary for a current to occur, touching a single "live" conductor has no effect if the body is isolated. However, if a person at the same time is in contact with a water pipe, or is standing on wet soil, or otherwise is grounded, a current will pass through his body. The resistance of the human body is of the order of magnitude of 1000 Ω, so contact via wet skin with a 120-V line will lead to a current in the neighborhood of $i = 120\ v/1000\ \Omega = 0.12\ A = 120\ mA$, which is extremely dangerous.

19-3 Resistivity

The resistance of a conductor that obeys Ohm's law depends upon three factors:

1. The material of which it is composed – the ability to carry an electric current varies more than almost any other physical property of matter;

2. Its length L – the longer the conductor, the greater its resistance;

3. Its cross-sectional area A – the thicker the conductor, the less its resistance.

Once again we note the correspondence to water flowing through a pipe: the longer the pipe, the more chance friction against the pipe wall has to slow down the water, and the wider the pipe, the larger the volume of water that can pass through per second when everything else is the same (Fig. 19-5).

The simple formula

$$R = \rho \frac{L}{A} \tag{19-4}$$

has been found to hold for the resistance of a conductor that obeys Ohm's law. The quantity ρ (the Greek letter *rho*) is called the *resistivity* of the material from which the conductor is made. Given the nature of a conductor and its dimensions, the value of R can be computed at once. In the SI system lengths are measured in m and areas in m², and the unit of resistivity is accordingly the Ω-m.

Resistivity

The resistivities of nearly all substances vary with temperature. In general, metals increase in resistivity with an increase in temperature while nonmetals decrease in resistivity. If R is the resistance of a conductor at a particular temperature, then the change ΔR in its resistance when the temperature changes by ΔT is approximately proportional to both R and ΔT, so that

$$\Delta R = \alpha R \, \Delta T. \tag{19-5}$$

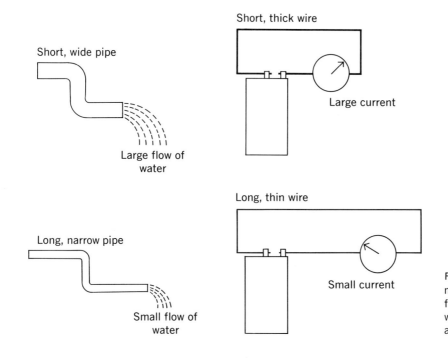

Fig. 19-5. The way in which the dimensions of a conductor affect the flow of charge in it is analogous to the way in which the dimensions of a pipe affect the flow of water in it.

Table 19-1. Approximate resistivities (at 20°C) and their temperature coefficients.

SUBSTANCE	$\rho(\Omega \cdot m)$	$\alpha(/°C)$
Conductors		
Aluminum	2.6×10^{-8}	0.0039
Constantan (60% Cu, 40% Ni)	49×10^{-8}	0.000002
Copper	1.7×10^{-8}	0.0039
Iron	12×10^{-8}	0.0050
Lead	21×10^{-8}	0.0043
Manganin (84% Cu, 12% Mn, 4% Ni)	44×10^{-8}	0.000000
Mercury	98×10^{-8}	0.00088
Platinum	11×10^{-8}	0.0036
Silver	1.6×10^{-8}	0.0038
Semiconductors		
Carbon	3.5×10^{-5}	−0.0005
Germanium	0.5	
Copper oxide (CuO)	1×10^3	
Insulators		
Glass	$10^{10} - 10^{14}$	
Quartz	7.5×10^{17}	
Sulfur	10^{15}	

The quantity α is the temperature coefficient of resistivity of the material. Table 19-1 lists the resistivities of various substances at room temperature (20°C) together with the temperature coefficients of a number of them. The temperature coefficient of carbon is labeled negative (−) because its resistivity decreases with increasing temperature.

Problem. A *resistance thermometer* makes use of the temperature variation of resistivity. (a) Find the resistance at 20°C of a coil that consists of 20 cm of platinum wire 0.05 mm in diameter. (b) When this coil is placed in a furnace, its resistance triples. What is the temperature of the furnace, assuming that α remains constant?

Solution. (a) Here $L = 0.2$ m and $A = \pi d^2/4 = \pi (5 \times 10^{-5}$ m$)^2/4 = 1.96 \times 10^{-9}$ m², and so, with the value of ρ given in Table 19-1,

$$R = \rho\frac{L}{A} = \frac{11 \times 10^{-8}\ \Omega \cdot \text{m} \times 0.2\ \text{m}}{1.96 \times 10^{-9}\ \text{m}^2} = 11\ \Omega.$$

(b) Since $R = 11\ \Omega$ and $\Delta R = 33\ \Omega - 11\ \Omega = 22\ \Omega$,

$$\Delta T = \frac{\Delta R}{\alpha R} = \frac{22\ \Omega}{0.0036/°C \times 11\ \Omega} = 556°C.$$

The temperature of the furnace is $T + \Delta T = 20°C + 556°C = 576°C.$

19-4 Electric Power

Electrical energy in the form of electric current is converted into heat in an electric stove, into radiant energy in a light bulb, into chemical energy when a storage battery is charged, and into mechanical energy in an electric motor. The widespread use of electrical energy is due as much to the ease with which it can be transformed into other kinds of energy as to the ease with which it can be transported through wires.

The work that must be done to take a charge q through the potential difference V is, by definition,

$$W = qV.$$

Since a current i carries the amount of charge $q = it$ in the time t, the work done is

$$W = iVt. \tag{19-6}$$

The energy input to a device of any kind through which the current i flows when the potential difference V is placed across it is equal to the product of the current, the potential difference, and the time span.

We recall from Chapter 7 that *power* is the term given to the rate at which work is being done, so that

Definition of power

$$P = \frac{W}{t} \tag{7-4}$$

$$\text{Power} = \frac{\text{work done}}{\text{time interval}}.$$

When the work is done by an electric current, $W = iVt$, and so

Electric power
$$P = iV \tag{19-7}$$

Electric power = current × potential difference.

The unit of power is the watt, and when i and V are in amperes and volts respectively, P will be in watts.

Problem.   How much current is drawn by a $\frac{1}{2}$ hp electric motor operated from a 120-V source of electricity? Assume that all of the electrical energy absorbed by the motor is turned into mechanical work.

Solution.   Since 1 hp = 746 W, we find from $P = iV$ that

$$i = \frac{P}{V}$$

$$= \frac{\frac{1}{2} \text{ hp} \times 746 \text{ W/hp}}{120 \text{ V}}$$

$$= 3.1 \text{ A.}$$

The power consumed by a resistance through which current passes may be expressed in the alternative forms

Formulas for electric power
$$P = iV, \tag{19-7}$$

$$P = i^2 R, \tag{19-8}$$

$$P = \frac{V^2}{R}. \tag{19-9}$$

Equation (19-7) holds regardless of the nature of the current-carrying device. Depending upon which quantities are known in a specific case, any of the above expressions for P may be used.

Problem.   Find the power consumed by a 240-Ω light bulb when the current through it is 0.5 A.

Solution.   The formula $P = i^2R$ is easiest to use here. We have

$$P = i^2 R$$

$$= (0.5 \text{ A})^2 \times 240 \ \Omega$$

$$= 60 \text{ W}.$$

Owing to the resistance that all conductors offer to the flow of charge through them, electrical power is dissipated whenever a current is maintained regardless of whether the current also supplies energy that is converted to some other form. Electrical resistance is analogous to friction, and so the power consumed in causing a current to flow is dissipated as heat. If too much current flows in a particular wire, it becomes so hot that it may start a fire or even melt. To prevent this from happening, nearly all electrical circuits are protected by fuses or circuit breakers which interrupt the current when i exceeds a safe value. For example, a 15-A fuse in a 120-V power line means that the maximum power that can be carried is

$$P = iV = 15 \text{ A} \times 120 \text{ V} = 1800 \text{ W}.$$

19-5 Resistors in Series

An electrical circuit in its simplest form consists of a source of electrical energy connected to a *load*, which may be a lamp, a motor, or any other device that absorbs electrical energy and converts it into some other form of energy or into work. In analyzing a circuit it is convenient to group its various components into individual *resistors* that are imagined to be joined together by resistanceless wires. The *equivalent resistance* of a set of interconnected resistors is the value of the single resistor that can be substituted for the entire set without affecting the current in the rest of the circuit.

Equivalent resistance

The symbol for a resistor, $-\wedge\!\wedge\!\wedge-$, represents any circuit component that has electrical resistance and obeys Ohm's law. Although there are an unlimited number of ways in which resistors can be put together, most of them are merely combinations of the basic *series* and *parallel* arrangements (Fig. 19-6). Resistors in series are connected consecutively, so that the same current is present in all of them. On the other hand, resistors in parallel have their ends connected together, so that the total current is split up among them.

Resistors in series and in parallel

The potential difference V across the ends of a series set of resistors is the sum of the potential differences $V_1, V_2, V_3, \ldots$, across each one. This statement follows from the principle of conservation of energy: if V is the work done per coulomb in "pushing" a charge through the set of resistors, then it must equal the sum $V_1 + V_2 + V_3 + \cdots$ of the amounts of work done

Fig. 19-6.   Resistors in series

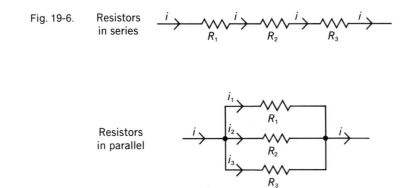

Resistors in parallel

per coulomb in pushing the charge through each resistor in turn. In the case of the three resistors in series of Fig. 19-7,

$$V = V_1 + V_2 + V_3.$$

Resistors in series have the same current

Since the same current i passes through all the resistors, the individual potential drops are

$$V_1 = iR_1, \qquad V_2 = iR_2, \qquad V_3 = iR_3.$$

If we denote the equivalent resistance of the set by R, the potential difference across it is

$$V = iR.$$

We therefore have

$$V = V_1 + V_2 + V_3,$$
$$iR = iR_1 + iR_2 + iR_3.$$

Fig. 19-7.   Resistors in series. The same current i passes through each resistor.

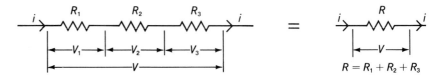

$$R = R_1 + R_2 + R_3$$

Dividing through by the current i yields

$$R = R_1 + R_2 + R_3.$$

In general, *the equivalent resistance of a set of resistors connected in series is equal to the sum of the individual resistances:*

$$R = R_1 + R_2 + R_3 + \cdots \qquad \text{Resistors in series} \quad (19\text{-}10)$$

Problem. A 5-Ω resistor and a 20-Ω resistor are connected in series and a potential difference of 100 volts is applied across them by means of a generator. Find the equivalent resistance of the circuit, the current that flows in it, the potential difference across each resistor, the power dissipated by each resistor, and the power dissipated by the entire circuit.

Solution. Successive stages in the solution of this problem are shown in Fig. 19-8.

(a) The equivalent resistance of the two resistors is

$$R = R_1 + R_2 = 5\ \Omega + 20\ \Omega = 25\ \Omega.$$

This resistance is more than that of either of the resistors.

(b) The current in the circuit is, from Ohm's law,

$$i = \frac{V}{R} = \frac{100\ \text{V}}{25\ \Omega} = 4\ \text{A}.$$

(c) The potential difference across R_1 is

$$V_1 = iR_1 = 4\ \text{A} \times 5\ \Omega = 20\ \text{V},$$

and the potential difference across R_2 is

$$V_2 = iR_2 = 4\ \text{A} \times 20\ \Omega = 80\ \text{V}.$$

We note that the sum of V_1 and V_2 is 100 V, the same as the impressed potential difference, as of course it must be.

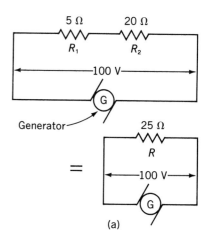

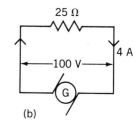

Generator

(a)

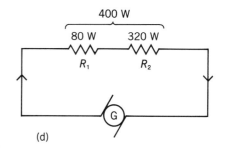

(b)

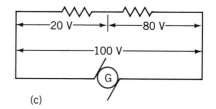

(c)

(d)

Fig. 19-8.

(d) The power dissipated by R_1 is

$$P_1 = iV_1 = 4 \text{ A} \times 20 \text{ V} = 80 \text{ W}.$$

We can also find P_1 from the alternative formulas $P = i^2R$ and $P = V^2/R$. In each case we get the same answer:

$$P_1 = i^2R_1 = (4 \text{ A})^2 \times 5 \text{ }\Omega = 80 \text{ W}, \; P_1 = \frac{V^2}{R} = \frac{(20 \text{ V})^2}{5 \text{ }\Omega} = 80 \text{ W}.$$

The power dissipated by R_2 is

$$P_2 = iV_2 = 4 \text{ A} \times 80 \text{ V} = 320 \text{ W}.$$

The power dissipated by the entire circuit is

$$P = iV = 4 \text{ A} \times 100 \text{ V} = 400 \text{ W}$$

and is equal to $P_1 + P_2$. The dissipated power appears as heat.

19-6 Resistors in Parallel

When two or more resistors are connected in parallel, the same potential difference is applied across all of them. As we shall find, the current that passes through each resistor is inversely proportional to its resistance: the less the resistance, the more the current.

Let us consider three resistors, R_1, R_2, and R_3, that are connected in parallel, as in Fig. 19-9. The total current i through the set is equal to the

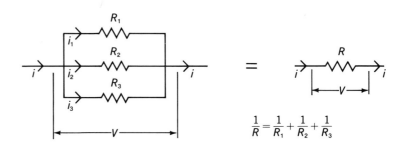

$$\frac{1}{R} = \frac{1}{R_1} + \frac{1}{R_2} + \frac{1}{R_3}$$

Fig. 19-9. Resistors in parallel. The same potential difference V exists across each resistor.

sum of the currents through the separate resistors, so that

$$i = i_1 + i_2 + i_3.$$

The potential difference V is the same across all the resistors, and by applying Ohm's law to each of them in turn we find that

Resistors in parallel have the same potential difference across them.

$$i_1 = \frac{V}{R_2}, \qquad i_2 = \frac{V}{R_2}, \qquad i_3 = \frac{V}{R_3}.$$

The smaller the resistance, the greater the proportion of the total current that flows through it.

The total current flowing through the set of three resistors is given in terms of their equivalent resistance R by

$$i = \frac{V}{R}.$$

Hence we have

$$i = i_1 + i_2 + i_3,$$

$$\frac{V}{R} = \frac{V}{R_1} + \frac{V}{R_2} + \frac{V}{R_3}.$$

The final step is to divide through by V, which yields

$$\frac{1}{R} = \frac{1}{R_1} + \frac{1}{R_2} + \frac{1}{R_3}.$$

In general, *the reciprocal of the equivalent resistance of a set of resistors connected in parallel is equal to the sum of the reciprocals of the individual resistances*:

Parallel resistors

$$\frac{1}{R} = \frac{1}{R_1} + \frac{1}{R_2} + \frac{1}{R_3} + \cdots \qquad \textit{Resistors in parallel} \quad (19\text{-}11)$$

The formulas for the equivalent resistances of series and parallel arrangements of resistors are in accord with the basic formula

$$R = \rho \frac{L}{A}$$

for the resistance of a conductor. When several resistors are connected in series, the effect is the same as increasing the length L, whereas when they are connected in parallel, the effect is the same as increasing the cross-sectional area A. Thus a series set of resistors lets through *less* current than any of the individual resistors, and a parallel set of resistors lets through *more* current than any of the individual resistors, in each case assuming the same potential difference.

In the case of two resistors in parallel, the equivalent resistance is

$$\frac{1}{R} = \frac{1}{R_1} + \frac{1}{R_2}.$$

The lowest common denominator of the right-hand side of this formula is R_1R_2, which enables us to write

$$\frac{1}{R} = \frac{R_1 + R_2}{R_1R_2}.$$

Taking the reciprocal of both sides of this equation yields the convenient result

$$R = \frac{R_1R_2}{R_1 + R_2}. \qquad\qquad \textit{Two resistors in parallel} \quad (19\text{-}12)$$

Problem. A 5-ohm resistor and a 20-ohm resistor are connected in parallel and a potential difference of 100 volts is applied across them by means of a generator. Find the equivalent resistance of the circuit, the current that flows in each resistor and in the circuit as a whole, the power dissipated by each resistor, and the power dissipated by the entire circuit.

Solution. Successive stages in the solution of this problem are shown in Fig. 19-10.

(a) The equivalent resistance of the resistors is

$$R = \frac{R_1R_2}{R_1 + R_2} = \frac{5\ \Omega \times 20\ \Omega}{5\ \Omega + 20\ \Omega} = 4\ \Omega.$$

(b) The current that flows in R_1 is

$$i_1 = \frac{V}{R_1} = \frac{100\ V}{5\ \Omega} = 20\ A,$$

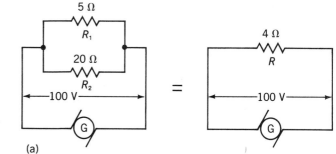

Fig. 19-10.

(a)

and the current that flows in R_2 is

$$i_2 = \frac{V}{R_2} = \frac{100 \text{ V}}{20 \text{ }\Omega} = 5 \text{ A}.$$

The total current in the circuit is

$$i = \frac{V}{R} = \frac{100 \text{ V}}{4 \text{ }\Omega} = 25 \text{ A},$$

and is equal to $i_1 + i_2$.

(c) The power dissipated by R_1 is

$$P_1 = i_1 V = 20 \text{ A} \times 100 \text{ V} = 2000 \text{ W},$$

and that dissipated by R_2 is

$$P_2 = i_2 V = 5 \text{ A} \times 100 \text{ V} = 500 \text{ W}.$$

The power dissipated by the entire circuit is

$$P = iV = 25 \text{ A} \times 100 \text{ V} = 2500 \text{ W},$$

and is equal to $P_1 + P_2$.

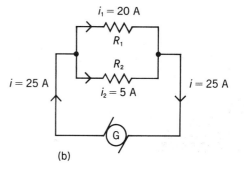

(b)

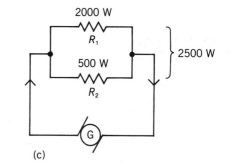

(c)

We recall from the previous problem that, when a 100-V potential difference is applied across the same two resistors connected in series, the power dissipated is only 400 watts. The difference is due to the lower equivalent resistance of the parallel combination (4 Ω instead of 25 Ω), which permits more current to flow. Since $P = iV$ and V is the same in both cases, the greater current in the parallel circuit leads to a greater power dissipation.

Fig. 19-11.

$R_1 = 3\ \Omega$
$R_2 = 12\ \Omega$
$R_3 = 8\ \Omega$
$R_4 = 6\ \Omega$

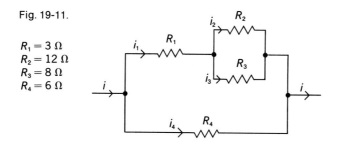

Problem. Find the equivalent resistance of the set of resistors shown in Fig. 19-11.

Solution. The procedure to follow here is shown in Fig. 19-12, where the original complex circuit is decomposed into its series and parallel elements. The first step is to find the equivalent resistance R' of the parallel resistors R_2 and R_3:

$$R' = \frac{R_2 R_3}{R_2 + R_3} = \frac{12\ \Omega \times 8\ \Omega}{12\ \Omega + 8\ \Omega} = 4.8\ \Omega.$$

Fig. 19-12. Successive steps in determining the equivalent resistance R of the resistor network shown in Fig. 19-11.

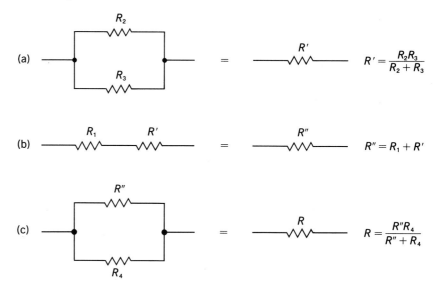

This pair of resistors is in series with R_1, and the equivalent resistance R'' of the upper branch of the circuit is therefore

$$R'' = R' + R_1 = 4.8 \ \Omega + 3 \ \Omega = 7.8 \ \Omega.$$

The upper and lower branches of the circuit are in parallel, which means that the equivalent resistance R of the entire circuit is

$$R = \frac{R''R_4}{R'' + R_4} = \frac{7.8 \ \Omega \times 6 \ \Omega}{7.8 \ \Omega + 6 \ \Omega} = 3.4 \ \Omega.$$

Problem.   A potential difference of 12 V is applied across the set of resistors shown in Fig. 19-11. Find the current that flows through each resistor.

Solution.   The full potential difference is applied across R_4, and so, by Ohm's law, the current i_4 in it is

$$i_4 = \frac{V}{R_4} = \frac{12 \ \text{V}}{6 \ \Omega} = 2.0 \ \text{A}.$$

The current i_1 that flows through R_1 also flows through the entire upper branch of the circuit. Since the equivalent resistance of this branch is R'',

$$i_1 = \frac{V}{R''} = \frac{12 \ \text{V}}{7.8 \ \Omega} = 1.54 \ \text{A}.$$

The potential difference V' across the parallel resistors R_2 and R_3 is equal to the current i_1 through the equivalent resistance R' multiplied by R':

$$V' = i_1 R' = 1.54 \ \text{A} \times 4.8 \ \Omega = 7.4 \ \text{V}.$$

Another way to find V' is to subtract the potential difference

$$V_1 = i_1 R_1 = 1.54 \ \text{A} \times 3 \ \Omega = 4.6 \ \text{V}$$

from the total of 12 volts to obtain

$$V' = V - V_1 = 12 \ \text{V} - 4.6 \ \text{V} = 7.4 \ \text{V}.$$

Hence the currents i_2 and i_3 through resistors R_2 and R_3 are respectively

$$i_2 = \frac{V'}{R_2} = \frac{7.4 \ \text{V}}{12 \ \Omega} = 0.62 \ \text{A}.$$

$$i_3 = \frac{V'}{R_3} = \frac{7.4 \text{ V}}{8 \text{ }\Omega} = 0.92 \text{ A.}$$

19-7 Electromotive Force

The potential difference that exists across a battery, generator, or other source of electrical energy when it is not connected to any external circuit **Emf** is called its *electromotive force*. Electromotive force is usually referred to simply as emf, and its symbol is $\mathcal{E}$.

As charges pass through a source of electrical energy, work is done on them, and the emf of the source is the work done per coulomb on the charges. The emf of an automobile storage battery is 12 V, which means that 12 J of work are done on each coulomb of charge that passes through the battery. In the case of a battery, chemical energy is converted into electrical energy by means of the work done on the charges in transit through it; in a generator, mechanical energy is converted into electrical energy; in a thermocouple, heat energy is converted into electrical energy; and so on. The emf of an electrical source bears a relationship to its power output analogous to that of applied force to mechanical power in a machine, which is the reason for its name.

When a source of electrical energy is part of a complete circuit, a current i flows, and the potential difference across the terminals of the source **Internal resistance** is always *less* than its emf owing to its *internal resistance*. Every electrical source has a certain amount of internal resistance r, which means that a potential drop ir occurs *within* the source. Hence the actual terminal voltage V across a source of emf $\mathcal{E}$ and internal resistance r is

$$V = \mathcal{E} - ir, \tag{19-13}$$

Terminal voltage = emf − potential drop within source.

If the source is disconnected, no current flows, and $V = \mathcal{E}$; the existence of a current lowers the value of V by an amount proportional to i.

Figure 19-13 shows a battery of emf $\mathcal{E}$ connected to an external circuit whose equivalent resistance is R. The total resistance in the entire circuit is R plus the internal resistance r of the battery, so that the current i that flows is

$$i = \frac{\mathcal{E}}{R + r}. \tag{19-14}$$

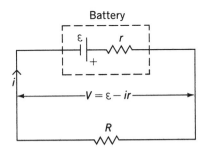

Fig. 19-13.   The terminal voltage V of a battery is always less than its emf $\mathcal{E}$ because of the potential drop ir within the battery itself.

The actual potential difference across the battery is given by Eq. (19-13).

Problem.   A battery of emf $\mathcal{E}$ and internal resistance r is connected to an external resistance R. What should the value of R be in order that it dissipate the maximum amount of power?

Solution.   The current in the circuit is $i = \mathcal{E}/(R + r)$ as in Eq. (19-14). The power dissipated in the external resistance is therefore

$$P = i^2 R = \frac{\mathcal{E}^2 R}{(R + r)^2}.$$

This formula is plotted in Fig. 19-14, from which it is clear that P is a maximum when $R = r$.

Impedance matching

The above result is an example of the phenomenon of *impedance matching:* when energy is being transferred from one system to another (here from the battery to the external resistance), the efficiency is greatest when both systems have the same impedance, which is a general term for resistance to the flow of energy in whatever form it may take in a particular case. We encountered impedance matching earlier, the first time in the Special Topic to Chapter 8. There we saw that when a moving object strikes a stationary one, the maximum transfer of energy takes place when both have the same mass. In this situation the inertia of each object, as measured by its mass, represents its impedance. The curve of relative energy transfer versus mass ratio in Fig. 8-14 is identical in form to that of Fig. 19-14.

Another example of impedance matching occurs when a pulse or wave moves down a stretched string, as discussed in Section 13-2. When two

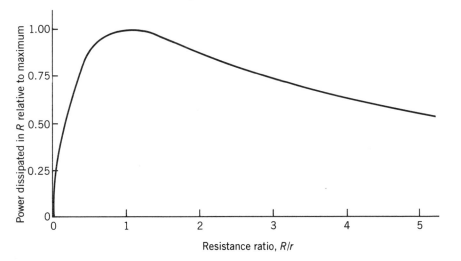

Fig. 19-14.   The power transferred to an external resistance R is a maximum when R equals the internal resistance r of the source.

strings having the same mass per unit length are joined together, a pulse in one passes to the other with no reflection. However, if the second string has either a greater or a smaller mass per unit length, reflection occurs at the junction and not all the energy of the pulse is transmitted (Figs. 13-6 and 13-7). Though best known for its application to electrical circuits, impedance matching is an important concept in many other branches of physics and engineering as well.

19-8 Kirchhoff's Rules

It is often difficult or impossible to determine the currents that flow in the various branches of a complex network merely by computing equivalent resistances. Two rules formulated by Gustav Kirchhoff (1824-1887) make it possible to find the current in each part of a direct-current circuit, no matter how complicated, if we are given the emf's of the sources of potential difference and the resistances of the various circuit elements. These rules apply to *junctions*, which are points where three or more wires come together, and to *loops*, which are closed conducting paths that are part of the circuit (Fig. 19-15).

Junctions and loops

Kirchhoff's first rule follows from the conservation of electric charge. Charge is never found to accumulate at any point in a circuit, nor can it be created there, and so there cannot be any net current into or out of a junction. Hence the first rule:

Kirchhoff's first rule

1. The sum of the currents flowing into a junction is equal to the sum of the currents flowing out of the junction.

The second rule is a consequence of the conservation of energy. The sum of the emf's in a loop equals the amount of work done per coulomb by the sources of emf *on* a charge that moves once around the loop. The work done per coulomb *by* the charge as it moves around the loop equals the sum of the *iR* potential drops along the way. To conserve energy, the work done on the charge must be the same as the work done by the charge, and so we have Kirchhoff's second rule:

Kirchhoff's second rule

2. The sum of the emf's around a loop is equal to the sum of the *iR* potential drops around the loop.

A definite procedure must be followed when Kirchhoff's rules are applied to a network. First, a direction is assumed for the current through each resistor. If we have guessed correctly, the solution of the problem will

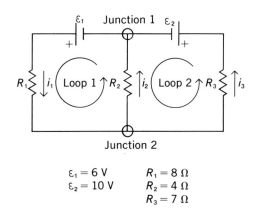

Fig. 19-15. A *junction* in an electrical network is a point where three or more wires come together, and a *loop* is any closed conducting path in the network. In the network shown, the internal resistances of the batteries are included in R_1 and R_3. The directions of the currents i_1, i_2, i_3 are chosen arbitrarily; when the problem is solved, the value of i_3 turns out to be negative, which means that its actual direction is opposite to the one shown.

$$\mathcal{E}_1 = 6 \text{ V} \qquad R_1 = 8 \text{ } \Omega$$
$$\mathcal{E}_2 = 10 \text{ V} \qquad R_2 = 4 \text{ } \Omega$$
$$R_3 = 7 \text{ } \Omega$$

give a positive value for the current; if we have guessed wrong, a negative value will indicate that the actual current is in the reverse direction. Second, in going around a loop to apply the second rule, we must follow a consistent path, either clockwise or counterclockwise. An emf is reckoned positive if we meet the negative terminal of its source first; if instead we meet the positive terminal first, the emf is reckoned negative. An iR drop is considered positive if the current in the resistor is in the same direction as the path we are following, and negative if its direction is opposite to that of our path.

How the second rule is applied

Problem.   Find the current in each of the resistors in the network shown in Fig. 19-15.

Solution.   The assumed directions of the unknown currents i_1, i_2, and i_3 are shown in the figure. Applying Kirchhoff's first rule to junction 1, we obtain

$$i_1 = i_2 + i_3.$$

This rule applied to junction 2 yields the same result.

In applying Kirchhoff's second rule to the two loops we shall follow counterclockwise routes. For loop 1,

$$\mathcal{E}_1 = i_1 R_1 + i_2 R_2,$$

and for loop 2,

$$-\mathcal{E}_2 = -i_2 R_2 + i_3 R_3.$$

In loop 2 we consider $\mathcal{E}_2$ as negative because we encounter its positive terminal first, and i_2 as negative because its direction is opposite to our counterclockwise path.

There is also a third loop, namely the outside one in Fig. 19-15, which similarly must obey Kirchhoff's second rule. Proceeding counterclockwise yields

$$-\mathcal{E}_2 + \mathcal{E}_1 = i_1 R_1 + i_3 R_3.$$

We note that this last equation is just the sum of the two preceding loop equations. Thus we may use the junction equation and any two of the loop equations to solve for the unknown currents; nothing will be gained by using all three loop equations.

There are three unknown currents, and we have three independent equations relating them, which means that the problem can be solved by routine algebra. As an example of how this might be done, we shall start by substituting $i_2 + i_3$ for i_1 in the first loop equation to obtain

$$\mathcal{E}_1 = i_2 R_1 + i_3 R_1 + i_2 R_2.$$

We now solve both this equation and the second loop equation for i_3:

$$i_3 = \frac{\mathcal{E}_1 - i_2 R_1 - i_2 R_2}{R_1}, \qquad i_3 = \frac{-\mathcal{E}_2 + i_2 R_2}{R_3}.$$

Setting the two expressions for i_3 equal enables us to solve for i_2 as follows:

$$\frac{\mathcal{E}_1 - i_2 R_1 - i_2 R_2}{R_1} = \frac{-\mathcal{E}_2 + i_2 R_2}{R_3},$$

$$\mathcal{E}_1 R_3 - i_2 R_1 R_3 - i_2 R_2 R_3 = -\mathcal{E}_2 R_1 + i_2 R_1 R_2,$$

$$i_2 (R_1 R_3 + R_2 R_3 + R_1 R_2) = \mathcal{E}_2 R_1 + \mathcal{E}_1 R_3,$$

$$i_2 = \frac{\mathcal{E}_2 R_1 + \mathcal{E}_1 R_3}{R_1 R_3 + R_2 R_3 + R_1 R_2}.$$

Finally we insert the values of $\mathcal{E}_1$, $\mathcal{E}_2$, R_1, R_2, and R_3 given in Fig. 19-15 to obtain

$$i_2 = 1.05 \text{ A}.$$

With the value of i_2 known we can find the value of i_3 from the second loop equation:

$$i_3 = \frac{-\mathcal{E}_2 + i_2 R_2}{R_3} = -0.83 \text{ A}.$$

The minus sign means that the direction of i_3 is opposite to that shown in Fig. 19-15. The junction equation finally provides us with the value of i_1.

$$i_1 = i_2 + i_3 = 0.22 \text{ A}.$$

The same results would have been obtained had we chosen other directions for the currents or taken other routes around the loops in applying Kirchhoff's second rule. The important thing is not what choice of directions is made but to follow that choice consistently in working out the problem.

Special Topic

Ammeters, Voltmeters, and Ohmmeters

The basic instrument for measuring electrical quantities is the *galvanometer*, whose principle of operation is described in Chapter 20. The deflection of a galvanometer's pointer is proportional to the current that passes through it. A galvanometer can be used to measure potential difference as well as current. It has a certain resistance, and according to Ohm's law the current in it is directly proportional to the potential difference across its terminals. Hence the meter's scale can be calibrated equally well in volts as in amperes. A meter which has a full-scale reading of, say, 1 mA (10^{-3} A) and a resistance of 50 Ω can also be used to measure potential differences of up to

$$V_{max} = i_{max} R_{coil} = 10^{-3} \text{ A} \times 50 \text{ }\Omega = 0.05 \text{ V}.$$

A galvanometer employed to measure current is called an *ammeter*. The inherent range of a galvanometer (0 to 1 mA in the case of the above meter) can be extended by connecting a low resistance shunt in parallel with it to carry the bulk of the current, leaving a known fraction of the current to be registered by the meter itself.

As an example, let us convert the above galvanometer to measure currents of up to 1 A. What we must do is place a resistor in parallel with the meter that will divert

$$1.000 \text{ A} - 0.001 \text{ A} = 0.999 \text{ A}$$

of the maximum current, leaving 0.001 A to give a full-scale reading on the meter (Fig. 19-16). To obtain the value of R_{shunt}, we note that since the potential difference V is the same across each branch of a parallel circuit,

$$V = i_{shunt} R_{shunt} = i_{galv} R_{galv}$$

and so

$$R_{shunt} = R_{galv} \frac{i_{galv}}{i_{shunt}} = 50 \text{ }\Omega \times \frac{0.001 \text{ A}}{0.999 \text{ A}} = 0.050 \text{ }\Omega.$$

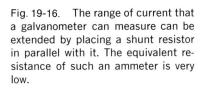

Fig. 19-16.   The range of current that a galvanometer can measure can be extended by placing a shunt resistor in parallel with it. The equivalent resistance of such an ammeter is very low.

An ideal ammeter has a very low equivalent resistance so that its presence in any circuit alters the properties of the circuit as little as possible. The above ammeter has an equivalent resistance of very nearly 0.050 Ω; unless the equivalent resistance of a circuit in which it is inserted is large compared with 0.050 Ω, the meter reading will not be an accurate measure of the current in the absence of the ammeter.

A galvanometer employed to measure potential difference is called a *voltmeter*. The inherent voltage range of a galvanometer can be extended by inserting a resistor in series with it in order to hold the maximum current down to whatever figure is appropriate for the meter's movement.

We might wish to convert the above 0 to 1 mA galvanometer to a voltmeter with a full-scale reading of 1 V. This means that when a potential difference of 1 V is across the voltmeter's terminals, a current of 1 mA flows in the meter's coil. The equivalent resistance of the voltmeter must therefore be

$$R = R_{series} + R_{galv} = \frac{V}{i}$$

and so

$$R_{series} = \frac{V}{i} - R_{galv} = \frac{1 \text{ V}}{0.001 \text{ A}} - 50 \text{ }\Omega = 950 \text{ }\Omega.$$

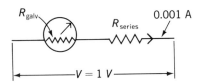

Fig. 19-17.   The range of potential difference a galvanometer can measure can be extended by placing a resistor in series with it. The equivalent resistance of such a voltmeter is very high.

A 950-Ω resistor connected in series with the galvanometer converts it to a voltmeter whose range is 0 to 1 V (Fig. 19-17).

Placing a voltmeter across a circuit element is the same thing as connecting a resistor in parallel with it, which changes the properties of the circuit and so leads to an incorrect potential difference measurement. For this reason an ideal voltmeter has a very high resistance.

A galvanometer can be used to measure an unknown resistance R with the help of a battery and a series resistor, as in Fig. 19-18. Such a device is called an *ohmmeter*. The meter's scale is calibrated backwards, so that a full-scale deflection corresponds to $R = 0$, when the current is a maximum; no deflection corresponds to $R = \infty$, when the current is 0. The purpose of the series resistor is to have $i = i_{max}$ for the galvanometer when the only resistance in the circuit is that of the resistor, that of the meter, and that of the battery. In the case of a 0 to 1-mA, 50-Ω galvanometer used with a 1.5-V, 0.05-Ω battery,

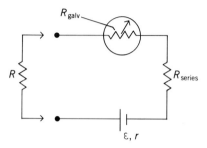

Fig. 19-18. How a galvanometer can be connected to measure resistance.

$$R_{series} + R_{galv} + r_{bat} = \frac{\mathcal{E}}{i}$$

$$R_{series} = \frac{\mathcal{E}}{i} - R_{galv} - r_{bat} = \frac{1.5\text{ V}}{0.001\text{ A}} - 50\ \Omega - 0.05\ \Omega = 1450\ \Omega.$$

An ohmmeter is not a precision instrument, but is handy and often sufficient. A Wheatstone bridge (problem 19) offers greater accuracy.

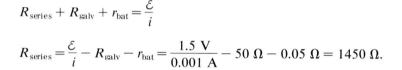

Important Terms

A flow of electric charge from one place to another is called an **electric current**. The unit of electric current is the **ampere**, which is equal to a flow of 1 coulomb/second.

Ohm's law states that the current in a metallic conductor is proportional to the potential difference between its ends. The **resistance** of a conductor is the ratio between the potential difference across its ends and the current that flows. The unit of resistance is the **ohm**, which is equal to 1 volt/amp. The resistance of a conductor is proportional to its length and to the **resistivity** ρ of the material of which it is made, and inversely proportional to its cross-sectional area.

The **equivalent resistance** of a set of interconnected resistors is the value of the single resistor that can be substituted for the entire set without affecting the current that flows in the rest of any circuit of which it is a part. Resistors in **series** are connected consecutively so that the same current flows through all of them, while resistors in **parallel** have their terminals connected together so that the total current is split up among them.

The **electromotive force** (emf) of a battery, generator, or other source of electrical energy is the potential difference across its terminals when no current flows. When a current is flowing, the terminal voltage is less than the emf owing to the potential drop in the **internal resistance** of the source.

Kirchhoff's rules for network analysis are: (1) The sum of the currents flowing into a junction of three or more wires is equal to the sum of the currents flowing out of the junction; (2) The sum of the emf's around a closed conducting loop is equal to the sum of the iR potential drops around the loop.

Important Formulas

Electric current:

$$i = \frac{q}{t}$$

Resistance:

$$R = \frac{V}{i}$$

Ohm's law:

$$i = \frac{V}{R} \qquad [holds\ for\ most\ metals]$$

Resistance of ohmic conductor:

$$R = \rho\frac{L}{A}$$

Power:

$$P = iV$$

$$= i^2R = \frac{V^2}{R} \qquad [ohmic\ conductor]$$

Resistors in series:

$$R = R_1 + R_2 + R_3 + \cdots$$

Resistors in parallel:

$$\frac{1}{R} = \frac{1}{R_1} + \frac{1}{R_2} + \frac{1}{R_3} + \cdots$$

$$R = \frac{R_1R_2}{R_1 + R_2} \qquad [two\ resistors]$$

Emf and terminal voltage:

$$V = \mathcal{E} - iR$$

Multiple Choice

1. The resistance of a conductor does not depend on its
   a. mass.
   b. length.
   c. cross-sectional area.
   d. resistivity.

2. A certain wire has a resistance R. The resistance of another wire, identical with the first except for having twice its diameter, is
   a. $\frac{1}{4}R$.
   b. $\frac{1}{2}R$.
   c. $2R$.
   d. $4R$.

3. A certain piece of copper is to be shaped into a conductor of minimum resistance. Its length and cross-sectional area
   a. should be, respectively, L and A.
   b. should be, respectively, $2L$ and $\frac{1}{4}A$.
   c. should be, respectively, $\frac{1}{2}L$ and $2A$.
   d. don't matter, since the volume of copper remains the same.

4. The temperature of a copper wire is raised. Its resistance
   a. decreases.
   b. remains the same.
   c. increases.
   d. any of the above, depending upon the temperatures involved.

5. The unit of emf is the
   a. ohm.
   b. ampere.
   c. volt.
   d. watt.

6. Of the following combinations of units, the one that is not equal to the watt is the
   a. J/s.
   b. AV.
   c. A²Ω.
   d. Ω²/V.

7. Which of the following is neither a basic physical law nor derivable from one?
   a. Coulomb's law
   b. Ohm's law
   c. Kirchhoff's first law
   d. Kirchhoff's second law

8. A battery is connected to an external circuit. The potential drop within the battery is proportional to
   a. the emf of the battery.
   b. the equivalent resistance of the circuit.
   c. the current in the circuit.
   d. the power dissipated in the circuit.

9. The power dissipated in a circuit is not equal to
   a. iV.
   b. i^2R.
   c. V^2/R.
   d. iR/V.

10. A flow of 1000 electrons/s constitutes a current of
    a. 1.6×10^{-16} A.   b. 2.7×10^{-18} A.
    c. 2.7×10^{-19} A.   d. 4.5×10^{-20} A.

11. An electric iron draws a current of 15 A when connected to a 120-V power source. Its resistance is
    a. 0.125 Ω.   b. 8 Ω.
    c. 16 Ω.   d. 1800 Ω.

12. The power rating of an electric motor which draws a current of 3 A when operated at 120 V is
    a. 40 W.   b. 360 W.
    c. 540 W.   d. 1080 W.

13. When a 100-W, 240-V light bulb is operated at 200 V, the current that flows in it is
    a. 0.35 A.   b. 0.42 A.
    c. 0.50 A.   d. 0.58 A.

14. A 50-V battery is connected across a 10-Ω resistor and a current of 4.5 A flows. The internal resistance of the battery is
    a. 0.   b. 0.5 Ω.
    c. 1.1 Ω.   d. 5 Ω.

15. The equivalent resistance of a network of three 2-Ω resistors cannot be
    a. 0.67 Ω.   b. 1.5 Ω.
    c. 3 Ω.   d. 6 Ω.

16. A 12-V potential difference is applied across a series combination of four 6-Ω resistors. The current in each resistor is
    a. 0.5 A.   b. 2 A.
    c. 8 A.   d. 18 A.

17. A 12-V potential difference is applied across a parallel combination of four 6-Ω resistors. The current in each resistor is
    a. 0.5 A.   b. 2 A.
    c. 8 A.   d. 18 A.

Exercises

1. Would you expect bends in a wire to affect its electrical resistance?

2. It is sometimes said that an electrical appliance "uses up" electricity. What does such an appliance actually use in its operation?

3. Why are two wires used to carry electric current instead of a single one?

4. How should two identical batteries be connected in order to obtain the maximum emf from the combination? Why?

5. Why is it undesirable to connect cells of different emf in parallel?

6. Currents of 3 A flow through two wires, one which has a potential difference of 60 V across its ends and another which has a potential difference of 120 V across its ends. Compare the rates at which charge passes through each wire.

7. A current of 3 A flows through a wire whose ends are at a potential difference of 12 V. How much charge flows through the wire per minute?

8. How many electrons flow through the filament of a 120-V, 60-W electric light bulb per second?

9. It is possible to measure the passage of as few as 60 electrons per second with a certain sensitive device. To what current does this correspond?

10. A certain 12-V storage battery is rated at 80 Ah, which means that when it is fully charged it can deliver a current of 1 A for 80 h, 2 A for 40 h, 80 A for 1 h, etc. (a) How many coulombs of charge can this battery deliver? (b) How much energy is stored in it?

11. A mercury cell with a capacity of 1.5 Ah and an emf of 1.35 V is to be used to power a cardiac pacemaker. If the power required is 0.1 mW, how long will the cell last? What will be the average current?

12. When a certain 1.5-V battery is used to power a 3-W flashlight bulb, it is exhausted after an hour's use. (a) How much charge has passed through the bulb in this period of time? (b) If the battery costs $0.30, find the cost of a kilowatt-hour of electric energy obtained in this way. How does this compare with the cost of the electric energy supplied to your home?

13. What is the power rating of an electric motor that draws a current of 5 A when operated at 240 V?

14. The starting motor of a certain car requires a current of 100 A from a 12-V battery to turn over

the engine. How many horsepower does this represent?

15. What is the current in a 60-W, 120-V light bulb when it is operated at 80 V?

16. A 120-V electric toaster draws a current of 9 A. Find its resistance.

17. A silver wire 2 m long is to have a resistance of 0.5 Ω. What should its diameter be?

18. A copper rod 1 m long and 1 cm in diameter is drawn out into a wire 1 mm in diameter. (a) What is the length of the wire? (b) Compare the resistance of the rod with the resistance of the wire.

19. A 40-Ω resistor is to be wound from platinum wire 0.1 mm in diameter. How much wire is needed?

20. A copper wire 1 mm (10^{-3} m) in diameter carries a current of 12 A. Find the potential difference between two points in the wire which are 100 m apart.

21. If the resistance of a carbon resistor is 500 Ω at 20°C, find its resistance at 100°C.

22. An iron wire has a resistance of 2.00 Ω at 0°C and a resistance of 2.46 Ω at 45°C. Find the temperature coefficient of resistivity of the wire.

23. When a metal object is heated, both its dimensions and its resistivity increase. Is the increase in resistivity likely to be a consequence of the increase in length?

24. Two 60-W, 240-V light bulbs are connected in series to a 240-V power source. (a) What is the current in each bulb? (b) How much power is dissipated by the combination?

25. Two 60-W, 240-V light bulbs are connected in parallel to a 240-V power source. (a) What is the current in each bulb? (b) How much power is dissipated by the combination?

26. In a certain bus, forty 15-W, 30-V light bulbs are connected in parallel to a 30-V power source. (a) What is the current provided by the source? (b) What is the current in each bulb? (c) How much power is provided by the source?

27. A set of Christmas tree lights consists of twelve bulbs connected in series to a 120-V power source. Each bulb has a resistance of 5 Ω. (a)

What is the current in the circuit? (b) How much power is dissipated in the circuit?

28. A 12-V battery of internal resistance 1.5 Ω is connected to an 8-Ω resistor. Find the total power produced by the battery and the percentage of this power dissipated as heat within it.

29. The brightness of a light bulb depends upon the power dissipated by its filament. As a dry cell ages, its internal resistance increases while its emf remains approximately unchanged at 1.5 V. A fresh No. 6 dry cell might have an internal resistance of 0.05 Ω and an old one an internal resistance of 0.20 Ω. Find the ratio between the powers dissipated in a 0.25-Ω bulb when it is connected to a fresh and an old dry cell.

Problems

1. Approximately 10^{20} electrons/cm participate in conducting electric current in a certain wire. (That is, 10^{20} electrons in each centimeter of the wire are in motion when a current is being carried by the wire.) What is the average speed of the electrons when there is a current of 1 A in the wire?

2. Aluminum wires are sometimes used to transmit electric power instead of copper wires. What is the ratio between the masses of an aluminum and a copper wire of the same length whose cross-sectional areas are such that they have the same resistance? (The density of aluminum is 2.70 × 10^3 kg/m³ and that of copper is 8.89 × 10^3 kg/m³.)

3. A common alloy used in constructing heating elements is nichrome (60% nickel, 26% iron, 12% chromium, 2% manganese). The resistivity of nichrome is 1.0 × 10^{-6} Ω · m. How much nichrome wire 0.70 mm in diameter is required for a 1000-W, 120-V electric heater?

4. List the resistances that can be obtained by combining three 100-Ω resistors in all possible ways.

5. In a Van de Graaff generator an insulating belt is used to carry charges to a large metal sphere. In a typical generator of this kind, the potential difference between the sphere and the source of the charges is 5 × 10^6 V. (a) If the belt carries charge

to the sphere at a rate of 10^{-3} A, how much power is required? (b) How much energy in eV will an electron have if it is accelerated by such a potential difference? (c) Express the answer to (b) in joules.

6. An electric water heater has a resistance of 12 Ω and is operated from a 120-V power line. If no heat escapes from it, how much time is required for it to raise the temperature of 40 kg of water from 15°C to 80°C?

7. A 12-volt storage battery with an internal resistance of 0.012 Ω delivers 80 A when used to crank a gasoline engine. If the battery mass is 20 kg and it has an average specific heat of 0.2 kcal/kg·°C, what is its rise in temperature during 1 min of cranking the engine?

8. A certain 16-cell "32-volt" storage battery has an emf of 33.5 V when charged to 75% of its 250 ampere-hour capacity. The internal resistance of the battery is 0.1 Ω. It is desired to charge the battery to its full capacity, when its emf will be 34.3 V. (a) What potential difference must be applied to the battery if it is to be charged at the initial rate of 40 A? (b) If this potential difference is held constant, what will be the rate of charge at the end of the process? (c) The emf of the battery arises from the conversion of chemical to electrical energy; the higher potential difference is required for charging in order to pass a current through the battery and thereby produce chemical changes that store energy. Find the proportion of the power supplied during the charging process that is stored as chemical energy and the proportion that is dissipated as heat. (Assume an average emf during charging of 33.9 V.)

9. (a) Find the equivalent resistance of the circuit of Fig. 19-19. (b) What is the current in the 8-Ω

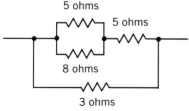

5 ohms

5 ohms

8 ohms

3 ohms

 Fig. 19-19.

resistor when a potential difference of 12 V is applied to the circuit?

10. (a) Find the equivalent resistance of the circuit of Fig. 19-20. (b) What is the current in the 12-Ω resistor when a potential difference of 100 V is applied to the circuit?

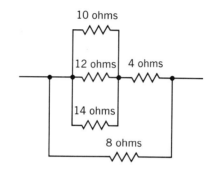

10 ohms

12 ohms   4 ohms

14 ohms

8 ohms

Fig. 19-20.

11. Three identical 1.5-V dry cells with internal resistances of 0.15 Ω are connected in parallel with an external 0.5-Ω resistor. How much current flows through the resistor?

12. Each of the resistors in Fig. 19-21 can safely dissipate 10 watts. What is the maximum power the entire circuit can dissipate?

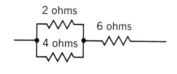

2 ohms

6 ohms

4 ohms

Fig. 19-21.

13. A 5-Ω and a 10-Ω resistor are connected in parallel. This combination is connected in series with another pair of parallel resistors whose resistances are both 8 Ω. (a) What is the equivalent resistance of the network? (b) The network is connected to a 24-V battery whose internal resistance is 1.5 Ω. Find the current in each of the resistors.

14. Find the current in the 20-Ω resistor in the circuit shown in Fig. 19-22.

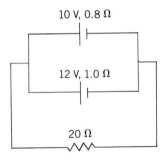

10 V, 0.8 Ω

12 V, 1.0 Ω

20 Ω

Fig. 19-22.

15. Find the currents in each of the resistors of the circuit shown in Fig. 19-23. The internal resistances of the batteries are included in the values of the resistors.

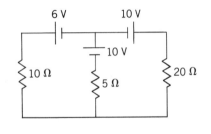

Fig. 19-23.

16. (a) Find the current in the 5-Ω resistor in the circuit of Fig. 19-24. (b) Find the potential difference between the points a and b

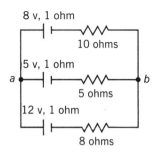

Fig. 19-24.

17. Find the potential differences between a and b and between a and c in the circuit shown in Fig. 19-25.

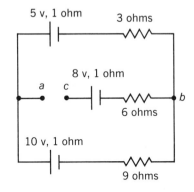

Fig. 19-25.

18. If a and c are connected in Fig. 19-25, find the potential difference between a and b.
19. A Wheatstone bridge (Fig. 19-26) provides a convenient means for measuring an unknown resistance R in terms of the known resistances A and B and the calibrated variable resistance C. The resistance of C is varied until no current flows through the galvanometer, in which case the bridge is said to be *balanced*. Show that $R = AC/B$ when the bridge is balanced.

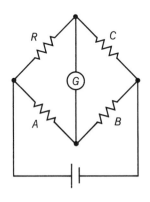

Fig. 19-26.

Answers to Multiple Choice

1. a	7. b	13. a
2. a	8. c	14. c
3. c	9. d	15. b
4. c	10. a	16. a
5. c	11. b	17. b
6. d	12. b	

20

Magnetic Field

Magnetism and electricity are both manifestations of the same basic interaction between electric charges. Charges at rest relative to an observer appear to him to exert only electric forces upon one another. When the charges are in motion relative to the observer, however, he finds that the forces acting between them are different from before, and these differences are traditionally attributed to "magnetic" forces. In reality, magnetic forces represent the modifications to electric forces that arise because of the motions of the charges involved. It is convenient to consider magnetic and electric forces separately, and to think in terms of separate magnetic and electric fields, but we should keep in mind that these distinctions are artificial.

20-1 The Nature of Magnetism

Electric charges in motion exert forces upon one another quite different from those they exert while at rest. For instance, if we place a current-carrying wire parallel to another current-carrying wire, with the currents in the same direction, we find that the wires attract each other (Fig. 20-1). If the currents are in opposite directions, the forces on the wires are repulsive.

These observations cannot be accounted for unless the motion of the charges is taken into account. Gravitational forces cannot be responsible since they are never repulsive, and the electrical forces discussed earlier cannot be responsible since there is no net charge on a wire when a current is present in it. The forces that come into being when electric currents interact are called *magnetic forces*. All magnetic effects can ultimately be traced to currents or, more exactly, to moving electric charges. Of course, the word "magnetic" suggests ordinary magnets and their familiar attraction for iron objects, but, as we shall see, this is but one aspect of the whole subject of magnetism.

Magnetic forces arise from the interaction of moving charges

The gravitational force between two masses and the electrical force between two charges at rest are both *fundamental forces* in the sense that they cannot be accounted for in terms of anything else. On the other hand, the force a bat exerts on a ball is not fundamental because it can be traced to the electrical forces between the atomic electrons of the bat and the atomic electrons of the ball.

What about magnetic forces? It is an important fact that whatever it is in nature that manifests itself as an electric force between stationary charges *must*, according to the theory of relativity, also manifest itself as a magnetic force between moving charges. It is impossible, even in principle, to have one without the other. There is only a single interaction between charges, the *electromagnetic interaction*. The distinction we make between electric and magnetic forces is an artificial one for the sake of convenience only.

Electromagnetic interaction

The theory of relativity is taken up in Chapter 27, and the nature of

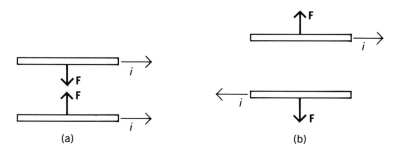

Fig. 20-1. (a) Parallel electric currents in the same direction attract each other. (b) When the currents are in opposite directions, they repel each other.

magnetic forces is further discussed there. In the meantime, it is sufficient for us to note that it is always possible to separate the force on a charge into an electric part, which is independent of its motion, and a magnetic part, which is proportional to its velocity relative to the observer. These forces are additive so that, for example, it is entirely possible for a magnetic force on a moving charge to exactly cancel an electric force on it under the proper circumstances. Because fields are always defined in terms of the forces they exert, the additive character of electric and magnetic forces permits a similar separation of an electromagnetic field into electric and magnetic parts.

20-2 Magnetic Field

We recall from Chapter 18 that the electric field **E** at a given location is defined in terms of the force **F** the field exerts on a stationary positive charge q placed there. Because the electric force on a charge is always found to be proportional to q, the magnitude of **E** is appropriately specified by the ratio

$$E = \frac{F}{q}.$$

Electric field magnitude

The direction of **E** is taken as the same as the direction of **F**. Once we know **E**, we can readily find the magnitude and direction of the electric force on *any* charge at that location.

The symbol for *magnetic field is* **B**; this quantity is also known as *magnetic induction* and as *magnetic flux density*. Magnetic forces only act on moving charges, so it is natural to use the magnetic force **F** exerted on a positive charge q whose velocity is **v** to define **B**. Experiment and theory both show that *every* magnetic force on a moving charge, regardless of the details of how the force arises, is proportional in magnitude to qv. However, we cannot simply proceed by analogy with the case of **E** to define the magnitude of **B** as F/qv because it is found that, in a given field **B**, the magnitude and direction of **F** both depend upon the direction in which the charge is moving. Hence we begin our definition of **B** by choosing its direction at any point as that in which a moving charge would experience no magnetic force. If θ is the angle between **B** and **v** (Fig. 20-2), the magnitude of **B** is then specified as

$$B = \frac{F}{qv \sin \theta}.$$

Magnetic field magnitude   (20-1)

Magnetic field

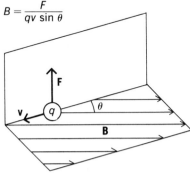

Fig. 20-2.   The magnitude of a magnetic field **B** is defined in terms of the force **F** it exerts on a moving charge.

The unit of magnetic field is, from the above definition, the newton/ampere-meter (N/A-m), since the units of qv are coulomb-meter/second = ampere-meter. The name *tesla*, abbreviated T, has been given to this unit: The tesla

$$1 \text{ T} = 1 \frac{\text{N}}{\text{A-m}}.$$

Thus a force of 1 N will be exerted on a charge of 1 C when it is moving at 1 m/s perpendicular to a magnetic field whose magnitude is 1 T. (When v is perpendicular to **B**, $\theta = 90°$ and $\sin \theta = 1$.)

The tesla is also referred to as the *weber/m²*. Another unit of **B** in common use is the *gauss*, where   The gauss

1 gauss $= 10^{-4}$ T,

1 T $= 10^4$ gauss.

Figure 20-3 contains some representative values of magnetic field which may help in acquiring a feeling for the magnitude of the tesla.

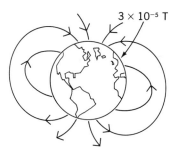

3×10^{-5} T

The magnitude of the earth's magnetic field at sea level is about 3×10^{-5} T.

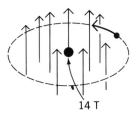

14 T

The magnetic field produced at the nucleus of a hydrogen atom by the electron circling around it is about 14 T.

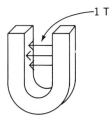

1 T

The magnetic field near a strong permanent magnet is about 1 T.

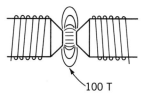

100 T

The most powerful magnetic fields achieved in the laboratory have magnitudes in the neighborhood of 100 T.

Fig 20-3.   Some representative values of magnetic field.

20-3 Magnetic Field of a Current

Now that both the direction and magnitude of magnetic field **B** have been defined in terms of procedures for finding them, we can go on to the nature of the magnetic field produced by various electric currents.

Unfortunately the calculations needed to determine **B** in a given situation are usually fairly difficult. To see why, let us consider a short length of wire Δl long that carries the current i, as in Fig. 20-4. (We ignore for the moment the rest of the circuit.) The field $\Delta \mathbf{B}$ due to the current element at a point P a distance r away has its magnitude given by *Biot's law*:

$$\Delta B = \frac{\mu}{4\pi} \frac{i \Delta l \sin \theta}{r^2}.$$

Biot's law   (20-2)

Here θ is the angle between the direction of the current and that of the vector **r**.

Permeability

The constant μ is called the *permeability* of the medium in which the magnetic field exists. In free space,

$$\mu_0 = 4\pi \times 10^{-7} \text{ T-m/A} = 1.257 \times 10^{-6} \text{ T-m/A},$$

so that

$$\frac{\mu_0}{4\pi} = 10^{-7} \text{ T-m/A}.$$

Fig. 20-4.   The direction of the magnetic field of a current element is given by the right-hand rule; its magnitude is given by the Biot law.

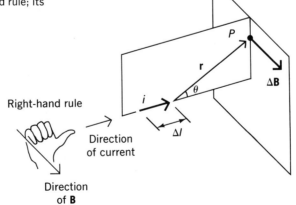

Right-hand rule

Direction of current

Direction of **B**

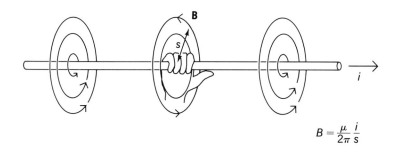

$$B = \frac{\mu}{2\pi} \frac{i}{s}$$

Fig. 20-5.   The lines of force of the magnetic field around a long, straight current consist of concentric circles. The sense of the field is given by the right-hand rule.

The value of μ in air is very close to μ_0; they will be assumed to be the same here. The unit of permeability is sometimes expressed in other ways, for instance as N/A².

The direction of $\Delta\mathbf{B}$ is, as shown in Fig. 20-4, perpendicular to the plane formed by $\Delta\mathbf{l}$ and $\mathbf{r}$. The *right-hand rule* specifies the sense of $\Delta\mathbf{B}$:

Grasp the wire with the right hand so the thumb points in the direction of the current; the curled fingers of that hand point in the direction of the magnetic field.

Right-hand rule for magnetic field direction

To apply Biot's law in an actual situation, we must compute the value of $\Delta\mathbf{B}$ at some point P for each of the tiny lengths of wire Δl that make up the complete circuit, and then add up the results to find the total field $\mathbf{B}$ at P. It is necessary to take the directions of the various $\Delta\mathbf{B}$ contributions into account, since they generally are not parallel to one another. With few exceptions such calculations involve calculus, so only the results in the most important cases will be given here.

Figure 20-5 shows the configuration of the magnetic field around a long, straight wire that carries the current i. The lines of force take the form of a series of concentric circles with the current at the center. The magnitude of the field a distance s from the wire is given by

$$B = \frac{\mu}{2\pi} \frac{i}{s}.$$ *Long, straight current*   (20-3)

The greater the current and the closer we are to it, the stronger the magnetic field.

Problem.   Find the magnetic field in air 1 cm from a wire that carries a current of 1 A.

Solution.   Since $1\ \text{cm} = 10^{-2}$ m, we have (Fig. 20-6)

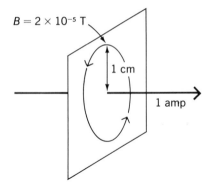

$B = 2 \times 10^{-5}$ T

1 cm

1 amp

Fig. 20-6.

$$B = \frac{\mu_0}{2\pi} \frac{i}{s}$$

$$= \frac{4\pi \times 10^{-7} \text{ T-m/A}}{2\pi} \times \frac{1 \text{ A}}{10^{-2} \text{ m}}$$

$$= 2 \times 10^{-5} \text{ T}.$$

This is only a little smaller than the magnitude of the earth's magnetic field. Hence great care is taken aboard ships to keep current-carrying wires away from magnetic compasses.

The magnetic field around a circular current loop has the configuration shown in Fig. 20-7. At the center of the loop **B** is perpendicular to the plane of the loop and has the magnitude

$$B = \frac{\mu}{2} \frac{i}{r}, \qquad \qquad \textit{Center of current loop} \quad (20\text{-}4)$$

Direction of **B** inside current loop

where i is the current in the loop and r is its radius. The direction of **B** is given by another right-hand rule:

Grasp the loop so the curled fingers of the right hand point in the direction of the current; the thumb of that hand then points in the direction of B.

Fig. 20-7.   (a) The magnetic field around a circular current loop. (b) At the center of the loop, **B** is perpendicular to the plane of the loop and its direction is given by the right-hand rule shown. (c) If there are N loops, the magnetic fields of the individual loops add up to give a field N times stronger than each one produces by itself.

In the case of a flat coil of more than one loop, as in Fig. 20-7(c), the magnetic fields of each individual loop add up to give a proportionately stronger field. If there are N turns, then,

$$B = \frac{\mu}{2} \frac{Ni}{r}. \qquad \qquad \textit{Center of flat coil} \quad (20\text{-}5)$$

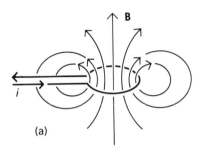

(a)

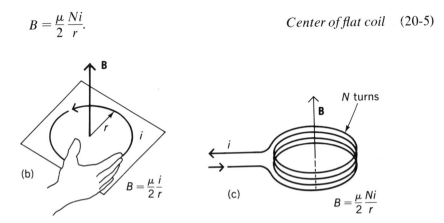

(b) $B = \frac{\mu}{2} \frac{i}{r}$

(c) N turns   B $B = \frac{\mu}{2} \frac{Ni}{r}$

A *solenoid* is a coil of wire in the form of a helix (Fig. 20-8). If the turns are close together and the solenoid is long relative to its diameter, then the magnetic field within it is uniform and parallel to its axis except near the ends. The direction of the field inside a solenoid is given by the same right-hand rule that gives the direction of **B** inside a current loop. The magnetic field in the interior of a solenoid L long that has N turns of wire and carries the current i has the magnitude

Solenoid

$$B = \mu \frac{N}{L} i. \qquad \qquad \textit{Interior of solenoid} \quad (20\text{-}6)$$

The diameter of the solenoid does not matter, provided it is small compared with the length L.

Problem. A solenoid 20 cm long and 4 cm in diameter with an air core has a total of 800 turns wound in four layers in such a manner that the current is in the same direction in all of them. What should the current be if the magnetic field in the center of the solenoid is to be 0.1 T?

Solution. In air we let $\mu = \mu_0$ and so, from Eq. (20-6),

Fig. 20-8.   (a) The magnetic field inside a solenoid is uniform except near its ends if the solenoid is long relative to its diameter and if its turns are close together. (b) An expanded view of a solenoid showing how the magnetic fields of the individual turns add together to yield a uniform field inside it.

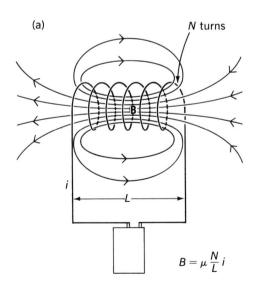

(a) N turns

$$B = \mu \frac{N}{L} i$$

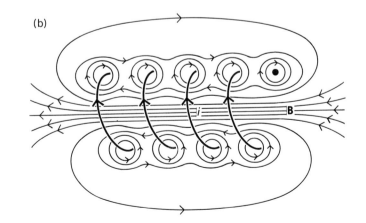

(b)

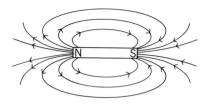

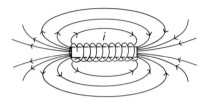

Fig. 20-9. The magnetic fields of a bar magnet and of a solenoid are the same.

$$i = \frac{BL}{\mu_0 N} = \frac{0.1 \text{ T} \times 0.2 \text{ m}}{1.26 \times 10^{-6} \text{ T-m/A} \times 8 \times 10^2} = 20 \text{ A.}$$

The solenoid diameter has no significance here except as a check that the solenoid is long relative to its diameter.

The magnetic field of a bar magnet is identical with that of a solenoid, which is not surprising since all permanent magnets owe their character to an alignment of atomic current loops no different in principle from the alignment of the current loops in a solenoid (Fig. 20-9). The behavior of permanent magnets is discussed in the Special Topic to this chapter.

20-4 Magnetic Properties of Matter

Ferromagnetism increases B greatly

The magnetic field produced by a current-carrying solenoid is changed in strength when a rod of almost any material is inserted in it. Some materials increase B (for instance, aluminum), others decrease B (for instance, bismuth), but in almost all cases the difference is very small. However, a few substances yield a dramatic increase in B when placed in a solenoid—the new field may be hundreds or thousands of times greater in magnitude than before. Such substances are called *ferromagnetic*; iron is the most familiar example (Fig. 20-10).

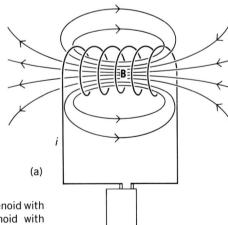

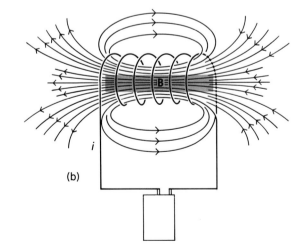

Fig. 20-10. (a) Solenoid with no core. (b) Solenoid with ferromagnetic core.

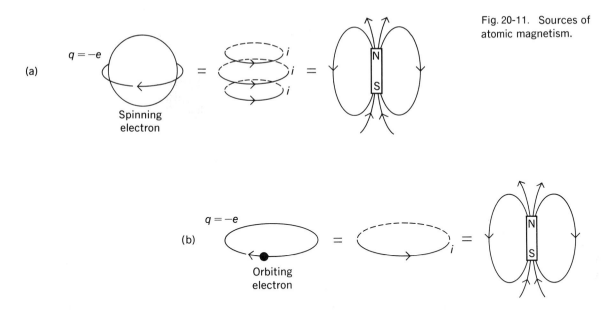

Fig. 20-11. Sources of atomic magnetism.

The magnetic properties of matter can be traced almost entirely to the electrons every atom contains; the contribution of the nucleus is very minor. There are two sources of the magnetic behavior of an atomic electron. First, an electron in certain respects resembles a spinning charged sphere, which we may imagine as a series of ultraminute current loops (Fig. 20-11a). Hence every electron has the magnetic field of a tiny bar magnet due to its spin.

Second, when it is part of an atom, an electron may be thought of as revolving around the nucleus much like a planet revolving around the sun. (This is a crude approximation of the actual situation, but it is adequate for many purposes.) An electron circulating in an orbit is a current loop, and so the result is again the magnetic field of a tiny bar magnet (Fig. 20-11b).

When any material whatever is placed in a magnetic field, the orbital motions of the electrons are affected by the field and, by Lenz's law (described in Section 21-3) the result is that the alterations in the orbital magnetic fields tend to oppose the external magnetic field. This effect is called *diamagnetism*. In most substances whose atoms or molecules have an even number of electrons, the various magnetic fields of the electrons cancel each other out in pairs when there is no external magnetic field. (If we shake an even number of small bar magnets together in a box, they will also end up paired off with opposite poles together.) In an external field, the orbital magnetic fields of the electrons will then reduce the magnetic field inside the material.

Diamagnetism decreases B slightly

In other substances there are one or more electrons per atom or molecule whose spin magnetic fields are not cancelled out. In an external magnetic field the fields of these electrons tend to line up to enhance the external field. This effect is called *paramagnetism*. The increase in *B* due to paramagnetism (when it is present) is always greater than the decrease due to diamagnetism (which is always present), so the net result is an increase in *B*. The increase is usually small, however, because the constant thermal agitation of the atoms or molecules prevents complete alignment of the spin magnetic fields with the external magnetic field. At low temperatures, as we might expect, there is less random thermal motion, so the alignment of the electron spins is more complete and the increase in *B* is correspondingly greater. The diamagnetic decrease in *B* described in the preceding paragraph is not affected by temperature.

Paramagnetism increases B slightly

In a ferromagnetic material, the unpaired electrons in each atom interact strongly with their counterparts in adjacent atoms, which causes the unpaired spin magnetic fields in all the atoms to be locked together. Atoms in a ferromagnetic material are accordingly grouped together in assemblies called *domains*, each about 5×10^{-5} m across and just visible in a microscope. In an unmagnetized sample, the directions of magnetization of the domains are randomly oriented, though within each domain the unpaired electron spins are parallel. When such a sample is placed in an external magnetic field, either the spins within the domains turn to line up with the field or, in pure and homogeneous materials, the domain walls change so that those domains lined up with the field grow at the expense of the others (Fig. 20-12). The former process requires stronger fields to take place than the latter. Hence good "permanent" magnets are irregular in structure—for instance steel rather than pure iron—and once magnetized, cannot change their magnetization by the easy process of domain wall motion. When all the unpaired spins in a ferromagnetic sample are lined up, no further increase in *B* is possible, and the sample is said to be *saturated*.

Magnetic domains

Above a certain temperature (760°C in the case of iron), the atoms in a domain acquire enough kinetic energy to overcome the interatomic forces that hold their spins in alignment. The spins and their magnetic fields then become randomly oriented, and the ferromagnetic material loses its special magnetic properties. Thus heating a "permanent magnet" sufficiently will cause it to become demagnetized. Hammering a permanent magnet also tends to disturb the alignment of spins, though some ferromagnetic materials are able to retain their magnetization despite almost any mechanical disturbance.

Iron, unlike steel (which is an alloy, or mixture, of iron with carbon and other elements), tends to lose its magnetization when an external magnetic field is removed. Hence if an iron rod is placed inside a solenoid, we have a

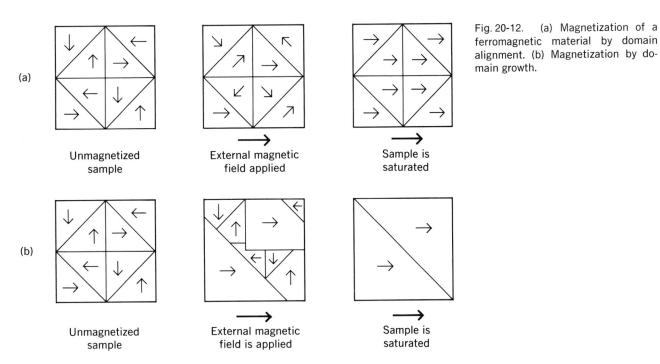

Fig. 20-12.   (a) Magnetization of a ferromagnetic material by domain alignment. (b) Magnetization by domain growth.

very strong magnet that can be turned on and off just by switching the current in the solenoid on and off. Such an *electromagnet* is much stronger than the solenoid itself and can be stronger than a permanent magnet as well; also, unlike a permanent magnet, its field can be controlled at will by adjusting the current in the solenoid. Electromagnets are among the most widely-used electrical devices. They range in size from the tiny one in a telephone receiver that causes a steel plate to vibrate and thus produce the sounds we hear to the giant electromagnets used to pick up automobiles in scrap yards.

Electromagnets

20-5 Force on a Moving Charge

The defining property of a magnetic field is its ability to exert a force on an electric current, whether it is a current in a wire, a moving charged particle, or an atomic current as in an iron bar. This property has been exploited both technologically, as in the electric motor, and scientifically, as in such research tools as the mass spectrometer and various kinds of particle accelerators. The law that governs the magnetic force on a current element in a magnetic field is a straightforward one, and in the remainder of this chapter we shall see how it is applied in a number of situations.

According to Eq. (20-1), the force on a particle of charge q and velocity **v** in the magnetic field **B** has the magnitude

$$F = qvB \sin\theta, \qquad\qquad \textit{Force on moving charge} \quad (20\text{-}7)$$

where θ is the angle between **v** and **B**.

The direction of the force **F** is perpendicular to both **v** and **B** and is given by still another right-hand rule (Fig. 20-13): *Open your right hand so the fingers are together and the thumb sticks out. When your thumb is in the direction of* **v** *and your fingers are in the direction of the magnetic field, your palm faces in the same direction as the force acting on the charge.* This rule holds only for a positive charge; for a negative charge, the force is in the opposite direction.

The work done by a force on a body upon which it acts depends upon the component of the force in the direction the body moves. Because the force on a charged particle in a magnetic field is perpendicular to its direction of motion, the force does no work on it. Hence the particle keeps the same velocity magnitude v and energy it had when it entered the field, even though it is deflected. On the other hand, the velocity magnitude and energy of a charged particle in an *electric* field are always affected by the interaction between the field and the particle, unless **v** is perpendicular to **B**.

A particle of charge q and velocity **v** that is moving in a uniform magnetic field so that **v** is perpendicular to **B** experiences a force of magnitude

$$F = qvB \qquad [\mathbf{v} \perp \mathbf{B}] \qquad\qquad (20\text{-}8)$$

since $\sin 90° = 1$. This force is directly perpendicular to both **v** and **B**, so the particle travels in a circular path (Fig. 20-14).

To find the radius R of the circular path of the charged particle, we note that the magnetic force qvB provides the particle with the centripetal force mv^2/R that keeps it moving in a circle. Equating the magnetic and centripetal forces yields

$$F_{\text{magnetic}} = F_{\text{centripetal}},$$

$$qvB = \frac{mv^2}{R}$$

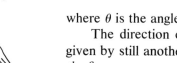

Positive charge

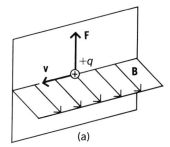

(a)

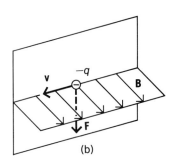

(b)

Fig. 20-13. (a) Right-hand rule for the direction of the magnetic force on a moving positively-charged particle. (b) When the particle is negatively charged, the force is in the opposite direction.

and so, solving for R, we obtain

$$R = \frac{mv}{qB}. \qquad (20\text{-}9)$$

The radius of a charged particle's orbit in a uniform magnetic field is directly proportional to its momentum mv and inversely proportional to its charge and to the magnitude of the field. The greater the momentum, the larger the circle, and the stronger the field, the smaller the circle.

A charged particle moving parallel to a magnetic field experiences no force and is not deflected; the same particle moving perpendicular to the field follows a circular path. Hence a charged particle whose direction of motion is oblique with respect to **B** follows a helical (corkscrew) path. If we call $v_\parallel$ the component of the particle's velocity **v** that is parallel to **B** and $v_\perp$ the component of **v** perpendicular to **B**, then the motion of the particle is the resultant of a forward motion at the velocity $v_\parallel$ and a circular motion perpendicular to this whose radius is $mv_\perp/qB$ (Fig. 20-15).

An extremely interesting phenomenon occurs when a charged particle moving in a magnetic field approaches a region where the field becomes stronger. The magnetic field that describe such a field converge, since their spacing is always proportional to the magnitude of the field they describe. The force the particle experiences now has a backward component as well as the inward component that leads to its helical path, as shown in Fig. 20-16. The backward force may be strong enough and extend over a long enough distance to reverse the particle's direction of motion. A converging magnetic field can thus act as a *magnetic mirror*.

Magnetic mirrors are found both in the laboratory and in nature. In the laboratory a pair of them are used as a "magnetic bottle," as in Fig. 20-17, to

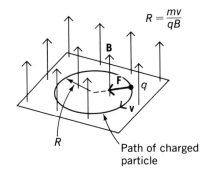

$$R = \frac{mv}{qB}$$

Path of charged particle

Fig. 20-14. The path of a charged particle moving perpendicular to a uniform magnetic field is a circle.

Magnetic mirror

Fig. 20-15. A charged particle which has velocity components both parallel and perpendicular to a magnetic field follows a helical path in the field.

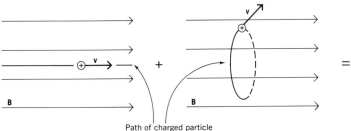

Path of charged particle

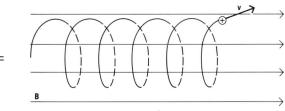

Fig. 20-16. The principle of the magnetic mirror.

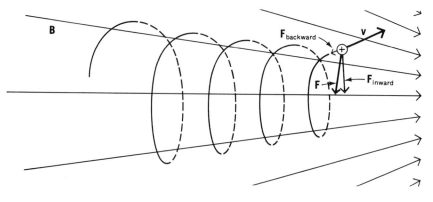

contain a hot plasma (highly-ionized gas) in research on thermonuclear fusion. If a solid container were used, contact with its walls would cool the plasma and the ions would not have enough energy to interact. Magnetic bottles of this kind are somewhat leaky, because ions moving along the axis of a magnetic mirror experience no backward force and hence are able to escape.

Magnetosphere

The earth's magnetic field traps electrons and protons from space in the *magnetosphere*, a giant doughnut-shaped magnetic bottle that surrounds the earth and extends from about 600 mi above the equator out to perhaps 40,000 mi. The magnetosphere contains large numbers of particles with relatively high energies (100 MeV, for instance). Figure 20-18 shows a typical particle trajectory in the magnetosphere.

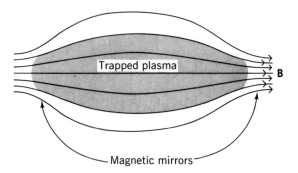

Fig. 20-17.   A "magnetic bottle".

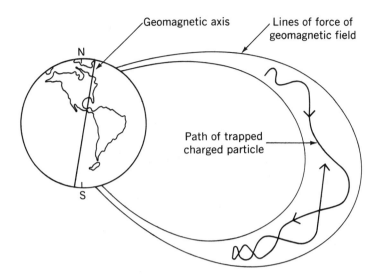

Geomagnetic axis

Lines of force of geomagnetic field

N

S

Path of trapped charged particle

Fig. 20-18. Protons and electrons are trapped by the earth's magnetic field.

20-6 The Mass Spectrometer

The mass of an atom is one of its most characteristic properties and an accurate knowledge of atomic masses provides considerable insight into nuclear phenomena. A variety of instruments with the generic name of *mass spectrometers* have been devised to measure atomic masses, and we shall consider the operating principles of the particularly simple one shown in Fig. 20-19.

The first step in the operation of this spectrometer is to produce ions of the substance under study. If the substance is a gas, ions can be formed readily by electron bombardment; if it is a solid, it is often convenient to incorporate it into an electrode that is used as one terminal of an electric arc discharge. The ions emerge from their source through a slit with the charge $+e$ and are then accelerated by an electric field. (Ions with other charges are sometimes present but are easily taken into account.)

When the ions enter the spectrometer, as a rule they are traveling in slightly different directions with slightly different speeds. A pair of slits serves to collimate the beam, that is, to eliminate those ions not moving in the desired direction. Then the beam passes through a *velocity selector*. The velocity selector consists of uniform electric and magnetic fields that are perpendicular to each other and to the beam of ions. The electric field $\mathbf{E}$ exerts the force $F_{\text{electric}} = eE$ on the ions to the right, whereas the magnetic field $\mathbf{B}$ exerts the force $F_{\text{magnetic}} = evB$ on them to the left. In order for an ion

Velocity selector

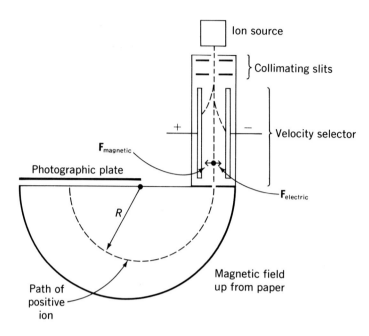

Fig. 20-19. A simple mass spectrometer. Modern instruments use electrical ion detectors and more complicated fields.

to reach the slit at the far end of the velocity selector it must suffer no deflection inside the selector, which means that the condition for escape is

$$F_{electric} = F_{magnetic},$$

$$eE = evB.$$

Hence the ions that escape all have the velocity

$$v = \frac{E}{B}.$$

Once past the velocity selector the ions enter a uniform magnetic field
and follow circular paths whose radius is given by Eq. (20-9). Since v, e, and
B are known, a measurement of R yields a value for m, the ion mass.

Problem. The velocity selector of a mass spectrometer consists of an
electric field of $E = 40{,}000$ V/m perpendicular to a magnetic field of $B =$
0.0800 T. The same magnetic field is used to deflect the ions that have
passed through the velocity selector. Ions of a certain isotope of lithium are
found to have radii of curvature in the magnetic field of 39.0 cm. What is
their mass?

Solution.   The velocity of the ions is

$$v = \frac{E}{B} = \frac{4.00 \times 10^4 \text{ V/m}}{8.00 \times 10^{-2} \text{ T}}$$

$$= 5.00 \times 10^5 \text{ m/s}.$$

From Eq. (20-9) we obtain

$$m = \frac{1.60 \times 10^{-19} \text{ C} \times 8.00 \times 10^{-2} \text{ T} \times 0.390 \text{ m}}{5.00 \times 10^5 \text{ m/s}}$$

$$= 9.98 \times 10^{-27} \text{ kg}.$$

20-7 Force on a Current

Since an electric current is a flow of charge, we would expect a current-carrying wire to be affected by a magnetic field in a manner similar to that of a moving charged particle. According to Eq. (20-1), the force on a charge q whose velocity is **v** when it is in the magnetic field **B** has the magnitude

$$F = qvB \sin \theta, \tag{20-1}$$

where θ is the angle between **v** and **B**. What we must do to find an expression for the force on a current is to replace the qv of the above formula with the quantity appropriate for a current.

Figure 20-20(a) shows a particle of charge q and velocity v. In the time t the particle travels the distance

$$\Delta l = vt,$$

and while it does so it is equivalent to a current of

$$i = \frac{q}{t}.$$

Hence

$$v = \frac{\Delta l}{t} \quad \text{and} \quad q = it,$$

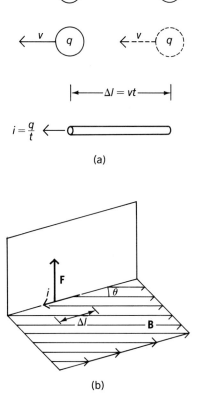

Fig. 20-20. The force on a charge q moving with the velocity v in a magnetic field is the same as that on a wire Δl long carrying the current i, where $i\Delta l = qv$.

so that

$$qv = i\ \Delta l.$$

We conclude that the force on an element Δl long of current i when it is in a magnetic field **B** has the magnitude

$$F = i\ \Delta l\ B\ \sin\theta,\qquad\qquad \textit{Force on current element}\quad(20\text{-}10)$$

where θ is the angle between the direction of i and that of **B** (Fig. 20-20b).

There are two simple ways to determine the direction of the force on a current element in a magnetic field. Both give the same result, of course, and deciding which one to use in a particular case is largely a matter of personal preference. The first is essentially the same as the right-hand rule used for a moving charge in Section 20-1, except that now the thumb points in the direction of the current. This version of the rule is illustrated in Fig. 20-21.

Another method for finding the direction of **F** is based upon the pattern of lines of force around a current in a magnetic field. Figure 20-22(a) shows the lines of force of a uniform field **B** in the absence of a current, and (b) shows the lines of force around a wire carrying a current i in the absence of a magnetic field of external origin. Since the current is into the paper, the lines of force are concentric circles in the clockwise sense. When field (a) is added

Right-hand rule for force on current

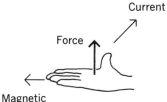

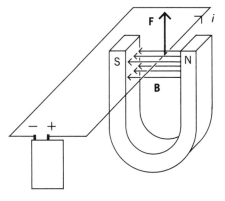

Fig. 20-21. The right-hand rule for the direction of the force on a current-carrying wire in a magnetic field.

Fig. 20-22. The direction of the force on a current element in a magnetic field is from the region of strong field to the region of weak field.

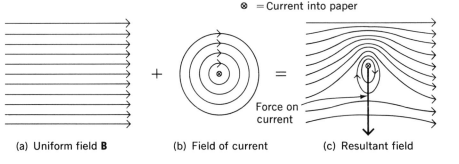

$\otimes$ = Current into paper

(a) Uniform field **B**   (b) Field of current   (c) Resultant field

vectorially to field (b), the resulting pattern of lines of force is like that shown in (c): the lines are closer together in the region above the wire where the field of the current is in the same direction as **B** and farther apart under the wire where the field of the current is opposite to **B**. *The direction of the force on the current element is from the region of strong field to the region of weak field*, as though the lines of force were rubber bands that try to straighten out when distorted by the presence of the current.

Lines-of-force rule for force on current

While the pictorial method for establishing the direction of **F** is easy to use and appeals to the intuition, it must be kept in mind that lines of force do not in fact exist but are only a device for visualizing the magnitude and direction of a force field. It is the field itself that exists in space as a continuous property of the region it occupies, not a series of strings. However, despite the artificial nature of lines of force, they can be extremely convenient in representing various aspects of the interaction between magnetic fields and electric currents, and we shall make use of them for this purpose whenever appropriate.

Problem. A wire carrying a current of 100 A due west is suspended between two towers 50 m apart. The lines of force of the earth's magnetic field enter the ground there in a northerly direction at a 45° angle; the magnitude of the field at that location is 3×10^{-5} T. Find the force on the wire exerted by the earth's field.

Solution. The wire is perpendicular to **B**, and so the magnitude of the force is

$$F = i \, \Delta l \, B \sin \theta$$
$$= 100 \text{ A} \times 50 \text{ m} \times 3 \times 10^{-5} \text{ T} \times \sin 90°$$
$$= 0.15 \text{ N}.$$

Since 1 N = 0.225 lb, the force is 0.034 lb, about $\frac{1}{2}$ oz. By either of the methods described above, the force acts downward at a 45° angle with the ground toward the south (Fig. 20-23).

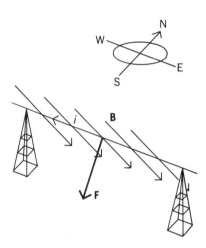

Fig. 20-23.

20-8 Force Between Two Currents

Every current is surrounded by a magnetic field, and because of this nearby currents exert forces upon one another. The forces are magnetic in origin; a current-carrying wire has no net electric charge, and hence cannot interact electrically with another such wire.

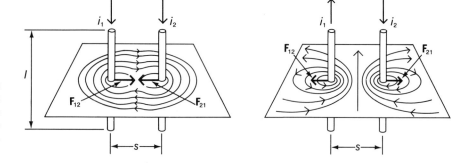

Fig. 20-24.   Equal and opposite forces are exerted by parallel currents on each other. The forces are attractive when the currents are in the same direction, repulsive when they are in opposite directions.

Figure 20-24 shows two parallel wires a distance s apart that carry the currents i_1 and i_2 respectively. The magnetic fields a distance s from each of the wires are, from Eq. (20-3),

$$B_1 = \frac{\mu}{2\pi} \frac{i_1}{s},$$

$$B_2 = \frac{\mu}{2\pi} \frac{i_2}{s}.$$

The fields are perpendicular to the wires, which means that $\theta = 90°$ and $\sin \theta = 1$, and therefore the force F_{12} on a length l of current 1 exerted by the magnetic field of current 2 is

$$F_{12} = i\,\Delta l B\, \sin \theta = \frac{\mu}{2\pi} \frac{i_1 i_2}{s}\, l. \tag{20-11}$$

The force F_{21} on current 2 that is exerted by the magnetic field of current 1 has exactly the same magnitude, and so we may express both of them in the alternative form

$$\frac{F}{l} = \frac{\mu}{2\pi} \frac{i_1 i_2}{s}, \qquad \textit{Force between parallel currents} \quad (20\text{-}12)$$

where F/l is the *force per unit length* each wire exerts on the other by virtue of its magnetic field. From the pattern of lines of force around each wire, it is clear that the forces are always opposite in direction and hence obey Newton's third law of motion, as they must. The forces are attractive when the currents are in the same direction and repulsive when they are in opposite directions.

Problem. The cables that connect the starting motor of a car with its battery are 1 cm apart for a distance of 40 cm. Find the forces between the cables when the current in them is 300 A.

Solution. The currents in the cables are opposite in direction, and hence the forces are repulsive. Their magnitudes, taking $\mu = \mu_0$, are

$$F = \frac{\mu_0}{2\pi} \frac{i^2}{s} l = \frac{4\pi \times 10^{-7} \text{ T-m/A} \times (300 \text{ A})^2 \times 0.4 \text{ m}}{2\pi \times 10^{-2} \text{ m}}$$

$$= 0.72 \text{ N},$$

which is 0.16 lb, a perceptible amount.

Equation (20-12) is used to define the ampere: An ampere is that current   Definition of ampere
in each of two parallel wires 1 m apart in free space which produces a force on each wire of exactly 2×10^{-7} N per meter of length. (Thus $\mu_0 = 4\pi \times 10^{-7}$ N-m/A.) In turn, the coulomb is defined in terms of the ampere as that amount of charge transferred per second by a current of 1 A. The ampere is chosen as the primary electrical unit instead of the coulomb because it can be defined in terms of a more direct and unambiguous experiment than would be possible with the coulomb.

20-9 Torque on a Current Loop

A straight current-carrying wire is acted upon by a force when it is in a magnetic field, provided that it is not parallel to the direction of **B**. A loop of current in a uniform magnetic field experiences no net force, but instead a torque occurs that tends to rotate the loop so that its plane is perpendicular to **B**. This is the principle that underlies the operation of all electric motors, from the tiniest one in a clock to the many-thousand-horsepower giant in a locomotive.

Let us examine the forces on each side of a rectangular current-carrying wire loop whose plane is parallel to a uniform magnetic field **B**, as in Fig. 20-25(a). The sides A and C of the loop are parallel to **B** and so there is no magnetic force on them. Sides B and D are perpendicular to **B**, however, and each therefore experiences a force. To find the directions of the forces on B and D we can use the right-hand rule: with the fingers of the right hand in line with **B** and the outstretched thumb in line with **i,** the palm faces the same way as **F**. What we find is that F_B is opposite in direction to F_D. The same conclusion can be obtained by examining the pattern of lines of force.

The forces F_B and F_D are the same in magnitude, so there is no net force on the current loop. But F_B and F_D do not act along the same line, and hence

Fig. 20-25.   (a) A current-carrying wire loop whose plane is parallel to a magnetic field experiences a torque. (b) If the plane of the loop is perpendicular to the magnetic field, there is no torque on the loop. In both cases there is no net force on the loop.

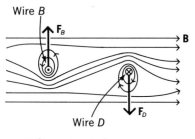

Wire B

Wire D

⊙ Current out of paper
⊗ Current into paper

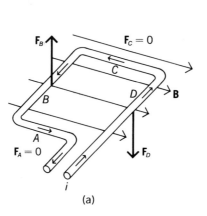

(a)

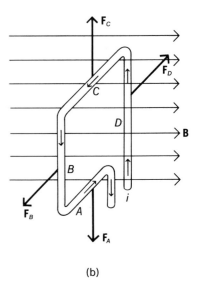

(b)

they exert a torque on the loop that tends to turn it. This is a perfectly general conclusion that holds for a current loop of any shape in a magnetic field.

If the plane of the loop is perpendicular to the magnetic field instead of parallel to it, there is neither a net force nor a net torque on it. This is easy to verify from Fig. 20-25(b), bearing in mind the right-hand rule for the direction of the force on each side of the loop. Evidently $\mathbf{F}_A$ and $\mathbf{F}_C$ cancel each other out, and $\mathbf{F}_B$ and $\mathbf{F}_D$ also cancel each other out. There is no torque force now because $\mathbf{F}_A$ and $\mathbf{F}_C$ have the same line of action, and $\mathbf{F}_B$ and $\mathbf{F}_D$ have the same line of action.

Current loop in magnetic field

The above results can be summarized by saying that *a current loop in a magnetic field always tends to turn so that it is perpendicular to the field.*

We have seen that a current-carrying wire loop tends to rotate in a magnetic field. The *galvanometer* capitalizes upon this behavior to furnish a means for measuring current. Figure 20-26 shows the basic construction of a galvanometer. A U-shaped permanent magnet is used to provide a magnetic field, and between its poles is a small coil wound on an iron core to enhance the torque developed when the unknown current is passed through it. The coil assembly is held in place by two bearings that permit it to rotate, and a pair of hairsprings keeps the pointer at 0 when there is no current in the coil.

Galvanometer

When a current flows, there is a torque on the coil because of the inter-action between the current and the magnetic field, and the coil rotates as far as it can against the opposing torque of the springs. The more the current, the stronger the torque, and the farther the coil turns. The restoring torque of the hairsprings is proportional to the angle through which they are twisted, and as a result the deflection of the pointer is directly proportional to the current i in the coil.

Galvanometers of the above type can be constructed that are able to respond to currents of as little as 0.1 microampere (10^{-7} A), though ordinary commercial meters are less sensitive. Even greater sensitivity can be attained if the moving coil is suspended by a thin wire to which a small mirror is attached: bearing friction is avoided in this way, and the mirror deflects a light beam so that the "pointer" may be a meter or more long instead of a few centimeters. Laboratory galvanometers like this can be used to measure currents of 10^{-10} A.

The torque which a magnetic field exerts on a current loop disappears when the loop turns so that its plane is perpendicular to the field direction. If the loop swings past this position, the torque on it will be in the opposite sense and will return the loop to the perpendicular orientation. In order to construct a motor capable of continuous rotation, then, the current in the loop must be automatically reversed each time it turns through 180°. The method by which this reversal is accomplished is shown in Fig. 20-27. The current is led to the loop by means of graphite rods called *brushes* which press against a split ring called a *commutator*. As the loop rotates, the cur-

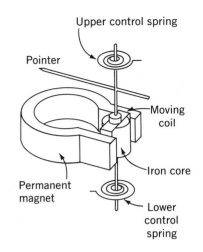

Fig. 20-26.   The construction of a common type of galvanometer.

Electric motor

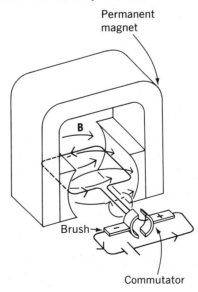

Fig. 20-27.   A simple dc electric motor. The commutator automatically reverses the current in the rotating loop twice per rotation so that the torque will stay in the same direction.

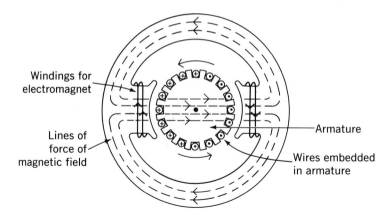

Windings for electromagnet

Lines of force of magnetic field

Armature

Wires embedded in armature

Fig. 20-28.   Actual dc electric motors employ various means to increase the available torque.

rent is reversed twice per turn as the commutator segments make contact alternately with the brushes. The torque is always in the same direction, except at the moments of switching when it is zero because the loop is perpendicular to the field. However, the angular momentum of the loop carries it past this point, and it can continue to turn indefinitely.

While actual direct-current electric motors are the same in principle as the simple device of Fig. 20-27, they employ a number of stratagems to increase the available torque. Electromagnets rather than permanent magnets provide the field, and there are six or more different coils with many turns each on a slotted iron core called an *armature*, instead of a single loop (Fig. 20-28). A commutator with a pair of segments for each coil is provided so that only those coils parallel to the magnetic field receive current at any time, which means that maximum torque is developed continuously.

Special Topic

Magnetic Poles

It may seem strange that there has been no mention until now of the "magnetic poles" that figure so prominently in naive discussions of magnetism. The reason is that all magnetic fields, including those of permanent

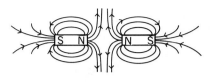

Like poles repel

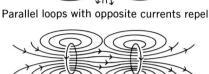

Parallel loops with opposite currents repel

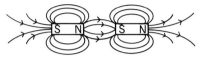

Unlike poles attract

Parallel loops with similar currents attract

Fig. 20-29.   Interactions between magnets can be traced to interactions between current loops.

magnets, originate in electric currents (or, more precisely, in moving charges); and all magnetic forces arise from interactions between currents and magnetic fields. To understand electromagnetic phenomena of any kind, it is necessary to start directly from these fundamental concepts.

The magnetic field of a bar magnet is identical with that of a solenoid, as we saw in Fig. 20-9, because in a permanent magnet atomic current loops are aligned by their mutual interactions. Hence we can use the ideas of this chapter to understand the behavior of permanent magnets.

Because the external magnetic field of a bar magnet seems to originate in its ends, these are by custom called its *poles*. At one time it was believed that the poles are "magnetic charges" analogous to electric charges, and that the field of the magnet is due to these poles. This belief was reinforced by the repulsion of like poles and the attraction of unlike ones, phenomena that have their true explanation in the forces between parallel and antiparallel currents (Fig. 20-29).

An important difference between magnetic poles and electric charges is that the former invariably come in pairs of equal strength and opposite polarity: if a magnet is sawed in half, the poles are not separated but instead two new magnets are created, as in Fig. 20-30. Magnetic poles are therefore only superficially like electric charges, and all effects that can be attributed to them can be explained in terms of the behavior of current-carrying solenoids.

Measurements of the earth's magnetic field show that it is very much like the field that would be produced by a powerful current loop whose center is a few hundred miles from the earth's center and whose plane is tilted by 11° from the plane of the earth's equator (Fig. 20-31). On the basis of geological evidence the earth is thought to have a core of molten iron 3470 km (2160 mi) in radius, a little over half the earth's radius, and there is little doubt today that electric currents in this core are responsible for the ob-

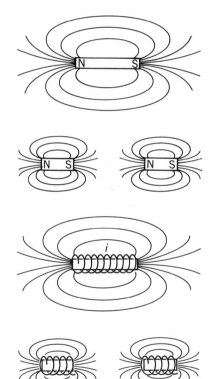

Fig. 20-30.   Cutting a magnet in half produces two new magnets.

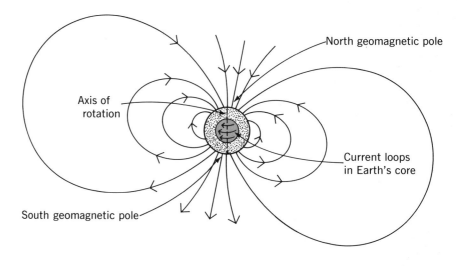

Fig. 20-31. The earth's magnetic field originates in currents in its core of molten iron. The magnetic axis is tilted by 11° from the axis of rotation.

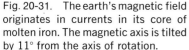
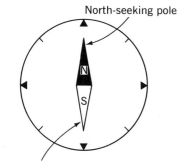

Fig. 20-32. A magnetic compass.

served geomagnetic field. The details of how these currents came into being and how they are maintained are still uncertain, but there seems little difficulty in accounting for their existence.

As we saw, a current loop tends to rotate in a magnetic field until its axis is parallel to the field. A bar magnet, too, tends to rotate in a magnetic field until it is aligned with the field direction. Because of the earth's magnetic field, a magnet suspended by a string turns so as to line up in an approximately north-south direction. A compass consists of a pivoted magnetized iron needle together with a card that permits directions relative to magnetic north to be determined (Fig. 20-32). The end of a freely-swinging magnet that points toward the north is called its north-seeking pole, usually shortened to just *north pole*, and the other end is its south-seeking pole, or *south pole*. Magnetic lines of force leave the north pole of a magnet and enter its south pole. (The north geomagnetic pole is thus in reality a south pole, and the south geomagnetic pole is a north pole; this has been a source of confusion for several hundred years.)

The mechanism by which a magnet or a solenoid attracts an iron object (or an object of any other ferromagnetic material such as cobalt or nickel) is very similar to the way in which an electric charge attracts an uncharged object. First the presence of the magnet causes the atomic magnets in the iron object to line up with its field by one of the mechanisms shown in Fig. 20-12, and then the attraction of opposite poles leads to a force on the object that draws it toward the magnet, as in Fig. 20-33.

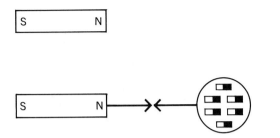

Ferromagnetic
substance

Fig. 20-33. How a permanent magnet attracts a ferromagnetic object.

Important Terms

Charged particles in motion relative to an observer exert forces upon one another that are different from the electric forces they exert when at rest. These differences are by custom said to arise from **magnetic forces.** In reality, magnetic forces represent modifications of electric forces due to the motion of the charges involved, as predicted by the theory of relativity.

A **magnetic field** exists wherever a magnetic force would act on a moving charged particle. The direction of a magnetic field **B** at a point is such that a charged particle would experience no force if it moves in that direction at the point. The magnitude of **B** is numerically equal to the force that would act on a charge of 1 C moving at 1 m/s perpendicular to **B.** The unit of magnetic field is the **tesla** (T), equal to 1 N/A-m.

When different substances are inserted in a current-carrying wire coil, the magnetic field in its vicinity changes. Those substances that lead to a slight increase in B are called **paramagnetic**, those that lead to a great increase in B are called **ferromagnetic**, and those that lead to a slight decrease in B are called **diamagnetic.** A **permanent magnet** is an object composed of ferromagnetic material whose atomic current loops have been aligned by an external current.

The **permeability** of a medium is a measure of its magnetic properties. Paramagnetic and ferromagnetic substances have higher permeabilities than free space and diamagnetic ones have lower permeabilities.

Important Formulas

Magnetic field:

$$B = \frac{F}{qv \sin \theta}$$

Field around long, straight current:

$$B = \frac{\mu}{2\pi} \frac{i}{s}$$

Field at center of flat coil:

$$B = \frac{\mu}{2} \frac{Ni}{r}$$

Field in interior of solenoid:

$$B = \mu \frac{N}{L} i$$

Force on moving charge:

$$F = qvB \sin \theta$$

Force on current element:

$$F = i\Delta l\, B \sin \theta$$

Force between parallel currents:

$$\frac{F}{l} = \frac{\mu}{2\pi} \frac{i_1 i_2}{s}$$

Multiple Choice

1. All magnetic fields originate in
   a. iron atoms.
   b. permanent magnets.
   c. magnetic domains.
   d. moving electric charges.

2. An observer moves past a stationary electron. His instruments measure
   a. an electric field only.
   b. a magnetic field only.
   c. both electric and magnetic fields.
   d. any of the above, depending upon his velocity.

3. The unit of magnetic field is the
   a. tesla.   b. weber.
   c. ampere/meter.   d. ampere-meter.

4. A drawing of the lines of force of a magnetic field provides information on
   a. the direction of the field only.
   b. the magnitude of the field only.
   c. both the direction and magnitude of the field.
   d. the source of the field.

5. In a drawing of magnetic lines of force, the stronger the field is,
   a. the closer together the lines of force are.
   b. the farther apart the lines of force are.
   c. the more nearly parallel the lines of force are.
   d. the more divergent the lines of force are.

6. The magnetic field near a strong permanent magnet might be
   a. 10^{-9} T.   b. 10^{-5} T.
   c. 1 T.   d. 100 T.

7. A typical value for the earth's magnetic field at sea level is
   a. 3×10^{-9} T.   b. 3×10^{-5} T.
   c. 3×10^{5} T.   d. 3×10^{9} T.

8. When a paramagnetic substance is inserted in a current-carrying solenoid, the magnetic field is
   a. slightly decreased. b. greatly decreased.
   c. slightly increased. d. greatly increased.

9. When a diamagnetic substance is inserted in a current-carrying solenoid, the magnetic field is
   a. slightly decreased. b. greatly decreased.
   c. slightly increased. d. greatly increased.

10. When a ferromagnetic substance is inserted in a current-carrying solenoid, the magnetic field is
    a. slightly decreased. b. greatly decreased.
    c. slightly increased. d. greatly increased.

11. Permanent magnets are made from
    a. diamagnetic substances.
    b. paramagnetic substances.
    c. ferromagnetic substances.
    d. any of the above.

12. The magnetic field a distance d from a long, straight wire is proportional to
    a. d.   b. d^2.
    c. $1/d$.   d. $1/d^2$.

13. Inside a solenoid the magnetic field
    a. is zero.
    b. is uniform.
    c. increases with distance from the axis.
    d. decreases with distance from the axis.

14. A current is flowing east along a power line. If we neglect the earth's field, the direction of the magnetic field below it is
    a. north.   b. east.
    c. south.   d. west.

15. The magnetic field 2 cm from a long, straight wire is 10^{-6} T. The current in the wire is
    a. 0.01 A.   b. 0.1 A.
    c. 1 A.   d. 10 A.

16. The magnetic field inside a 100-turn solenoid 2 cm long that carries a 10-A current is
    a. 0.00063 T.   b. 0.00126 T.
    c. 0.063 T.   d. 0.126 T.

17. Magnetic fields do not interact with
    a. stationary electric charges.
    b. moving electric charges.
    c. stationary permanent magnets.
    d. moving permanent magnets.

18. A charged particle moves perpendicularly through a magnetic field. The effect of the field is to change the particle's
    a. charge.   b. mass.
    c. velocity.   d. energy.

19. A current-carrying loop in a magnetic field always tends to rotate until the plane of the loop is
    a. parallel to the field.

b. perpendicular to the field.

c. either parallel or perpendicular to the field, depending on the direction of the current.

d. at a 45° angle with the field.

20. A current-carrying wire is in a uniform magnetic field with the direction of the current the same as that of the field.
    a. There is a force on the wire that tends to move it parallel to the field.
    b. There is a force on the wire that tends to move it perpendicular to the field.
    c. There is a torque on the wire that tends to rotate it until it is perpendicular to the field.
    d. There is neither a force nor a torque on the wire.

21. A particle of charge q, mass m, and velocity v moves perpendicular to a magnetic field B. The force on the particle is equal to
    a. qvB.　　　　b. q^2vB.
    c. $qmvB$.　　　d. qvB/m.

22. The right-hand rule for the direction of the force on a charged particle in a magnetic field applies
    a. only to positive charges.
    b. only to negative charges.
    c. to both positive and negative charges.
    d. only when the particle is moving parallel to the field.

Exercises

1. What is the relationship between electric and magnetic phenomena? Why are all of them considered to arise from a single electromagnetic interaction between charges? What property of the electromagnetic interaction makes it convenient to separate its effects into electric and magnetic parts?

2. A current is flowing north along a power line. What is the direction of the magnetic field above it? Below it?

3. A current-carrying wire is in a magnetic field. (a) What angle should the wire make with **B** for the force on it to be zero? (b) What should the angle be for the force to be a maximum?

4. A current is passed through a loop of highly flexible wire. What shape does the loop assume? Why?

5. A stream of protons is moving parallel to a stream of electrons. Is the force between the two streams necessarily attractive? Explain.

6. A beam of protons, initially moving slowly, is accelerated to higher and higher speeds. What happens to the diameter of the beam during this process?

7. A weak magnetic field exists in interplanetary space between the earth and the sun. What effect would you expect this field to have on the transit times of fast protons emitted by the sun which strike the earth's atmosphere to produce auroras?

8. The electron beam in a cathode-ray tube is aimed parallel to a nearby wire carrying a current in the same direction. Is the electron beam deflected? If so, in what direction?

9. A charged particle moving in a magnetic field follows a curved path, which means that its linear momentum is changing. How can this observation be reconciled with the principle of conservation of linear momentum?

10. A current-carrying wire loop is in a uniform magnetic field. Under what circumstances, if any, will there be no torque on the loop? No net force? Neither torque nor net force?

11. A power line 10 m above the ground carries a current of 5000 A. Find the magnetic field of the current on the ground directly under the cable.

12. What should the current be in a long, straight wire if the magnetic field 10 cm from it is to be 0.02 T?

13. At what distance from a long, straight wire carrying a current of 12 A does the magnetic field equal that of the earth, approximately 3×10^{-5} T?

14. A 10-turn circular coil of radius 2 cm carries a current of 0.5 A. Find the magnetic field at its center.

15. What should the current be in a wire loop 1 m in diameter if the magnetic field at the center of the loop is to be 1 T?

16. A long solenoid is wound with 30 turns/inch. What is the magnetic field inside the solenoid when the current is 3.8 A?

17. A solenoid 0.3 m long is wound with 1000 turns of wire. The solenoid is aligned with its axis parallel to the direction of the earth's magnetic field at a place where the latter is 2.3×10^{-5} T in magnitude. What current should be passed through the solenoid to exactly cancel the earth's field in the solenoid's interior?

18. In a certain electric motor, wires carrying currents of 4 A are perpendicular to a 1.2-T magnetic field. Find the force on each centimeter of these wires.

19. A horizontal north-south wire 5 m long is in a 0.02 T magnetic field whose direction is northeast. (a) What is the magnitude and direction of the force on the wire when a 4-A current flows north in it? (b) When the same current flows south in it?

20. A vertical wire 2 m long is in a 10^{-2} T magnetic field whose direction is northeast. (a) What is the magnitude and direction of the force on the wire when a 5-A current flows upward in it? (b) When the same current flows downward in it?

21. A certain size of copper wire has a mass of 5 kg per 100 m. A length of this wire is placed horizontally in a 1-T magnetic field whose lines of force are horizontal and perpendicular to the direction of the wire. What should the current in the wire be for it to be supported against gravity by the magnetic force on it?

22. A copper wire whose linear density is 10 g/m is stretched horizontally perpendicular to the direction of the horizontal component of the earth's magnetic field at a place where the magnitude of that component is 2×10^{-5} T. What must the current in the wire be for its weight to be supported by the magnetic force on it? What do you think would happen to such a wire if this current were passed through it?

23. What is the radius of the path of an electron whose velocity is 10^7 m/s in a magnetic field of 0.02 T when the electron's path is perpendicular to the field?

24. Compute the maximum radius of curvature in the earth's magnetic field at sea level, assuming $B = 3 \times 10^{-5}$ T, of (a) a proton whose velocity is 2×10^7 m/sec, and (b) an electron of the same velocity.

25. An electron in a television picture tube travels at a velocity of 3×10^7 m/s and is acted upon both by gravity and by the earth's magnetic field. Which exerts the greater force on the electron?

26. A certain electric transmission line consists of two wires 4 m apart that carry currents of 10^4 A. If the towers supporting the wires are 200 m apart, how much force does each current exert on the other between the towers?

27. The parallel wires in a lamp cord are 2.0 mm apart. What is the force per meter between them when the cord is used to supply power to a 120-V, 200 W light bulb?

Problems

1. Two parallel wires 10 cm apart carry currents in the same direction of 8 A. Find the magnetic field midway between the wires.

2. Two parallel wires 10 cm apart carry currents in opposite directions of 8 A. Find the magnetic field midway between the wires.

3. Two parallel wires 20 cm apart carry currents in the same direction of 5 A. Find the magnetic field between the wires 5 cm from one of them and 15 cm from the other.

4. Two parallel wires 20 cm apart carry currents in opposite directions of 5 A. Find the magnetic field between the wires 5 cm from one of them and 15 cm from the other.

5. A tube 20 cm long is wound with 400 turns of wire in one layer and with 200 turns in a second layer in the opposite sense to the first. A current of 0.1 A is passed through the coils. What is the magnetic field in the interior of the tube?

6. A solenoid is wound with 18 turns/cm and carries a current of 1.2 A. Another layer of turns is wound over the solenoid with 10 turns/cm and a

current of 5.0 A is passed through the new coil, opposite to the direction of the current in the solenoid. What is the magnetic field in the interior of the solenoid?

7. A current of 0.5 A is passed through a solenoid wound with 20 turns/cm. A single loop of wire 1 cm in radius is bent around the middle of the solenoid. What should the current in this loop be in order for the magnetic field at its center to cancel out the field of the solenoid there?

8. Two long, parallel wires a distance s apart each carry the current i in opposite directions. Verify that the magnetic field at a point equidistant from the wires and x away from the plane in which they lie is given by $2\mu_0 is/\pi(4x^2 + s^2)$.

9. Prove that the time required for a charged particle in a magnetic field to make a complete revolution is independent of its speed and the radius of its orbit.

10. A charge of $+10^{-6}$ C is moving at 500 m/s along a path parallel to a long, straight wire and 0.1 m from it. The wire carries a current of 2 A in the same direction as that of the charge. What is the magnitude and direction of the force on the charge?

11. A long, straight wire carries a current of 100 A. (a) What is the force on an electron traveling parallel to the wire, in the opposite direction to the current, at a velocity of 10^7 m/s when it is 10 cm from the wire? (b) Find the force on the electron under the above circumstances when it is traveling perpendicularly toward the wire.

12. A charge of $+2 \times 10^{-6}$ C is moving at 10^3 m/s at a distance of 12 cm away from a straight wire carrying a current of 4 A. Find the magnitude and direction of the force on the charge when it is moving parallel to the wire (a) in the same direction as the current, and (b) in the opposite direction.

13. An electron is traveling at 6×10^7 m/s at a distance of 5 cm from a long, straight wire carrying a current of 40 A. Find the magnitude and direction of the force on the electron when it is moving parallel to the wire (a) in the same direction as the current, and (b) in the opposite direction to the current.

14. The charge of Problem 12 is moving in a direction perpendicular to the same wire. Find the magnitude and direction of the force on the charge when it is moving (a) toward the wire, and (b) away from the wire.

15. The electron of Problem 13 is moving in a direction perpendicular to the same wire. Find the magnitude and direction of the force on the electron when it is moving (a) toward the wire, and (b) away from the wire.

16. An electron moving through an electric field of 500 V/m and a magnetic field of 0.1 T experiences no force. The two fields and the electron's direction of motion are all mutually perpendicular. What is the speed of the electron?

17. A particle of charge q is moving with a velocity $\mathbf{v}$ perpendicular to an electric field $\mathbf{E}$. What should be the magnitude and direction of a magnetic field in the same region that will exactly cancel out the effects of the electric field on the moving charge? What if $\mathbf{v}$ is parallel to $\mathbf{E}$?

18. A velocity selector uses a magnet to produce a 0.05-T magnetic field and a pair of parallel metal plates 1 cm apart to produce a perpendicular electric field. What potential difference should be applied to the plates to permit singly charged ions of velocity 5×10^6 m/s to pass through the selector?

19. A mass spectrometer employs a velocity selector consisting of a magnetic field of 0.0400 T perpendicular to an electric field of 50,000 V/m. The same magnetic field is then used to deflect the ions. Find the radius of curvature of singly charged lithium ions of mass 1.16×10^{-26} kg in this spectrometer.

Answers to Multiple Choice

1. d	7. b	13. b	19. b
2. c	8. c	14. a	20. d
3. a	9. a	15. b	21. a
4. c	10. d	16. c	22. a
5. a	11. c	17. a	
6. c	12. c	18. c	

21

Electromagnetic Induction

We have seen that, by causing a current to flow, an electric field is able to produce a magnetic field. Is there any way in which a magnetic field can produce an electric field? This problem was unsuccessfully tackled by many of the early workers in electricity and magnetism. Finally, in 1831, Michael Faraday in England and Joseph Henry in the United States independently discovered the phenomenon of electromagnetic induction. Two familiar applications of electromagnetic induction are the electric generator, which is the source of most electric power, and the transformer, which enables the emf of an alternating current to be increased or decreased easily.

21-1 Electromagnetic Induction

There is no current in a stationary wire in a static magnetic field when the wire is not connected to a source of emf. What Faraday and Henry found is that *moving* the wire produces a current (Fig. 21-1). It does not matter whether the wire or the source of the magnetic field is being moved, provided that a component of the motion is perpendicular to the field; the origin of the current lies in the *relative motion* between a condutor and a magnetic field. This effect is known as *electromagnetic induction.*

Faraday generalized his observations with the help of the notion of lines of force:

An electromotive force is produced in a conductor whenever it cuts across magnetic lines of force.

Electromagnetic induction

It is this electromotive force that leads to the current that flows whenever there is relative motion between a conductor and a magnetic field. In fact, it is not even necessary for there to be actual motion of either a wire or of a source of magnetic field, because a magnetic field that changes in strength has moving lines of force associated with it.

Moving a wire *parallel* to a magnetic field does not give rise to a current: electromagnetic induction only occurs when there is a component of the wire's velocity perpendicular to the lines of force.

Electromagnetic induction has a straightforward explanation in terms of the force exerted by a magnetic field on a moving charge. As we saw in the previous chapter, a wire carrying a current is pushed sideways in a magnetic field because of the forces exerted on the moving electrons. In Faraday's and Henry's experiments electrons are also moved through a magnetic

Fig. 21-1. (a) There is no current in a stationary wire in a magnetic field. (b) When the wire is moved across the field, a current is produced. Reversing the direction of motion also reverses the direction of the current.

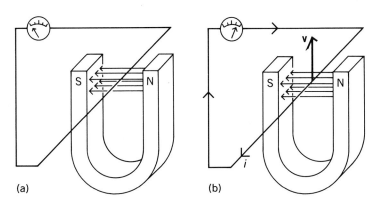

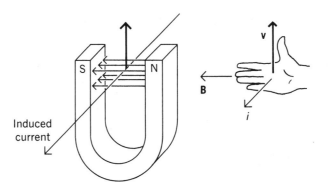

Fig. 21-2. Right-hand rule for the direction of an induced current.

field, but now by shifting the entire wire. As before, the electrons are pushed to the side, and in consequence move along the wire. The motion of these electrons along the wire is the electric current we measure.

Direction of induced current

The right-hand rule for the force on a moving positive charge in a magnetic field can also be used to give the direction of the induced current in a wire moving across a magnetic field. Hold your right hand so that the fingers point in the direction of **B** and the outstretched thumb is in the direction of **v**. The palm then faces in the same direction as the force on the positive charges in the wire, and hence in the direction of the conventional current (Fig. 21-2).

21-2 Moving Wire in a Magnetic Field

When a wire moves through a magnetic field **B** with a velocity **v** that is not parallel to **B**, an electromotive force $\mathcal{E}$ comes into being between the ends of the wire. It is not hard on the basis of what we already know to determine the magnitude of $\mathcal{E}$ in terms of **B**, **v**, and the length l of the wire. We shall assume that the wire, its direction of motion, and the magnetic field are all perpendicular to one another as in the previous diagrams.

According to Eq. (20-8) the force on a charge q in a moving wire when the wire has the velocity component $\mathbf{v}_\perp$ perpendicular to a magnetic field **B** has the magnitude

$$F = qv_\perp B.$$

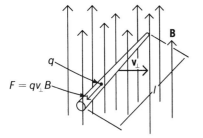

Fig. 21-3. The force on a charge q within a conductor when the conductor is moved through a magnetic field is $qv_\perp B$, where $v_\perp$ is the component of the conductor's velocity perpendicular to **B**.

The direction of this force is along the wire, as in Fig. 21-3.

Let us consider a charge q that moves from one end of the wire to the other under the influence of the magnetic force $qv_\perp B$. The work done by this force on the charge is

Work = force × distance,

$$W = Fl = qv_\perp Bl,$$

since the wire is l long. By definition the potential difference between two points is the work done in moving a unit charge between these points, so the potential difference between the ends of the wire is W/q or

$$\mathcal{E} = Blv_\perp. \qquad\qquad \textit{Motional emf} \quad (21\text{-}1)$$

This potential difference is the emf induced in the wire by its motion through the magnetic field.

How large will the current that flows in the wire be? The answer is given by Ohm's law, since an induced emf is like any other emf in its ability to cause a current to flow. If the total resistance in the circuit is R, the resulting current will be equal to $i = \mathcal{E}/R$ as long as the emf exists.

Problem. Compute the potential difference between the wing tips of a jet airplane induced by its motion through the earth's magnetic field. The total wing span of the airplane is 40 m and its speed is 300 m/s in a region where the vertical component of the earth's field has the magnitude 3×10^{-5} T (Fig. 21-4).

Solution. Substituting the given values of $v_\perp$, l, and B in Eq. (21-1) for the induced emf yields

$$\mathcal{E} = Blv_\perp = 300\ \frac{\text{m}}{\text{s}} \times 3 \times 10^{-5}\ \text{T} \times 40\ \text{m} = 0.36\ \text{V}$$

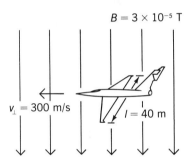

Fig. 21-4.

21-3 Faraday's Law

A convenient approach to electromagnetic induction makes use of the notion of *magnetic flux*. If the area of a wire loop is A and it is perpendicular to a magnetic field **B**, the total magnetic flux Φ (Greek letter *phi*) through the loop is defined as

Magnetic flux

$$\Phi = BA. \qquad\qquad \textit{Magnetic flux} \quad (21\text{-}2)$$

The loop can have any shape, but its area must be taken from its projection on a plane perpendicular to **B**; see Fig. 21-5. The unit of flux is the *weber* (Wb), where 1 Wb = 1 T-m².

Faraday's law

Faraday found that the electromotive force $\mathcal{E}$ in such a wire loop is *equal to the rate of change of the flux through it*. That is,

$$\mathcal{E} = -\frac{\Delta\Phi}{\Delta t},$$

Induced emf (21-3)

where $\Delta\Phi$ is the change in the flux Φ that takes place during a period of time Δt. The values of the magnetic induction B and the loop area A are, in themselves, irrelevant; only the speed with which either or both of them changes is important. Equation (21-3) is called Faraday's law of electromagnetic induction. The reason for the minus sign is given below.

The induced emf in a wire loop is equal to the potential difference that would be found between the ends of an identical *open* wire loop. The notion of emf is useful here because it is meaningless to speak of the potential difference around a closed circuit; the emf in a closed circuit is the potential difference that would be found between the ends of the circuit if it were cut anywhere.

If a coil of N turns replaces the single loop of Fig. 21-5, the induced emf's in the turns are added together, and the total emf in the entire coil is

$$\mathcal{E} = -N\frac{\Delta\Phi}{\Delta t}.$$

(21-4)

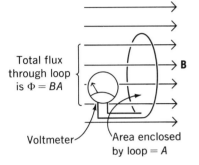

Total flux through loop is $\Phi = BA$

B

Voltmeter

Area enclosed by loop = A

Fig. 21-5. The emf induced in the wire loop is equal to the rate at which the flux Φ it encloses is changing.

The minus sign in Eqs. (21-3) and (21-4) is a consequence of the law of conservation of energy. If the sign of $\mathcal{E}$ were the same as that of $\Delta\Phi/\Delta t$, the induced electric current would be in such a direction that its own magnetic field would *add* to that of the external field **B**; this additional changing field would then augment the existing rate of change of the flux Φ, and more and more current would flow even if the external contribution to Φ were to stay constant. But energy is associated with every current, and no current can increase without external energy being supplied to it. The only possibility, then, is that $\mathcal{E}$ be opposite in sign to $\Delta\Phi/\Delta t$, which means that

Lenz's law

The direction of an induced current is always such that its own magnetic field opposes the changes in flux responsible for producing it.

This observation is known as *Lenz's law*. An example of Lenz's law is illustrated in Fig. 21-6.

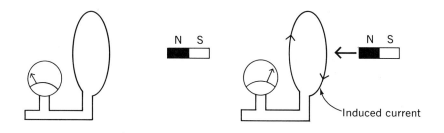

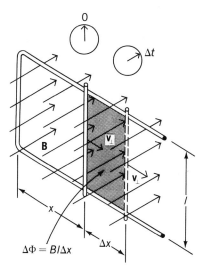

Fig. 21-6. The current induced in the wire loop is such that its own magnetic field is opposite to the field of the permanent magnet, in accord with Lenz's law.

Problem. A 12-turn coil 10 cm in diameter has its axis parallel to a magnetic field of 0.5 T which is produced by a nearby electromagnet. The current in the electromagnet is cut off, and as the field collapses an average emf of 8 V is induced in the coil. What was the length of time required for the field to disappear?

Solution. The change in the flux through the coil is

$$\Delta\Phi = BA = B\pi r^2 = 0.5 \text{ T} \times \pi (0.05 \text{ m})^2$$
$$= 0.0039 \text{ Wb}.$$

If we assume that the flux drops to zero at a uniform rate,

$$\Delta t = \frac{N \Delta\Phi}{\mathcal{E}} = \frac{12 \times 0.0039 \text{ Wb}}{8 \text{ V}} = 0.0059 \text{ s}.$$

Lenz's law has no significance in this problem.

$\Delta\Phi = Bl\Delta x$

Fig. 21-7.

Derivation of Faraday's law

It is not hard to derive Faraday's law of electromagnetic induction from the formula for the emf induced in a wire moved across a magnetic field. Figure 21-7 shows a moving wire that slides across the legs of a U-shaped metal frame, so that the frame completes the loop. At the moment (say $t = 0$) when the moving wire is the distance x from the closed end of the frame, the flux enclosed by the loop is

$$\Phi = BA = Blx.$$

At the later time $t = \Delta t$ the moving wire is $x + \Delta x$ from the closed end of the frame, so the flux now enclosed by the loop is

$$\Phi + \Delta\Phi = Blx + Bl\Delta x.$$

Hence the increase in the enclosed flux during the time Δt is

$$\Delta\Phi = Bl\Delta x.$$

The velocity of the moving wire is $v_\perp$, and so

$$\Delta x = v_\perp \Delta t,$$

and

$$\Delta\Phi = Blv_\perp \Delta t,$$

$$\frac{\Delta\Phi}{\Delta t} = Blv_\perp.$$

But $Blv_\perp$ is the emf induced in the moving wire. Hence we have, inserting a minus sign to remind ourselves of Lenz's law,

$$\mathcal{E} = -\frac{\Delta\Phi}{\Delta t},$$

which is Faraday's law.

21-4 The Generator

Electromagnetic induction is of immense practical importance since it is the means whereby nearly all the world's electric power is produced. In a generator a coil of wire is rotated in a magnetic field so that the flux through the coil changes constantly. The resulting potential difference across the ends of the coil causes a current to flow in an external circuit, and this current can be transmitted by a suitable system of wires for long distances from its origin.

Despite its name, a generator does not *create* electrical energy; what it does is *convert* mechanical energy into electrical energy, just as a battery converts chemical energy into electrical energy.

The construction of a simple generator is shown in Fig. 21-8. A wire loop is rotated in a magnetic field, and the ends of the loop are connected to two *slip rings* on the shaft. Brushes pressing against the slip rings permit the loop to be connected to an external circuit. Only sides B and D of the loop contribute to the induced emf. Because the sides of the loop reverse their direction of motion through the magnetic field twice per rotation, the induced current is also reversed twice per rotation and for this reason is

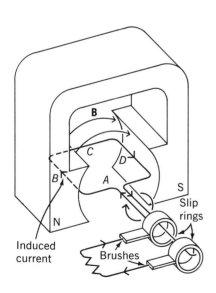

Fig. 21-8.   A simple ac-generator.

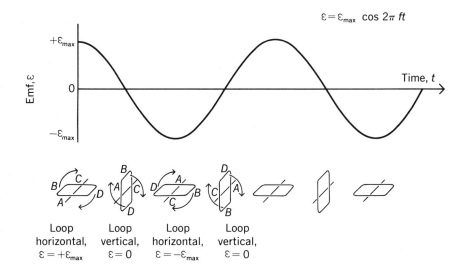

$$\varepsilon = \varepsilon_{max} \cos 2\pi \, ft$$

Fig. 21-9 The variation with time of
the emf of a simple ac-generator.

called an *alternating current*. The variation with time of the emf of this   Alternating current
generator is shown in Fig. 21-9.

The ac-output of a simple generator can be changed to dc by substituting
a commutator for the slip rings that connect the rotating coil with the external circuit (Fig. 21-10). The emf of such a dc-generator varies with time as
shown in Fig. 21-11; the emf is always in the same direction, but it rises to
the maximum and drops to zero twice per complete rotation.

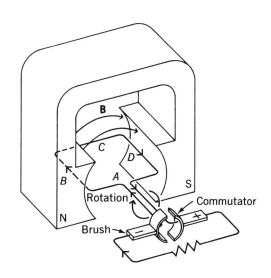

Fig. 21-10. A simple dc-generator.
The current in the brushes is always in
the same direction.

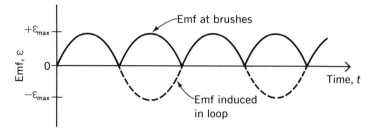

Fig. 21-11.   The variation with time of the emf of a simple dc-generator.

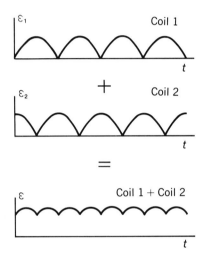

Fig. 21-12.   The output of a dc-generator whose armature contains two perpendicular coils has only a slight ripple.

The emf of Fig. 21-11, though unidirectional, pulsates too much for most applications. A steadier emf can be obtained by using a pair of coils on the armature set perpendicular to each other. When one of the coils has its maximum emf, the other has an emf of zero. Figure 21-12 shows the emf's of each coil and also what happens when the outputs of the two are connected in series: the resulting emf has only a moderate ripple. Commercial generators have many coils in their armatures to produce very nearly constant emf's.

In nearly all actual generators, electromagnets produce the magnetic field. (If a permanent magnet is used, the generator is called a *magneto*.) In an ac-generator, or *alternator*, the armature and field windings are often exchanged so that, in effect, the field rotates inside a stationary armature. This makes no difference to the physics involved since the essential condition is that there be relative motion of a conductor and a magnetic field. The advantage of a stationary armature is that a high-voltage output can be taken directly from it without any moving electrical contacts, since the field current can be supplied at a modest voltage.

21-5 Back Emf

In an electric motor a torque is produced through the interaction of the current in its armature with a magnetic field. As we have just learned, a coil rotated in a magnetic field is a source of emf, which implies that there is an emf generated in the rotating armature of an electric motor. This induced emf exists simply because wires are being moved through a magnetic field: the presence of a current of external origin in the armature at the same time, which is what causes it to turn in the first place, does not alter the situation. Every electric motor is also a generator.

By Lenz's law all induced emf's are such as to oppose the changes that bring them into being, which means that the induced emf in a motor armature is opposite in direction to the external voltage applied to the armature. The effect of this *back emf* (or *counter emf*) is to reduce the current in the armature to

$$i = \frac{V - \mathcal{E}_b}{R},$$ (21-5)

where V is the external voltage, $\mathcal{E}_b$ the back emf, and R the resistance of the armature coils.

Because the back emf is proportional to the speed of the armature, it is zero when the motor is turned on and increases as the speed increases. The current is therefore a maximum when the motor starts and drops afterward. The existence of back emf in a motor helps to keep its speed constant. At the normal speed,

Back emf and motor speed

$$V = \mathcal{E}_b + iR,$$ (21-6)

Impressed voltage = back emf + potential drop in armature.

If the speed falls, $\mathcal{E}_b$ decreases and i goes up, and the greater current means more torque on the armature to restore the speed to its original value. If the speed rises above normal, $\mathcal{E}_b$ increases and i goes down, with the result that the torque is reduced and the motor loses speed until it assumes its normal value.

Problem. A 120-V dc-electric motor has its armature and field windings connected in parallel. (Such a motor is said to be "shunt wound." When armature and field are in series, the motor is said to be "series wound.") The armature resistance is 2.0 Ω and the field resistance is 200 Ω. When the motor is at its operating speed of 1800 rpm, the total current is 3.0 A. Find (a) the back emf of the motor at its operating speed, (b) the power output of the motor, and (c) the current in the motor at the moment its switch is turned on.

Solution.   (a) The current in the field winding is

$$i_f = \frac{V}{R_f} = \frac{120 \text{ V}}{200 \text{ }\Omega} = 0.6 \text{ A}.$$

Since the total current is 3.0 A, the armature current is

$$i_a = i - i_f = 3.0 \text{ A} - 0.6 \text{ A} = 2.4 \text{ A}.$$

The magnitude of the back emf is therefore

$$\mathcal{E}_b = V - i_a R_a = 120 \text{ V} - 2.4 \text{ A} \times 2.0 \text{ }\Omega = 115 \text{ V}.$$

The direction of the back emf is, of course, opposite to that of the impressed potential difference V.

(b) The mechanical power output of the motor (ignoring friction) is equal to the product $\mathcal{E}_b i_a$ of the back emf and the armature current. Hence

$$P = \mathcal{E}_b i_a = 115 \text{ V} \times 2.4 \text{ A} = 276 \text{ W}.$$

(c) At the moment the motor is started, there is no back emf because the armature is still stationary. The total resistance of the armature and field windings is

$$R = \frac{R_a R_f}{R_a + R_f} = \frac{2.0 \text{ }\Omega \times 200 \text{ }\Omega}{2.0 \text{ }\Omega + 200 \text{ }\Omega} = 1.98 \text{ }\Omega.$$

The current is therefore

$$i = \frac{V}{R} = \frac{120 \text{ V}}{1.98 \text{ }\Omega} = 61 \text{ A}.$$

The starting current is more than 20 times greater than the operating current!

21-6 The Transformer

Earlier it was mentioned that a current can be induced in a wire or other conductor by a changing magnetic field as well as by relative motion between a wire and a constant magnetic field. The essential condition for an induced emf is that magnetic lines of force cut across the wire (or vice versa), and it does not matter exactly how this comes about.

A change in the current in a wire loop is accompanied by a corresponding change in its magnetic field, and if there is another wire loop nearby, an emf will therefore be induced in it. If an alternating current flows in the first loop, the magnetic field around it will vary periodically and an alternating emf will be induced in the second loop. A *transformer* is a device based upon this effect which is used to produce an emf in a secondary ac-circuit larger or smaller than the emf in a primary circuit.

With the help of Fig. 21-13 we can see how a changing current in a wire loop induces a changing current in another wire loop not connected to it:

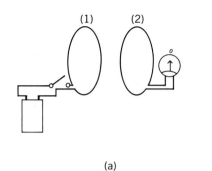

(a)

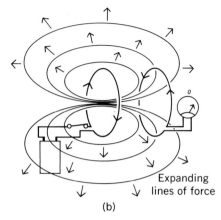

Expanding
lines of force

(b)

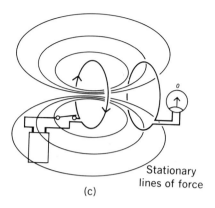

Stationary
lines of force

(c)

(a) The switch connecting loop (1) with the battery is open, and no current is present in either loop.

(b) The switch has just been closed, and the expanding lines of force resulting from the increasing current in loop (1) cut across loop (2), inducing a current in it. The current in loop (2) is in the *opposite* direction to that in loop (1) because of Lenz's law: the direction of an induced current is always such that its own magnetic field opposes the change in flux that is inducing it.

(c) Now there is a constant current in loop (1), and since the flux through it does not change, no current is induced in loop (2).

(d) The switch has just been opened, and the contracting lines of force resulting from the decreasing current in loop (1) cut across loop (2), inducing a current in it. The current in loop (2) is now in the *same* direction as that in loop (1), since it is the decreasing current in (1) that leads to the current in (2) and Lenz's law requires that an induced current oppose the change in flux that brings it about.

(e) Finally the current in loop (1) disappears, and no current is present in either loop.

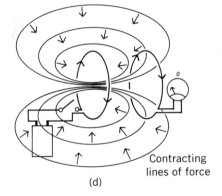

Contracting
lines of force

(d)

Fig. 21-13. When the current in a wire loop is turned on or off, the magnetic field around it changes. The changing field can induce a current in another nearby loop, even though both loops are stationary.

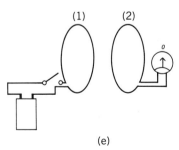

(e)

A transformer operates on alternating current

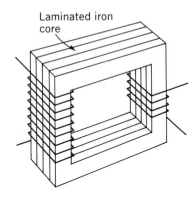

Laminated iron core

Fig. 21-14.   A simple transformer.

Instead of being switched on and off, the current input to a transformer is alternating current which reverses its direction regularly. The construction of a simple transformer is shown in Fig. 21-14. Coils are used instead of single loops, and they are wound on a common iron core so that the alternating magnetic flux set up by an alternating current in one coil links with the other coil. This alternating flux induces an alternating emf in the latter coil. The *primary winding* of a transformer is the coil which is fed with an alternating current, and the *secondary winding* is the coil to which power is transferred via the changing magnetic flux. The designations of primary and secondary depend only upon how the transformer is connected; either winding may be the primary, and the other is then the secondary.

Let us consider an ideal transformer in which there is no "leakage" of flux outside the iron core and no losses within it, so that $\Delta\Phi/\Delta t$ is the same for each of the turns of both windings. The emf *per turn* is therefore the same in both windings, and the total emf in each winding is proportional to the number of turns it contains. Hence

$$\frac{\mathcal{E}_1}{\mathcal{E}_2} = \frac{N_1}{N_2} \tag{21-7}$$

$$\frac{\text{Primary emf}}{\text{Secondary emf}} = \frac{\text{total primary turns}}{\text{total secondary turns}}.$$

Step-up and step-down transformers

When the secondary winding has more turns than the primary winding, $\mathcal{E}_2$ is greater than $\mathcal{E}_1$ and the result is a *step-up transformer*; when the reverse is the case and $\mathcal{E}_2$ is less than $\mathcal{E}_1$, the result is a *step-down transformer*.

In an ideal transformer the power output is equal to the power input. If the primary current is i_1 and the secondary current is i_2, then

$$\mathcal{E}_1 i_1 = \mathcal{E}_2 i_2$$

Power input = power output;

and

$$\frac{i_1}{i_2} = \frac{\mathcal{E}_2}{\mathcal{E}_1} = \frac{N_2}{N_1}. \qquad\qquad \textit{Transformer} \tag{21-8}$$

Increasing the voltage with a step-up transformer means a proportionate decrease in the current, while dropping the voltage leads to a greater current.

Eddy currents

A transformer core is itself an electrical conductor, and the changing magnetic flux within it leads to internal currents called *eddy currents*. Eddy currents are wasteful partly because of the power lost as heat and partly because the flux they establish interferes with the proper operation of the transformer. To minimize eddy currents, actual transformer cores are usually laminated, with many thin sheets arranged parallel to the flux. The laminations are insulated from each other by natural oxide layers on their surfaces or by varnish coatings; only very weak eddy currents are induced in a core of this kind. In small transformers an alternative technique is to use a core of compressed powdered iron, so that each grain of iron is fairly well insulated from the others, although magnetically the core behaves as if it were solid.

There are still other sources of power loss in a transformer. The windings themselves have a certain amount of resistance, and part of the energy that flows through them is dissipated as heat. Also, in all magnetic materials some energy is lost as heat per cycle in the atomic rearrangements that accompany flux changes. Leakage of flux from the core of Fig. 21-14 between the separated primary and secondary windings is more easily remedied. Instead, transformers normally have concentric windings as in Fig. 21-15, and in this arrangement the flux through each winding is virtually the same. Efficiencies of as much as 99% are achieved in large transformers.

Alternating current owes its wide use largely to the ability of transformers to change the voltage at which it is transmitted from one value to another. Electricity generated at perhaps 11,000 volts is stepped up by transformers at the power station to as much as 750,000 volts (or even more) for transmission, and subsequently is stepped down by transformers at substations near the point of consumption to less than a thousand volts. High voltages are desirable because, since $P = iV$, the higher the voltage, the smaller the current for a given amount of power; and since power is dissipated as heat at the rate i^2R, the smaller the current, the less the power lost due to resistance in the transmission line.

Problem. Find the power lost as heat when a 10 Ω cable is used to transmit 1 kW of electricity at 240 V and at 240,000 V.

Solution. The current in the cable in each case is

$$i_a = \frac{P}{V_a} = \frac{1000 \text{ W}}{240 \text{ V}} = 4.17 \text{ A,}$$

$$i_b = \frac{P}{V_b} = \frac{1000 \text{ W}}{240,000 \text{ V}} = 4.17 \times 10^{-3} \text{ A.}$$

The respective rates of heat production per kilowatt are therefore

$$i_a^2 R = (4.17 \text{ A})^2 \times 10 \text{ Ω} = 174 \text{ W,}$$

Why AC is widely used

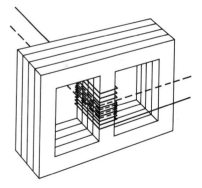

Fig. 21-15. Most transformers have concentric windings to minimize flux leakage.

$$i_b^2 R = (4.17 \times 10^{-3}\ \text{A})^2 \times 10\ \Omega = 1.74 \times 10^{-4}\ \text{W},$$

a difference of a factor of 10^6.

Special Topic

The Betatron

An interesting application of electromagnetic induction is the betatron. In a betatron (Fig. 21-16) electrons are accelerated to high speeds through the action of an increasing magnetic field. As the magnetic field **B** increases, the associated electric field contributes to the energy of an electron moving in a circular path of area A by an amount per revolution equal to its increase in energy when it moves through an emf of

$$\mathcal{E} = -\frac{\Delta \Phi}{\Delta t} = -A\,\frac{\Delta B}{\Delta t}.$$

That is, we can regard a circular path of this kind exactly as though it were a loop of wire, insofar as electromagnetic induction is concerned. As the electron goes faster and faster, gaining the energy $q\mathcal{E}$ in each revolution, it will require a stronger magnetic field **B** if it is to stay in the same orbit; it is

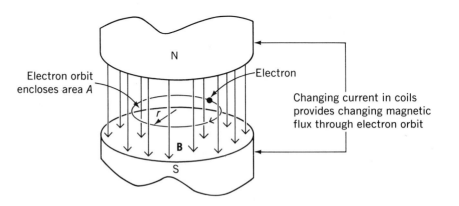

Fig. 21-16. Schematic diagram of a betatron.

not difficult to arrange matters so that the very increase in **B** that accelerates the electron in the first place exactly keeps pace with its increasing speed, thereby maintaining a constant orbit radius.

A typical betatron that accelerates electrons to 100 MeV might have an orbit radius of 1 m and a magnetic field changing at the rate of 100 T/s. The area enclosed by the orbit is

$$A = \pi R^2 = 3.1 \text{ m}^2,$$

and the emf per revolution is

$$\mathcal{E} = -A\,\frac{\Delta B}{\Delta t} = -3.1 \text{ m}^2 \times 100 \text{ T/s} = 310 \text{ V}.$$

An electron in this betatron acquires 310 eV each time it makes a complete circle; to acquire 100 MeV it must make

$$\frac{100 \times 10^6 \text{ eV}}{310 \text{ eV/rev}} = 3.23 \times 10^5 \text{ revolutions.}$$

Important Terms

Electromagnetic induction refers to the production of an electric field in a conductor whenever magnetic lines of force move across it.

Lenz's law states that the direction of an induced current must be such that its own magnetic field opposes the changes in flux that are inducing it.

A **generator** is a device that converts mechanical energy into electrical energy.

The direction of an **alternating current** reverses itself periodically.

A **back** (or **counter**) **emf** is induced in the rotating coils of an electric motor and is opposite in direction to the external voltage applied to them.

An alternating current flowing in the primary coil of a **transformer** induces another alternating current in the secondary coil. The ratio of the emfs is proportional to the ratio of turns in the coils.

Important Formulas

Motional emf:
$$\mathcal{E} = Blv$$

Magnetic flux:
$$\Phi = BA$$

Induced emf:
$$\mathcal{E} = -\frac{\Delta \Phi}{\Delta t}$$

Transformer:
$$\frac{i_1}{i_2} = \frac{\mathcal{E}_2}{\mathcal{E}_1} = \frac{N_2}{N_1}$$

Multiple Choice

1. The appearance of an electric field in a conductor when it is moved across a magnetic field is an example of
   a. Lenz's law.
   b. Ohm's law.
   c. electromagnetic induction.
   d. magnetic flux.

2. The emf produced in a wire by its motion across a magnetic field does *not* depend on
   a. the length of the wire.
   b. the diameter of the wire.
   c. the orientation of the wire.
   d. the flux density of the field.

3. Electromotive force is most closely related to
   a. electric field.
   b. magnetic field.
   c. potential difference.
   d. mechanical force.

4. The unit of electromotive force is the
   a. newton.
   b. joule.
   c. volt.
   d. ampere.

5. Lenz's law is a consequence of the law of conservation of
   a. charge.
   b. momentum.
   c. lines of force.
   d. energy.

6. A dynamo is often said to "generate electricity." It actually acts as a source of
   a. charge
   b. emf.
   c. electrons.
   d. magnetism.

7. The unit of magnetic flux is the weber, where
   1 Wb =

   a. 1 T-m².
   b. 1 T/m².
   c. 1 A-m².
   d. 1 A/m².

8. When a wire loop is rotated in a magnetic field, the direction of the induced emf changes once in each
   a. $\frac{1}{4}$ revolution.
   b. $\frac{1}{2}$ revolution.
   c. 1 revolution.
   d. 2 revolutions.

9. The emf produced by an ac-generator varies between $\mathcal{E}_{max}$ and
   a. 0.
   b. $\frac{1}{2} \mathcal{E}_{max}$.
   c. $- \mathcal{E}_{max}$.
   d. the back emf.

10. When the speed of a dc-motor drops because of an increased load, there is also a drop in the
    a. impressed voltage.
    b. back emf.
    c. current.
    d. armature resistance.

11. The ac-output of a simple generator can be changed to dc by connecting the output of the rotating coil to the external circuit with
    a. slip rings.
    b. a commutator.
    c. a resistor.
    d. a transformer.

12. The alternating current in the secondary coil of a transformer is induced by
    a. a varying electric field.
    b. a varying magnetic field.
    c. the iron core of the transformer.
    d. motion of the primary coil.

13. The ratio between primary and secondary currents in a transformer does *not* depend on the
    a. ratio of turns in the two windings.

b. the resistance of the windings.

c. the nature of the core.

d. the primary voltage.

14. The primary winding of a transformer has 200 turns and its secondary winding has 50 turns. If the current in the secondary winding is 40 A, the current in the primary is
    a. 10 A.
    b. 80 A.
    c. 160 A.
    d. 8000 A.

15. The primary winding of a transformer has 200 turns and its secondary winding has 50 turns. If an ac-emf of 12 V is applied to the primary, the secondary emf will be
    a. 3 V.
    b. 6 V.
    c. 24 V.
    d. 48 V.

16. A 100% efficient transformer has 100 turns in its primary winding and 300 turns in its secondary. If the power input to the transformer is 60 W, the power output is
    a. 20 W.
    b. 60 W.
    c. 180 W.
    d. 540 W.

Exercises

1. A bar magnet held vertically with its north pole downward is dropped through a wire loop whose plane is horizontal. Is the current induced in the loop just before the magnet enters it clockwise or counterclockwise as seen by an observer above? What is the direction of the current when the magnet passes through the center of the loop? Just after it leaves the loop?

2. One end of a bar magnet is thrust into a coil, and the induced current is clockwise as seen from the front of the coil. (a) Was the end of the magnet its north or south pole? (b) What will be the direction of the induced current when the magnet is withdrawn?

3. A car is traveling from New York to Florida. Which of its wheels have a positive charge and which a negative charge?

4. Why is it easy to turn the shaft of a generator when it is not connected to an outside circuit, but much harder when such a connection is made?

5. *Magnetic storms* occur when swarms of charged particles from the sun enter the earth's magnetic field. Explain why such storms always produce a decrease in the magnetic field at sea level.

6. Why does an electric motor require more current when it is turned on than when it is running continuously?

7. A loop of copper wire is rotated in a magnetic field about an axis along a diameter. (a) Why does the loop resist this rotation? (b) If the loop were made of aluminum wire, would the resistance to rotation be different?

8. Why are transformer cores made from thin sheets of steel? Would there be any advantage in making permanent magnets in this way?

9. Why must alternating current be used in a transformer?

10. In a "series-wound" generator the magnetic field is produced by an electromagnet whose windings are connected in series with the armature windings. What happens to the emf in such a generator when the current drawn by the external circuit increases? When the external circuit current decreases? Why?

11. In a "shunt-wound" generator the electromagnet windings are connected in parallel with the armature windings. What happens to the emf in such a generator when the current drawn by the external circuit increases? When the current decreases? Why?

12. A train is traveling at 80 mi/hr in a region where the vertical component of the earth's magnetic field is 3×10^{-5} T. The railway is standard gauge (that is, its rails are 4 ft $8\frac{1}{2}$ in. apart). Find the potential difference between the wheels on each axle of the train's cars.

13. An automobile has a velocity of 30 m/s on a road where the vertical component of the earth's magnetic field is 3×10^{-5} T. What is the potential difference between the ends of its axles, which are 2 m long?

14. A potential difference of 1.8 V is found between the ends of a 2-m wire moving in a direction perpendicular to a magnetic field at a speed of 12 m/s. What is the magnitude of the field?

15. A rectangular wire loop 5 cm × 10 cm in size is perpendicular to a magnetic field of 10^{-3} T. (a) What is the flux through the loop? (b) If the magnetic field drops to zero in 3 sec, what is the potential difference induced between the ends of the loop during that period?

16. A wire loop is 5 cm in diameter and is situated so that its plane is perpendicular to a magnetic field. How rapidly should the field change if a potential difference of 1 V is to appear across the ends of the loop?

17. The magnetic flux through a 20-turn coil drops from 0.3 weber to 0 in 1 sec. What is the induced potential difference across the ends of the coil?

18. A square wire loop 10 cm on a side is situated so that its plane is perpendicular to a magnetic field. The resistance of the loop is 5 ohms. How rapidly should the magnetic field change if a current of 2 A is to flow in the loop?

19. A coil is rotated 100 times per second in a magnetic field. What time interval separates successive instants when the induced emf is zero?

20. The primary winding of a transformer has 120 turns and the secondary winding has 1800 turns. A current of 10 A flows in the primary when a potential difference of 550 V is applied across it. What is the potential difference across the secondary coil and what is the current in it? (Assume 100% efficiency in Exercises 20–23.)

21. An electric welding machine employs a current of 400 A. The device uses a transformer whose primary winding has 400 turns and which draws 2 A from a 220-V power line. (a) How many turns are there in the secondary winding of the transformer? (b) What is the potential difference across the secondary?

22. A transformer has 600 turns in its primary winding and 90 turns in its secondary. What are the secondary voltage and current when the primary draws 3.4 A at 115 V?

23. A transformer is used to couple an 11,000-V transmission line to a 230-V distribution circuit. If 5 kW of power is delivered to the 230-V circuit, find the primary and secondary currents in the transformer.

Problems

1. The armature of a dc-generator has a resistance of 0.20 Ω. The terminal potential difference of the generator is 120 V with no load and 115 V on full load. How much power is delivered at full load?

2. A 1.5 kW dc generator has a full-load potential difference of 35 V and a no-load potential difference of 38 V. How much power is dissipated as heat in the armature at full load?

3. A shunt-wound 120-V, 3-kW generator has an armature resistance of 0.1 Ω and a field resistance of 120 Ω. Find its efficiency.

4. A series-wound 24-V dc electric motor has an armature resistance of 0.1 Ω and a field resistance of 0.4 Ω. The motor draws 20 A at its normal operating speed. Find the power output, power input, and efficiency of the motor.

5. A shunt-wound 120-V dc electric motor has an armature resistance of 1.0 Ω and a field resistance of 200 Ω. The motor has a power input of 420 W at 2000 rpm. (a) Find the power output and efficiency of the motor. (b) What is the starting current? (c) What series resistance is required if the starting current is to be limited to 25 A?

6. Find the power output, power input, and efficiency of the motor of Problem 5 when it is operating at 1800 rpm.

Answers to Multiple Choice

1. c	7. a	13. d
2. b	8. b	14. a
3. c	9. c	15. a
4. c	10. b	16. b
5. d	11. b	
6. b	12. b	

22

Capacitance and Inductance

An electric field requires the expenditure of energy to be brought into existence, and this energy is available for doing work under suitable circumstances. A capacitor is a device that stores energy in the form of an electric field. Energy is also associated with a magnetic field, and is thus present in the vicinity of every electric current. Devices called inductors are designed to exploit this property of currents. When a charged capacitor is connected to an inductor, energy flows back and forth between them in the same way that KE and PE are interchanged in a harmonic oscillator. As in the case of a harmonic oscillator, the alternations in energy flow in a given capacitor-inductor combination occur at a certain natural frequency, a property that is made use of in practically every branch of electronics.

22-1 Electric Field Energy

Work must be done to create an electric field, since work must be done to separate positive and negative charges against the Coulomb forces attracting them together. Where does the work go? The answer is that it is stored as potential energy, which we can think of as residing in the electric field between the charges.

Let us consider a pair of parallel metal plates of area A that are the distance d apart, as in Fig. 22-1. We imagine that we create an electric field between them by bringing electrons from one plate to the other until a total charge of Q has been transferred. When we are finished, there will be a potential difference of V between the plates.

At the beginning, the potential difference between the plates is small, and little work is needed to move each electron. As the charge builds up, the potential difference becomes greater, and more work is needed per electron. The average potential difference $\bar{V}$ during the charge transfer is

$$\bar{V} = \frac{V_{final} + V_{initial}}{2}$$

$$= \frac{V + 0}{2}$$

$$= \tfrac{1}{2}V.$$

Since the total charge transferred is Q, the work W that was done is the product of Q and the average potential difference $\bar{V}$, namely

$$W = Q\bar{V} = \tfrac{1}{2}QV. \tag{22-1}$$

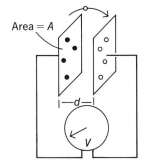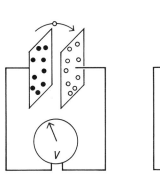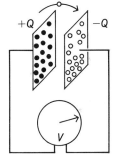

• +e   ○ −e

Fig. 22-1. As a capacitor is charged, the potential difference between its plates builds up and more work must be done to transfer each successive charge.

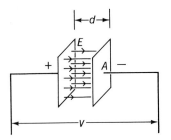

Fig. 22-2. The energy density of the electric field in a region is the total energy in the region divided by its volume.

Because the metal plates are good conductors, the $+$ and $-$ charges on them are spread out evenly, and the field between the plates (except near the edges) is uniform with the magnitude

$$E = 4\pi k \frac{Q}{A}.$$ (22-2)

The derivation of this formula, which follows from the considerations discussed in Chapter 18, is not given here because it involves calculus. We can rewrite Eq. (22-2) in the form

$$Q = \frac{AE}{4\pi k}.$$ (22-3)

The potential difference between the plates is equal to

$$V = Ed.$$ (22-4)

Hence the electric potential energy of the system of two charged plates is

$$W = \tfrac{1}{2}QV$$

$$= \frac{Ad}{8\pi k} E^2.$$ (22-5)

Electric energy density

We note that Ad, the product of the area of each plate and the distance between them, is the volume occupied by the electric field $\mathbf{E}$ (Fig. 22-2). If we define the *energy density* w of an electric field as the electric potential energy per unit volume associated with it, we have here

$$w = \frac{W}{Ad}$$

$$= \frac{E^2}{8\pi k}.$$ *Electric energy density* (22-6)

This important formula states that the energy density of an electric field is directly proportional to the square of its magnitude E. Even though we derived this formula for a special situation, it is a completely general result.

Problem. Dry air is an insulator provided the electric field in it does not exceed about 3×10^6 V/m. What energy density does this correspond to?

Solution. Here $E = 3 \times 10^6$ V/m, and so

$$w = \frac{E^2}{8\pi k} = \frac{(3 \times 10^6 \text{ V/m})^2}{8\pi \times 9 \times 10^9 \text{ N·m}^2/\text{C}^2}$$

$$= 40 \text{ J/m}^3.$$

22-2 Capacitance

A system of conductors that stores energy in the form of electric field is called a *capacitor*. The pair of parallel metal plates considered above is an example of a capacitor.

Capacitor

The potential difference V across a capacitor is always directly proportional to the charge Q on either of its plates (Fig. 22-3): the more the charge, the stronger the electric field between the plates, and the greater the potential difference. The ratio between Q and V is therefore a constant for any capacitor and is known as its *capacitance* (symbol C):

Capacitance

$$C = \frac{Q}{V} \qquad\qquad (22\text{-}7)$$

$$\text{Capacitance} = \frac{\text{charge on either conductor}}{\text{potential difference between conductors}}.$$

The unit of capacitance is the *farad*, abbreviated F, where

The farad

$$1 \text{ farad} = 1\ \frac{\text{coulomb}}{\text{volt}}.$$

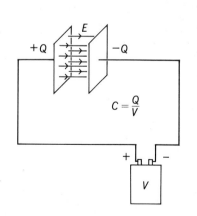

Fig. 22-3. The ratio between the charge on a particular capacitor and the potential difference across it is a constant called its *capacitance*.

The farad is so large a unit that, for practical purposes, it is usually replaced by the *microfarad* (μF) or *picofarad* (pF), whose values are

$$1\,\mu\text{F} = 10^{-6}\text{F}, \qquad 1\,\text{pF} = 10^{-12}\,\text{F}.$$

The capacitance of a pair of separated conductors depends solely upon their geometry and upon the material between them. In the case of a parallel-plate capacitor in vacuum, from Eqs. (22-3) and (22-4) we have

$$
\begin{aligned}
C &= \frac{Q}{V} \\[4pt]
&= \frac{AE/4\pi k}{Ed} \\[4pt]
&= \frac{1}{4\pi k}\frac{A}{d}.
\end{aligned}
$$

Parallel-plate capacitor   (22-8)

To a good degree of approximation the same formula can be used for such a capacitor in air as well.

Problem.   The plates of a parallel-plate capacitor are 10 cm square and 1 mm apart. Find its capacitance in air and the charge each plate will acquire when a potential difference of 100 V is applied.

Solution.   Here $A = 10^{-2}$ m^2 and $d = 10^{-3}$ m. In air the capacitance is

$$C = \frac{1}{4\pi \times 9 \times 10^9 \text{ N·m}^2/\text{C}^2} \times \frac{10^{-2} \text{ m}^2}{10^{-3} \text{ m}} = 8.85 \times 10^{-11} \text{ F} = 88.5 \text{ pF}.$$

If a potential difference of 100 V is placed across the plates of this capacitor, they will acquire charges of

$$
\begin{aligned}
Q &= CV = 8.85 \times 10^{-11} \text{ F} \times 100 \text{ V} \\[4pt]
&= 8.85 \times 10^{-9} \text{ C}.
\end{aligned}
$$

The formula $\tfrac{1}{2}QV$ for the potential energy of a charged capacitor derived in the previous section can be written in three equivalent ways:

$$W = \tfrac{1}{2} QV \tag{22-9}$$

$$= \tfrac{1}{2} CV^2 \tag{22-10}$$

$$= \frac{1}{2}\frac{Q^2}{C}. \tag{22-11}$$

Potential energy of charged capacitor   (22-11)

These equations hold for all capacitors, regardless of their construction.

Problem. A heart attack often leads to a condition called *fibrillation* in which the heart's actions lose their synchronization and it is unable to pump blood effectively. This condition can often be corrected by an electric shock to the heart which completely stops it for a moment; the heart may then start again spontaneously in its normal rhythm. An appropriate such shock can be provided by the discharge of a 10-μF capacitor that has been charged to a potential difference of 6000 V. (a) How much energy is released in the current pulse? (b) How much charge passes through the patient's body? (c) If the pulse lasts 5 ms, what is the average current?

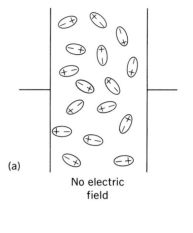

(a) No electric field

Solution. (a) $W = \frac{1}{2} CV^2 = \frac{1}{2} \times 10^{-5}$ F $\times$ (6000 V)$^2 = 180$ J.

(b) $Q = CV = 10^{-5}$ F $\times$ 6000 V $= 0.06$ C.

(c) $\bar{\imath} = \dfrac{Q}{t} = \dfrac{0.06 \text{ C}}{5 \times 10^{-3} \text{ s}} = 12$ A.

22-3 Dielectric Constant

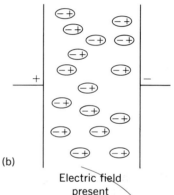

(b) Electric field present

Let us now examine what happens when a slab of an insulating material is placed between the plates of a capacitor.

Although an insulator cannot conduct electric current, it can respond to an electric field in another way. The molecules of all substances either normally have a nonuniform distribution of electric charge within them or assume such a distribution under the influence of an electric field. As we know, a molecule of the former kind is called a polar molecule, and behaves as though one end is positively charged and the other negatively charged. In an assembly of polar molecules when there is no external electric field, the molecules are randomly oriented as in Fig. 22-4(a). When an electric field is present, it acts to align the molecules opposite to the field, as in Fig. 22-4(b).

Fig. 22-4. An electric field tends to align polar molecules opposite to the field.

While nonpolar molecules ordinarily have symmetric charge distributions, an electric field is able to distort their inner structures so that an effective separation of charge takes place (Fig. 22-5). Again the molecules have their charged ends aligned opposite to the external field. In either case, then, the net electric field between the plates of the capacitor is *less* than it would be with nothing between them.

The *dielectric constant*, symbol K, of a substance is a measure of how

Dielectric constant

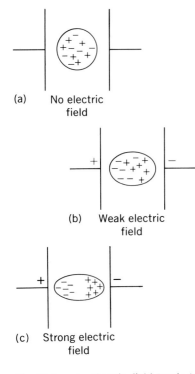

(a)   No electric
field

(b)   Weak electric
field

(c)   Strong electric
field

Fig. 22-5.   An electric field tends to
distort the charge distributions in
molecules which are ordinarily nonpolar.

Table 22-1. Dielectric constants at 20°C except where noted

Substance	K	Substance	K
Air	1.0006	Mica	2.5–7
Air, liquid (−191°C)	1.4	Neoprene	6.7
Alcohol, ethyl	26	Sulfur	3.9
Benzene	2.3	Teflon	2.1
Glass	5–8	Water	80
Ice (−2°C)	94	Waxed paper	2.2

effective it is in reducing an electric field set up across a sample of it. If the
capacitance of a capacitor is C_0 when there is a vacuum between its plates,
its capacitance will be

$$C = KC_0 \qquad\qquad (22\text{-}12)$$

when a substance of dielectric constant K is between the plates. Table 22-1
is a list of dielectric constants for various substances. Water and alcohol
molecules are highly polar, and the values of K for water, ice, and ethyl
alcohol are accordingly high.

Commercial capacitors consist of many interleaved plates to make
possible a high capacitance in a small unit (Fig. 22-6). Solid dielectrics are

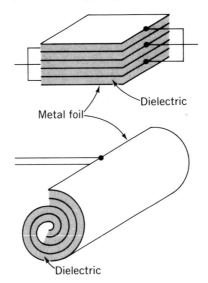

Dielectric

Metal foil

Dielectric

Fig. 22-6.   Most fixed capacitors are
made of interleaved sheets of metal
foil separated by layers of dielectric
material.

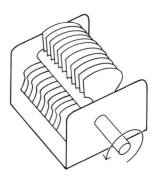

Fig. 22-7.   Variable capacitors of this
type are widely used to tune radio receivers.

usually used, both to increase C and to maintain a fixed distance between the plates. With sheets of a solid dielectric such as waxed paper or mica between them, the plates can be of inexpensive metal foil, without the rigidity required to prevent accidental contact if only air separated them. In an *electrolytic capacitor* the plates have extremely thin dielectric layers formed on their surfaces by chemical action. A conducting paste between these plates constitutes the other electrode of the capacitor. The very small thickness of the dielectrics in electrolytic capacitors permits them to have capacitances of as much as 50 μf without excessive bulk.

Variable capacitors normally have two sets of rigid aluminum plates   Variable capacitors
that are interleaved with air as the dielectric, as in Fig. 22-7. One of the sets is mounted on a shaft, and by rotating the shaft the amount of overlap between the plates can be adjusted. Since the overlapped area of the plates is the chief contributor to their capacitance, turning the shaft varies C.

22-4 Capacitors in Combination

The equivalent capacitance of two or more capacitors connected together can be determined in a manner analogous to that used in the case of resistors in combination, though the results are different.

Figure 22-8 shows three capacitors in series. Each has charges of the   Capacitors in series
same magnitude Q on its plates, in agreement with the principle of conservation of charge. Hence the potential differences across the capacitors are respectively

$$V_1 = \frac{Q}{C_1}, \qquad V_2 = \frac{Q}{C_2}, \qquad V_3 = \frac{Q}{C_3},$$

and so, if C is the equivalent capacitance of the set, we have

$$V = V_1 + V_2 + V_3$$

$$\frac{Q}{C} = \frac{Q}{C_1} + \frac{Q}{C_2} + \frac{Q}{C_3}$$

$$\frac{1}{C} = \frac{1}{C_1} + \frac{1}{C_2} + \frac{1}{C_3}.$$

For any number of capacitors in series,

$$\frac{1}{C} = \frac{1}{C_1} + \frac{1}{C_2} + \frac{1}{C_3} + \cdots \qquad \textit{Capacitors in series} \quad (22\text{-}13)$$

Fig. 22-8.   Capacitors in series. A charge of the same magnitude Q is present on all the plates of the capacitors.

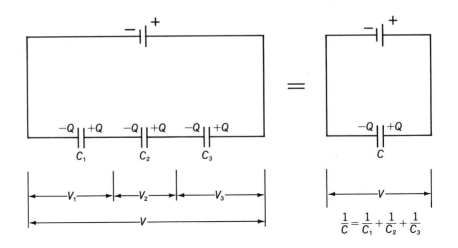

$$\frac{1}{C} = \frac{1}{C_1} + \frac{1}{C_2} + \frac{1}{C_3}$$

The reciprocal of the equivalent capacitance of a series arrangement of capacitors is equal to the sum of the reciprocals of the individual capacitors. Evidently C is smaller than the capacitance of any of the individual capacitors.

Two capacitors in series   If there are only two capacitors in series, Eq. (22-13) becomes

$$C = \frac{C_1 C_2}{C_1 + C_2}. \tag{22-14}$$

Capacitors in parallel   Figure 22-9 shows three capacitors connected in parallel. The same potential difference is across all of them, so that the charges on their plates have the respective magnitudes

$$Q_1 = C_1 V, \qquad Q_2 = C_2 V, \qquad Q_3 = C_3 V.$$

The total charge $Q_1 + Q_2 + Q_3$ on either the positive or negative plates of the capacitors is equal to the charge Q on the corresponding plate of the equivalent capacitor, and hence

$$Q = Q_1 + Q_2 + Q_3,$$
$$CV = C_1 V + C_2 V + C_3 V,$$
$$C = C_1 + C_2 + C_3.$$

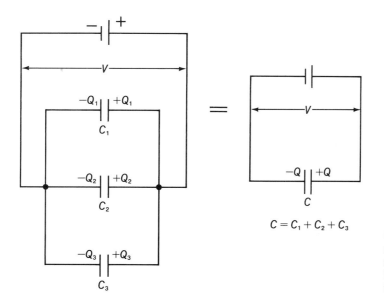

$$C = C_1 + C_2 + C_3$$

Fig. 22-9. Capacitors in parallel. The same potential difference V is across all of them, but each has a charge on its plates whose magnitude is proportional to its capacitance.

This result can be generalized to

$$C = C_1 + C_2 + C_3 + \cdots \qquad \textit{Capacitors in parallel} \quad (22\text{-}15)$$

More energy is stored in a set of capacitors when they are connected in parallel across a given potential difference than when they are connected in series, because the electric fields in them are stronger in the former case.

22-5 Magnetic Field Energy

To appreciate why energy is associated with a magnetic field, we can think of what happens when a wire loop is connected to a battery. As the current starts to flow, a magnetic field begins to build up. But this changing magnetic field induces an emf in the *same* wire loop whose current causes the field in the first place. By Lenz's law the *self-induced emf* is such as to oppose the change in flux that is responsible for it. The battery must therefore push electrons through the wire loop against the self-induced emf due to the increasing magnetic field, which means that work has to be done in order to produce the magnetic field. (The work needed to overcome the effect of the self-induced emf has nothing to do with the resistance of the wire; it is a consequence of electromagnetic induction only.)

Self-induced emf

When the current in the wire loop reaches its final value, there is no more self-induced emf. The work performed to establish the magnetic field has become magnetic energy.

Let us determine the amount of energy contained in a solenoid when a steady current i flows through it. This energy is equal to the work W that had to be done against the self-induced emf $\mathcal{E}$ in order to establish the current starting from $i = 0$ (Fig. 22-10).

If we suppose that the current rises at a uniform rate from $i = 0$ to $i = i$ in a time interval Δt, then the average current $\bar{\imath}$ during Δt is

$$\bar{\imath} = \frac{i_{\text{final}} + i_{\text{initial}}}{2}$$

$$= \frac{i + 0}{2}$$

$$= \tfrac{1}{2}\,i.$$

Thus the total charge Q that passes through the coil while the current is building up to its final value i is

$$Q = \bar{\imath}\,\Delta t = \tfrac{1}{2}\,i\,\Delta t.$$

The work that was done during the buildup of the current is the product of the total charge Q and the emf $\mathcal{E}$:

Work done to create field
$$W = Q\mathcal{E} = \tfrac{1}{2}\,i\mathcal{E}\,\Delta t. \tag{22-16}$$

To find expressions for i and $\mathcal{E}$, we begin with the formula

$$B = \mu i\,\frac{N}{l} \tag{20-6}$$

for the magnetic field in the interior of an N-turn solenoid l long when it carries the current i. This formula can be rewritten to give an expression for i in terms of B, namely

$$i = \frac{Bl}{\mu N}. \tag{22-17}$$

The magnitude of the emf is given by Faraday's law (the minus sign is irrelevant here) as

Self-induced emf
$$\mathcal{E} = N\,\frac{\Delta\Phi}{\Delta t}.$$

The final flux Φ through the solenoid is BA, where A is the cross-sectional area (see Fig. 22-10). Since the initial flux is $\Phi = 0$,

$$\frac{\Delta\Phi}{\Delta t} = \frac{BA}{\Delta t}$$

and

$$\mathcal{E} = \frac{NAB}{\Delta t}. \qquad (22\text{-}18)$$

By substituting Eqs. (22-17) and (22-18) in Eq. (22-16) we obtain for the energy contained in the magnetic field of the solenoid

Magnetic field energy

$$W = \tfrac{1}{2} i \mathcal{E} \, \Delta t$$

$$= \tfrac{1}{2} \times \frac{Bl}{\mu N} \times \frac{NAB}{\Delta t} \times \Delta t$$

$$= \frac{lA}{2\mu} B^2. \qquad (22\text{-}19)$$

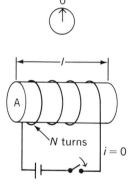

The quantity lA is the volume occupied by the magnetic field $\mathbf{B}$, since it is the product of the length of the solenoid and its cross-sectional area. (Since we are considering an ideal solenoid whose length is large relative to its diameter, we can ignore the magnetic field escaping from the ends.) The *magnetic energy density* of the field is therefore

$$w = \frac{W}{lA} = \frac{B^2}{2\mu}. \qquad \textit{Magnetic energy density} \quad (22\text{-}20)$$

The energy density of a magnetic field is directly proportional to the square of its magnitude B. This formula holds for all magnetic fields, not just for those inside solenoids. We recall from Eq. (22-6) that the energy density of an electric field $\mathbf{E}$ is similarly proportional to E^2.

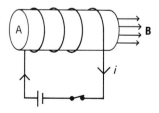

Problem. The earth's magnetic field in a certain region has the magnitude 6×10^{-5} T. Find the magnetic energy per cubic mile in this region.

Solution. The magnetic energy density corresponding to $B = 6 \times 10^{-5}$ T is, since here $\mu = \mu_0$,

$$w = \frac{B^2}{2\mu_0} = \frac{(6 \times 10^{-5} \text{ T})^2}{2 \times 1.26 \times 10^{-6} \text{ T-m/A}} = 1.4 \times 10^{-3} \text{ J/m}^3.$$

Fig. 22-10. A self-induced emf opposes the establishment of a current in a solenoid. The work done against this emf becomes magnetic energy.

A cubic mile is equal to

$$1 \text{ mi}^3 \times \left(1.61 \frac{\text{km}}{\text{mi}}\right)^3 \times \left(10^3 \frac{\text{m}}{\text{km}}\right)^3 = 4.2 \times 10^9 \text{ m}^3,$$

so that the total magnetic energy in a cubic mile is

$$W = wV = 1.4 \times 10^{-3} \frac{\text{J}}{\text{m}^3} \times 4.2 \times 10^9 \text{ m}^3 = 6 \times 10^6 \text{ J}.$$

This amount of energy is enough to raise the 365,000-ton Empire State Building by 1.8 mm.

22-6 Inductance

A circuit element in which a self-induced emf accompanies a changing current is called an *inductor*. A solenoid is an example of an inductor.

Inductor

Normally the energy content W of an inductor is most conveniently expressed in terms of the current i present in it. Since the magnetic field B in an inductor is proportional to i in the absence of ferromagnetic materials, W is proportional to i^2. The constant of proportionality is written as $L/2$, where L is called the *inductance* of the inductor. Thus

$$W = \tfrac{1}{2} Li^2. \qquad\qquad \textit{Potential energy of inductor} \quad (22\text{-}21)$$

The henry

The unit of inductance is the *henry* (H), where

$$1 \text{ H} = 1 \frac{\text{J}}{A^2} = 1 \frac{V\text{-}s}{A}.$$

The henry, like the farad, is usually too large a unit for convenience. Inductances are accordingly often expressed in *millihenries* (mH) or *microhenries* (μH), where

$$1 \text{ mH} = 10^{-3} \text{ H}, \qquad 1 \text{ }\mu\text{H} = 10^{-6} \text{ H}.$$

Factors influencing inductance

The inductance of an inductor depends upon its geometry and upon the presence of a core with magnetic properties. A coil consisting of many turns has a greater inductance than one consisting of few turns because its magnetic field is stronger for a given current (Fig. 22-11). The longer the coil and the greater its cross-sectional area, the greater its inductance be-

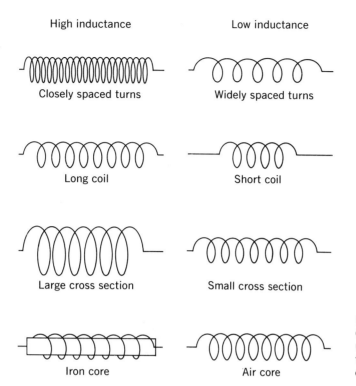

High inductance   Low inductance

Closely spaced turns   Widely spaced turns

Long coil   Short coil

Large cross section   Small cross section

Iron core   Air core

Fig. 22-11. The inductance of a coil depends upon how closely wound it is, upon its length, upon its cross-sectional area, and upon the nature of its core.

cause it contains a larger volume of magnetic field. A core in a coil changes its inductance by changing the flux through the coil (see Section 20-4). Thus a diamagnetic core decreases the inductance slightly from its value in free space or air, a paramagnetic core increases it slightly, and a ferromagnetic core increases it considerably — by a factor of as much as 10^4 in some cases — but by an amount that is not constant.

Let us calculate the inductance of a solenoid. From Eqs. (22-19) and (22-21), which are different ways to express the energy of a solenoid when it carries the current i, we have

Inductance of solenoid

$$\frac{lA}{2\mu} B^2 = \tfrac{1}{2} Li^2,$$

$$L = \frac{lA}{\mu i^2} B^2.$$

Since the magnetic field within the solenoid is

$$B = \mu i \frac{N}{l},$$

we see that

$$L = \frac{lA}{\mu i^2} \times \left(\mu i \frac{N}{l}\right)^2,$$

$$= \mu N^2 \frac{A}{l}. \qquad\qquad \textit{Inductance of solenoid} \quad (22\text{-}22)$$

Problem. Find the inductance in air of a 1000-turn solenoid 10 cm long that has a cross-sectional area of 20 cm². How much energy is stored in the magnetic field of the solenoid when it carries a 0.01-A current?

Solution. Since 10 cm $= 10^{-1}$ m, 20 cm² $= 2 \times 10^{-3}$ m², and $\mu = \mu_0$ in air (very nearly),

$$L = \mu_0\, N^2 \frac{A}{l} = 1.26 \times 10^{-6}\,\frac{\text{T-m}}{\text{A}} \times (10^3)^2 \times \frac{2 \times 10^{-3}\ \text{m}^2}{10^{-1}\ \text{m}}$$

$$= 2.5 \times 10^{-2}\ \text{H} = 25\ \text{mH}.$$

The energy stored in the solenoid's magnetic field when $i = 0.01$ A $= 10^{-2}$ A is

$$W = \tfrac{1}{2} L i^2 = \tfrac{1}{2} \times 2.5 \times 10^{-2}\ \text{H} \times (10^{-2}\ \text{A})^2 = 1.3 \times 10^{-6}\ \text{J}.$$

22-7 Self-induced Emf

It is easy to show that the inductance L of an inductor determines the magnitude of the self-induced emf $\mathcal{E}$ that accompanies a changing current in the inductor. From Eqs. (22-16) and (22-21) we have

$$\tfrac{1}{2} L i^2 = \tfrac{1}{2} i \mathcal{E}\ \Delta t,$$
$$\mathcal{E} = L \frac{i}{\Delta t}.$$

Since the current in the inductor rose from 0 to i in the time Δt, we can write $\Delta i / \Delta t$ in place of $i / \Delta t$. Also, Lenz's law requires that $\mathcal{E}$ be opposite in sign to $\Delta i / \Delta t$. Hence we have the important formula

$$\mathcal{E} = -L \frac{\Delta i}{\Delta t}.$$

Self-induced emf (22-23)

A high value of L means a large $\mathcal{E}$ for a given rate of change $\Delta i/\Delta t$ and a low value of L means a small $\mathcal{E}$.

Inductance and self-induced emf

Problem. What is the average self-induced emf in a 25-mH coil when the current in it falls from 0.3 A to 0 in 0.01 s?

Solution. From Eq. (22-23)

$$\mathcal{E} = -L \frac{\Delta i}{\Delta t} = -2.5 \times 10^{-2} \text{ H} \times \frac{0.3 \text{ A}}{10^{-2} \text{ s}} = -0.75 \text{ V}$$

22-8 Electrical Oscillations

Let us connect a charged capacitor to an inductor, as in Fig. 22-12. The following sequence of events occurs:

(a) Initially the capacitor's plates each possess the charge Q and its electric energy is $\frac{1}{2}Q^2/C$, while the inductor has no energy since $i = 0$.

(b) The capacitor is partially discharged, and the current in the inductor leads to a magnetic field whose energy is $\frac{1}{2}Li^2$.

(c) No charge is left on the capacitor's plates, and all the energy of the circuit is magnetic energy $\frac{1}{2}LI^2$.

(d) Now charge begins to build up on the capacitor's plates with polarities the reverse of those originally present, and the electric energy of the capacitor grows at the expense of the magnetic energy of the inductor as the current i drops.

(e) At the instant the capacitor is fully charged again, the current is 0 and the energy of the circuit is once more entirely electric.

(f) Next the capacitor begins to discharge, and eventually the entire circuit returns to its original state at (a). The cycle will continue to repeat itself indefinitely if no resistance is present.

The above sequence is wholly analogous to the behavior of a harmonic oscillator. As we learned in Chapter 12, the basic process involved in the operation of a harmonic oscillator is the continual interchange of energy between KE and PE; here the same sort of interchange occurs, and we can regard the electrostatic energy of a charged capacitor as corresponding to PE and the magnetic energy of an inductor through which a current flows as corresponding to KE.

Electrical oscillator is analog of harmonic oscillator

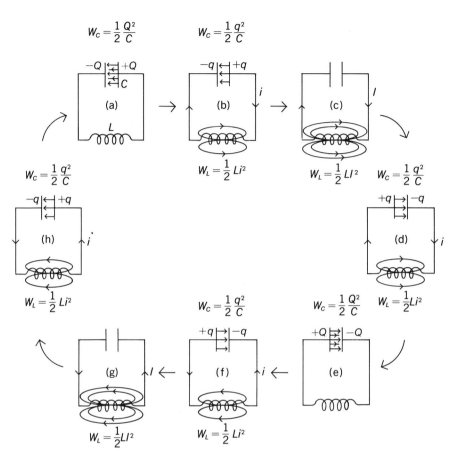

Fig. 22-12. The various stages in the oscillation of a circuit containing the inductance L and the capacitance C. The maximum charge on the capacitor is Q, and the maximum current in the circuit is I.

A harmonic oscillator whose mass is m and whose spring constant is k has the potential and kinetic energies

$$\text{PE} = \tfrac{1}{2}ks^2$$

$$\text{KE} = \tfrac{1}{2}mv^2$$

where s is the displacement of the object from its equilibrium position and v is its velocity. The corresponding formulas for the electric and magnetic energies in an electrical oscillator are

$$W_e = \frac{1}{2}\frac{1}{C}q^2,$$

$$W_m = \frac{1}{2} Li^2.$$

We note that $1/C$ is analogous to k, since both determine the amount of energy present when nothing is moving in the respective oscillators. Similarly L is analogous to m. Both are measures of inertia—a large inductance tends to slow down changes in current, and a large m tends to slow down changes in velocity. The frequency of a harmonic oscillator is

$$f = \frac{1}{2\pi} \sqrt{\frac{k}{m}},$$ Harmonic oscillator

and so it is not surprising that the frequency of an electrical oscillator is

$$f = \frac{1}{2\pi\sqrt{LC}}.$$ *Oscillator frequency*   (22-24)   Electrical oscillator

The larger the inductance L and the capacitance C, the lower the frequency.

   The parallel between electrical and mechanical oscillations is a very close one: just as every mechanical system has certain specific natural frequencies of vibration which depend upon its properties, so every electrical circuit has a natural frequency of oscillation which depends upon L and C. Just as the energy of every actual mechanical vibration is eventually dissipated as heat owing to the inevitable presence of friction, so the energy of every electrical oscillation is eventually dissipated as heat owing to the inevitable presence of resistance in the wires of the circuit. Electrical resonance is much like mechanical resonance: if a source of potential which alternates at the natural frequency of a circuit is connected to it, energy is fed into the oscillations that can maintain or increase their amplitude. At any other frequency the energy absorbed is negligible.

   Radio communication is made possible by the ability of an L-C circuit to oscillate only at the frequency given by Eq. (22-24). Such a circuit is used to "tune" a transmitter so that only a single frequency of radio waves is produced, and another similar circuit is used to "tune" a receiver to that same frequency. Electrical oscillations are discussed further in the next chapter.

Problem.   In the antenna circuit of a radio receiver, the inductance is fixed at 4 mH but the capacitance can be varied. What should the capacitance be in order to receive a 750-kHz radio signal?

Solution.   Since $f = \frac{1}{2}\pi\sqrt{LC} = 7.5 \times 10^5$ Hz,

$$C = \frac{1}{(2\pi f)^2 L} = \frac{1}{(2\pi \times 7.5 \times 10^5 \text{ Hz})^2 \times 4 \times 10^{-3} \text{ H}}$$

$$= 1.13 \times 10^{-11} \text{ F} = 11.3 \text{ pF}.$$

Special Topic

Time Constants

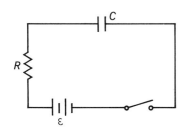

Fig. 22-13.   A circuit that contains a capacitor, a resistor, and a source of emf in series. When the switch is closed, the charge on the capacitor increases gradually to its ultimate value of $C\varepsilon$.

Circuits that contain resistance as well as capacitance or inductance do not respond instantaneously to changes in the applied emf. For instance, when a capacitor is connected to a battery, as in Fig. 22-13, it does not immediately become fully charged. At first the only limit to the current that flows to the capacitor is the resistance R in the circuit, so that the initial current is $i = \varepsilon/R$, where ε is the emf of the battery. As the capacitor becomes charged, however, a potential difference appears across it, whose polarity is such as to tend to oppose the further flow of current. When the charge on the capacitor has built up to some value q, this opposing potential difference is $V = q/C$. Hence the net potential difference is $\varepsilon - q/C$, and the current is

$$i = \frac{\varepsilon - q/C}{R}.$$

As q increases, then, its *rate* of increase drops. Figure 22-14 is a graph showing how q varies with time when a capacitor is being charged; the capacitor is connected to the battery at $t = 0$.

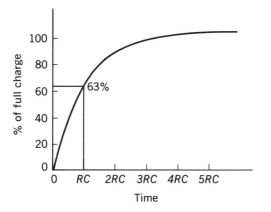

Fig. 22-14.   The growth of charge in a capacitor.

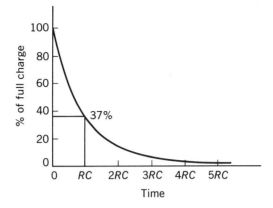
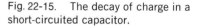

Fig. 22-15. The decay of charge in a short-circuited capacitor.

A mathematical analysis of the above equation shows that after a time interval of RC (the product of the resistance R in the circuit and the capacitance C of the capacitor), the charge on the capacitor reaches 63% of its ultimate value of $Q = C\mathcal{E}$. The time RC is therefore a convenient measure of how rapidly the capacitor becomes charged and is accordingly called the *time constant* of the circuit. In principle the capacitor acquires its ultimate charge Q only after an infinite time has elapsed, but, as we can see from Fig. 22-14 this value is very nearly reached after only 3 or 4 RC.

If a capacitor with an initial charge is discharged through a resistance, its charge decreases with time as shown in Fig. 22-15. After the time RC the charge on the capacitor is reduced to 37% of its original value, a drop of 63%.

When a circuit containing inductance is connected to a battery, the current in the circuit does not rise instantly to its ultimate value $i = \mathcal{E}/R$, where $\mathcal{E}$ is the emf of the battery and R is the total resistance in the circuit. At the moment the switch in Fig. 22-16 is closed, the current starts to grow, and as a result the induced emf $-L(\Delta i/\Delta t)$ comes into being in the opposite direction to the battery emf $\mathcal{E}$. The net emf acting to establish current in the circuit is therefore $\mathcal{E} - L(\Delta i/\Delta t)$, and the current reaches its ultimate value of $\mathcal{E}/R$ in a gradual manner.

The graph in Fig. 22-17 shows how i varies with time when a current is being established in a circuit containing inductance. A mathematical analysis shows that, after a time interval of L/R, the current reaches 63% of its final value V/R. The time L/R is therefore a convenient measure of how rapidly a current rises in a circuit containing inductance, and, like RC in a circuit containing capacitance, is called the *time constant* of the circuit.

When the battery in Fig. 22-16 is short-circuited by a wire, the current i drops slowly in the manner shown in Fig. 22-18, since the induced emf now

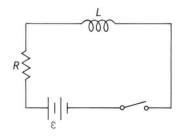

Fig. 22-16. A circuit that contains an inductor, a resistor, and a source of emf in series. When the switch is closed, the current increases gradually to its ultimate value of $\mathcal{E}/R$ because of the opposing self-induced emf in the inductor.

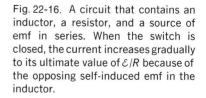

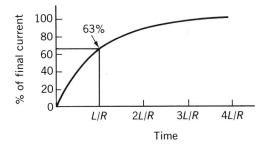

Fig. 22-17. The growth of current in a circuit containing inductance and resistance.

tends to maintain the existing current. The current falls to 37% of its original value in the time L/R after the battery is short-circuited.

A coil of inductance L has an energy of $\frac{1}{2}Li^2$ stored in it when a current i is present in it. When the potential difference across the coil that is responsible for the current is removed, the energy $\frac{1}{2}Li^2$ is what powers the self-induced emf that retards the drop in current. The gradual rise of current in a circuit containing inductance may be thought of as the result of the initial absorption of $\frac{1}{2}Li^2$ of potential energy by the circuit, and its gradual drop may be thought of as the result of the restoration of the potential energy to the circuit.

Fig. 22-18. The decay of current in a circuit containing inductance and resistance when the battery is short-circuited.

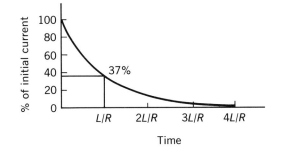

Important Terms

The **energy density** of an electric field is the electric potential energy per unit volume associated with it. The **energy density** of a magnetic field is the magnetic energy per unit volume associated with it.

A **capacitor** is a device that stores electric energy in the form of an electric field. The ratio between the charge on either plate of a capacitor and the potential difference between the plates is called its **capacitance**. The unit of capacitance is the **farad**, which is equal to 1 coul/volt.

The **dielectric constant** K of a particular material is a measure of how effective it is in reducing an electric field set up across a sample of it.

The **inductance** L of a circuit is the ratio between the magnitude of the self-induced emf $\mathcal{E}$ due to a changing current in it and the rate of change $\Delta i/\Delta t$ of the current. The unit of inductance is the **henry**.

Important Formulas

Energy density of electric field:

$$w = \frac{E^2}{8\pi k}$$

Capacitance:

$$C = \frac{Q}{V}$$

Parallel-plate capacitor:

$$C = \frac{\mathrm{K}}{4\pi k}\frac{A}{d} \qquad [\mathrm{K} = \text{dielectric constant}]$$

Potential energy of charged capacitor:

$$W = \tfrac{1}{2}QV = \tfrac{1}{2}CV^2 = \frac{Q^2}{2\,C}$$

Capacitors in series:

$$\frac{1}{C} = \frac{1}{C_1} + \frac{1}{C_2} + \frac{1}{C_3} + \cdots$$

Capacitors in parallel:

$$C = C_1 + C_2 + C_3 + \cdots$$

Magnetic energy density:

$$w = \frac{B^2}{2\mu}$$

Inductance of solenoid:

$$L = \mu N^2 \frac{A}{l}$$

Potential energy of inductor:

$$W = \tfrac{1}{2}Li^2$$

Self-induced emf:

$$\mathcal{E} = -L\frac{\Delta i}{\Delta t}$$

Oscillator frequency:

$$f = \frac{1}{2\pi\sqrt{LC}}$$

Multiple Choice

1. The energy content of a charged capacitor resides in its
   a. plates.   b. potential difference.
   c. charge.   d. electric field.

2. When a slab of insulating material is placed between the plates of a charged capacitor, the electric field there relative to what it was before is
   a. less.
   b. the same.
   c. more.
   d. any of the above, depending on the circumstances.

3. The farad is not equivalent to which of the following combination of units?
   a. C^2/J b. C/V
   c. $C\cdot V^2$ d. J/V^2

4. Magnetic fields invariably contain
   a. a ferromagnetic material.
   b. inductance.
   c. electric current.
   d. energy.

5. The unit of inductance is the henry, where $1\ \mathrm{H} =$
   a. $1\ \mathrm{J\text{-}A^2}$.
   b. $1\ \mathrm{J/A^2}$.
   c. $1\ \mathrm{V\text{-}A}$.
   d. $1\ \mathrm{V/A}$.

6. A wire coil carries the current i. The potential energy of the coil does not depend upon

a. the value of i.
b. the number of turns in the coil.
c. whether the coil has an iron core or not.
d. the resistance of the coil.

7. The inductance of a solenoid that contains a total of N turns is proportional to
   a. N.
   b. N^2.
   c. $1/N$.
   d. $1/N^2$.

8. A large increase in the inductance of a coil can be achieved by using a core that is
   a. diamagnetic.
   b. paramagnetic.
   c. ferromagnetic.
   d. polar.

9. A capacitor acquires a charge of 0.002 C when connected across a 50-V battery. Its capacitance is
   a. 1 μF.　　　　b. 2 μF.
   c. 4 μF.　　　　d. 40 μF.

10. A 50-μF capacitor has a potential difference of 8 V across it. Its charge is
    a. 4×10^{-3} C.
    b. 4×10^{-4} C.
    c. 6.25×10^{-5} C.
    d. 6.25×10^{-6} C.

11. The plates of a parallel-plate capacitor of capacitance C are brought together to one-third their original separation. The capacitance is now
    a. $\frac{1}{9}C$.　　　　b. $\frac{1}{3}C$.
    c. $3C$.　　　　d. $9C$.

12. If a 20-μF capacitor is to have an energy content of 2.5 J, it must be placed across a potential difference of
    a. 150 V.　　　　b. 350 V.
    c. 500 V.　　　　d. 250,000 V.

13. A parallel-plate capacitor has a capacitance of 50 pF in air and 110 pF when immersed in turpentine. The dielectric constant of turpentine is
    a. 0.45.　　　　b. 0.55.
    c. 1.1.　　　　d. 2.2.

14. A parallel-plate capacitor with air between its plates is charged until a potential difference of V appears across it. Another capacitor, having hard rubber (dielectric constant = 3) between its plates but otherwise identical, is also charged to the same potential difference. If the energy of the first capacitor is W, that of the second is
    a. $\frac{1}{3} W$.　　　　b. W.
    c. $3 W$.　　　　d. $9 W$.

15. Two 50 μF capacitors are connected in series. The equivalent capacitance of the combination is
    a. 25 μF.
    b. 50 μF.
    c. 100 μF.
    d. 200 μF.

16. The capacitor combination of Question 15 is connected across a 100-V battery. The potential difference across each capacitor is
    a. 25 V.　　　　b. 50 V.
    c. 100 V.　　　　d. 200 V.

17. Two 50 μF capacitors are connected in parallel. The equivalent capacitance of the combination is
    a. 25 μF.
    b. 50 μF.
    c. 100 μF.
    d. 200 μF.

18. The capacitor combination of Question 17 is connected across a 100-V battery. The potential difference across each capacitor is
    a. 25 V.　　　　b. 50 V.
    c. 100 V.　　　　d. 200 V.

19. The energy contained in a cubic meter of space in which the magnetic induction is 1 T is
    a. 6.3×10^{-7} J.
    b. 3.97×10^{-5} J.
    c. 3.97×10^{5} J.
    d. 7.94×10^{5} J.

20. The self-induced emf in a 0.1-H coil when the current in it is changing at the rate of 200 A/s is
    a. 125 V.
    b. 20 V.
    c. 8×10^{-4} V.
    d. 8×10^{-5} V.

21. The current in a circuit falls to 0 from 16 A in 0.01 s. The average emf induced in the circuit during the drop is 64 V. The inductance of the circuit is
    a. 0.032 H.
    b. 0.04 H.
    c. 0.25 H.
    d. 4 H.

22. A 2-mH coil carries a current of 10 A. The energy stored in its magnetic field is
    a. 0.05 J.
    b. 0.1 J.
    c. 1.0 J.
    d. 100 J.

23. An L-C circuit has the initial capacitance C. In order to double its natural frequency of oscillation, the capacitance must be changed to
    a. $\frac{1}{4}C$.
    b. $\frac{1}{2}C$.
    c. $2C$.
    d. $4C$.

Exercises

1. Is there any kind of material that, when inserted between the plates of a capacitor, reduces its capacitance?

2. What effect does placing a slab of a material of dielectric constant K between the plates of a charged capacitor have on the energy content of the capacitor? (The capacitor is disconnected from the charging circuit before the dielectric is inserted.) If the energy is greater than before, where does the additional energy come from? If the energy is less than before, where does the lost energy go?

3. A parallel-plate capacitor of capacitance C is given the charge Q and then disconnected from the circuit. How much work is required to pull the plates of this capacitor to twice their original separation?

4. A sheet of mica whose dielectric constant is 5 is placed between the plates of a charged, isolated parallel-plate capacitor. How is the potential difference across the capacitor affected? How is the charge on the capacitor affected?

5. The greater its capacitance, the less energy is stored in a capacitor when it is given a certain charge. On the other hand, the greater its inductance, the more energy is stored in an inductor when a certain current is present in it. Explain the difference.

6. What becomes of the work done against the back emf in an inductive circuit when a current is being established in it?

7. What is the direction of the self-induced emf in a coil when the current in it increases? When the current decreases? What is the reason in each case?

8. A potential difference of 300 V is applied across a pair of parallel metal plates 1 cm apart. What is the energy density of the electric field between the plates?

9. A 25-μF capacitor is connected to a source of potential difference of 1000 V. What is the resulting charge on the capacitor? How much energy does it contain?

10. What is the potential difference between the plates of a 20-μF capacitor whose charge is 0.01 C? How much energy does it contain?

11. The plates of a parallel-plate capacitor are 50 cm^2 in area and 1 mm apart. (a) What is its capacitance? (b) When the capacitor is connected to a 45-V battery, what is the charge on either plate? (c) What is the energy of the charged capacitor?

12. The space between the plates of the capacitor of the previous problem is filled with sulfur. Answer questions (a), (b), and (c) for this case.

13. A capacitor with air between its plates is connected to a battery and each of its plates receives a charge of 10^{-4} C. While still connected to the battery the capacitor is immersed in oil, and a

further charge of 10^{-4} C is added to each plate. What is the dielectric constant of the oil?

14. Find the equivalent capacitance of a 20-μF capacitor and a 50-μF capacitor that are connected in series.

15. Find the equivalent capacitance of a 20-μF capacitor and a 50-μF capacitor that are connected in parallel.

16. The strongest magnetic fields that have been produced in the laboratory have been about 10^2 T. (a) How much energy is contained in a 10^3 cm^3 volume of such a field? (b) What intensity would an electric field require to have the same energy density?

17. A 1 mH inductor is to be made by winding fine copper wire on a tube 2 cm in diameter and 10 cm long. How many turns are needed?

18. A 50-mH coil carries a current of 2 A. How much energy is stored in its magnetic field?

19. What is the self-induced emf in a 0.4-H coil when the current in it is changing at a rate of 500 A/s?

20. The current in a circuit drops from 10 A to 2 A in 0.1 s. An average of 32 V is induced in the circuit while this happens. What is the inductance of the circuit?

21. Find the natural frequency of an *L-C* circuit in which $L = 12$ mH and $C = 5$ μF.

22. What inductance is needed in a circuit in which $C = 60$ μF if its natural frequency is to be 30 Hz?

23. What capacitance is needed in a circuit in which $L = 2$ H if its natural frequency is to be 200 Hz?

24. The frequencies used in commercial radio broadcasting range from 550 to 1600 kHz. What range of capacitance should a variable capacitor have if it is connected to a coil of inductance 1 mH in a circuit designed to respond to frequencies in this band?

Problems

1. A potential difference of 100 V is applied across a pair of parallel metal plates 5 cm square and 1 mm apart. (a) What is the force between the plates? (b) Is the force attractive or repulsive? (c) What is the energy density in the region between the plates?

2. A capacitor is charged by connecting it through a resistance to a battery. Verify that half the work done by the battery is dissipated as heat.

3. Three capacitors whose capacitances are 5, 10, and 50 μF are connected in series across a 12-V battery. Find the charge on each capacitor and the potential difference across it.

4. Three capacitors whose capacitances are 2, 4, and 5 μF are connected in series across a 100-V battery. Find the charge on each capacitor and the potential difference across it.

5. The three capacitors of problem 3 are connected in parallel across the same battery. Find the charge on each capacitor and the potential difference across it.

6. The three capacitors of problem 4 are connected in parallel across the same battery. Find the charge on each capacitor and the potential difference across it.

7. A 1-μF capacitor and a 2-μF capacitor are each charged across a potential difference of 1200 V. The capacitors are then connected with terminals of opposite sign together. What is the final charge of each capacitor?

8. A solenoid 20 cm long and 2.4 cm in diameter is wound with 1200 turns of wire whose total resistance is 40 Ω. The solenoid is connected to a 12-V battery whose internal resistance is 2 Ω. Find the inductance of the solenoid, the final current in it, and its energy content when the final current flows.

9. The solenoid of the previous problem is immersed in liquid oxygen whose relative perme-

ability is 1.0049. Find the inductance of the solenoid, the final current in it, and its energy content when the final current flows.

10. A 1-μF capacitor is charged by being connected to a 10-V battery. The battery is then removed and the capacitor connected to a 10-mH coil. (a) Find the frequency of the resulting oscillations. (b) Find the maximum value of the charge on the capacitor. (c) Find the maximum current that flows through the inductor.

11. Verify that the time required for the energy in an *L-C* circuit to be equally divided between the electric and magnetic fields starting from the moment when the capacitor is fully charged is $1/8f$. (If $q = Q$ at $t = 0$, then $q = Q \cos 2\pi ft$ and $i = I \sin 2\pi ft$.)

Answers to Multiple Choice

1. d	9. d	17. c
2. a	10. b	18. c
3. c	11. c	19. c
4. d	12. d	20. b
5. b	13. d	21. b
6. d	14. c	22. b
7. b	15. a	23. a
8. c	16. b	

23

Alternating Current

Nearly all the world's electrical energy is carried by alternating current. The preference for ac is chiefly due to the economy of high-voltage transmission which minimizes i^2R heat losses; with step-up transformers at the production end and step-down transformers at the consumption end, the transmission voltage is limited only by insulation and atmospheric discharge problems. Alternating current is also preferred in industry because ac-electric motors are, as a class, cheaper, more durable, and less in need of maintenance than dc-motors; although dc-motors have certain characteristics, such as better speed regulation, that make them more suitable for specialized duties. Alternating currents are involved in all aspects of modern communication: the electrical equivalent of a sound wave of a certain frequency is an alternating current of that frequency, and radio waves are produced by antennas fed with high-frequency ac. Alternating current behaves in a circuit in a very different way from direct current, and some knowledge of ac-circuit behavior is necessary to understand much of modern technology.

23-1 Effective Current and Voltage

A direct current is described in terms of its direction and magnitude. In a simple circuit there might be a current of 6 A that flows from the positive terminal of a battery through a resistance network to its negative terminal. An alternating current has neither a constant direction nor a constant magnitude. How shall we speak of it quantitatively?

Since alternating current flows back and forth in a circuit, it has no "direction" in the same sense as a direct current has. However, the fluctuations have a certain frequency in each case, and the value of this frequency — that is, how many times per second the current goes through a complete cycle — forms part of the description of the current.

An alternating current varies with time in the manner shown in Fig. 23-1, and it would seem natural to specify its maximum value, i_{max}, as well as its frequency. The flaw in doing this is that i_{max} is not a measure of the ability of a current to do work or produce heat. A 6-amp direct current is not equivalent to an alternating current in which $i_{max} = 6$ amp. A better procedure is to define an *effective current* i_{eff} such that a direct current of this magnitude produces heat in a resistor at the same rate as the alternating current.

The variation of an alternating current with time obeys the formula

$$i = i_{max} \sin 2\pi ft, \tag{23-1}$$

where f is the frequency of the current and $i = 0$ and is increasing when $t = 0$. Figure 23-1 is a graph of this formula. The rate at which heat is dissipated in a resistance R by an alternating current is, at any time t, given by

$$i^2R = i_{max}^2R \sin^2 2\pi ft.$$

The average value of i^2R over a complete cycle is

$$[i^2R]_{av} = i_{eff}^2R = i_{max}^2R[\sin^2 2\pi ft]_{av}. \tag{23-2}$$

What we must find is the average value of $\sin^2 2\pi ft$ over a complete cycle. (The average value of i over a cycle is 0, since i is positive for half the cycle and negative for the other half. However, i^2 is always positive, and its average is a positive quantity. See Fig. 23-2.)

We begin with the trigonometric identity

$$\sin^2 \theta = \tfrac{1}{2}(1 - \cos 2\theta).$$

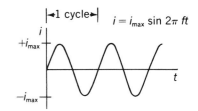

Fig. 23-1.   The variation of an alternating current with time. The frequency of the current is the number of cycles that occur per second.

Effective current

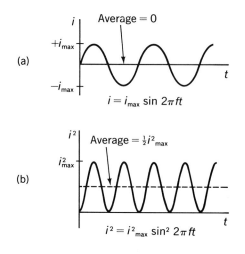

(a)

(b)

Fig. 23-2. (a) The average value of i in an ac-circuit is zero. (b) The average value of i^2 in an ac-circuit is $\frac{1}{2}i^2_{max}$.

The average value of $\sin^2 \theta$ over a complete cycle is therefore

$$[\sin^2 \theta]_{av} = \frac{1}{2}[1 - \cos 2\theta]_{av}$$
$$= \frac{1}{2} - \frac{1}{2}[\cos 2\theta]_{av}$$
$$= \frac{1}{2},$$

since, over a complete cycle, the average value of $\cos 2\theta$ is 0 by the same reasoning as in the case of $\sin \theta$. Hence we see that

$$i^2_{eff}R = i^2_{max}R \ [\sin^2 2\pi ft]_{av}$$
$$= \frac{1}{2} i^2_{max}R,$$

and

Effective current and voltage are 70.7% of maximum values

$$i_{eff} = \frac{i_{max}}{\sqrt{2}} = 0.707 i_{max}. \qquad \qquad \textit{Effective current} \quad (23\text{-}3)$$

The effective magnitude of an alternating current is 70.7% of its maximum value.

In a similar way the effective voltage in an ac-circuit turns out to be

$$V_{eff} = \frac{V_{max}}{\sqrt{2}} = 0.707 V_{max}. \qquad \qquad \textit{Effective voltage} \quad (23\text{-}4)$$

It is customary to express currents and voltages in ac-circuits in terms of their effective values. Thus the potential difference across a "120-volt, 60 cycle/s" power line actually varies from

$$+V_{max} = +\frac{V_{eff}}{0.707} = +\frac{120 \text{ volts}}{0.707} = +170 \text{ volts}$$

through 0 to -170 volts and back to $+170$ volts a total of 60 times per second.

In what follows, when current, potential difference, and emf values are given for an ac-circuit without other qualification, they will refer to the effective magnitudes of these quantities.

By analogy with circular motion, angular frequency ω (in radians/s) is often used instead of frequency f (in Hz) in discussing alternating currents, where

Angular frequency of alternating current

$$\omega = 2\pi f.$$

In this notation, which will not be used here, the instantaneous current in an ac-circuit is written

$$i = i_{max} \sin \omega t.$$

What are called here "effective" values of current and voltage are elsewhere sometimes called "root-mean-square" or "rms" values from their definitions as the square roots of the average values of i^2 and V^2.

23-2 Phase Relations

All actual circuits exhibit resistance, capacitance, and inductance to some degree. When direct current flows through a circuit, only its resistance is significant, but all three properties of the circuit affect the flow of alternating current.

We shall first consider a pure-resistance ac-circuit, an idealized circuit whose capacitance and inductance are negligible. The instantaneous values of the voltage and current are *in phase* at all times in such a circuit. This statement means that corresponding variations in V and i occur simultaneously: both V and i are 0 at the same time, both V and i pass through their maximum values at the same time, and so on (Fig. 23-3).

In a resistor, V is in phase with i

Now let us look at a pure-inductance ac-circuit, an idealized circuit whose resistance and capacitance are negligible. There is no iR potential drop across the inductance, and the potential difference across it is therefore proportional to $\Delta i/\Delta t$, the rate of change of the current (see Section 22-7). In

Fig. 23-3. Current and voltage are in phase in a pure resistance.

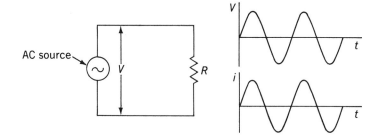

Fig. 23-4. The voltage across a pure inductance leads the current in the inductance by $\frac{1}{4}$ cycle.

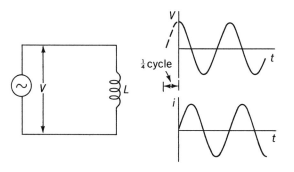

this situation V and i cannot be in phase with each other: i changes most rapidly when $i = 0$, so that $V = \pm V_{max}$ when $i = 0$, while $\Delta i/\Delta t = 0$ at $i = i_{max}$, so that $V = 0$ when $i = \pm i_{max}$. As shown in Fig. 23-4,

In an inductor, V leads i

The voltage across a pure inductor leads the current in the inductor by $\frac{1}{4}$ cycle.

That is, the variations in the voltage occur $\frac{1}{4}$ cycle *earlier than* the corresponding variations in the current.

Another way to express the phase relationship between i and V in an ac-inductive circuit is based on the sinusoidal variation of these quantities. On a graph both i and V appear as sine curves, and a sine curve repeats itself every 360°. It is therefore natural to think of a complete cycle as the equivalent of 360°, and so we see that a lead of $\frac{1}{4}$ cycle can be regarded as a lead of 90°.

The potential difference across a capacitor depends upon the amount of charge stored on its plates. The charge is a maximum at each moment that

$i = 0$, which is when the current is about to reverse direction and carry away the stored charge. The potential difference across a capacitor is

$$V = \frac{Q}{C},$$

and so $V = \pm V_{max}$ when $i = 0$. The stored charge is 0 at each moment that $i = \pm i_{max}$, because at these times the former stored charge is all gone and charge of the opposite sign is about to build up. Hence $V = 0$ when $i = \pm i_{max}$. As shown in Fig. 23-5,

The voltage across a pure capacitor lags behind the current into and out of the capacitor by $\frac{1}{4}$ cycle.

In a capacitor, V lags behind i

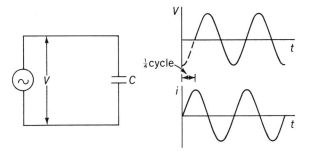

Fig. 23-5. The voltage across a pure capacitor lags behind the current into and out of the capacitor by $\frac{1}{4}$ cycle.

That is, the variations in the voltage occur $\frac{1}{4}$ cycle *later than* the corresponding variations in the current. In terms of angles, we can say that the voltage in a pure-capacitance circuit lags behind the current by 90°.

In a pure-capacitance circuit the phase relationship between current and voltage is different from what it is in pure-resistance and pure-inductance circuits. While alternating current does not flow *through* a capacitor, it does flow *into one plate and out of the other* since changes in the amount of charge stored on one of the plates are mirrored by changes in the charge stored on the other plate. A flow of $+Q$ into one plate means that $+Q$ flows out of the other plate to leave the latter with a net charge of $-Q$. If direct current were involved, eventually enough charge would accumulate to stop the arrival of any more, and the current would cease. In the case of an alternating current, however, the current always stops and reverses itself periodically anyway, so the presence of a series capacitor does not prevent ac from flowing in the circuit.

23-3 Inductive Reactance

Resistors, inductors, and capacitors all impede the flow of alternating current in a circuit. The effect of resistance is to dissipate part of the electrical energy that passes through it into heat. In those conductors in which Ohm's law is valid for direct current, it is valid for alternating current as well, and

$$i = \frac{V_R}{R}. \tag{23-5}$$

Here V_R represents the effective potential difference across the resistance R, and i is the effective current through it.

Origin of inductive reactance The opposition an inductor offers to the flow of alternating current arises from the self-induced back emf produced in it by the changing current. The back emf represents a potential drop across the inductor, and the current in the circuit is correspondingly reduced.

The *inductive reactance* X_L of an inductor is a measure of its effect on an alternating current passing through it. The effective current i in an inductor is related to the effective potential difference V_L across it and the inductive reactance X_L by

$$i = \frac{V_L}{X_L}. \tag{23-6}$$

The unit of inductive reactance is the ohm. Equation (23-6) is analogous to Ohm's law, but there is a basic distinction between reactance and resistance in that there is no power loss in an inductor while power is dissipated as heat in a resistor.

The inductive reactance of an inductor is given by the formula

$$X_L = 2\pi f L, \qquad\qquad\qquad \textit{Inductive reactance} \quad (23-7)$$

where f is the frequency of the current in Hz (cycles/sec) and L is the inductance in henries. The direct dependence of X_L on f and L is reasonable: the self-induced back emf, which is what opposes the current, is proportional to both $\Delta i/\Delta t$ and L, and so the more rapidly the current changes and the larger the value of L, the greater the back emf.

Problem. What is the current in a coil of negligible resistance and inductance 0.40 H when it is connected to a 120-V, 60-Hz power line?

Solution. The reactance of the coil is

$$X_L = 2\pi f L = 2\pi \times 60 \text{ Hz} \times 0.4 \text{ H} = 150 \ \Omega,$$

and the current in it is accordingly

$$i = \frac{V_L}{X_L} = \frac{120 \text{ V}}{150 \ \Omega} = 0.80 \text{ A}$$

Both the 120 V and 0.80 A figures represent effective values (Fig. 23-6).

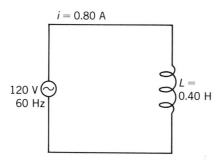

Fig. 23-6.

23-4 Capacitive Reactance

The extent to which a capacitor opposes the flow of alternating current depends upon its *capacitive reactance* X_C. If V_C is the effective potential difference across a capacitor whose reactance is X_C, the effective current into and out of the capacitor is

$$i = \frac{V_C}{X_C}. \tag{23-8}$$

The capacitive reactance of a capacitor is given by the formula

$$X_C = \frac{1}{2\pi f C}. \qquad\qquad \textit{Capacitive reactance} \quad (23\text{-}9)$$

If f is in Hz and the capacitance C in farads, the unit of X_C is the ohm.

Origin of capacitive reactance   A capacitor impedes the flow of alternating current by virtue of the reverse potential difference that appears across it as charge builds up on its plates. Thus there is a potential drop across a capacitor in an ac-circuit that affects the current just as the potential drop in a resistor does.

The inverse dependence of X_C upon f and C follows from the way in which a capacitor responds to alternating current. When the frequency is high, each cycle is brief, and less charge is deposited on the capacitor plates in each cycle; hence there is a smaller opposing potential difference, and correspondingly more current flows into and out of the capacitor. When the capacitance C is large, the opposing potential difference is less for a given amount of charge on the capacitor plates, and hence more current can flow in the circuit.

Problem.   A capacitor whose reactance is 80 Ω at 50 Hz is used in a 60 Hz circuit. What is its reactance in the latter circuit?

Solution.   The capacitance of the capacitor, from Eq. (23-9), is

$$C = \frac{1}{2\pi f X_C} = \frac{1}{2\pi \times 50 \text{ Hz} \times 80 \text{ } \Omega} = 4 \times 10^{-5} \text{ F},$$

which is 40μF. Its reactance at 60 Hz is

$$X_C = \frac{1}{2\pi f C} = \frac{1}{2\pi \times 60 \text{ Hz} \times 4 \times 10^{-5} \text{ F}} = 66 \text{ } \Omega.$$

In the limit of $f = 0$, which means direct current, $X_L = 0$ and $X_C = \infty$. When the current does not vary, there is no self-induced back emf in an inductor, and no inductive reactance to impede current. On the other hand, a capacitor completely obstructs direct current because the charge that builds up on its plates remains there instead of surging back and forth as it does when an ac-potential is applied.

23-5 Impedance

How to add instantaneous voltages   A series circuit that contains resistance, inductance, and capacitance can be represented as in Fig. 23-7, where each of these circuit properties is considered as lumped into a single resistor, inductor, and capacitor. If an ac-source of emf is connected to the circuit, at any instant the applied voltage V is equal to the sum of the voltage drops across the various circuit elements:

$$V = V_R + V_L + V_C. \qquad \qquad \textit{Instantaneous voltage} \quad (23\text{-}10)$$

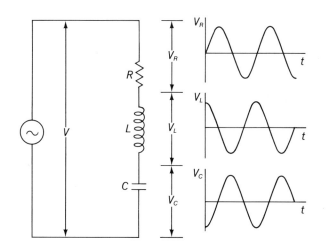

Fig. 23-7. The instantaneous value of the voltage V applied to a series RLC circuit is equal to the sum of the instantaneous values of V_R, V_L, and V_C. This relationship does *not* hold for the effective values of the various voltages because of the phase differences among them.

However, V_R, V_L, and V_C are *out of phase with one another.* The voltage across the resistor, V_R, is always in phase with the current i, but V_L is $\frac{1}{4}$ cycle ahead of i, and V_C is $\frac{1}{4}$ cycle behind i. This situation is shown in Fig. 23-7. While Eq. (23-10) is always correct when V, V_R, V_L, and V_C refer to the instantaneous values of the various voltages, it is *not* correct when effective (or maximum) values are involved, and we must use a vectorial approach to take the phase differences into account.

Figure 23-8 is a polar diagram in which the effective values of V_R, V_L, and V_C are plotted as vectors with the phase differences between them appearing as angles. By convention $\mathbf{V}_L$ is drawn in the $+y$-direction, $\mathbf{V}_R$ in the $+x$-direction, and $\mathbf{V}_C$ in the $-y$-direction. The vector sum $\mathbf{V}$ of $\mathbf{V}_R$, $\mathbf{V}_L$, and $\mathbf{V}_C$

How to add effective voltages

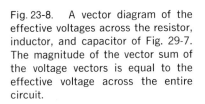

$$V = \sqrt{V_R^2 + (V_L - V_C)^2}$$

Fig. 23-8. A vector diagram of the effective voltages across the resistor, inductor, and capacitor of Fig. 29-7. The magnitude of the vector sum of the voltage vectors is equal to the effective voltage across the entire circuit.

represents the effective voltage across the terminals of the circuit, and its magnitude is

$$V = \sqrt{V_R^2 + (V_L - V_C)^2}. \qquad \textit{Effective voltage} \quad (23\text{-}11)$$

Phase angle The angle ϕ between **V** and $\mathbf{V}_R$ is called the *phase angle* because it is a measure of how much the voltage in the circuit leads or lags behind the current.

The above vector procedure is based upon the fact that the projection on an arbitrary axis of a uniformly rotating radius vector varies sinusoidally with time (this is discussed in the Special Topic for Ch. 12). We are therefore to imagine the diagrams of Fig. 23-8 as rotating counterclockwise f times per second, so that the horizontal component of **V** at any time is proportional to the instantaneous magnitude of the voltage across the resistor. The rotating vector **V** is sometimes called a *phasor* or a *rotor*.

In Fig. 23-8, V_L is greater than V_C, and $(V_L - V_C)$ is a positive quantity. If instead V_L is the smaller quantity, $(V_L - V_C)$ is negative and the vector **V** is below the *x*-axis (Fig. 23-9). However, Eq. (23-11) still applies, since $(V_L - V_C)^2 = (V_C - V_L)^2$.

Because

$$V_R = iR, \qquad V_L = iX_L, \qquad \text{and} \qquad V_C = iX_C,$$

we can rewrite Eq. (23-11) in the form

$$V = i\sqrt{R^2 + (X_L - X_C)^2}.$$

The quantity

$$Z = \sqrt{R^2 + (X_L - X_C)^2} \qquad \textit{Impedance} \quad (23\text{-}12)$$

Fig. 23-9. The procedure for finding **V** is the same whether V_L is greater than V_C or not.

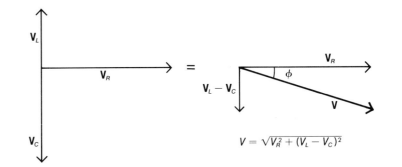

$$V = \sqrt{V_R^2 + (V_L - V_C)^2}$$

is known as the *impedance* of a series circuit containing resistance, inductance, and capacitance. The unit of impedance is evidently the ohm. Impedance in an ac-circuit plays the same role that resistance does in a dc-circuit, and in an ac-circuit

$$i = \frac{V}{Z} \qquad\qquad \text{Current in ac-circuit} \quad (23\text{-}13)$$

is the effective current that flows when the effective voltage V is applied. It is important to keep in mind that Z does not depend only upon the circuit parameters R, L, and C but also varies with the frequency f.

Because the current i is the same in all parts of the circuit at all times and

$$Z = \frac{V}{i}, \qquad\qquad R = \frac{V_R}{i},$$

$$X_L = \frac{V_L}{i}, \qquad \text{and} \qquad X_C = \frac{V_C}{i},$$

the vector voltage diagram of Fig. 23-8 can be replaced by the vector impedance diagram of Fig. 23-10. The phase angle ϕ is the same in both cases, of course, and can be calculated from the relation

$$\tan \phi = \frac{X_L - X_C}{R}. \qquad\qquad \text{Phase angle} \quad (23\text{-}14)$$

Problem. Analyze in detail a series circuit that consists of a 10-mH inductor, a 10-μF capacitor, and a 30-Ω resistor connected to a source of 100-V, 400-Hz alternating current.

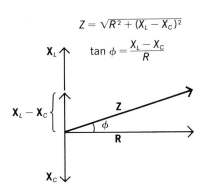

$$Z = \sqrt{R^2 + (X_L - X_C)^2}$$

$$\tan \phi = \frac{X_L - X_C}{R}$$

Fig. 23-10.   The vector impedance diagram that corresponds to the vector voltage diagram shown in Fig. 29-8.

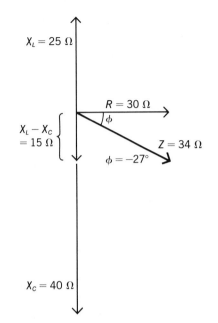

Fig. 23-11. A negative phase angle occurs when X_C is greater than X_L and signifies that the voltage in the circuit lags behind the current.

Solution. The reactances of the inductor and capacitor at 400 Hz are

$$X_L = 2\pi f L = 2\pi \times 400 \text{ Hz} \times 0.01 \text{ H} = 25 \text{ }\Omega,$$

$$X_C = \frac{1}{2\pi f C} = \frac{1}{2\pi \times 400 \text{ Hz} \times 10^{-5} \text{ F}} = 40 \text{ }\Omega.$$

The vector impedance diagram for this circuit is shown in Fig. 23-11. The impedance Z is

$$Z = \sqrt{R^2 + (X_L - X_C)^2} = \sqrt{(30 \text{ }\Omega)^2 + (40 \text{ }\Omega - 25 \text{ }\Omega)^2}$$
$$= 34 \text{ }\Omega.$$

The phase angle ϕ is found as follows:

$$\tan \phi = \frac{X_L - X_C}{R} = -\frac{15 \text{ }\Omega}{30 \text{ }\Omega} = -0.50,$$

$$\phi = -27°.$$

A negative phase angle signifies that the voltage lags behind the current. Here the lag is 27°, which is $\frac{27}{360}$ or 0.075 of a complete cycle.

The current in the circuit is

$$i = \frac{V}{Z} = \frac{100 \text{ V}}{34 \text{ }\Omega} = 2.94 \text{ A.}$$

The effective potential difference across each of the circuit elements is

$$V_L = iX_L = 2.94 \text{ A} \times 25 \text{ }\Omega = 74 \text{ V,}$$
$$V_C = iX_C = 2.94 \text{ A} \times 40 \text{ }\Omega = 118 \text{ V,}$$
$$V_R = iR = 2.94 \text{ A} \times 30 \text{ }\Omega = 88 \text{ V.}$$

The arithmetic sum of these potential differences is 270 V, but their vector sum, which takes into account the phase differences among them, is

$$V = \sqrt{V_R^2 + (V_L - V_C)^2} = \sqrt{(88 \text{ V})^2 + (74 \text{ V} - 118 \text{ V})^2}$$
$$= 100 \text{ V.}$$

Evidently the effective voltage across an inductor or capacitor in an ac-circuit can exceed the effective voltage applied to the entire circuit.

23-6 Resonance

When an ac-voltage is applied to a series circuit, the current that flows depends upon the frequency. The greatest current flows when the impedance Z is a minimum. Since

$$Z = \sqrt{R^2 + (X_L - X_C)^2},$$

the condition for minimum impedance in a given circuit is that the frequency be such that the inductive and capacitive reactances are equal. When this is true,

$$2\pi f_0 L = \frac{1}{2\pi f_0 C},$$

and

$$f_0 = \frac{1}{2\pi\sqrt{LC}}. \qquad\qquad \textit{Resonant frequency} \quad (23\text{-}15)$$

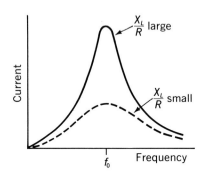

Fig. 23-12.   The variation of the current in a series ac-circuit with frequency. The maximum current occurs at the resonant frequency f_0.

Sharpness of tune depends upon X_L/R ratio

Current and voltage are in phase at resonance

When the impressed voltage has this frequency, the current is a maximum and is limited only by the resistance R. The frequency f_0 is called the *resonant frequency* of the circuit, and *resonance* occurs in a series circuit when the impressed voltage oscillates with the resonant frequency.

We note that Eq. (23-15) is the same as Eq. (22-24) for the "natural" frequency at which an LC-circuit will oscillate if the capacitor is initially charged. When a circuit is in resonance with an applied voltage, the energy alternately stored and discharged from the capacitor is precisely equal to the energy stored and discharged from the inductor. At other frequencies the energy contents of the capacitor and inductor are different, which interferes with the back-and-forth flow of power and gives rise to an impedance that exceeds the resistance R. The current that flows in an RLC-circuit at resonance is

$$i = \frac{V}{R},$$

just as in a dc-circuit.

Figure 23-12 shows how the current in a series circuit varies with frequency. The less the resistance R relative to the inductive reactance X_L, the sharper the peak in the curve. The antenna circuit of a radio receiver is tuned to respond to a particular frequency of radio waves by adjusting a variable capacitor or inductor until the resonant frequency of the circuit is equal to the signal frequency. A circuit in which X_L/R is large has a narrow response curve and can separate signals from two stations very close together in frequency.

At resonance, $X_L = X_C$, and there is no phase difference between current and voltage. Another way to specify the condition for resonance, then, is to require that current and voltage be in phase at all times, just as they are in a pure-resistance circuit.

Problem.   The antenna circuit of a radio receiver consists of a 10-mH coil and a variable capacitor; the resistance in the circuit is 50 Ω. An 880-kHz (kilohertz) radio wave produces a potential difference of 10^{-4} V across the circuit. Find the capacitance required for resonance and the current at resonance.

Solution.   The capacitance required for resonance at 880 kHz is

$$C = \frac{1}{(2\pi f)^2 L} = \frac{1}{(2\pi \times 8.8 \times 10^5 \text{ Hz})^2 \times 10^{-2} \text{ H}}$$

$$= 3.2 \times 10^{-12} \text{ F} = 3.2 \text{ pF}.$$

At resonance, inductive and capacitive reactances cancel each other out, and the current that flows is

$$i = \frac{V}{R} = \frac{10^{-4}\ \text{V}}{50\ \Omega} = 2 \times 10^{-6}\ \text{A}.$$

23-7 Power in AC-Circuits

No power is consumed in a pure inductor or capacitor in an ac-circuit, since these elements act merely as temporary reservoirs of energy and return whatever energy they absorb in one quarter of a cycle to the circuit in the next quarter of that cycle. No power is therefore needed to maintain an ac-current in the inductive and capacitive parts of a circuit. The resistance in the circuit, however, dissipates power as heat at the rate $P = iV_R$, where, as usual, i and V_R are effective values. From Fig. 23-8 we see that

$$V_R = V \cos \phi, \tag{23-16}$$

since $V \cos \phi$ is the component of $\mathbf{V}$ that is in phase with the current i. Hence the effective power absorbed in an ac-circuit is

$$P = iV \cos \phi. \qquad \textit{Power in ac-circuit} \quad (23\text{-}17)$$

The quantity $\cos \phi$ is called the *power factor* of the circuit. The power factor is equal to 1 only at resonance, when current and voltage are in phase; at resonance, $\phi = 0$ and $\cos \phi = 1$. Under other circumstances the voltage is not in phase with the current, and the actual power in the circuit is less than iV.

From Fig. 23-10 we see that

$$\cos \phi = \frac{R}{Z} = \frac{R}{\sqrt{R^2 + (X_L - X_C)^2}}. \qquad \textit{Power factor} \quad (23\text{-}18)$$

The power factor of a circuit is the ratio between its resistance and its impedance. Often power factors are expressed as percentages rather than as decimals or fractions. Thus a phase angle of 60° means a power factor of

$$\cos 60° = 0.50 = 50\%.$$

Instruments called wattmeters have been devised which respond directly to the effective product of V and i. An ac-wattmeter connected in a circuit gives a lower value for the power than the product of the effective values of V and i obtained from a separate voltmeter and ammeter in the same circuit (except at resonance), because the separate meters are not affected by the phase difference between current and voltage. Alternating-current generators, transformers, and power lines are usually rated in

The volt-ampere

volt·amperes, the product of effective voltage and current without regard to actual power, because higher values of V and/or i must be supplied to a circuit whose power factor is less than 1 than is reflected in its power consumption. It is convenient to think of the power factor as having the unit of watts/volt·ampere.

Alternating-current devices of various kinds, for example ac electric motors and fluorescent lamps, may have net inductive or capacitive reactances and consequently have power factors of less than 100%. A power factor of 70%, for instance, means that 1 kVA (kilovolt·ampere) of apparent power must be supplied for every 700 W of power actually consumed. This is an uneconomical situation because of the additional generator capacity required as well as because of the additional heat losses in the transmission lines as the unused power circulates between the generator and the device. The remedy is to introduce capacitors or inductors into the power line to increase the power factor to an acceptable figure.

23-8 Impedance Matching

In Section 19-7 we saw that the maximum power transfer between two dc-circuits occurs when their resistances are equal. The same conclusion applies to ac-circuits with the impedances of the circuits rather than their resistances as the quantities to be matched. Ac-circuits have the advantage that a transformer can be used to correct impedance mismatches. A common use of a transformer for this purpose occurs in an audio system where the amplifier circuit might have an impedance of several thousand ohms whereas the voice coil of the loudspeaker has an impedance of only a few ohms. Connecting the voice coil directly to the amplifier would be grossly inefficient.

Let us calculate the ratio of turns N_1/N_2 between the primary and secondary windings of a transformer needed to couple a circuit of impedance Z_1 to a circuit of impedance Z_2. If the voltages and currents in the circuits are respectively V_1, i_1 and V_2, i_2, then from Eq. (21-8)

$$\frac{V_1}{V_2} = \frac{N_1}{N_2} \quad \text{and} \quad \frac{i_2}{i_1} = \frac{N_1}{N_2}.$$

Since $Z_1 = V_1/i_1$ and $Z_2 = V_2/i_2$,

$$\frac{Z_1}{Z_2} = \frac{V_1 i_2}{V_2 i_1} = \left(\frac{N_1}{N_2}\right)^2$$

and the ratio of the turns is

$$\frac{N_1}{N_2} = \sqrt{\frac{Z_1}{Z_2}}.$$ *Impedance matching*   (23-19)

Thus if a loudspeaker whose voice coil has an impedance of 10 Ω is to be used with an amplifier whose load impedance is 9,000 Ω, a transformer should be used whose ratio of turns is

$$\frac{N_1}{N_2} = \sqrt{\frac{Z_1}{Z_2}} = \sqrt{\frac{9,000\ \Omega}{10\ \Omega}} = 30.$$

Special Topic

Parallel AC-Circuits

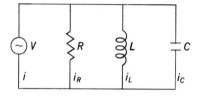

Fig. 23-13. A parallel *RLC* circuit. The potential differences across the circuit elements are the same.

When a resistor, an inductor, and a capacitor are connected in parallel, as in Fig. 23-13, the potential difference is the same across each circuit element:

$$V = V_R = V_L = V_C.$$

The total instantaneous current is the sum of the instantaneous currents in each branch, as in the case of a dc-parallel circuit, but this is not true of the total effective current because the branch currents are not in phase. The current i_R is always in phase with the voltage V, but i_C leads V by 90° and i_L lags behind V by 90°. A vector diagram of the situation is shown in Fig. 23-14. The magnitudes of the currents in the branches are

$$i_R = \frac{V}{R}, \qquad i_C = \frac{V}{X_C}, \qquad i_L = \frac{V}{X_L},$$

and their vector sum is

$$i = \sqrt{i_R{}^2 + (i_C - i_L)^2}.$$

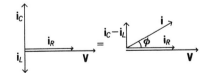

Fig. 23-14. A vector diagram of the effective currents in the resistor, inductor, and capacitor of Fig. 23-13. The magnitude i of the vector sum of the current vectors is equal to the effective current through the entire circuit.

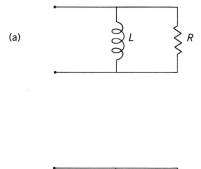

(a)

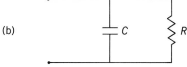

(b)

Fig. 23-15. (a) A high-pass filter. The higher the frequency, the more the reactance X_L, and the greater the proportion of the current that passes through the load. (b) A low-pass filter. The lower the frequency, the more the reactance X_C, and the greater the proportion of the current that passes through the load.

The phase angle ϕ, specified by

$$\tan \phi = \frac{i_C - i_L}{i_R},$$

is the angle between the current and the voltage. A positive phase angle means that the current leads the voltage, a negative one means that the current lags behind the voltage.

Parallel ac-circuits have various uses. For instance, an inductor connected in parallel across a resistive load, as in Fig. 23-15(a), acts as a "high-pass filter" to permit only high-frequency currents to reach the load when more than one frequency is present. Since $X_L = 2\pi fL$, the higher the frequency, the less the current through the inductor and the greater the current through the load R. Low-frequency currents are diverted through the inductor and little of them reach the load. Similarly, a capacitor in parallel across a resistive load, as in Fig. 23-15(b), serves as a "low-pass filter." Since $X_C = \frac{1}{2}\pi fC$, the lower the frequency, the less the current through the capacitor and the greater the current through R.

In a series RLC circuit, the impedance is a minimum at resonance and increases at higher and lower frequencies than f_0. At resonance, $X_L = X_C$, $Z = R$, and $i = V/R$. In a parallel RLC circuit, resonance again corresponds to $X_L = X_C$ and $i = V/R$, but now the impedance is a maximum at f_0 since at higher and lower frequencies some current can pass through the inductor and capacitor as well as through the resistor. Thus a series circuit can be used as a selector to favor a particular frequency, and a parallel circuit with the same L and C can be used as a selector to discriminate against the same frequency.

Important Terms

The **effective value** of an alternating current is such that a direct current of this magnitude produces heat in a resistor at the same rate as the alternating current.

The **phase relationships** between the instantaneous voltage and instantaneous current in ac-circuit components are as follows: the voltage across a pure resistor is in phase with the current; the voltage across a pure inductor leads the current by $\frac{1}{4}$ cycle; the voltage across a pure capacitor lags behind the current by $\frac{1}{4}$ cycle.

The **inductive reactance** X_L of an inductor is a measure of its effect on an alternating current. The **capacitive reactance** X_C of a capacitor is a measure of its effect on an alternating current. Both X_L and X_C vary with the frequency of the current.

The **impedance** Z of an ac-circuit is analogous to the resistance of a dc-circuit. The **resonant frequency** of an ac-circuit is that frequency for which the impedance is a minimum.

The **power factor** of an ac-circuit is the ratio between the power consumed in the circuit and the product of the effective current and voltage there; this ratio is equal to that between the resistance and the impedance of the circuit, and is less than 1, except at resonance. The unit of the apparent power $V_{eff}i_{eff}$ is the **volt·ampere,** as distinct from the watt, which is the unit of consumed power.

Impedance-matching transformer:

$$\frac{N_1}{N_2} = \sqrt{\frac{Z_1}{Z_2}}$$

Important Formulas

Effective current:

$$i_{eff} = \frac{i_{max}}{\sqrt{2}} = 0.707 \, i_{max}$$

Effective voltage:

$$V_{eff} = \frac{V_{max}}{\sqrt{2}} = 0.707 \, V_{max}$$

Inductive reactance:

$$X_L = 2\pi f L$$

Capacitive reactance:

$$X_C = \frac{1}{2\pi f C}$$

Impedance:

$$Z = \sqrt{R^2 + (X_L - X_C)^2}$$

Current in ac-circuit:

$$i = \frac{V}{Z}$$

Phase angle:

$$\tan \phi = \frac{X_L - X_C}{R}, \quad \cos \phi = \frac{R}{Z}$$

Resonant frequency:

$$f_0 = \frac{1}{2\pi \sqrt{LC}}$$

Power in ac-circuit:

$$P = iV \cos \phi$$

Multiple Choice

1. The effective voltage in an ac-circuit is equal to
   a. $0.5 \, V_{max}$.
   b. $0.707 \, V_{max}$.
   c. $V_{max}/0.707$.
   d. $i_{max}/0.707$.

2. The reactance of a capacitor is proportional to
   a. $\sqrt{f}$.   b. f.
   c. f^2.   d. $1/f$.

3. The reactance of an inductor is proportional to
   a. $\sqrt{f}$.   b. f.
   c. f^2.   d. $1/f$.

4. The unit of inductive reactance is the
   a. henry.   b. tesla.
   c. weber.   d. ohm.

5. In an ac-circuit, the voltage
   a. leads the current.
   b. lags the current.
   c. is in phase with the current.
   d. any of the above, depending on the circumstances.

6. When voltage and current are in phase in an ac-circuit, the
   a. impedance is 0.
   b. reactance is 0.
   c. resistance is 0.
   d. phase angle is 90°.

7. The current in an ac-circuit varies between
   a. 0 and i_{eff}.
   b. 0 and i_{max}.
   c. $-i_{eff}$ and $+i_{eff}$.
   d. $-i_{max}$ and $+i_{max}$.

8. The power dissipated in an ac-circuit depends on its

a. resistance.   b. reactance.

c. impedance.   d. phase angle.

9. A coil of inductance L has an inductive reactance of X_L in an ac-circuit in which the effective current is i. The coil is made from a superconducting material and has no resistance. The rate at which power is dissipated in the coil is

a. 0.   b. iX_L.

c. i^2X_L.   d. iX_L^2.

10. The power factor of a circuit is equal to

a. RZ.   b. R/Z.

c. X_L/Z.   d. X_C/Z.

11. At resonance, it is *not* true that

a. $R = Z$.   b. $X_L = 1/X_C$.

c. $P = iV$.   d. $i = V/R$.

12. The impedance of a circuit does not depend on

a. i.   b. f.

c. R.   d. C.

13. The power factor of a certain circuit in which the voltage lags behind the current is 80%. To increase the power factor to 100%, it is necessary to add to the circuit additional

a. resistance.

b. capacitance.

c. inductance.

d. impedance.

14. The capacitive reactance of a 5-μF capacitor in a 20-kHz circuit is

a. 0.63 Ω.   b. 1.6 Ω.

c. 5 Ω.   d. 16 Ω.

15. The inductive reactance of a 1-mH coil in a 5-Hz circuit is

a. 3.1 Ω.   b. 6.3 Ω.

c. 10 Ω.   d. 31 Ω.

16. In an ac-circuit $R = 10\ \Omega$, $X_L = 8\ \Omega$, and $X_C = 6\ \Omega$ when the frequency is f. The impedance at this frequency is

a. 10.2 Ω.   b. 12 Ω.

c. 24 Ω.   d. 104 Ω.

17. The phase angle in the circuit of Problem 16 is

a. 0.2°.   b. 2°.

c. 11°.   d. 45°.

18. The resonant frequency of the circuit of Problem 16 is

a. less than f.

b. equal to f.

c. more than f.

d. any of the above, depending on the applied voltage.

19. The power factor of a circuit in which $X_L = X_C$

a. is 0.

b. is 1.

c. depends on the ratio X_L/X_C.

d. depends on the value of R.

20. A voltmeter across an ac-circuit reads 50 V and an ammeter in series with the circuit reads 5 A. The power consumption of the circuit

a. is less than or equal to 250 W.

b. is exactly equal to 250 W.

c. is equal to or more than 250 W.

d. may be less than, equal to, or more than 250 W.

Exercises

1. What properties of an ac-circuit are described by its resistance, capacitive reactance, inductive reactance, and impedance? What are the similarities and differences among these quantities?

2. What is the reactance of a coil of inductance L in a dc-circuit? The reactance of a capacitor of capacitance C?

3. An alternating potential difference whose frequency is higher than the resonant frequency of a series RLC-circuit is applied to it. Does the voltage in the circuit lead or lag the current?

4. The frequency of the alternating potential difference applied to a series RLC-circuit is doubled. What happens to the resistance, the inductive reactance, and the capacitive reactance of the circuit? What further information is needed to establish what happens to the impedance of the circuit?

5. What is the significance of the power factor of a circuit? Is it independent of frequency? Under what circumstances (if any) is it zero? Under what circumstances (if any) is it 100%?

6. How many times per second does 400-Hz alternating current reverse its direction?

7. In an ac-circuit, how many times per second does $i^2 = i^2_{\text{eff}}$?

8. An ammeter in series with an ac-circuit reads 4 A and a voltmeter across the circuit reads 85 V. (a) What is the maximum current in the circuit? (b) What is the maximum potential difference across the circuit? (c) What additional information is required in order to determine the power consumed in the circuit?

9. An ammeter in series with an ac-circuit reads 10 A and a voltmeter across the circuit reads 60 V. (a) What is the maximum current in the circuit? (b) What is the maximum potential difference across the circuit? (c) Does the maximum current necessarily occur at the same moments as the maximum voltage?

10. What is the reactance of a 5-μF capacitor at 10 Hz? At 10 kHz?

11. The reactance of a capacitor is 50 Ω at 200 Hz. What is its capacitance?

12. What is the reactance of a 5-mH inductor at 10 Hz? At 10 kHz?

13. The reactance of an inductor is 1000 Ω at 200 Hz. What is its inductance?

14. The current in a resistor is 2 A when it is connected across a 240-V, 50-Hz line. How much capacitance should be connected in series with the resistor to reduce the current to 1 A?

15. How much inductance should be connected in series with the resistor of Example 14 to reduce the current to 1 A?

16. A 3.0-H inductor of negligible resistance is connected to a 40-V, 50-Hz power source. Find (a) the reactance of the inductor, and (b) the current that flows in it.

17. A current of 0.20 A flows through a 0.15-H inductor of negligible resistance that is connected to an 80-V source of ac. What is the frequency of the source?

18. A current of 5.5 A flows through a 0.3-H inductor of negligible resistance that is connected to a source of 120-Hz ac. What is the potential difference across the source?

19. A 5-μF capacitor is connected to a 12-V, 600-Hz power source. Find (a) the reactance of the capacitor, and (b) the current that flows.

20. A 60-mA current flows into and out of a 12-μF capacitor connected to a source of 200-Hz ac. Find the potential difference across the source.

21. A 5-μF capacitor is connected in series with a 300-Ω resistor and a 120-V, 50-Hz potential difference is applied. Find the current in the circuit and the power dissipated.

22. A 0.1-H, 30-Ω inductor is connected to a 50-V, 100-Hz power source. Find the current in the inductor and the power dissipated in it.

23. A pure capacitor, a pure inductor, and a pure resistor are connected in series across a 60-V ac-power source. The potential difference across the capacitor is 60 V, and that across the inductor is also 60 V. What is the potential difference across the resistor?

24. A pure capacitor, a pure inductor, and a pure resistor are connected in series across an ac-power source. A voltmeter placed in turn across each circuit element reads 7 V, 12 V, and 12 V respectively. What is the potential difference of the source?

25. A series circuit has a resistance of 20 Ω, and inductive reactance of 20 Ω, and a capacitive reactance of 20 Ω when connected to an ac-source. Find (a) the impedance of the circuit, (b) the phase angle, and (c) the potential difference required for a current of 20 A to flow.

26. A series circuit has a resistance of 40 Ω, an inductive reactance of 25 Ω, and a capacitive reactance of 60 Ω when connected to a certain ac-

source. Find (a) the impedance of the circuit, (b) the phase angle, and (c) the potential difference required for a current of 1 A to flow.

27. A 10-kW electric motor has an inductive power factor of 70%. What minimum rating in kVA must the power line have? A capacitor is connected in series with the motor to increase the power factor to 100%. How does this affect the required rating of the power line?

Problems

1. The potential difference across a source of alternating current varies sinusoidally with time with $V_{max} = 100$ V. Find the value of the instantaneous potential difference (a) $\frac{1}{8}$ cycle, (b) $\frac{1}{4}$ cycle, (c) $\frac{3}{8}$ cycle, and (d) $\frac{1}{2}$ cycle after $V = 0$.

2. A certain 50-Hz ac-power source has a maximum potential difference of 60 V. Find the value of the instantaneous potential difference (a) 0.0025 s, (b) 0.005 s, (c) 0.010 s, (d) 0.0175 s, and (e) 0.020 s after $V = 0$.

3. A circuit that contains inductance and resistance has an impedance of 50 Ω at 100 Hz and an impedance of 100 Ω at 500 Hz. What are the values of the inductance and the resistance of the circuit?

4. A circuit that contains capacitance and resistance has an impedance of 20 Ω at 100 Hz and an impedance of 5 Ω at 500 Hz. What are the values of the capacitance and the resistance of the circuit?

5. An inductor of negligible resistance whose reactance is 120 Ω at 200 Hz is connected to a 240-V, 60-Hz power line. What is the current in the inductor?

6. A capacitor of unknown capacitance is found to have a reactance of 120 Ω at 200 Hz. The capacitor is connected to a 240-V, 60-Hz power line. (a) What is the current in the circuit? (b) Why is this current not the same as that of Problem 5?

7. The voltage leads the current in a certain ac-circuit by 30°. The effective current in the circuit is 5 A. (a) Is the capacitive reactance greater

than or less than the inductive reactance? (b) What is the value of i when $V = 0$?

8. A capacitor of unknown capacitance is connected in series with a 64-Ω resistor. The combination is found to draw 1.6 A when connected to a 240-V, 240-Hz power source. Find (a) the capacitance of the capacitor, (b) the power dissipated in the capacitor, and (c) the power dissipated in the resistor.

9. A coil of unknown inductance and resistance is observed to draw 8 A when a 40-V dc-potential difference is applied, and 5 A when a 40-V, 60 Hz ac potential difference is applied. (a) Find the inductance and resistance of the coil. (b) Find the power dissipated in the coil in each situation.

10. A coil of unknown inductance and resistance is observed to draw 42 mA when a dc-potential difference of 6 V is applied. When the coil is connected to a source of 400-Hz ac, however, a potential difference of 10 V is required to yield the same current. (a) Find the inductance and resistance of the coil. (b) Find the power dissipated in the coil in each situation.

11. A 60-μF capacitor, a 0.3-H inductor, and a 50-Ω resistor are connected in series with a 120-V, 60-Hz power line. Find (a) the impedance of the circuit, (b) the current that flows in it, (c) the power factor, (d) the power dissipated, and (e) the minimum volt·amp rating of the power line.

12. A 5-μF capacitor, a 40-mH inductor, and a 12-Ω resistor are connected in series with a 20-V, 400-Hz power line. Find (a) the impedance of the circuit, (b) the current that flows in it, (c) the power dissipated, and (d) the minimum volt·amp rating of the power line.

13. (a) Find the resonant frequency of the circuit of problem 11. (b) What current will flow if the circuit is connected to a 120-V power line whose frequency is equal to the resonant frequency?

14. (a) Find the resonant frequency of the circuit of problem 12. (b) What current will flow if the circuit is connected to a 20-V power line whose frequency is equal to the resonant frequency?

15. A 10-μF capacitor, a 0.10-H inductor, and a 60-Ω resistor are connected in series across a 120-V, 60-Hz power line. Find (a) the current in the circuit, (b) the power dissipated in it, and (c) the potential difference across each of the circuit elements.

16. The circuit of problem 15 is connected across a 120-V, 25-Hz power line. Answer the same questions for this situation.

17. An inductor dissipates 75 W of power when it draws 1.0 A from a 120-V, 60-Hz power line. (a) What is its power factor? (b) What capacitance should be connected in series with it to increase its power factor to 100%? (c) How much current would the circuit then draw? (d) How much power would it then dissipate? (e) What should the minimum volt·amp rating of the power line be then?

18. A circuit consisting of a capacitor in series with a resistor draws 3.6 A from a 50-V, 100-Hz power line. The circuit dissipates 120 W. (a) What is the precise phase relationship between voltage and current in the circuit? (b) What series inductance should be inserted in the circuit if the current and voltage are to be in phase? (c) How much current would the circuit then draw? (d) How much power would it then dissipate?

Answers to Multiple Choice

1. b	8. a	15. d
2. d	9. a	16. a
3. b	10. b	17. c
4. d	11. b	18. a
5. d	12. a	19. b
6. b	13. c	20. a
7. d	14. b	

24

Light

Among the most noteworthy achievements of nineteenth-century science was the realization that light consists of electromagnetic waves. Electromagnetic waves themselves were hypothesized by James Clerk Maxwell in 1864 on the basis of his theory of electric and magnetic fields. The speed Maxwell calculated for these waves turned out to be the same as the speed of light. He then concluded that, since both were transverse waves, they were the same phenomenon, a conclusion that has since been verified in every detail. Although every aspect of the wave behavior of light can be determined from Maxwell's electromagnetic theory of light, such calculations are quite complicated. A less comprehensive but simpler and more intuitive approach to optical phenomena was devised in 1678 by Christian Huygens; it can be applied to all kinds of waves, not just light waves. In this chapter the reflection and refraction of light at plane surfaces will be considered with the help of Huygens' principle. We should note in passing that in certain important respects light has the character of a stream of particles rather than that of a series of waves. This duality is examined in later chapters.

24-1 Maxwell's Hypothesis

In electromagnetic induction, a changing magnetic field induces an emf in a nearby wire loop or other conducting path. Thus a changing magnetic field is equivalent in its effects to an electric field. The converse is also true: a changing electric field is equivalent in its effects to a magnetic field. This is true even in empty space, where electric currents cannot flow. No simple experiment can directly demonstrate the latter equivalence (unlike electromagnetic induction, which is very easy to exhibit), and it was first proposed by Maxwell on the basis of an indirect argument. Electromagnetic waves occur as a consequence of these two effects—a changing magnetic field produces an electric field, and a changing electric field produces a magnetic field. The two constantly-varying fields are coupled together as they travel through space.

When Maxwell expressed mathematically what was known in his time about electricity and magnetism, he found that the resulting equations became particularly symmetrical if he assumed that a magnetic field arises from a varying electric field as well as from an actual electric current. This assumption is actually less arbitrary than it may seem. Suppose we attach two metal rods to the opposite terminals of a battery, as in Fig. 24-1. At first currents flow in the wires, and electric charge of opposite sign accumulates on the rods. Ultimately enough charge is present for the potential difference between the rods to equal the emf of the battery, and the currents cease.

While charge is building up on the rods, the electric field between them changes accordingly. The greater the current, the more rapid the change of the field; the less the current, the slower the change of the field. The behavior of the current is thus mirrored in the *rate of change* of the electric field.

When he assumed that a varying electric field produces a magnetic field, then, Maxwell was in effect extending the above correspondence between an electric current and a varying electric field to include magnetic effects. That such an extension also simplified the equations of electricity and magnetism encouraged Maxwell to take this notion seriously, because fundamental physical principles usually (though not always) can be expressed in simple form.

It is not surprising that electromagnetic induction became known many years before its converse was suspected. Even a slight electric field causes a current to flow in a conductor, and, if the resistance of the conductor is low enough, the current may be sufficiently large to detect despite the feebleness of the field itself. However, electric current has no magnetic counterpart because single magnetic poles do not exist; the opposite poles of a magnet

A changing electric field is equivalent to a magnetic field

Weak electric fields can be detected

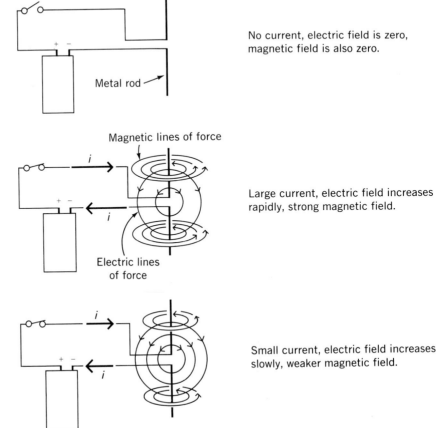

No current, electric field is zero, magnetic field is also zero.

Large current, electric field increases rapidly, strong magnetic field.

Small current, electric field increases slowly, weaker magnetic field.

Fig. 24-1.

Weak magnetic fields are difficult to detect

cannot be separated from one another the way opposite electric charges can. Unlike weak electric fields, weak magnetic fields are therefore hard to measure no matter where they occur; and those due to changing electric fields are seldom strong. Maxwell's notion that an electric field that varies with time gives rise to a magnetic field accordingly did not originate in an observation, but instead developed from an intuitive feeling for order in the natural world.

One of the tenets of the scientific method of inquiry is that all hypotheses must be capable of being experimentally verified, directly or indirectly, if they are to mean anything at all. Maxwell's next step was to seek a phenomenon that was a unique consequence of his hypothesis.

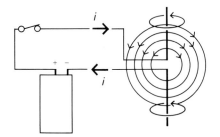

Very small current, electric field increases very slowly, very weak magnetic field.

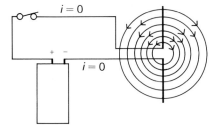

No current, electric field is constant, no magnetic field.

Fig. 24-1. Continued.

24-2 Electromagnetic Waves

It follows from electromagnetic induction that whenever there is a change in a magnetic field, an electric field is produced, and it follows from Maxwell's hypothesis that whenever there is a change in an electric field, a magnetic field is produced. Evidently it is impossible to have either effect occur alone. The electric field that arises from a change in a magnetic field is in itself a change in the pre-existing electric field (which might have had any original value, including zero), and therefore causes another magnetic field. The latter magnetic field, too, represents a change, and from the change an electric field is in turn produced. The process continues indefinitely, with a definite coupling between the fluctuating electric and magnetic fields. On the basis of his hypothesis, together with the other principles of electricity and magnetism, Maxwell was able to develop a detailed picture of how these field fluctuations travel through space.

Em waves consist of coupled electric and magnetic fields

The first idea that emerged from Maxwell's analysis was that the field fluctuations spread out in space from an initial disturbance in the same manner that waves spread out from a disturbance in a body of water: hence the name *electromagnetic waves* to describe them. If we throw a stone into a pond or otherwise alter the state of the water surface at some point, oscillations occur in which energy is continually interchanged between the kinetic energy of moving water and the potential energy of water higher than its normal level. These oscillations begin where the stone lands, and spread out as waves across the surface of the pond. The wave speed depends upon the properties of the pond water, varying with temperature, impurity content, and so on, but it is independent of the wave amplitude. As we saw in Chapter 13, this is typical wave behavior. When electromagnetic waves spread out from an electric or magnetic disturbance, their energy is constantly being interchanged between the fluctuating electric field and the fluctuating magnetic field of the waves.

Formation of an em wave

Let us connect the pair of metal rods of Fig. 24-1 to a source of alternating emf, as in Fig. 24-2. (Such a source is called an oscillator, as we know.) For clarity we will imagine that there is only a single charge in each rod at any time.

(a) When the oscillator is switched on, a positive charge in the upper rod begins to move upward and a negative charge in the lower rod begins to move downward. The electric lines of force around the charges are indicated by the black lines, and the magnetic lines of force due to the motion of the charges (which are concentric circles perpendicular to the paper) are indicated by crosses when their direction is into the paper and by dots when their direction is out of the paper. (The dots represent arrowheads and the crosses represent the tail feathers of arrows.)

(b) The charges have reached the limit of their motion and have stopped, so that they cease to produce a magnetic field. The outer magnetic lines of force do not disappear instantly because of the finite speed at which changes in electric and magnetic fields travel.

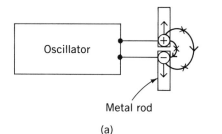

Fig. 24-2.   A pair of metal rods connected to an electrical oscillator emit electromagnetic waves.

Oscillator

Metal rod

(a)

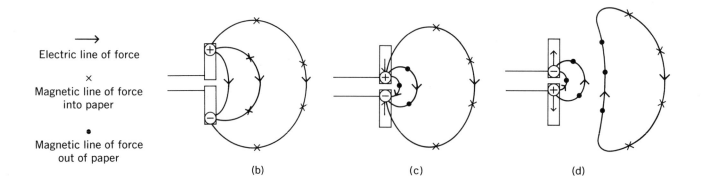

Electric line of force

×

Magnetic line of force into paper

•

Magnetic line of force out of paper

(b)   (c)   (d)

(c) The emf of the oscillator now begins to decrease, and the charges move toward each other. The result is a magnetic field in the opposite direction to the earlier field. The electric field is in the same direction as before.

(d) The emf has passed through 0 and begun to increase in the opposite sense. Consequently there is a negative charge in the upper rod and a positive one in the lower rod which begin to move apart. The electric field is therefore opposite in direction to the earlier field, but the magnetic field is in the same direction since magnetically a positive charge moving downward is equivalent to a negative charge moving upward.

Owing to this sequence of changes in the fields, the outermost electric and magnetic lines of force respectively form into closed loops. These loops of force, which lie in perpendicular planes, are divorced from the oscillating charges that gave rise to them and continue moving outward, constituting an electromagnetic wave. As the charges continue oscillating back and forth, further associated loops of electric and magnetic lines of force are emitted, forming an expanding pattern of loops.

Figure 24-3 shows the configuration of the electric and magnetic fields that spread outward from a pair of oscillating charges. The actual fields are present in three dimensions, so that the magnetic lines of force form loops in planes perpendicular to the line joining the charges.

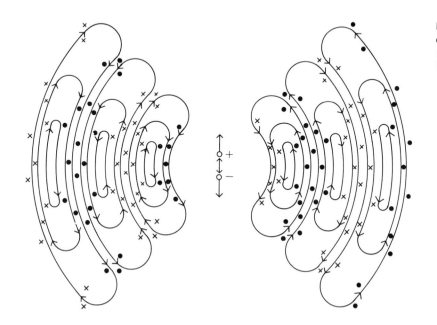

Fig. 24-3. The configuration of the electric and magnetic fields that spread outward from a pair of oscillating charges.

Properties of em waves

Three properties of electromagnetic waves are worth noting:

(1) The variations occur simultaneously in both fields (except close to the oscillating charges), so that the electric and magnetic fields have maxima and minima at the same times and in the same places.

(2) The directions of the electric and magnetic fields are perpendicular to each other and to the direction in which the waves are moving. Light waves are therefore transverse.

(3) The speed of the waves depends only upon the electric and magnetic properties of the medium they travel in, and not upon the amplitudes of the field variations.

Figure 24-4 is an attempt at portraying (1) and (2) of the above in terms of lines of force of **E** and **B** a long distance from a source of electromagnetic waves. Closer to the source the lines of force are curved, as in Fig. 24-3. It is worth keeping in mind that, unlike the other types of waves considered in Chapter 13—waves in a stretched string, water waves, sound waves—nothing material moves in the path of an electromagnetic wave. The only changes are in electric and magnetic fields.

In the preceding discussion we considered a special kind of electromagnetic wave source. Actually, *all* accelerated charges radiate electromagnetic waves, regardless of the manner in which the acceleration occurs.

The **E** and **B** fields of an electromagnetic wave have the important property that their magnitudes fall off with distance r as $1/r$. This is in contrast to the **E** and **B** fields around a charge whose velocity is constant, which fall off as $1/r^2$, a much more rapid variation with distance. A charge is

All accelerated charges emit em waves

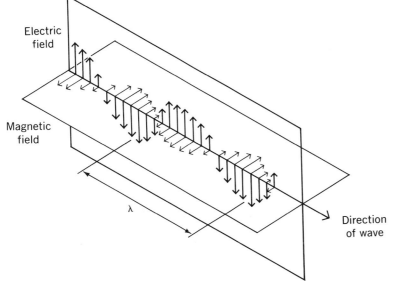

Electric field

Magnetic field

λ

Direction of wave

Fig. 24-4. The electric and magnetic fields in an electromagnetic wave vary simultaneously. The field directions are perpendicular to each other and to the direction of propagation.

severely limited in the range of its interaction with other charges by the inverse-square decrease in its normal **E** and **B** fields, but no such limitation applies to the electromagnetic waves produced when it is accelerated. For this reason electromagnetic signals may be detected for quite remarkable distances from their sources—light and radio waves reach us from galaxies at the outer limit of the universe—even though static electric and magnetic fields produced by these sources would be imperceptible not very far away.

Maxwell's theory of electromagnetic waves showed that their velocity c in vacuum depends solely upon ϵ_0 and μ_0, the permittivity and permeability of free space. Maxwell found that the velocity c is given by

Velocity of light

$$c = \frac{1}{\sqrt{\epsilon_0 \mu_0}} \qquad\qquad \textit{Velocity of light} \quad (24\text{-}1)$$

$$= \frac{1}{\sqrt{8.85 \times 10^{-12} \text{ C}^2/\text{N·m}^2 \times 1.26 \times 10^{-6} \text{ T-m/A}}}$$

$$= 3.00 \times 10^8 \text{ m/s},$$

which is the same velocity that had been experimentally measured for light waves in free space! The correspondence was too great to be accidental, and, as further evidence became known, the electromagnetic nature of light found universal acceptance. To five significant figures the value of c is 2.9979×10^8 m/s.

24-3 Types of Electromagnetic Waves

Light is not the only example of an electromagnetic wave. While all electromagnetic waves share certain basic properties, other features of their behavior depend upon their frequencies. Light waves themselves span a brief frequency interval, from about 4.2×10^{14} Hz for red light to 7.9×10^{14} Hz for violet light. Electromagnetic waves with frequencies between these limits are the only ones that the eye responds to, and specialized instruments of various kinds are required to detect waves with higher and lower frequencies. Figure 24-5 shows the electromagnetic wave *spectrum* from the low frequencies used in radio communication to the high frequencies found in x-rays and gamma rays (which are considered in later chapters). The wavelengths corresponding to the various frequencies are also shown.

There are many kinds of em waves but all have same basic nature

As we recall, the product of the frequency f of a wave and its wavelength λ is just the wave speed, here c. Hence, given the wavelength or frequency of a particular electromagnetic wave, we can immediately find the complementary quantity.

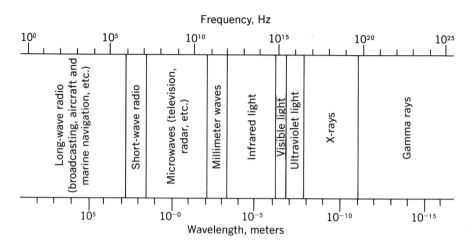

Fig. 24-5.   The electromagnetic wave spectrum. The boundaries of the various categories are not sharp except in the case of visible light.

Problem.   Find the wavelengths of yellow light whose frequency is 5×10^{14} Hz and of radio waves whose frequency is 1 mHz. (1 mHz = 1 megahertz = 10^6 Hz.)

Solution.   The wavelength of the yellow light is

$$\lambda = \frac{c}{f} = \frac{3 \times 10^8 \text{ m/s}}{5 \times 10^{14} \text{ Hz}} = 6 \times 10^{-7} \text{ m} = 600 \text{ nm},$$

which is less than 1/1000 of a millimeter. (1 nm = 1 nanometer = 10^{-9} m. The nanometer is often used for expressing the wavelengths found in light.) By the same procedure we find that a 1-mHz radio wave has a wavelength of 300 m.

Electromagnetic waves provide a means for transmitting information from one place to another without wires or other material links between them. There are two principal methods of incorporating information in an
Amplitude modulation   electromagnetic wave, *amplitude modulation* and *frequency modulation*.

In amplitude modulation, a "carrier wave" of constant frequency is varied in amplitude, with the variations constituting the signal. The simplest example is a flashlight switched on and off to give a series of dots and dashes. Another example is the radiotelegraph. Coded sequences of dots and dashes are used to represent letters of the alphabet, numbers, or other specific information; the Morse code is an example (Fig. 24-6).

A carrier wave can also be modulated in such a way that the amplitude variations correspond to sound waves, as shown schematically in Fig. 24-7.

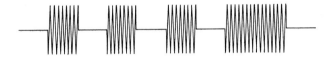

Fig. 24-6. In radiotelegraphy, a constant-frequency electromagnetic wave is switched on and off in a coded sequence to transmit information. The letter "v," which is ···– in Morse code, is shown.

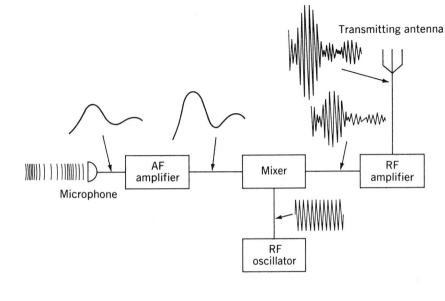

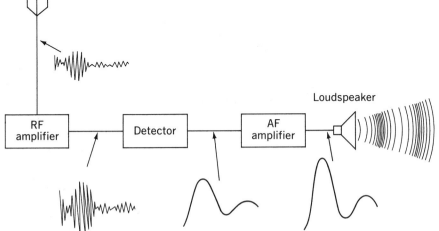

AF = Audio frequency
RF = Radio frequency

Fig. 24-7. In amplitude modulation, the amplitude of a constant-frequency carrier wave is varied in accordance with an audio signal. The transmitting and receiving systems shown are highly simplified.

Fig. 24-8.   The information content of a frequency-modulated wave resides
in its frequency variations rather than in its amplitude variations.

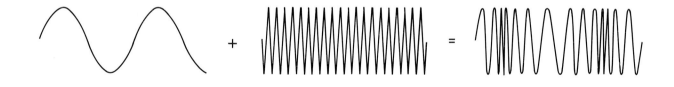

Frequency modulation

A microphone converts sound waves to an equivalent electrical signal, which is then amplified and combined with the carrier. The final wave is broadcast from an antenna, and at the receiving station the carrier is removed to leave an audio signal. The latter is then amplified and used to generate sound waves in a loudspeaker.

One difficulty with amplitude modulation is that such sources of random electromagnetic waves as electrical storms and electrical machinery can interfere with the broadcast waves to cause "static." In frequency modulation the frequency, not the amplitude, of the carrier is varied in accordance with the audio signal (Fig. 24-8). The information content of a frequency-modulated electromagnetic wave is virtually immune to disturbance.

In television the desired scene is focussed on the light-sensitive screen of a special tube which is then scanned in a zig-zag fashion by an electron beam. The signal extracted from the screen varies with the image brightness at each successive point that is scanned, and this signal is then used to modulate a carrier wave for broadcasting. At the receiver, suitable circuits demodulate the wave, and in the picture tube the image is reproduced with the help of an electron beam that moves in synchronism with the electron beam in the camera tube.

24-4 Huygens' Principle

Wavefront

Huygens' method of analysis concerns *wavefronts*. A wavefront is an imaginary surface that joins points where all of the waves involved are in the same phase of oscillation. As in Fig. 24-9, waves from a point source spread out in a succession of spherical wavefronts. (In the case of water waves, the wavefronts that result when a stone is dropped in a lake are circular.) At a long distance from a point source, the curvature of the wavefronts is so small that they can be considered as a succession of planes.

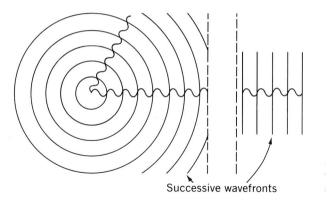

Successive wavefronts

Fig. 24-9. The spherical wave fronts of a point source become plane wave fronts at a long distance from the source.

Huygens' principle states that

Every point on a wavefront can be considered as a point source of secondary wavelets which spread out in all directions with the wave speed of the medium. The wavefront at any time is the envelope of these wavelets.

Huygens' principle

Figure 24-10 shows how Huygens' principle is applied to the propagation of plane and spherical wavefronts in a uniform medium. It does not matter just

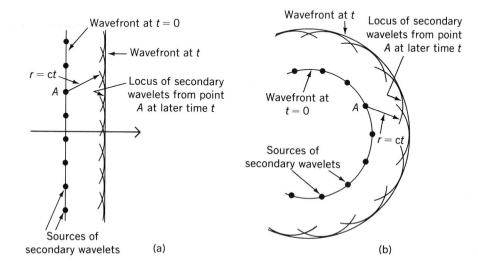

Fig. 24-10. Huygens' principle applied to the propagation of (a) plane and (b) spherical wavefronts.

where on the initial wavefront we imagine the sources of the secondary wavelets to be.

Despite its name, Huygens' principle is *not* in the same category as such fundamental principles as those of conservation of mass energy, momentum, and electric charge, but is rather a convenient means for studying wave motion in a geometrical manner. We shall find Huygens' principle useful in understanding a variety of optical phenomena.

Light rays

While we shall use the notion of wavefronts for the actual analysis of wave propagation, the results are commonly represented in terms of *rays*. A light ray is simply an imaginary line in the direction in which the wavefronts advance, and so it is perpendicular to the wavefronts. The picture most of us have of a light ray is a narrow pencil of light, which is perfectly legitimate. However, an approach based exclusively on rays does not reveal such characteristic wave behavior as diffraction (the bending of waves around an obstacle into the "shadow" region); hence we must remember that the motion of wavefronts is what is really significant, although it is both proper and convenient to use rays to summarize our conclusions.

24-5 Reflection

All real objects reflect a certain proportion of the light falling upon them, and it is this reflected light that enables us to see them. In most cases the surface of the object has irregularities that spread out an initially parallel beam of light in all directions to produce *diffuse reflection* (Fig. 24-11a). A surface so smooth that any irregularities in it are small relative to the wavelength of the light falling upon it behaves differently; when a parallel beam of light is directed at such a surface, it is *specularly reflected* in only one direction (Fig. 24-11b). Reflection from a brick wall is diffuse, while reflection from a mirror is specular. Often a mixture of both kinds of reflection occurs,

Diffuse reflection

Specular reflection

Fig. 24-11.

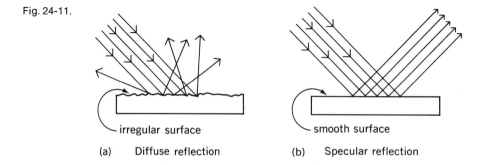

(a)   Diffuse reflection   (b)   Specular reflection

as in the case of a surface coated with varnish or glossy enamel. Our concern here is with specular reflection only.

When we see an object, what enters our eyes are light waves reflected from its surface. What we perceive as the object's color therefore depends upon two things, the kind of light falling on it and the nature of its surface. If white light is used to illuminate an object that absorbs all colors other than red, the object will appear red. If green light is used instead, the object will appear black because it absorbs green light. A white object reflects light of all wavelengths equally well, and its apparent color depends entirely upon the color of the light reaching it. A black object, on the other hand, absorbs light of all wavelengths, and it appears black no matter what color light reaches it.

Let us examine the specular reflection of a series of plane wavefronts, which corresponds to a parallel beam of light. At $t = 0$ in Fig. 24-12 the wavefront AB just touches the mirror, and secondary wavelets start to spread out from A. After a time t the end B of the wavefront reaches the mirror at C. The secondary wavelets from A are now at the point D. The wavefront which was AB at $t = 0$ is therefore CD at the later time t.

The angle i between an approaching wavefront and the reflecting surface is called the *angle of incidence*, and the angle r between a receding wavefront and the reflecting surface is called the *angle of reflection*. Here

Angles of incidence and reflection

$$\sin i = \frac{BC}{AC} \qquad \text{and} \qquad \sin r = \frac{AD}{AC}.$$

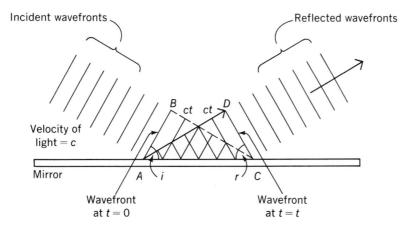

Fig. 24-12. The behavior of successive wave fronts during reflection from a plane mirror. The angle of incidence i and the angle of reflection r are equal.

The wavelet that originates at A takes the time t to reach D. Hence $AD = ct$, where c is the velocity of light. The point B of the wavefront AB also takes the time t to reach C, and so $BC = ct$ as well. Therefore $AD = BC$ and $\sin i = \sin r$, from which we conclude that the angles i and r are equal. Thus we have the basic law of reflection:

Law of reflection

The angle of reflection of a plane wavefront with a plane mirror is equal to the angle of incidence.

In the ray model of light propagation, the angles of incidence and reflection are measured with respect to the *normal* to the reflecting surface at the point where the light strikes it (Fig. 24-13). The normal is a line drawn perpendicular to the surface at that point. In this representation, too, the angles of incidence and reflection are equal. The incident ray, the reflected ray, and the normal all lie in the same plane.

The normal to a surface is perpendicular to it

The image of an object we see in a plane mirror appears to be the same size as the object and as far behind the mirror as the object is in front of it. Let us try to figure out how this situation arises.

In Fig. 24-14, three typical light rays from a point object at A impinge on a mirror and are reflected, in each case with an angle of reflection equal to the angle of incidence. To the eye, the three diverging rays apparently come from the point A' behind the mirror. The rays that seem to come from the image do not actually pass through it, and for this reason the image is said to be *virtual*. (A *real image* is formed by light rays that pass through it.) From the geometry of the figure it is clear that $AM = A'M$, so that the object and the image are the same distance on either side of the mirror.

Real and virtual images

In Fig. 24-15, an object of finite size is being reflected in a plane mirror. Again simple geometry shows that the object and its virtual image have the same dimensions. Every point in the image is directly behind the corre-

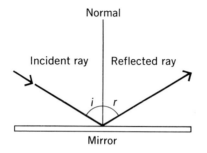

Fig. 24-13. The ray representation of reflection. The angles of incidence and reflection are measured with respect to the normal to the reflecting surface.

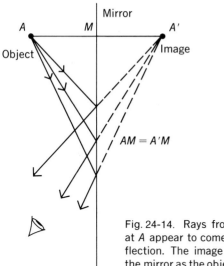

Fig. 24-14. Rays from a point object at *A* appear to come from *A'* after reflection. The image is as far behind the mirror as the object is in front of it.

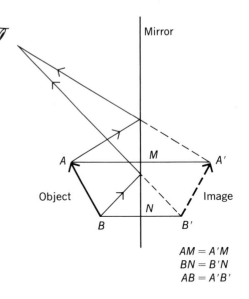

$AM = A'M$
$BN = B'N$
$AB = A'B'$

Fig. 24-15. Image formation by a plane mirror. The image is erect and is the same size and shape as the object.

sponding point on the object, and so the orientation of the image is the same as that of the object; the image is therefore *erect*. However, left and right are interchanged, as in Fig. 24-16: a "mirror image" of anything is the same size and shape as the original but its transverse features are reversed. A printed page appears backward in a mirror, and what seems to be one's left hand in a mirror is actually one's right hand.

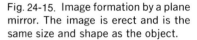

A plane mirror reverses left and right

24-6 Refraction

It is a matter of experience that a beam of light passing obliquely from one medium to another, say from air to water, is deflected at the surface between the two media. The bending of a light beam when it passes from one medium to another is called *refraction*, and it is responsible for such familiar phenomena as the apparent distortion of objects partially submerged in water.

Several important aspects of refraction are illustrated in Fig. 24-17. When a light beam goes from air into water along the normal to the surface between them, it simply continues along the same path, but when it enters the water at any other angle, it is bent toward the normal. The paths are

Fig. 24-16. Left and right are interchanged in reflection.

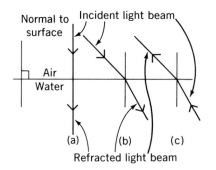

Fig. 24-17. The bending of a light beam when it passes from one medium to another is called *refraction*.

reversible; thus a light beam emerging from the water is bent *away* from the normal as it enters the air.

Origin of refraction

Refraction occurs because light travels at different velocities in the two media. Let us see what happens to the plane wavefront AB in Fig. 24-18 when it passes obliquely from a medium in which its velocity is v_1 to a medium in which its speed is v_2, where v_2 is less than v_1. At $t = 0$ the wavefront AB just comes in contact with the interface between the two media. After a time t, the end B of the wavefront reaches the interface at C, and the secondary wavelets from A are now at D. Since v_2 is less than v_1, the secondary wavelets generated at the interface travel a shorter distance in the same time interval than do wavelets in the first medium, and the distance AD is shorter than BC. The refracted wavefronts accordingly move in a different direction from that of the incident wavefronts.

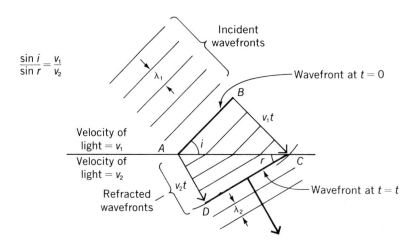

Fig. 24-18. The behavior of successive wavefronts during refraction. The velocity of light v_1 in the first medium is greater than that in the second medium.

The angle i between an approaching wavefront and the interface between two media is called the *angle of incidence* of the wavefront, and the angle r between a receding wavefront and the interface it has passed through is called the *angle of refraction*. From Fig. 24-18 we see that

$$\sin i = \frac{BC}{AC} \qquad \text{and} \qquad \sin r = \frac{AD}{AC}.$$

Because $BC = v_1 t$ and $AD = v_2 t$,

$$\frac{\sin i}{\sin r} = \frac{BC}{AD} = \frac{v_1 t}{v_2 t},$$

and so we have

$$\frac{\sin i}{\sin r} = \frac{v_1}{v_2}. \tag{24-2}$$

This useful result is known as *Snell's law* after its discoverer, the 17th-century Dutch astronomer Willebrord Snell. It states that

> **The ratio of the sines of the angles of incidence and refraction is equal to the ratio of the velocities of light in the two media.**   Snell's law

Refraction, like reflection, can be described in terms of the ray model of light. As in Fig. 24-19 the angles i and r are taken with respect to the normal to the interface at the point where the rays meet it.

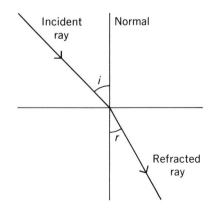

Fig. 24-19. Ray representation of refraction.

24-7 Index of Refraction

The ratio between the velocity of light c in free space and its velocity v in a particular medium is called the *index of refraction* of the medium. The greater the index of refraction, the greater the extent to which a light beam is deflected upon entering or leaving the medium. The symbol for index of refraction is n, so that

$$n = \frac{c}{v}.$$
Index of refraction (24-3)

Table 24-1 is a list of the values of n for a number of substances.

Table 24-1. Indexes of refraction.

Substance	n	Substance	n
Air	1.0003	Glass, flint	1.63
Benzene	1.50	Glycerin	1.47
Carbon disulfide	1.63	Ice	1.31
Diamond	2.42	Quartz	1.46
Ethyl alcohol	1.36	Water	1.33
Glass, crown	1.52	Zircon	1.92

It is easy to rewrite Snell's law in terms of the indexes of refraction n_1 and n_2 of two successive media. In these media light has the respective velocities

$$v_1 = \frac{c}{n_1} \quad \text{and} \quad v_2 = \frac{c}{n_2},$$

and so Snell's law becomes

$$\frac{\sin i}{\sin r} = \frac{v_1}{v_2} = \frac{c/n_1}{c/n_2} = \frac{n_2}{n_1}.$$

This is usually written in the form

$$n_1 \sin i = n_2 \sin r.$$ *Snell's law* (24-4)

Problem. A beam of parallel light enters a block of ice at an angle of incidence of 30° (Fig. 24-20). What is the angle of refraction in the ice?

Solution. The indexes of refraction of air and ice are respectively 1.00 and 1.31. From Snell's law,

$$\sin r = \frac{n_1}{n_2} \sin i = \frac{1.00}{1.31} \sin 30° = 0.382,$$

$$r = 22°.$$

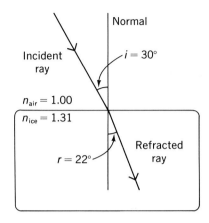

Fig. 24-20.

The index of refraction of a medium depends to some extent upon the frequency of the light involved, with the highest frequencies having the highest values of n. In ordinary glass the index of refraction for violet light is about one percent greater than that for red light, for example. Since a different index of refraction means a different degree of deflection when a light beam enters or leaves a medium, a beam containing more than one frequency is split into a corresponding number of different beams when it is refracted. This effect, called *dispersion*, is illustrated in Fig. 24-21, which shows the result of directing a narrow pencil of white light at one face of a glass prism. The initial beam separates into beams of various colors, from which we conclude that white light is actually a mixture of light of these different colors. The band of colors that emerges from the prism is known as a *spectrum*.

Dispersion

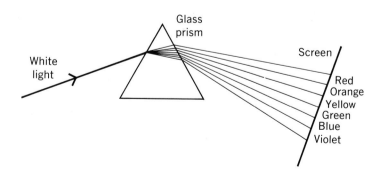

Fig. 24-21. Dispersion by a prism.

24-8 Apparent Depth

Submerged objects seem closer to surface

A familiar example of refraction is the apparent reduction in the depth of an object submerged in water or other transparent liquid. Thus an oar dipped in a lake seems bent upward where it enters the water because its submerged portion appears to be closer to the surface than it actually is. Figure 24-22 shows how this effect comes about. Light leaving the tip of the oar at B is bent away from the normal upon entering the air. To an observer above, who instinctively interprets what he sees in terms of the straightline propagation of light, the tip of the oar is at C, and the submerged part of the oar seems to be AC and not AB.

It is not difficult to relate the apparent and actual depths of a submerged object. Figure 24-23 shows a fish at the point F, an actual depth h below the surface of a body of water, whereas to an observer in the air the fish is at F', only h' below the surface. If we restrict ourselves to rays that are nearly vertical,

$$\frac{\sin i}{\sin r} \approx \frac{\tan i}{\tan r}, \tag{24-5}$$

since $\sin \theta \approx \tan \theta$ when θ is small. From Fig. 24-23 we see that

$$\tan i = \frac{x}{h} \qquad \text{and} \qquad \tan r = \frac{x}{h'}$$

where x is the horizontal distance between the position of the fish and the point O where the ray under consideration leaves the water. Hence

$$\frac{\tan i}{\tan r} = \frac{x/h}{x/h'} = \frac{h'}{h},$$

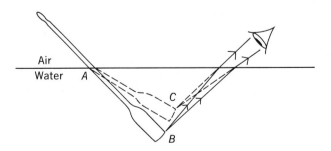

Air

Water A

C

B

Fig. 24-22.   An oar appears bent when partly immersed in water because of refraction.

and so, from Eq. (24-5) and Snell's law,

$$\frac{h'}{h} = \frac{n_2}{n_1}. \qquad \text{\textit{Apparent depth}} \quad (24\text{-}6)$$

Problem. To a sailor standing on the deck of a yacht, the water depth appears to be 10 ft. If this estimate is correct in terms of what he sees, what is the true depth?

Solution. Here, with $n_1 = 1.34$ and $n_2 = 1.00$ (the indexes of refraction of water and air respectively),

$$\frac{h'}{h} = \frac{1.00}{1.33} = 0.752;$$

the apparent depth h' is only about three-quarters of the true depth h. Hence

$$h = \frac{h'}{0.752} = \frac{10 \text{ ft}}{0.752} = 13.3 \text{ ft.}$$

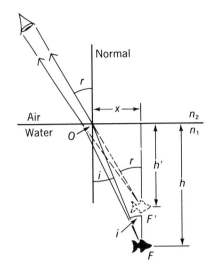

Fig. 24-23. Submerged bodies seem closer to the surface than they actually are.

24-9 Total Internal Reflection

An interesting phenomenon known as *total internal reflection* can occur when light passes from one medium to another which has a *lower* index of refraction, for instance from water or glass to air. In this case the angle of refraction is greater than the angle of incidence, and a light ray is bent *away* from the normal (ray 2 in Fig. 24-24). As the angle of incidence is increased, a certain *critical angle* i_c is reached for which the angle of refraction is 90°. The "refracted" ray now travels along the interface between the two media and cannot escape (ray 3). A ray approaching the boundary at an angle of incidence exceeding the critical angle is reflected back into the medium it comes from, with the angle of reflection being equal to the angle of incidence as in any other instance of reflection (ray 4).

To find the value of the critical angle, we set $i = i_c$ and $r = 90°$ in Snell's law, and we obtain

$$n_1 \sin i_c = n_2 \sin 90°,$$

$$\sin i_c = \frac{n_2}{n_1}. \qquad (24\text{-}7)$$

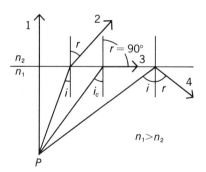

Fig. 24-24. Total internal reflection occurs when the angle of refraction of a light ray going from one medium to another of lower index of refraction equals or exceeds 90°.

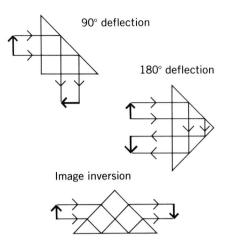

90° deflection

180° deflection

Image inversion

Fig. 24-25.   Three applications of total internal reflection.

For a ray going from water to air,

$$\sin i_c = \frac{n_{\text{air}}}{n_{\text{water}}} = \frac{1.00}{1.33} = 0.752,$$

$$i_c = 49°.$$

What does a person (or a fish) who is underwater see when he looks upward? By reversing all the rays of light in Fig. 24-24 (the paths taken by light rays are always reversible), it is clear that light from everywhere above the water's surface reaches his eyes through a circle on the surface. The rays are all concentrated in a cone whose angular width is twice i_c, which is 98°.

The sharpness and brightness of a light beam are better preserved by total internal reflection than by reflection from an ordinary mirror, and optical instruments accordingly utilize the former in preference to the latter whenever light is to be changed in direction. Figure 24-25 shows three types of totally reflecting prisms in common use.

Total internal reflection makes it possible to "pipe" light from one place to another with a rod of glass or transparent plastic. As in Fig. 24-26, successive internal reflections occur at the surface of the rod, and nearly all the light entering at one end emerges at the other. If a cluster of narrow glass fibers is used instead of a single thick rod, an image can be transferred from one end to the other since each fiber carries intact a part of the image.

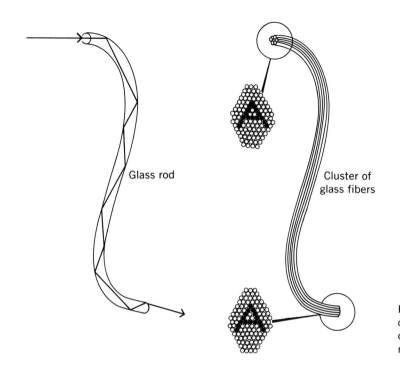

Glass rod

Cluster of
glass fibers

Fig. 24-26.   Light can be "piped" from
one place to another by means of successive internal reflections in a glass
rod.

Special Topic

Illumination

Visible light represents only part of the electromagnetic radiation emitted by most light sources, perhaps 10 percent in the case of an ordinary light
bulb. Also, the eye does not respond with equal sensitivity to light of different colors; the sensitivity is a maximum for green light and it decreases
toward both ends of the visible spectrum (see Section 25-5). Simply stating
the power output of a light source in watts therefore does not convey much
information about the illumination it actually provides, and a separate system of units has been devised for this purpose that is more closely related to
the visual response of the eye.

The brightness of a light source is referred to as its *luminous intensity,* for which the *candela* (cd) is the SI unit. The candela is defined in terms of the light emitted by a small pool of platinum at its melting point. A candle has a luminous intensity of about 1 cd, and in fact the former standard of this quantity was an actual candle of specified composition and dimensions.

The *luminous flux* emitted by a light source refers to the total amount of light it gives off; a small bright source may give off less light than a large dim source, just as a small hot object may give off less heat than a large warm one. The unit of luminous flux is the *lumen* (lm), which equals the luminous flux that falls on each m^2 of a sphere 1 m in radius at whose center is a 1-candela light source that radiates equally well in all directions. Since the area of a sphere of radius *r* is $4\pi r^2$, such a sphere has an area of 4π m^2, and the total luminous flux radiated by a 1-cd source is thus 4π lm.

The luminous flux emitted per watt of power input by a light source is its *luminous efficiency.* Ordinary tungsten-filament lamps increase in efficiency with their power, because the higher the power, the higher the filament temperature and the greater the proportion of visible light in the total radiation. A 10-W lamp, for instance, has a luminous efficiency of about 8 lm/W, whereas it is about 22 lm/W for a 1000-W lamp. Fluorescent lamps have higher efficiencies, from 40 to 75 lm/W.

The *illumination* of a surface is the luminous flux per unit area that reaches it. In the SI system the unit of illumination is the lumen/m^2, or *lux;* an illumination of at least 200 luxes is recommended for reading.

Important Terms

Maxwell's hypothesis was that a changing electric field produces a magnetic field.

Electromagnetic waves consist of coupled electric and magnetic field oscillations. Radio waves, microwaves, light waves, x-rays, and gamma rays are all electromagnetic waves differing only in their frequency.

In **amplitude modulation**, information is contained in variations in the amplitude of a constant-frequency carrier wave. In **frequency modulation**, information is contained in variations in the frequency of the carrier wave.

A **wavefront** is an imaginary surface that joins points where all the waves from a source are in the same phase of oscillation. According to **Huygens' principle**, every point on a wavefront can be considered as a point source of secondary wavelets which spread out in all directions with the wave velocity of the medium. The wavefront at any time is the envelope of these wavelets.

In **diffuse reflection** an incident beam of parallel light is spread out in many directions, while in **specular reflection** the angle of reflection is equal to the angle of incidence. A **mirror** is a specular reflecting surface of regular form that can produce an image of an object placed before it.

A **real image** of an object is formed by light rays

that pass through the image; the image would therefore appear on a properly placed screen. A **virtual image** can only be seen by the eye because the light rays that seem to come from the image actually do not pass through it.

The bending of a light beam when passing from one medium to another is called **refraction**. The quantity that governs the degree to which a light beam will be deflected in entering a medium is its **index of refraction**, defined as the ratio between the velocity of light in free space and its value in the medium. **Dispersion** refers to the splitting up of a beam of light containing different frequencies by passage through a substance whose index of refraction varies with frequency.

Snell's law states that the ratio between the sine of the angle of incidence of a light ray upon an interface between two media and the sine of the angle of refraction is equal to the ratio of the speeds of light in the two media.

Important Formulas

Index of refraction:

$$n = \frac{c}{v}$$

Snell's law:

$$n_1 \sin i = n_2 \sin r$$

Apparent depth:

$$\frac{h'}{h} = \frac{n_2}{n_1}$$

Critical angle:

$$\sin i_c = \frac{n_2}{n_1}$$

Multiple Choice

1. According to Maxwell's hypothesis, a changing electric field gives rise to
   a. an electric current.
   b. an emf.
   c. a magnetic field.
   d. radiation pressure.

2. Electromagnetic waves can travel
   a. only through a vacuum.
   b. only through gases.
   c. only through electric and magnetic fields.
   d. through all of the above.

3. The energy of an electromagnetic wave resides in its
   a. frequency.
   b. wavelength.
   c. velocity.
   d. electric and magnetic fields.

4. Electromagnetic waves transport
   a. wavelength.   b. frequency.
   c. charge.   d. energy.

5. Compared with the electric and magnetic fields of a charge moving at constant velocity, those of an electromagnetic wave decrease in magnitude with distance
   a. less rapidly.
   b. the same.
   c. more rapidly.
   d. any of the above, depending upon the medium.

6. The direction of the magnetic field in an electromagnetic wave is
   a. parallel to the electric field.
   b. perpendicular to the electric field.
   c. parallel to the direction of propagation.
   d. random.

7. Which of the following are *not* electromagnetic in nature?
   a. infrared rays   b. ultraviolet rays
   c. radar waves   d. sound waves

8. In a vacuum, the velocity of an electromagnetic wave

a. depends upon its frequency.
b. depends upon its wavelength.
c. depends upon its Poynting vector.
d. is a universal constant.

9. Reflection from a mirror is said to be
   a. specular.
   b. diffuse.
   c. real.
   d. virtual.

10. All real images
    a. are erect.
    b. are inverted.
    c. can appear on a screen.
    d. cannot appear on a screen.

11. When an object is reflected in a plane mirror, the image is always
    a. real.
    b. inverted.
    c. enlarged.
    d. left-right reversed.

12. When a beam of light enters one medium from another, a quantity that never changes is its
    a. direction.
    b. velocity.
    c. frequency.
    d. wavelength.

13. The bending of a beam of light when it passes from one medium to another is known as
    a. refraction.   b. reflection.
    c. diffraction.   d. dispersion.

14. The index of refraction of a material medium
    a. is always less than 1.
    b. is always equal to 1.
    c. is always greater than 1.
    d. may be less than, equal to, or greater than 1.

15. Relative to the angle of incidence, the angle of refraction
    a. is smaller.
    b. is the same.
    c. is larger.
    d. may be any of the above.

16. A light ray enters one medium from another along the normal. The angle of refraction
    a. is 0.
    b. is 90°.
    c. equals the critical angle.
    d. depends upon the index of refraction of the two media.

17. Dispersion is the term used to describe
    a. the splitting of white light into its component colors in refraction.
    b. the propagation of light in straight lines.
    c. the bending of a beam of light when it goes from one medium to another.
    d. the bending of a beam of light when it strikes a mirror.

18. The index of refraction of benzene is 1.5. The velocity of light in benzene is
    a. 1.5×10^8 m/s.
    b. 2×10^8 m/s.
    c. 3×10^8 m/s.
    d. 4.5×10^8 m/s.

19. According to Snell's law,
    a. $\sin i = \sin r$.
    b. $v_1 \sin i = v_2 \sin r$.
    c. $v_2 \sin i = v_1 \sin r$.
    d. $\sin i / \sin r = v_1 v_2$.

20. The depth of an object submerged in a transparent liquid
    a. always seems less than its actual depth.
    b. always seems more than its actual depth.
    c. may seem less or more than its actual depth, depending on the index of refraction of the liquid.
    d. may seem less or more than its actual depth, depending on the angle of view.

21. Total internal reflection can occur when light passes from one medium to another
    a. which has a lower index of refraction.
    b. which has a higher index of refraction.
    c. which has the same index of refraction.
    d. at less than the critical angle.

22. When a light ray approaches a glass-air interface from the glass side at the critical angle, the angle of refraction is
    a. 0.
    b. 45°.
    c. 90°.
    d. equal to the angle of incidence.

Exercises

1. Why was electromagnetic induction discovered much earlier than its converse, the production of a magnetic field by a varying electric field?

2. Under what circumstances does a charge radiate electromagnetic waves?

3. Why are light waves able to travel through a vacuum whereas sound waves cannot?

4. Light is said to be a transverse wave phenomenon. What is it that varies at right angles to the direction in which a light wave travels?

5. In an electromagnetic wave, what is the relationship, if any, between the variations in the electric and magnetic fields?

6. A radio transmitter has a vertical antenna. Does it matter whether the receiving antenna is vertical or horizontal? Explain.

7. Huygens' principle is only useful for investigating the propagation of light in certain simple situations, whereas the electromagnetic theory of light has a much broader scope. However, Huygens' principle has one major advantage over the electromagnetic theory of light. What is it?

8. What is a wavefront?

9. What is the relationship between a light ray and the wavefronts whose motion it is used to describe?

10. What is the difference between a real image and a virtual image?

11. Do electromagnetic waves have the same velocity in all transparent media? If not, how do their velocities in a transparent medium compare with their velocity in free space?

12. Under what circumstances do the incident ray, the reflected ray, and the normal all lie in the same plane?

13. What advantage has a prism over a mirror when a light beam is to be deflected? Can you think of any disadvantages?

14. Under what circumstances is it appropriate to consider light as a wave phenomenon and under what circumstances as a ray phenomenon?

15. Flint glass and carbon disulfide have almost the same index of refraction. How does this explain the fact that a flint-glass rod immersed in carbon disulfide is nearly invisible?

16. Why is a beam of white light not dispersed into its component colors when it passes perpendicularly through a pane of glass?

17. Explain why a cut diamond held in white light shows flashes of color. What would happen if it were held in red light?

18. Fermat's principle states that light travels between two points in such a manner that the transit time is a minimum. Is refraction in accord with this principle?

19. A nanosecond is 10^{-9} s. (a) What is the frequency of an electromagnetic wave whose period is 1 nanosecond? (b) What is its wavelength?

20. An atom that has been excited to a higher energy state than its normal one divests itself of the excess energy in one or more bursts of radiation. If the frequency of a particular burst is 6×10^{14} Hz and it lasts for 10^{-9} s, find the length of the wavetrain and the number of wavelengths it consists of.

21. A plane mirror is mounted on the back of a truck which is traveling at 20 mi/hr. How fast does the image of a man standing in the road behind the truck seem to be moving away from him?

22. What is the height of the smallest mirror in which a man 6 ft tall can see himself at full length?

23. Using the indexes of refraction given in Table 24-1 and taking the speed of light in free space as 3×10^8 m/s, find the speed of light in (a) air, (b) diamond, (c) crown glass, and (d) water.

24. What is the angle of refraction of a beam of light that enters the surface of a lake at an angle of incidence of 45°?

25. A flashlight is frozen into a block of ice. If its beam strikes the surface of the ice at an angle of incidence of 37°, what is the angle of refraction?

26. A beam of light enters a liquid of unknown composition at an angle of incidence of 20° and is deflected by 10° from its original path. Find the index of refraction of the liquid.

27. A beam of light strikes a pane of glass at an angle of incidence of 60°. If the angle of refraction is 35°, find the index of refraction of the glass.

28. The critical angle for total internal reflection in dense flint glass is 37°. Find its index of refraction.

29. Find the critical angle for total internal reflection in a diamond when the diamond is (a) in air, and (b) immersed in water.

Problems

1. The frequency of the light in a particular beam depends solely upon its source, whereas the wavelength depends upon the velocity of light in the medium it travels through. Find the ratio between the wavelengths of a light beam that passes from one medium to another in terms of their indexes of refraction.

2. A ray of light passes through a plane boundary separating two media whose indexes of refraction are $n_1 = 1.5$ and $n_2 = 1.3$. (a) If the ray goes from medium 1 to medium 2 at an angle of incidence of 30°, what is the angle of refraction? (b) If the ray goes from medium 2 to medium 1 at the same angle of incidence, what is the angle of refraction?

3. A ray of light strikes a glass plate at an angle of incidence of 60°. If the reflected and refracted rays are perpendicular to each other, find the index of refraction of the glass.

4. A ray of light is incident at the angle θ on one side of a glass plate of thickness d whose sides are plane and parallel. Verify that the ray that emerges from the other side of the plate is parallel to the incident ray but displaced sideways from it.

5. A light bulb is on the bottom of a swimming pool 2 m below the water surface. A person on a diving board above the pool sees a circle of light. What is its diameter?

6. The index of refraction of ordinary crown glass for red light is 1.51 and for violet light is 1.53. A beam of white light falls on a plate of thin glass at an incident angle of 40°. What is the difference between the angles of refraction of the red and the violet light?

7. The olive in a Martini cocktail ($n = 1.35$) appears to be 2 in. below the surface. What is the actual depth of the olive?

8. A glass paperweight in the form of a 3 in. cube is placed on a letter. How far underneath the top of the cube does the letter appear if the index of refraction of the glass is 1.5?

9. A barrel of ethyl alcohol is 0.80 m high. How high does it appear to somebody looking into it from above?

10. Prisms are used in optical instruments such as binoculars instead of mirrors because total internal reflection better preserves the sharpness and brightness of a light beam. Find the minimum index of refraction of the glass to be used in a prism that is meant to change the direction of a light beam by 90°.

Answers to Multiple Choice

1. c	9. a	17. a
2. d	10. c	18. b
3. d	11. d	19. c
4. d	12. c	20. a
5. a	13. a	21. a
6. b	14. c	22. c
7. d	15. d	
8. d	16. a	

25

Lenses

A lens refracts light from an object that passes through it in such a way as to be able to form an image of the object. The image may be real or virtual, erect or inverted, and larger, smaller, or the same size as the object. In this chapter we shall investigate not only how individual lenses produce images but also how this is done by such systems of lenses as the microscope and the telescope. The discussion will be limited to lenses whose thicknesses are small relative to their diameters.

25-1 Lenses

A *lens* is a transparent object of regular form that alters the shape of wavefronts of light that pass through it. A simple way to appreciate how a lens produces its effects is to begin with a pair of prisms, as in Fig. 25-1(a). When the prisms have their bases together, parallel light that approaches is deviated so that the various rays intersect. However, they do not intersect at a single focal point because each prism merely changes the direction of the rays without affecting the shape of the wavefronts. If properly curved rather than flat surfaces are employed, as in Fig. 25-1(b), the object is called a *converging lens,* and it acts to bring an incoming parallel beam of light to a single focal point.

A *diverging lens* may similarly be thought of as a development of a pair of prisms with their apexes together (Fig. 25-2). The prisms deviate incoming parallel light outward, but the diverging rays cannot be projected back to a single point. The corresponding lens has curved surfaces so shaped that the diverging rays have a single virtual focal point.

Figure 25-3 shows how lenses affect wavefronts. The speed of light in glass (or other suitable transparent material) is less than in air, and light takes more time to pass through a certain thickness of glass than through the same thickness of air. Light that enters the center of a converging lens is

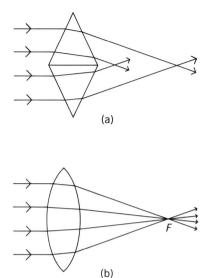
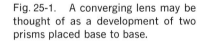

Fig. 25-1.   A converging lens may be thought of as a development of two prisms placed base to base.

Fig. 25-2.   A diverging lens may be thought of as a development of two prisms placed apex to apex.

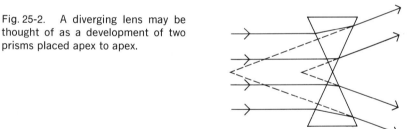
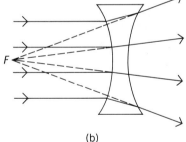

(a)   (b)

Fig. 25-3.   How converging and diverging lenses affect plane wavefronts.

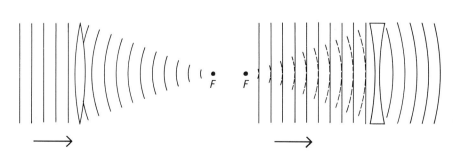

accordingly retarded more than light that enters toward the edge, with the result that the wavefronts converge after passing through the lens. Exactly the opposite effect occurs with a diverging lens.

A lens with spherical surfaces, or one plane and one spherical surface, has the important characteristic that, provided it is relatively thin, it produces an undistorted image of something placed in front of it. Such lenses are also the easiest to manufacture. *Aspherical* (nonspherical) lens surfaces are used in special applications to avoid certain kinds of distortion that occur with spherical lenses. The various forms of simple lenses are shown in Fig. 25-4; converging lenses are always thickest in the center, whereas diverging lenses are always thinnest in the center. A meniscus lens has one concave and one convex surface.

The distance from a lens to its focal point is its focal length

The focal length f of a lens depends both upon the index of refraction n of its material relative to that of the medium it is in, and upon the radii of curvature R_1 and R_2 of its surfaces. In the case of a lens whose thickness is small compared with R_1 and R_2, these quantities are related by the *lensmaker's equation*:

$$\frac{1}{f} = (n-1)\left(\frac{1}{R_1} + \frac{1}{R_2}\right). \qquad \textit{Lensmaker's equation} \quad (25\text{-}1)$$

In using this equation, it does not matter which surface of the lens is considered as 1 and which as 2. However, the sign given to each radius of curvature *is* important. We shall reckon a radius as $+$ if the surface is convex (curved outward), and as $-$ if the surface is concave (curved inward). A plane surface has, in effect, an infinite radius of curvature, and $1/R$ for a plane surface is therefore 0. In Fig. 25-4(a), both radii are $+$; in (b), the first radius is infinite and the second is $+$; in (c), the first radius is $-$ and the second is $+$; in (d), both radii are $-$; and so on.

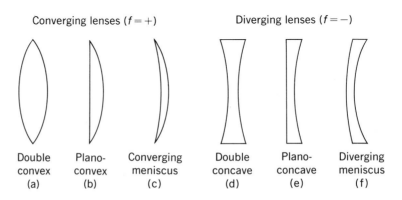

Converging lenses ($f = +$)   Diverging lenses ($f = -$)

Double convex (a)   Plano-convex (b)   Converging meniscus (c)   Double concave (d)   Plano-concave (e)   Diverging meniscus (f)

Fig. 25-4.   Some simple lenses. The focal length of a converging lens is reckoned positive, that of a diverging lens is reckoned negative.

Depending upon its shape, a lens may have a positive or a negative focal length. A positive focal length signifies a converging lens, and a negative one signifies a diverging lens.

Problem.   A meniscus has a convex surface whose radius of curvature is 25 cm and a concave surface whose radius of curvature is 15 cm. The index of refraction is 1.52. Find the focal length of the lens and whether it is converging or diverging.

Solution.   Here $R_1 = +25$ cm and $R_2 = -15$ cm in accord with the sign convention. From the lensmaker's equation,

$$\frac{1}{f} = (n-1)\left(\frac{1}{R_1} + \frac{1}{R_2}\right) = (1.52 - 1.00)\left(\frac{1}{25 \text{ cm}} - \frac{1}{15 \text{ cm}}\right)$$

$$= (0.52)(0.040 - 0.067) \text{ cm}^{-1} = -0.014 \text{ cm}^{-1}$$

and so

$$f = -71 \text{ cm}.$$

The negative focal length indicates a diverging lens.

Problem.   The above lens is placed in water, whose index of refraction is 1.33. Find its focal length there.

Solution.   The index of refraction of the glass relative to water is

$$n' = \frac{\text{index of refraction of glass}}{\text{index of refraction of water}} = \frac{1.52}{1.33} = 1.14.$$

From the lensmaker's equation, since R_1 and R_2 are the same in both air and water, the ratio between the focal length f' of the lens in water and its focal length f in air is

$$\frac{f'}{f} = \frac{n-1}{n'-1} = \frac{1.52 - 1}{1.14 - 1} = 3.7.$$

Hence

$$f' = 3.7f = 3.7 \times (-71 \text{ cm}) = -263 \text{ cm}.$$

The focal length of *any* lens made of this glass is 3.7 times longer in water than in air.

25-2 Image Formation

A scale drawing provides a convenient way to determine the size and position of an image formed by a lens. The procedure is to consider two different light rays that originate at a certain point on an object and to trace their paths until they (or their extensions) come together again after being refracted by the lens. In this connection it is worth noting that a lens has *two* focal points, one on each side of the lens the distance *f* from its center. We shall call the focal point on the side of the lens from which the light comes the *near focal point,* and the one on the other side of the lens the *far focal point.*

Three rays that are particularly easy to trace are shown in Fig. 25-5. They are:

(1) A ray that leaves the object parallel to the lens axis. When this ray is refracted by the lens, it passes through the far focal point of a converging lens or seems to come from the near focal point of a diverging lens;

(2) A ray that leaves the object and passes through the near focal point of a converging lens or is directed toward the far focal point of a diverging lens. When this ray is refracted, it proceeds parallel to the axis;

(3) A ray that leaves the object and proceeds through the center of the lens. If the lens is thin relative to the radii of curvature of its surfaces, this ray is not deviated.

Near and far focal points

Rays for finding an image

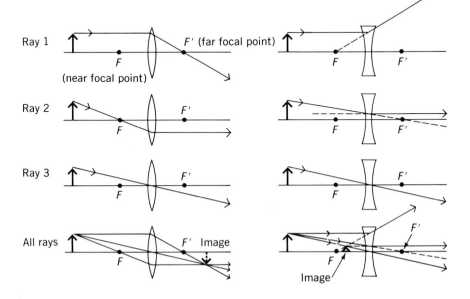

Fig. 25-5. The position and size of an image produced by a thin lens can be determined by tracing any two of the rays shown. In tracing the rays, any deviations produced by the lens are assumed to occur at its central plane.

Ordinarily only two of these rays are required to establish the image of an object. Figure 25-6 shows how the properties of the image produced by a converging lens depend upon the position of the object. As usual, solid lines represent the actual paths taken by light rays, and dashed lines represent virtual paths. When the object is closer to the lens than F, the near focal point, the image is erect, enlarged, and virtual. The image seems to be behind the lens because the refracted rays diverge as though coming from a point behind it. No image is formed of an object precisely at the focal point because the refracted rays are all parallel and hence never intersect; the image in this case may be said to be at infinity. An object farther from the lens than F always has an inverted real image which may be smaller than,

Images produced by converging lens

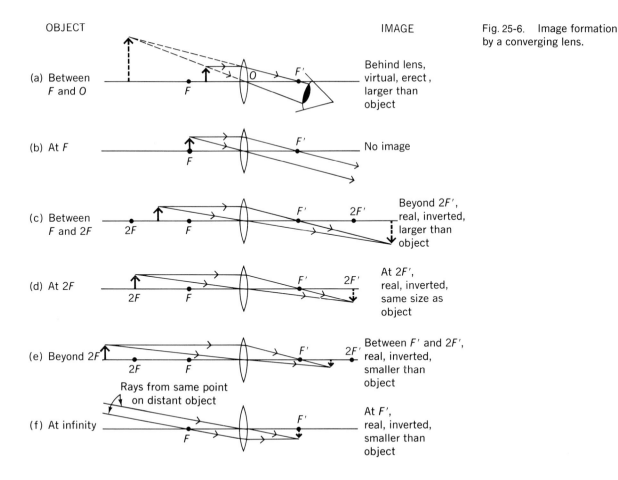

Fig. 25-6. Image formation by a converging lens.

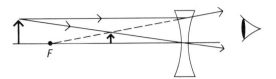

Fig. 25-7. The image of a real object formed by a diverging lens is always virtual, erect, and smaller than the object.

Images produced by diverging lens

the same size as, or larger than the object, depending upon whether the object distance is between f and $2f$, equal to f, or greater than $2f$.

In contrast to the diversity of image sizes, natures, and locations produced by a converging lens, the images of real objects formed by a diverging lens are always virtual, erect, and smaller than the object (Fig. 25-7).

25-3 The Lens Equation

There is a simple formula that relates the positions of the image and the object of a thin lens to the lens's focal length f. In terms of the symbols

$p =$ distance of object from lens

$q =$ distance of image from lens

$f =$ focal length of lens

this formula is

$$\frac{1}{p} + \frac{1}{q} = \frac{1}{f}. \qquad \text{\textit{Lens equation}} \quad (25\text{-}2)$$

When any two of the three quantities are known, the third can be calculated. The sign conventions to be observed when using the lens equation are:

Sign conventions for lenses

(1) The focal length f is considered $+$ for a converging lens, $-$ for a diverging lens;

(2) Object distance p and image distance q are considered $+$ for real objects and images, $-$ for virtual objects and images.

The lens equation can be derived geometrically with the help of Fig. 25-8. We observe that the triangles ABO and $A'B'O$ are similar, which means that corresponding sides of these triangles are proportional. Hence

$$\frac{A'B'}{AB} = \frac{A'O}{AO} = \frac{q}{p}. \qquad (25\text{-}3)$$

Fig. 25-8.   A ray diagram for
deriving the lens equation.

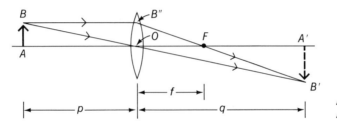

$$\frac{1}{p} + \frac{1}{q} = \frac{1}{f}$$

$\triangle ABO$ and $\triangle A'B'O$ are similar
$\triangle OB''F$ and $\triangle A'B'F$ are similar

The triangles $OB''F$ and $A'B'F$ are also similar, and

$$\frac{A'B'}{OB''} = \frac{B'F}{OF} = \frac{q-f}{f} = \frac{q}{f} - 1. \qquad (25\text{-}4)$$

Now, it is clear that

$$OB'' = AB,$$

since the light ray BB'' is parallel to the axis. Therefore

$$\frac{A'B'}{OB''} = \frac{A'B'}{AB},$$

which permits us to set the right-hand sides of Eqs. 25-3 and 25-4 equal:

$$\frac{q}{f} - 1 = \frac{q}{p}.$$

Dividing each term of this equation by q yields

$$\frac{1}{f} - \frac{1}{q} = \frac{1}{p},$$

and, rearranging terms, we obtain

$$\frac{1}{p} + \frac{1}{q} = \frac{1}{f}.$$

The above derivation involved a converging lens with an object distance greater than the focal length, but the resulting formula is valid for diverging lenses as well and for any object distance.

It is often convenient to solve the lens equation for p or q before using it in a calculation. We find that

$$\frac{1}{p} = \frac{1}{f} - \frac{1}{q} = \frac{q-f}{qf},$$

$$p = \frac{qf}{q-f}, \tag{25-5}$$

and, in a similar way, that

$$q = \frac{pf}{p-f} \tag{25-6}$$

$$f = \frac{pq}{p+q}. \tag{25-7}$$

Problem. A camera consists of a converging lens that forms images on a light-sensitive photographic film, as in Fig. 25-9. The lens of a certain camera that employs 35-mm film has a focal length of 50 mm. What range of motion should the lens have if the camera is to be capable of photographing objects as close as 2 m from the lens?

Solution. Here the image distance q for an object distance of $p = 2.00$ m, since $f = 50$ mm $= 0.05$ m, is

$$q = \frac{pf}{p-f} = \frac{2.00 \text{ m} \times 0.05 \text{ m}}{2.00 \text{ m} - 0.05 \text{ m}} = 0.0513 \text{ m} = 51.3 \text{ mm}.$$

An object at infinity is brought to a sharp focus on the film when the image distance is equal to the focal length, which in this case is 50 mm. Hence a range of adjustment of 1.3 mm will permit the camera to photograph objects from 2 m to infinity.

Problem. A double-concave lens whose focal length is -12 in. is used to examine an object 8 in. behind it. Where does the image seem to be located?

Solution. From the thin lens equation,

$$q = \frac{pf}{p-f} = \frac{8 \text{ in.} \times (-12 \text{ in.})}{8 \text{ in.} - (-12 \text{ in.})}$$

$$= -4.8 \text{ in.}$$

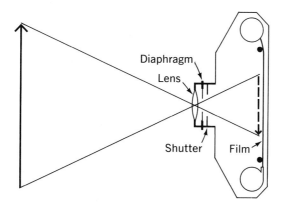

Fig. 25-9   A camera. The light-sensitive film is exposed by opening the shutter for a fraction of a second. The adjustable diaphragm permits the intensity of the light entering the camera to be varied to suit the shutter speed and film used.

The image is located 4.8 in. behind the lens. The negative image distance signifies a virtual image, which is always on the same side of a lens as its object. This situation is pictured in Fig. 25-7.

25-4 Magnification

The *linear magnification m* of an optical system of any kind is the ratio between the size of the image and that of the object. A magnification of exactly 1 means that the image and the object are the same size; a magnification of more than 1 means that the image is larger than the object; and a magnification of less than 1 means that the image is smaller than the object. By "size" is meant any transverse linear dimension, for instance height or width. Thus we can write

$$m = \frac{h'}{h},$$

$$\text{Linear magnification} = \frac{\text{height of image}}{\text{height of object}},$$

where h = height of object and h' = height of image.

Figure 25-10 shows a converging lens that forms the image $A'B'$ of the object AB. The object height is $AB = h$, and that of the image is $A'B' = h'$. The image is inverted and therefore its height is considered negative. The triangles ABO and $A'B'O$ are similar, and their corresponding sides are proportional. Hence

$$\frac{A'B'}{AB} = \frac{A'O}{AO}.$$

Fig. 25-10. The magnification of a lens is equal to minus the ratio of image and object distances. A positive magnification means an erect image, while a negative magnification means an inverted image.

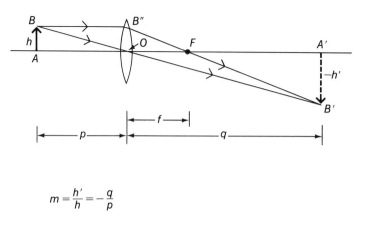

$$m = \frac{h'}{h} = -\frac{q}{p}$$

Because $A'B' = -h'$, $AB = h$, $A'O = q$, and $AO = p$, we find that

$$\frac{h'}{h} = -\frac{q}{p},$$

and the magnification produced by the lens is

$$m = \frac{h'}{h} = -\frac{q}{p}. \qquad\qquad \textit{Linear magnification of lens} \quad (25\text{-}8)$$

This formula is a general one that holds for diverging as well as converging lenses and for any object distance.

A useful feature of Eq. (25-8) is that it automatically indicates whether an image is erect (m positive) or inverted (m negative). Table 25-1 summarizes the various sign conventions we have been using.

Table 25-1. Sign conventions for lenses.

Quantity	Positive	Negative
Focal length f	Converging lens	Diverging lens
Object distance p	Real object	Virtual object
Image distance q	Real image	Virtual image
Magnification m	Erect image	Inverted image

Problem. A "magnifying glass" is a converging lens held less than its focal length from an object being examined (see Fig. 25-6a). How far should a double-convex lens whose focal length is 6 in. be held from an object to produce an erect image three times larger?

Solution. The required magnification is $+3$ since an erect image is required. From Eq. (25-8),

$$m = -\frac{q}{p},$$

$$q = -mp = -3p.$$

We also know from the thin lens equation that

$$q = \frac{pf}{p-f}.$$

By equating these expressions for q, with $f = +6$ in., we obtain

$$\frac{pf}{p-f} = -3p,$$

$$pf = -3p(p-f),$$

$$f = -3p + 3f,$$

$$p = \tfrac{2}{3}f = \tfrac{2}{3} \times 6 \text{ in.}$$

$$= 4 \text{ in.}$$

When the lens is held 4 in. from the object, the image will be virtual, erect, and enlarged three times.

Problem. A slide projector is to be used with its lens 6 m from a screen. If a projected image 1.5 m square of a slide 5 cm square is desired, what should the focal length of the lens be?

Solution. Projectors use converging lenses to produce enlarged, inverted, real images of transparencies which are illuminated from behind (Fig. 25-11). In the present case, the image distance is given as $q = 6$ m, and the required magnification is

$$m = \frac{h'}{h} = \frac{-1.5 \text{ m}}{0.05 \text{ m}} = -30.$$

Fig. 25-11.   An optical projection system. The condenser causes the slide to be evenly illuminated, thereby making possible an image of uniform brightness on the screen.

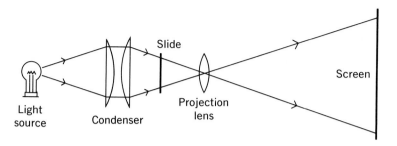

The image height is reckoned negative as it is inverted. From Eq. (25-8),

$$p = -\frac{q}{m} = -\frac{6 \text{ m}}{-30} = 0.2 \text{ m}.$$

From Eq. (25-7) we find that the lens should have a focal length of

$$f = \frac{pq}{p+q} = \frac{0.2 \text{ m} \times 6 \text{ m}}{0.2 \text{ m} + 6 \text{ m}} = 0.194 \text{ m} = 19.4 \text{ cm}.$$

25-5 The Eye

Structure of the eye

The structure of the human eye is shown in Fig. 25-12. Its diameter is typically a little less than 3 cm, and its approximately spherical shape is maintained by internal pressure. Incoming light first passes through the *cornea,* the transparent outer membrane, and then enters a liquid called the *aqueous humor.* The *lens* of the eye is a jellylike assembly of minute transparent fibers that slide over one another when the shape of the lens is altered by the *ciliary muscle* to which it is attached by ligaments. In front of the lens is the colored *iris* whose aperture is the *pupil.* Behind the lens is a cavity filled with another liquid, the *vitreous humor,* and the lining of this cavity is the *retina.*

The retina

The retina contains millions of tiny structures called *rods* and *cones* that are sensitive to light; the cones are responsible for vision in bright conditions, the rods for vision in dim conditions. Because the sensation of color is produced by the cones, colors are difficult to distinguish in a faint light. The rods and cones respond to the image formed by the lens on the retina and transmit the information to the brain through the *optic nerve.* There are no

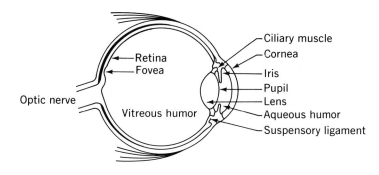

Fig. 25-12. The human eye slightly larger than life size.

photoreceptors near the optic nerve and that region is accordingly called the "blind spot." The blind spot covers a field of view about 8° high and 6° wide. Its existence is not conspicuous for two reasons: the blind spots of the two eyes obscure different fields of vision, and the eyes are never completely at rest but are in constant scanning motion. Vision is most acute for the image focused on the *fovea,* a small region near the center of the retina, while the rest of the field of view appears less distinct. The fovea contains only cones, hence it is easier to see something in a dim light by looking a bit to one side instead of directly at it.

In bright light the pupil contracts to reduce the amount of light reaching the retina, and it dilates in faint light. A fully opened pupil admits about 16 times as much light as a fully contracted one, since the respective pupil diameters are about 2 and 8 mm for a ratio of 4:1. The retina itself is also able to cope with a considerable range of brightnesses. The sensitivity of the eye varies with wavelength, so that light of the same intensity but of different wavelengths will give different impressions of brightness. Maximum sensitivity in bright light occurs in the middle of the visible spectrum at about $\lambda = 555$ nm, which corresponds to green. Maximum sensitivity in dim light, when the rods rather than the cones are chiefly involved, occurs at about 510 nm, which is closer to the blue end of the spectrum and accounts for the reduced ability of the eye to respond to red light against a dark background. For this reason red illumination is provided when accommodation to darkness must be preserved, for instance on the bridge of a ship at night.

Refraction at the cornea is responsible for most of the focusing power of the eye; the lens is less effective because its index of refraction is not very different from those of the aqueous and vitreous humors on both sides of it. The shape of the lens is controlled by the ciliary muscle. When this relaxes, the lens brings objects at infinity to a sharp focus on the retina. To permit a closer object to be viewed, the ciliary muscle forces the lens into a more convex shape whose focal length is appropriate for the object distance in-

Accommodation

volved. The image distance in this situation, which is the lens-retina distance, is fixed, and focusing is thus accomplished by changing the focal length of the lens. The latter process is called *accommodation*. In the case of a camera, on the other hand, the focal length of the lens is fixed, and focusing is done by changing the image distance.

Underwater vision

The limited range of accommodation of the lens is not sufficient to make up for the loss of refractive power that occurs when the eye is immersed in water, which is considerable since the index of refraction of the cornea (1.38) is close to that of water (1.33). Hence underwater objects cannot be brought to a sharp focus unless goggles are used to keep water away from the cornea. Fish are able to see clearly when submerged because the lenses of their eyes are spherical and have high indexes of refraction. Focusing in most fish eyes is carried out by shifting the lens closer or farther from the retina, in the same way a camera lens is focused.

25-6 Defects of Vision

Myopia and hyperopia

Two common defects of vision are *myopia* (nearsightedness) and *hyperopia* (farsightedness). In myopia, the eyeball is too long, and light from an object at infinity comes to a focus in front of the retina (Fig. 25-13b). Accommodation permits nearby objects to be seen clearly, but not more distant ones. A diverging lens of the proper focal length can correct this condition. In hyperopia, the eyball is too short, and light from an object at infinity does not come to a focus within the eyeball at all. Its power of accommodation permits a hyperopic eye to focus on distant objects, but the range of accommodation is not enough for nearby objects to be seen clearly. The correction for hyperopia is a converging lens, as in Fig. 25-13(c).

Fig. 25-13.   Myopia and hyperopia are common defects of vision that can be remedied with corrective lenses.

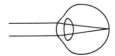

(a) Normal eye

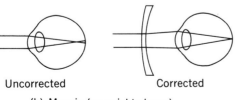

Uncorrected   Corrected

(b) Myopia (nearsightedness)

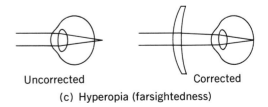

Uncorrected   Corrected

(c) Hyperopia (farsightedness)

Opticians often use the *power P* of a lens, expressed in *diopters*, in place of its focal length. If the focal length f is given in meters, then

$$P \text{ (diopters)} = \frac{1}{f} \tag{25-9}$$

Thus a converging lens of $f = +50$ cm $= +0.50$ m could also be described as having a power of $+2$ diopters.

Problem. A certain nearsighted eye cannot see objects distinctly when they are more than 25 cm away. Find the power in diopters of a correcting lens that will enable this eye to see distant objects clearly.

Solution. The purpose of the lens is to form an image 25 cm in front of the eye of an object that is infinitely far away. Since the image is to be on the same side of the lens as the object, the image must be virtual, and so $q = -25$ cm. The object distance is $p = \infty$, hence

$$\frac{1}{f} = \frac{1}{p} + \frac{1}{q} = \frac{1}{\infty} - \frac{1}{25 \text{ cm}} = 0 - \frac{1}{25 \text{ cm}},$$

$$f = -25 \text{ cm} = -0.25 \text{ m}.$$

The minus sign means that the lens is diverging. Its power in diopters is

$$P = \frac{1}{f} = \frac{-1}{0.25 \text{ m}} = -4 \text{ diopters}.$$

Problem. A certain farsighted eye cannot see objects distinctly when they are closer than 1 m away. Find the power in diopters of a correcting lens that will enable this eye to read a letter 25 cm away.

Solution. The purpose of the lens is to form an image 1 m in front of the eye of an object that is 25 cm away. Since the image is to be on the same side of the lens as the object, the image must be virtual, and so $q = -1$ m. The object distance is $p = 25$ cm $= 0.25$ m, hence

$$f = \frac{pq}{p + q} = \frac{0.25 \text{ m} \times (-1 \text{ m})}{0.25 \text{ m} + (-1 \text{ m})} = +0.33 \text{ m}.$$

The plus sign means that the lens is converging. Its power in diopters is

$$P = \frac{1}{f} = \frac{1}{0.33 \text{ m}} = +3 \text{ diopters}.$$

Presbyopia

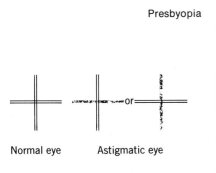

Fig. 25-14. How a cross is seen by a normal and by an astigmatic eye.

Astigmatism

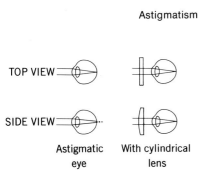

Fig. 25-15. How a cylindrical lens can improve the image formed by an astigmatic eye.

Objective and eyepiece

The range of accommodation decreases with age as the lens hardens, a condition known as *presbyopia*. This does not affect distant vision, but the "near point" of the eye (the closest position at which objects can be seen distinctly) gets increasingly far away. From perhaps 4 in. at age 20, the near point recedes until it is often 6 ft or more at age 60. The correction for presbyopia is a converging lens; if the range of accommodation is severely limited, more than one set of corrective lenses may be required.

The closer an object is to the eye, the larger it seems and the more the detail that can be made out. Since the eye cannot focus on objects closer than the near point, this sets a limit to the magnification of which the eye is capable. By convention, the distance of most distinct vision is taken as 25 cm in metric units or 10 in. in British units, which is a comfortable object distance for most people. For purposes of calculation, an optical instrument such as a microscope or a telescope is therefore assumed to form a virtual image 25 cm (or 10 in.) behind the lens (or lens system) nearest the eye.

Astigmatism is a defect of vision caused by the cornea (or sometimes the lens) having different curvatures in different planes. When light rays that lie in one plane are in focus on the retina of an astigmatic eye, those in other planes will be in focus either in front or in back of the retina. As a result only one of the bars of a cross can be in focus at the same time (Fig. 25-14). Astigmatism is a source of eyestrain because the mechanism of accommodation continually varies the focus of the lens in an effort to produce a completely sharp image. The remedy is a corrective lens which has a cylindrical curvature, as in Fig. 25-15.

25-7 The Microscope

In many applications a system of two or more lenses is superior to a simple lens. An ordinary magnifying glass, for instance, cannot produce images enlarged beyond 3× or so without severe distortion. Greater magnifications of nearby objects can be satisfactorily obtained by using instead two converging lenses arranged as a *microscope*. The optical system of a microscope is shown in Fig. 25-16. The *objective* is a lens of short focal length that forms an enlarged real image of the object. This image is further enlarged by the *eyepiece*, which acts as a simple magnifier to form a virtual final image. The image produced by the objective is thus the object of the eyepiece, and the final image has been magnified twice.

The total magnification produced by a microscope is the product of the magnification m_1 of the objective and the magnification m_2 of the eyepiece:

$$m = m_1 m_2. \tag{25-10}$$

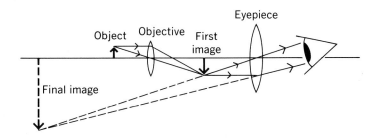

Typical objectives yield magnifications of 10× to 100×, and standard eyepiece magnifications are 5×, 10×, and 15×. Total magnifications of up to 1500× are therefore possible with a good laboratory microscope. Actual microscopes use compound lenses that consist of two to six elements in place of the single lenses shown in Fig. 25-16, but their optical behavior is the same.

Problem.   A microscope has an objective of focal length 4 mm and an eyepiece of focal length 20 mm. If the image distance of the objective is 160 mm and that of the eyepiece is 250 mm (which are the usual figures for these quantities), find the magnification produced by each lens and by the entire microscope. What is the distance between the objective and the eyepiece?

Solution.   (a) The object distance of the objective is

$$p_1 = \frac{q_1 f_1}{q_1 - f_1} = \frac{160 \text{ mm} \times 4 \text{ mm}}{160 \text{ mm} - 4 \text{ mm}} = 4.10 \text{ mm},$$

and so its magnification is

$$m_1 = -\frac{q_1}{p_1} = -\frac{160 \text{ mm}}{4.10 \text{ mm}} = -39.$$

The minus sign means the image is inverted.

(b) With $q_2 = -250$ mm and $f_2 = 20$ mm the same procedure yields for the eyepiece

$$p_2 = \frac{q_2 f_2}{q_2 - f_2} = \frac{-250 \text{ mm} \times 20 \text{ mm}}{-250 \text{ mm} - 20 \text{ mm}} = 18.5 \text{ mm},$$

$$m_2 = \frac{q_2}{p_2} = -\frac{-250 \text{ mm}}{18.5 \text{ mm}} = 13.5.$$

(c) The magnification of the microscope is

$$m = m_1 \times m_2 = -39 \times 13.5 = -527,$$

and the distance between the objective and the eyepiece (see Fig. 25-16) is

$$L = q_1 + p_2 = 160 \text{ mm} + 18.5 \text{ mm} = 178.5 \text{ mm}.$$

25-8 The Telescope

A telescope is a lens system used to examine distant objects. As in a microscope, two lenses are involved, with an eyepiece to enlarge the image produced by the objective. A telescope objective, however, has a long focal length, while that of a microscope is very short.

Figure 25-17 shows a simple telescope. The image produced by the objective is real, inverted, and smaller than the object. The eyepiece then acts as a simple magnifier to form an enlarged virtual image whose object is the initial image. In practice, the object distance is usually very long relative to the focal length of the objective, and the final image is smaller than the object itself. However, the image seen by the eye is larger than it would be without the telescope, so the effect is the same as if the object were closer to the eye than it actually is.

Angular magnification

Telescopes are described in terms of the *angular magnification* they produce. This quantity is the ratio between the angle β subtended at the eye by the image and the angle α subtended at the eye by the object seen directly. For distant objects *(p >> f)* the angular magnification of a telescope is simply the ratio between the focal lengths of its objective and eyepiece:

Fig. 25-17.   Ray diagram of a simple telescope. The rays used to locate the final image are not the same as those used to locate the first image.

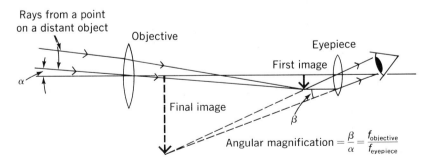

Angular magnification $= \dfrac{\beta}{\alpha} = \dfrac{f_{\text{objective}}}{f_{\text{eyepiece}}}$

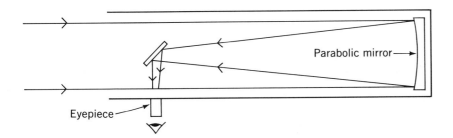

Fig. 25-18.   One type of reflecting telescope.

Parabolic mirror

Eyepiece

$$\text{Angular magnification} = \frac{f_{\text{objective}}}{f_{\text{eyepiece}}} = \frac{\beta}{\alpha}. \qquad (25\text{-}11)$$

The higher the magnification, the greater in diameter the objective must be in order to gather in enough light for the image to be visible. This sets a limit on the size of a telescope since a large glass lens tends to distort under its own weight. The largest refracting telescope in the world is the 40 in. diameter instrument at Yerkes Observatory in Wisconsin, whose objective has a focal length of 65 ft. Modern astronomical telescopes invariably use concave parabolic mirrors as their objectives, since such a mirror produces a real image of a distant object and, furthermore, can be adequately supported from behind. A small secondary mirror reflects the image outside of the telescope for viewing or, more often, for photographing (Fig. 25-18). The largest reflecting telescope is at Mt. Palomar in California and has a mirror 200 in. in diameter.

Reflecting telescope

A telescope can be constructed that produces an erect image if a third lens is introduced between the objective and the eyepiece (Fig. 25-19). The additional lens merely inverts the initial image, but it has the disadvantage of lengthening the telescope tube. A better scheme employs a pair of prisms which serve both to invert the image so that it is erect and to shorten the instrument length. The latter method is employed in prism binoculars (Fig. 25-20).

Problem.   A telescope with an objective of focal length 60 cm and an eyepiece of focal length 1.5 cm is used to examine a hummingbird 5 cm long that is 20 m away. If the image distance of the eyepiece is 25 cm (the distance of most distinct vision), what is the apparent length of the hummingbird?

Solution.   The angular magnification of the telescope is

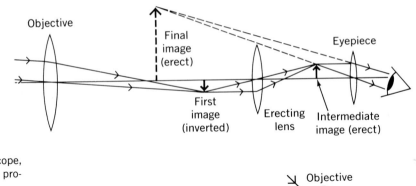

Fig. 25-19. In a terrestrial telescope, an intermediate lens is used to produce an erect image.

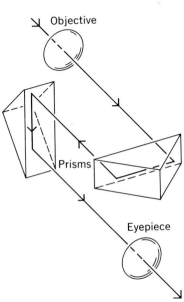

Fig. 25-20. A pair of prisms is used to erect the image in each half of a prism binocular.

$$m_{\text{ang}} = \frac{f_{\text{objective}}}{f_{\text{eyepiece}}} = \frac{60 \text{ cm}}{1.5 \text{ cm}} = 40.$$

The angle α subtended by the bird from the location of the telescope is

$$\alpha = \frac{\text{object length}}{\text{object distance}} = \frac{0.05 \text{ m}}{20 \text{ m}} = 0.0025 \text{ radian}.$$

If L is the length of the bird's image as seen through the telescope, the angle β this image subtends is

$$\beta = \frac{\text{image length}}{\text{image distance}} = \frac{L}{25 \text{ cm}}.$$

Since the angular magnification of the telescope is 40, from Eq. (25-11) we have

$$\beta = m_{ang}\alpha,$$

$$\frac{L}{25 \text{ cm}} = 40 \times 0.0025 \text{ radian}$$

$$L = 2.5 \text{ cm}.$$

The hummingbird seems to be 2.5 cm long, half its actual length, and to be located 25 cm from the viewer's eye.

25-9 Lens Aberrations

The image formed by a single lens is never a perfect replica of its object. Of the variety of aberrations such an image is subject to, perhaps the most familiar is the presence of fringes of color around whatever is being viewed. This *chromatic aberration* is a consequence of the variation with wavelength of the index of refraction of glass (Fig. 24-21). Because of this variation, the focal length of a lens is slightly different for light of different colors, and the fringes are the result. The remedy for chromatic aberration is to combine a converging and a diverging lens made of different glass so that the dispersion produced by one is canceled by the other while leaving a net converging or diverging power. Such an *achromatic lens* is illustrated in Fig. 25-21.

The lens equation was derived on the basis of light rays that made only small angles with the axis. When a simple lens is used to form images of objects some distance from the axis, a variety of aberrations arise even if the light used is monochromatic. One of them is *spherical aberration*, in which rays passing near the lens rim come to a focus closer to the lens than rays near the axis (Fig. 25-22). Another is distortion of the image because the magnification of a simple lens varies with the distance of an object from the axis. By using several lenses of different types of glass and different curvatures to replace a single lens, the most conspicuous of the aberrations can be minimized. Compound lenses that consist of two or more elements are invariably used in high-performance optical systems such as those in microscopes, prism binoculars, and quality cameras.

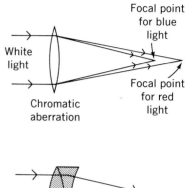

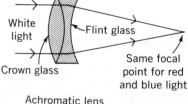

Fig. 25-21.   A compound lens made of different types of glass can correct for chromatic aberration.

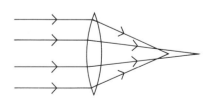

Fig. 25-22.

Special Topic

Spherical Mirrors

Mirrors with spherical reflecting surfaces form images in much the same way that lenses do. Figure 25-23(a) shows how a concave mirror converges a parallel beam of light to a real focal point, and Fig. 25-23(b) shows how a convex mirror diverges such a beam so that the reflected rays seem to originate in a virtual focal point behind the mirror. If R is the radius of the reflecting surface in each case, the focal lengths of these mirrors are

$$f = \frac{R}{2}, \qquad\qquad\qquad \textit{Concave mirror}$$

$$f = -\frac{R}{2}. \qquad\qquad\qquad \textit{Convex mirror}$$

The position, size, and nature of the image produced by a spherical mirror of an object in front of it can be determined with the help of a scale drawing. As in the case of a lens, the procedure is to consider two different light rays that originate at a certain point on the object and to trace their paths until they (or their extensions) come together again after reflection. There are three rays that are especially useful because they are so easily traced (Fig. 25-24):

(1) A ray that leaves the object parallel to the mirror axis. When this ray is reflected, it passes through the focal point of a concave mirror or seems to come from the focal point of a convex mirror;

(2) A ray that leaves the object and passes through the focal point of a concave mirror, or is directed toward the focal point of a convex mirror. When this ray is reflected, it proceeds parallel to the axis;

(3) A ray that leaves the object along a radius of the mirror. When this ray is reflected, it returns along its original path.

In any given situation only two of these rays are needed to locate the image of a reflected object. Figure 25-25 shows how the properties of the image produced by a concave mirror vary with the position of the object. Solid lines represent the actual paths taken by light rays, and dashed lines represent virtual paths.

When the object is closer to the mirror than the focal point F, the image is erect (right side up), enlarged, and virtual. The image *seems* to be behind

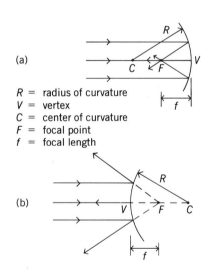

(a)

R = radius of curvature
V = vertex
C = center of curvature
F = focal point
f = focal length

(b)

Figure 25-23. (a) A concave spherical mirror converges a parallel beam of incident light. (b) A convex spherical mirror diverges a parallel beam of incident light.

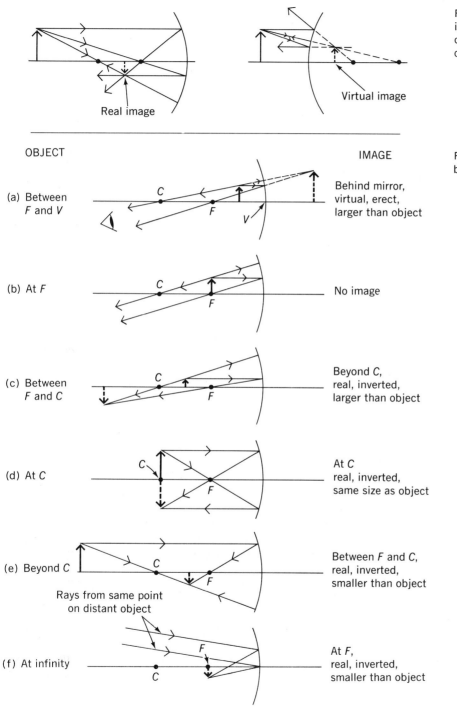

Fig. 25-24. The position and size of an image produced by a spherical mirror can be determined by tracing any two of the rays shown.

Real image

Virtual image

OBJECT

IMAGE

Fig. 25-25. Image formation by a concave mirror.

(a) Between F and V

Behind mirror, virtual, erect, larger than object

(b) At F

No image

(c) Between F and C

Beyond C, real, inverted, larger than object

(d) At C

At C real, inverted, same size as object

(e) Beyond C

Between F and C, real, inverted, smaller than object

Rays from same point on distant object

(f) At infinity

At F, real, inverted, smaller than object

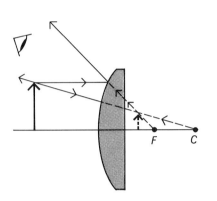

Fig. 25-26. Images formed by convex mirrors are always virtual, erect, and smaller than their objects.

the mirror because the reflected light rays from the mirror diverge as though coming from a point behind it.

When the object is precisely at the focal point, no image is formed because the reflected rays are all parallel and so do not intersect; the image in this case is sometimes said to be at infinity.

An object between the focal point F and the center of curvature C has an inverted, enlarged image that is real; the image would appear on a screen placed at its position.

When the object is at C its image is at the same place and is the same size but is inverted, while an object past C has a real image that is reduced in size and is inverted.

The image formed by a convex mirror of a real object is always erect, smaller than the object, and virtual (Fig. 25-26). The field of view of a convex mirror is wider than that of a plane mirror, which accounts for a number of its applications.

The object and image distances p and q of a spherical mirror are related to its focal length by the same formula that holds for a thin lens:

$$\frac{1}{p} + \frac{1}{q} = \frac{1}{f}. \qquad \textit{Mirror equation}$$

The linear magnification of a mirror also follows the same formula as for a lens:

$$m = \frac{h'}{h} = -\frac{q}{p}. \qquad \textit{Magnification of a mirror}$$

The sign conventions for a mirror are the same as those for a lens (Table 25-1) as well.

Mirrors do not suffer from chromatic aberration but do exhibit spherical aberration. In the case of a concave mirror, rays reflected from the outer part of the mirror cross the axis closer to the vertex than rays reflected from the central part. When a high-quality image is required, as in the case of an astronomical telescope (Fig. 25-18), a concave mirror whose cross-sectional shape is a parabola is the answer. However, despite the presence of spherical aberration, spherical mirrors are common because they are the easiest to manufacture.

Important Terms

A **lens** is a transparent object of regular form that can produce an image of an object placed before it. A **converging lens** brings parallel light to a single real focal point, while a **diverging lens** deviates parallel light outward as though it originated at a single virtual focal point. The distance from a lens to its focal point is its **focal length.**

A **microscope** is a lens system used to produce enlarged images of nearby objects. A **telescope** is a lens system used to produce larger images of distant objects than would be seen by the unaided eye, though the image itself may be smaller than the actual object.

Important Formulas

Lensmaker's equation:

$$\frac{1}{f} = (n-1)\left(\frac{1}{R_1} + \frac{1}{R_2}\right) \quad [R = + \text{ for convex,} \\ - \text{for concave surface}]$$

Lens equation:

$$\frac{1}{p} + \frac{1}{q} = \frac{1}{f}$$

Alternate forms of lens equation:

$$p = \frac{qf}{q-f}$$

$$f = \frac{pq}{p+q}$$

$$q = \frac{pf}{p-f}$$

Magnification:

$$m = -\frac{q}{p}$$

$$m = m_1 \times m_2$$

Multiple Choice

1. In order to calculate the focal length of a glass lens, it is not necessary to know
   a. the index of refraction of the glass.
   b. the index of refraction of the medium in which the lens is located.
   c. the radii of curvature of the lens surfaces.
   d. the diameter of the lens.

2. A converging lens may not have
   a. a positive focal length.
   b. a negative focal length.
   c. one plane surface.
   d. one concave surface.

3. An object farther from a converging lens than its focal point always has an image that is
   a. inverted.
   b. virtual.
   c. the same in size.
   d. smaller in size.

4. An object closer to a converging lens than its focal point always has an image that is
   a. inverted.
   b. virtual.
   c. the same in size.
   d. smaller in size.

5. Images of real objects formed by diverging lenses are always
   a. real.   b. virtual.
   c. inverted.   d. enlarged.

6. A positive magnification signifies an image that is
   a. erect.
   b. inverted.
   c. smaller than the object.
   d. larger than the object.

7. A negative image distance signifies an image that is
   a. real.   b. virtual.
   c. erect.   d. inverted.

8. The image produced by a converging lens of an object that is infinitely far away is always
   a. real.   b. virtual.
   c. erect.   d. larger than the object.

9. A term that does not describe a defect of vision is
   a. accommodation.
   b. astigmatism.
   c. presbyopia.
   d. hyperopia.

10. The pupil of the eye controls
    a. the focal length of the eye.
    b. the range of accommodation of the eye.
    c. the distance of most distinct vision.
    d. the amount of light reaching the eye.

11. The human eye has a focal length of about 5 cm. Its power in diopters is
    a. 0.05.   b. 0.2.
    c. 5.   d. 20.

12. An object is located 10 in. from a converging lens of focal length 12 in. The image distance is
    a. +5.45 in.
    b. −5.45 in.
    c. +60 in.
    d. −60 in.

13. An object is located 12 in. from a converging lens of focal length 10 in. The image distance is
    a. +5.45 in.
    b. −5.45 in.
    c. +60 in.
    d. −60 in.

14. The image of an object 10 cm from a lens is located 10 cm behind the object. The focal length of the lens is
    a. +6.7 cm.   b. −6.7 cm.
    c. +20 cm.   d. −20 cm.

15. A pencil 10 cm long is placed 70 cm in front of a lens of focal length +50 cm. The image is
    a. 4 cm long and erect.
    b. 4 cm long and inverted.
    c. 25 cm long and erect.
    d. 25 cm long and inverted.

16. A pencil 10 cm long is placed 100 cm in front of a lens of focal length +50 cm. The image is
    a. 5 cm long and erect.
    b. 5 cm long and inverted.
    c. 10 cm long and erect.
    d. 10 cm long and inverted.

17. A pencil 10 cm long is placed 175 cm in front of a lens of focal length +50 cm. The image is
    a. 4 cm long and erect.
    b. 4 cm long and inverted.
    c. 25 cm long and erect.
    d. 25 cm long and inverted.

18. A magnifying glass is to be used at the fixed object distance of 1 in. If it is to produce an erect image magnified 5 times, its focal length should be
    a. +0.2 in.
    b. +0.8 in.
    c. +1.25 in.
    d. +5 in.

19. Four lenses with the listed focal lengths are being considered for use as a microscope objective. The one that will produce the greatest magnification with a given eyepiece has the focal length
    a. −5 mm.
    b. +5 mm.
    c. −5 cm.
    d. +5 cm.

20. Four lenses with the listed focal lengths are being considered for use as a telescope objective. The one that will produce the greatest magnification with a given eyepiece has the focal length
    a. −1 ft.   b. +1 ft.
    c. −10 ft.   d. +10 ft.

Exercises

1. Is there any way in which a diverging lens, used by itself, can form a real image of a real object? Is there any way in which a converging lens, used by itself, can form a virtual image of a real object?

2. Under what circumstances, if any, is a light ray that passes through a converging lens not deflected? Under what circumstances, if any, is a light ray that passes through a diverging lens not deflected?

3. Is the mercury column in a thermometer wider or narrower than it appears?

4. A fortune-teller's crystal ball is 6 in. in diameter. (a) Would you expect the lensmaker's equation to hold for the focal length of this ball? (b) Would you expect the ball to form undistorted images?

5. A slide is in sharp focus on a screen. The screen is then moved farther away from the projector. Should the projector's lens be moved closer to or farther from the slide to bring it back into focus on the screen?

6. What are the characteristics of the image formed on the retina by the lens of the eye?

7. Under what circumstances, if any, will a diverging lens form an inverted image of a real object? Under what circumstances, if any, will a converging lens form an erect image?

8. (a) Plot a graph of magnification versus object distance for a converging lens as p is varied from ∞ to 0. (b) Plot a similar graph for a diverging lens. In each case indicate when the image is real and when it is virtual.

9. (a) What can you say about the properties of an image when the magnification is less than -1? (b) Between -1 and 0? (c) Between 0 and $+1$? (d) Greater than $+1$?

10. A double-concave lens has surfaces whose radii of curvature are both 50 cm. The lens is made from flint glass whose index of refraction is 1.60. Find its focal length.

11. A converging meniscus lens has surfaces whose radii of curvature are 20 cm and 30 cm. The lens is made from crown glass whose index of refraction is 1.50. Find its focal length.

12. A diverging meniscus lens has surfaces whose radii of curvature are 20 cm and 30 cm. The lens is made from crown glass whose index of refraction is 1.50. Find its focal length.

13. A plano-concave lens of focal length 9 in. is to be ground from quartz of index of refraction 1.55. Find the required radius of curvature.

14. A double-convex lens whose focal length is 13.3 in. has surfaces whose radii of curvature are 10 in. and 20 in. respectively. Find the index of refraction of the glass.

15. A lens is to be used to focus sunlight on a piece of paper to ignite it. What kind of lens should be used? If the focal length of the lens is 8 cm, how far should it be held from the paper?

16. A candle is placed with its flame 10 cm from a lens whose focal length is +15 cm. Find the location of the image of the flame. Is the image larger or smaller than the actual flame? What is the character of the image; that is, is it erect or inverted, real or virtual?

17. The flame of the candle of Exercise 16 is 15 cm from the lens. Answer the same questions for this case.

18. The flame of the candle of Exercise 16 is 25 cm from the lens. Answer the same questions for this case.

19. The flame of the candle of Exercise 16 is 30 cm from the lens. Answer the same questions for this case.

20. The flame of the candle of Exercise 16 is 50 cm from the lens. Answer the same questions for this case.

21. A lens held 20 cm from an object produces a real, inverted image of it 30 cm on the other side. What is the focal length of the lens? Is it converging or diverging?

22. A camera whose lens has a focal length of 10 cm is used to photograph a person 3 m away. (a) How far in front of the film should the lens be placed? (b) If the film is 4 cm square, what is the area of the subject that it covers?

23. A camera whose lens has a focal length of 6 in. is used to photograph a person 8 ft away. (a) How far in front of the film should the lens be placed? (b) If the film is 4 in. × 5 in. in size, what is the area of the subject that it covers?

24. An aerial camera whose lens has a focal length of 30 in. is used to photograph a military base from an altitude of 20,000 ft. How long is the image on the film of a tank 30 ft long?

25. The focal length of a thin double-convex lens is 3 in. The lens is used as a magnifying glass to examine a spider egg 0.02 in. long. If the lens is

held 2.3 in. from the egg, find the apparent size of the egg to the observer.

26. A magnifying glass of 10 cm focal length is held 8 cm from a stamp. What is the actual length of a feature of the stamp that appears to be 1 cm long?

Problems

1. A converging lens made from glass of $n = 1.60$ has a focal length in air of 6.5 in. (a) Is it converging or diverging when immersed in water? (b) What is its focal length in water?

2. A diverging lens made from glass of $n = 1.52$ has a focal length of -10 cm in air. (a) Is it converging or diverging when immersed in carbon disulfide? (b) What is its focal length in carbon disulfide?

3. Motion-picture directors often use diverging lenses to examine scenes before they are filmed to see how they will ultimately appear on the screen. What focal length should such a lens have if it is to achieve a reduction of 1/10 when held 20 ft from a scene?

4. The actual size of each frame of 16 mm motion picture film is 7.5 mm × 10.5 mm. A projector whose lens has a focal length of 1 in. is to be used with a screen 5 ft wide. How far from the projector should the screen be located?

5. A photographic enlarger is being designed to produce prints 16 in. × 20 in. from negatives 4 in. × 5 in. (a) If the maximum distance from negative to print paper is to be 2 ft, what should the focal length of the lens be? (b) How far would a 4 in. × 5 in. negative be from the print paper using the same lens if the enlargement size were 8 in. × 10 in.?

6. An object should be about 25 cm from a normal eye for maximum distinctness of vision. (a) Find a formula for the magnification of a converging lens when it is used as a magnifying glass (see Fig. 25-6a) with an image distance of -25 cm. (b) Find the magnification of a lens for which $F = +6$ cm.

7. A farsighted eye has a near point of 60 cm. What is its near point when a correcting lens of $+3.33$ diopters is used?

8. A nearsighted person whose eyes have far points of 60 cm is lent a pair of glasses whose power is -1.5 diopters. How far can he see clearly with these glasses?

9. A certain presbyopic eye has a near point 1.0 m away. What type of lens is needed to permit the eye to see clearly objects 25 cm away, and what focal length should such a lens have?

10. A certain myopic eye cannot bring to a focus objects farther than 30 cm away. What type of lens is needed to permit this eye to see clearly objects at infinity, and what focal length should such a lens have?

11. In a certain microscope an objective of 4 mm focal length is used with a 10× eyepiece. (a) What is the magnification of the microscope? (b) How far should the objective be from the specimen being examined?

12. In a certain microscope an objective of 10 mm focal length is used with a 15× objective. (a) What is the magnification of the microscope? (b) How far should the objective be from the specimen being examined?

13. A telescope with an objective of focal length 1.0 m and an eyepiece of focal length 5.0 cm is used to examine a penguin 40 cm high. Find the apparent height of the penguin when it is (a) 50 m, and (b) 5 m away from the telescope. Assume the image to be 25 cm in front of the eyepiece.

14. A telescope has an objective of 30 in. focal length and an eyepiece of 1.0 in. focal length. The telescope is focussed on a distant albatross. (a) What is the angular magnification? (b) How far apart are the lenses? Assume that the final image is located 10 in. in front of the eyepiece.

15. Verify that the effective focal length f of two thin lenses of focal lengths f_1 and f_2 that are in contact is given by

$$\frac{1}{f} = \frac{1}{f_1} + \frac{1}{f_2}.$$

(To do this, let the image produced by the first lens be the object of the second.)

Answers to Multiple Choice

1. d	8. a	15. d
2. b	9. a	16. d
3. a	10. d	17. b
4. b	11. d	18. c
5. b	12. d	19. b
6. a	13. c	20. d
7. b	14. c	

26

Physical Optics

In the previous chapter we had little need to invoke the wave character of light. But, while the properties of lenses and mirrors are more readily analyzed in terms of rays that in terms of waves, there are other optical phenomena in which the wave nature of light is directly involved. The study of the latter phenomena is called physical optics, while the study of those aspects of light behavior that can be understood using a ray treatment is geometrical optics. Since the basic laws of refraction and reflection follow from Huygens' principle, geometrical optics is evidently an approximation of physical optics whose usefulness comes from its simplified view of light propagation. In this chapter we shall find the wave approach of physical optics necessary for explaining interference, diffraction, and polarization.

26-1 Interference of Light

Light was recognized as a wave phenomenon well before its electromagnetic character became known a century ago. The problem of the nature of light was an old one: Newton felt sure light consists of a stream of tiny particles, whereas his contemporary Christian Huygens (1629-1695) thought it to be a succession of waves. Neither man offered any hypothesis as to what kind of particle or wave is involved. Eventually interference, diffraction, and polarization were discovered in light, and these phenomena could only be explained on the basis that light consists of transverse waves. But Newton was not completely wrong. As we shall learn in Chapter 28, light has certain distinctly particle properties in addition to its wave properties, and one of the central problems of contemporary physics has been the resolution of this apparent paradox.

When light waves from one source are mixed with those from another source, the two wave trains are said to *interfere*. We have already met the principle of superposition, which governs interference: When two or more waves of the same nature travel past a point at the same time, the amplitude at that point is the sum of the instantaneous amplitudes of the individual waves. Constructive interference refers to the reinforcement of waves in phase ("in step") with one another, and destructive interference refers to the partial or complete cancellation of waves out of phase with one another (Fig. 26-1).

Principle of superposition

Anyone with a pan of water can see for himself how interference between water waves can lead to a water surface disturbed in a variety of characteristic patterns. Two people who hum fairly pure tones slightly different in frequency will hear beats as the result of interference in the sound waves. But if we shine light from two flashlights at the same place on a screen, there is no evidence of interference: the region of overlap is merely uniformly bright.

There are two reasons for the difficulty of observing interference in light. First, light waves have extremely short wavelengths—the visible part of the spectrum extends only from 3.8×10^{-7} m for violet light to 7×10^{-7} m for red light. Second, every natural source of light emits light waves only as short trains of random phase, so that any interference that occurs is usually averaged out during even the briefest period of observation by the eye or photographic film unless special procedures are used. Interference in light is nevertheless just as real a phenomenon as interference in water or sound waves, and there is one example of it familiar to everybody—the bright colors of a thin film of oil spread out on a water surface.

Two sources of waves are said to be *coherent* if there is a fixed phase relationship between the waves they emit during the time the waves are

Coherent sources

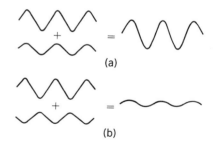

Fig. 26-1.   (a) Constructive interference. (b) Destructive interference.

being observed. It does not matter whether the waves are exactly in step when they leave the sources, or exactly out of step, or anything in between; the important thing is that the phase relationship stay the same. If the sources shift back and forth in relative phase while the observation is made, the phase differences average out, and there will be no interference pattern. The latter sources are *incoherent*.

Incoherent sources

The question of coherence is especially significant for light waves because an excited atom radiates for 10^{-8} s or less, depending upon its environment. Therefore a monochromatic light source such as a gas discharge tube (a neon sign is an example) does not emit a continuous wave train as a radio antenna does but instead a series of individual wave trains whose phases are random. The light from such a tube actually comes from a great many individual, uncoordinated sources, namely the individual atoms, and these individual sources are in effect being switched on and off rapidly and irregularly (Fig. 26-2).

Suppose we have two point sources of monochromatic light, for instance a discharge tube with a shield that has two pinholes close together. Different atoms are behind each pinhole, so they are independent sources. Therefore we have at most 10^{-8} s to observe the interference of waves from the two sources. If our detecting instruments are fast enough, as some modern electronic devices are, interference can be demonstrated and the sources can be considered coherent. If we are limited to the eye and to photographic film, which average arriving light signals over times far greater than 10^{-8} s, no interference can be observed in the light from the two sources, and they must then be considered incoherent. Like beauty, coherence lies in the eye of the beholder.

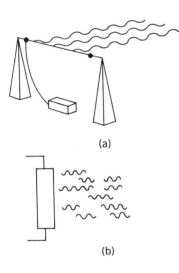

Fig. 26-2.   (a) Radio waves from an antenna are coherent. (b) Light waves from a gas discharge tube are incoherent.

Methods of producing coherent light sources

Does the brief lifetime of an excited atom mean that interference patterns can never be literally seen but can only be recorded by instruments? Hardly. There are three ways to construct separate sources of light coherent for long enough periods of time to produce visible interference patterns. These are:

1. Illuminate two (or more) slits with light from one slit behind them. Then the light waves from the secondary slits are automatically coordinated.
2. Obtain coherent virtual sources from a single source by reflection or refraction. This is how interference is produced by thin films of oil.
3. Coordinate the radiating atoms in each separate source so that they always have the same phase even though different atoms are radiating at successive instants. This is done in the *laser*.

We shall examine the first two of these methods here. The laser is discussed in a later chapter.

26-2 Double Slit

The interference of light waves was demonstrated in 1801 by Thomas Young, who used an arrangement similar to that shown in Fig. 26-3. A source of monochromatic light (that is, light consisting of only a single wavelength) is placed behind a narrow slit S in an opaque screen, and another screen with two similar slits A and B is placed on the other side. Light from S passes through both A and B and then to the viewing screen. If light were not a wave phenomenon, we would expect to find the viewing screen completely dark, since no light ray can reach it from the source along a straight path. What actually happens is that each slit acts as a source of secondary wavelets—we recall Huygens' principle from Section 24-4—so that almost the entire screen is illuminated. Even the point S', separated from S by the opaque barrier between the slits A and B, turns out to be bright rather than dark (Fig. 26-4).

Fig. 26-3. Young's double-slit experiment. In accord with Huygens' principle, each slit acts as a source of secondary wavelets.

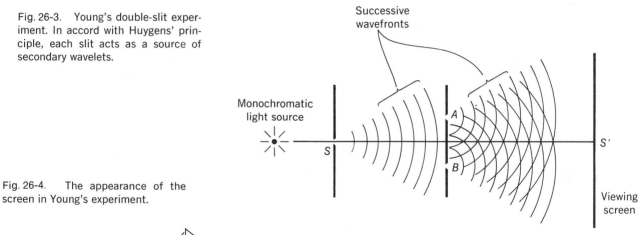

Fig. 26-4. The appearance of the screen in Young's experiment.

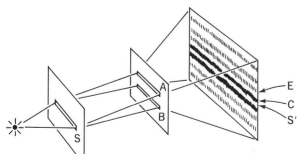

An interference pattern consists
of light and dark lines

Owing to interference the screen is not evenly illuminated but shows a pattern of alternate bright and dark lines. Light waves from slits A and B are exactly in phase, since A and B are the same distance from S. The centerline S' of the screen is equally distant from A and B, so light waves from these slits interfere constructively there to produce a bright line.

Let us next see what happens at the position C on the screen located to one side of S'. The distance BC is longer than the distance AC by the amount BD, which is equal to exactly half a wavelength of the light being used. That is,

$$BD = \tfrac{1}{2}\lambda.$$

When a crest from A reaches C, this difference in path length means that a trough from B arrives there at the same time, since $\tfrac{1}{2}\lambda$ separates a crest and a trough in the same wave. The two cancel each other out, the light intensity at C is zero, and a dark line results on the screen there. At S' the equality of path length gives rise to constructive interference; at C the difference of $\tfrac{1}{2}\lambda$ in path length gives rise to destructive interference (Fig. 26-5).

If we go past C on the screen we will come to a point E such that the distance BE is greater by exactly one wavelength than the distance AE. That is, the difference BF between BE and AE is

$$BF = \lambda.$$

Consequently, when a crest from A reaches E, a crest from B also arrives there, although the latter crest left B earlier than that from A owing to the longer path it had to cover. Because $BF = \lambda$, waves arriving at E from both

Fig. 26-5.   Origin of the double-slit interference pattern.

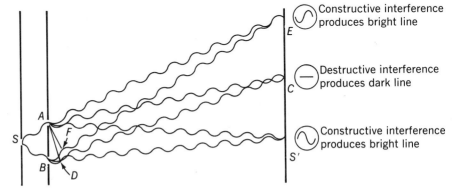

slits are always in the same part of their cycles, and they constructively interfere to produce a bright line at E.

By continuing the same analysis, we find that the alternate light and dark lines actually observed on the screen correspond respectively to locations where constructive and destructive interference occurs. Waves reaching the screen from A and B along paths that are equal or differ by a whole number of wavelengths (λ, 2λ, 3λ, and so on) reinforce, while those whose paths differ by an odd number of half wavelengths ($\frac{1}{2}\lambda$, $\frac{3}{2}\lambda$, $\frac{5}{2}\lambda$, and so on) cancel. At intermediate locations on the screen the interference is only partial, so that the light intensity on the screen varies gradually between the bright and dark lines.

26-3 Wavelength of Light

Interference patterns like those produced by the double slit permit us to determine the wavelength of the light used. To illustrate the procedure, we shall consider the double-slit experiment quantitatively. Figure 26-6 is a diagram of the experiment: d is the separation between the slits, L is the distance from the slits to the screen, and y is the distance from the central point S' on the screen to the point Q whose illumination we are observing.

The waves that travel from slit B to Q must travel s farther than those from slit A. As we have seen, when the path difference s is

Conditions for bright and dark lines

Bright lines $\qquad s = 0, \lambda, 2\lambda, 3\lambda, \ldots ,$ $\qquad\qquad$ (26-1)

where λ is the wavelength of the light from the source, waves from A and B arrive at Q in the same stage of their cycles and reinforce one another to produce a bright line there. When the path difference s is

Dark lines $\qquad s = \frac{1}{2}\lambda, \frac{3}{2}\lambda, \frac{5}{2}\lambda, \ldots ,$ $\qquad\qquad$ (26-2)

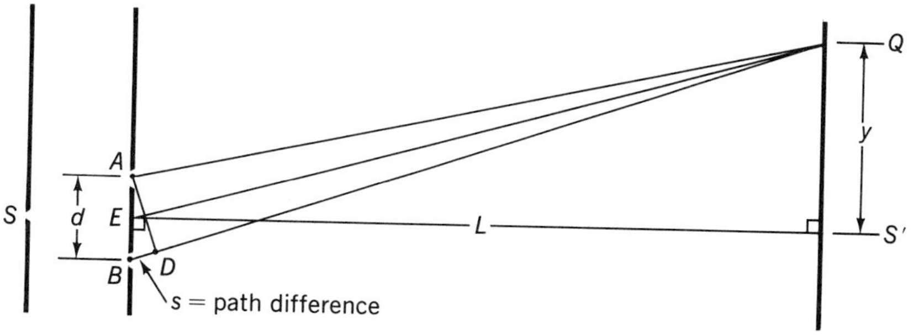

Fig. 26-6. A diagram of the double-slit experiment.

on the other hand, waves from A and B arrive at Q in the opposite stages of their cycles and cancel one another out to produce a dark line there. The triangles ABD and QES' are similar, since each is a right triangle and two sides of each are perpendicular to two sides of the other. Corresponding sides of similar triangles are proportional, and so

$$\frac{s}{y} = \frac{d}{EQ}. \qquad (26\text{-}3)$$

In an actual experiment, y is very much smaller than L (their ratio is usually less than 1:1000; in Fig. 26-6, y is exaggerated for clarity), which means that EQ is almost exactly equal to L. At the expense of introducing a negligible error, we may substitute L for EQ in Eq. (26-3), with the result that

$$\frac{s}{y} = \frac{d}{L},$$

$$s = \frac{dy}{L}. \qquad (26\text{-}4)$$

Combining Eqs. (26-1) and (26-2) with Eq. (26-4) yields the conditions for bright and dark lines to occur at Q:

Bright lines $y = 0, \quad \dfrac{L\lambda}{d}, \quad \dfrac{2L\lambda}{d}, \quad \dfrac{3L\lambda}{d}, \quad \dots, \qquad (26\text{-}5)$

Dark lines $y = \dfrac{L\lambda}{2d}, \quad \dfrac{3L\lambda}{2d}, \quad \dfrac{5L\lambda}{2d}, \quad \dots. \qquad (26\text{-}6)$

According to Eq. (26-5), there is a bright line when $y = 0$, corresponding to the center of the screen S', a bright line on either side of S' a distance $L\lambda/d$ from it, another bright line on either side a distance $2L\lambda/d$ from S', and so on (Fig. 26-4). Similarly, Eq. (26-6) states that there is a dark line on either side of S' a distance $L\lambda/2d$ from it, another dark line on either side a distance $3L\lambda/2d$ from it, and so on.

In an actual experiment, L and d are known initially, and the distance y from the center of the screen to any particular light or dark line can be measured. The wavelength of the light may then be computed by using the equation corresponding to the particular line.

Problem.   Monochromatic yellow light from a sodium-vapor lamp illuminates two narrow slits 1 mm apart. The viewing screen is 1 m from the slits,

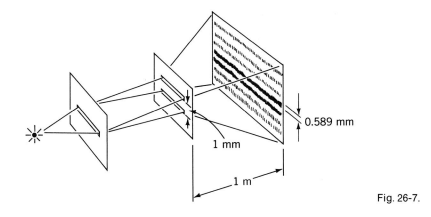

0.589 mm

1 mm

1 m

Fig. 26-7.

and the distance from the central bright line to the bright line nearest it is found to be 0.589 mm (Fig. 26-7). Find the wavelength and frequency of the light.

Solution. For the first bright line,

$$y = \frac{L\lambda}{d},$$

and so

$$\lambda = \frac{yd}{L} = \frac{5.89 \times 10^{-4} \text{ m} \times 10^{-3} \text{ m}}{1 \text{ m}} = 5.89 \times 10^{-7} \text{ m} = 589 \text{ nm}.$$

26-4 Diffraction Grating

There are two difficulties in using a double slit for measuring wavelengths. First, the "bright" lines on the screen are actually extremely faint and an intense light source is therefore required; second, the lines are relatively broad and it is hard to locate their centers accurately. A *diffraction grating* that consists, in essence, of a large number of parallel slits, overcomes both of these difficulties. Gratings are made by ruling grooves on a glass or metal plate with a diamond; the clear bands between the grooves are the "slits." Replica gratings, made by allowing a transparent liquid plastic to harden in contact with an original grating, are ordinarily used in practice. Replica gratings are often given a thin coating of silver or aluminum and produce their characteristic diffraction patterns by the interference of

A diffraction grating uses interference to measure wavelengths of light

reflected rather than transmitted light. A long-playing phonograph record held at a glancing angle acts as a reflection grating by virtue of its closely spaced grooves.

Gratings are ruled with from 2,000 to 10,000 lines/m, and a lens is used to focus the light from the slits between them on a screen. The effect of phase differences among the rays from the various slits is accentuated by their great number, and as a result the intensity of light on the screen falls rapidly on either side of the center of each bright line. The bright lines are therefore sharp, and, because there are so many slits, they are also bright in a literal sense. Very accurate wavelength determinations can be made with the help of a grating, and wavelengths that are close together can be resolved. The analysis of a light beam in terms of the particular wavelengths it contains is today almost invariably carried out by using a grating.

Figure 26-8 shows a diffraction grating that forms a bright line on a screen at a deviation angle of θ from the original beam of light. The condition for a bright line is that $s = n\lambda$, where $n = 1, 2, 3$, and so on. Since $s = d \sin \theta$, where d is the spacing of the slits, bright lines occur at those angles for which

$$\sin \theta = n \frac{\lambda}{d}, \qquad n = 1, 2, 3, \ldots \qquad \textit{Bright lines} \quad (26\text{-}7)$$

When polychromatic light is incident on a grating, a series of spectra is formed on each side of the original beam corresponding to $n = 1$, $n = 2$, and Spectral orders   so on. The *first-order spectrum* contains bright lines for which $n = 1$, the

Fig. 26-8.   The plane diffraction grating.

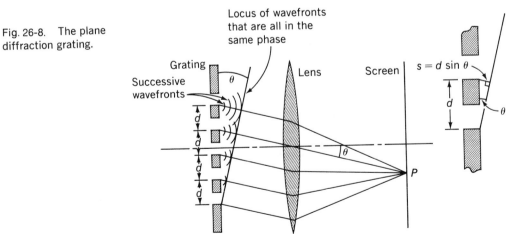

second-order spectrum contains bright lines for which $n = 2$, and so on. In some gratings the higher-order spectra overlap, so that, for example, the blue end of the third-order spectrum may be deviated by less than the red end of the second-order spectrum. According to Eq. (26-7), the angle θ increases with wavelength, so that blue light is deviated least and red light most, which is the reverse of what occurs when a prism is used to form a spectrum.

Problem. Visible light includes wavelengths from approximately 4×10^{-7} m (blue light) to 7×10^{-7} m (red light). Find the angular width of the first-order spectrum produced by a grating ruled with 8000 lines/cm.

Solution. The slit spacing d that corresponds to 8000 lines/cm is

$$d = \frac{10^{-2} \text{ m/cm}}{8 \times 10^3 \text{ lines/cm}} = 1.25 \times 10^{-6} \text{ m.}$$

Since $n = 1$ for a first-order spectrum, the angular deviations of blue and red light respectively are given by

$$\sin \theta_b = \frac{\lambda_b}{d} = \frac{4 \times 10^{-7} \text{ m}}{1.25 \times 10^{-6} \text{ m}} = 0.32,$$

$$\theta_b = 19°,$$

and

$$\sin \theta_r = \frac{\lambda_r}{d} = \frac{7 \times 10^{-7} \text{ m}}{1.25 \times 10^{-6} \text{ m}} = 0.56,$$

$$\theta_r = 34°.$$

The total width of the spectrum is therefore $34° - 19° = 15°$.

A legitimate question is why spectra are not formed whenever specular reflection (see Section 24-5) occurs. Reflection of light waves takes place when electrons at the surface of the reflecting material are caused to oscillate by the incident electromagnetic waves, and the reradiated waves have their central maximum ($n = 0$) at an angle of reflection equal to the angle of incidence, as Fig. 24-12 shows. Because atoms in a solid or liquid are only about 10^{-10} m apart, λ/d is equal to several thousand for visible light, which means that spectra corresponding to $n = 1$ or more do not occur.

26-5 Thin Films

We have all seen the marvelous rainbow colors that appear in soap bubbles and thin oil films. Some of us may also have observed the patterns of light and dark bands that occur when two glass plates are almost (but not quite) in perfect contact. Both phenomena owe their origins to a combination of reflection and interference.

Let us consider a beam of monochromatic light that strikes a thin film of soapy water. Figure 26-9 shows a ray picture of what happens. We notice that some reflection takes place at both the air-soap and soap-air interfaces. This is a general result: waves are always partially reflected when they go from one medium to another in which their speed is different. (See the discussion in Section 13-2 of the reflection of pulses in a string under similar circumstances.)

A light ray actually consists of a succession of wave fronts. Figure 26-10 is the same diagram with the wave fronts drawn in. In (a) the two reflected wave trains are out of phase and they interfere destructively to partially or completely cancel out. Most or all of the light reaching this part of the soap bubble therefore passes right through.

Another part of the soap film may have a different thickness. When the film is a little thinner than in (a), the waves in the two reflected trains are exactly in phase, and they interfere constructively to reinforce one another (Fig. 26-10b). Light reaching this part of the soap bubble is strongly reflected. Shining monochromatic light on a soap bubble therefore yields a pattern of light and dark which results from the varying thickness of the bubble.

When white light is directed at a soap bubble, light waves of each wavelength present pass through the soap film without reflection at those places where the film is exactly the right thickness for the two reflected rays to destructively interfere. Light waves of the other wavelengths are reflected

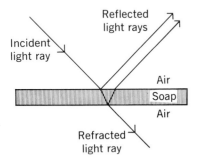

Fig. 26-9. Reflection occurs at both surfaces of a soap film.

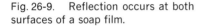

Why thin films appear colored

Fig. 26-10. (a) Destructive and (b) constructive interference in a thin film.

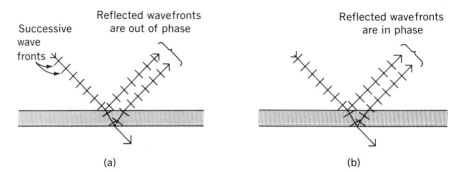

to at least some extent, and give rise to the vivid colors seen. The varying thickness of the bubble means that the color of the light reflected from the bubble changes from place to place. Exactly the same effect is responsible for the coloration of thin oil films. Generally speaking, soap or oil films whose thickness is comparable with the wavelengths in visible light give rise to the most striking color effects.

A thin film of air between two sheets of glass or transparent plastic also yields a pattern of colored bands when illuminated with white light. A notable example is *Newton's rings*, which occur when a slightly curved lens is placed on a flat glass plate (the curvature is exaggerated in Fig. 26-11). Again reflection takes place at both the top and bottom of the film, and again the result is constructive or destructive interference, depending upon the film thickness. Because the thickness of the air film increases with distance from the central point of contact, the pattern of light and dark bands consists of concentric circles.

Newton's rings

The dark spot at the center of a set of Newton's rings is not what we might expect to find (Fig. 26-12). At the center, where the air film between the pieces of glass is minute, the path difference between the waves reflected from the upper and lower surfaces of the film is negligible. Hence there ought to be constructive interference and reinforcement of the light to yield a bright spot. What this analysis overlooks is the fact that a wave reflected at the surface of a new medium in which its speed is less (in optical terms, a medium of higher index of refraction) is shifted by half a wavelength. That is, a positive displacement of the wave variable is reflected as a negative one, and vice versa. The same effect was noted in the discussion of pulses in a stretched string and is shown in Fig. 13-6. Thus a $\frac{1}{2}\lambda$ shift occurs when light waves are reflected in going from air to glass, but not in going from glass to air. In consequence the two wave trains reflected at the center of a

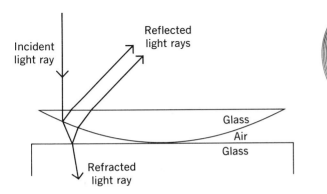

Fig. 26-11.   Newton's rings.

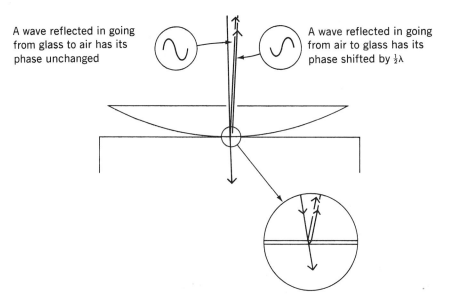

A wave reflected in going from glass to air has its phase unchanged

A wave reflected in going from air to glass has its phase shifted by $\frac{1}{2}\lambda$

Fig. 26-12. Origin of dark spot at center of Newton's rings.

Newton's ring pattern exactly cancel each other out to yield the dark spot actually observed.

About 4 per cent of the light striking a glass-air interface is reflected. This is not a lot, but there may be many glass-air interfaces in an optical instrument and the total amount of light lost through reflection may be considerable. There are 10 such interfaces in each of the optical systems in a pair of binoculars, for instance, so that only about $\frac{2}{3}$ of the incoming light actually gets through to the observer's eyes. Even in a camera, where there are fewer glass-air interfaces, reflections are a nuisance because they may lead to secondary images that blur the picture.

Coated lenses   To reduce reflections at a glass-air interface, the glass can be coated with a very thin layer of a transparent substance (usually magnesium fluoride) whose index of refraction is intermediate between those of glass and of air. If the layer is exactly $\frac{1}{4}\lambda$ thick, light reflected at its bottom will have traveled $\frac{1}{2}\lambda$ farther when it rejoins light reflected at the top of the layer, and the two will cancel out exactly. (Light waves reflected at either surface are shifted in phase by $\frac{1}{2}\lambda$, so these shifts have no effect on the cancellation.)

But the cancellation described above is exact only for a particular wavelength λ, while white light contains a range of wavelengths. What is therefore done is to choose a wavelength in the middle of the visible spectrum, which corresponds to green light, so that at least partial cancellation occurs over a wide range. The red and violet ends of the spectrum are accordingly least affected and the light reflected from a coated lens is a mixture of these colors, a purplish hue. The average reflectivity of a glass surface coated in this way is only a little over 1 percent. Multiple coatings are sometimes used to reduce reflection even further; a triple coating brings the average reflectivity below 0.5 percent. Despite their lack of perfection at suppressing reflections, coated lenses transmit appreciably more light than uncoated ones, and they are universally used in fine optical instruments.

26-6 Diffraction

Waves are able to bend around the edge of an obstacle in their path, a property called *diffraction*. We all have heard sound that originated around the corner of a building from where we were standing, for example. These sound waves cannot have traveled in a straight line from their source to our ears, and refraction cannot account for their behavior. Water waves, too, diffract, as the simple experiment illustrated in Fig. 26-13 shows. The waves on the far side of the gap spread out into the geometrical "shadow" of the gap's edges, though with reduced amplitude. The diffracted waves spread out as though they originated at the gap, in accord with Huygens' principle.

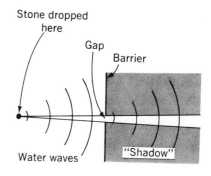

If we look very carefully at the edge of the shadow cast by an obstruction in the path of the light spreading out from a pinhole or other point source, we will see that it is not sharp but smeared out (Fig. 26-14). The fuzzy edges of shadows are not easy to observe because the wavelengths in visible light are so short, less than 10^{-6} m, and the extent of diffraction into the shadow zone is correspondingly small. (In contrast, a typical audible sound wave might have a wavelength of 1 m and a typical wave in a pan of water might have a wavelength of 10 cm, and it is easy to observe

Fig. 26-13.   Diffraction in water waves.

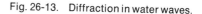

Shadows do not have sharp edges because of diffraction

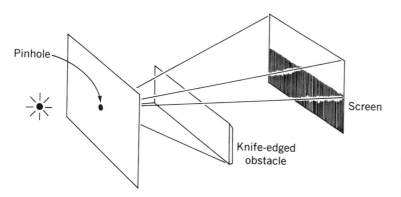

Fig. 26-14.   Even under ideal conditions, the edge of a shadow is never completely sharp.

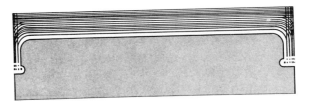

Fig. 26-15.   The shadow of a razor blade.

diffraction effects with such waves.) In fact, because he was not able to perceive any diffraction with his relatively crude apparatus, Newton felt sure that light must be corpuscular in nature and not consist of waves at all.

A broad light source such as a light bulb or the sun does not produce sharp shadows for another reason. In this case light from different parts of the source passes the edge of the obstacle at different angles, which is not true of light from a point source.

We can refine our observation of diffraction further by using a monochromatic light source and a sheet of photographic film instead of a screen. When we enlarge the developed image on the film, we find a pattern of light and dark fringes at the edge of the shadow. Figure 26-15 shows what the shadow caused by a razor blade looks like. Patterns like this are the result of interference between secondary wavelets from different parts of the same wavefront, not from different sources as in Young's double-slit experiment. The wavefronts in a beam of unobstructed light produce secondary wavelets that interfere in such a way as to produce new wavefronts exactly like the old ones. By obstructing part of the wavefronts, points in the shadow region

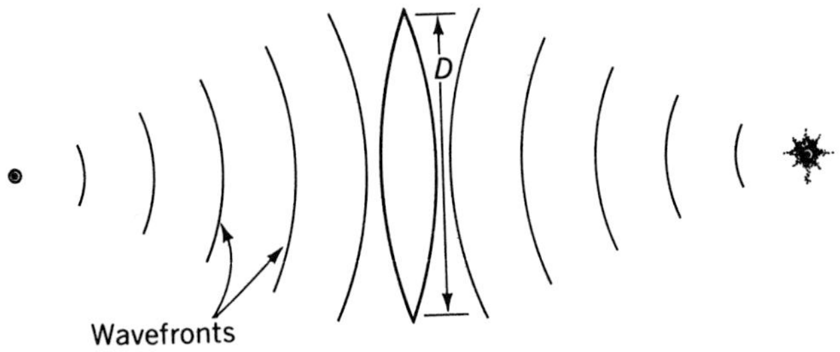

Fig. 26-16.   The image a lens produces of a point source of light is always a tiny disc with bright and dark fringes around it. The smaller the lens, the larger the image.

Wavefronts

Large lens

Small lens

Very small lens

Fig. 26-17.   A large lens or mirror is better able to resolve nearby objects than a small one.

are not reached by secondary wavelets from the entire initial wavefronts but only from part of them, and the result is an interference pattern.

Diffraction sets a limit to the useful magnification of an optical system such as that of a telescope or microscope. Diffraction occurs whenever wavefronts of light are obstructed, and the light that enters a lens (or mirror) is affected by the finite opening which admits only part of each incident wavefront. No matter how perfect a lens is, the image of a point source of light it produces is always a tiny disc of light with bright and dark fringes around it (Fig. 26-16); only if the lens has an infinite diameter can a point source give rise to a point image. The smaller the lens, the larger the image of a point source. The angular width of the radius of this disc of light is about

$$\theta_0 = 1.22 \frac{\lambda}{D} \tag{26-8}$$

in radians, where λ is the wavelength of the light used and D is the lens diameter.

Two objects separated by less than θ_0 cannot be *resolved*, that is, distinguished apart, no matter how high the magnification employed, because their images will overlap (Fig. 26-17). Hence there is no advantage in using a higher magnification than will just reveal features that subtend the angle θ_0 at the position of the lens. While Eq. (26-8) was derived for the image of a point source of light an infinite distance from a lens, it is a reasonable approximation of the resolving power of a telescope or microscope when D is taken as the diameter of the objective lens.

Diffraction limits magnification of optical system

Resolution

If two objects d apart are the distance L from an observer, the angle between them, in radians, is

$$\theta = \frac{d}{L}.$$

Hence Eq. (26-8) can be rewritten

$$d_0 = 1.22 \, \frac{\lambda L}{D}, \qquad\qquad \textit{Resolving power} \quad (26\text{-}9)$$

where d_0 is the minimum separation of objects that can be resolved.

Problem. The pupils of a person's eyes under ordinary conditions of illumination are about 3 mm in diameter, and the distance of most distinct vision is 25 cm for most people. What is the resolving power of the eye at this distance under the assumption that it is limited only by diffraction?

Solution.   Using $\lambda = 550$ nm, which is in the middle of the visible spectrum,

$$d_0 = 1.22 \, \frac{\lambda L}{D} = \frac{1.22 \times 5.5 \times 10^{-7} \text{ m} \times 0.25 \text{ m}}{0.003 \text{ m}}$$

$$= 5.6 \times 10^{-5} \text{ m} = 0.056 \text{ mm}.$$

The photoreceptors in the retina are not quite close enough together to permit this degree of resolution, and 0.1 mm is a more realistic figure under ideal conditions.

26-7 Polarization

A *polarized* beam of transverse waves is one whose vibrations occur in only a single direction perpendicular to the direction in which the beam travels, so that the entire wave motion is confined to a plane called the *plane of polarization* (Fig. 26-18). When many different directions of polarization are present in a beam of transverse waves, vibrations occur equally often in all directions perpendicular to the direction of motion, and the beam is then said to be *unpolarized*. Since the vibrations that constitute longitudinal waves can only take place in one direction, namely that in which the waves travel, longitudinal waves cannot be polarized.

Light waves are transverse, and it is possible to produce and detect polarized light. To clarify the ideas involved, let us first consider the behavior of transverse waves in a stretched string. If the string passes through

Plane of polarization

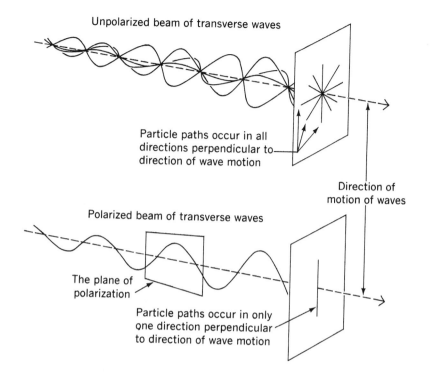

Unpolarized beam of transverse waves

Particle paths occur in all directions perpendicular to direction of wave motion

Direction of motion of waves

Polarized beam of transverse waves

The plane of polarization

Particle paths occur in only one direction perpendicular to direction of wave motion

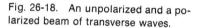

Fig. 26-18. An unpolarized and a polarized beam of transverse waves.

a tiny hole in a fence, as in Fig. 26-19(a), waves traveling down the string are stopped since the string cannot vibrate there. When the hole is replaced by a vertical slot, waves whose vibrations are vertical can get through the fence, but waves with vibrations in other directions cannot (b and c). In a situation in which several waves vibrating in different directions move down the string, the slot stops all but vertical vibrations (d): an initially unpolarized series of waves has become polarized.

The above approach can be used to determine whether a particular wave phenomenon can be polarized or not. In the case of a stretched string, what we do is erect another fence a short distance from the first, as in Fig. 26-19(e). If the slot in the new fence is also vertical, those waves that can get through the first fence can also get through the second. If the slot in the new fence is horizontal, however, it will stop all waves that reach it from the first fence (f).

Should longitudinal waves (say in a spring) go through the fence, it is possible that their amplitudes might decrease in passing through the slots, but the relative alignments of the slot would not matter (Fig. 26-19g and h).

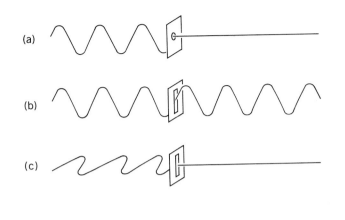

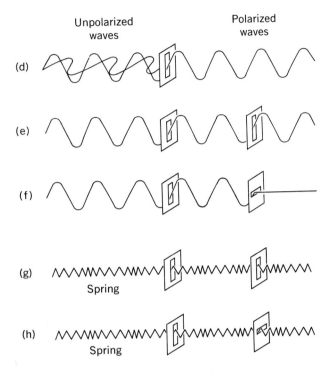

Fig. 26-19. Mechanical analogies of fundamental polarization phenomena.

On the other hand, the alignment of the slots is the critical factor in the case of transverse waves.

The preceding chain of reasoning made it possible for the polarization of light waves to be demonstrated in the last century. A number of substances, for instance quartz, calcite, and tourmaline, have different indexes of refraction for light with different planes of polarization relative to their crystal structures, and prisms can be made from them that transmit light in only a single plane of polarization. When a beam of unpolarized light is incident upon such a prism, only those of its waves whose planes of polarization are parallel to a particular plane in the prism emerge from the other side. The remainder of the waves are absorbed or deflected.

Polaroid is an artificially made polarizing material in wide use that only transmits light with a single plane of polarization. To exhibit the transverse nature of light waves, we first place two Polaroid discs in line so that their axes of polarization are parallel, Fig. 26-20(a), and note that all light passing through one disc also passes through the other. Then we turn one disc until its axis of polarization is perpendicular to that of the other, Fig. 26-20(b), and note that all light passing through one disc is now *stopped* by the other.

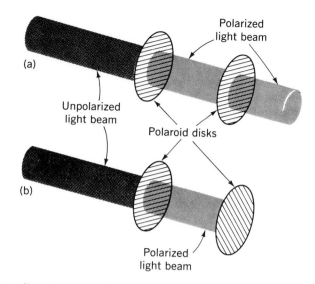

Fig. 26-20. Experiment showing the transverse nature of light waves.

Just what is it whose vibrations are aligned in a beam of polarized light? As discussed in Chapter 24, light waves actually consist of oscillating electric and magnetic fields perpendicular to each other. Because it is the electric fields of light waves whose interactions with matter produce nearly all common optical effects, the plane of polarization of a light wave is considered to be that in which both the direction of its electric field and its direction of motion lie (Fig. 26-21). Even though nothing material moves during the passage of a light wave, it is possible to establish its transverse nature and identify its plane of polarization.

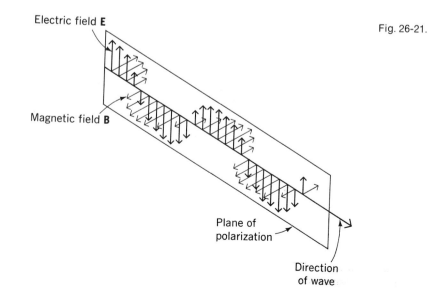

Fig. 26-21.

Special Topic

Numerical Aperture

The *numerical aperture* NA of a microscope objective is a measure of its resolving power. If α is the half-angle of the cone of light rays that reach an objective from an object, as in Fig. 26-22, and n is the index of refraction of the medium between the objective and the object, then by definition

$$NA = n \sin \alpha.$$ *Numerical aperture*

The resolving power of the objective is

$$d_0 = 1.22 \frac{\lambda}{2NA}.$$

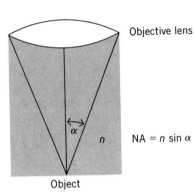

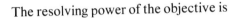

Fig. 26-22. The numerical aperture of a microscope objective depends upon the angular width of the cone of light that reaches it from the object and also upon the index of refraction of the medium between the objective and the object.

The greater the diameter of an objective relative to its focal length and the higher the index of refraction n, the greater the NA and the finer the detail that can be resolved.

For an objective used in air, the angle α has a maximum possible value of 90°, which would mean NA = 1 and $d_0 = 0.61 \lambda$. In practice, α seldom exceeds 60°, which corresponds to an NA of 0.86. However, if a liquid of high index of refraction is used between the objective and the object, the NA can be significantly increased. An immersion oil of n = 1.55 together with a lens for which $\alpha = 60°$ gives an NA of 1.34. The higher the magnification of an objective, the higher the NA it requires in order that the enlarged details be sharp enough to be distinguished. An NA of 0.25 is sufficient for a 10× objective, for instance, whereas an NA of about 1.3 is usual for a 100× objective.

The eyepiece of a microscope provides the user with an image 25 cm in front of his eye, and so the limit to the detail he can see clearly is about 0.1 mm, as mentioned at the end of Section 26-6. The finest detail the microscope itself can resolve is d_0, which for $\lambda = 550$ nm and NA = 1.3 is

$$d_0 = 1.22 \frac{\lambda}{2NA} = \frac{1.22 \times 5.5 \times 10^{-7} \text{ m}}{2 \times 1.3} = 2.6 \times 10^{-7} \text{ m}.$$

The ratio between these figures is

$$M = \frac{10^{-4} \text{ m}}{2.6 \times 10^{-7} \text{ m}} = 385,$$

which represents the maximum useful magnification of the microscope in the sense that higher magnifications will give larger images but will not reveal any more detail. A larger image is nevertheless useful when examining small objects in order to reduce eyestrain, but magnifications much beyond 1000× serve no purpose. For really high magnifications an electron microscope must be used; the electrons in such a device behave in certain respects like waves with very short wavelengths (see Chapter 28) and so greater resolving powers can be attained than are available with an optical microscope.

Important Terms

Two sources of waves are **coherent** if there is a fixed phase relationship between the waves they emit during the time the waves are being observed. Interference can be observed only in waves from coherent sources.

The ability of waves to bend around the edges of obstacles in their paths is called **diffraction**. A **diffraction grating** is a series of parallel slits that produces a spectrum through the interference of light that is diffracted by them.

The **resolving power** of an optical system refers to its ability to produce separate images of nearby objects; resolving power is limited by diffraction, and the larger the objective lens of an optical system, the greater its resolving power.

A **polarized** beam of transverse waves is one whose vibrations occur in only a single direction perpendicular to the direction in which the beam travels, so that the entire wave motion is confined to a plane called the **plane of polarization**. An **unpolarized** beam of transverse waves is one whose vibrations occur equally often in all directions perpendicular to the direction of motion.

Important Formulas

Double slit:

$$y = 0, \frac{L\lambda}{d}, \frac{2L\lambda}{d}, \frac{3L\lambda}{d}, \ldots \quad \text{[bright lines]}$$

Diffraction grating:

$$\sin \theta = n\frac{\lambda}{d} \quad \text{[bright lines]}$$

Resolving power :

$$d_0 = 1.22\,\frac{\lambda L}{D}$$

Multiple Choice

1. For two light beams to interfere, their sources must be
   a. coherent.
   b. incoherent.
   c. lasers.
   d. slits.

2. Coherent electromagnetic waves are not emitted by
   a. two antennas connected to the same radio transmitter.
   b. two pinholes in an opaque shield over a sodium-vapor lamp.
   c. a pinhole in an opaque shield over a sodium-vapor lamp and its reflection in a mirror.
   d. two lasers.

3. In a double-slit experiment, the maximum intensity of the first bright line on either side of the central one occurs on the screen at locations where the arriving waves differ in path length by
   a. $\lambda/4$ b. $\lambda/2$
   c. λ d. 2λ

4. Wavelength determinations cannot be made with the help of
   a. a glass prism.
   b. a pair of narrow parallel slits.
   c. a diffraction grating.
   d. a pair of Polaroid discs.

5. An interference pattern is produced whenever
   a. reflection occurs.
   b. refraction occurs.
   c. diffraction occurs.
   d. polarization occurs.

6. Thin films of oil and soapy water owe their brilliant colors to a combination of reflection and
   a. refraction.
   b. interference.
   c. diffraction.
   d. polarization.

7. A characteristic property of the spectra produced by a diffraction grating is the
   a. sharpness of the bright lines.
   b. diffuseness of the bright lines.
   c. absence of bright lines.
   d. absence of dark lines.

8. The minimum separation of two features of a distant object that can be discerned by a telescope does *not* depend on
   a. the diameter of the objective lens.
   b. the focal length of the objective lens.
   c. the wavelength of the light being used.
   d. the distance to the object.

9. The wavelength of light plays no role in
   a. interference.
   b. diffraction.
   c. resolving power.
   d. polarization.

10. An unpolarized beam of transverse waves is one whose vibrations
    a. are confined to a single plane.
    b. occur in all directions.
    c. occur in all directions perpendicular to their direction of motion.
    d. have not passed through a Polaroid disk.

11. Longitudinal waves do not exhibit
    a. refraction.   b. reflection.
    c. diffraction.   d. polarization.

12. The greater the number of lines that are ruled on a grating of given width,
    a. the shorter the wavelengths that can be diffracted.
    b. the longer the wavelengths that can be diffracted.
    c. the narrower the spectrum that is produced.
    d. the broader the spectrum that is produced.

13. Monochromatic green light of wavelength 5×10^{-7} m illuminates a pair of narrow slits 1 mm apart. The separation of bright lines on the interference pattern formed on a screen 2 m away is
    a. 0.1 mm.   b. 0.25 mm.
    c. 0.4 mm.   d. 1.0 mm.

14. Monochromatic light is used to illuminate a pair of narrow slits 0.3 mm apart, and the interference pattern is observed on a screen 0.9 m away. The second dark band appears 3 mm from the center of the pattern. The wavelength of the light is
    a. 2.2×10^{-7} m.   b. 3.3×10^{-7} m.
    c. 6.7×10^{-7} m.   d. 1.3×10^{-7} m.

15. A pair of binoculars designated "7 × 50" has a magnification of 7 and an objective lens diameter of 50 mm. The smallest detail that in principle can be perceived with such an instrument when view-

ing an object 1 km away in light of wavelength 5×10^{-7} m has a linear dimension of approximately

a. 1 mm.
b. 1 cm.
c. 10 cm.
d. 1 m.

Exercises

1. Can light from incoherent sources interfere? If so, then why is a distinction made between coherent and incoherent sources?

2. The waves used to carry television signals cannot reach receivers beyond the visual horizon of their transmitting antennas, while ordinary radio waves readily travel beyond the visual horizon of their transmitting antennas. Can you think of a reason for this contradictory behavior?

3. Radio waves diffract pronouncedly around buildings, while light waves, which are also electromagnetic waves, do not. Why?

4. In Young's double-slit experiment, which effects are due to diffraction and which to interference?

5. What do diffraction and interference have in common? How do they differ?

6. Which of the following can occur in (a) transverse waves and (b) longitudinal waves: refraction, dispersion, interference, diffraction, polarization?

7. Explain the peculiar appearance of a distant light source when seen through a piece of finely woven cloth.

8. What advantages has a diffraction grating over a double slit for determining wavelengths of light?

9. What is the difference between the first-order and the second-order spectra produced by a grating? Which is wider? Does a prism produce spectra of different orders?

10. What governs the angular width of the first-order spectrum of white light produced by a grating?

11. What becomes of the energy of the light waves whose destructive interference leads to dark lines in an interference pattern?

12. As a soap bubble is blown up, its wall becomes thinner and thinner. Just before the bubble breaks, the thinnest part of its wall turns black. Why?

13. Give two advantages a large-diameter telescope objective has over a small-diameter one.

14. A camera can be made by using a pinhole instead of a lens. What happens to the sharpness of the picture if the hole is too large? If it is too small?

15. Since light consists of transverse waves, why is not every light beam polarized?

16. What is the relationship between the plane of polarization of a transverse wave and its direction of propagation?

17. How can a single sheet of Polaroid be used to show that sky light is partially polarized?

18. In Young's experiment, how does the spacing of the slits affect the pattern of bright and dark lines?

19. Light of unknown wavelength is used to illuminate two parallel slits 1 mm apart. Adjacent bright lines on the interference pattern that results on a screen 1.5 m away are 0.65 mm apart. What is the wavelength of the light?

20. Two parallel slits 0.12 mm apart are illuminated by light of wavelength 6×10^{-7} m. A viewing screen is 1.5 m from the slits. (a) How far from the central bright line is the next bright line? (b) How far is the first dark line? (c) How far is the fifth dark line?

21. Two parallel slits 0.1 mm apart are illuminated by light of wavelength 5.46×10^{-7} m. A viewing screen is 0.8 m from the slits. (a) How far from the central bright line is the second bright line? (b) How far is the third bright line? (c) How far is the third dark line?

22. A 5000 line/cm diffraction grating produces an image deviated by 30° in the second order. Find the wavelength of the light.

23. Light of wavelength 7.5×10^{-7} m is shined on a grating ruled with 4000 lines/cm. What is the angular deviation of this light in (a) the first order, and (b) the third order?

24. A radar operating at a wavelength of 3 cm has a resolving power of 10 m at a range of 1 km. What is the width of its antenna?

25. A radar has a resolving power of 30 yards at a range of $\frac{1}{2}$ mile. What is the minimum width of its antenna in inches if its operating frequency is 9500 mHz?

26. An astronaut circles the earth in a satellite at an altitude of 150 km. If the diameter of his pupils is 2 mm and the average wavelength of the light reaching his eyes is 550 nm, is it conceivable that he can distinguish sports stadiums on the earth's surface? Private houses? Cars?

27. The largest telescope in the world, the Hale telescope at Mt. Palomar in California, has a concave mirror 5 m in diameter. How many meters apart must two features of the moon's surface be in order to be resolved by this telescope? Take the distance from the earth to the moon as 386,000 km and use 500 nm for the wavelength of the light.

28. The index of refraction of a material used to provide an antireflection coating on a lens is 1.25. How thick should the coating be for maximum cancellation at 560 nm?

Problems

1. Two parallel slits are illuminated by light of two wavelengths, one of which is 5.8×10^{-7} m. On a viewing screen an unknown distance from the slits the fourth dark line of the light of the known wavelength coincides with the fifth bright line of the light of the unknown wavelength. Find the unknown wavelength.

2. Two parallel slits 0.2 mm apart are illuminated by light of two wavelengths, 5×10^{-7} m and 6×10^{-7} m. A viewing screen is 2 m away from the slits. How far are the bright lines of one wavelength from the bright lines of the other?

3. How many diffracted images are formed on either side of the central image when radiation of wavelength 6×10^{-7} m falls on a 4000 line/cm grating?

4. Light containing wavelengths of 5.0 and 5.7×10^{-7} m is directed at a 2000 line/cm grating. How far apart are the lines formed by these wavelengths on a screen 5 m away in the second order?

5. White light that contains wavelengths from 4 to 7×10^{-7} m is directed at a 3000 line/cm grating. How wide is the first-order spectrum on a screen 2 m away?

6. White light that contains wavelengths from 4 to 7×10^{-7} m is directed at a 5000 line/cm grating. Do the first- and second-order spectra overlap? The second- and third-order spectra? If there is an overlap, will the use of a grating with a different number of lines/cm change the situation?

7. The Jodrell Bank radiotelescope has a parabolic reflecting "dish" 76 m in diameter. (a) What is the angular diameter in degrees of a point source of radio waves of wavelength 21 cm as seen by this telescope? (b) How does the above figure compare with the angular width of the moon, whose diameter is 2160 miles and whose average distance from the earth is 240,000 mi?

8. At night, the pupils of a man's eyes are 8 mm in diameter. (a) How many km away from a car facing him will he be able to distinguish its headlights from one another? (b) If his pupils were 4 mm in diameter (say at twilight), how far away from the car could he distinguish its headlights from one another? Assume the headlights are 1.5 m apart and that the average wavelength of their light is 600 nm.

9. The smaller the aperture of a camera lens, the greater the depth of field. However, a small aperture means reduced resolution. The criterion for an enlarged print to show sharp detail from a small negative is that the image of a point object on it be no more than about 0.01 mm across. Find the maximum $f/$ number (this is the ratio between the

focal length and diameter of a lens) of a camera lens in order that this criterion be met for a distant object in 550-nm light.

10. A certain laser produces a beam of monochromatic light whose wavelength is 550 nm and whose initial diameter is 1 mm. (a) In what distance will diffraction have caused the beam to double its diameter? (b) If the initial diameter of the beam were 1 cm, would the doubling distance be different? If so, what would it be?

Answers to Multiple Choice

1. a	6. b	11. d
2. b	7. a	12. d
3. c	8. b	13. d
4. d	9. d	14. c
5. c	10. c	15. b

27

Relativity

Few physical theories represent so drastic an assault on traditional habits of thought as the theory of relativity. Relativity links time and space, matter and energy, electricity and magnetism —and for all the seeming magic of its conclusions, most of them can be reached with the simplest of mathematics. The theory of relativity was proposed in 1905 by Albert Einstein, and little of physical science since then has remained unaffected by his ideas.

27-1 Special Relativity

Thus far no special point was raised about how measurements of such physical quantities as length, time, and mass are carried out. It was simply assumed that these quantities could be determined in some way, and that, since standard units have been established for each of them, it doesn't matter who makes a particular determination—everybody ought to get the same figure. There is certainly no question of principle associated with, say, finding the length of an airplane on the ground: all we need do is place one end of a tape measure at the airplane's nose and note the number on the tape at the airplane's tail.

But what if the airplane is in flight and we are on the ground? It is not hard to find the length of a distant object with the help of a surveyor's transit to measure angles, a tape measure to establish a base line, and a knowledge of trigonometry. When the object is moving, however, things become more complicated because now we must take into account the fact that light does not travel instantaneously from one place to another but does so at a definite, fixed velocity—and light is the means by which information is carried from a distant object to our measuring instruments. When a careful analysis is made of the problem of measuring physical quantities when there is relative motion between the measuring instruments and whatever is being observed, many surprising results emerge.

In Chapter 2 the notion of *frame of reference* was introduced. When we observe something moving, what we actually detect is that its position relative to something else is changing (Fig. 27-1). A passenger moves relative to a train; the train moves relative to the earth; the earth moves relative to the sun; the sun moves relative to the galaxy of stars (the "Milky Way") of which it is a member; and so on. In each case a frame of reference is part of the description of the motion: it is meaningless to say that something is moving unless it is understood with respect to what the motion occurs.

There is no universal frame of reference that can be used everywhere. If we see something changing its position with respect to us at constant velocity, we have no way of knowing whether *it* is moving or *we* are moving. If we were isolated from the rest of the universe, we would be unable to find out if we were moving at constant velocity or not—indeed, the question would make no sense. All motion is relative to the observer, and there is no such thing as "absolute motion."

The theory of relativity is concerned with the physical consequences of the absence of a universal frame of reference. The special theory of relativity, published in 1905 by Albert Einstein, is confined to problems involving the motion of frames of reference at constant velocity (that is,

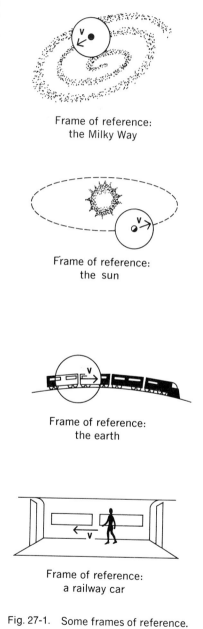

Frame of reference:
the Milky Way

Frame of reference:
the sun

Frame of reference:
the earth

Frame of reference:
a railway car

Fig. 27-1. Some frames of reference.

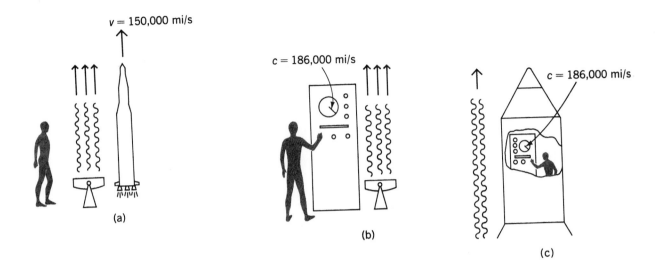

Fig. 27-2. The velocity of light is the same to all observers.

both constant speed and constant direction) with respect to one another; the general theory of relativity, published 10 years later by Einstein, deals with problems involving frames of reference accelerated with respect to one another. The special theory has had an enormous impact on all of physics, and its chief conclusions will be examined here.

Two principles are fundamental to the special theory of relativity. The principle of relativity states that

Principle of relativity

The laws of physics are the same in all frames of reference moving at constant velocity with respect to one another.

This principle follows directly from the absence of a universal frame of reference. If the laws of physics were different for different observers in relative motion, they could infer from these differences which of them were "stationary" in space and which were "moving." But such a distinction does not exist in nature, and the principle of relativity is an expression of this fact.

Thus experiments of any kind performed, for instance, in an elevator which is ascending at a constant velocity yield exactly the same results as the same experiments performed when the elevator is at rest or is descending at a constant velocity. On the other hand, an isolated observer *can* detect accelerations, as any elevator passenger can verify.

The second principle states that

The velocity of light in free space has the same value for all observers, regardless of their state of motion.

At first glance the constancy of the velocity of light may not seem so very extraordinary, which is a misleading impression. Let us examine a hypothetical experiment in essence no different from actual experiments that have been performed in a number of ways.

Suppose I turn on a searchlight at the same moment you take off in a spacecraft at a speed of 150,000 mi/s (Fig. 27-2). We both measure the speed of the light waves from the searchlight using identical instruments. From the ground I find their speed to be 186,000 mi/s, as usual. "Common sense" tells me you ought to find a speed of (186,000 − 150,000) mi/s or only 36,000 mi/s for the same light waves. But you also find their speed to be 186,000 mi/s, even though to me you seem to be moving parallel to the waves at 150,000 mi/s. As so often, common sense is wrong.

There is only one way to account for the apparent discrepancy between the above results without violating the principle of relativity, and that is to conclude that measurements of space and time are not absolute but depend upon the relative motion of the observer and that which is observed. If your clock ticks more slowly than it did on the ground and your meter stick is shorter in the direction of motion of the spacecraft, then you will find 186,000 mi/s for a velocity that I think should be only 36,000 mi/s. To you, your clock and meter stick are the same as they were on the ground before you took off, but to me they are different because of the relative motion. Time intervals and lengths are relative quantities, not absolute ones.

27-2 The Relativity of Time

Measurements of time intervals are affected by relative motion between an observer and what he observes. As a result, all moving clocks tick more slowly than clocks at rest do, and all natural processes (including those of life) that involve regular time intervals occur more slowly when they take place in a moving frame of reference.

We begin by considering the operation of the particularly simple clock shown in Fig. 27-3. In this clock a pulse of light is reflected back and forth

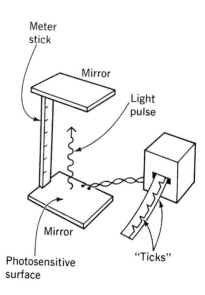

Fig. 27-3. A light-pulse clock.

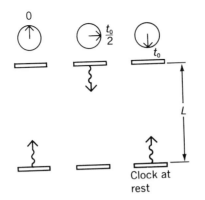

Fig. 27-4. Light-pulse clock in the laboratory.

between two mirrors. Whenever the light strikes the lower mirror, an electrical signal is produced that is registered as a mark on the recording tape. Each mark corresponds to the tick of an ordinary clock.

Let us consider two of these clocks, one of them at rest in a laboratory and another in a spaceship moving at the velocity v relative to the laboratory. An observer in the laboratory watches both clocks: does he find that they tick at the same rate?

Figure 27-4 shows the laboratory clock in operation. The mirrors are L apart, and the time interval between ticks is t_0. Hence the time needed for the light pulse to travel the distance L between the mirrors at the velocity

Fig. 27-5. Light-pulse clock in the spaceship as seen by an observer in the laboratory.

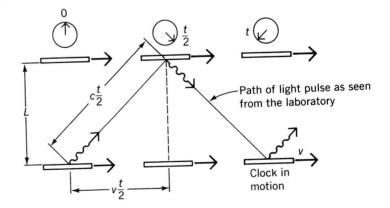

c is $t_0/2$, and so

$$L = c\left(\frac{t_0}{2}\right), \qquad t_0 = \frac{2L}{c}.$$

Figure 27-5 shows the moving clock as seen from the laboratory. The time interval between ticks is t. Because the clock is moving, the pulse of light follows a zigzag path in which it travels the distance $ct/2$ in going from one mirror to the other in the time $t/2$. From the Pythagorean theorem,

Light in the moving clock follows a zigzag path

$$\left(\frac{ct}{2}\right)^2 = L^2 + \left(\frac{vt}{2}\right)^2.$$

How is t related to t_0? To find out, we first solve the preceding equation for t:

$$\left(\frac{ct}{2}\right)^2 = L^2 + \left(\frac{vt}{2}\right)^2, \qquad \frac{t^2}{4}(c^2 - v^2) = L^2,$$

$$t^2 = \frac{4L^2}{(c^2 - v^2)} = \frac{(2L)^2}{c^2(1 - v^2/c^2)},$$

$$t = \frac{2L/c}{\sqrt{1 - v^2/c^2}}.$$

But the quantity $2L/c$ is the time interval t_0 between ticks in the laboratory clock, as we saw. Hence

$$t = \frac{t_0}{\sqrt{1 - v^2/c^2}}. \qquad\qquad \textit{Time dilation (27-1)}$$

Here is a reminder of what the symbols in this important formula represent:

t_0 = time interval on clock at rest,

t = time interval on clock in relative motion as determined by outside observer,

v = velocity of relative motion,

c = velocity of light.

Because the quantity $\sqrt{1 - v^2/c^2}$ is always smaller than 1 for a moving object, t is always greater than t_0. *A clock moving with respect to an observer*

A moving clock ticks more slowly than one at rest

ticks more slowly than a clock that is stationary with respect to the same observer. This effect is referred to as *time dilation* (to *dilate* is to become larger).

Now let us turn the situation around and ask what an observer in a spacecraft finds when he compares his clock with one on the ground. The only change needed in the preceding derivation is the direction of motion: if the man on the ground sees the spacecraft moving to the east, the man in the spacecraft sees the laboratory on the ground moving to the west. To the man in the spacecraft the light pulse of the ground clock follows a zigzag path that requires a total time per round trip of

$$t = \frac{t_0}{\sqrt{1 - v^2/c^2}},$$

whereas the light pulse in his own clock takes t_0 for the round trip. Thus to the man in the spacecraft the clock on the ground ticks at a slower rate than his own clock does. A clock moving relative to an observer *always* is slower than a clock at rest relative to him, regardless of where the observer is located.

A light clock is a rather more exotic timepiece than most of us are accustomed to. What if a stationary cuckoo clock and a moving one are compared: do we again find that the moving clock runs more slowly?

The principle of relativity makes it easy for us to predict the outcome of the experiment. Suppose cuckoo clocks tick at exactly the same rate to all observers, whether there is relative motion or not. We put a cuckoo clock and a light-pulse clock (which *does* tick more slowly when in motion) on a spacecraft. On the ground they show the same time.

In flight, the two clocks show different times to an observer on the ground, since the light-pulse clock ticks slower whereas the cuckoo clock (by hypothesis) does not. To an observer in the spacecraft, however, the two clocks agree, since to him the clocks are stationary and it is the ground which is moving away from him. Therefore the laws of physics which govern the operation of the clocks must be different on the spacecraft from what they are on the ground—which contradicts the principle of relativity. *All moving clocks tick more slowly than clocks at rest.*

Time dilation is independent of the nature of the clock

It is important to keep in mind that the slowing down of a moving clock is significant only at relative velocities not far from the velocity of light. Such velocities are readily attained by elementary particles, and most of the experiments that have confirmed time dilation have employed such particles. Today's spacecraft are far too slow to exhibit time dilation. For instance, the highest velocity reached by the Apollo 11 spacecraft on its way to the moon was only about 10,840 m/s, or 0.0036 percent of the velocity of light. At this velocity, clocks on the spacecraft differ from those on the earth by less than 1 part in 10^9.

27-3 Muon Decay

A good illustration of time dilation occurs in the decay of the unstable elementary particles called *muons*.

Muons have masses 207 times that of the electron and may have positive or negative electric charges. A muon decays into an electron an average of 2.0×10^{-6} s (2.0 μs) after it comes into being. Muons are created in the atmosphere thousands of meters above sea level as the ultimate result of collisions between air atoms and fast cosmic-ray particles (largely protons) that reach the earth from space. A muon passes through each cm² of the earth's surface a little more often than once a minute. The muon velocities are observed to be about 2.994×10^8 m/s, or 0.998 c where c is the velocity of light. But in $t_0 = 2.0 \times 10^{-6}$ s, the average muon lifetime, they can travel a distance of only

A muon at rest decays into an electron in 2 microseconds

$$vt_0 = 2.994 \times 10^8 \frac{m}{s} \times 2.0 \times 10^{-6} \text{ s}$$

$$= 600 \text{ m,}$$

whereas they actually come into being at elevations ten or more times greater than this.

The key to resolving this paradox is to note that the average muon lifetime of 2 μs is what an observer at rest with respect to a muon would find. If we could collect some muons at the instant of their creation and time their decays when they are at rest, we would find an average of $t_0 = 2$ μs.

However, when we are on the ground and the muons are hurtling toward us at the considerable speed of 0.998 c, we find instead that their lifetimes have been extended by time dilation to

$$t = \frac{t_0}{\sqrt{1 - v^2/c^2}}$$

$$= \frac{2 \times 10^{-6} \text{ s}}{\sqrt{1 - \frac{(0.998\ c)^2}{c^2}}} = 31.6 \times 10^{-6} \text{ s.}$$

Moving muons have longer lifetimes

The fast muons have lifetimes almost 16 times longer than those at rest. In a time interval of $t = 31.6 \times 10^{-6}$ s, a muon whose velocity is 0.998 c can cover the distance

$$vt = 2.994 \times 10^8 \frac{\text{m}}{\text{s}} \times 31.6 \times 10^{-6} \text{ s} = 9500 \text{ m.}$$

Despite its brief lifespan of $t_0 = 2$ μs in its own frame of reference, a muon is able to reach the ground from considerable altitudes because in the frame of reference in which these altitudes are measured, the muon lifetime is $t = 31.6$ μs.

27-4 The Lorentz Contraction

As we saw, the arrival of cosmic-ray muons at sea level from high altitudes is not in conflict with the brevity of their lives (in their frames of reference) since these lives are increased 16-fold (in our frame of reference) by their relative motion. But what if somebody could accompany the muons downward at the same velocity of 0.998 c, so that to him the muons are at rest? Both the muons and the observer are now in the same frame of reference, and the muon lifetime is only 2 μs in this frame. The question is, does the moving observer find that the muons reach the ground, or does he find that they decay beforehand?

The principle of relativity states that the laws of physics are the same in all frames of reference moving at constant velocity with respect to one another. An observer on the ground and an observer moving with the muons are in relative motion at a constant velocity, namely 0.998 c, and if we on the ground find that the muons reach our apparatus before they decay, then the moving observer must also find the same thing. Though the appearance of an event may be different to different observers, the fact of the event's occurrence is not subject to dispute.

Distances are shorter to a moving object

The only way an observer in a muon's frame of reference can reconcile its arrival at sea level with the lifetime of 2 μs he finds is if the distance the muon travels is shortened by virtue of its motion. The principle of relativity enables us to infer at once the extent of the shortening — it must be by the same factor of $\sqrt{1 - v^2/c^2}$ that the muon lifetime is extended from the point of view of a stationary observer. Thus a distance we on the ground measure to be L_0 will appear to the muon as the abbreviated distance

$$L = L_0 \sqrt{1 - v^2/c^2}$$

In our frame of reference, the average distance a muon can go at the velocity $v = 0.998\ c$ before it decays is $L_0 = 9500$ m. The corresponding distance in the muon's frame of reference is

$$L = L_0 \sqrt{1 - v^2/c^2}$$

$$= 9500 \text{ m} \times \sqrt{1 - \frac{(0.998\ c)^2}{c^2}} = 600 \text{ m}.$$

This is precisely how far a muon traveling at $0.998\ c$ can go in 2 μs. Both points of view—from the frame of reference of someone on the ground, to whom the muon lifetime is dilated, and from the frame of reference of the muon itself, to which the distance to the ground is contracted—give the same result.

The relativistic shortening of distances is an example of the general *Lorentz contraction* of lengths:

$$L = L_0 \sqrt{1 - v^2/c^2} \qquad \text{\textit{Lorentz contraction}} \quad (27\text{-}2)$$

The symbols in this formula have these meanings:

$L_0 =$ length measured when the object is at rest,

$L =$ length measured when the object is in relative motion,

$v =$ velocity of relative motion,

$c =$ velocity of light.

The length of an object in motion with respect to an observer is measured by the observer to be shorter than when it is at rest with respect to him. This shortening works both ways; to a man in a spacecraft, measurements indicate that objects on the earth are shorter than they were when he was on the ground, and someone on the ground finds that the spacecraft is shorter in flight than when it was at rest (Fig. 27-6). (To the man in the spacecraft, its length is the same whether on the ground or in flight, since it is always at rest with respect to him.) The length of an object is a maximum when determined in a reference frame in which it is stationary, and its length is less when determined in a reference frame in which it is moving.

Moving objects are shorter than when at rest

Only lengths in the direction of motion undergo contractions. Thus to the outside observer a spacecraft is shorter in flight than on the ground, but it is not narrower.

The Lorentz contraction occurs only in direction of motion

The relativistic length contraction is negligible for ordinary speeds, but is an important effect at speeds close to the speed of light. A speed of 1000

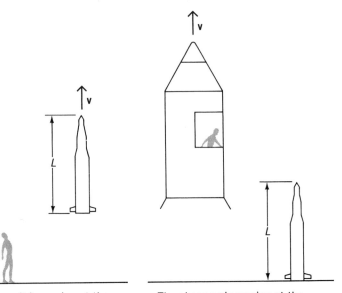

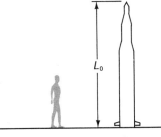

An observer and a spacecraft are at rest on the ground. The observer finds the spacecraft's length to be L_0.

The spacecraft is moving at the velocity v. The observer on the ground finds its length to be $L = L_0\sqrt{1 - v^2/c^2}$.

The observer is moving at the velocity v and the spacecraft is on the ground. The observer finds its length to be $L = L_0\sqrt{1 - v^2/c^2}$.

Fig. 27-6.

mi/s seems enormous to us, yet it results in a shortening in the direction of motion to only

$$\frac{L}{L_0} = \sqrt{1 - (v^2/c^2)} = \sqrt{1 - \frac{(1000 \text{ mi/s})^2}{(186{,}000 \text{ mi/s})^2}}$$

$$= 0.999985 = 99.9985\%$$

of the length at rest. On the other hand, something traveling at nine-tenths the speed of light is shortened to

$$\frac{L}{L_0} = \sqrt{1 - \frac{(0.9c)^2}{c^2}} = 0.436 = 43.6\%$$

of its length at rest, a significant change.

27-5 Origin of Magnetic Forces

The theory of relativity provides the connection between electric and magnetic phenomena. It is hardly obvious that, when we pick up a nail with a magnet, we are witnessing a consequence of relative motion. Most relativistic effects are imperceptible in everyday life because the speeds of the objects around us are so small compared with the speed of light. Even though experiments show that there are moving electrons in the atoms of the nail and the magnet, their speeds are nowhere near that of light. The puzzle is underscored when we consider that the effective speeds of the electrons that carry a current in a wire are less than 1 mm/s — slower than a caterpillar — yet current-carrying wires do give rise to appreciable magnetic effects, as anyone who has seen an electric motor in operation can testify.

If we think about the matter for a moment, though, the idea that electricity and magnetism are connected via relativity becomes less implausible. For one thing, electric forces are extremely strong, so even a small alteration in their character due to relative motion (which is what magnetic forces represent) may have large consequences. As we saw in Chapter 17 the electric force between the electron and proton in a hydrogen atom is more than 10^{39} times greater than the gravitational force between them. Second, though the individual charges involved in magnetic forces usually do move slowly, there may be such enormous numbers of them that the total effect is not negligible; for example, even a modest current in a wire involves the motion of 10^{20} electrons in each centimeter of the wire.

As an illustration of how relativity accounts for the magnetic forces between moving charges, let us look into how the forces between two parallel currents come into being. In doing so, we must keep in mind that, like the velocity of light, electric charge is relativistically invariant, so that a charge whose magnitude is found to be q in one frame of reference will be found to be q in all other frames of reference regardless of their relative velocities.

Figure 27-7(a) shows two parallel conductors when no current is present. They contain equally-spaced positive and negative charges that are at rest. The conductors are electrically neutral.

In Fig. 27-7(b) we see the same conductors when they carry currents in the same direction. The positive charges move to the right at the velocity u and the negative charges move to the left at the same velocity u, as seen from the laboratory frame of reference. The spacing of the charges is smaller than before by the factor $\sqrt{1 - v^2/c^2}$ because of the relativistic length contraction. Since the charges of both signs have the same velocity in the

Magnetism is a relativistic effect

Electric charge is relativistically invariant

idealized situation we are considering, the contractions in their spacings are the same, and the conductors are still neutral to an observer in the laboratory frame of reference. There is an attractive force between the conductors: How does it arise?

We begin by looking at conductor 2 from the frame of reference of one of the negative charges in conductor 1 (Fig. 27-7c). To this charge, the negative charges in conductor 2 are at rest, since they are (as we see the situation from the outside) all moving at the same velocity u as it is. The spacing of the negative charges is not contracted, as it is to an observer in the laboratory, so they are farther apart than in the previous diagram. However, in this frame the positive charges in conductor 2 are moving at the velocity $2u$, and their spacing accordingly exhibits a greater contraction. Conductor 2 therefore appears positively charged to the negative charge in conductor 1, and there is an attractive electric force on this charge in its own frame of reference.

From the frame of reference of one of the positive charges in conductor 1, the positive charges in conductor 2 are at rest and their spacing, in the absence of any relativistic length contraction, is greater than we find in the laboratory (Fig. 27-7d). The negative charges in conductor 2 have the velocity $2u$ and they are accordingly closer together than in the laboratory frame of reference. There is a net negative charge on conductor 2 as seen

Origin of forces on charges in conductor 1

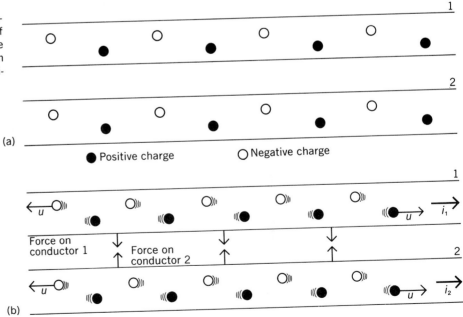

Fig. 27-7. (a) Two idealized conductors that contain equal numbers of positive and negative charges. (b) The conductors attract each other when they carry currents in the same direction.

by a positive charge in conductor 1, and it is attracted electrically to conductor 2.

An identical argument shows that both the negative and positive charges in conductor 2 are attracted to conductor 1. To any of the charges in either conductor, the force on it is an "ordinary" electric force that occurs because the charges of opposite sign in the other conductor are closer together than the charges of the same sign, yielding a net attractive force. To an observer in the laboratory, both conductors are electrically neutral, and he therefore finds it natural to ascribe the force to a special "magnetic" interaction between the currents in the conductors.

As in everything else where there is relative motion, the frame of reference from which a phenomenon is viewed is an essential part of the description of the phenomenon. Although for many purposes it is convenient to think of magnetic forces as something different from electric ones, it is worth keeping in mind that both are manifestations of a single electromagnetic interaction that occurs between charges.

A similar approach accounts for the repulsive force between parallel currents in opposite directions. Again, the "magnetic force" turns out to be an inevitable consequence of Coulomb's law, charge invariance, and the principles of special relativity.

Repulsive magnetic forces

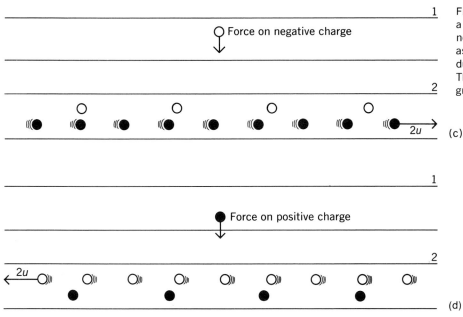

Fig. 27-7.   (c) Conductor 2 as seen by a negative charge in conductor 1 has a net positive charge. (d) Conductor 2 as seen by a positive charge in conductor 1 has a net negative charge. The various length contractions are greatly exaggerated here.

Actual currents in metal wires consist of flows of electrons only, with the positive ions remaining in place. The advantage of considering the idealized currents above, which are electrically equivalent to actual currents, is that they are easier to analyze; the results are exactly the same in both cases.

As we have seen, a current-carrying conductor which is electrically neutral in one frame of reference might not be neutral in another frame. But this observation does not apply to the *entire* circuit of which the conductor is a part. Every electric circuit in which a current exists more than momentarily is a closed circuit, so for every current element in one direction that a moving observer finds to have a positive charge, there must be another current element in the opposite direction, which the same observer finds to have a negative charge. Hence we would expect magnetic forces to occur between different parts of a circuit, which is experimentally observed, even though all observers agree on the electrical neutrality of the circuit as a whole. The latter agreement is required by charge invariance.

27-6 Relativity of Mass

Another important finding of the theory of relativity is that the mass of a body is not the same to all observers but depends upon the body's velocity with respect to each observer who measures its mass. The variation of mass with velocity obeys the formula

$$m = \frac{m_0}{\sqrt{1 - v^2/c^2}},$$ *Mass increase*   (27-3)

whose symbols have these meanings:

m_0 = mass measured when object is at rest ("rest mass"),

m = mass measured when object is in relative motion,

v = velocity of relative motion,

c = velocity of light.

A moving object is more massive than when it is at rest

Since the denominator of Eq. (27-3) is always less than one, an object will always appear more massive when in relative motion than when at rest. This mass increase is reciprocal; to a measuring device on the rocket ship in flight its twin ship on the ground also appears to have a mass m greater than its own mass m_0.

Relativistic mass increases are significant only at velocities approaching that of light (Fig. 27-8). The rest mass of the Apollo 11 spacecraft (apart from its launch vehicle, which dropped away after accelerating the space-craft) was about 63,070 kg (nearly 70 tons). On its way to the moon, the spacecraft's velocity was about 10,840 m/s (6,730 miles/s) and so its mass in flight, as measured by an observer on the earth, increased to

$$m = \frac{m_0}{\sqrt{1 - \dfrac{v^2}{c^2}}} = \frac{63{,}070 \text{ kg}}{\sqrt{1 - \dfrac{(1.084 \times 10^4 \text{ m/s})^2}{(3 \times 10^8 \text{ m/s})^2}}} = 63{,}070.000041 \text{ kg.}$$

This is not much of a change.

Smaller objects can be given much higher velocities. It is not very difficult to accelerate electrons (rest mass 9.109×10^{-31} kg) to velocities of, say, $0.9999c$. The mass of such an electron is

$$m = \frac{m_0}{\sqrt{1 - \dfrac{v^2}{c^2}}} = \frac{9.109 \times 10^{-31} \text{ kg}}{\sqrt{1 - \dfrac{(0.9999c)^2}{c^2}}} = 644 \times 10^{-31} \text{ kg.}$$

The electron's mass is 71 times its rest mass! The mass increases predicted by the relativistic mass formula, even such remarkable ones as this, have been experimentally verified without exception.

Equation (27-3) has something interesting to say about the greatest velocity an object can have. The closer v approaches c, the closer v^2/c^2 approaches one, and the closer $\sqrt{1 - (v^2/c^2)}$ approaches zero. As the de-nominator of Eq. (27-3) becomes smaller, the mass m becomes larger, so that if the relative velocity v actually were equal to the velocity of light, the object's mass would be infinite. The concept of an infinite mass anywhere in the universe is, of course, nonsense on many counts; it would have re-quired an infinite force to have accelerated it to the velocity of light, its length in the direction of motion would be zero by Eq. (27-1) so that its volume would be zero, and it would exert an infinite gravitational force on all other bodies in the universe. Hence we interpret Eq. (27-3) to mean that no material body can ever equal or exceed the velocity of light.

The most famous conclusion of the theory of relativity is the equiva-lence of mass and energy according to the formula $E = mc^2$, which was discussed in Section 7-7.

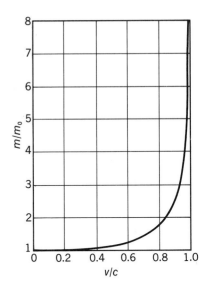

Fig. 27-8. The relativity of mass.

No object can move faster than light

Special Topic

General Relativity

The special theory of relativity shows us how to interpret what we observe in frames of reference that move at constant velocity with respect to us. The laws of physics are valid in all such frames of reference, but measurements of some quantities—notably time intervals, lengths, and masses—are affected by relative motion, whereas measurements of others—notably the velocity of light and electric charge—are not. The *general theory of relativity* explores the effects of accelerated motion on what we observe. Published by Einstein in 1915, it provides insights into gravitational phenomena as profound and far-reaching as those of the special theory into the relationships between mass and energy and between electricity and magnetism.

One of the basic ideas of general relativity is the *principle of equivalence:* An observer in a closed laboratory cannot distinguish between the effects produced by a gravitational field and those produced by an acceleration of the laboratory. This principle is another way to express the experimental finding that the inertial mass of an object, which determines its acceleration when a force acts on it, is always equal to its gravitational mass, which determines the gravitational pull another object exerts on it. (The two masses are actually proportional; the constant of proportionality is set equal to 1 by an appropriate choice of the gravitational constant G.)

An immediate consequence of the principle of equivalence is that light should be subject to gravity. We can come to this conclusion in two ways. First, light carries energy, and since $E = mc^2$, inertial mass must be associated in some way with light waves. Gravitational mass is inseparable from inertial mass, hence light waves are affected by gravity. Alternatively, we can start by considering the passage of a light beam across an accelerated laboratory. The light beam will pursue a curved path relative to the laboratory, which to an observer there is exactly the path that would be taken if it were subject to the same gravitational field the acceleration is equivalent to.

To be sure, the gravitational mass of a pulse of light is not very great, but its existence has been confirmed in a variety of experiments. The first was a comparison of the positions of stars in the sky that appeared near the sun during an eclipse, when the sun's disk was obscured by the moon, with their positions at other times when the light from them did not have to pass close to the sun. The observed deviation of about 0.0005° for light grazing the sun was in agreement with the prediction of general relativity. More recently laboratory experiments based on the quantum theory of light (Chapter 28) have independently verified that light is indeed subject to gravity.

Gas molecules are held in the atmospheres of the earth and other planets by gravity despite their considerable average speeds. Is it possible that an astronomical object both massive enough and small enough to have an enormously strong gravitational field could similarly prevent light from escaping from it? The answer seems to be yes. Such an object is called a "black hole" because, besides emitting no light of its own, it sucks in light from other sources that happens to pass nearby. Objects that behave in the manner predicted for black holes have been identified by astronomers.

The Twin Paradox

Another conclusion of general relativity — also verified by experiment — is that a clock of any kind ticks more slowly in a strong gravitational field than in a weak one. A clock whose ticks are t_0 apart when it is far away from all other matter will have an interval between ticks of

$$t = \frac{t_0}{\sqrt{1 - 2GM/Rc^2}}$$

when it is R away from a body of mass M, where G is the constant of gravitation. This finding enables us to understand more fully the famous relativistic phenomenon called the twin paradox.

The twin paradox concerns the slowing down of a moving clock, which was discussed in Section 27-2. Since life processes occur with regular rhythms, a person constitutes a biological clock and must behave in the same way as any other clock when in motion relative to an observer. There is no difference in principle between heartbeats and the ticks of a clock. This means that the life processes of a person in a moving spacecraft occur at a slower rate than they do on the ground, so he ages more slowly than somebody on the ground does.

The celebrated case of the twin brothers Alphonse and Gaston illustrates the consequences of time dilation in space travel. Alphonse is 20 years old when he embarks on a space voyage at a speed of 2.97×10^8 m/s, which is 99% of the velocity of light. To Gaston, who has stayed behind, the pace of Alphonse's life is slower than his own by a factor of

$$\sqrt{1 - v^2/c^2} = \sqrt{1 - (0.99c)^2/c^2} = 0.141 \approx 1/7.$$

Alphonse's heart beats only once for every 7 beats of Gaston's heart; Alphonse takes only one breath for every 7 of Gaston's; Alphonse thinks only one thought for every 7 of Gaston's. Eventually Alphonse returns after 70 years have elapsed by Gaston's calendar—but Alphonse is only 30 years old, whereas Gaston, the stay-at-home twin, is 90 years old.

Gaston is baffled by the youth of his astronaut brother. "After all," he argues, "according to the principle of relativity, *my* life processes should have appeared 7 times slower to Alphonse, so by the same reasoning I ought to be 30 and Alphonse ought to be 90."

But advanced age has dulled Gaston's powers of reasoning. The two situations are not at all interchangeable. Alphonse, in his spacecraft, had experienced enormous and prolonged accelerations in reaching the velocity of 99% of the velocity of light, in turning around, and finally in coming in to a landing. In each of these accelerations Alphonse changed from one frame of reference moving at constant velocity relative to the earth to a succession of others. Gaston, on the ground, had not been accelerated and hence had stayed in the same frame of reference all the time. What Gaston measured is in accord with the formulas of special relativity. Alphonse's accelerations affected his own measuring instruments, and indeed his own life processes, and the general theory of relativity must be applied to his journey. According to this theory, a large acceleration produces effects indistinguishable from those produced by a strong gravitational field—and clocks tick more slowly in strong gravitational fields. The conclusion is that Alphonse is indeed younger than Gaston on his return, and by the amount expected on the basis of Gaston's measurements. Of course, Alphonse's lifespan has not been extended to *him,* since regardless of his brother's experience of the passing of time, he has spent only 10 years on the journey by his own reckoning.

Important Terms

The **special theory of relativity** relates measurements made on an object or phenomenon from frames of reference moving at constant velocity with respect to one another.

The relativistic **time dilation** refers to the fact that a clock moving with respect to an observer appears to tick less rapidly than it does to an observer traveling with the clock.

The **Lorentz contraction** refers to the decrease in the measured length of an object when it is moving relative to an observer.

The **relativity of mass** refers to the increase in the measured mass of an object when it is moving relative to an observer.

Important Formulas

Lorentz contraction:

$$L = L_0\sqrt{1 - v^2/c^2}$$

Time dilation:

$$t = \frac{t_0}{\sqrt{1 - v^2/c^2}}$$

Mass increase:

$$m = \frac{m_0}{\sqrt{1 - v^2/c^2}}$$

Multiple Choice

1. According to the principle of relativity, the laws of physics are the same in all frames of reference
   a. at rest with respect to one another.
   b. moving toward or away from one another at constant velocity.
   c. moving parallel to one another at constant velocity.
   d. all of the above.

2. The lifetime of a muon in motion relative to an observer appears to him to be
   a. shorter than its lifetime at rest.
   b. the same as its lifetime at rest.
   c. longer than its lifetime at rest.
   d. any of the above, depending upon the relative velocity.

3. An observer in a closed laboratory wishes to determine whether the laboratory is at rest or in motion at constant velocity.
   a. He can find out by measuring the apparent velocity of light in the laboratory.
   b. He can find out by measuring his mass.
   c. He can find out by comparing two different clocks in the laboratory over a period of time.
   d. He cannot find out.

4. A spacecraft has left the earth and is moving toward Mars. An observer on the earth finds that, relative to its value when at rest, the spacecraft's
   a. length is greater.
   b. mass is smaller.
   c. clocks tick faster.
   d. momentum is greater.

5. Of the following quantities, which one always has the same value to all observers?
   a. The length of an object.
   b. The velocity of an object.
   c. The velocity of light.
   d. The duration of a time interval.

6. The theory of relativity shows that Newtonian mechanics is
   a. wholly incorrect.
   b. correct only for velocities up to the velocity of light.
   c. approximately correct for all velocities.
   d. approximately correct only for velocities much smaller than that of light.

7. An electron is moving through a body of water. The greatest velocity it can possibly have is
   a. the velocity of water waves.
   b. the velocity of sound waves in water.
   c. the velocity of light waves in water.
   d. the velocity of light waves in vacuum.

8. When an object whose length is 1 m when at rest approaches the velocity of light, its length approaches
   a. 0.   b. 0.5 m.
   c. 2 m.   d. ∞.

9. When an object whose rest mass is 1 kg approaches the velocity of light, its mass approaches
   a. 0.   b. 0.5 kg.
   c. 2 kg.   d. ∞.

10. Which of the following velocities must an object have if its mass is to be double its rest mass?
    a. $c/2$ b. $\sqrt{3}c/2$
    c. c d. $2c$

11. A man 6 ft tall lies along the axis of a space vehicle traveling at 0.9 c. His height as measured by a stationary observer is
    a. 1.9 ft.   b. 2.6 ft.
    c. 6.0 ft.   d. 14 ft.

12. The velocity of an electron whose mass is ten times its rest mass is
    a. 2×10^8 m/s.   b. 2.98×10^8 m/s.
    c. 4×10^8 m/s.   d. 3×10^9 m/s.

Exercises

1. Can an observer in a windowless laboratory in principle determine if the earth is (a) moving through space with a uniform velocity; (b) moving through space with a uniform linear acceleration; (c) rotating on its axis?

2. The electron beam in a television picture tube can move across the screen faster than the speed of light. Why does this not violate the special theory of relativity?

3. If the speed of light were smaller than it is, would relativistic phenomena be more or less conspicuous than they are now?

4. Is the mass of an object the same whether it is moving toward an observer or away from him at the same velocity? Is it the same whether it moves toward the observer or the observer moves toward it at the same velocity? What is the connection between the answers to these questions and the principle of relativity?

5. The length of a rod is measured by several observers, one of whom is stationary with respect to the rod. What must be true of the figure obtained by the latter observer?

6. An object moving in a circle has an average velocity v of zero. Would you expect it to have a relativistic mass increase?

7. The hydrogen molecule contains two protons and two electrons. The electrons are moving considerably faster than the protons. The most accurate measurements to date indicate that the hydrogen molecule is electrically neutral to at least one part in 10^{10}. What significance has this result with respect to the question of whether electric charge is an invariant quantity, like the velocity of light in free space, or depends upon relative motion, like mass, length, and the duration of a time interval?

8. Does a laboratory at rest on the earth's surface constitute a nonaccelerated frame of reference? If not, where on the surface is the acceleration greatest? Where is it least? Is it zero anywhere in the earth?

9. Is the density of a moving object less than, the same as, or more than it appears to an observer when the object is at rest?

10. A spacecraft in flight is observed to be 99% of its length when it was at rest on the earth. What is its velocity relative to the earth?

11. An airplane 100 ft long on the ground is flying at 400 mi/hr (180 m/s). How much shorter does it appear to somebody on the ground?

12. A meter stick is moving so fast that its length appears contracted to only 50 cm. What is its velocity?

13. How fast would a rocket ship have to go for each year on the ship to correspond to two years on the earth?

14. A certain process requires 10^{-6} s to occur in an atom at rest in the laboratory. How much time will this process require to an observer in the laboratory when the atom is moving at a speed of 5×10^7 m/s?

15. What velocity must an object have if its mass is to triple?

16. Find the mass of an object whose rest mass is 1000 g when it is traveling at 10%, 90%, and 99% of the speed of light.

17. A man has a mass of 100 kg on the ground. When he is in a spacecraft in flight, an observer on the earth measures his mass to be 101 kg. How fast is the spacecraft moving?

Problems

1. An airplane is flying at 300 m/s (672 mi/hr). How much time must elapse before a clock in the airplane and one on the ground differ by 1 s?

2. Two observers, A on earth and B in a rocket ship whose velocity is 2×10^8 m/s, both set their watches to the same time when the ship is abreast of the earth. (a) How much time must elapse by A's reckoning before the watches differ by 1 s? (b) To A, B's watch seems to run slow. To B, does A's watch seem to run fast, run slow, or keep the same time as his own watch?

Answers to Multiple Choice

1. d	5. c	9. d
2. c	6. d	10. b
3. d	7. d	11. b
4. d	8. a	12. b

28

Particles and Waves

Particles and waves are entirely separate concepts in everyday life: it is impossible to confuse a stone, for instance, with the waves it produces when it is dropped into a lake. But in the realm of the atom, the situation is very different. Electromagnetic waves—which, as we know, exhibit such characteristic wave phenomena as interference—in many ways behave exactly as particles do. And electrons—whose particle nature is amply shown in the operation of the television picture tube—nevertheless behave in many ways exactly as waves do. On the microscopic level, then, a wave-particle duality replaces the distinction between waves and particles so evident on a macroscopic level. This duality turns out to be the key to understanding the structure of atoms and why they behave as they do.

28-1 Photoelectric Effect

The formulation of the theory of relativity and that of the quantum theory of light, both of which took place early in this century, profoundly altered our ways of thinking about the physical world. We have already examined some of the remarkable consequences of relativity, and now we come to the realm of quanta, which will be no less remarkable. But such a subjective term is really not justified, because it is merely based upon the limits to our imagination that are imposed by our expereience; if we were in a world where we were about the same size as an electron, relativistic and quantum phenomena would be familiar (though then most macroscopic phenomena would not).

Toward the end of the nineteenth century a number of experiments were performed that revealed the emission of electrons from a metal surface when light (particularly ultraviolet light) falls on it (Fig. 28-1). This phe-nomenon is known as the *photoelectric effect*. It is not, at first glance, any-thing to surprise us, for light waves carry energy, and some of the energy absorbed by the metal may somehow concentrate on individual electrons and reappear as kinetic energy. Upon closer inspection of the data, however, we find that the photoelectric effect can hardly be explained in so straight-forward a manner.

Photoelectric effect

The first peculiarity of the photoelectric effect is that, even when the metal surface is only faintly illuminated, the emitted electrons (which are called *photoelectrons*) leave the surface immediately. But according to the electromagnetic theory of light, the energy content of light waves is spread

Photoelectrons

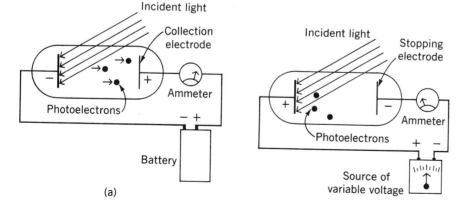

(a)

(b)

Fig. 28-1.   (a) A method of detecting the photoelectric effect. The photo-electrons ejected from the metal plate being irradiated are attracted to the positive collection electrode at the other end of the tube, and the current that results is measured with the am-meter (b) A method of detecting the maximum energy of the photoelec-trons. As the stopping electrode is made more negative, the slower photo-electrons are repelled before they can reach it. Finally a voltage will be reached at which no photoelectrons whatever are received at the stopping electrode, as indicated by the current dropping to zero, and this voltage corresponds to the maximum photo-electron energy.

out across the width of the wavefronts of the light beam involved. Calculations show that a definite period of time—about *a year* in the case of a beam of very low intensity—must elapse before any individual electrons accumulate enough energy to leave the metal. Instead, the electrons are found to be emitted as soon as the light is turned on.

Another unexpected discovery is that the energy of the photoelectrons does not depend upon the intensity of the light. A bright light yields more electrons than a dim one, but their average energy remains the same. This behavior contradicts the electromagnetic theory of light, which predicts that the energy of photoelectrons should depend upon the intensity of the light beam responsible for them.

Photoelectron energies vary with frequency of light

The energies of the photoelectrons emitted from a given metal surface turn out, most surprisingly of all, to depend upon the *frequency* of the light employed. At frequencies below a certain critical one (which is characteristic of the particular metal), no electrons whatever are given off. Above this threshold frequency, the photoelectrons have a range of energies from zero to a certain maximum value, and *this maximum energy increases with increasing frequency* (Fig. 28-2). High-frequency light yields high maximum

Fig. 28-2.   The variation of maximum photoelectron energy with the frequency of the incident light for two target metals. No photoelectrons are emitted for frequencies less than f_0^A in the case of metal *A* and less than f_0^B in the case of metal *B*. In both cases, however, the angle between the experimental line and either axis is the same. Hence we may write the equation of the lines as $KE_{max} = h(f - f_0)$, where h has the same value in all cases, but where f_0, the minimum frequency required for photoelectric emission to occur, depends upon the nature of the target metal.

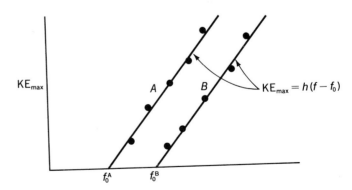

Frequency of incident light, *f*

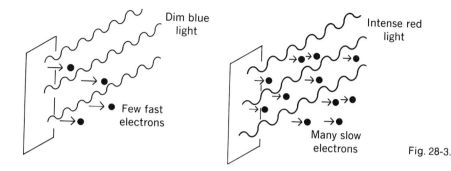

Dim blue light

Few fast electrons

Intense red light

Many slow electrons

Fig. 28-3.

photoelectron energies; low-frequency light yields low maximum photo-electron energies. Thus dim blue light produces electrons with more energy than those produced by intense red light, although the latter results in a greater number of them (Fig. 28-3).

28-2 Quantum Theory of Light

Aware that the electromagnetic theory of light, despite its notable success in accounting for other optical phenomena, failed to explain the photoelectric effect, Albert Einstein in 1905 sought some other basis for interpreting it. He found what he needed in a novel assumption that Max Planck, a German physicist, had had to make a few years earlier in order to understand the origin of the radiation given off by an object so hot that it is luminous (a poker thrust in a fire, for instance). Planck found that the accepted physical laws of the time predicted the observed characteristics of this radiation *provided* that the radiation is considered as though emitted in little bursts of energy, rather than continuously. These bursts of energy are called *quanta*.

Planck showed that the energy E of each quantum had to be related to the light frequency f by the formula

$$E = hf,$$

Quantum energy   (28-1)

where h is a constant, known today as Planck's constant, whose value is

$$h = 6.63 \times 10^{-34} \text{ joule-second.}$$

Planck's constant

Although the energy radiated by a heated object must be regarded as coming out intermittently, in order for theory and experiment to agree, Planck held to the conventional view that it nevertheless travels through space as continuous waves.

Einstein saw that Planck's idea could be used to interpret the photoelectric effect if light not only is emitted a quantum at a time but also propagates as separate quanta. Then the h of the photoelectric effect equation (Fig. 28-2) is the same as the h of the formula $E = hf$, and the significance of the former equation becomes clear when it is rewritten.

$$hf = KE_{max} + hf_0.$$ *Photoelectric effect* (28-2)

What this equation states is that

Quantum energy = maximum electron energy + energy required to eject an electron.

The reason for a threshold frequency f_0 is clear: it corresponds to the energy required to dislodge an electron from the metal surface. (There must be such a minimum energy, or electrons would leave metals all the time.) And there are several plausible reasons why not all photoelectrons have the same energy even though a single frequency of light is used. For instance, not all the quantum energy hf may be transferred to a single electron, and an electron may lose some of its initial energy in collisions with other electrons within the metal before it actually emerges from the surface.

Problem. The threshold frequency for copper is 1.1×10^{15} Hz. When ultraviolet light of frequency 1.5×10^{15} Hz is shone on a copper surface, what is the maximum energy of the photoelectrons?

Solution. From Eq. (28-2),

$$KE_{max} = hf - hf_0 = h(f - f_0)$$
$$= 6.63 \times 10^{-34} \text{ J·s} \times (1.5 - 1.1) \times 10^{15} \text{ Hz}$$
$$= 2.7 \times 10^{-19} \text{ J}.$$

Since $1 \text{ eV} = 1.6 \times 10^{-19}$ J, the maximum photoelectron energy is

$$\frac{2.7 \times 10^{-19} \text{ J}}{1.6 \times 10^{-19} \text{ J/eV}} = 1.7 \text{ eV}.$$

Photons are quanta of light

Einstein's notion that light travels as a series of little packets of energy (sometimes referred to as quanta, sometimes as *photons*) is in complete contradiction with the wave theory of light (Fig. 28-4). And the wave theory, as we know, has some powerful observational evidence on its side. There is no other way to explain interference effects, for example. According to the wave theory, light spreads out from a source in a manner analogous to the spreading out of ripples on the surface of a lake when a stone is dropped into it, with the energy of the light distributed continuously throughout the wave pattern. According to the quantum theory, light spreads out from a source as a succession of localized packets of energy, each sufficiently small to permit its being absorbed by a single electron. Yet, despite the particle picture of light that it presents, the quantum theory requires a knowledge of the light frequency f, a wave quantity, in order to determine the energy of each quantum.

On the other hand, the quantum theory of light is able to explain the photoelectric effect. It predicts that the maximum photoelectron energy should depend upon the frequency of the incident light and not upon its intensity, precisely the opposite of what the wave theory suggests, and it is able to explain why even the feeblest light can lead to the immediate emission of photoelectrons. The wave theory can give no reason why there should be a threshold frequency below which no photoelectrons are observed, no matter how strong the light beam, something that follows naturally from the quantum theory.

Which theory is correct? The history of physics is filled with examples of physical ideas that required revision or even replacement when new empirical data conflicted with them, but this is the first occasion in which two completely different theories are both required to explain a single physical phenomenon. In thinking about this, it is important for us to note that, in a particular situation, light behaves *either* as though it has a wave nature *or* a particle nature. While light sometimes assumes one guise and sometimes the other, there is no physical process in which both are simultaneously exhibited. The same light beam can diffract around an obstacle and then impinge on a metal surface to eject photoelectrons, but these two processes occur separately.

The electromagnetic theory of light and the quantum theory of light complement each other; by itself, each theory is "correct" in certain experiments, and there are no relevant experiments which neither can account for. Light must be thought of as a phenomenon that incorporates both particle and wave characters. Although we cannot visualize its "true nature," these complementary theories of light are able to account for its behavior, and we have no choice but to accept them both.

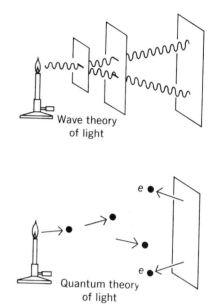

Fig. 28-4. The wave theory of light is necessary to explain diffraction and interference phenomena, which the quantum theory cannot explain. The quantum theory of light is necessary to explain the photoelectric effect, which the wave theory cannot explain.

The wave and quantum theories of light complement each other

28-3 X-Rays

If photons of light can give up their energy to electrons, can the kinetic energy of moving electrons be converted into photons? The answer is that such a transformation is not only possible, but had in fact been discovered (though not understood) prior to the work of Planck and Einstein. In 1895 Roentgen found that a mysterious, highly penetrating radiation is emitted when high-speed electrons impinge on matter. The x-rays (so called because their nature was then unknown) caused phosphorescent substances to glow, exposed photographic plates, traveled in straight lines, and were not affected by electric or magnetic fields. The more energetic the electrons, the more penetrating the x-rays, and the greater the number of electrons, the greater the density of the resulting x-ray beam.

X-rays are high-frequency em waves

After over 10 years of study, it was finally established that x-rays exhibit, under certain circumstances, both interference and polarization effects, leading to the conclusion that they are electromagnetic waves. From the interference experiments their frequencies were found to be very high, above those in ultraviolet light.

Figure 28-5 is a diagram of an x-ray tube. Battery *A* sends a current through the filament, heating it until it emits electrons. These electrons are then accelerated toward a metallic target by the potential difference *V* provided by battery *B*. The tube is evacuated to permit the electrons to reach the target unimpeded. The impact of the electrons causes the evolution of x-rays from the target.

How x-rays are produced

What is the physical process involved in the production of x-rays? It is known that charged particles emit electromagnetic waves whenever they are accelerated, and so we may reasonably identify x-rays as the radiation accompanying the slowing down of fast electrons when they strike a metal. The great majority of the incident electrons, to be sure, lose their kinetic energy too gradually for x-rays to be evolved, and merely act to heat the

Fig. 28-5. An x-ray tube.

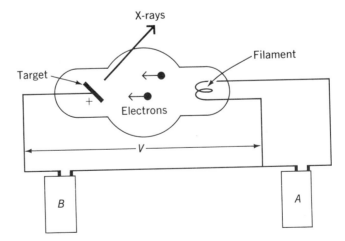

target. (Consequently the targets in x-ray tubes are made of metals with high melting points, and a means for cooling the target is often provided.) A few electrons, however, lose much or all their energy in single collisions with target atoms, and this is the energy that appears as x-rays. In other words, we may regard x-ray production as an inverse photoelectric effect.

Since the threshold energy hf_0 needed to remove an electron from a metal is only a few eV whereas the accelerating potential V in an x-ray tube usually exceeds 10,000 volts, we can neglect hf_0 here. (The electron volt is discussed in Section 18-7). The highest frequency f_{max} found in the x-rays emitted from a particular tube should therefore correspond to a quantum energy of hf_{max}, where hf_{max} equals the kinetic energy

$$KE = eV$$

of an electron that has been accelerated through a potential difference of V. We conclude that

$$hf_{max} = eV \qquad\qquad\text{X-ray energy} \quad (28\text{-}3)$$

in the operation of an x-ray tube.

Problem. Find the highest frequency present in the radiation from an x-ray machine whose operating potential is 50,000 volts.

Solution. From the formula $hf_{max} = eV$ we find that

$$f_{max} = \frac{eV}{h}$$

$$= \frac{1.6 \times 10^{-19}\text{ C} \times 5.0 \times 10^4\text{ V}}{6.6 \times 10^{-34}}$$

$$= 1.2 \times 10^{19}\text{ Hz.}$$

28-4 Matter Waves

As we have seen, electromagnetic waves under certain circumstances have properties indistinguishable from those of particles. It requires no greater stretch of the imagination to speculate whether what we normally think of as particles might not have wave properties, too. This speculation was first made by Louis de Broglie in 1924. Soon afterward de Broglie's idea was taken up and developed by a number of other physicists (notably Heisenberg, Schrödinger, Born, Pauli, and Dirac) into the elaborate, mathematically difficult—but very beautiful—theory called *quantum mechanics*.

Quantum mechanics

The advent of quantum mechanics did more than provide a supremely accurate and complete description of atomic phenomena; it also altered the way in which the physicist approaches nature, so that he now thinks in terms of probabilities instead of in terms of certainties. The universe is closer in many respects to a roulette wheel than to a clock—but it is a roulette wheel that obeys certain rules, and the laws of physics are these rules.

De Broglie started with the formula for the linear momentum of a photon, which is

Photon momentum

$$p = \frac{hf}{c}.$$

Since $\lambda f = c$, the momentum of a photon can be expressed in terms of its wavelength as

$$p = \frac{h}{\lambda}.$$

Hence for a photon

Photon wavelength

$$\lambda = \frac{h}{p}.$$

De Broglie suggested that this equation for wavelength is a perfectly general one, applying to material objects as well as to photons. In the case of an object

$$p = mv,$$

and so its *de Broglie wavelength* is

$$\lambda = \frac{h}{mv}. \qquad \textit{De Broglie wavelength} \quad (28\text{-}4)$$

The more momentum an object has, the shorter its wavelength. The relativistic formula $m = m_0/\sqrt{1 - v^2/c^2}$ must usually be used for m in computing de Broglie wavelengths.

Matter waves exhibit interference

How can de Broglie's hypothesis be verified? Perhaps the most striking example of wave behavior is a diffraction pattern, which depends upon the ability of waves both to bend around obstacles and to interfere constructively and destructively with one another. Several years after de Broglie's work, Davisson and Germer, in the United States, and G. P. Thomson in England independently demonstrated that streams of electrons are diffracted when they are scattered from crystals. The diffraction patterns they observed were in complete accord with the electron wavelengths predicted by de Broglie's formula.

In certain aspects of its behavior, a moving object resembles a wave, and in other aspects it resembles a particle. Which type of behavior is most conspicuous depends upon how the object's de Broglie wavelength compares with its dimensions and with the dimensions of whatever it interacts with. Two examples will help us appreciate this statement.

Criterion for type of behavior

In one of their experiments, Davisson and Germer aimed a beam of 54-eV electrons at a nickel crystal (Fig. 28-6). The momentum of a 54-eV electron can be calculated nonrelativistically. Since

$$KE = \tfrac{1}{2}mv^2$$

and

$$KE = 54 \text{ eV} \times 1.6 \times 10^{-19} \text{ J/eV} = 8.6 \times 10^{-18} \text{ J},$$

we have

$$mv = \sqrt{2m\,KE}$$
$$= \sqrt{2 \times 9.1 \times 10^{-31}\text{kg} \times 8.6 \times 10^{-18} \text{ J}}$$
$$= 4.0 \times 10^{-24} \frac{\text{kg-m}}{\text{s}}.$$

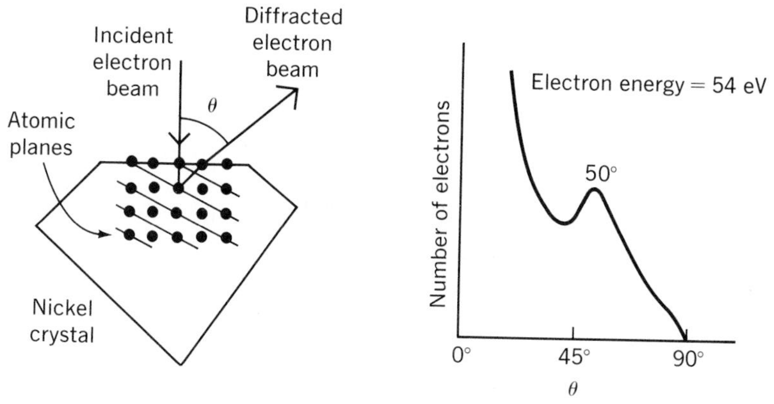

Fig. 28-6. The Davisson-Germer experiment. The peak in the number of diffracted 54-eV electrons at $\theta = 50°$ is in agreement with de Broglie's formula for the wavelength of a moving particle.

The electron wavelength is therefore

Electron wavelength

$$\lambda = \frac{h}{mv}$$

$$= \frac{6.63 \times 10^{-34} \text{ J-s}}{4.0 \times 10^{-24} \text{ kg-m/s}}$$

$$= 1.7 \times 10^{-10} \text{ m.}$$

This is the same order of magnitude as the spacing of the atoms in the nickel crystal, and we recall that diffraction is prominent only when the wavelength of the waves involved is comparable with the spacing of the scattering centers. Davisson and Germer found that the scattered 54-eV electrons were concentrated in just the direction ($\theta = 50°$) predicted by the theory of diffraction for waves of $\lambda = 1.7 \times 10^{-10}$ m, instead of the more even distribution expected for purely billiard-ball scattering. They interpreted the result as support for de Broglie's hypothesis.

On the other hand, a 1500-kg car whose velocity is 30 m/s has a de Broglie wavelength of

Car wavelength

$$\lambda = \frac{h}{mv}$$

$$= \frac{6.63 \times 10^{-34} \text{ J-s}}{1.5 \times 10^3 \text{ kg} \times 30 \text{ m/s}}$$

$$= 1.5 \times 10^{-38} \text{ m.}$$

The car's wavelength is so small relative to its dimensions that no wave behavior can be detected.

The above examples are extreme ones. More ambiguous is the case of a moving atom or molecule. In Chapter 15 we saw how a model of a gas in which molecules are considered as particles was very successful in explaining the properties of gases. But at 0°C the average wavelength of helium atoms, to give a specific illustration, is 7.6×10^{-11} m, which is of the same order of magnitude as atomic dimensions. Such atoms may exhibit either particle or wave characteristics, depending upon the situation.

In water waves, the physical quantity that varies periodically is the height of the water surface. In sound waves, the variable quantity is pressure in the medium the waves travel through. In light waves, the variable quantities are the electric and magnetic fields. What is it that varies in the case of matter waves?

Wave function of moving object

The quantity whose variations constitute the matter waves of a moving object is known as its *wave function*. The symbol for wave function is ψ, the Greek letter *psi*.

The value of ψ^2 for a particular object at a certain place and time is proportional to the probability of finding the object at that place at that time.

For this reason the quantity ψ^2 is called the *probability density* of the object. A large value of ψ^2 signifies that the body is likely to be found at the specified place and time; a small value of ψ^2 signifies that the body is unlikely to be found at that place and time. Thus matter waves may be regarded as waves of probability.

A group or packet of matter waves is associated with every moving object. The packet travels with the same velocity as the object does. The waves in the packet have the average wavelength $\lambda = h/mv$ given by de Broglie's formula (Fig. 28-7). Even though we cannot visualize what is meant by ψ and so cannot form a mental image of matter waves, the agreement between theory and experiment signifies that the notion of matter waves is a meaningful way to describe moving objects.

28-5 Particle in a Box

When an object is confined to a certain region of space instead of being able to move freely, its wave properties lead to certain remarkable consequences. Let us see what these consequences are in the simplest possible case—that of a particle trapped in a box L wide whose walls are infinitely hard, so that the particle does not lose energy as it bounces back and forth (Fig. 28-8). We shall assume that the particle's velocity v is sufficiently small so that relativistic considerations can be ignored.

The wave equivalent of a particle in a box is a standing de Broglie wave. The reason for this is the same as in the analogous phenomenon of standing waves in a stretched string (Section 13-6): the wave variable—transverse displacement in the case of a string, wave function ψ in the case of a moving particle—must be 0 at the walls of the box in order that the waves be reflected there. Hence the possible de Broglie wavelengths depend upon the

Probability density

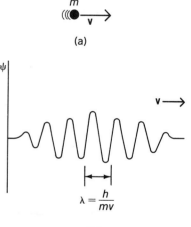

(a)

(b)

Fig. 28-7.   (a) Particle description of moving object. (b) Wave description of moving object.

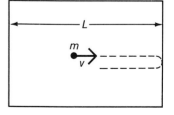

Fig. 28-8.   A moving particle in a box L wide.

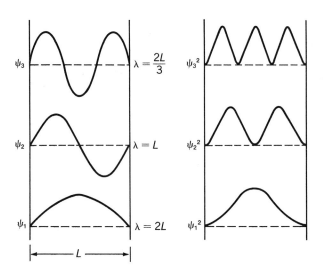

Fig. 28-9. Wave functions and probability densities of a particle in a box.

width L of the box (Fig. 28-9). The longest possible wavelength is $\lambda = 2L$, next is $\lambda = L$, then $\lambda = 2L/3$, and so on; in general, the permitted wavelengths are given by

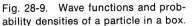
Trapped-particle wavelengths

$$\lambda_n = \frac{2L}{n}, \qquad n = 1,2,3, \ldots \tag{28-5}$$

The wavelength of a particle of mass m and velocity v is

$$\lambda = \frac{h}{mv},$$

so the restrictions on λ imposed by the size of the box amount to restrictions on the velocity v the particle can have and, in turn, to restrictions on its energy. The kinetic energy of a moving particle of wavelength λ is

$$\text{KE} = \tfrac{1}{2}mv^2 = \frac{1}{2m}\,(mv)^2 = \frac{h^2}{2m\lambda^2}.$$

Our trapped particle has only kinetic energy, and because the only wavelengths it can have are $\lambda_n = 2L/n$, the only energies it can have are

Trapped-particle energies

$$E_n = \frac{n^2 h^2}{8mL^2} \qquad n = 1,2,3, \ldots \tag{28-6}$$

Each permitted energy is called an *energy level,* and the integer *n* that corresponds to a given energy level is called its *quantum number.*

We can draw three important general conclusions from the preceding analysis. These conclusions apply to *any* particle confined to a certain region of space, such as an atomic electron held captive by the attraction of the positively-charged nucleus, and not just to the artificial situation of a particle in a rigid-walled box.

1. A trapped particle can possess only certain specific energies and no others. The energy of such a particle is said to be *quantized,* and the magnitudes of the energy levels depend upon the manner in which the particle's motion is restricted.

Energy quantization

2. A trapped particle cannot have zero energy; it must have a certain minimum amount E_1. Since the de Broglie wavelength of a particle is $\lambda = h/mv$, a velocity of $v = 0$ means an infinite wavelength. But there is no way to reconcile an infinite wavelength with a trapped particle, so such a particle must possess at least some kinetic energy. This is the origin of the "zeropoint" energy mentioned in Section 15-6.

Minimum energy

3. Because Planck's constant is so small—only 6.63×10^{-34} J-s— quantization of energy is conspicuous only when m and L are also small. For example, an electron in a box 10^{-10} m wide, which is the order of magnitude of atomic dimensions, is restricted to the energies

When energy quantization is significant

$$E_n = \frac{n^2 h^2}{8mL^2}$$

$$= \frac{n^2 \times (6.63 \times 10^{-34} \text{ J-s})^2}{8 \times 9.1 \times 10^{-31} \text{ kg} \times (10^{-10} \text{ m})^2}$$

$$= 6.0 \times 10^{-18} \, n^2 \text{ J}$$

$$= 38 \, n^2 \text{ eV} \qquad n = 1, 2, 3, \ldots$$

The energies the electron can have are therefore 38 eV, 152 eV, 342 eV, 608 eV, and so on. If such a box existed, the quantization of a trapped electron's energy would be a prominent feature of the system. (And, indeed, energy quantization *is* prominent in the case of an atomic electron, as we shall find in Chapter 29.)

On the other hand, the energy levels of a 10-g marble in a box 10 cm wide are revealed by a similar calculation to be

$$E_n = 5.5 \times 10^{-64} \, n^2 \text{ J} \qquad n = 1, 2, 3, \ldots$$

Here energy quantization is hardly likely to be detected. When $n = 1$, the

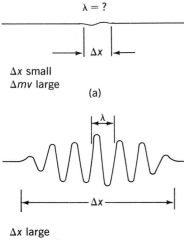

Fig. 28-10. (a) A narrow wave packet means a small uncertainty Δx in the position of a moving object but a large uncertainty in its wavelength and hence momentum. (b) A wide wave packet means a large uncertainty in the position of a moving object but a small uncertainty in its wavelength and hence momentum. It is impossible to have a narrow wave packet with a well-defined wavelength.

marble's velocity is 3.3×10^{-31} m/s, a velocity no experiment could distinguish from zero, and the spacing between the higher energy levels is too minute for measurement. On a macroscopic scale, then, quantum effects are unobservable, which is why classical physics is adequate on this scale. But on the scale of the atom, quantum effects are dominant, and the concepts and principles of classical physics must be replaced by others of a more sophisticated character. In fact, classical physics turns out to be just an approximation of quantum physics, which is perfectly general in its range of application.

28-6 Uncertainty Principle

To regard a moving object as a wave packet raises the problem of the ultimate accuracy with which such "particle" properties as its position and momentum can be determined. In the macroscopic world, where the wave aspects of matter are insignificant, the limit to how accurately position and momentum can be measured depends solely upon our instruments, and there is, in principle, no absolute limit at all. But in the microscopic world, where the wave aspects of matter are very significant indeed, these wave aspects set a fundamental limit to the accuracy of measurements of position and momentum. Of course, poor instruments will give poor results, but even if perfect instruments were available, they would not yield exact results. The *uncertainty principle* is the physical law which expresses quantitatively the basic indeterminacy which the wave nature of matter imposes on measurements of moving objects.

Figure 28-10 shows how the wave nature of matter leads to the uncertainty principle. A narrow wave packet means a small uncertainty in the position of the corresponding object, and a wide wave packet means a large uncertainty in position. On the other hand, the wavelength of the waves in a wide packet is fairly well defined, and the momentum of the object has only a small uncertainty Δmv. The wavelength of a narrow wave packet is imprecise, and the momentum of the object therefore has a large uncertainty Δmv. We conclude that

1. If an object has a well-defined position at a certain time, its momentum must have a large uncertainty;
2. If an object has a well-defined momentum at a certain time, its position must have a large uncertainty.

Evidently a reciprocal relationship of some kind exists between the uncertainty Δx in an object's position and the uncertainty Δmv in its momentum. The details of this relationship can be found from the mathematical nature of the wave packet that corresponds to a moving object. Such a packet can be regarded as being formed by the superposition of many trains of sinusoidal waves, as discussed in the Appendix to Chapter 13, with each train having a different wavelength (Fig. 28-11). The narrower the wave packet, the greater the spread of wavelengths involved and hence the greater the uncertainty Δmv. A smaller range of wavelengths is needed to construct a wide wave packet, but now Δx is necessarily large. The advantage of this approach is that it permits a precise analysis of the uncertainty principle, with the result that

$$\Delta x\,\Delta mv \geqslant \frac{h}{2\pi}. \qquad\qquad \textit{Uncertainty principle} \quad (28\text{-}7)$$

In words, the uncertainty principle states that

$$\begin{pmatrix}\text{Uncertainty}\\ \text{in position}\end{pmatrix} \times \begin{pmatrix}\text{uncertainty}\\ \text{in momentum}\end{pmatrix} \begin{array}{c}\textit{is equal to or}\\ \textit{greater than}\end{array} \frac{\text{Planck's constant}}{2\pi}.$$

The uncertainty principle makes sense from the point of view of the particle properties of waves as well as from the point of view of the wave properties of particles. Suppose we wish to measure the position and momentum of something at a certain moment. To do so, we must touch it with something else that will carry the required information back to us; that is, we must poke it with a stick, shine light on it, or perform some similar act. The measurement process itself thus requires that the body be interfered with in some way, and if we consider this interference in detail, we are led to the same uncertainty principle as before even without taking into account the wave nature of moving objects.

Particle approach to uncertainty principle

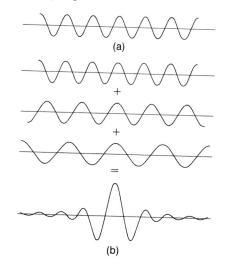

(a)

+

+

=

(b)

Fig. 28-11.   (a) If a moving object can be represented by a wave with a single wavelength, its position cannot be established at all. (b) A wave packet is the result of superposing waves of different wavelengths; the greater the range of wavelengths, the narrower the packet.

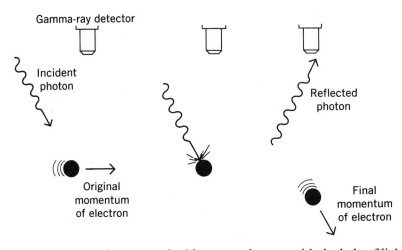

Fig. 28-12.   An example of the uncertainty principle: it is impossible to accurately determine the position and momentum of an object at the same time.

Let us imagine we are looking at an electron with the help of light whose wavelength is λ, as in Fig. 28-12. We cannot use ordinary visible light, because the electron's diameter is perhaps a hundred million times smaller than the wavelengths found in visible light, but there is no reason in principle why gamma rays of ultrashort wavelength could not be employed instead. Each photon of this light has the momentum h/λ.

When one of these photons bounces off the electron (which must occur if we are to "see" it), the electron's original momentum will be changed. The exact amount of the change Δp cannot be predicted, but it will be of the same order of magnitude as the photon momentum h/λ. Hence

$$\Delta p \approx \frac{h}{\lambda}$$

The *larger* the wavelength, the smaller the uncertainty in momentum.

Because light is a wave phenomenon as well as a particle phenomenon, we cannot expect to determine the electron's location with perfect accuracy even with the best of instruments. A reasonable estimate of the irreducible uncertainty Δx in the measurement might be one wavelength. That is,

$$\Delta x \geqslant \lambda.$$

The *smaller* the wavelength, the smaller the uncertainty in location. Hence if we use light of short wavelength to increase the accuracy of our position measurement, there will be a corresponding decrease in the accuracy of our momentum measurement, whereas light of long wavelength will yield an accurate momentum but an inaccurate position.

By combining the two formulas above, we find that

$$\Delta x \, \Delta mv \geqslant h.$$

A more elaborate calculation shows that somewhat better accuracy is in principle possible, so that the limit of the product $\Delta x \, \Delta mv$ is $h/2\pi$ instead of just h.

To repeat, the uncertainty principle follows from the wave character of moving objects, and has nothing to do with inaccuracies in the instruments used in making measurements. On a macroscopic scale, since Planck's constant h is such a minute quantity, the limitation imposed upon measurements by the uncertainty principle is negligible, but on a microscopic scale the uncertainty principle dominates many phenomena and probability considerations are correspondingly important.

The uncertainty principle is important in atomic physics

Special Topic

Atomic Electrons

A significant application of the uncertainty principle is to the question of whether electrons are present within atomic nuclei. Experiments show that nuclei are about 10^{-14} m across. If an electron is to be confined within a nucleus, the uncertainty Δx in its position cannot exceed about 10^{-14} m. Hence the uncertainty in the electron's momentum must be

$$\Delta mv \geqslant \frac{h}{2\pi \Delta x} \geqslant \frac{6.63 \times 10^{-34} \text{ J-s}}{2\pi \times 10^{-4} \text{ m}} \geqslant 1.1 \times 10^{-20} \, \frac{\text{kg-m}}{\text{s}}.$$

This momentum uncertainty corresponds to an energy uncertainty of over 20 MeV (a relativistic calculation is needed here). If this is the uncertainty in the electron's kinetic energy, the energy itself must be at least comparable in magnitude. However, the electrons associated with atoms, even unstable ones, never have much more than 10 percent of this amount of energy, and we therefore conclude that electrons are not present inside atomic nuclei.

How much energy must an electron have if it is confined to an atom? An atom is much larger in size than its nucleus, as discussed in Chapter 17, so the electron energy can be smaller than in the case of the nucleus. The hydrogen atom, to take a specific example, has a radius of about 5×10^{-11} m, which means that the uncertainty Δx in the location of its single electron cannot exceed this figure. The momentum uncertainty of such an electron is

$$\Delta mv \geq \frac{h}{2\pi\Delta x} \geq \frac{6.63 \times 10^{-34} \text{ J-s}}{2\pi \times 5 \times 10^{-11} \text{ m}} \geq 2.1 \times 10^{-24} \frac{\text{kg-m}}{\text{s}}.$$

This momentum uncertainty corresponds to a kinetic energy uncertainty of only 15 eV (the calculation here can be made using Newtonian mechanics, with $\text{KE} = [mv]^2/2m$). Electrons in hydrogen atoms actually do have kinetic energies of this order of magnitude.

Important Terms

The **photoelectric effect** is the emission of electrons from a metal surface when light is shone on it.

The **quantum theory of light** states that light travels in tiny bursts of energy called **quanta** or **photons**. The quantum theory of light is required to account for the photoelectric effect.

X-rays are high-frequency electromagnetic waves emitted when fast electrons impinge on matter.

A moving body behaves as though it has a wave character. The waves representing such a body are **matter waves**, also called **de Broglie waves**. The wave variable in a matter wave is its **wave function**, whose square is the **probability density** of the body. The value of the probability density of a particular body at a certain place and time is proportional to the probability of finding the body at that place at that time. Matter waves may thus be regarded as waves of probability.

Because of its wave nature, a particle restricted to a definite region of space can have only certain specific energies, each of which is called an **energy level** and corresponds to a **quantum number.**

The **uncertainty principle** is an expression of the limit set by the wave nature of matter on finding the position and state of motion of a moving body.

Important Formulas

Quantum energy:

$$E = hf$$

Photoelectric effect:

$$hf = \text{KE}_{\text{max}} + hf_0$$

X-ray energy:

$$hf_{\text{max}} = eV$$

De Broglie wavelength:

$$\lambda = \frac{h}{mv}$$

Uncertainty principle:

$$\Delta x \Delta mv \geq \frac{h}{2\pi}$$

Multiple Choice

1. Photoelectrons are emitted by a metal surface only when the light directed at it exceeds a certain minimum
   a. wavelength.
   b. frequency.
   c. velocity.
   d. charge.

2. When light is shone on a metal surface, the energies of the emitted electrons
   a. vary with the intensity of the light.
   b. vary with the frequency of the light.
   c. vary with the speed of the light.
   d. are random.

3. The photoelectric effect can be understood on the basis of
   a. the electromagnetic theory of light.
   b. the special theory of relativity.
   c. the principle of superposition.
   d. none of the above.

4. Modern physical theories indicate that
   a. all particles exhibit wave behavior.
   b. only moving particles exhibit wave behavior.
   c. only charged particles exhibit wave behavior.
   d. only uncharged particles exhibit wave behavior.

5. The wave packet that corresponds to a moving particle
   a. has the same size as the particle.
   b. has the same velocity as the particle.
   c. has the velocity of light.
   d. consists of X-rays.

6. The de Broglie wavelength of a particle is
   a. proportional to its momentum.
   b. proportional to its energy.
   c. inversely proportional to its momentum.
   d. inversely proportional to its energy.

7. A photon and an electron have the same wavelength.
   a. The photon has the greater momentum.
   b. The electron has the greater momentum.
   c. They have the same momentum.
   d. Either may have the greater momentum, depending on the wavelength.

8. A photon, an electron, and an automobile all have the same wavelength. The one with the most energy
   a. is the photon.
   b. is the electron.
   c. is the automobile.
   d. depends on the wavelength and properties of the various bodies.

9. When a beam of light is used to determine the position of an object, the greatest accuracy is obtained if the light
   a. is polarized.
   b. has a short wavelength.
   c. has a long wavelength.
   d. has a low intensity.

10. Heisenberg's uncertainty principle states that
    a. moving bodies exhibit both particle and wave properties.
    b. moving charged particles resemble electromagnetic waves in their behavior.
    c. neither the position nor the momentum of a particle can ever be precisely determined.
    d. the position and momentum of a particle cannot both be precisely determined.

11. Light of wavelength 5×10^{-7} m consists of photons whose energy is
    a. 1.1×10^{-48} J.
    b. 1.3×10^{-27} J.
    c. 4×10^{-19} J.
    d. 1.7×10^{-15} J.

12. The de Broglie wavelength of an electron whose speed is half that of light is
    a. 3.6×10^{-12} m.
    b. 4.2×10^{-12} m.
    c. 4.9×10^{-12} m.
    d. 1.2×10^{-11} m.

13. The lowest energy possible for a certain particle trapped in a certain box is 2 eV. The next highest energy the particle can have is
    a. 3 eV.
    b. 4 eV.
    c. 6 eV.
    d. 8 eV.

14. An electron is confined to a box 5×10^{-11} m across, which is approximately the size of a hydrogen atom. The electron's momentum is uncertain by about
    a. 3.5×10^{-25} kg-m/s.
    b. 2.2×10^{-24} kg-m/s.
    c. 3.8×10^{-23} kg-m/s.
    d. 5×10^{-11} kg-m/s.

Exercises

1. Why do you think the wave aspect of light was discovered earlier than its particle aspect?

2. If Planck's constant were smaller than it is, would quantum phenomena be more or less conspicuous than they are now in everyday life?

3. A photon strikes a free electron in space and some of the photon's energy and momentum are transferred to the electron. Why is it impossible for all of the photon's energy and momentum to be given to the electron even though this can happen in the photoelectric effect when the target electron is part of a metal?

4. A proton and a photon have the same wavelength. How are their momenta related, if at all? Their energies?

5. Must a particle have an electric charge in order for matter waves to be associated with its motion?

6. Can the rest mass of a moving particle be determined by measuring its de Broglie wavelength?

7. What is the simplest experimental procedure that can distinguish between a gamma ray whose wavelength is 10^{-11} m and an electron whose de Broglie wavelength is also 10^{-11} m?

8. The uncertainty principle applies to all objects, yet its consequences are only significant for such minute particles as electrons, protons, and neutrons. Why?

9. The eye can detect as little as 10^{-18} J of electromagnetic energy. How many photons of $\lambda = 6 \times 10^{-7}$ m does this energy represent?

10. Yellow light has a wavelength of 6×10^{-7} m. How many photons are emitted per second by a yellow lamp radiating at a power of 10 W?

11. A radio transmitter operates at a frequency of 880 kHz and a power of 10 kW. How many photons per second does it emit?

12. What is the lowest-frequency light that will cause the emission of photoelectrons from a surface whose nature is such that 1.9 eV is required to eject an electron?

13. Electrons are accelerated in television tubes through potential differences of about 10,000 V. Find the highest frequency of the electromagnetic waves that are emitted when these electrons strike the screen of the tube. What type of waves are these?

14. A target is bombarded by 8×10^4 eV electrons. What is the highest frequency present in the emitted x-rays?

15. What is the de Broglie wavelength of a 1-mg grain of sand blown by the wind at a velocity of 20 m/s?

16. What is the de Broglie wavelength of a proton whose velocity is 10^7 m/s? Use a nonrelativistic calculation.

17. (a) An electron is confined in a box 10^{-9} m in length. What is the uncertainty in its velocity? (b) A proton is confined in the same box. What is the uncertainty in its velocity?

18. What is the approximate momentum imparted to a proton initially at rest by a measurement which locates its position to 10^{-11} m?

Problems

1. Photoelectrons are emitted with a maximum speed of 7×10^5 m/s from a surface when light of frequency 8×10^{14} Hz is shone on it. What is the threshold frequency for this surface?

2. Light of wavelength 5×10^{-7} m falls on a potassium surface whose nature is such that 2 eV is needed to eject an electron. What is the maximum kinetic energy in eV of the photoelectrons that are emitted?

3. Light from the sun reaches the earth at the rate of about 1.4×10^3 W/m² of area perpendicular to the direction of the light. Assume sunlight is monochromatic with a frequency of 5×10^{14} Hz. (a) How many photons fall per second on each m² of the earth's surface directly facing the sun? (b)

How many photons are present in each m³ near the earth on the sunlit side?

4. What potential difference must be applied across an x-ray tube for it to emit x-rays with a minimum wavelength of 10^{-11} m?

5. Find the energy and momentum of an x-ray photon whose wavelength is 2×10^{-11} m.

6. Find the energy and momentum of an x-ray photon whose frequency is 5×10^{18} Hz.

7. Calculate the de Broglie wavelength of a proton whose kinetic energy is 1 MeV. This calculation may be made nonrelativistically.

8. Calculate the de Broglie wavelength of (a) an electron whose velocity is 1×10^8 m/s, and (b) an electron whose velocity is 2×10^8 m/s. Use relativistic formulas.

9. The position and momentum of a 1000-eV electron are determined at the same time. If the position is found to within 10^{-10} m, what is the percentage of uncertainty in the momentum?

10. At a certain time a measurement establishes the position of an electron with an accuracy of $\pm 10^{-11}$ m. Calculate the uncertainty in the electron's momentum and, from this, the uncertainty in its position 1 s later.

11. A typical atomic nucleus is 10^{-14} m in diameter. Find the lowest energy of a neutron in a box 10^{-14} m wide.

12. From the formula for the possible energies of a particle trapped in a box, find the minimum momentum the particle can have and compare it with the prediction of the uncertainty principle.

13. Verify that the uncertainty principle can be expressed in the form $\Delta L \Delta \theta \geq h/2\pi$, where ΔL is the uncertainty in the angular momentum of a body and $\Delta \theta$ is the uncertainty in its angular position. To do this, consider a particle of mass m moving in a circle.

Answers to Multiple Choice

1. b	6. c	11. c
2. b	7. c	12. b
3. d	8. d	13. d
4. b	9. b	14. b
5. b	10. d	

29

The Hydrogen Atom

A hydrogen atom consists of a proton with an electron circling it. In the Bohr model of this atom, the particle and wave properties of the electron are combined: its orbital motion is such that the centripetal force needed is provided by the electric force exerted by the proton, and the circumference of its orbit is exactly one de Broglie wavelength. This model predicts that the hydrogen atom can have certain specific energies only, and no others. This prediction agrees remarkably well with experiment. Despite this success, however, the Bohr theory has certain deficiencies, and has been superseded by the more general (and more complicated) quantum theory of the atom which is the subject of Chapter 30.

29-1 Standing Waves in the Atom

The picture of the atom that emerged from Rutherford's work, as we learned in Section 17-5, consists of a tiny, massive, positively-charged nucleus surrounded by enough negatively-charged electrons to leave the atom as a whole electrically neutral. These electrons cannot be stationary, since the electric attraction of the nucleus would pull them in at once. If the electrons are in motion around the nucleus, however, stable orbits like those of the planets about the sun would seem to be possible.

A hydrogen atom, the simplest of all, has a single electron and a nucleus that consists of a single proton. Experiments indicate that 13.6 eV of work must be performed to break apart a hydrogen atom into a proton and electron that go their separate ways. With the help of calculations based upon what we already know about mechanics and electricity this figure leads to an orbital radius of

Mechanical model of hydrogen atom

$$r = 5.3 \times 10^{-11} \text{ m}$$

and an orbital velocity of

$$v = 2.2 \times 10^6 \text{ m/s}$$

for the electron in the hydrogen atom. The proton mass is 1836 times the electron mass, so we can consider the proton as stationary with the electron revolving around it in such a way that the required centripetal force is provided by the electric force exerted by the proton (Fig. 29-1).

What the preceding analysis overlooks is that, according to electromagnetic theory, all accelerated electric charges radiate electromagnetic waves—and an electron moving in a circular path is certainly accelerated.

Mechanical model is inconsistent with electromagnetic theory

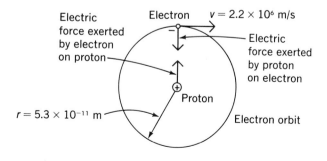

Fig. 29-1. The hydrogen atom consists of an electron circling a proton. The electric force exerted by the proton on the electron provides the centripetal force required to hold it in a circular path. The proton is nearly 2000 times heavier than the electron, and so its motion under the influence of the electric force which the electron exerts is relatively insignificant."

Thus an atomic electron circling its nucleus to keep from falling into it by electrostatic attraction cannot help radiating away energy, which means that it must spiral inward until it is swallowed up by the nucleus. Clearly the ordinary laws of physics cannot account for the stability of the hydrogen atom, the simplest atom of all, whose electron must be whirling around the nucleus to keep from being pulled into it and yet must be radiating electromagnetic energy continuously. However, since other phenomena similarly impossible to understand—the photoelectric effect, for instance—find complete explanation in terms of quantum concepts, it is appropriate for us to inquire whether this might also be true for the atom.

In the discussion that follows of the Bohr model of the atom, the initial argument is somewhat different from that of Bohr who did not, in 1913, have the notion of matter waves to guide his thinking. The results are exactly the same as Bohr obtained, though.

Wave properties of atomic electron

Let us begin by looking into the wave properties of the electron in the hydrogen atom. The de Broglie wavelength λ of an object of mass m and velocity v is $\lambda = h/mv$ where h is Planck's constant. Now the velocity of the electron in a hydrogen atom is

$$v = 2.2 \times 10^6 \, \frac{\text{m}}{\text{s}}$$

and so the wavelength of its matter waves is

$$\lambda = \frac{h}{mv} = \frac{6.63 \times 10^{-34} \, \text{J·s}}{9.1 \times 10^{-31} \, \text{kg} \times 2.2 \times 10^6 \, \text{m/s}} = 3.3 \times 10^{-10} \, \text{m}.$$

This is a most exciting result, because the electron's orbit has a circumference of exactly

$$2\pi r = 2\pi \times 5.3 \times 10^{-11} \, \text{m} = 3.3 \times 10^{-10} \, \text{m}.$$

The electron orbit is one wavelength in circumference

We therefore conclude that *the orbit of the electron in a hydrogen atom corresponds to one complete electron wave joined on itself* (Fig. 29-2).

The fact that the electron orbit in a hydrogen atom is one electron wavelength in circumference is just the clue we need to construct a theory of the atom. If we examine the vibrations of a wire loop, as in Fig. 29-3, we see that their wavelengths always fit an integral number of times into the loop's circumference, each wave joining smoothly with the next. In the absence of dissipative effects, such vibrations would persist indefinitely. Why are these the only vibrations possible in a wire loop? A fractional number of wavelengths cannot be fitted into the loop and still allow each wave to join

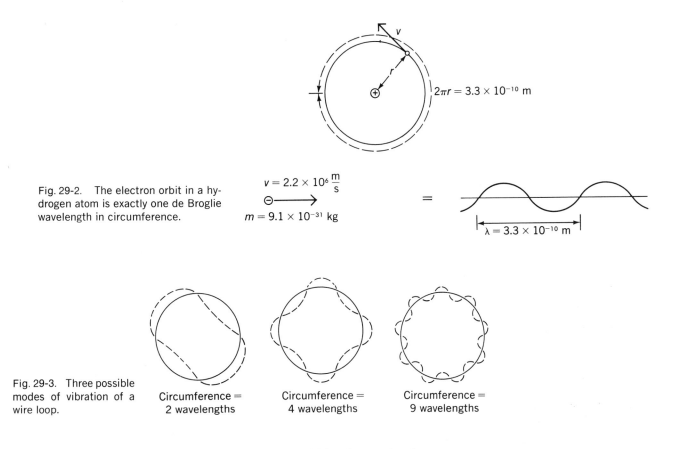

Fig. 29-2. The electron orbit in a hydrogen atom is exactly one de Broglie wavelength in circumference.

$$v = 2.2 \times 10^6 \, \frac{m}{s}$$

$$m = 9.1 \times 10^{-31} \, kg$$

$$2\pi r = 3.3 \times 10^{-10} \, m$$

$$\lambda = 3.3 \times 10^{-10} \, m$$

Fig. 29-3. Three possible modes of vibration of a wire loop.

Circumference = 2 wavelengths

Circumference = 4 wavelengths

Circumference = 9 wavelengths

smoothly with the next (Fig. 29-4); the result would be destructive interference as the waves travel around the loop, and the vibrations would die out rapidly.

By considering the behavior of electron waves in the hydrogen atom as analogous to the vibrations of a wire loop, then, we may postulate that

An electron can circle an atomic nucleus only if its orbit is an integral number of electron wavelengths in circumference.

The above postulate is the decisive one in our understanding of the atom. We note that it combines both the particle and wave characters of the electron into a single statement; although we can never observe these antithetical characters at the same time in an experiment, they are inseparable in nature.

It is easy to express in a formula the condition that an integral number

Condition for orbital stability

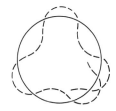

Fig. 29-4. Unless a whole (integral) number of wavelengths fits into the wire loop, destructive interference causes the vibrations to die out rapidly.

of electron wavelengths fit into the electron's "orbit." The circumference of a circular orbit of radius r is $2\pi r$, and so the condition for orbit stability is

$$n\lambda = 2\pi r_n, \qquad n = 1, 2, 3, \ldots \qquad \textit{Condition for orbit stability} \quad (29\text{-}1)$$

where r_n designates the radius of the orbit that contains n wavelengths. The quantity n is called the *quantum number* of the orbit.

A straightforward calculation (given in Section 29-4) shows that the stable electron orbits are those whose radii are given by the formula

$$r_n = n^2 r_1, \qquad n = 1, 2, 3, \ldots \qquad \textit{Orbital radii in Bohr atom} \quad (29\text{-}2)$$

where

$$r_1 = 5.3 \times 10^{-11} \text{ m}$$

is the radius of the innermost orbit (Fig. 29-5).

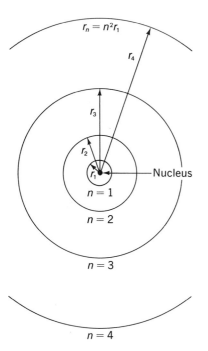

Fig. 29-5. Electron orbits in the hydrogen atom according to the Bohr model.

29-2 Energy Levels

The total energy of a hydrogen atom is not the same in the various permitted orbits. The energy E_n of a hydrogen atom whose electron is in the nth orbit is given by

$$E_n = \frac{E_1}{n^2}, \qquad n = 1, 2, 3, \ldots \qquad \textit{Energy levels of hydrogen atom} \quad (29\text{-}3)$$

where $E_1 = -13.6$ eV $= -2.18 \times 10^{-18}$ J is the energy corresponding to the innermost orbit. The energies specified by the above formula are called the *energy levels* of the hydrogen atom.

The energy levels of the hydrogen atom are all less than zero, which signifies that the electron does not have enough energy to escape from the atom. The lowest energy level E_1, corresponding to the quantum number $n = 1$, is called the *ground state* of the atom; the higher levels E_2, E_3, E_4, and so on are called *excited states* (Fig. 29-6).

As the quantum number n increases, the energy E_n approaches closer and closer to zero. In the limit of $n = \infty$, $E_\infty = 0$ and the electron is no longer bound to the proton to form an atom. An energy greater than zero signifies an unbound electron which, since it has no closed orbit that must satisfy quantum conditions, may have any positive energy whatever.

The work needed to remove an electron from an atom in its ground

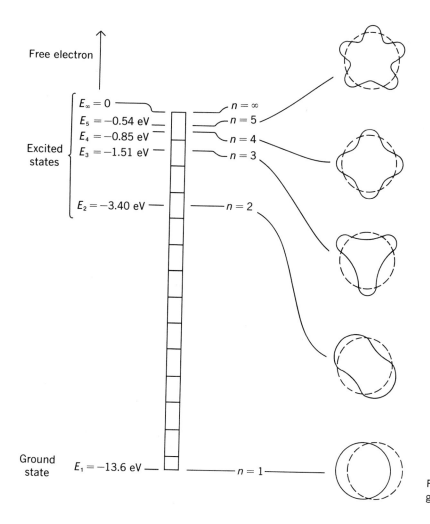

Fig. 29-6.   Energy levels of the hydrogen atom.

state is called its *ionization energy*. The ionization energy is therefore equal to the amount of energy that must be provided to raise an electron from its ground state to an energy of $E = 0$, when it is free. In the case of hydrogen, the ionization energy is 13.6 eV, since the ground-state energy of the hydrogen atom is -13.6 eV.

The presence of definite energy levels in an atom—which is true for all atoms, not just the hydrogen atom—is another example of the fundamental graininess of physical quantities on a microscopic scale. In the everyday world, matter, electric charge, energy, and so on seem continuous and capable of being cut up, so to speak, into parcels of any size we like. In the

Ionization energy

world of the atom, however, matter consists of elementary particles of fixed masses which join together to form atoms of fixed masses; electric charge always comes in multiples of $+e$ and $-e$; energy in the form of electromagnetic waves of frequency f always comes in separate photons of energy hf; and stable systems of particles, such as atoms, can have only certain energies and no others.

Other quantities in nature are also grainy, or *quantized*, and it has turned out that this graininess is the key to understanding how the properties of matter we are familiar with in everyday life originate in the interactions of elementary particles. In the case of the atom, the quantization of energy is a consequence of the wave nature of moving bodies: in an atom the electron wave functions can only occur in the form of standing waves, much as a violin string can vibrate only at those frequencies that give rise to standing waves.

Quantization of characteristic of many quantities in nature

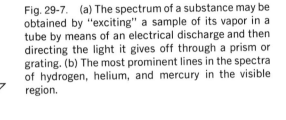

Slit

Prism

Lens

Screen

Rarefied gas or vapor "excited" by electrical discharge

(a)

Fig. 29-7. (a) The spectrum of a substance may be obtained by "exciting" a sample of its vapor in a tube by means of an electrical discharge and then directing the light it gives off through a prism or grating. (b) The most prominent lines in the spectra of hydrogen, helium, and mercury in the visible region.

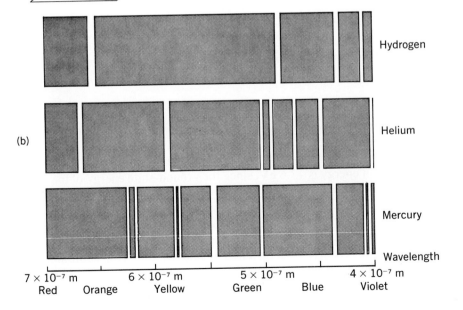

(b)

Hydrogen

Helium

Mercury

Wavelength

| 7×10^{-7} m | | 6×10^{-7} m | | 5×10^{-7} m | | 4×10^{-7} m |
| Red | Orange | Yellow | | Green | Blue | Violet |

29-3 Atomic Spectra

The Bohr theory is strikingly confirmed in its successful explanation of atomic spectra. When a gas or vapor at somewhat less than atmospheric pressure is "excited" by the passage of an electric current through it, light whose spectrum consists of a limited number of individual wavelengths is emitted (Fig. 29-7). The characteristic red-orange color of a neon sign is an example of this phenomenon.

Atomic spectra are called *line spectra* from their appearance. Every element exhibits a unique line spectrum when a sample of it is suitably excited, and the presence of any element in a substance of unknown composition can be ascertained by the appearance of its characteristic wavelengths in the spectrum of the substance. Spectral lines are found in the infrared and ultraviolet as well as in the visible region.

It is worth noting that whereas unexcited gases and vapors do not radiate their characteristic spectral lines, they do *absorb* light of certain of those wavelengths when white light is passed through samples of them. In other words, the *absorption spectrum* of an element is closely related to its *emission spectrum*. Emission spectra consist of bright lines on a dark background; absorption spectra consist of dark lines on a bright background. Figure 29-8 shows the absorption and emission spectra of sodium vapor.

The wavelengths present in atomic spectra fall into definite series. The spectral series of hydrogen are shown in Fig. 29-9. The presence of a sequence of definite, discrete energy levels in the hydrogen atom suggests a connection with line spectra. Let us assert that when an electron in an excited state drops to a lower state, the difference in energy between the states is emitted as a single photon of light. Because electrons cannot, according to our model, exist in an atom except in certain specific energy levels, a rapid "jump" from one level to the other, with the energy difference being given off all at once in a photon rather than in some gradual manner, fits in well with this model.

If the quantum number of the initial (higher energy) state is n_i and the quantum number of the final (lower energy) state is n_f, what we assert is that

$$E_i - E_f = hf \qquad \text{\textit{Origin of spectral lines}} \quad (29\text{-}4)$$

Initial energy − final energy = quantum energy

where f is the frequency of the emitted photon and h is Planck's constant. This formula agrees exactly with the experimental data on the spectrum of hydrogen.

Fig. 29-8.   (a) Absorption spectrum of sodium vapor. (b) Emission spectrum of sodium vapor. Each dark line in the absorption spectrum corresponds to a bright line in the emission spectrum.

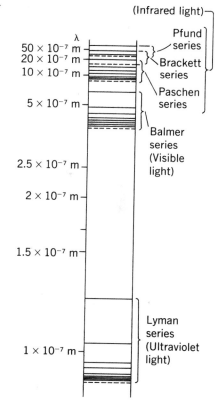

Fig. 29-9.   The line spectrum of hydrogen with the various series of spectral lines indicated. The wavelength scale is not linear in order to cover the entire spectrum.

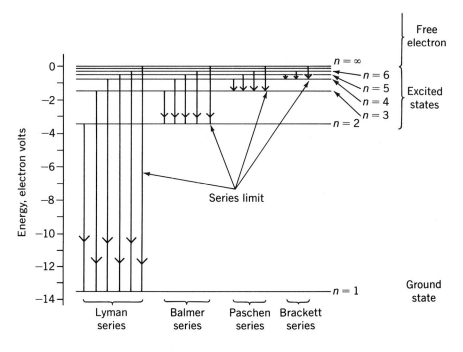

Fig. 29-10.   Energy-level diagram of the hydrogen atom with some of the transitions that give rise to spectral lines indicated.

Figure 29-10 is an energy-level diagram for the hydrogen atom showing the possible transitions from initial quantum states to final ones. Each transition — or "jump" — involves a characteristic amount of energy, and hence a photon of a certain characteristic frequency. The larger the energy difference between initial and final energy levels, as indicated by the lengths of the arrows, the higher the frequency of the emitted photon. The origins of the various series of spectral lines are indicated on the diagram.

29-4 The Bohr Model

In this section the various results of the Bohr theory of the hydrogen atom will be derived.

We start with the classical picture of the hydrogen atom in which the electron is assumed to circle the proton in a stable orbit. If the orbit radius is r, the centripetal force $F_c = \dfrac{mv^2}{r}$ on the electron is provided by the electric force $F_e = k\dfrac{e^2}{r^2}$ between the electron and the proton. Hence

$$F_c = F_e, \qquad \frac{mv^2}{r} = k\frac{e^2}{r^2},$$

Force balance

and the electron velocity is related to its orbit radius r by the formula

$$v = e\sqrt{\frac{k}{mr}}. \qquad (29\text{-}5)$$

Because the proton is nearly 2000 times heavier than the electron, we are justified in ignoring the proton's motion under the influence of the electric force the electron exerts on it.

The total energy E of a hydrogen atom is the sum of the electron's kinetic energy $KE = \frac{1}{2}mv^2$ and the electric potential energy of the system of electron and proton, which is $PE = -k\frac{e^2}{r}$. The latter formula was discussed in Section 18-5. Hence

$$E = KE + PE = \frac{mv^2}{2} - k\frac{e^2}{r}.$$

and so, in view of Eq. (29-5),

$$E = k\left(\frac{e^2}{2r} - \frac{e^2}{r}\right) = -\frac{k}{2}\frac{e^2}{r}. \qquad (29\text{-}6)$$

The total energy of the hydrogen atom is negative, which expresses the fact that the electron is bound to the proton. If E were greater than zero, the electron would not be able to remain in a closed orbit about the proton.

Experiments indicate that 13.6 eV of work must be performed to break a hydrogen atom apart into a proton and an electron which go their separate ways. In other words, the energy of the hydrogen atom in its ground state is $E = -13.6$ eV. We can find the orbital radius of the electron from Eq. (29-6) for E. Since 13.6 eV $= 2.18 \times 10^{-18}$ J, and the value of the electric constant k is 9.0×10^9 N·m²/C², we have

$$r = -\frac{k}{2}\frac{e^2}{E}$$

Orbit radius

$$= -\frac{9.0 \times 10^9 \text{ N·m}^2/\text{C}^2 \times (1.6 \times 10^{-19} \text{ C})^2}{2 \times (-2.18 \times 10^{-18} \text{ J})}$$

$$= 5.3 \times 10^{-11} \text{ m}.$$

The electron's velocity is, from Eq. (29-5),

Electron velocity

$$v = e \sqrt{\frac{k}{mr}}$$

$$= 1.6 \times 10^{-19} \text{ C} \sqrt{\frac{9.0 \times 10^9 \text{ N} \cdot \text{m}^2/\text{C}^2}{9.1 \times 10^{-31} \text{ kg} \times 5.3 \times 10^{-11} \text{ m}}}$$

$$= 2.2 \times 10^6 \text{ m/s}.$$

This velocity is well below the velocity of light ($c = 3 \times 10^8$ m/s), and so the above nonrelativistic calculation is justified.

To derive a formula for the radii r_n of the possible electron orbits in the hydrogen atom, we start from the quantum condition of Eq. (29-1),

$$n\lambda = 2\pi r_n \qquad n = 1,2,3,\ldots$$

The formula for the de Broglie wavelength of the electron is

$$\lambda = \frac{h}{mv}.$$

Using Eq. (29-5) for v with r_n in place of r yields

$$\lambda = \frac{h}{me} \sqrt{\frac{mr_n}{k}} = \frac{h}{e} \sqrt{\frac{r_n}{mk}}.$$

Hence the quantum condition becomes

$$n\lambda = \frac{nh}{e} \sqrt{\frac{r_n}{mk}} = 2\pi r_n.$$

Squaring both sides and solving for r_n yields

$$\frac{n^2 h^2}{e^2} \frac{r_n}{mk} = 4\pi^2 r_n^2$$

$$r_n = \frac{h^2 n^2}{4\pi^2 k m e^2} \qquad n = 1,2,3,\ldots \tag{29-7}$$

which we can write as

$$r_n = r_1 n^2 \qquad n = 1,2,3,\ldots$$

where r_1 is the radius of the first (innermost) orbit. This result was given as Eq. (29-2).

The total energy E_n of a hydrogen atom whose electron is in the nth orbit depends only upon the radius r_n of that orbit, since

$$E = -\frac{k}{2}\frac{e^2}{r}$$

for all orbits. Hence the possible energies of a hydrogen atom are limited to

$$E_n = -\frac{k}{2}\frac{e^2}{h^2n^2/4\pi^2kme^2}$$ Electron energies

$$= -\frac{2\pi^2k^2me^4}{h^2n^2} \qquad n = 1,2,3,\ldots \tag{29-8}$$

which we can write as

$$E_n = \frac{E_1}{n^2} \qquad n = 1,2,3,\ldots$$

where E_1 is the energy of the innermost orbit. This result was given as Eq. (29-3).

It is worth noting that the formula for E_n contains only the directly-measurable quantities k, m, e, and h, all of which are constants of nature. The formula is in complete agreement with the experimentally-determined values of E_n.

29-5 Atomic Excitation

There are two principal mechanisms by which an atom may be excited   Excitation by collision
to an energy level above that of its ground state and thereby become capable of radiating. One of these mechanisms is a collision with another atom during which part of their kinetic energy is transformed into electron energy within either or both of the participating atoms. An atom excited in this way will then lose its excitation energy by emitting one or more photons in the course of returning to its ground state (Fig. 29-11). In an electric discharge in a rarefied gas, an electric field accelerates electrons and charged atoms and molecules (whose charge arises from either an excess or a deficiency in the electrons required to neutralize the positive charge of their nuclei) until their kinetic energies are sufficient to excite atoms with which they happen to collide. A neon sign is a familiar example of how applying a strong electric field between electrodes in a gas-filled tube leads to the

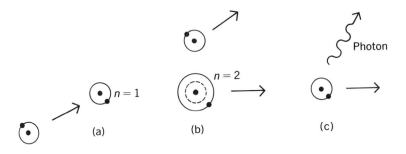

$n = 1$

$n = 2$

Photon

(a)   (b)   (c)

Fig. 29-11.   Excitation by collision. In (a) both atoms are in their ground states. During the collision some kinetic energy is transformed into excitation energy, and in (b) the target atom is in an excited state. In (c) it has returned to its ground state by emitting a photon.

emission of the characteristic spectral radiation of that gas, which happens to be orange light in the case of neon.

Excitation by absorption   Another excitation mechanism is the absorption by an atom of a photon of light whose energy is just the right amount to raise it to a higher energy level. A photon of wavelength 121.7 nm is emitted when a hydrogen atom in the $n = 2$ state drops to the $n = 1$ state; hence the absorption of a photon of wavelength 121.7 nm by a hydrogen atom initially in the $n = 1$ state will bring it up to the $n = 2$ state. This process explains the origin of absorption spectra (Fig. 29-12).

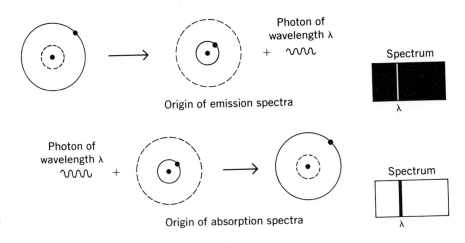

Photon of wavelength λ

Spectrum

Origin of emission spectra

Photon of wavelength λ

Spectrum

Origin of absorption spectra

Fig. 29-12.   The origins of emission and absorption spectra.

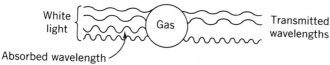

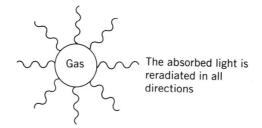

Fig. 29-13. The dark lines in an absorption spectrum are never totally dark.

When white light (in which all wavelengths are present) is passed through hydrogen gas, photons of those wavelengths that correspond to transitions between hydrogen energy levels are absorbed. The resulting excited hydrogen atoms reradiate their excitation energy almost at once, but these photons come off in random directions, not all in the same direction as in the original beam of white light (Fig. 29-13). The dark lines in an absorption spectrum are therefore never totally dark, but only appear so by contrast with the bright background of transmitted light. We would expect the lines in the absorption spectrum of a particular substance to be the same as lines in its emission spectrum, which is what is found. Only a few of the emission lines appear in absorption, however, since reradiation is so rapid that essentially all the atoms in an absorbing gas or vapor are initially in their lowest ($n = 1$) energy states and therefore take up and re-emit only photons that represent transitions to and from the $n = 1$ state.

29-6 The Laser

The *laser* is a device for producing a light beam with a number of remarkable properties:

1. The beam is extremely intense, more intense by far than the light from any other source. To achieve an energy density equal to that in some laser beams, a hot object would have to be at a temperature of 10^{30} K.

Properties of a laser beam

2. There is very little divergence, so that a laser beam from the earth that was reflected by a mirror left on the moon during the Apollo 11 expedition remained collimated enough to be detected upon its return to the earth—a total journey of nearly a half-million miles.
3. The light is essentially monochromatic.
4. The light is coherent, with the waves all exactly in step with one another. It is possible to obtain interference patterns not only by merely placing two slits in a laser beam but also by using beams from two separate lasers.

The term "laser" stands for *l*ight *a*mplification by *s*timulated *e*mission of *r*adiation. Let us look into how a laser functions.

Left to itself, an atom always remains in its ground state, the quantum state in which it has the least possible energy. The atom can be raised to an excited state by acquiring energy in a collision with an electron or another atom or by absorbing a photon of exactly the right frequency. An excited atom normally falls to its ground state almost at once, sometimes first dropping to an intermediate excited state, with the emission of a photon during each transition. The lifetime of most excited states is only about

Metastable state 10^{-8} s, but certain states are *metastable* (temporarily stable). An atom may remain in a metastable state for 10^{-3} second or more before radiating, provided it does not undergo a collision in the meantime in which its excitation energy would be lost (Fig. 29-14). The operation of lasers depends upon the existence of metastable states in atoms.

Processes of absorption and emission Three kinds of transition involving electromagnetic radiation are possible between two energy levels in an atom whose energy difference is $E_2 - E_1 = hf$. In *induced absorption* the atom when in the lower level absorbs a photon of energy hf and is elevated to the upper level (29-15).

Fig. 29-14. An atom can exist in a metastable energy level for a longer time before radiating than it can in an ordinary energy level.

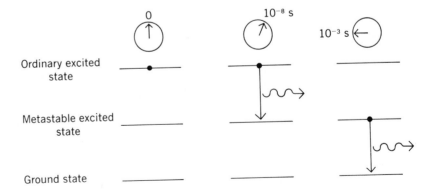

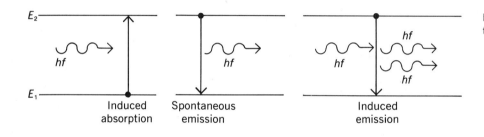

Fig. 29-15. The three types of transition between energy levels in an atom.

In *spontaneous emission* the atom when in the upper level emits a photon of energy *hf* and falls to the lower one. There is also a third possibility, *induced emission*, in which the presence of radiation of frequency *f* causes a transition from the upper level to the lower one. In induced emission, the emitted light waves are exactly in step with the incident ones, so that the result is an enhanced beam of coherent light.

Now let us suppose we have an assembly of atoms of some kind which have metastable states of excitation energy *hf*. If we can somehow raise a majority of the atoms to the metastable level and then shine light of frequency *f* on the assembly, there will be more induced emission than induced absorption and the result will be an amplification of the original light. This is the concept that underlies the operation of the laser. The term *population inversion* is given to an assembly of atoms in which a majority are in excited states, because under normal circumstances the ground states are the more highly populated.

There are a number of ways in which population inversions can be produced. One of them, called "optical pumping," employs an external light source some of whose photons have the right frequency to raise the atoms to excited states that decay into the desired metastable ones. This is the method used in the ruby laser, in which the chromium ions Cr^{+++} in a ruby crystal are excited by light from a xenon-filled flash lamp (Fig. 29-16). (A ruby is a crystal of Al_2O_3 a small number of whose Al^{+++} ions are replaced by Cr^{+++} ions, which are responsible for the reddish color. Such ions are Cr atoms that have lost three electrons each.)

The Cr^{+++} ions are raised in the pumping process to a level E_3 from which they decay to the metastable level E_2 by losing energy to other atoms in the crystal. This metastable level has a lifetime of about 0.003 s. Because there are comparatively few Cr^{+++} ions in the ruby rod and a xenon flash lamp emits a great deal of light, most of the Cr^{+++} ions can be pumped to the

Optical pumping

Operation of ruby laser

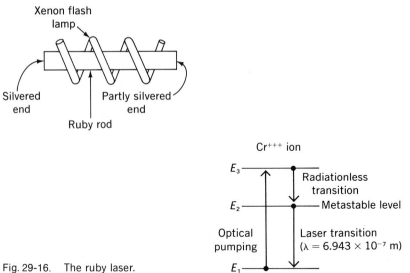

Fig. 29-16.   The ruby laser.

E_2 level in this way. A few Cr^{+++} ions in the E_2 level then spontaneously fall to the E_1 level, and some of the resulting photons bounce back and forth between the reflecting ends of the ruby rod. The presence of light of exactly the right frequency now stimulates the other Cr^{+++} ions in the E_2 level to radiate, and the result is an avalanche process that produces a large pulse of red light in a few microseconds that emerges from the partly-silvered end of the rod. The length of the rod is made to be exactly an integral number of half-wavelengths long, so the radiation trapped in it forms an optical standing wave. (The rod constitutes a "resonant cavity.") Since the induced emissions are stimulated by the standing wave, their waves are all in step with it, which is why the final pulse of light is both coherent and collimated.

Operation of helium-neon laser

The helium-neon gas laser achieves a population inversion in a different way. A mixture of about 7 parts of helium and 1 part of neon at a low pressure (1 mm of mercury) is placed in a glass tube that has parallel mirrors at both ends. An electric discharge is then produced in the gas by electrodes connected to a source of high-frequency alternating current, and collisions with electrons from the discharge excite He and Ne atoms to metastable states respectively 20.61 and 20.66 eV above their ground states (Fig. 29-17). Some of the excited He atoms transfer their energy to ground-state Ne atoms in collisions, with the 0.05 eV additional energy being provided by

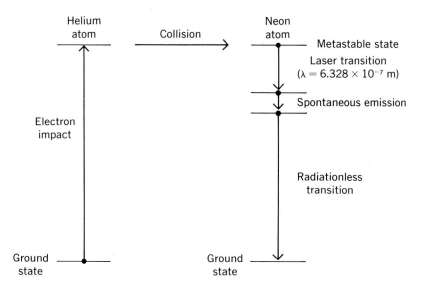

Fig. 29-17.   Energy levels involved in the operation of the helium-neon laser.

the kinetic energy of the atoms. The purpose of the He atoms is thus to assist in producing a population inversion in the Ne atoms.

An excited Ne atom emits a photon of wavelength $\lambda = 6.328 \times 10^{-7}$ m in the transition to another excited state 18.70 eV above the ground state; this is one of the transitions that leads to laser action. Then another photon is spontaneously emitted in a transition to a lower metastable state (this transition produces only incoherent radiation because it is spontaneous rather than induced), and the remaining excitation energy is lost in collisions with the tube walls. Because the electron impacts that produce a population inversion in the Ne atoms are not intermittent but occur all the time, a He-Ne laser operates continuously.

Many other types of laser have been constructed, including solid-state ones. Among the most interesting is the "chemical laser," in which molecules in metastable excited states are produced by chemical reactions. In this way the overall efficiency of the system is considerably improved, since elaborate and energy-wasting methods of producing a population inversion are not necessary.

Important Terms

An **emission spectrum** consists of the various wavelengths of light emitted by an excited substance. An **absorption spectrum** consists of the various wavelengths of light absorbed by a substance when white light is passed through it.

According to the **Bohr theory of the atom**, an electron can circle an atomic nucleus indefinitely without radiating energy if its orbit is an integral number of electron wavelengths in circumference. The number of wavelengths that fit into a particular permitted orbit is called the **quantum number** of that orbit. The electron energies corresponding to the various quantum numbers constitute the **energy levels** of the atom, of which the lowest is the **ground state** and the rest are **excited states**. When an electron in an excited state drops to a lower state, the difference in energy between the states is emitted as a single photon of light; when a photon of the same wavelength is absorbed, the electron goes from the lower to the higher state. The above two processes account for the properties of **emission** and **absorption spectra** respectively.

A **laser** is a device for producing a narrow, monochromatic, coherent beam of light. The terms stands for **l**ight **a**mplification by **s**timulated **e**mission of **r**adiation.

Important Formulas

Condition for orbit stability:

$$n\lambda = 2\pi r_n$$

Energy levels of hydrogen atom:

$$E_n = \frac{E_1}{n^2}, \qquad n = 1, 2, 3, \ldots$$

Origin of spectral lines:

$$E_i - E_f = hf$$

Multiple Choice

1. The bright-line spectrum produced by the excited atoms of an element contains wavelengths that
   a. are the same for all elements.
   b. are characteristic of the particular element.
   c. are evenly distributed throughout the entire visible spectrum.
   d. are different from the wavelengths in its dark-line spectrum.

2. A neon sign does *not* produce
   a. a line spectrum.
   b. an emission spectrum.
   c. an absorption spectrum.
   d. photons.

3. The electron in the Bohr model of the hydrogen atom is pictured as revolving around the nucleus in order for it to
   a. emit photons.
   b. possess energy.
   c. keep from being pulled into the nucleus.
   d. keep from being repelled by the nucleus.

4. An electron can rotate around an atomic nucleus indefinitely without radiating energy if its orbit
   a. is a perfect circle.
   b. is sufficiently far from the nucleus.
   c. is less than a de Broglie wavelength in circumference.
   d. is 1 de Broglie wavelength in circumference.

5. An atom emits a photon when one of its electrons
   a. collides with another of its electrons.
   b. exchanges quantum states with another of its electrons.
   c. undergoes a transition to a quantum state of lower energy.
   d. undergoes a transition to a quantum state of higher energy.

6. Which of the following transitions in a hydrogen atom emits the photon of highest frequency?
   a. $n = 1$ to $n = 2$
   b. $n = 2$ to $n = 1$

c. $n = 2$ to $n = 6$

d. $n = 6$ to $n = 2$

7. Which of the following transitions in a hydrogen atom absorbs the photon of highest frequency?

    a. $n = 1$ to $n = 2$

    b. $n = 2$ to $n = 1$

    c. $n = 2$ to $n = 6$

    d. $n = 6$ to $n = 2$

8. Which of the following transitions in a hydrogen atom emits the photon of lowest frequency?

    a. $n = 1$ to $n = 2$

    b. $n = 2$ to $n = 1$

    c. $n = 2$ to $n = 6$

    d. $n = 6$ to $n = 2$

9. A hydrogen atom is in its ground state when its orbital electron

    a. is within the nucleus.

    b. has escaped from the atom.

    c. is in its lowest energy level.

    d. is stationary.

10. With increasing quantum number, the energy difference between adjacent energy levels

    a. decreases.

    b. remains the same.

    c. increases.

    d. sometimes decreases and sometimes increases.

11. Most excited states of an atom have lifetimes of about

    a. 10^{-8} s.

    b. 10^{-3} s.

    c. 1 s.

    d. 10 s.

12. The operation of the laser is based upon

    a. the uncertainty principle.

    b. diffraction.

    c. induced emission of radiation.

    d. spontaneous emission of radiation.

13. The light produced by a laser is not

    a. incoherent.

    b. monochromatic.

    c. in the form of a narrow beam.

    d. electromagnetic.

Exercises

1. How are the Bohr and Rutherford models of the hydrogen atom related?

2. In the Bohr model of the hydrogen atom, why is the electron pictured as revolving around the nucleus?

3. In the Bohr theory of the hydrogen atom, the electron is in constant motion. How can such an electron have a negative amount of energy?

4. Is the Bohr theory of the atom compatible with the uncertainty principle? If not, why not?

5. Explain why the spectrum of hydrogen has many lines, although a hydrogen atom contains only one electron.

6. Would you expect the fact that the atoms of an excited gas are in rapid random motion to have any effect on the sharpness of the spectral lines they produce?

7. What kind of spectrum is observed in (a) light from the hot filament of a light bulb; (b) light from a sodium-vapor highway lamp; (c) light from an electric-light bulb that has passed through cool sodium vapor?

8. When radiation with a continuous spectrum is passed through a volume of hydrogen gas whose atoms are all in the ground state, which spectral series will be present in the resulting absorption spectrum?

9. The three kinds of transition involving electromagnetic radiation that can occur between two energy levels in an atom are induced absorption, spontaneous emission, and induced emission. Why is spontaneous absorption impossible?

10. Why is the length of the optical cavity of a laser so important?

11. A beam of electrons is used to bombard gaseous hydrogen. What is the minimum energy in eV the electrons must have if the first line of the Balmer series, corresponding to a transition from the $n = 3$ state to the $n = 2$ state, is to be emitted?

12. What is the shortest wavelength present in the Paschen series of spectral lines?

13. What is the shortest wavelength present in the Brackett series of spectral lines?

14. A proton and an electron, both at rest initially, combine to form a hydrogen atom in the ground state. A single photon is emitted in this process. What is its wavelength?

15. How much energy in eV is required to remove the electron from a hydrogen atom when it is in the $n = 5$ state?

16. A beam of electrons whose energy is 13 eV is used to bombard gaseous hydrogen. What series of wavelengths will be emitted?

Problems

1. (a) Calculate the de Broglie wavelength of the earth. (b) What is the quantum number that characterizes the earth's orbit about the sun? (The earth's mass is 6.0×10^{24} kg, its orbital radius is 1.5×10^{11} m, and its orbital speed is 3×10^4 m/s.)

2. Calculate the average kinetic energy per molecule in a gas at room temperature (20°C), and show that this is much less than the energy required to raise a hydrogen atom from its ground state ($n = 1$) to its first excited state ($n = 2$).

3. To what temperature must a hydrogen gas be heated if the average molecular kinetic energy is to equal the binding energy of the hydrogen atom?

4. (a) Derive a formula for the frequency of revolution of an electron in the nth orbit of the Bohr atom. (b) Find the frequencies of revolution of the electron when it is in the $n = 1$ and $n = 2$ orbits. (c) An electron spends about 10^{-8} sec in an ex-

cited state before it drops to a lower state by giving up energy in the form of a photon. How many revolutions does an electron in the $n = 2$ state of the Bohr atom make before dropping to the $n = 1$ state? How does this compare with the number of revolutions the earth has made around the sun in the 4.5×10^9 years of its existence?

5. A hydrogen atom emits a photon of wavelength λ in going from a state of quantum number n_i to a state of quantum number n_f. Verify that

$$\frac{1}{\lambda} = \frac{E_1}{ch}\left(\frac{1}{n_i^2} - \frac{1}{n_f^2}\right)$$

(Note that E_1 is a negative quantity.)

6. The electron in a hydrogen atom may be thought of as confined to a region a radius of 5×10^{-11} m from the proton. (a) Use the uncertainty principle to calculate the momentum of the electron, and from this calculate its kinetic energy. (b) What keeps the electron in this region despite its kinetic energy?

7. Repeat the derivation of the Bohr theory for a one-electron ion whose nuclear charge is Ze and show that the energy of the electron is proportional to Z^2.

8. Calculate the radius and speed of an electron in the ground state of doubly-ionized lithium and compare them with the radius and speed of the electron in the ground state of the hydrogen atom. (Li^{++} has a nuclear charge of $3e$.)

9. A negative muon ($m = 207\ m_e$, $q = -e$) can be captured by a proton to form a *mesic atom*. What is the radius of the first Bohr orbit of such an atom? What is the ionization energy of such an atom? Assume that the proton remains stationary as the muon revolves around it; in reality, both revolve around a common center of mass.

10. Use Eq. (29-8) to calculate the ground-state energy E_1 of the hydrogen atom.

Answers to Multiple Choice

1. b	6. b	11. a
2. c	7. a	12. c
3. c	8. d	13. a
4. d	9. c	
5. c	10. a	

30

Quantum Theory of the Atom

The Bohr theory of the atom is indeed impressive in its agreement with experiment, but it has certain serious limitations. These limitations are absent from the quantum theory of the atom, which was developed a decade after Bohr's work, and whose refinement and application to new problems continues to the present day. Not only has the quantum theory of the atom provided the theoretical framework for understanding the structure of the atom itself, but it has also furnished key insights which explain how and why atoms join together to form molecules, solids, and liquids.

30-1 Bohr Theory Versus Quantum Theory

The Bohr theory is unable to account for many important atomic phenomena. While correctly predicting the wavelengths of the spectral lines in hydrogen, which has but a single atomic electron, the Bohr theory fails when attempts are made to apply it to more complex atoms. Even with hydrogen, it is not possible to calculate from the Bohr theory the relative probabilities of the various transitions between energy levels, for instance, whether it is more likely that an atom in the $n = 3$ quantum state will go directly to the $n = 1$ state or instead first drop to the $n = 2$ state. (In other words, we cannot find from the Bohr theory which of the spectral lines of hydrogen will show up brightest in an emission spectrum and which will be faint.) For another thing, the careful study of spectral lines shows that many of them actually consist of two or more separate lines that are close together, something that the Bohr theory cannot account for.

Deficiencies of Bohr theory

In cataloging these objections to the Bohr theory, the intent is not to detract from its eminence in the history of science, which is certainly secure, but instead to emphasize that a more general approach capable of wider application is necessary. Such an approach was developed in the 1920's by Schrödinger, Heisenberg, and others, under the name of the quantum theory of the atom. Instead of trying to visualize an atomic electron as a kind of hybrid of a particle and a wave and thinking of it as occupying one of various possible orbits, the quantum theory of the atom avoids all reference to anything not capable of direct measurement and restricts itself only to such observable quantities as photon energies, the mass and charge of the electron, and so on.

In the Bohr theory we compute the radius of an electron orbit from a knowledge of its de Broglie wavelength, which depends upon the electron's momentum; however, according to the uncertainty principle, the position (and hence orbital radius) and momentum of an electron can never simultaneously have well-defined values, and so even in principle the Bohr theory cannot be subjected to an experimental test. The quantum theory sacrifices such easily pictured notions as that of electrons circling a nucleus like planets around the sun in favor of an abstract mathematical formulation, each of whose statements can be verified. Freed from the necessity of working in terms of a single model, the quantum theory is able to tackle successfully a broad range of atomic problems.

The traditional laws of mechanics and electromagnetism do not "work" when applied to the atomic world, as we have seen—these laws lead to predictions that do not agree with experimental findings. But quantum theory works on all scales of size, and it turns out that the traditional laws of physics are merely approximate versions of quantum theory that are

Classical physics is approximation of quantum physics

successful for large-scale phenomena because on a large scale all that is apparent is the average behavior of a great many separate particles. In dealing with individual particles, however, no averaging is possible, and only quantum theory can be applied.

Unfortunately the quantum theory of the hydrogen atom involves advanced mathematics and so cannot be developed here. The conclusions of the theory, however, are fairly straightforward.

In the Bohr model of the hydrogen atom, the motion of the orbital electron is essentially one-dimensional. There the electron is regarded as being confined to a definite circular orbit, and the only quantity that changes as it revolves around its nucleus is its position on this fixed circle. A single quantum number is all that is required to specify the physical state of such an electron.

In the more general quantum theory of the atom, the electron is not restricted to a specific orbit, and three quantum numbers turn out to be necessary instead of the single one of the Bohr theory. These are:

Quantum numbers of atomic electron

1. The principal quantum number n;
2. The orbital quantum number l;
3. The magnetic quantum number m_l.

The value of n is the chief factor which governs the total energy of an electron bound to a nucleus. The energy levels of a hydrogen atom are the same in the Bohr and quantum theories, and are specified by the formula

$$E = \frac{E_1}{n^2} \qquad n = 1, 2, 3, \ldots \tag{30-1}$$

where $E_1 = -13.6$ eV is the energy of the ground state.

30-2 Angular Momentum Quantization

Orbital quantum number

The orbital quantum number l governs the magnitude L of the electron's angular momentum about the nucleus. Even though the quantum theory discards the picture of an atomic electron as circling the nucleus in favor of a certain distribution of probability density ψ^2, angular momentum can be a property of the atom. The possible angular momenta are restricted to

$$L = \sqrt{l(l+1)}\, \frac{h}{2\pi} \qquad l = 0, 1, 2, \ldots, (n-1). \tag{30-2}$$

There are several interesting things about this result. First, there is the quantization of angular momentum itself, which is a universal phenomenon: like mass, electric charge, and the energy of a trapped particle, the angular momentum of a particle or system of particles can have only certain specific values. The quantity $h/2\pi$ is the natural unit of angular momentum. Because h is so small, the quantization of angular momentum is only perceptible in the physics of atoms and molecules, where it plays an important role.

The orbital quantum number l can be 0 or any integer up to $n-1$, where n is the principal quantum number of the electron's quantum state. If $n=3$, for instance, l can be 0, 1, or 2. When $l=0$, the angular momentum $L=0$ also. Here is a significant difference from the Bohr theory, since an electron revolving in a circle *must* possess angular momentum. No such requirement emerges from the quantum theory of the atom.

The magnetic quantum number m_l governs the *direction* of the electron's angular momentum L. Angular momentum is a vector quantity with both direction and magnitude, so it is natural (though perhaps surprising) that the direction of L should also be restricted in some way. Figure 30-1 shows the right-hand rule for the direction of L.

What can be the significance of the direction of L? And with respect to what is this direction reckoned? The answers to these questions become clear when we consider that an electron with angular momentum behaves like a loop of electric current and interacts with an external magnetic field in a manner similar to that of a bar magnet. The greater the angular momentum L, the stronger the equivalent bar magnet. The potential energy of a bar magnet in a magnetic field B varies with the strength of the magnet and with its direction relative to B; the energy is least when the magnet is parallel to the field, and is most when it is antiparallel (Fig. 30-2).

The magnetic quantum number m_l determines the angle between L and B and hence governs the extent of the magnetic contribution to the total energy of the atom when the atom is in a magnetic field. An atomic electron characterized by a certain value of m_l will assume a certain corresponding

Magnetic quantum number

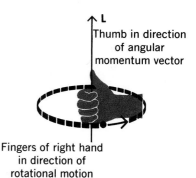

Fig. 30-1.   The right-hand rule for the direction of angular momentum L.

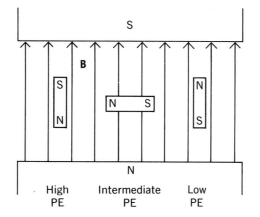

Fig. 30-2.   The potential energy of a bar magnet in a magnetic field depends upon the orientation of the magnet relative to B.

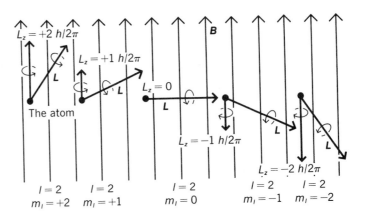

Fig. 30-3. Possible orientations of the angular momentum vector **L** of an atomic electron in a magnetic field when the orbital quantum number of the electron is $l = 2$.

orientation of its angular momentum **L** relative to a magnetic field when placed in such a field.

The direction of **L** relative to **B** is specified in terms of the component of **L** parallel to the magnetic field **B**. By convention this component is called L_z, and its possible values are also in units of $h/2\pi$:

$$L_z = m_l \frac{h}{2\pi}. \tag{30-3}$$

The magnetic quantum number m_l can be any integer from $-l$ through 0 to $+l$. An atomic electron for which $l = 2$, for instance, could have a magnetic quantum number of -2, -1, 0, $+1$, or $+2$. The angular momentum vector **L** of the atom can never be exactly parallel to **B**, even when $m_l = l$ since L_z is always smaller than L (Fig. 30-3).

Why L cannot have a fixed direction

The uncertainty principle enables us to understand why the angular momentum **L** of an atomic electron can never be exactly parallel to **B**. If **L** were parallel to **B**, or if **L** pointed in any other definite direction in space, it would mean that the electron is always present in a specific plane perpendicular to **B**. But if there is such precision in the z coordinate of the electron, then the electron's momentum component in the z direction must be infinitely uncertain. An infinite uncertainty in the momenta of atomic electrons cannot be reconciled with the existence of stable atoms. However, the problem does not in fact arise because $L > L_z$ for all values of m_l. As Fig. 30-4 shows, **L** may point anywhere along the circle at its tip and still have the component $m_l h/2\pi$ in the direction of **B**. Thus there is a built-in uncertainty

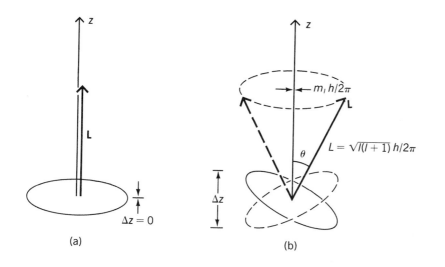

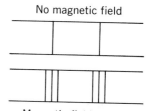

Fig. 30-4. (a) If **L** were parallel to a given direction in space, for instance that of a magnetic field, then the electron would be confined to a definite plane. (b) In reality, L is always greater than L_z, and so there is an intrinsic uncertainty in the electron's position in three dimensions.

in the electron's position, and its momentum uncertainty in consequence does not exceed an amount compatible with its presence in an atom.

Although the interpretation of the magnetic quantum number m_l was discussed in terms of an external magnetic field, actually the component of **L** in *any* direction must be $m_l h/2\pi$. The point is that, for an isolated atom, an external magnetic field provides a definite, experimentally meaningful reference direction. But when the atom is part of a molecule, for instance, there are then other experimentally meaningful reference directions, namely those defined by the relative positions of the various atoms in the molecule. Thus a line joining the two H atoms in the hydrogen molecule H_2 is a quite specific reference direction, and along this line the components of the angular momenta of the two H atoms are fixed by their m_l values.

Because the magnetic quantum number m_l usually has several possible values, the presence of an external magnetic field splits the energy levels of a particular atom into two or more sublevels. The emission spectrum of an element in a magnetic field accordingly differs from its ordinary spectrum in that the spectral lines of the latter now have several components whose spacing varies with B (Fig. 30-5). This phenomenon is known as the *Zeeman effect*. For example, a field of 1 T leads to an energy difference of 5.8×10^{-4} eV between adjacent sublevels of different m_l, which is observed as a wavelength difference between the split components of a given spectral line of order of magnitude 0.1%.

No magnetic field

Magnetic field present

Fig. 30-5. An example of the Zeeman effect.

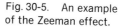
Zeeman effect

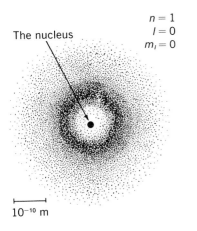

$n = 1$
$l = 0$
$m_l = 0$

The nucleus

10^{-10} m

Fig. 30-6. Probability cloud for the ground state of the hydrogen atom.

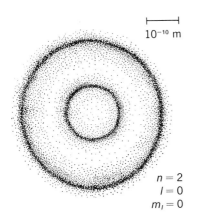

10^{-10} m

$n = 2$
$l = 0$
$m_l = 0$

Fig. 30-7. The probability cloud for the $n = 2$, $l = 0$ state of the hydrogen atom.

Standing waves in 3 dimensions

30-3 Probability Clouds

In place of the picture of a hydrogen atom as an electron circling a proton in a definite orbit, the quantum theory of the atom provides a description in terms of the probability density of the electron. We recall from Section 28-4 what is meant by the probability density ψ^2: The value of ψ^2 for a particular object at a certain place and time is proportional to the probability of finding the object at that place at that time. A large value of ψ^2 signifies that the object is likely to be found; a small value of ψ^2 signifies that the object is unlikely to be found.

The calculation of probability densities starts from *Schrödinger's equation*, an equation as central to atomic physics as Newton's laws of motion are to classical mechanics, but far more complicated. The results show that the electron probability density distribution is different for each quantum state of a hydrogen atom, that is, for each set of quantum numbers n, l, and m_l. We might call the ψ^2 distribution that corresponds to a particular quantum state a *probability cloud*. In the Bohr model, an atomic electron travels in a specific orbit; in the quantum theory, an atomic electron moves about within a probability cloud that forms a certain pattern in space.

There is only a single probability cloud possible for a ground-state hydrogen atom, since when $n = 1$, $l = 0$ and $m_l = 0$. This cloud is spherically symmetric—that is, the likelihood of finding the electron at a given distance from the nucleus is the same in all directions. We can picture ψ^2 for an $n = 1$ electron by a shaded drawing in which the darker the shading, the greater the value of ψ^2. As we can see in Fig. 30-6, the probability cloud does not have a sharp boundary. The electron spends most of the time near the nucleus, but there is a certain probability, which becomes vanishingly small at large distances, that it may be found anywhere.

An electron with the principal quantum number $n = 2$ may exist in one of four quantum states:

$$n = 2 \begin{cases} l = 0 & m_l = 0 \\ l = 1 & \begin{cases} m_l = +1 \\ m_l = 0 \\ m_l = -1. \end{cases} \end{cases}$$

The $n = 2$, $l = 0$ probability cloud is spherically symmetric (indeed, *all* $l = 0$ clouds are spherically symmetric), but it does not fade out uniformly with distance as does the $n = 1$, $l = 0$ cloud (Fig. 30-7). Instead there is a gap at a radius of 1.06×10^{-10} m; the electron is *never* exactly this far from the nucleus. This gap corresponds to a node in a standing-wave pattern—in fact, we can think of the entire probability cloud as a kind of standing-wave pattern in three dimensions. The dense part of the cloud, which covers the region in which the electron is nearly always to be found, is four times as far

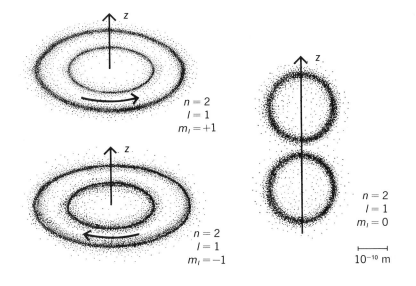

$n = 2$
$l = 1$
$m_l = +1$

$n = 2$
$l = 1$
$m_l = -1$

$n = 2$
$l = 1$
$m_l = 0$

10^{-10} m

Fig. 30-8. Probability clouds for the three possible $n = 2$, $l = 1$ states of the hydrogen atom.

across as the $n = 1$ cloud. This is in accord with the Bohr theory, in which an electron's orbital radius is proportional to n^2. We can think of the cloud in the drawing as representing a fuzzy sphere inside a fuzzy spherical shell.

The sizes of the $n = 2$, $l = 1$ probability clouds are roughly the same as that of the $n = 2$, $l = 0$ cloud, but they have different shapes (Fig. 30-8). The $m_l = +1$ cloud has the form of a doughnut centered on the z axis. Because $m_l = +1$, the cloud possesses angular momentum whose component in the z direction is up (that is, in the $+z$ direction). This is indicated in the figure by the arrow, which is in accord with the right-hand rule of Fig. 30-1. If we like, we can think of the cloud as a rotating distribution of charge, with the rotation furnishing both the observed angular momentum and the observed magnetic behavior. In probability terms, if we were to determine the electron's direction of motion within the atom in a series of experiments, more often than not we would find it to be moving counterclockwise around the z axis.

The $m_l = -1$ probability cloud is exactly the same as the $m_l = +1$ cloud except that the angular momentum is in the opposite direction. The rotational motion associated with the cloud is now in the clockwise sense around the z axis.

The $m_l = 0$ probability cloud has two concentrations around the z axis on either side of the nucleus. These concentrations are symmetric about the z axis—the pattern is the same on all sides. Because $m_l = 0$, there is no angular momentum about the z axis. A series of measurements of the electron's direction of motion would show it going as often clockwise as counterclockwise. But since $l = 1$, the atom nevertheless has angular momentum. Evidently the rotational motion that corresponds to this angular momentum occurs in such a way that **L** is always perpendicular to the z axis; then **L** has no component along this axis.

The probability clouds for quantum states whose principal quantum number n is greater than 2 are more complicated than those for $n = 1$ and $n = 2$. However, when $l = 0$, the clouds are always spherically symmetric, and when $l = 1$, the $m_l = 0, \pm 1$ clouds are similar in form to the corresponding ones for $n = 2$, $l = 1$ depicted above.

30-4 Electron Spin

We mentioned earlier that one of the many problems facing the atomic theorist is the observed splitting of many spectral lines into several components close together. For example, the first line of the Balmer series of hydrogen, which both theory and experiment place at a wavelength of 656.280 nm, actually consists of a pair of lines 0.014 nm apart.

An electron behaves much like a spinning charged sphere

In 1925 it was pointed out that the fine structure of spectral lines would occur if the electron behaves like a charged sphere spinning on its axis rather than like a single point charge; several years later the British physicist P. A. M. Dirac was able to show on the basis of a relativistic version of quantum theory that the electron *must* have the spin attributed to it. A spinning electron is in effect a tiny bar magnet, and it interacts with the magnetic field produced by its own motion in an atom as well as with any magnetic fields originating outside the atom.

Spin magnetic quantum number

Electron spin is described by the *spin magnetic quantum number*, m_s, whose values are $+\frac{1}{2}$ and $-\frac{1}{2}$. The magnitude of the angular momentum associated with its spin is the same for every electron; the quantum number m_s refers to the *direction* of the spin. Figure 30-9 shows how electrons with $m_s = +\frac{1}{2}$ and $m_s = -\frac{1}{2}$ align themselves in an external magnetic field. The component L_{sz} of spin angular momentum in the direction of **B** is $m_s h / 2\pi$.

Fig. 30-9. The spin magnetic quantum number m_s of an atomic electron has two possible values, $+\frac{1}{2}$ and $-\frac{1}{2}$.

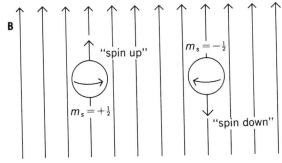

Table 30-1. Quantum numbers of an atomic electron.

Name	Symbol	Possible Values	Quantity Determined
Principal	n	1, 2, 3, . . .	Electron energy
Orbital	l	0, 1, 2, . . . , $n-1$	Magnitude of angular momentum
Magnetic	m_l	$-l, . . . , 0, . . . , +l$	Direction of angular momentum
Spin magnetic	m_s	$-\frac{1}{2}, +\frac{1}{2}$	Direction of electron spin

The notion of an electron as a spinning charged sphere is simply a model to make it easier to understand what is going on. No one knows what an electron really looks like, and in fact it is unlikely that any picture based on our perceptions of the macroscopic world can be relevant to elementary particles. But we *can* say that the possession of angular momentum and the magnetic behavior that goes with it are properties as fundamental to an electron as its rest mass and charge.

In its own frame of reference, an atomic electron "sees" the positively charged nucleus moving around *it,* and the electron accordingly experiences a magnetic field. Because the energy of an electron is different in each orientation of its spin relative to the magnetic field it experiences, the various energy levels of an atom are split into sublevels. The presence of these sublevels of energy is responsible for the observed fine structure of spectral lines. The introduction of electron spin into the theory of the atom means that a total of four quantum numbers, n, l, m_l, and m_s, is required to describe each possible state of an atomic electron. These are listed in Table 30-1.

There are $2n^2$ possible quantum states for each value of the principal quantum number n. For example, when $n=3$ there are 18 states, as shown in Table 30-2.

Origin of fine structure in spectra

Table 30-2. Quantum states of an electron of principal quantum number $n = 3$.

	$m_l=0$	$m_l=-1$	$m_l=+1$	$m_l=-2$	$m_l=+2$	
$l=0$:	⇕					
						↑ $m_s=+\frac{1}{2}$
$l=1$:	⇕	⇕	⇕			↓ $m_s=-\frac{1}{2}$
$l=2$:	⇕	⇕	⇕	⇕	⇕	

30-5 The Exclusion Principle

The probability clouds of atoms that contain more than one electron are similar in shape to the corresponding ones in the hydrogen atom. Normally the electron in a hydrogen atom is in its ground state of $n = 1$. What about the electrons in a more complex atom? Are they all in the same $n = 1$ state, jammed together in a single probability cloud, or are they distributed in some special way among the various other quantum states? The answer to this question is fundamental to chemistry because it shows how the properties of the various elements arise and explains the mechanisms by which elements combine to form compounds.

The arrangement of the electrons in a complex atom is governed by the *exclusion principle* discovered by Wolfgang Pauli. This principle states that

Exclusion principle   **No two electrons in an atom can exist in the same quantum state.**

That is, each electron in a complex atom must have a different set of the quantum numbers n, l, m_l, and m_s. The exclusion principle can be generalized to refer to the electrons in *any* small region of space, regardless of whether or not they constitute an atom. We shall make use of this generalization in the next chapter.

Pauli was led to the exclusion principle by a study of atomic spectra. It is possible to determine the quantum states of an atom empirically by analyzing its spectrum, since states with different quantum numbers differ in energy (even if only slightly) and the various wavelengths present correspond to transitions between these states. Pauli found several lines missing from the spectra of every element except hydrogen; the missing lines correspond to transitions to and from atomic states having certain sets of quantum numbers. Every one of these absent states has two or more electrons with identical sets of quantum numbers, a result that is expressed in the exclusion principle. Hydrogen, with a single electron, naturally has no absent states.

30-6 Families of Elements

Certain elements resemble one another so closely in their physical and chemical properties that it is natural to think of them as forming a "family." Three particularly striking examples of such families are the *halogens*, the *inert gases*, and the *alkali metals*. The members of these groups of elements are listed in Table 30-3 together with the number of electrons the respective atoms contain. This number is called the *atomic number* (symbol Z) of the element.

Atomic number

Table 30-3. Three families of elements. The atomic numbers of the elements are in parentheses.

Halogens	Inert gases	Alkali metals
	(2) Helium	(3) Lithium
(9) Fluorine	(10) Neon	(11) Sodium
(17) Chlorine	(18) Argon	(19) Potassium
(35) Bromine	(36) Krypton	(37) Rubidium
(53) Iodine	(54) Xenon	(55) Cesium
(85) Astatine	(86) Radon	(87) Francium

The halogens are nonmetallic elements with a high degree of chemical activity. At room temperature fluorine and chlorine are gases, bromine is a liquid, and iodine and astatine are solids. The halogens have valences of -1, and form diatomic molecules in the vapor state. The inert gases, as their name suggests, are inactive chemically: they form virtually no compounds with other elements, and their atoms do not join together into molecules. The inert gases have valences of 0. The alkali metals, like the halogens, are chemically very active, but they are active as reducing agents rather than as oxidizing agents. They are soft, not very dense, and have low melting points (only lithium is solid above 100°C). The alkali metals have valences of $+1$.

A curious feature of the three groups listed above is that, while the atomic numbers of the member elements of each group bear no obvious relation to one another, each inert gas is preceded in atomic number by a halogen (except in the case of helium) and followed by an alkali metal. Thus fluorine, neon, and sodium have the atomic numbers 9, 10, and 11, respectively, a sequence that persists through astatine (85), radon (86), and francium (87). If we list *all* the elements in order of their atomic number, *elements with similar properties recur at regular intervals.* This observation, first formulated in detail by Dmitri Mendeleev about 1869, is known as the *periodic law.*

Periodic law

A periodic table is a listing of the elements according to atomic number in a series of rows such that elements with similar properties form vertical columns. Table 30-4 is a simple form of periodic table; the number above the symbol of each element is its atomic mass, and the number below the symbol is its atomic number. The elements whose atomic masses appear in parentheses are radioactive and are not found in nature, but have been prepared in nuclear reactions. The atomic mass in each such case is the mass number of the longest-lived isotope of the element.

Table 30-4. Periodic classification of the elements

Group\Period	I	II											III	IV	V	VI	VII	VIII
1	1.008 H 1																	4.00 He 2
2	6.94 Li 3	9.01 Be 4											10.82 B 5	12.01 C 6	14.01 N 7	16.00 O 8	19.00 F 9	20.18 Ne 10
3	22.99 Na 11	24.31 Mg 12											26.98 Al 13	28.09 Si 14	30.98 P 15	32.06 S 16	35.46 Cl 17	39.95 Ar 18
4	39.10 K 19	40.08 Ca 20	44.96 Sc 21	47.90 Ti 22	50.94 V 23	52.00 Cr 24	54.94 Mn 25	55.85 Fe 26	58.93 Co 27	58.71 Ni 28	63.54 Cu 29	65.37 Zn 30	69.72 Ga 31	72.59 Ge 32	74.92 As 33	78.96 Se 34	79.91 Br 35	83.8 Kr 36
5	85.47 Rb 37	87.62 Sr 38	88.91 Y 39	91.22 Zr 40	92.91 Nb 41	95.94 Mo 42	(99) Tc 43	101.1 Ru 44	102.91 Rh 45	106.4 Pd 46	107.87 Ag 47	112.40 Cd 48	114.82 In 49	118.69 Sn 50	121.76 Sb 51	127.61 Te 52	126.90 I 53	131.30 Xe 54
6	132.91 Cs 55	137.34 Ba 56	* 57–71	178.49 Hf 72	180.95 Ta 73	183.85 W 74	186.2 Re 75	190.2 Os 76	192.2 Ir 77	195.09 Pt 78	196.97 Au 79	200.59 Hg 80	204.37 Tl 81	207.19 Pb 82	208.98 Bi 83	(209) Po 84	(210) At 85	(222) Rn 86
7	(223) Fr 87	226.05 Ra 88	† 89–104														Halogens	Inert gases
	Alkali metals																	

* Rare earths	138.91 La 57	140.12 Ce 58	140.91 Pr 59	144.27 Nd 60	(147) Pm 61	150.35 Sm 62	151.96 Eu 63	157.25 Gd 64	158.92 Tb 65	162.50 Dy 66	164.93 Ho 67	167.26 Er 68	168.93 Tm 69	173.04 Yb 70	174.97 Lu 71
† Actinides	(227) Ac 89	232.04 Th 90	(231) Pa 91	238.03 U 92	(244) Np 93	(244) Pu 94	(243) Am 95	(247) Cm 96	(247) Bk 97	(251) Cf 98	(254) Es 99	(257) Fm 100	(256) Mv 101	(254) No 102	(257) Lw 103

Elements created in the laboratory

Groups The columns in the periodic table are called *groups*. We recognize group I as the alkali metals plus hydrogen, group VII as the halogens, and group VIII as the inert gases. In addition to the elements forming the eight principal groups, there are a number of *transition elements* falling between groups II and III. The transition elements are metals which share certain general properties: most are hard and brittle, have high melting points, and exhibit several different valences. The rows in the periodic table are called *periods*. **Periods** Each period starts with an active alkali metal and proceeds through less active metals to weakly active nonmetals to an active halogen and an inactive inert gas. The transition elements in each period may be very much alike; the rare earths and actinides are so much alike that they are usually considered as separate categories.

For nearly a century the periodic law has been a mainstay of the chemist by permitting him to predict the behavior of undiscovered elements and by providing a framework for organizing his knowledge. It is one of the triumphs of the quantum theory of the atom that it enables us to account for the periodic law in complete detail without invoking any new assumptions or postulates.

30-7 Atomic Structures

Two basic principles determine the structures of atoms with more than one electron:

1. An atom is stable when its total energy is a minimum. This means that in the normal configuration of an atom its various electrons are present in the lowest energy states available to them;
2. Only one electron can exist in each quantum state of an atom. This is the exclusion principle.

The notion of shells and subshells of electrons in an atom is a useful one. The electrons in an atom that share the same principal quantum number n are said to occupy the same *shell*. These electrons average about the same distance from the nucleus and have comparable, though not identical, energies as a rule. Electrons in a given shell that also have the same orbital quantum number l are said to occupy the same *subshell*. In complex atoms the various subshells of the same shell vary in energy because electrons with different angular momenta have different probability density distributions and hence different average distances from the nucleus. The higher the value of l, the higher the energy, in the same shell.

Electron shells and subshells

We can account for the periodic table of the elements with the help of the above considerations. Our procedure will be to investigate the status of a new electron added to an existing electronic structure. (Of course, the nuclear charge must also increase by $+e$ each time this is done.) In the simplest case we add an electron to a hydrogen atom ($Z = 1$) to give a helium atom ($Z = 2$). Both electrons in helium fall into the same $n = 1$ shell. Since $l = 0$ is the only value l can have when $n = 1$, both electrons have $l = m_l = 0$. The exclusion principle is not violated here since one electron can have the spin magnetic quantum number $m_s = +\frac{1}{2}$ while the other has $m_s = -\frac{1}{2}$. It is customary to describe this situation by saying that the electrons have *opposite spins*, that is, that they behave as though they rotate in opposite directions.

Closed shells and subshells

Because no more than two electrons can occupy the $n = 1$ shell, helium atoms have *closed shells*. The characteristic properties of closed shells and subshells are that the orbital and spin angular momenta of their constituent electrons cancel out independently and that their effective electric charge distributions are perfectly symmetrical. As a result atoms with closed shells do not tend to interact with other atoms, which we know to be true of helium.

Lithium

Lithium, with $Z = 3$, has one more electron than helium; it is the lightest of the alkali metals. There is no room left in the $n = 1$ shell, and so the additional electron goes into the $l = 0$ subshell of the $n = 2$ shell. The outer electron is relatively far from the nucleus in this shell, and is much less tightly bound. The chemical activity of lithium is a consequence of the low binding energy of this electron, which is readily lost to other atoms.

Beryllium

The next element, beryllium, has two electrons of opposite spin in the $l = 0$ subshell of the $n = 2$ shell. The nuclear charge is $+4e$, and so these outer electrons are more tightly held than the single outer electron in lithium; beryllium is accordingly less reactive than lithium.

Boron

Boron, with $Z = 5$, has an electron in the $l = 1$ subshell as well as two in the $l = 0$ one. The $l = 1$ subshell can contain a total of six electrons, corresponding to two electrons of opposite spin in the $m_l = +1$, $m_l = 0$, and $m_l = -1$ states. This subshell is closed (and the $n = 2$ shell is also closed) in neon, whose atomic number is 10. We therefore expect neon to be chemically inert, as indeed it is.

Fluorine

Fluorine, the element just before neon in the periodic table and the lightest of the halogens, lacks but one electron of having a closed outer shell. Just as lithium tends to lose its single outermost electron in interacting with other elements, thereby leaving it with a closed shell configuration, fluorine tends to gain a single electron to close its outer shell. The very different behaviors of the alkali metals, the inert gases, and the halogens thus find explanation in terms of their respective atomic structures. Further analysis of the preceding kind is able to account for all the other relationships revealed in the periodic table.

Figure 30-10 is a highly schematic representation of the electron structures of the elements in the first three periods of the periodic table. A more realistic picture of each atom would show the probability clouds associated with each electron having a particular set of quantum numbers n, l, and m_l (two electrons of opposite spin can have the same probability cloud in an atom.)

Table 30-5 shows the electron configurations of the 36 lightest elements. We note that the sequence of electron addition becomes irregular with potassium, which has an electron in its $n = 4$ shell even though its $n = 3$ shell is

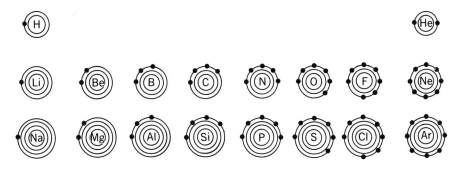

Fig. 30-10. Schematic representation of the electron structures of the elements in the first three periods of the periodic table. The electrons in the filled inner shells of the second and third period elements are not shown.

incomplete. The origin of this apparent anomaly is that electrons in the $n = 4, l = 0$ subshell have less energy, and therefore are more tightly bound, than those in the $n = 3, l = 2$ subshell, since the probability clouds of the former electrons have concentrations that are closer to the nucleus than those of the latter electrons. The energy difference between the $n = 3, l = 2$ and $n = 4, l = 0$ subshells is actually quite small, as we can see from the configurations of chromium ($Z = 24$) and copper ($Z = 29$). In both of these elements an additional electron is present in the $n = 3, l = 2$ subshell at the expense of the $n = 4, l = 0$ subshell, leaving a vacancy in the latter that is filled in the succeeding element.

The ferromagnetic properties of iron, cobalt, and nickel are a consequence of their unfilled $n = 3, l = 2$ subshells; without violating the exclusion principle, the electrons in these subshells do *not* pair off, and their spins do not cancel out. In the $n = 3, l = 2$ subshell of iron, for instance, five of the six electrons have parallel spins, leaving each iron atom with a strongly magnetic character, due to its four unpaired electrons. The reason atomic electrons in the same subshell have parallel spins whenever possible can be traced to their mutual electrostatic repulsion. A consequence of this repulsion is that the more widely separated electrons are, the lower the energy of the system to which they belong (everything else being equal). Electrons in the same subshell which have parallel spins must have different values of m_l and therefore different probability clouds. Hence such electrons are farther apart on the average than paired electrons with the same m_l values, and this arrangement is the more stable one since it represents the lowest possible energy.

As we can see, the quantum theory of the atom is the most powerful and fruitful approach yet devised for understanding the properties of matter.

Origin of ferromagnetism

Table 30-5. Electron configurations of the thirty-six lightest elements

Atomic number	Symbol	Element	Electron configuration of atom			
			$n = 1$	$n = 2$	$n = 3$	$n = 4$
			$l = 0$	$l = 0, l = 1$	$l = 0, l = 1, l = 2$	$l = 0, l = 1, \ldots$
1	H	Hydrogen	1			
2	He	Helium	2	*Inert gas*		
3	Li	Lithium	2	1 *Alkali metal*		
4	Be	Beryllium	2	2		
5	B	Boron	2	2 1		
6	C	Carbon	2	2 2		
7	N	Nitrogen	2	2 3		
8	O	Oxygen	2	2 4		
9	F	Fluorine	2	2 5	*Halogen*	
10	Ne	Neon	2	2 6	*Inert gas*	
11	Na	Sodium	2	2 6	1 *Alkali metal*	
12	Mg	Magnesium	2	2 6	2	
13	Al	Aluminum	2	2 6	2 1	
14	Si	Silicon	2	2 6	2 2	
15	P	Phosphorus	2	2 6	2 3	
16	S	Sulfur	2	2 6	2 4	
17	Cl	Chlorine	2	2 6	2 5 *Halogen*	
18	Ar	Argon	2	2 6	2 6 *Inert gas*	

Important Terms

In the **quantum theory of the atom** only experimentally measurable quantities are considered, and no use is made of models inconsistent with the uncertainty principle. According to this theory, each electron in an atom is described by its **principal quantum number** n, which governs its energy, its **orbital quantum number** l, which governs the magnitude of its angular momentum, and its **magnetic quantum number** m_l, which governs the orientation of its angular momentum. To each set of quantum numbers there is a corresponding **probability cloud** that governs the likelihood the electron thus described will be found in any particular location in an atom.

Every electron has a certain intrinsic amount of angular momentum called its **spin**. The **spin magnetic**

Table 30-5. Electron configurations of the thirty-six lightest elements (continued)

Atomic number	Symbol	Element	Electron configuration of atom								
			$n = 1$	$n = 2$		$n = 3$			$n = 4$		
			$l = 0$	$l = 0,$	$l = 1$	$l = 0,$	$l = 1,$	$l = 2$	$l = 0,$	$l = 1, \ldots$	
19	K	Potassium	2	2	6	2	6		1		*Alkali metal*
20	Ca	Calcium	2	2	6	2	6		2		
21	Sc	Scandium	2	2	6	2	6	1	2		
22	Ti	Titanium	2	2	6	2	6	2	2		
23	V	Vanadium	2	2	6	2	6	3	2		
24	Cr	Chromium	2	2	6	2	6	5	1		
25	Mn	Manganese	2	2	6	2	6	5	2		
26	Fe	Iron	2	2	6	2	6	6	2		*Ferro-*
27	Co	Cobalt	2	2	6	2	6	7	2		*magnetic*
28	Ni	Nickel	2	2	6	2	6	8	2		*metals*
29	Cu	Copper	2	2	6	2	6	10	1		
30	Zn	Zinc	2	2	6	2	6	10	2		
31	Ga	Gallium	2	2	6	2	6	10	2	1	
32	Ge	Germanium	2	2	6	2	6	10	2	2	
33	As	Arsenic	2	2	6	2	6	10	2	3	
34	Se	Selenium	2	2	6	2	6	10	2	4	
35	Br	Bromine	2	2	6	2	6	10	2	5	*Halogen*
36	Kr	Krypton	2	2	6	2	6	10	2	6	*Inert gas*

quantum number m_s of an atomic electron has two possible values, $+\frac{1}{2}$ and $-\frac{1}{2}$. Owing to its spin, every electron acts like a tiny bar magnet.

According to the Pauli **exclusion principle**, no two electrons in an atom can exist in the same quantum state.

The **periodic law** of chemistry states that if we list the elements in order of atomic number, elements with similar properties recur at regular intervals. The quantum theory of the atom together with the exclu-sion principle is able to explain the origin of the periodic law.

The electrons in an atom that have the same total quantum number n are said to occupy the same **shell**. Electrons in a given shell which have the same orbital quantum number l are said to occupy the same **sub-shell**. Shells and subshells containing the maximum number of electrons permitted by the exclusion principle are **closed**. Atoms whose subshells are all closed possess unusual stability.

Multiple Choice

1. The quantum theory of the atom
   a. is based upon the Bohr theory.
   b. is more comprehensive but less accurate than the Bohr theory.
   c. cannot be reconciled with Newton's laws of motion.
   d. is not based upon a mechanical model and considers only observable quantities.

2. The energy of a quantum state of an atom depends chiefly upon its
   a. principal quantum number n.
   b. orbital quantum number l.
   c. magnetic quantum number m_l.
   d. spin magnetic quantum number m_s.

3. The angular momentum of an atomic electron
   a. is not quantized.
   b. is quantized in magnitude only.
   c. is quantized in direction only.
   d. is quantized in both magnitude and direction.

4. The splitting of the spectral lines of an element when it radiates in a magnetic field is known as
   a. the Schrödinger effect.
   b. the Zeeman effect.
   c. induced emission.
   d. population inversion.

5. A large value of the probability density ψ^2 of an atomic electron at a certain place and time signifies that the electron
   a. is likely to be found there.
   b. is certain to be found there.
   c. has a great deal of energy there.
   d. has a great deal of charge there.

6. The quantum number which is not involved in describing the probability cloud of an atomic electron is the
   a. principal quantum number.
   b. orbital quantum number.
   c. magnetic quantum number.
   d. spin magnetic quantum number.

7. The probability cloud of an atomic electron is spherically symmetric when
   a. $n = 0$ b. $l = 0$
   c. $m_l = 0$ d. $m_s = 0$

8. The natural unit of angular momentum in atomic physics is the quantity
   a. $2h$ b. $2\pi h$
   c. $h/2$ d. $h/2\pi$

9. The number of possible orientations of an atomic electron in a magnetic field is
   a. 1   b. 2
   c. 4   d. 2π

10. The number of possible m_l values of an atomic electron for which $l = 3$ is
    a. 3   b. 6
    c. 7   d. 14

11. Pauli's exclusion principle states that no two electrons in an atom can
    a. be present in the same probability cloud.
    b. have the same spin.
    c. have the same quantum numbers.
    d. interact with each other.

12. The two electrons in a helium atom
    a. occupy different shells.
    b. occupy different subshells.
    c. have opposite spins.
    d. have parallel spins.

13. Each vertical column of the periodic table contains elements whose atoms have
    a. similar chemical properties.
    b. different chemical properties.
    c. the same number of electrons.
    d. the same atomic mass.

14. Each horizontal row of the periodic table contains elements whose atoms have
    a. similar chemical properties.
    b. different chemical properties.
    c. the same number of electrons.
    d. the same atomic mass.

15. In the periodic table, elements are listed in order of
    a. atomic number.   b. atomic mass.
    c. density.
    d. ability to form covalent bonds.

16. At room temperature and atmospheric pressure, most elements are
    a. metals.
    b. nonmetals.
    c. gases.
    d. liquids.

17. An alkali metal atom
    a. has a closed outer shell.
    b. is one electron short of having a closed outer shell.
    c. has one electron in its outer shell.
    d. has two electrons in its outer shell.

18. A halogen atom
    a. has a closed outer shell.
    b. is one electron short of having a closed outer shell.
    c. has one electron in its outer shell.
    d. has two electrons in its outer shell.

19. Atoms of the inert gases
    a. form diatomic molecules.
    b. combine chemically only with the alkali metals and halogens.
    c. exhibit chemical behavior like that of hydrogen.
    d. are almost never found in molecules.

Exercises

1. The quantum theory of the atom is consistent with what fundamental principle that is violated by the Bohr theory?

2. What quantity is governed by each of the quantum numbers n, l, m_l, and m_s?

3. In the quantum theory of the atom, an atomic electron can have zero angular momentum. Is this possible in the Bohr theory?

4. What are the possible values for the orbital and magnetic quantum numbers of an atomic electron with the principal quantum number $n = 3$?

5. Why is the probability cloud for an atomic electron with no orbital angular momentum the same in all directions?

6. Verify that an atomic electron of principal quantum number n might have any one of n^2 possible probability clouds depending upon its l and m_l values.

7. According to the Bohr model, the radius of an atomic electron's orbit is proportional to n^2. Is there any aspect of the probability-cloud model that might correspond to this relationship?

8. What is the minimum orbital angular momentum of an atomic electron according to the Bohr theory? According to the quantum theory?

9. Under what circumstances do electrons exhibit spin? Do electrons in inner shells spin faster than those in outer shells?

10. What significance does electron spin have in atomic structure?

11. What is true in general of the chemical characteristics of elements in each vertical column of the periodic table?

12. What is true in general of the chemical characteristics of elements in each horizontal row of the periodic table?

13. How many electrons do the elements in group II of the periodic table have in their outermost shells?

14. How can you account for the fact that fluorine and chlorine exhibit similar chemical behavior?

15. How can you account for the fact that lithium and and sodium exhibit similar chemical behavior?

16. Under what circumstances can two electrons share the same probability cloud in an atom?

17. State the quantum numbers of the electrons in the sodium atom ($Z = 11$).

18. State the quantum numbers of the electrons in the argon atom ($Z = 18$).

19. How many elements would there be if atoms with occupied electron shells up through $n = 6$ could exist?

Answers to Multiple Choice

1. d	5. a	9. b	13. a	17. c
2. a	6. d	10. c	14. b	18. b
3. d	7. b	11. c	15. a	19. d
4. b	8. d	12. c	16. a	

31

Atoms
in Combination

Individual atoms are rare on the earth and in its lower atmosphere only the inert gas atoms occur by themselves. All other atoms are found linked in small groups called molecules or in larger ones as liquids or solids. Some of these groups consist exclusively of atoms of the same element, others consist of atoms of different elements, but in every case the arrangement is favored over separate atoms by virtue of interactions between the atoms that reduce the total energy of the system. The quantum theory of the atoms is able to account for these interactions in a natural way, with no special assumptions, which is further testimony to the power of this approach.

31-1 The Hydrogen Molecule

A molecule is a group of atoms that stick together strongly enough to act as a single particle. A molecule always has a certain definite structure; hydrogen molecules always consist of two hydrogen atoms each, for instance, and water molecules always consist of one oxygen atom and two hydrogen atoms each. A piece of iron is not a molecule because, even though its atoms stay together, any number of them do so to form an object of any size or shape.

A molecule of a given kind is complete in itself with little tendency to gain or lose atoms. If one of its atoms is somehow removed or another atom is somehow attached, the result is a molecule of a different kind with different properties. A liquid or a solid, on the other hand, can gain or lose additional atoms of the kinds already present without changing its character.

The chief mechanism that bonds atoms together to form molecules involves the sharing of electrons by the atoms involved. As the shared electrons circulate around the atoms, they spend more time between the atoms than they do on the outside, which produces an attractive force. To see why this force should occur, we may think of the atoms that are sharing the electrons as positive ions, so the presence of the shared electrons between them means a negative charge that holds the positive ions together. This is especially easy to see in the case of the hydrogen molecule, H_2, which consists of two hydrogen atoms.

Electron sharing

In the hydrogen molecule, the two protons are 7.42×10^{-11} m apart and the two electrons, one contributed by each atom, belong to the entire molecule rather than to their parent nuclei. Figure 31-1(a) is a picture of the H_2 molecule in terms of electron orbits. A more realistic picture of this molecule is provided by the quantum theory, in which the notion of electrons with definite, predictable positions and velocities at every moment is replaced by the notion of probability clouds. Figure 31-1(b) shows how the electron probability clouds of two hydrogen atoms join to form the probability cloud of a hydrogen molecule.

Because the electrons spend more time on the average between the protons than they do on the outside, there is effectively a net negative charge between the protons. The attractive force this charge exerts on the protons is more than enough to counterbalance the direct repulsion between them. If the protons are too close together, however, their repulsion becomes dominant and the molecule is not stable. The balance between attractive and repulsive forces occurs at a separation of 7.42×10^{-11} m, where

Origin of the bonding force

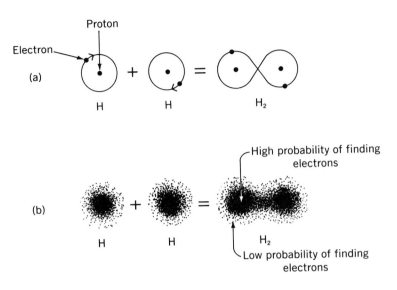

Fig. 31-1. (a) Orbit model of the hydrogen molecule. (b) Probability cloud model of the hydrogen molecule.

the total energy of the H_2 molecule is -4.5 eV. Hence 4.5 eV of work must be done to break a H_2 molecule into two H atoms:

$$H_2 + 4.5 \text{ eV} \rightarrow H + H.$$

By comparison, the binding energy of the hydrogen atom is 13.6 eV:

$$H + 13.6 \text{ eV} \rightarrow p^+ + e^-.$$

This is an example of the general rule that it is easier to break up a molecule than to break up an atom.

An interesting argument based on the uncertainty principle makes it possible, even without a detailed calculation starting from Schrödinger's equation, to understand why a hydrogen molecule should have less energy (and hence more stability) than two separate hydrogen atoms. According to the uncertainty principle, the smaller the uncertainty in a particle's position, the greater the uncertainty in its momentum. Therefore the smaller the region of space in which an electron is confined, the greater its momentum and hence energy must be. An electron shared by two protons has more room in which to move about than an electron bound to a single proton to form a hydrogen atom, so its energy is less.

According to the Pauli exclusion principle, the two electrons in the H_2 molecule must have different spin magnetic quantum numbers, with one electron having $m_s = +\frac{1}{2}$ and the other $m_s = -\frac{1}{2}$. That is, the two electrons

The electrons in H_2 have opposite spins

must have opposite spins. If the spins were the same, the electrons would have to spend most of their time near the far ends of the H_2 molecule where they interact least. The result would be both a direct repulsion between the nuclei and a repulsion induced by the outward attractions of the electrons, making a stable molecule impossible. Figure 31-2 shows the total energy of two hydrogen atoms plotted versus their separation. Bonding is evidently possible only when the electron spins are antiparallel. The minimum of the lower curve corresponds to the H_2 molecule, which is a stable configuration since work must be done to move the H atoms closer together or farther apart than the distance $a = 7.42 \times 10^{-11}$ m.

Why do only two hydrogen atoms join together to form a molecule? Why not three, or four, or a hundred?

The basic reason for the limited size of molecules is the exclusion principle, which prohibits more than one electron in an atomic system from having the same set of quantum numbers. If a third H atom were to be brought up to an H_2 molecule, its electron would have to leave the $n = 1$ shell and go to the $n = 2$ shell in order for an H_3 molecule to be formed, since only two electrons can occupy the $n = 1$ shell. But an $n = 2$ electron in the hydrogen atom has over 10 eV more energy than an $n = 1$ electron, and the binding energy in H_2 is only 4.5 eV. Hence an H_3 molecule, if it could somehow be put together, would immediately break apart into $H_2 + H$ with the release of energy. Similar considerations hold for other molecules.

The ability of atoms to join with only a limited number of others to form molecules is referred to as *saturation*.

The exclusion principle is also responsible for the fact that the inert gases—helium, neon, argon, and so on—do not occur as molecules. The electron shells of the inert gas atoms are all filled to capacity, so in order for two of them to share an electron pair, one of the electrons would have

The exclusion principle and molecular size

The inert gases

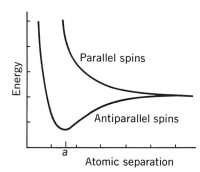

Fig. 31-2. The total energy of two hydrogen atoms plotted as a function of their separation. Binding is only possible when their electron spins are antiparallel; the minimum of the lower curve corresponds to the H_2 molecule, which is a stable configuration since work must be done to move the H atoms closer together or farther apart than the distance a.

to go into an empty shell of higher energy. The increase in energy involved would be more than the energy decrease produced by sharing the electrons, and therefore no such molecules as He_2 or Ne_2 occur.

31-2 Covalent Bond

A shared pair of electrons constitutes a covalent bond

The mechanism by which electron sharing holds atoms together to form molecules is known as *covalent bonding*. Often it is convenient to think of the atoms as being held together by *covalent bonds*, with each shared pair of electrons constituting a bond.

More complex atoms than hydrogen also join together to form molecules by sharing electrons. Depending upon the electron structures of the atoms involved, there may be one, two, or three covalent bonds—shared electron pairs—between the atoms. For example, in the oxygen molecule O_2 there are two bonds between the O atoms, and in the nitrogen molecule N_2 there are three bonds between the N atoms. Covalent bonds are represented either by a pair of dots or a single dash for each shared pair of electrons. Thus the H_2, O_2, and N_2 molecules can be represented as follows:

$$H:H \quad \text{or} \quad H—H,$$
$$O::O \quad \text{or} \quad O=O,$$
$$N:::N \quad \text{or} \quad N\equiv N.$$

Covalent bonds are not limited to atoms of the same element nor to only two atoms per molecule. Here are two examples of more complicated molecules, with a line representing each covalent bond:

$$\text{Water, } H_2O \qquad \begin{matrix} H \\ | \\ O—H \end{matrix} \qquad \text{Ammonia, } NH_3 \qquad \begin{matrix} H \\ | \\ N—H \\ | \\ H \end{matrix}$$

We notice that oxygen participates in two bonds in H_2O and nitrogen in three bonds in NH_3, while H participates in only one bond in both cases. This behavior is consistent with the fact that the hydrogen, oxygen, and nitrogen molecules respectively have one, two, and three bonds between their atoms.

A carbon atom forms four covalent bonds

Carbon atoms tend to form four covalent bonds at the same time, since they have four electrons in their outer shells and these shells lack four electrons for completion. Various distributions of these bonds are possible, including bonds between adjacent carbon atoms in a complex molecule. The structures of the common covalent molecules methane, carbon dioxide, and acetylene illustrate the different bonds in which carbon atoms can participate to form molecules.

$$
\begin{array}{ccc}
\quad\;\; H & & \\
\quad\;\; | & & \\
H—C—H & O{=}C{=}O & H—C{\equiv}C—H \\
\quad\;\; | & & \\
\quad\;\; H & \text{carbon dioxide} & \\
\text{methane} & & \text{acetylene}
\end{array}
$$

Carbon atoms are so versatile in forming covalent bonds with each other as well as with other atoms that literally millions of carbon compounds are known, some whose molecules contain tens of thousands of atoms. Such compounds were once thought to originate only in living things, and their study is accordingly known even today as organic chemistry.

Organic chemistry is the study of carbon compounds

When atoms join together to form a molecule, their inner, complete electron shells undergo little change. In molecules whose atoms are the same, such as H_2, O_2, and N_2, the atoms share their outer electrons evenly, and on the average have uniform distributions of electric charge.

In a molecule composed of different atoms, the shared electrons favor one or another of the atoms, depending upon which elements are involved. The resulting molecule has a nonuniform distribution of electric charge, with some parts of the molecule being positively charged and other parts being negatively charged. A molecule of this kind, as we learned earlier, is called a *polar molecule*, whereas a molecule whose charge distribution is symmetric is called a *nonpolar molecule*. Thus the covalent molecule HCl is polar because a chlorine atom has greater attraction for an electron than a hydrogen atom, a situation we can represent by placing the pair of dots that indicate a covalent bond closer to the Cl atom:

Polar and nonpolar molecules

Hydrochloric acid molecule   H   :Cl.

The water molecule is highly polar because the two O—H bonds are 104.5° apart (not 180° apart) and the two shared electron pairs spend more time near the O atom than near the H atoms:

Water molecules are highly polar

Water molecule   H   :O
                              ..
                      H

31-3 Solids and Liquids

The same covalent bonds that can tie several atoms together into a molecule can in certain circumstances also tie an unlimited number of them together into a solid or a liquid. Other bonding mechanisms are found in solids and liquids as well, one of which is responsible for the ability of metals to conduct electric current. While only a minute proportion of the universe is in the solid state and even less is in the liquid state, matter in these forms constitutes much of the physical world of our experience, and modern technology is to a large extent based upon the exploitation of the unique characteristics of various solid materials.

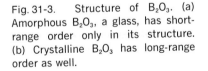

Most solids are crystalline in nature, with their constituent atoms or molecules arranged in regular, repeated patterns. A crystal is thus characterized by the presence of *long-range order* in its structure.

Other solids lack the definite arrangements of atoms and molecules so conspicuous in crystals. They can be thought of as liquids whose stiffness is due to an exaggerated viscosity. Examples of such *amorphous* ("without form") solids are pitch, glass, and many plastics. The structures of amorphous solids exhibit *short-range order* only.

Some substances, for instance B_2O_3, can exist in either crystalline or amorphous forms. In both cases each boron atom is surrounded by three larger oxygen atoms, which represents a short-range order. In a B_2O_3 crystal a long-range order is also present, as shown in a two-dimensional representation in Fig. 31-3, whereas amorphous B_2O_3, a glassy material, lacks this additional regularity.

Fig. 31-3. Structure of B_2O_3. (a) Amorphous B_2O_3, a glass, has short-range order only in its structure. (b) Crystalline B_2O_3 has long-range order as well.

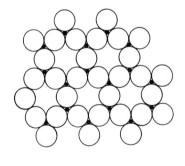

○ Oxygen
• Boron

(b)

(a)

The lack of long-range order in amorphous solids means that the various bonds vary in strength. When an amorphous solid is heated, the weakest bonds are ruptured at lower temperatures than the others and the solid softens gradually, while in a crystalline solid the bonds break simultaneously, and melting is a sudden process.

Liquids have more in common with solids than with gases, even though liquids share with gases the ability to flow from place to place. Because the density of a given liquid is usually approximately the same as that of the same substance in solid form, we infer that the bonding mechanism is similar in both cases. When a solid is heated to its melting point, its atoms or molecules acquire enough energy to shift the bonds holding them together so that they form different groupings, but not until the liquid is heated to its vaporization point are the atoms or molecules able to break loose completely and form a gas. This interpretation of the liquid state is confirmed by x-ray studies that reveal definite groupings of atoms or molecules in a liquid (that is, short-range order like that in amorphous solids), but the groupings are constantly shifting their arrangements unlike the permanent arrangements in a solid.

Liquids

Few crystals are perfect in structure, with completely regular arrangements of atoms. The defects in the structure of a crystal—atoms in the "wrong" place, missing atoms, extra atoms, impurity atoms, and so on—are extremely important in determining its mechanical strength.

A common type of crystal defect is a missing line of atoms. Such a defect is called a *dislocation*. The existence of dislocations makes it possible to understand why a solid can be deformed permanently by bending, squeezing, or stretching without breaking. We might think that such plastic deformations occur when a force is applied to a solid by the sliding of layers of atoms over one another, but calculations show that such sliding—which involves the breaking of millions of bonds between atoms simultaneously—would require forces about a thousand times stronger than those actually able to produce the deformations experimentally.

Dislocations

Figure 31-4(a) shows a crystal with a missing row of atoms. When equal and opposite forces are applied to the crystal along different lines of action, as in Fig. 31-4(b), the dislocation shifts to the right as the atoms in the layer containing the missing row shift their bonds, one row at a time, with the atoms in the layer above. In Fig. 31-4(c) the dislocation has reached the end of the crystal, which is now permanently deformed. Because the bonds were shifted one row at a time, instead of all at once, much less force was needed than would have been required to deform a perfect solid.

To increase the hardness of a solid, it is necessary to impede the motion of dislocations in its structure. There are two chief ways to do this in a metal. One is to increase the number of dislocations by hammering the metal or

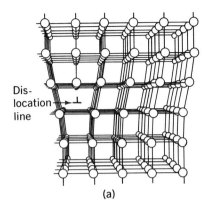

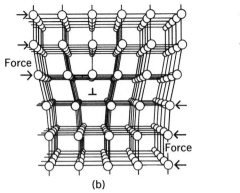

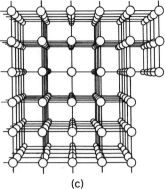

(a)   (b)   (c)

Fig. 31-4. The motion of a dislocation in crystal under stress results in a permanent deformation.

squeezing it between rollers. The dislocations then become so numerous and tangled together that they interfere with each other's motion. This effect is called *work hardening*. Another approach is to add foreign atoms to the metal that act as roadblocks to the progress of dislocations. Thus the addition of small amounts of carbon, chromium, manganese, and other elements to iron turns it into the much stronger steel by impeding the ability of dislocations to move through the material.

Work hardening

31-4 Covalent and Ionic Solids

Molecular bonds are all essentially covalent in character, even though in a few highly polar molecules the shared electrons stay so close to one of the partner atoms that there is effectively a transfer rather than a sharing of them. In solids there are four bonding mechanisms that occur, depending upon the nature of the solid: covalent, ionic, van der Waals, and metallic.

The covalent bonds between atoms that are responsible for the formation of molecules also act to hold certain crystalline solids together. Figure 31-5 shows the array of carbon atoms in a diamond crystal, with each carbon atom sharing electron pairs with the four other carbon atoms adjacent to it. All the electrons in the outer shells of the carbon atoms participate in the bonding, and it is therefore not surprising that diamond is extremely hard and must be heated to over 3500°C before its crystal structure is disrupted and it melts. Other covalent crystals are those of silicon, silicon carbide ("Carborundum"), and germanium.

In covalent bonding, two atoms share one or more pairs of electrons. In *ionic bonding*, one or more electrons from one atom transfer to another atom, producing a positive ion and a negative ion that attract each other.

Ionic bonding

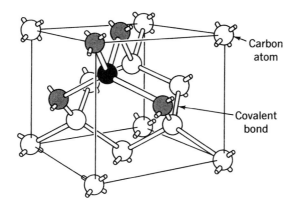

Fig. 31-5.   Covalent structure of diamond. Each carbon atom shares electron pairs with four other carbon
atoms.

Let us see how ionic bonding functions in the case of NaCl, whose crystals
constitute ordinary table salt. We shall consider a hypothetical NaCl "molecule" for convenience; salt crystals are aggregates of Na and Cl atoms
which, although they do not pair off into individual molecules, do interact
through ionic bonds whose nature can be most easily examined in terms of
a single Na-Cl unit.

The work that must be done to remove an electron from an atom is
called the ionization energy of that atom. An atom with just one electron in   Ionization energy
its outer shell has a relatively low ionization energy which reflects the
shielding effect of the inner electrons that cancel out most of the charge of
the nucleus (Fig. 31-6). Such an atom tends to lose the outer electron and
thereby assume an electron configuration composed solely of closed shells.
At the other extreme, an atom lacking just one electron for the completion
of its outer shell has a relatively high ionization energy which reflects the
greater attractive force exerted by the nucleus when the nuclear charge is
less effectively shielded by the intervening electrons. Thus sodium, with
filled $n = 1$ and $n = 2$ shells and 1 electron in its $n = 3$ shell, has an ionization
energy of 5.1 eV, while chlorine, also with filled $n = 1$ and $n = 2$ shells but
with 7 electrons in its $n = 3$ shell, has an ionization energy of 13.0 eV.

Another atomic property relevant to ionic binding is *electron affinity*.   Electron affinity
The electron affinity of an atom is the amount of energy released when it
acquires an additional electron. Since this amount of work must be done to
remove the additional electron after it is in place, electron affinity is a
measure of the tendency of an atom to attract electrons in excess of its
normal complement. As we would expect, atoms with high ionization

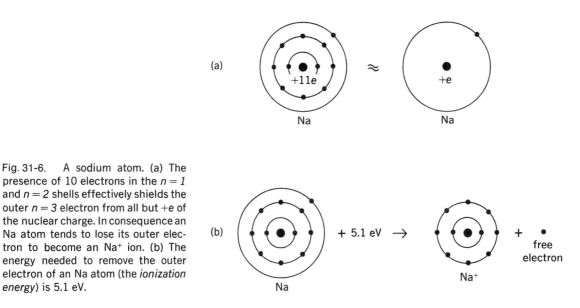

Fig. 31-6.   A sodium atom. (a) The presence of 10 electrons in the $n = 1$ and $n = 2$ shells effectively shields the outer $n = 3$ electron from all but $+e$ of the nuclear charge. In consequence an Na atom tends to lose its outer electron to become an Na⁺ ion. (b) The energy needed to remove the outer electron of an Na atom (the *ionization energy*) is 5.1 eV.

energies have high electron affinities, while those with low ionization energies have low electron affinities. Atoms of the former kind, such as chlorine, tend to form negative ions by picking up electrons (Fig. 31-7), and atoms of the latter kind, such as sodium, tend to form positive ions by losing

Fig. 31-7. A chlorine atom. (a) The inner electrons leave unshielded $+7e$ of the nuclear charge. In consequence a Cl atom tends to pick up another electron to become a Cl⁻ ion. (b) The energy released when an electron becomes attached to a Cl atom (the *electron affinity*) is 3.6 eV.

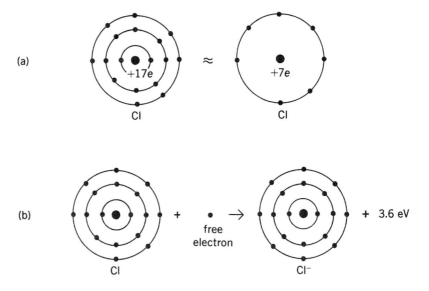

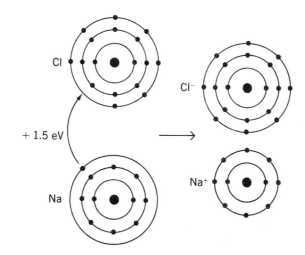

Fig. 31-8.   Electron transfer in NaCl. The resulting ions attract each other electrically.

electrons. An ionic bond between an atom of each kind occurs when one or more electrons pass from the atom with the low ionization energy to the one with the high electron affinity, since the resulting ions, both with complete outer shells, then attract each other electrically (Fig. 31-8). The bond is stable if more energy is needed to separate the ions than can be supplied by the return of the transferred electron or electrons.

As was said, the ionization energy of sodium is 5.1 eV, so that 5.1 eV of work is required to remove its single outer electron. The electron affinity of chlorine is 3.6 eV, so that 3.6 eV of energy is liberated when an electron is added to fill the vacancy in its outer shell. Transferring an electron from a sodium atom to a chlorine atom to form Na^+ and Cl^- ions therefore involves a net energy expenditure of 5.1 eV − 3.6 ev or 1.5 eV:

$$Na + Cl + 1.5\ ev \rightarrow Na^+ + Cl^-.$$

An Na^+ and a Cl^- ion attract each other. The closer they get, the stronger the force between them, and the more negative their mutual electric potential energy becomes. When the ions are 9.6×10^{-10} m apart, their electric potential energy is −1.5 eV, so the total energy of the system is 0 (Fig. 31-9). The work done *on* the system Na + Cl to transfer an electron from Na to Cl is equal to the work done *by* the system $Na^+ + Cl^-$ as the ions come to the distance 9.6×10^{-10} m from each other. When the ions are still closer together, their total energy is negative, and external energy is needed to convert the system $Na^+ + Cl^-$ back to Na + Cl.

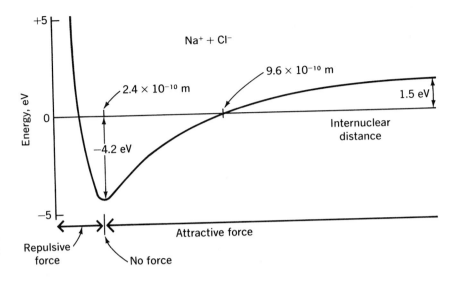

Fig. 31-9. The variation with internuclear distance of the energy of the system of a Na⁺ and a Cl⁻ ion. The system is stable at an internuclear distance corresponding to the lowest point of the curve.

What keeps ions apart

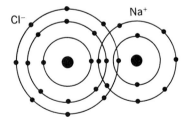

Fig. 31-10. Strong repulsive forces whose origin is associated with the exclusion principle prevent atoms from coming so close together that their electron structures mesh.

If the Na⁺ and Cl⁻ ions were to come very close to each other, their electron structures would mesh together (Fig. 31-10). Two different effects prevent this from occurring. First, when such meshing takes place, the Na⁺ and Cl⁻ nuclei cease being shielded by their electrons and repel electrically. Second but more important, such meshing means that the inner electrons of the Na⁺ and Cl⁻ ions no longer constitute separate atomic systems but instead constitute a single one. According to the Pauli exclusion principle, no two electrons in the same system can be in the same quantum state, and so some of the electrons must go into unoccupied states of higher energy. The result of both effects is that work must be done to push the ions together to make their electron structures overlap. The force between the ions is now repulsive.

As the Na⁺ and Cl⁻ ions approach each other, then, at first the only force between them is an attractive electric one and the energy of the system decreases. When they are in the neighborhood of 4×10^{-10} m apart, a repulsive force whose origin is the exclusion principle comes into being, and the energy of the system begins to level out. At a separation of 2.4×10^{-10} m the attractive and repulsive forces are in balance, and the system is stable. If the ions are closer together, the energy of the system increases. To break apart a stable Na⁺ + Cl⁻ system, 4.2 eV of work must be done.

The structure of an NaCl crystal is shown in Fig. 31-11. Each ion behaves essentially like a point charge and thus tends to attract to itself as

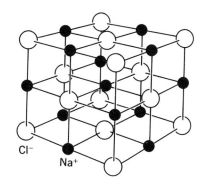

Cl⁻

Na⁺

Fig. 31-11. Ionic structure
of NaCl crystal.

many ions of opposite sign as can fit around it. The latter ions, of course,
repel one another, and so the resulting crystal is an equilibrium configuration
in which the various attractions and repulsions balance out. In a NaCl
crystal each Na^+ ion is surrounded by six Cl^- ions and vice versa. In crystals
having different structures the number of "nearest neighbors" around each
ion may be 3, 4, 6, 8, or 12. Ionic bonds are usually fairly strong, and consequently ionic crystals are strong, hard, and have high melting points.

Many crystalline bonds are partly ionic and partly covalent in origin.
An example of such mixed bonding is quartz, SiO_2.

31-5 van der Waals' Bonds

A number of molecules and nonmetallic atoms exist whose electronic
structures do not lend themselves to either of the above kinds of binding.
The inert gas atoms, which have filled outer electron shells, and molecules
such as methane,

whose valence electrons are fully involved in the molecular bond itself, fall
into this category. However, even these virtually noninteracting substances
condense into solids and liquids at low enough temperatures through the
action of what are known collectively as *van der Waals' forces.*

The electric attraction between polar molecules that was discussed in
Section 17-7 is an example of a van der Waals force. These molecules have
asymmetric distributions of charge, with one end positive and the other

Polar-polar attraction

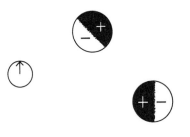

Fig. 31-12.   Polar molecules attract each other electrically.

negative. When one such molecule is near another, the ends of opposite polarity attract each other to hold the molecules together (Fig. 31-12).

Polar-nonpolar attraction

A somewhat similar effect occurs when a polar molecule is near a nonpolar one. The electric field of the polar molecule distorts the initially symmetric charge distribution of the nonpolar one, as in Fig. 31-13, and the two then attract each other in the same way as any other pair of polar molecules. This phenomenon is similar to that involved in the attraction of bits of paper by a charged rubber comb shown in Fig. 17-8.

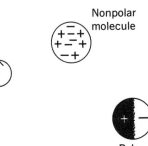

Nonpolar molecule

Polar molecule

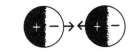

Fig. 31-13.   A polar molecule attracts a nonpolar one by first distorting the latter's originally symmetric charge distribution.

Nonpolar molecules attract one another in much the same way that polar molecules attract nonpolar ones. In a nonpolar molecule, the electrons are distributed symmetrically *on the average*, but *at any moment* one part of the molecule contains more electrons than usual and the rest of the molecule contains fewer. Thus *every* molecule (and atom) behaves as though it is polar, though the direction and magnitude of the polarization vary constantly. The fluctuations in the charge distributions of nearby nonpolar molecules keep in step through the action of electric forces, and these forces also hold the molecules together (Fig. 31-14).

Nonpolar-nonpolar attraction

In general, van der Waals' bonds are considerably weaker than ionic, covalent, and metallic bonds; only about 1% as much energy is needed to remove an atom or molecule from a van der Waals solid as is required in the case of ionic or covalent crystals. As a result the inert gases and compounds with symmetric molecules liquify and vaporize at rather low temperatures. Thus the boiling point of argon is −186°C and the boiling point of

van der Waals' bonds are weak

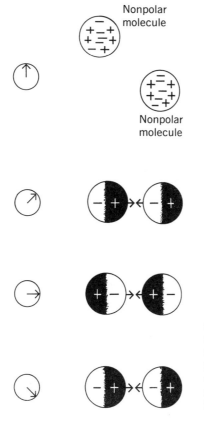

Fig. 31-14. Nonpolar molecules have, on the average, symmetric charge distributions, but at any instant the distributions are not necessarily symmetric. The fluctuations in the charge distributions of adjacent nonpolar molecules keep in step, which leads to an attractive force between them.

methane is −161°C. Molecular crystals, whose lattices consist of individual molecules held together by van der Waals' forces, generally lack the mechanical strength of other kinds of crystals.

31-6 Metallic Bond

A fourth important type of cohesive force in crystalline solids is the *metallic bond*, which has no molecular counterpart.

A characteristic property of all metal atoms is the presence of only a few electrons in their outer shells, and these electrons can be detached relatively easily to leave behind positive ions. According to the theory of the metallic bond, a metal in the solid state consists of an assembly of atoms that have given up their outermost electrons to a common "gas" of freely moving electrons that pervades the entire metal. The electric interaction between the positive ions and the negative electron gas holds the metal together.

An electron "gas" bonds metals

This theory has many attractive features. The high electrical and thermal conductivity of metals follows from the ability of the free electrons to migrate through their crystal structures, while all the electrons in ionic and covalent crystals are bound to particular atoms or pairs of atoms. The absence of specific atom-to-atom bonds permits metals to be readily deformed without fracture, unlike crystals of other kinds. While certain solids that are amorphous rather than crystalline in structure can also be deformed readily, they lack the strength and hardness conferred by the metallic bond. Furthermore, since the atoms in a metal interact through the medium of a common electron gas, the properties of mixtures of different metal atoms should not depend critically on the relative proportions of each kind of atom, provided that their sizes are similar. This prediction is fulfilled in the observed behavior of alloys, in contrast to the specific atomic proportions characteristic of ionic and covalent solids.

Why metals are opaque

The opacity and metallic luster exhibited by metals can also be traced to the gas of free electrons that pervades them. When light of any frequency shines on a metal, the free electrons are set in vibration by the oscillating electromagnetic fields and thereby absorb the light. The oscillating electrons themselves then act as sources of light, sending out electromagnetic waves of the original frequency in all directions. Those waves that happen to be directed back toward the metal surface are able to escape, and their emergence gives the metal its lustrous appearance. If the metal surface is smooth, the reradiated waves appear to us as a reflection of the original incident light.

Table 31-1 summarizes the properties of the four types of crystalline solids that have been discussed here.

Table 31-1. Types of crystalline solids.

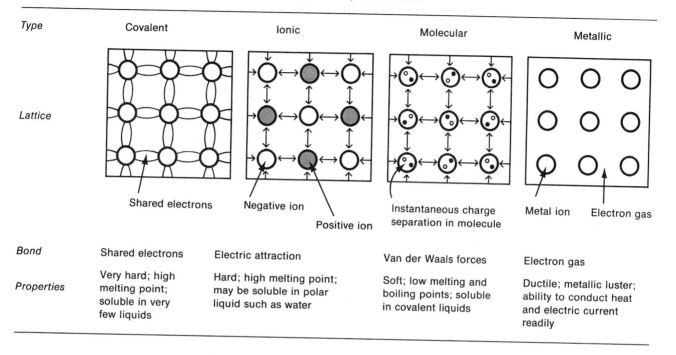

Type	Covalent	Ionic	Molecular	Metallic
Lattice	Shared electrons	Negative ion / Positive ion	Instantaneous charge separation in molecule	Metal ion Electron gas
Bond	Shared electrons	Electric attraction	Van der Waals forces	Electron gas
Properties	Very hard; high melting point; soluble in very few liquids	Hard; high melting point; may be soluble in polar liquid such as water	Soft; low melting and boiling points; soluble in covalent liquids	Ductile; metallic luster; ability to conduct heat and electric current readily

31-7 Energy Bands

The notion of energy bands provides a useful framework for understanding the electrical behavior of solids. The nature and properties of semiconductors in particular are clarified with the help of an energy-band analysis.

Origin of energy bands

When atoms are brought as close together as those in a crystal, they interact with one another to such an extent that their outer electron shells constitute a single system of electrons common to the entire array of atoms. The Pauli exclusion principle prohibits more than two electrons (one with each spin) in any energy level of a system. This principle is obeyed in a crystal because the energy levels of the outer electron shells of the various atoms are all slightly altered by their mutual interaction. (The inner shells do not interact and therefore do not undergo a change.) As a result of the shifts in the energy levels, an *energy band* exists in a crystal in place of each sharply defined energy level of its component atoms (Fig. 31-15). While these bands are actually composed of a multitude of individual energy levels, as many as there are atoms in the crystal, the levels are so near one another as to form a continuous distribution.

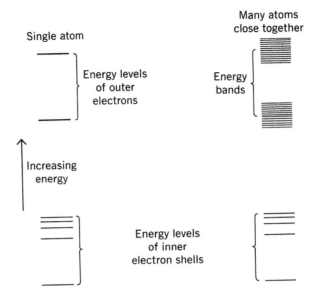

Fig. 31-15. Energy bands replace energy levels of outer electrons in an assembly of atoms that are close together.

The energy bands in a crystal correspond to energy levels in an atom, and an electron in a crystal can only have an energy that falls within one of these bands. The various energy bands in a crystal may or may not overlap, depending upon the composition of the crystal (Fig. 31-16). If they do not overlap, the gaps between them represent energy values which electrons in the crystal cannot have. The gaps are accordingly known as *forbidden bands*.

Forbidden bands

The energy bands we have been speaking of contain all the possible energies that can be possessed by electrons. The electrical properties of a crystalline solid depend upon both its energy-band structure and the way in which the bands are normally occupied by electrons. We shall examine a few specific cases to see how the energy-band approach accounts for the observed electrical behavior of such solids.

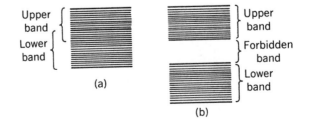

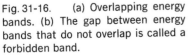

Fig. 31-16. (a) Overlapping energy bands. (b) The gap between energy bands that do not overlap is called a forbidden band.

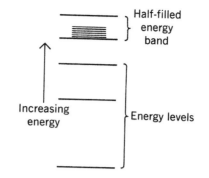

Fig. 31-17. Energy levels and bands in solid sodium, which is a metal. (Not to scale.)

Figure 31-17 shows the energy bands of solid sodium. A sodium atom has only one electron in its outer shell. This means that the upper energy band in a sodium crystal is only half filled with electrons, since each level within the band, like each level in the atom, is capable of containing *two* electrons. When an electric field is established in a sodium crystal, electrons readily acquire the small additional energy they need to move up in their energy band. The additional energy is in the form of kinetic energy, and the moving electrons constitute an electric current. Sodium is therefore a good conductor, as are other crystalline solids with energy bands that are only partially filled. Such solids are metals.

Figure 31-18 shows the energy-band structure of diamond. The two lower energy bands are completely filled, and there is a gap of 6 eV between the top of the higher of these bands and the empty band above it. Hence a minimum of 6 eV of additional energy must be given to an electron in a diamond crystal if it is to be capable of free motion, since it cannot have an energy lying in the forbidden band. Such an energy increment cannot readily be imparted to an electron in a crystal by an electric field. Diamond, like other solids with similar energy band structures, is therefore an electrical insulator.

Silicon has a crystal structure similar to that of diamond, and, as in diamond, a gap separates the top of a filled energy band from an empty higher band. However, whereas the gap is 6 eV wide in diamond, it is only 1.1 eV wide in silicon. At very low temperatures silicon is hardly better than diamond as a conductor, but at room temperature a small proportion of its electrons can possess enough kinetic energy of thermal origin to exist in the higher band. These few electrons are sufficient to permit a limited amount of current to flow when an electric field is applied. Thus silicon has a resistivity intermediate between those of conductors (such as sodium) and those of insulators (such as diamond), and is classified as a *semiconductor*.

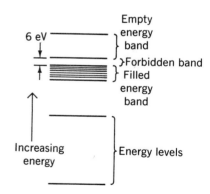
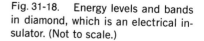

Fig. 31-18. Energy levels and bands in diamond, which is an electrical insulator. (Not to scale.)

Special Topic

Semiconductor Devices

The conductivity of semiconductors is markedly affected by slight amounts of impurity. Suppose that we add several arsenic atoms to a silicon crystal. Arsenic atoms have five electrons in their outermost shells, while silicon atoms have four. When an arsenic atom replaces a silicon atom in a silicon crystal, four of its electrons participate in covalent bonds with its nearest neighbors. The remaining electron needs very little energy to be detached and move about freely in the crystal. Such a solid is called an *n-type* semiconductor because electric current in it is carried by the motion of negative charges toward the positive end of a sample of it. In an energy-band diagram, the effect of arsenic as an impurity in silicon is to supply occupied energy levels just below an empty energy band. These levels are called *donor levels.*

If we add gallium atoms to a silicon crystal, a different effect occurs. Gallium atoms have only three electrons in their outer shells, and their presence leaves vacancies called *holes* in the electron structure of the crystal. An electron requires little energy to move into a hole, but as it does so it leaves a new hole in its previous location. When an electric field is applied to a silicon crystal in which gallium is present as an impurity, electrons move toward the positive electrode by successively filling holes. The flow of current here is best described in terms of the motion of the holes, which behave as though they are positive charges since they move toward the negative electrode. A material of this kind is therefore called a *p-type* semiconductor. In an energy band diagram, the effect of gallium as an impurity is to provide energy levels, called *acceptor levels,* just above the highest filled band; electrons that enter these levels leave behind unoccupied levels in the formerly filled band which make possible the conduction of current.

The significance of semiconductors in technology arises from the degree of control of electric current that can be accomplished by combining *n-* and *p-*type semiconductors in various ways. Most semiconductor devices depend for their action on the properties of junctions between *n-* and *p-*type materials. Perhaps the most important of these properties is that ordinarily current can flow in only one direction through a $p-n$-junction. Figure 31-19 shows a crystal so manufactured that part exhibits *n-*type conductivity (current carried by electrons) and the rest exhibits *p-*type conductivity (current carried by holes). When a potential difference is applied across the

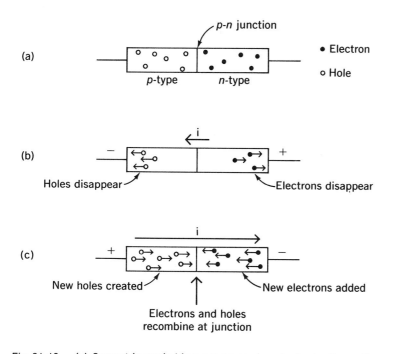

Fig. 31-19. (a) Current is carried in a *p*-type semiconductor by the motion of holes, and in an *n*-type semiconductor by the motion of electrons. (b) Reverse bias. (c) Forward bias.

crystal so that the *p*-end is negative and the *n*-end positive, the holes in the *p*-region migrate to the left and the electrons in the *n*-region migrate to the right. Only a limited number of holes and electrons are in the respective regions and new ones appear spontaneously at only a very slow rate; the current through the entire crystal is therefore negligible. This situation is called *reverse bias*.

Figure 31-19(c) shows the same crystal with the connections changed so that the *p*-end is positive and the *n*-end negative. Now new holes are created continuously by the removal of electrons at the *p*-end while new electrons are fed into the *n*-end of the crystal. The holes migrate to the right and the electrons to the left to produce a net positive current flowing from + to −. The electrons and holes meet at the junction between the *p*- and *n*-regions and recombine there: a hole is a missing electron, and when electrons and holes come together they disappear into the regular structure of the crystal and can no longer act as current carriers. Thus current can flow readily through a semiconductor junction from the *p*- to the *n*-region, but

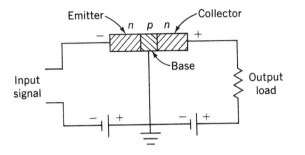

Fig. 31-20.   A simple transistor amplifier.

hardly at all in the opposite direction. Semiconductor rectifiers are widely used today.

Transistors are semiconductor devices that are able to amplify weak signals into strong ones. Figure 31-20 shows an *n-p-n*-junction transistor, one of a variety of such devices, which consists of a *p*-type material sandwiched between *n*-type materials. The *p*-type region is called the *base*, and the two *n*-type regions are the *emitter* and the *collector*. In the figure the transistor is connected as a simple amplifier. There is a forward bias across the emitter-base junction in this circuit, so electrons pass readily from the emitter to the base. Depending upon the rate at which holes are produced in the base, which in turn depends upon its potential relative to the emitter and hence upon the signal input, a certain proportion of the electrons from the emitter will recombine there. The rest of the electrons migrate across the base to the collector to complete the emitter-collector circuit. If the base is at a high positive potential relative to the emitter, many holes are produced there to recombine with electrons from the emitter, and little current can flow through the transistor. If the relative positive potential of the base is low, the number of electrons arriving from the emitter will exceed the number of holes being formed in the base, and the surplus electrons will continue across the base to the collector. A modest change in the emitter-base voltage can cause a large change in the emitter-collector current.

Important Terms

A **molecule** is a group of atoms that stick together strongly enough to act as a single particle. A molecule of a given compound always has a certain definite structure and is complete in itself with little tendency to gain or lose atoms.

In a **covalent bond** between atoms in a molecule or a solid, the atoms share one or more electron pairs.

Solids whose constituent atoms or molecules are arranged in regular, repeated patterns are called crystalline. When only short-range order is present, the solid is **amorphous.**

In an **ionic bond,** electrons are transferred from one atom to another, and the two then attract each other electrically. The **ionization energy** of an atom is the amount of work that must be done to remove an electron from it. The **electron affinity** of an atom is the amount of energy released when it acquires an additional electron.

Van der Waals' forces arise from the electrostatic attraction between asymmetrical charge distributions in atoms and molecules.

The **metallic bond** which holds metal atoms together in the solid state arises from a "gas" of freely moving electrons that pervades the entire metal.

Because the atoms in a crystal are so close together, the energy levels of their outer electron shells are altered slightly to produce **energy bands** characteristic of the entire crystal, in place of the individual sharply defined energy levels of the separate atoms. Gaps between energy bands represent energies forbidden to electrons in the crystal and are called **forbidden bands.**

Multiple Choice

1. The reason only two H atoms join to form a hydrogen molecule is a consequence of
   a. the uncertainty principle.
   b. the exclusion principle.
   c. the size of the H atom.
   d. the mass of the H atom.

2. In a diatomic molecule,
   a. only a single pair of electrons can be shared.
   b. more than one pair of electrons can be shared.
   c. one of the atoms must be a hydrogen atom.
   d. one of the atoms must be a carbon atom.

3. The number of covalent bonds a hydrogen atom forms when it combines chemically is
   a. 1.
   b. 2.
   c. 3.
   d. 4.

4. The number of covalent bonds a carbon atom usually forms when it combines chemically is
   a. 1.   b. 2.
   c. 3.   d. 4.

5. When two or more atoms join to form a stable molecule,
   a. energy is absorbed.   b. energy is given off.
   c. there is no energy change.
   d. any of the above, depending upon the circumstances.

6. Short-range order is never found in
   a. crystalline solids.
   b. amorphous solids.
   c. liquids.
   d. gases.

7. An amorphous solid is closest in structure to
   a. a covalent crystal.
   b. an ionic crystal.
   c. a semiconductor.
   d. a liquid.

8. The lowest melting points are usually found in solids held together by
   a. covalent bonds.
   b. ionic bonds.
   c. metallic bonds.
   d. van der Waals bonds.

9. A crystalline solid always
   a. is transparent.
   b. is held together by covalent bonds.
   c. is held together by ionic bonds.
   d. has long-range order in its structure.

10. The work that must be done to remove an electron from an atom is called its
    a. electron affinity.
    b. ionization energy.
    c. energy band.
    d. heat of vaporization.

11. The highest ionization energies are found in the
    a. inert gases.   b. alkali metals.
    c. halogens.   d. transition elements.

12. The highest electron affinities are found in the
    a. inert gases.   b. alkali metals.
    c. halogens.   d. transition elements.

13. The weakest crystals are
    a. covalent.   b. ionic.
    c. van der Waals.   d. metallic.

14. The least brittle crystals are
    a. covalent.   b. ionic.
    c. van der Waals.   d. metallic.

15. Van der Waals forces arise from
    a. electron transfer.
    b. electron sharing.
    c. symmetrical charge distributions.
    d. asymmetrical charge distributions.

16. The particles that make up the lattice of the van der Waals crystal of a compound are
    a. electrons.   b. atoms.
    c. ions.   d. molecules.

17. The particles that make up the lattice of an ionic crystal are
    a. electrons.   b. atoms.
    c. ions.   d. molecules.

18. The particles that make up the lattice of a covalent crystal are
    a. electrons.   b. atoms.
    c. ions.   d. molecules.

19. A property of metals that is not due to the electron "gas" that pervades them is their unusual ability to
    a. conduct electricity.
    b. conduct heat.
    c. reflect light.
    d. form oxides.

Exercises

1. Under what circumstances can two electrons share the same probability cloud in a molecule?

2. What is wrong with the model of a hydrogen molecule in which the two electrons are supposed to follow figure-of-eight orbits that encircle the two protons?

3. What property of carbon atoms enables them to form so many varied and complex molecules?

4. Why do the inert gas atoms almost never participate in covalent bonds?

5. The atoms in a molecule are said to share electrons, yet some molecules are polar. Explain.

6. Small "perfect crystals" have been prepared with very few structural defects. How would you expect their strength to compare with that of ordinary crystals of the same kind?

7. (a) An iron bar is hammered flat on an anvil, and becomes harder as a result. What change occurred in the structure of the iron? (b) The piece of iron is then strongly heated and allowed to cool slowly, and it becomes softer. Why do you think this happened?

8. The ionization energies of Li, Na, K, Rb, and Cs are, respectively, 5.4, 5.1, 4.3, 4.2, and 3.9 eV. All are in group I of the periodic table. Account for the decrease in ionization energy with increasing atomic number.

9. Why are electrons much more readily liberated from lithium when it is irradiated with ultraviolet light than from fluorine?

10. Why are Cl atoms more active chemically than Cl^- ions?

11. Why are Na atoms more active chemically than Na^+ ions?

12. What is the chief cause of the repulsive force that keeps atoms from meshing together despite any attractive forces that may be present?

13. Does the "gas" of freely-moving electrons in a metal include all the electrons present? If not, which electrons are members of the "gas"?

14. Van der Waals' forces can hold inert gas atoms together to form solids, but they cannot hold such atoms together to form molecules in the gaseous state. Why not?

15. Which of the four fundamental interactions is responsible for each of the principal bonding mechanisms in solids?

16. The temperature of a gas falls when it passes slowly from a full container to an empty one through a porous plug. Since the expansion is into a rigid container, no mechanical work is done. What is the origin of the fall in temperature?

17. The separation between Na$^+$ and Cl$^-$ ions in an NaCl crystal is 2.8×10^{-10} m, whereas it is 2.4×10^{-10} m in an NaCl molecule such as might exist in the gaseous state. Account for the fact that these separations are different.

18. Ionic and covalent crystals are supposed to have stronger interatomic bonds than metal crystals, but everyday experience indicates that metals are, as a class, stronger than most other substances. What reason can you think of that would explain why ionic and covalent solids usually cannot develop their theoretical strength while metals can?

19. What is the basic reason that energy bands rather than specific energy levels exist in a solid?

20. How does the energy band structure of a solid determine whether it is a conductor, a semiconductor, or an insulator of electricity?

21. What is the connection between the ability of a metal to conduct electricity and its ability to conduct heat?

22. Explain the following optical properties of solids with the help of the notion of energy bands. (a) Metals are opaque to light of all wavelengths. (b) Semiconductors are opaque to visible light but transparent to infrared light. (c) Most insulators are transparent to visible light.

23. The forbidden energy band in germanium that lies between the highest filled band and the empty band above it has a width of 0.7 e V. Compare the conductivity of germanium with that of silicon at (a) very low temperatures and (b) room temperature.

Problems

1. At what temperature would the average energy of the hydrogen molecules in a gas sample be equal to their binding energy? Compare the answer with the answer to Problem 3 of Chapter 29.

2. The ionization energy of potassium is 4.3 eV and the electron affinity of chlorine is 3.8 eV. (a) What is the net amount of energy required to form a K$^+$ and Cl$^-$ ion pair from a pair of the same atoms? (b) Considering them as point charges, how close together must a K$^+$ and a Cl$^-$ ion be if the total energy of the pair is to be zero?

3. The ionization energy of lithium is 5.4 eV and the electron affinity of bromine is 3.5 eV. (a) What is the net amount of energy required to form a Li$^+$ and Br$^-$ ion pair from a pair of the same atoms? (b) Considering them as point charges, how close together must a Li$^+$ and a Br$^-$ ion be if the total energy of the pair is to be zero?

Answers to Multiple Choice

1. b	6. d	11. a	16. d
2. b	7. d	12. c	17. c
3. a	8. d	13. c	18. b
4. d	9. d	14. d	19. d
5. b	10. b	15. d	

32

The Nucleus

Until now we have not had to regard the nucleus of an atom as anything but a tiny positively-charged lump whose sole functions are to provide the atom with most of its mass and to hold its several electrons in place. Since the behavior of atomic electrons is responsible for the behavior of matter in bulk, the properties of matter we have been exploring, save for mass, have nothing directly to do with atomic nuclei. Nevertheless, the nucleus turns out to be of supreme importance in the universe: the elements exist by virtue of the ability of nuclei to hold multiple electric charges, and the energy involved in nearly all natural processes has its ultimate origin in nuclear reactions and transformations.

32-1 Nuclear Structure

The nature and behavior of the electron structure of the atom was understood before even the composition of its nucleus was known. The reason is that the nucleus is held together as a unit by forces vastly stronger than the electric forces that hold the electrons to the nucleus, and it is correspondingly harder to break apart a nucleus to find out what is inside. Changes in the electron structure of an atom, such as those that occur in the emission of photons or in the formation of chemical bonds, involve energies of only several eV, whereas changes in nuclear structure involve energies of several MeV, a million times more. Let us first inquire into the composition of the nucleus: What is it that gives a nucleus its characteristic mass and charge?

The nucleus of the hydrogen atom consists of a single proton, whose charge is $+e$ and whose mass is

$$m_{\text{proton}} = 1.673 \times 10^{-27} \text{ kg.}$$

The proton mass is 1836 times that of the electron, so by far the major part of the hydrogen atom's mass resides in its nucleus. This is true of all other atoms as well.

Elements more complex than hydrogen have nuclei that contain *neutrons* as well as protons. The neutron, as its name suggests, is uncharged. The neutron mass is slightly more than that of the proton:

Atomic nuclei consist of protons and neutrons

$$m_{\text{neutron}} = 1.675 \times 10^{-27} \text{ kg.}$$

Neutrons and protons are jointly called *nucleons*.

Every atom contains the same number of protons and electrons; this number is the *atomic number* of the element involved. Except in the case of ordinary hydrogen atoms, the number of neutrons in a nucleus equals or exceeds the number of protons. The compositions of atoms of the four lightest elements are illustrated in Fig. 32-1.

Fig. 32-1.   The electronic and nuclear compositions of hydrogen, helium, lithium, and beryllium atoms.

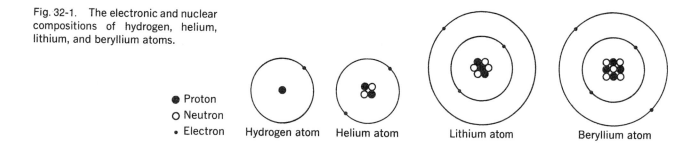

● Proton
○ Neutron
· Electron   Hydrogen atom   Helium atom   Lithium atom   Beryllium atom

The mass of an atom can be determined with the help of a mass spectrometer, such as the one described in Section 20-6. Atomic masses are conventionally expressed in terms of the *atomic mass unit,* abbreviated u, whose value is

Atomic mass unit

$$1 \, u = 1.660 \times 10^{-27} \, kg.$$

The proton and neutron masses in u are respectively

$$m_e = 0.00055 \, u, \qquad m_p = 1.007277 \, u,$$
$$m_n = 1.008665 \, u.$$

Because m_p and m_n are so close to 1, we would expect atomic masses to be very nearly whole numbers. This is often true. For example, the atomic mass of helium is 4.003, that of lithium is 6.939, and that of beryllium is 9.012. However, the chlorine found in nature has an atomic mass of 35.45, which does not fit in with this picture.

Isotopes

Chlorine is an example of an element whose nuclei do not all have the same composition. The several varieties of an element are called its *isotopes.* The number of protons is always equal to the atomic number Z of the element, of course, but the number of neutrons may be different. Thus chlorine consists of two isotopes, one whose nuclei contain 17 protons and 18 neutrons and another whose nuclei contain 17 protons and 20 neutrons. There are about three times as many nuclei of the former type as there are of the latter, which yields an average atomic mass of 35.45 (Table 32-1).

Table 32-1. The isotopes of hydrogen and chlorine.

ELEMENT	PROPERTIES OF ELEMENT		PROPERTIES OF ISOTOPE			
	Atomic number	Atomic mass, u	Protons in nucleus	Neutrons in nucleus	Atomic mass, u	Relative abundance
Hydrogen	1	1.008	1	0	1.008	99.985%
			1	1	2.014	0.015%
			1	2	3.016	very small
Chlorine	17	35.45	17	18	34.97	75.53%
			17	20	36.97	24.47%

All elements have isotopes, even hydrogen. The most abundant hydrogen isotope has nuclei that each consist of a single proton. Less common is the isotope *deuterium*, whose nuclei each consist of a proton and a neutron, and the isotope *tritium*, whose nuclei each consist of a proton and two neutrons. Deuterium is stable, but tritium is radioactive and a sample of it gradually changes to an isotope of helium. The flux of cosmic rays from space continually replenishes the earth's tritium by nuclear reactions in the atmosphere; there is only about 2 kg of tritium of natural origin on the earth's surface, nearly all of it in the oceans.

Deuterium and tritium

Because the chemical properties of an element depend upon the distribution of the electrons in its atoms, which in turn depends upon the nuclear charge, nuclear structure beyond the number of protons present has little significance for the chemist. The physical properties of an element, however, depend strongly on the nuclear structures of its isotopes, whose behavior may be very different from one another although chemically they are indistinguishable.

Isotopes of an element have similar chemical behavior

The conventional symbols for isotopes all follow the pattern $_Z^A X$, where

X = chemical symbol of the element.

Z = atomic number of the element

   = number of protons in the nucleus.

A = mass number of the isotope

   = number of protons and neutrons in the nucleus.

Hence ordinary hydrogen is designated $_1^1 H$, since its atomic number and mass number are both 1, while tritium is designated $_1^3 H$. The two isotopes of chlorine mentioned above are designated $_{17}^{35} Cl$ and $_{17}^{37} Cl$ respectively.

32-2 Nuclear Size and Composition

The Rutherford scattering experiment (Section 17-5) provides information on nuclear dimensions as well as on atomic structure. The observed distribution of scattering angles is consistent with a nucleus of infinitely small size provided the alpha particles are not too energetic. That is, below a certain particle energy the size of the nucleus is small compared with the minimum distance to which the incident alpha particles approach it. At higher energies, discrepancies occur between theory and data which suggest that the particles have come so close to the nucleus that it no longer can be thought of as a point charge, and from the energy at which these discrepancies appear an estimate can be made of nuclear dimensions.

More recent experiments that employ high-energy electrons, protons, and neutrons yield more precise figures for nuclear sizes. Nuclear radii are found to range from about 1.1×10^{-15} m for the proton to about 7.4×10^{-15} m for the $^{235}_{92}U$ nucleus. These experiments are in essence observations of the diffraction of the matter waves of the fast particles by the nuclei of the target atoms; the result is a diffraction pattern formed by the scattered particles, from which the size of the "obstacle"—the nucleus—can be inferred. The observations show that the diffraction patterns are almost, but not quite, the same as those that would be produced by a black disk. The nucleus does not have a sharp boundary, but, like the atom itself, a fuzzy one.

The density of nuclear matter is about 2×10^{17} kg/m^3, which is 3 billion tons per cubic inch. "White dwarf" stars consist of atoms whose electron

Measuring nuclear size

White dwarf stars

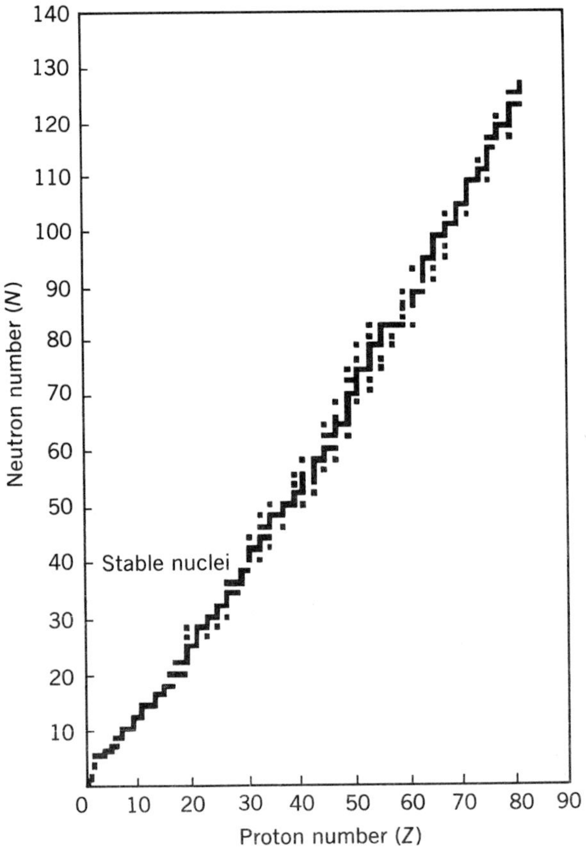

Fig. 32-2. The number of neutrons versus the number of protons in stable nuclei. The larger the nucleus, the greater the proportion of neutrons.

structures have collapsed because of immense pressures, and the densities of these stars approach the density of nuclear matter.

Only certain combinations of protons and neutrons form stable nuclei. In light nuclei there are about as many neutrons as protons, while in heavier ones there are somewhat more neutrons than protons. Figure 32-2 is a plot of neutron number N versus Z for stable nuclei. Evidently the number of neutrons in the nuclei of a given element is a fairly critical quantity. Let us see why.

There are two opposing tendencies in a nucleus. The first is a tendency for N to equal Z, which arises from the existence of quantum states of different energy in a nucleus. Just like an electron in an atom, a nucleon in a nucleus can possess only certain specific energies. Because neutrons and protons obey the exclusion principle, at most two of each kind of nucleon (one whose spin is "up" and one whose spin is "down") can occupy each quantum state. Again as in the case of atomic energy levels, nuclear energy levels are filled in sequence to achieve nuclei of minimum energy and hence maximum stability. Thus the boron isotope $^{12}_{5}B$ has more energy than the carbon isotope $^{12}_{6}C$ because one of its neutrons is in a higher energy level, and it is accordingly unstable (Fig. 32-3). If created in a nuclear reaction, a $^{12}_{5}B$ nucleus changes by beta decay into a $^{12}_{6}C$ nucleus in a fraction of a second (See Section 32-5).

Nuclear energy levels

The other tendency in a nucleus is for the number of neutrons to exceed the number of protons. This is a consequence of the strong electric repulsion exerted by the protons upon one another, which must be balanced by the attractive nuclear forces that act between nucleons. The repulsive electric forces increase more rapidly with Z than the attractive nuclear forces increase with A, the total number of nucleons, and so a greater proportion of neutrons, which produce only attractive forces owing to their electrical neutrality, is necessary for stability in large nuclei.

In heavy nuclei, N exceeds Z

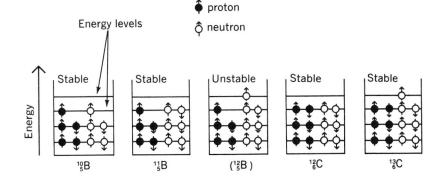

Fig. 32-3. Each nuclear energy level can contain two protons of opposite spins and two neutrons of opposite spins. In the light nuclei, energy levels are filled in sequence so the resulting configuration is one of minimum energy.

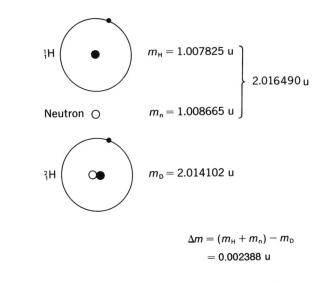

$m_H = 1.007825$ u

Neutron ○ $m_n = 1.008665$ u

2.016490 u

Fig. 32-4.   The mass of every atom is less than the total of the masses of its constituent neutrons, protons, and electrons. This phenomenon is illustrated here for the deuterium atom.

$m_D = 2.014102$ u

$\Delta m = (m_H + m_n) - m_D$

$= 0.002388$ u

32-3 Binding Energy

It was stated above that the nucleus of a deuterium atom consists of a proton and a neutron. Thus we would expect that the mass of a deuterium atom, $^2_1$H, should be equal to the mass of an ordinary hydrogen atom, $^1_1$H, plus the mass of a neutron. However, it turns out that the mass of $^2_1$H is 0.002388 u *less* than the combined masses of a $^1_1$H atom and a neutron (Fig. 32-4).

The preceding result is an example of a general observation: stable atoms always have less mass than the combined masses of their constituent particles. The energy equivalent of the "missing" mass of a nucleus is called its *binding energy*. In order to break a nucleus apart into its constituent nucleons, an amount of energy equal to its binding energy must be supplied either in a collision with another particle or by the absorption of a sufficiently energetic photon. Binding energies are due to the action of the nuclear forces that hold nuclei together, just as ionization energies of atoms, which must be supplied to remove electrons from them, are due to the action of the electric forces that hold them together.

The binding energy of a nucleus is the energy needed to break it up

The difference between the $^2_1$H atomic mass and the combined masses of $^1_1$H and a neutron is 0.002388 u. Since the energy equivalent of 1 u is 931 MeV, the binding energy of the deuteron (as the deuterium nucleus is called) is therefore

$$0.002388 \text{ u} \times 931 \frac{\text{MeV}}{\text{u}} = 2.22 \text{ MeV}.$$

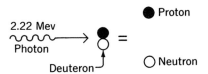

This figure is confirmed by experiments that show that the minimum energy a photon must have in order to disrupt a deuteron is 2.22 MeV (Fig. 32-5).

The binding energy per nucleon in a nucleus is equal to the total binding energy (calculated from the mass deficiency of the nucleus) divided by the number of neutrons and protons it contains. A very interesting curve results when we plot binding energy per nucleon versus mass number, as in Fig. 32-6. Except for an anomalously high peak for $^4_2$He, the curve is a quite regular one. Nuclei of intermediate size have the highest binding energies per nucleon, which means that their nucleons are held together more securely than the nucleons in both heavier and lighter nuclei. The maximum in the curve is 8.8 MeV/nucleon at $A = 56$, which corresponds to the iron nucleus $^{56}_{26}$Fe.

The binding-energy curve is one of the most significant in all of science

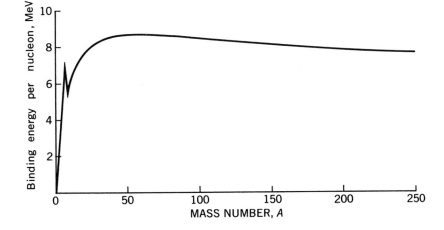

Fig. 32-6.   The binding energy per nucleon versus mass number. The higher the binding energy per nucleon, the more stable the nucleus. When a heavy nucleus is split into two lighter ones, a process called *fission*, the greater binding energy of the latter causes the liberation of energy. When two very light nuclei join to form a heavier one, a process called *fusion*, the greater binding energy of the latter again causes the liberation of energy.

Nuclear fission

A remarkable feature of nuclear structure is illustrated by this curve. Suppose that we split the nucleus $^{235}_{92}U$, whose binding energy is 7.6 MeV/nucleon, into two fragments. Each fragment will be the nucleus of a much lighter element, and therefore will have a higher binding energy per nucleon than the uranium nucleus. The difference is about 0.8 MeV/nucleon, and so, if such *nuclear fission* were to take place, an energy of

$$0.8 \; \frac{\text{MeV}}{\text{nucleon}} \times 235 \; \text{nucleons} = 190 \; \text{MeV}$$

would be given off per splitting. This is a truly immense amount of energy to be produced in a single atomic event. As a comparison, chemical processes involve energies of the order of magnitude of one electron volt per reacting atom, 10^{-8} the energy involved in fission.

Nuclear fusion

Figure 32-6 also shows that if two of the extremely light nuclei are combined to form a heavier one, the higher binding energy of the latter will also result in the evolution of energy. For instance, if two deuterons were to join to make a $^4_2$He nucleus, over 23 MeV would be released. This process is known as *fusion*, and, together with fission, promises to be the source of more and more of the world's energy as reserves of fossil fuels are depleted.

32-4 Strong Nuclear Interaction

The existence of stable nuclei can only be explained on the basis of a special interaction between nucleons. This interaction cannot be electrical, because neutrons are uncharged and the positive charges of protons lead to repulsive forces only. It cannot be gravitational, because gravitational forces are far too weak to be able to counterbalance the repulsive electric forces between protons. Thus we have a third fundamental interaction, the *strong nuclear interaction*, which is responsible for the existence of atomic nuclei more complex than that of $^1_1$H, which is a single proton.

Properties of strong nuclear interaction

Two properties of the strong interaction stand out. First, it is by far the strongest of all the fundamental interactions, as we can tell from the magnitude of nuclear binding energies. To pull apart the neutron and proton in a deuterium nucleus takes 2.22 MeV, but to pull the electron in a deuterium atom away from tne nucleus takes only 13.6 eV, over a hundred thousand times less work (Fig. 32-7).

The second noteworthy aspect of the strong interaction is its short range. Electric and gravitational forces fall off with distance as $1/r^2$, and are

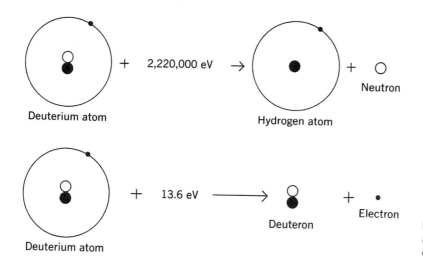

Fig. 32-7. Nuclear binding energies are much greater than atomic binding energies.

effective at considerable separations between the interacting objects — the planet Pluto is 6×10^{12} m from the sun, yet is kept in orbit by the gravitational attraction of the sun. But nuclear forces are effective only over a range of a few nucleon diameters. Up to a separation of about 3×10^{-15} m, the nuclear attraction between two protons is about 100 times stronger than the electric repulsion between them, but beyond this distance the nuclear force dies out rapidly. The nuclear interactions between protons and protons, between protons and neutrons, and between neutrons and neutrons appear to be identical.

The short range of nuclear forces is responsible for the restricted number of stable elements. The larger a nucleus, the stronger the electric repulsive forces that act on each of its protons, but the attractive nuclear forces on each nucleon cannot increase indefinitely because only a limited number of other nucleons are close enough to interact with it. The largest stable nucleus is the bismuth isotope $^{209}_{83}\text{Bi}$, and nuclei larger than the uranium isotope $^{238}_{92}\text{U}$ are too unstable to have survived on earth since its formation.

Why the number of stable elements is limited

32-5 Radioactivity

Not all atomic nuclei are stable. At the beginning of the 20th century it became known, as the result of research by Becquerel, the Curies, and

Radiations from unstable nuclei

others, that some nuclei exist which spontaneously transform themselves into other nuclear species with the emission of radiation. Such nuclei are said to be *radioactive*.

Unstable nuclei emit three kinds of radiation:

1. *Alpha particles*, which are the nuclei of $^4_2$He atoms;
2. *Beta particles*, which are electrons;
3. *Gamma rays*, which are photons of high energy.

The early experimenters identified these radiations with the help of a magnetic field. Figure 32-8 shows a radium sample in a magnetic field directed into the paper: the positively charged alpha particles are deflected to the left and the negatively charged beta particles are deflected to the right. Gamma rays carry no charge and are not affected by the magnetic field.

To understand why alpha, beta, and gamma decays take place, let us return to Fig. 32-2, which is a plot showing the number of neutrons versus the number of protons in stable nuclei. For light stable nuclei the number of neutrons and protons are approximately the same, while for heavier nuclei slightly more neutrons than protons are required for stability. It is evident that an element of a given atomic number has only a very narrow range of possible numbers of neutrons if it is to be stable.

Suppose now that a nucleus exists which has too many neutrons for

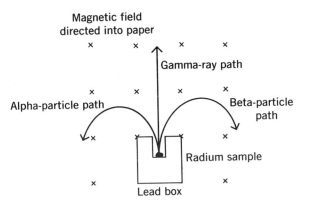

Fig. 32-8.   The radiations from a radium sample may be analyzed with the help of a magnetic field. In the figure the direction of the field is into the paper; hence the positively charged alpha particles (which are helium nuclei) are deflected to the left and the negatively charged beta particles (which are electrons), to the right. Gamma rays (which are energetic photons) carry no charge and are not affected by the magnetic field.

stability relative to the number of protons present. If one of the excess neutrons transforms itself into a proton, this will simultaneously reduce the number of neutrons while increasing the number of protons. To conserve electric charge, such a transformation requires the emission of a negative electron, and we may write it in equation form as

Negative beta decay

$$n^0 \rightarrow p^+ + e^-. \qquad \textit{Beta decay} \quad (32\text{-}1)$$

The electron leaves the nucleus, and is detectable as a "beta particle." The residual nucleus may be left with some extra energy as a consequence of its shifted binding energy, and this energy is given off in the form of gamma rays. Sometimes more than one such *beta decay* is required for a particular unstable nucleus to reach a stable configuration.

Should the nucleus have too few neutrons, the inverse reaction

Positive beta decay

$$p^+ \rightarrow n^0 + e^+, \qquad \textit{Positron emission} \quad (32\text{-}2)$$

in which a proton becomes a neutron with the emission of a positron, may take place. This is also called beta decay, since it resembles the emission

A positron is a positively-charged electron

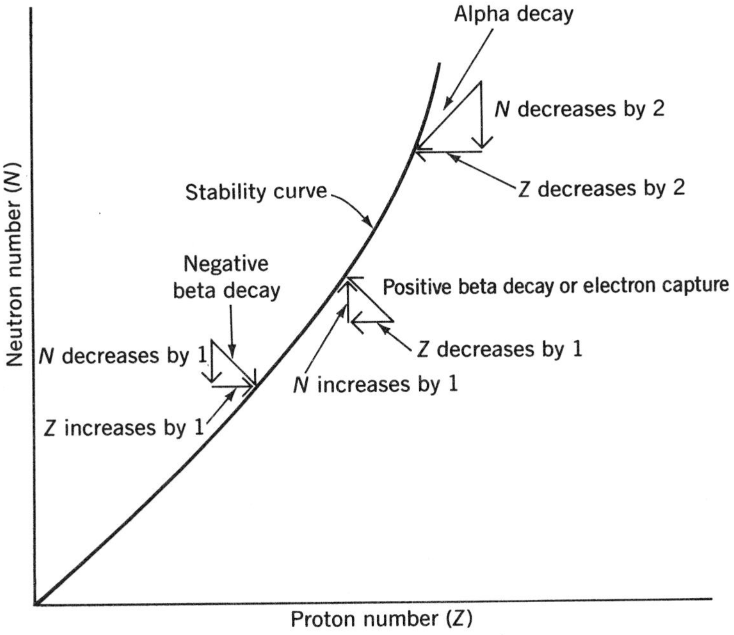

Fig. 32-9. How alpha and beta decays tend to bring an unstable nucleus to a stable configuration.

Electron capture

of negative electrons from an unstable nucleus in every way save for the difference in charge.

A process that competes with positron emission is the capture by a nucleus with too small a neutron/proton ratio of one of the electrons in its innermost atomic shell. The electron is absorbed by a nuclear proton which becomes a neutron in so doing. This process can be expressed as

$$p^+ + e^- \to n^0 \qquad\qquad \textit{Electron caputre} \quad (32\text{-}3)$$

Electron capture does not lead to the emission of a particle, but can be detected by the x-ray photon that is produced when one of the atom's outer electrons falls into the vacancy left by the absorbed electron.

Alpha decay

Another way of altering its structure to achieve stability may involve a nucleus in *alpha decay*, in which an alpha particle consisting of two neutrons and two protons is emitted. Thus negative beta decay increases the number of protons by one and decreases the number of neutrons by one; positive beta decay and electron capture decrease the number of protons by one and increases the number of neutrons by one; and alpha decay decreases both the number of protons and the number of neutrons by two. These processes are shown schematically in Fig. 32-9. Very often a succession of alpha and beta decays, with accompanying gamma decays to carry off excess energy, is required before a nucleus reaches stability.

32-6 The Weak Interaction

The strong nuclear interaction that holds nucleons together to form nuclei cannot account for the occurrence of beta decay. Another fundamental interaction turns out to be responsible: the *weak interaction*. The range of the latter is so short that it operates only *within* certain elementary particles and leads to their transformation into other particles.

Insofar as the structure of matter is concerned, the role of the weak interaction seems to be confined to causing beta decays in nuclei whose neutron/proton ratios are not appropriate for stability. This interaction also affects elementary particles that are not part of a nucleus. The name "weak interaction" arose because the other short-range force acting upon nucleons is extremely strong, as the high binding energies of nuclei attests. Actually the gravitational interaction is weaker than the weak interaction at distances where the latter is a factor.

The four fundamental interactions

The four fundamental interactions that govern the structure and behavior of the entire physical universe, from atoms to galaxies of stars, are therefore as follows:

1. *Gravitational*—acts between all masses, determines structures of planets, stars, galaxies;
2. *Electromagnetic*—acts between all charged particles, determines structures of atoms, molecules, and solids, and is an important factor in the astronomical universe;
3. *Strong nuclear*—acts between protons and neutrons, determines structures of atomic nuclei;
4. *Weak nuclear*—acts within elementary particles, helps determine compositions of atomic nuclei.

Recent theoretical studies, supported by experiment, indicate that the weak and electromagnetic interactions may be so closely related in certain fundamental respects as to represent different aspects of the same phenomenon. Some properties of the strong interaction seem to suggest an association between it and the weak and electromagnetic interactions as well; if true, this would represent a most remarkable unity. How the gravitational interaction would fit into an even grander scheme, however, remains unclear.

32-7 Alpha Decay

Nuclei that contain more than about 210 nucleons are so large that the short-range forces holding them together are barely able to counterbalance the long-range electric repulsive forces of their protons. Such a nucleus can reduce its bulk and thereby achieve greater stability by emitting an alpha particle, which decreases its mass number A by 4.

It is appropriate to ask why it is that only alpha particles are given off by excessively heavy nuclei, and not, for example, individual protons or $_2^3$He nuclei. The reason is a consequence of the high binding energy of the alpha particle, which means that it has significantly less mass than four individual nucleons. Because of this small mass, an alpha particle can be ejected by a heavy nucleus with energy to spare. Thus the alpha particle released in the decay of $_{92}^{232}$U has a kinetic energy of 5.4 MeV, while 6.1 MeV would have to be supplied from the outside to this nucleus if it is to release a proton, and 9.6 MeV supplied if it is to release a $_2^3$He nucleus.

Even though alpha decay may be energetically possible in a particular nucleus, it is not obvious just how the alpha particle is able to break away from the nuclear forces that bind it to the rest of the nucleus. Typically, an alpha particle has available about 5 MeV of energy with which to escape. However, an alpha particle located at a point near the nucleus but just outside the range of its nuclear forces has an electric potential energy of perhaps 25 MeV; that is, if released from this position it will have a kinetic energy of 25 MeV when it is an infinite distance away as a result of electric repulsion

Why alpha particles are emitted

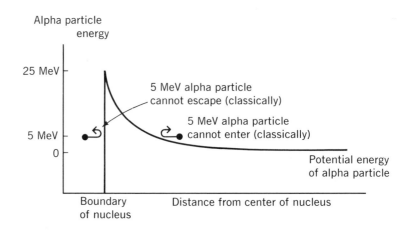

Fig. 32-10. The variation in alpha-particle potential energy near a typical heavy nucleus. The alpha particle in this nucleus has a kinetic energy of 5 MeV.

(Fig. 32-10). An alpha particle inside the nucleus therefore should require a minimum of 25 MeV in energy, five times more than is available, in order to break loose.

The alpha particle, then, is located in a box whose walls are of such a height that an energy of 25 Mev is needed to surmount them, while the particle itself has only 5 Mev for the purpose.

Quantum mechanics provides the answer to the paradox of alpha decay. Two assumptions are needed: (1) an alpha particle can exist as an individual entity within a nucleus, and (2) it is in constant motion there.

Quantum theory of alpha decay

According to quantum theory, a moving particle has a wave character, so that the proper classical analog of an alpha particle in a nucleus is a light wave trapped between mirrors and not a particle bouncing back and forth between solid walls. Now in order for a light wave to be reflected from a mirror, it must actually penetrate the reflecting surface for a short distance (Fig. 32-11). The intensity of the wave drops off quite rapidly inside the reflecting surface, to be sure, but it *must* penetrate to some extent. If the mirror is thick, all of the incident light is reflected. However, if the mirror is very thin, some of the incident light can pass right through the mirror, as shown. The formal theory of this partial transmission, with only minor changes, is able to account quantitatively for alpha decay. The very existence of alpha decay, in fact, is further confirmation of the validity of quantum ideas, since the principles of physics that follow from Newton's laws of motion prohibit such decay.

Of course, a 25-MeV energy barrier is not very "transparent" to

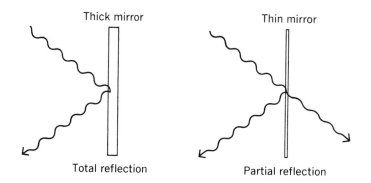

Fig. 32-11. In reflection, a wave penetrates the reflecting surface for a short distance and may pass through it if the mirror is sufficiently thin.

a 5-MeV alpha particle. An alpha-radioactive nucleus is usually about 1.5×10^{-14} m in diameter, and an alpha particle within it might oscillate back and forth with a speed of 2×10^{7} m/s. Hence the alpha particle strikes the confining potential-energy wall nearly 10^{21} times per second, but may nevertheless have to wait as much as 10^{10} years to escape from certain nuclei.

The quantum-mechanical phenomenon of barrier penetration is sometimes called the *tunnel effect*, because the particle escapes *through* the barrier and not over it.

Tunnel effect

32-8 Half-life

The rate at which a sample of a radioactive material decays is called its *activity*. The SI unit of activity is the *becquerel*, where 1 Bq = 1 event/s. The activities encountered in practice are usually so high that the MBq (10^{6}Bq) and GBq (10^{9}Bq) are more suitable. The traditional unit of activity, which is still in common use, is the *curie*. Originally the curie was defined as the activity of 1 g of radium ($^{226}_{88}$Ra), and its precise value accordingly changed as measuring techniques improved. For this reason the curie is now defined arbitrarily as

Activity

$$1 \text{ curie} = 3.70 \times 10^{10} \text{ events/s} = 37 \text{ GBq};$$

the activity of 1 g of radium is a few percent smaller. The *millicurie* (10^{-3} curie) and *microcurie* (10^{-6} curie) are frequently employed to supplement the curie. A luminous watch dial contains several microcuries of $^{226}_{88}$Ra; ordinary potassium has an activity of about 1 millicurie/kg owing to the presence of the radioactive isotope $^{40}_{19}$K; "cobalt-60" sources of 1 or more curies are widely used in medicine for radiation therapy and industrially for the inspection of metal castings and welded joints.

One of the characteristics of all types of radioactivity is that the rate at which the nuclei in a given sample decay always follows a curve whose shape is like that shown in Fig. 32-12. If we start with a sample whose rate of decay is, say, 100 events/s, it will not continue to decay at that rate but instead fewer and fewer disintegrations will occur in each successive second.

Some isotopes decay faster than others, but in each case a certain definite time is required for half of an original sample to decay. This time is called the *half life* of the isotope. The element radon has a half life of 3.8 days, for instance. Should we start with 1 gram of radon in a closed container (since it is a gas), $\frac{1}{2}$ g will remain undecayed after 3.8 days; $\frac{1}{4}$ g will remain undecayed after 7.6 days; $\frac{1}{8}$ g will remain undecayed after 11.4 days; and so on (Fig. 32-13).

Problem. The hydrogen isotope tritium, ^3_1H, is radioactive and emits an electron with a half-life of 12.5 years. (a) What does ^3_1H become after beta decay? (b) What percentage of an original sample of tritium will remain 25 years after its preparation?

Solution. (a) When a nucleus emits an electron, its atomic number increases by 1 unit (corresponding to an increase in nuclear charge of $+e$) and its mass number is unchanged. Helium has the atomic number 2, and so

$$^3_1\text{H} \rightarrow {}^3_2\text{He} + e^-.$$

(b) Twenty-five years represents two half-lives of tritium; hence $\frac{1}{2} \times \frac{1}{2} = \frac{1}{4} = 25\%$ of the original sample remains undecayed.

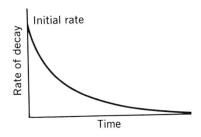

Half-life

Fig. 32-12. The rate at which a sample of radio-active substance decays is not constant, but varies with time in the manner shown in the curve.

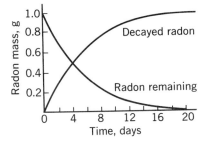

Fig. 32-13. The decay of radon, whose half-life is 3.8 days.

Half lives range from billionths of a second to billions of years. Samples of radioactive isotopes decay in the manner illustrated because a great many individual nuclei are involved, each having a certain probability of decaying. The fact that radon has a half life of 3.8 days signifies that every radon nucleus has a 50% chance of decaying in any 3.8-day period. Because a nucleus does not have a memory, this does *not* mean that a radon nucleus has a 100% chance of decaying in 7.6 days: the likelihood of decay of a given nucleus stays the same until it actually does decay. Thus a half life of 3.8 days means a 75% probability of decay in 7.6 days, an 87.5% probability of decay in 11.4 days, and so on, because in each interval of 3.8 days the probability is 50%.

The above discussion suggests that radioactive decay involves individual events that take place within individual nuclei, rather than collective processes that involve more than one nucleus in interaction. This idea is confirmed by experiments which show that the half life of a particular isotope is invariant under changes of pressure, temperature, electric and magnetic fields, and so on, which might, if strong enough, influence internuclear phenomena.

Three aspects of radioactivity are wholly remarkable from the point of view of classical (pre-relativity and pre-quantum theory) physics. First, the atomic number of a nucleus that undergoes alpha or beta decay changes, so that the nucleus becomes one characteristic of a different element. Elements *can* be transmuted into other elements, though hardly in a manner anticipated by alchemy. Second, radioactive decay liberates energy that can only come from *within* individual atoms. One gram of radium in a sealed container (to prevent the escape of radon, a gaseous product of its decay that is itself radioactive) evolves energy at the rate of 0.14 kcal/hr; this rate decreases so slowly that after 1600 years it has dropped by only 50%. Where does all this energy come from? Not until 1905, when Einstein proposed the equivalence of mass and energy, was this puzzle understood. Third, radioactivity is a statistical process. Every nucleus of a radioisotope has a certain likelihood of decaying, but, because the decay obeys the laws of chance, we have no way of predicting *which* nuclei will actually decay at a particular time. There is no cause-effect relationship in radioactivity as there is in all classical physics.

Classical physics cannot explain radioactivity

32-9 Nuclear Fission

When two nuclei approach close enough together, it is possible for a rearrangement of their constituent nucleons to occur with one or more new nuclei formed. Such a process is called a nuclear reaction, by analogy with chemical reactions in which two or more compounds may combine to form new ones.

Atoms and molecules are neutral, so it is easy for them to come together and react. Nuclei all have positive charges, as much as $+92e$ in nuclei found in nature, and the electric repulsion between them is sufficient to keep them beyond the range where they can interact unless they are moving very fast to begin with. In the sun and other stars, whose interiors are at temperatures of many millions of K, the nuclei present are moving fast enough on the average for nuclear reactions to be frequent, and indeed nuclear reactions provide the energy that maintains these temperatures.

Mutual repulsion hinders nuclear reactions

In the laboratory it is easy enough to produce nuclear reactions on a very small scale, either with alpha particles from radioactive substances or with protons, deuterons, or even heavier nuclei accelerated in cyclotrons and similar devices. But only one type of nuclear reaction has as yet proved to be a practical source of energy on the earth, namely the fission that occurs when neutrons strike the nuclei of certain very heavy nuclei.

Fission products are highly radioactive

In nuclear fission, which can take place only in certain very heavy nuclei such as $^{235}_{92}U$, the absorption of an incoming neutron causes the target nucleus to split into two smaller nuclei called *fission fragments* (Fig. 32-14). Because stable light nuclei have proportionately fewer neutrons than do heavy nuclei, the fragments are unbalanced when they are formed and at once release one or two neutrons each. Usually the fragments are still somewhat unstable and may undergo radioactive decay to achieve appropriate neutron-proton ratios. The products of fission, such as the fallout from a nuclear bomb burst, are accordingly highly radioactive.

Although a variety of nuclear species may appear as fission fragments, we might cite as a typical fission reaction

$$^{235}_{92}U + {}^{1}_{0}n \rightarrow {}^{236}_{92}U \rightarrow {}^{140}_{54}Xe + {}^{94}_{38}Sr + {}^{1}_{0}n + {}^{1}_{0}n + 200 \text{ MeV}. \tag{32-4}$$

The $^{236}_{92}U$ which is first formed lasts for only a small fraction of a second before it splits into two parts. About 84% of the total energy liberated during

Fig. 32-14. In nuclear fission, an absorbed neutron causes a heavy nucleus to split in two parts, with the emission of several neutrons and gamma rays.

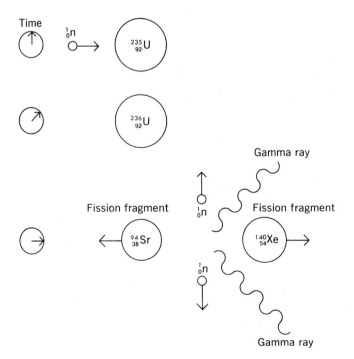

fission appears as kinetic energy of the fission fragments, about 2.5% as kinetic energy of the neutrons, and about 2.5% in the form of instantaneously emitted gamma rays, with the remaining 11% being given off in the decay of the fission fragments.

Because each fission event liberates two or three neutrons while only one neutron is required to initiate it, a rapidly multiplying sequence of fissions can occur in a lump of suitable material (Fig. 32-15). When uncontrolled, such a *chain reaction* evolves an immense amount of energy in a short time. If we assume that two neutrons emitted in each fission are able to induce further fissions (the average figure is lower in practice) and that 10^{-8} s elapses between the emission of a neutron and its subsequent absorption, a chain reaction starting with a single fission will release 2×10^{13} J of energy in less than 10^{-6} s! An uncontrolled chain reaction evidently can cause an explosion of exceptional magnitude.

When properly controlled so as to assure that exactly one neutron per fission causes another fission, a chain reaction occurs at a constant level of power output. A reaction of this kind makes a very efficient source of power: an output of about 1000 kW is produced by the fission of 1 g of a suitable isotope per day, as compared with the consumption of over 3 tons of coal per day per 1000 kW in a conventional power plant. A device in which a chain reaction can be initiated and controlled is called a *nuclear reactor*.

Chain reaction

Fission occurs at a steady rate in a nuclear reactor

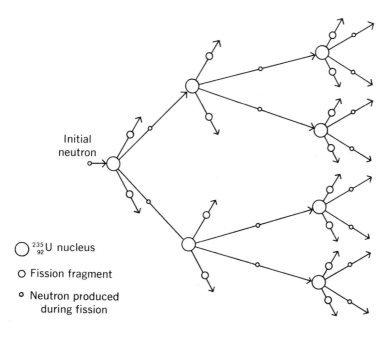

Initial
neutron

$\bigcirc$ $^{235}_{92}$U nucleus

O Fission fragment

° Neutron produced
during fission

Fig. 32-15. Simplified sketch of a chain reaction in $^{235}_{92}$U.

A number of basic problems must be overcome for a nuclear reactor to operate properly. Let us consider as an example a reactor fueled with natural uranium, which is composed of 0.7% of $^{235}_{92}U$ and 99.3% of $^{238}_{92}U$. The latter isotope absorbs rapidly moving neutrons readily, but fission does not occur as a result. Neutrons absorbed by $^{238}_{92}U$ are therefore wasted. However, $^{238}_{92}U$ has little ability to capture *slow* neutrons, while slow neutrons are more apt to cause fission in $^{235}_{92}U$ than fast ones. The neutrons liberated in fission are fast, and accordingly must be slowed down as soon as possible both to prevent them from being taken up by $^{238}_{92}U$ and also to render them more likely to cause further fissions in the $^{235}_{92}U$ content of the fuel.

Neutrons are slowed down by the moderator

In order to slow down the neutrons released in fission, the fuel in a uranium reactor is embedded in a suitable substance called a *moderator*. When a fast neutron collides with a nucleus, some of its initial energy is imparted to the target nucleus. Just how much energy is lost by the neutron depends upon the details of the interaction, but as a general rule the more nearly equal the masses of the particles involved, the greater the energy transferred (see Special Topic for Chapter 8). Hence hydrogen would seem the best moderator, since hydrogen nuclei are simply protons whose mass is almost identical to the neutron mass.

Enriched uranium

Unfortunately protons have a marked affinity for neutrons, and often a neutron colliding with a proton becomes attached to it to form a deuteron, ^2_1H. Deuterons themselves have much less tendency to pick up neutrons, and so make a better choice. "Heavy water" is H_2O whose hydrogen content is ^2_1H rather than ^1_1H, and is much superior to ordinary water as a moderator when natural uranium is the fuel. When *enriched uranium* whose $^{235}_{92}U$ content exceeds that in natural uranium is the fuel, ordinary water may be satisfactory. Early reactors employed graphite, a dense form of carbon, as the moderator, despite its inefficiency, because it does not capture neutrons to any appreciable extent. It was more readily available then than heavy water and, being a rigid material, was easier to incorporate in a reactor structure.

Merely bringing together a bit of uranium and some graphite or heavy water will not necessarily lead to a self-sustaining chain reaction. Neutrons escape from the surface of a reactor, which means that the reactor must have a certain minimum size (and fuel content), since a large one has less surface in proportion to its volume than a small one. When enough fissionable material—here $^{235}_{92}U$—is assembled in the presence of a moderator, a stray neutron, perhaps from a spontaneous fission, causes a $^{235}_{92}U$ nucleus to split, which releases two or three other neutrons. These neutrons are slowed down in the moderator in 0.001 s or so, and are then absorbed by other $^{235}_{92}U$ nuclei to cause further fissions. To control the resulting chain reaction, rods of cadmium or boron, which are good absorbers of slow neutrons, may be inserted in the reactor. The farther in they are placed, the feebler the reaction, until it ceases altogether when the rods pick up neutrons more rapidly than they are produced.

The energy liberated in a nuclear reactor appears as heat, and it is extracted by circulating a coolant liquid or gas. The hot coolant can then be used as the heat source of a conventional steam turbine which in turn may power a ship, a submarine, or an electric generator.

At the same time that $^{235}_{92}$U nuclei in a reactor are absorbing slow neutrons and undergoing fission as a result, $^{238}_{92}$U nuclei are absorbing fast neutrons to become $^{239}_{92}$U. This uranium isotope beta-decays with a half-life of about 23 min into $^{239}_{93}$Np, an isotope of the element *neptunium*. The neptunium isotope is also beta-radioactive, and decays with a half-life of a little over 2 days into $^{129}_{94}$Pu, an isotope of *plutonium* whose half-life against alpha decay is 24,000 yr. Both neptunium and plutonium are transuranic elements, none of which are found in nature because their half-lives against decay are too short for them to have survived even if they had been created when the other elements were some billions of years ago.

Plutonium

Like $^{235}_{92}$U, $^{239}_{94}$Pu is fissionable and can be used in nuclear reactors and weapons. Because plutonium is chemically different from uranium, it is not nearly so difficult to separate from the $^{238}_{92}$U that remains in a used reactor fuel element as it is to enrich natural uranium in the $^{235}_{92}$U isotope. Hence all nuclear reactors are a potential source of raw material for nuclear weapons, even in countries whose technology and resources are not equal to the task of uranium enrichment. "Breeder" reactors, now under development, are especially designed to produce more fissionable plutonium than the $^{235}_{92}$U they consume, so that, since there is no shortage of $^{238}_{92}$U (over 99% of natural uranium), their advent will ease the problem of nuclear fuel supply. Their advent will also complicate the problem of nuclear weapons control.

Breeder reactors

32-10 Nuclear Fusion

Virtually all the energy in the universe originates in the fusion of hydrogen nuclei into helium nuclei in stellar interiors, where hydrogen is the most abundant element. The *proton-proton* cycle of fusion reactions is the chief source of the sun's energy:

$$^1_1H + {}^1_1H \rightarrow {}^2_1H + e^+ + 0.4 \text{ MeV},$$

$$^1_1H + {}^2_1H \rightarrow {}^3_2He + 5.5 \text{ MeV}, \qquad \textit{Proton-proton cycle} \quad (32\text{-}5)$$

$$^3_2He + {}^3_2He \rightarrow {}^4_2He + 2\,{}^1_1H + 12.9 \text{ MeV}.$$

The first two of these reactions must each occur twice for every synthesis of $_2^4\text{He}$, so that the total energy produced per cycle is 24.7 MeV. The symbol e^+ represents a *positron* as we recall.

Thermonuclear energy

The energy liberated by nuclear fusion is often called *thermonuclear energy*. High temperatures and densities are necessary for fusion reactions to occur in such quantity that a substantial amount of thermonuclear energy is produced: the high temperature assures that the initial light nuclei have enough thermal energy to overcome their mutual electrostatic repulsion and come close enough together to react, while the high density assures that such collisions are frequent. A further condition for the proton-proton and similar cycles is a large reacting mass, such as that of a star, since a number of separate steps is involved in each cycle and much time may elapse between the initial fusion of a particular proton and its ultimate incorporation in an alpha particle.

On the earth, where any reacting mass must be very limited in size, an efficient thermonuclear process cannot involve more than a single step. Two reactions that appear promising as sources of commercial power involve the combination of two deuterons to form a triton and a proton

$$_1^2\text{H} + {}_1^2\text{H} \rightarrow {}_1^3\text{H} + {}_1^1\text{H} + 4.0 \text{ MeV}, \tag{32-6}$$

or their combination to form a $_2^3\text{H}$ nucleus and a neutron,

$$_1^2\text{H} + {}_1^2\text{H} \rightarrow {}_2^3\text{He} + {}_0^1\text{n} + 3.3 \text{ MeV}. \tag{32-7}$$

Both reactions have about equal probabilities. A major advantage of these reactions is that deuterium is relatively abundant on the earth, so that there should be no fuel problems in power plants operating on deuteron fusion. While there are many difficulties to surmount in achieving practical thermonuclear power, it will almost certainly become an eventual reality. The use of "magnetic bottles" to contain plasmas in the laboratory in which thermonuclear reactions can occur is discussed in Section 20-5.

Special Topic

Radiation Dosage

Radiation energetic enough to ionize matter in its path—which means x-rays as well as the various radiations from radioactive nuclei—is harmful to living tissue, and exposure to it should be kept to a minimum. In conflict with this aim is the desirability of many processes that involve ionizing radiation, from the use of x-rays in diagnosis and therapy to the operation of nuclear reactors. The most insidious aspect of the problem is the delay between an exposure and certain of the consequences it may have, among them leukemia in the case of the person irradiated and genetic defects in his or her descendants. Particularly unfortunate is the widespread abuse of diagnostic x-rays. For example, careful studies have shown that children whose mothers were x-rayed while pregnant develop cancer more often than those whose mothers were not, yet many pregnant women are still x-rayed as a matter of routine rather than solely for specific cause. Another example is the mass screening by x-ray of symptomless young women for breast cancer, which many experts feel has increased the overall rate of cancer mortality rather than decreased it. The only dosage of ionizing radiation that is unquestionably harmless is zero, and the benefits claimed for any irradiation, whether of a person or of the public in general, must be weighed against the hazard that accompanies it.

The SI unit of radiation dosage is the *gray*, where 1 Gy is equal to 1 J of energy absorbed per kg of target material. The gray is a large unit and the *rad*, equal to 0.01 Gy, is more widely used. The dosage in such units is insufficient as an index of the biological effectiveness of a certain irradiation, however: radiations of different kinds and different energies do not have the same effects on a given tissue, and different tissues respond differently to the same radiation. The relative biological effectiveness (RBE) of a particular target material is therefore an important quantity, even though hard to specify with precision. The RBE of 250 KeV x-rays is taken as 1, and a dosage of 1 rad of such x-rays to a person is considered 1 *rem* (rad equivalent man). The RBEs of other x-rays and of gamma and beta rays are close to 1, but the RBE may be as much as 20 for other radiations, for instance alpha particles.

A radiation dosage of 400 rem is considered lethal in the sense that a person receiving such a dosage in a short time has only a 50 percent likelihood of survival. At the other extreme, the average lifetime dosage from such unavoidable natural sources as cosmic rays and radioactive materials in the earth and in the body itself is less than 10 rem, or about 0.125 rem/year. The recommended maximum dosage rate for people who are exposed to radiation in the course of their work is 5 rem/year, and for the general public it is 0.5 rem/year. These figures, in particular the latter one, are not free from dispute — even if such dosage rates cannot be unambiguously associated with specific maladies, their genetic consequences are without question, and an acceleration of the mutation rate is not something to be embarked upon without considerable thought on the part of all sections of the community.

Important Terms

The **atomic number** of an element is the number of electrons in each of its atoms or, equivalently, the number of protons in each of its atomic nuclei.

The **neutron** is an electrically neutral particle, slightly heavier than the proton, which is present in nuclei together with protons. Neutrons and protons are jointly called **nucleons**. The **mass number** of a nucleus is the number of nucleons it contains.

Isotopes of an element have the same atomic number but different mass numbers. Isotopic symbols following the pattern

$$_Z^A X,$$

where X is the chemical symbol of the element, Z its atomic number, and A the mass number of the particular isotope.

The **binding energy** of a nucleus is the energy equivalent of the difference between its mass and the sum of the masses of its individual constituent nucleons. This amount of energy must be supplied to the nucleus if it is to be completely disintegrated.

Radioactive nuclei spontaneously transform themselves into other nuclear species with the emission of radiation. The radiation may consist of **alpha particles**, which are the nuclei of helium atoms, or **beta particles**, which are positive or negative electrons. The emission of **gamma rays**, which are energetic photons, enables an excited nucleus to lose its excess energy. Positive electrons are known as **positrons**.

The time required for half of a given sample of a radioactive substance to decay is called its **half-life**.

In **nuclear fission**, the absorption of neutrons by certain heavy nuclei causes them to split into smaller **fission fragments**. Because each fission also liberates several neutrons, a rapidly multiplying sequence of fissions called a **chain reaction** can occur if a sufficient amount of the proper material is assembled. A **nuclear reactor** is a device in which a chain reaction can be initiated and controlled.

In **nuclear fusion**, two light nuclei combine to form a heavier one with the emission of energy. The energy liberated in the process when it takes place on a large scale is called **thermonuclear energy**.

Multiple Choice

1. The chemical behavior of an atom is determined by its
   a. atomic number.
   b. mass number.
   c. binding energy.
   d. number of isotopes.

2. The atomic number of a nucleus is equal to the number of
   a. electrons it contains.
   b. protons it contains.
   c. neutrons it contains.
   d. nucleons it contains.

3. The mass number of a nucleus is equal to the number of
   a. electrons it contains.
   b. protons it contains.
   c. neutrons it contains.
   d. nucleons it contains.

4. The mass number of a nucleus is
   a. always less than its atomic number.
   b. always more than its atomic number.
   c. sometimes equal to its atomic number.
   d. sometimes less than and sometimes more than its atomic number.

5. Atoms whose atomic numbers are the same but whose mass numbers are different are called
   a. alpha particles.
   b. ions.
   c. radioactive.
   d. isotopes.

6. The number of neutrons in the nucleus of the aluminum isotope $^{27}_{13}$Al is
   a. 13.
   b. 14.
   c. 27.
   d. 40.

7. Relative to the sum of the masses of its constituent nucleons, the mass of a nucleus is
   a. greater.
   b. the same.
   c. smaller.
   d. sometimes greater and sometimes smaller.

8. The ionization energy of an atom relative to the binding energy of its nucleus is
   a. greater.
   b. the same
   c. smaller.
   d. sometimes greater and sometimes smaller.

9. The mass of a $^{7}_{3}$Li nucleus is 0.042 u less than the sum of the masses of 3 protons and 4 neutrons. The binding energy per nucleon in $^{7}_{3}$Li is
   a. 5.6 MeV
   b. 10 MeV.
   c. 13 MeV.
   d. 39 MeV.

10. The binding energy per nucleon is
    a. the same for all nuclei.
    b. greatest for very small nuclei.
    c. greatest for nuclei of intermediate size.
    d. greatest for very large nuclei.

11. In a stable nucleus, the number of neutrons is always
    a. less than the number of protons.
    b. less than or equal to the number of protons.
    c. more than the number of protons.
    d. more than or equal to the number of protons.

12. The strong nuclear interaction
    a. is effective only over short distances.
    b. is stronger between neutrons than between protons.
    c. is stronger between protons than between neutrons.
    d. binds electrons to nuclei to form atoms.

13. As a sample of a radioactive isotope decays, its half life
    a. decreases.   b. remains the same.
    c. increases.
    d. any of the above, depending upon the isotope.

14. After 2 hours, $\frac{1}{16}$ of the initial amount of a certain radioactive isotope remains undecayed. The half-life of the isotope is
    a. 15 min.   b. 30 min.
    c. 45 min.   d. 60 min.

15. When a nucleus undergoes radioactive decay, its new mass number is
    a. always less than its original mass number.
    b. always more than its original mass number.
    c. never less than its original mass number.
    d. never more than its original mass number.

16. A nucleus with an excess of neutrons may decay radioactively with the emission of
    a. a neutron.   b. a proton.
    c. an electron.   d. a positron.

17. A positron has the same mass as
    a. an electron.
    b. a proton.
    c. a neutron.
    d. an alpha particle.

18. The fundamental interaction with the least significance in nuclear physics is the
    a. gravitational interaction.
    b. electromagnetic interaction.
    c. strong nuclear interaction.
    d. weak interaction.

19. The weakest of the four fundamental interactions is the
    a. gravitational interaction.
    b. electromagnetic interaction.
    c. strong nuclear interaction.
    d. weak interaction.

20. The first nuclear reaction ever observed occurred when $^{14}_{7}$N was bombarded with alpha particles and protons were ejected. This reaction produces
    a. $^{17}_{7}$N.   b. $^{17}_{8}$O.
    c. $^{17}_{9}$F.   d. $^{17}_{10}$Ne.

21. A reaction between a proton and $^{11}_{5}$B that produces $^{11}_{6}$C must also liberate
    a. a proton.
    b. a neutron.
    c. an electron.
    d. an alpha particle.

22. The process by which a heavy nucleus splits into two lighter nuclei is known as
    a. fission.   b. fusion.
    c. alpha decay.   d. a chain reaction.

23. The sun's energy comes from
    a. nuclear fission.
    b. radioactivity.
    c. the conversion of hydrogen to helium.
    d. the conversion of helium to hydrogen.

24. By "chain reaction" is meant
    a. the joining together of protons and neutrons to form atomic nuclei.
    b. the joining together of light nuclei to form heavy ones.
    c. the successive fissions of heavy nuclei induced by neutrons emitted in the fissions of other heavy nuclei.
    d. the burning of uranium in a special type of furnace called a nuclear reactor.

25. Enriched uranium is a better fuel for nuclear reactors than natural uranium because it has a greater proportion of
    a. slow neutrons.   b. deuterium.
    c. $^{235}_{92}$U.   d. $^{238}_{92}$U.

26. From the point of view of a power-plant engineer, a nuclear reactor is a source of
    a. heat.   b. electric current.
    c. slow neutrons.   d. gamma rays.

Exercises

1. In what ways are the isotopes of an element similar to one another? In what ways are they different?

2. State the number of neutrons and protons in each of the following nuclei: $^{6}_{3}$Li; $^{13}_{6}$C; $^{31}_{15}$P; $^{94}_{40}$Zr; $^{137}_{56}$Ba.

3. State the number of neutrons and protons in each of the following nuclei: $^{10}_{5}$Be; $^{22}_{10}$Ne; $^{36}_{16}$S; $^{88}_{38}$Sr; $^{180}_{72}$Hf.

4. What advantage is there in using neutrons as bombarding particles to investigate nuclear reactions?

5. Where does the energy liberated in nuclear fission come from? In nuclear fusion?

6. Which nucleus would you expect to be more stable, $^{7}_{3}$Li or $^{8}_{3}$Li? $^{9}_{4}$Be or $^{10}_{4}$Be?

7. What limits the size of a stable nucleus?

8. Why do stable nuclei never have more protons than neutrons?

9. The gravitational interaction alone governs the motions of the planets about the sun. Explain why the other fundamental interactions are not significant in planetary motion.

10. Why are the conditions in the interior of a star favorable for nuclear fusion reactions?

11. What are the differences and similarities between nuclear fusion and nuclear fission?

12. What happens to the atomic number and mass number of a nucleus that emits a gamma ray? What happens to the actual mass of the nucleus?

13. Radium undergoes spontaneous decay into helium and radon. Why is radium regarded as an element rather than as a chemical compound of helium and radon?

14. How many disintegrations per second occur in a 25 millicurie sample of thorium?

15. Under what circumstances does a nucleus emit an electron? A positron?

16. The nuclei $^{14}_{8}O$ and $^{19}_{8}O$ both undergo beta decay in order to become stable nuclei. Which would you expect to emit a positron and which an electron?

17. The nucleus $^{11}_{6}C$ undergoes positive beta decay. What is the symbol of the resulting daughter nucleus?

18. The nucleus $^{233}_{90}Th$ undergoes two negative beta decays in becoming an isotope of uranium. What is the symbol of the isotope?

19. The nucleus $^{238}_{92}U$ decays into a lead isotope through the successive emissions of eight alpha particles and six electrons. What is the symbol of the resulting lead nucleus?

20. Sixty hours after a sample of the beta emitter $^{24}_{11}Na$ has been prepared, only 6.25% of it remains un-decayed. What is the half-life of this isotope?

21. The half-life of radium is 1600 years. How long will it take for $\frac{15}{16}$ of a given sample of radium to decay?

22. Every living thing on the earth has a certain very small proportion of the radioactive carbon isotope $^{14}_{6}C$ mixed with the stable isotope $^{12}_{6}C$. The radio-carbon is produced in the earth's atmosphere by the action of cosmic rays on nitrogen atoms. When a plant or animal dies, it no longer takes in radiocarbon atoms, while the radiocarbon atoms

it already has continually beta decay to $^{14}_{7}N$ with a half life of 5600 years. It is therefore possible to determine the time that has passed since the death of a plant or animal by measuring the ratio between its radiocarbon and ordinary carbon contents. How old is a piece of charcoal from the remains of an ancient campfire if its relative radiocarbon content is $\frac{1}{4}$ that of a modern specimen?

23. Complete the following nuclear reactions:

$$^{6}_{3}Li + ? \rightarrow ^{7}_{4}Be + ^{1}_{0}n$$
$$^{10}_{5}B + ? \rightarrow ^{7}_{3}Li + ^{4}_{2}He$$
$$^{35}_{17}Cl + ? \rightarrow ^{32}_{16}S + ^{4}_{2}He$$

24. A nucleus of $^{15}_{7}N$ is struck by a proton. A nuclear reaction takes place with the emission of (a) a neutron, or (b) an alpha particle. Give the atomic number, mass number, and chemical name of the remaining nucleus in each of the above cases.

25. A nucleus of $^{9}_{4}Be$ is struck by an alpha particle. A nuclear reaction takes place with the emission of a neutron. Give the atomic number, mass number, and chemical name of the resulting nucleus.

26. A reaction often used to detect neutrons occurs when a neutron strikes a $^{10}_{5}B$ nucleus, with the subsequent emission of an alpha particle. What is the atomic number, mass number, and chemical name of the remaining nucleus?

Problems

1. The mass of $^{4}_{2}He$ is 4.002603 u. What is its binding energy? What is its binding energy per nucleon?

2. The mass of $^{20}_{10}Ne$ is 19.99244 u. What is its binding energy? What is its binding energy per nucleon?

3. The mass of $^{35}_{17}$Cl is 34.96885 u. What is its binding energy? What is its binding energy per nucleon?

4. The atomic masses of $^{15}_{7}$N, $^{15}_{8}$O, and $^{16}_{8}$O are respectively 15.0001, 15.0030, and 15.9949 u. (a) Find the average binding energy per nucleon in $^{16}_{8}$O. (b) How much energy is needed to remove one proton from $^{16}_{8}$O? (c) How much energy is needed to remove one neutron from $^{16}_{8}$O? (d) Why are these figures different from one another?

5. The neutron decays in free space into a proton and an electron. What must be the minimum binding energy contributed by a neutron to a nucleus in order that the neutron not decay inside the nucleus? How does this figure compare with the observed binding energies per nucleon in stable nuclei?

6. The electric potential energy of two protons a distance r apart is ke^2/r. Compare the electric potential energy of two protons 6×10^{-15} m apart with the binding energy per nucleon of nuclei with $A \approx 60$. Such nuclei are about 6×10^{-15} m in radius.

7. The distance between the two protons in a $^3_2$He nucleus is roughly 1.7×10^{-15} m. (a) Calculate the electric potential energy of these protons. (b) Show that this energy is of the right order of magnitude to account for the difference in binding energy between $^3_1$H and $^3_2$He. What conclusion can be drawn from this result about the dependence of nuclear forces upon electric charge? The atomic mass of $^3_1$H is 3.016050 u and that of $^3_2$He is 3.016030 u.

8. Compare the density of the $^1_1$H atom, assuming it to be a sphere whose radius is equal to that of the first Bohr orbit, with the density of its nucleus. The volume of a sphere of radius r is $\frac{4}{3}\pi r^3$.

9. The nuclear reaction
$$^6_3\text{Li} + ^2_1\text{H} \rightarrow 2\ ^4_2\text{He}$$
evolves 22.4 MeV Calculate the mass of $^6_3$Li in u.

10. The isotope $^{232}_{92}$U (mass 232.0372 u) alpha decays into $^{228}_{90}$Th (mass 228.0287 u). (a) Find the amount of energy released in the decay. (b) The alpha particle emitted in the decay of $^{232}_{92}$U is observed to have a kinetic energy of 5.3 MeV. If this is not the same as the answer to (a), account for the difference. (c) Is it possible for $^{232}_{92}$U to decay into $^{231}_{92}$U (mass 231.0364 u) by emitting a neutron? Why? (d) Is it possible for $^{232}_{92}$U to decay into $^{231}_{91}$Pa (mass 231.0359 u) by emitting a proton? Why?

11. If each fission in $^{235}_{92}$U releases 200 MeV, how many fissions must occur per second to produce a power of 1 kW?

12. How much mass is lost per day by a nuclear reactor operated at a 10 megawatt (10×10^6 W) power level?

13. $^{235}_{92}$U loses about 0.1% of its mass when it undergoes fission. (a) How much energy is released when 1 kg of $^{235}_{92}$U undergoes fission? (b) One ton of TNT releases about 10^6 kcal when it is detonated. How many tons of TNT are equivalent in destructive power to a bomb that contains 1 kg of $^{235}_{92}$U?

14. A body of mass m_1 and velocity v_1 collides elastically with a stationary body of mass m_2. As discussed in the Special Topic of Chapter 8, the ratio between the energy imparted to m_2 and the original kinetic energy of m_1 is $4\ (m_2/m_1)/(1 + m_2/m_1)^2$. In a nuclear reactor the fast neutrons produced during fission are slowed down by elastic collisions with the atomic nuclei of the "moderator." As mentioned in the text, two substances often used as moderators are deuterium, whose nuclei have masses approximately double that of the neutron, and carbon, whose nuclei have masses approximately twelve times that of the neutron. (These substances are used because they have little tendency to absorb neutrons.) With the help of the above formula, find the percentage of the initial energy lost by a neutron colliding head-on

with (1) a stationary deuterium nucleus, and (2) a stationary carbon nucleus.

15. In certain stars three alpha particles join in succession to form a $^{12}_{6}$C nucleus. The mass of $^{12}_{6}$C is 12.0000 u. How much energy is evolved in this reaction?

16. Find the temperature of a deuterium plasma (ionized gas) such that the average kinetic energy of a deuteron is enough to bring it to within 3×10^{-15} m of another deuteron despite their electric repulsion, which is close enough for them to undergo a nuclear reaction.

Answers to Multiple Choice

1. a	10. c	19. a
2. b	11. d	20. b
3. d	12. a	21. b
4. c	13. b	22. a
5. d	14. b	23. c
6. b	15. d	24. c
7. c	16. c	25. c
8. c	17. a	26. a
9. a	18. a	

33

Elementary Particles

Ordinary matter is composed of neutrons, protons, and electrons. Another significant elementary particle is the neutrino, which is emitted along with the electron in beta decay. All these particles have antiparticle counterparts whose mass is the same but whose charge (and other properties) differ. When a particle and its antiparticle come together, they both disappear as their mass is turned into energy. A great many other elementary particles have been discovered in recent years, all of which are unstable and decay rapidly after being created in high-energy collisions between other elementary particles. These short-lived particles are not fully understood at present, though it seems probable that they should not be considered as true "elementary particles" but rather as composites of a small number of basic particles called quarks.

33-1 The Neutrino

In radioactive decay, as in all other natural processes, energy (including mass energy) is conserved. For this reason the total mass of the products of a particular decay must be less than the mass of the initial nucleus, with the missing mass appearing as photon energy in the case of gamma decay and as kinetic energy in the cases of alpha and beta decay. In gamma and alpha decay the liberated energy is indeed precisely equal to the energy equivalent of the lost mass, but in beta decay a strange effect occurs: instead of all having the same energy, the emitted electrons from a particular isotope exhibit a variety of energies. These energies range from zero up to a maximum figure equal to the energy equivalent of the missing mass in the transformation. This effect is illustrated in Fig. 33-1, which shows the spread in electron energy in the decay of $^{210}_{83}$Bi.

Momentum as well as energy is apparently not conserved in beta decay. When an object at rest disintegrates into two parts, they must move apart in opposite directions in order that the total momentum of the system remain zero. Experiments show, however, that the emitted electron and the residual nucleus do *not* in general travel in opposite directions after beta decay occurs, so that their momenta cannot cancel out to equal the initial momentum of zero.

A third difficulty concerns angular momentum. The spin angular momentum component in a given direction of the neutron, the proton, the electron, and the positron is in every case $\pm\frac{1}{2}h/2\pi$. The conversion of a neutron into a proton and an electron or of a proton into a neutron and a positron therefore leaves an angular momentum discrepancy of $\frac{1}{2}h/2\pi$.

The process of beta decay is the first one we have encountered in which the conservation laws of energy, linear momentum, and angular momentum

Energy, momentum, and angular momentum discrepancies in beta decay

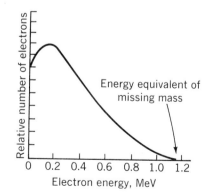

Fig. 33-1. The distribution of electron energies found in the beta decay of $^{210}_{83}$Bi. The maximum electron energy is equal to the energy equivalent of the mass lost by the decaying nucleus minus the electron mass.

The neutrino

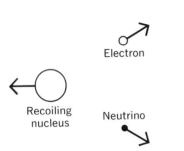

Electron

Recoiling
nucleus   Neutrino

Fig. 33-2.   An electron and a neutrino are simultaneously emitted in the beta decay of a nucleus, which makes possible the conservation of both energy and momentum in the process.

do not seem to hold. To account for the above discrepancies without abandoning three of the most fundamental and otherwise well-established physical principles, the existence of a new particle was postulated, the *neutrino*, symbol ν_e (Greek letter "nu"). The neutrino has no electric charge and no mass, but is able to possess both energy and momentum and has an intrinsic spin. (Lest this seem unlikely, we might reflect that the photon, also massless, has energy, momentum, and angular momentum. The neutrino is *not* a photon, however, but an entirely different entity.) According to the neutrino theory, an electron and neutrino are simultaneously emitted in beta decay, which permits energy and momentum to be conserved (Fig. 33-2).

For a quarter of a century the neutrino hypothesis was accepted despite the absence of any direct evidence in its support. This was a very unusual situation: here was a particle of a rather odd kind, which nobody had ever detected experimentally, yet practically no physicists doubted that it really existed. It was far from being a matter of blind faith, however. There were no theoretical objections to the neutrino hypothesis, whereas there were strong theoretical and experimental objections to dropping the principles of conservation of energy, linear momentum, and angular momentum. Furthermore, the neutrino, which lacks mass and charge and is not electromagnetic in nature as is the photon, interacts only feebly with matter, so it was not easy to think of a way to detect it.

Discovery of the neutrino

Finally, in 1956, an experiment was performed in which a nuclear reaction that, in theory, could only be caused by a neutrino, was actually found to take place. In the operation of a nuclear reactor, a great many beta decays take place, and as a result more than 10^{16} neutrinos per second may emerge from each square meter of the shielding around a reactor. A neutrino striking a proton has a small probability of inducing the reaction

$$\nu_e + \mathrm{p} \rightarrow \mathrm{n} + \mathrm{e}^+$$

in which a neutron and positron are created. By placing a sensitive detecting chamber containing hydrogen near a nuclear reactor, the simultaneous appearance of a neutron and a positron could be registered each time the above reaction occurred. Calculations were made initially of how many such reactions per second should occur based on the known properties of the detecting apparatus and on the theoretical properties of the neutrino. When this reaction rate was actually found, there was no doubt that neutrinos indeed exist.

Only the weak interaction affects neutrinos

Neutrinos are able to travel unimpeded through vast amounts of matter because they are limited to the weak interaction. On the average, a neutrino must traverse 130 light-years of solid iron before being absorbed—and a light-year, the distance light goes in a year, is 9.5×10^{15} m.

In the sequence of nuclear reactions by which hydrogen is converted to helium in the sun and other stars, two beta decays occur for each helium nucleus formed. A vast number of neutrinos is therefore produced in the sun at all times. Because neutrinos travel freely through matter, almost all of these neutrinos escape into space and ought to take with them 6 to 8% of the total energy generated by the sun. The flux of neutrinos from the sun is such that every cubic inch on the earth should contain perhaps 100 neutrinos at any moment, although this has not yet been confirmed by experiment. The energy carried by the neutrinos created in the sun and the other stars is apparently lost forever from the universe in the sense that it is no longer available for conversion into other forms of energy, such as matter.

33-2 Antiparticles

Another recent experimental discovery of a particle whose existence had been predicted theoretically decades ago is that of the negative proton, or *antiproton*, whose symbol is $\bar{p}$. This is a particle with the same properties as the proton except that it has a negative electric charge. The existence of antiprotons was predicted largely on the basis of symmetry arguments: since the electron has a positive counterpart in the positron, why should the proton not have a negative counterpart as well? Actually, as sophisticated theories show, this is an excellent argument, and few physicists were surprised when the antiproton was actually found.

The reason positrons and antiprotons are so difficult to find is that they are readily *annihilated* upon contact with ordinary matter (Fig. 33-3). When a positron is in the vicinity of an electron, they attract one another electrically, come together, and then both vanish simultaneously, with their missing mass appearing in the form of two gamma-ray photons:

$$e^+ + e^- \rightarrow \gamma + \gamma.$$

The total mass of the two particles is the equivalent of 1.02 MeV, and so each photon has an energy of 0.51 MeV. (Their energies must be equal and they must be emitted in opposite directions in order that momentum be conserved.)

While the similar reaction

$$p + \bar{p} \rightarrow \gamma + \gamma$$

can occur when a proton and antiproton undergo annihilation, it is more usual for the vanished mass to reappear in the form of several mesons, particles which we shall consider in the next section.

The antiproton

Annihilation of matter

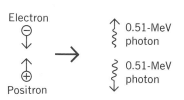

Fig. 33-3. The mutual annihilation of an electron and a positron.

Fig. 33-4. The production of (a) an electron-positron pair and (b) a proton-antiproton pair by the materialization of sufficiently energetic photons. Pair production can occur only in the presence of a nucleus.

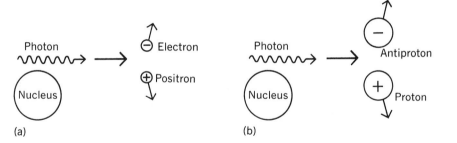

(a)   (b)

The reverse of annihilation can also take place, with the electromagnetic energy of a photon materializing into a positron and an electron or, if it is energetic enough, into a proton and an antiproton (Fig. 33-4). This phenomenon is known as *pair production,* and requires the presence of a nucleus in order that momentum as well as energy be conserved. Any photon energy in excess of the amount required to provide the mass of the created particles (1.02 MeV for a positron-electron pair, 1872 MeV for a proton-antiproton pair) appears as kinetic energy.

Pair production

Antineutrons (symbol ñ) and antineutrinos (symbol $\bar{\nu}_e$) have also been identified. Antineutrons can be detected through their mutual annihilation with neutrons, while more indirect, though equally definite, evidence supports the existence of antineutrinos. The antineutrino differs from the neutrino in that, while the spin axes of both are parallel to their directions of motion, the spin of the former is clockwise and that of the latter is counterclockwise when viewed from behind. A moving neutrino may be thought of as resembling a left-handed screw, and a moving antineutrino as resembling a right-handed screw. An antineutrino is released during a beta decay in which an electron is emitted, and a neutrino is released during a beta decay in which a positron is emitted. Thus the fundamental equations of beta decay are

Antineutrons and antineutrinos

$$p \rightarrow n + e^+ + \nu_e, \qquad (33\text{-}1)$$

$$n \rightarrow p + e^- + \bar{\nu}_e. \qquad (33\text{-}2)$$

Annihilation and pair production are consequences of the facts that matter is a form of energy and that conversions from matter to energy and from energy to matter are no more improbable than conversions from, say, gravitational potential energy to kinetic energy.

Ordinary atoms are composed of neutrons, protons, and electrons. There is apparently no reason why atoms composed of antineutrons, antiprotons, and positrons should not be stable and behave in every way like ordinary atoms. It is an attractive notion that equal amounts of matter and *antimatter*

Antimatter

came into being at the origin of the universe which became segregated into separate galaxies. The spectra of the light emitted by the members of anti-matter galaxies would be exactly the same as the spectra of the light emitted by the members of galaxies of ordinary matter, which gives us no way to distinguish between the two. Of course, if antimatter comes in contact with ordinary matter, their mutual annihilation occurs with the release of a great deal of energy, and it is possible that certain astrophysical phenomena derive their energy from this source.

33-3 Meson Theory of Nuclear Forces

In 1935 the Japanese physicist Yukawa suggested that the strong nuclear interaction could be regarded as the result of an interchange of certain particles between nucleons. Today these particles are called *pions*. Pions may be charged or neutral; those with charges of $+e$ or $-e$ have rest masses of 273 times the electron mass, while neutral pions have rest masses of 264 times the electron mass. Pions are members of a class of elementary particles collectively called *mesons*—the word pion is a contraction of the original name π-meson.

Pions

The crude analogy illustrated in Fig. 33-5 may help in understanding how meson exchange can lead to both attractive and repulsive forces between nuclei. Each child in the figure has a pillow. When the children exchange pillows by snatching them from each other's grasp, the effect is like that of a mutually attractive force. On the other hand, the children may also exchange pillows by throwing them at each other. Here conservation of momentum requires that the children move apart, just as if a repulsive force were present between them.

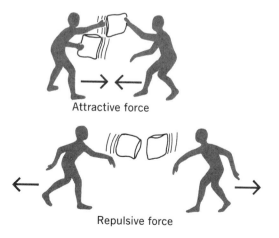

Attractive force

Repulsive force

Fig. 33-5. Particle exchange can lead to attractive or repulsive forces.

The uncertainty principle permits the temporary nonconservation of energy

According to Yukawa's theory, nearby nucleons constantly exchange mesons without themselves being altered. We note that the emission of a meson by a nucleon at rest which does not lose a corresponding amount of mass violates the law of conservation of energy. However, the law of conservation of energy, like all physical laws, deals only with measurable quantities. Because the uncertainty principle restricts the accuracy with which we can perform certain measurements, it limits the range of application of physical laws such as that of energy conservation.

In Chapter 28 the origin of the uncertainty principle in the form

$$\Delta x\ \Delta mv \geqslant \frac{h}{2\pi}$$

was discussed. This formula states that the product of the uncertainty in the position of a body and the uncertainty in its momentum cannot be less than $h/2\pi$, where h is Planck's constant. There is another form of the uncertainty principle that relates the uncertainty ΔE in an energy measurement with the uncertainty Δt in the time when the energy measurement is made. This form of the uncertainty principle states that

Alternate form of the uncertainty principle

$$\Delta E\ \Delta t \geqslant \frac{h}{2\pi}. \tag{33-3}$$

We conclude that a process can take place in which energy is *not* conserved by an amount ΔE *provided* that the time interval Δt in which the process takes place is not more than $h/2\pi\Delta E$. Thus the creation, transfer, and dis-

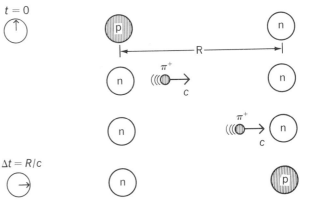

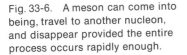

Fig. 33-6.   A meson can come into being, travel to another nucleon, and disappear provided the entire process occurs rapidly enough.

appearance of a meson do not conflict with the conservation of energy provided the sequence takes place fast enough. The latter condition provides a way to estimate the mass of the pion.

Let us assume that the temporary energy discrepancy ΔE is of the same magnitude as the rest energy mc^2 of the pion, and that the pion travels at very nearly the speed of light c as it goes from one nucleon to another. (These assumptions are crude because the kinetic energy of the pion is ignored, but all we are after is an approximate figure for m.) The time Δt the pion spends between its creation in one nucleon and its absorption in another cannot be greater than R/c, where R is the maximum distance that can separate interacting nucleons (Fig. 33-6).

We therefore have, using the symbol $\approx$ to indicate that the result is only a rough approximation,

$$\Delta E \Delta t \approx \frac{h}{2\pi},$$

$$mc^2 \times \frac{R}{c} \approx \frac{h}{2\pi},$$

$$m \approx \frac{h}{2\pi R c}. \tag{33-4}$$

Estimate of pion mass

The strong nuclear interaction responsible for the attractive forces between nucleons has a range of about 1.7×10^{-15} m. When we substitute

$$R = 1.7 \times 10^{-15} \text{ m}$$

in the above formula, we obtain $m \approx 2.1 \times 10^{-28}$ kg for the pion mass, which is about 230 electron masses. Of course, the preceding calculation is hardly a rigorous one, but if Yukawa's theory has any validity, the pions he postulated should have masses somewhere in this vicinity — as they do.

33-4 Pions and Muons

Not long after Yukawa's work, charged particles of about the right mass were experimentally discovered in the cosmic radiation. Their discovery was not unexpected because a sufficiently energetic nuclear collision should be able to liberate mesons by providing enough energy to create them without violating conservation of energy, and nuclear collisions between fast

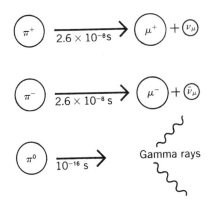

Fig. 33-7. Pion decay.

cosmic-ray protons from space and oxygen and nitrogen nuclei occur constantly in the atmosphere. However, these particles did not behave at all in the way they were expected to behave. Far from strongly interacting with nuclei, as Yukawa's mesons were supposed to do, they barely interacted at all. Instead of being absorbed in at most a meter of earth, they penetrated thousands of meters into the ground.

Finally, in 1947, the explanation for their peculiar behavior was found. The weakly interacting mesons, known as *muons* (a contraction of μ-meson, where μ is the Greek letter "mu") and found profusely in cosmic rays at sea level, are not the direct products of nuclear collisions, but are secondary particles that result from the decay of pions.

Outside a nucleus a charged pion decays in an average of 2.6×10^{-8} s in a muon and a neutrino (or antineutrino) plus kinetic energy (Fig. 33-7). That is,

Charged pion decay
$$\pi^+ \rightarrow \mu^+ + \nu_\mu, \tag{33-5}$$

$$\pi^- \rightarrow \mu^- + \bar{\nu}_\mu. \tag{33-6}$$

A small proportion of charged pions also decay directly into an electron or positron plus a neutrino or antineutrino. Neutral pions, whose masses of 264 m_e are a little less than the charged pion mass of 273 m_e, decay in about 10^{-16} s into a pair of gamma-ray photons:

Neutral pion decay
$$\pi^0 \rightarrow \gamma + \gamma. \tag{33-7}$$

The neutrinos involved in pion decay have been denoted ν_μ, whereas those

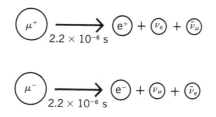

Fig. 33-8. Muon decay.

involved in beta decay have been denoted ν_e. Until 1962 it had been thought
that there is only a single kind of neutrino. In that year an experiment was
performed in which pions were produced by bombarding a metal target with
high-energy protons from an accelerator. The pion decays liberated neutrinos, and the interactions of these neutrinos with matter was studied. The
only inverse reactions found led to the production of muons; no electrons
whatever were created. Hence the neutrinos set free in pion decay are different from those set free in beta decay.

Two kinds of neutrino

The muon, whose mass is 207 m_e, is less well understood than the pion.
The relatively long muon lifetime of 2.2×10^{-6} s and their feeble interaction
with matter account for the penetrating ability of these particles. In contrast to the pion, which fits in well with the theory of nuclear forces, the
muon seems to have no particular function in the scheme of things; why such
a particle should exist is an unsolved problem. The positive and negative
muons decay into positrons and electrons respectively together with a
neutrino-antineutrino pair in each case (Fig. 33-8).

33-5 Categories of Elementary Particles

Several hundred elementary particles have been discovered in recent
years in addition to the sixteen we have thus far discussed. Such abundance
where scarcity had been expected stimulated an intense research effort, with
the result that a number of suggestive regularities have been found in the
properties of the various particles. But as yet there is no theory of elementary particles comparable in ability to account for their existence and behavior to, say, the quantum theory of the atom. The search for such a theory
continues — perhaps more difficult, frustrating, and exciting than any other in
the history of science.

Table 33-1. The 37 longest-lived elementary particles. Particles subject to the strong nuclear interaction are called hadrons; the various mesons and baryons are hadrons.

Name	Particle	Anti-particle	Mass, in electron masses	Stability and chief decay mode	Average lifetime, s	Category
Photon	γ	(γ)	0	Stable		Photon
Neutrino	ν_e	$\bar{\nu}_e$	0	Stable		Leptons
	ν_μ	$\bar{\nu}_\mu$	0	Stable		
Electron	e^-	e^+	1	Stable		
Muon	μ^-	μ^+	207	Unstable; decays into electron plus two neutrinos	2.2×10^{-6}	
Pion	π^+	π^-	273	Unstable; decays into muon plus neutrino	2.6×10^{-8}	Mesons
	π^0	(π^0)	264	Unstable; decays into two gamma rays	8.9×10^{-17}	
K meson	K^+	K^-	966	Unstable; decays into muon plus neutrino or into two or three pions	1.2×10^{-3}	
	K_1^0	$\overline{K_1^0}$	974	Unstable; decays into two pions	8.7×10^{-11}	
	K_2^0	$\overline{K_2^0}$	974	Unstable; decays into three pions or into pion and neutrino plus muon or electron	5.3×10^{-8}	
Eta meson	η^0	(η^0)	1073	Unstable; decays into three pions or two gamma rays	10^{-18}	
Proton	p^+	p^-	1836	Stable		Baryons
Neutron	n^0	$\overline{n^0}$	1839	Unstable in free space; decays into electron, proton, and neutrino	1×10^3	
Lambda hyperon	Λ^0	$\overline{\Lambda^0}$	2182	Unstable; decays into pion plus neutron or proton	2.5×10^{-10}	
Sigma hyperon	Σ^+	$\overline{\Sigma^+}$	2328	Unstable; decays into pion plus neutron or proton	8×10^{-11}	
	Σ^-	$\overline{\Sigma^-}$	2341	Unstable; decays into neutron plus pion	1.6×10^{-10}	
	Σ^0	$\overline{\Sigma^0}$	2332	Unstable; decays into lambda hyperon plus gamma ray	10^{-14}	
Xi hyperon	Ξ^-	$\overline{\Xi^-}$	2583	Unstable; decays into lambda hyperon plus pion	1.7×10^{-10}	
	Ξ^0	$\overline{\Xi^0}$	2571	Unstable; decays into lambda hyperon plus pion	2.9×10^{-10}	
Omega hyperon	Ω^-	$\bar{\Omega}^-$	3290	Unstable; decays into lambda hyperon plus K meson or into xi hyperon plus pion	10^{-10}	

Table 33-1 is a list of the 37 longest-lived elementary particles, together with some of their properties. The other known particles have much briefer lifetimes, in most cases so short (10^{-22}–10^{-23} s) that their existence is inferred on the basis of indirect evidence. The particles in the table, together with the unlisted ones, fall into four categories: *photons, leptons, mesons,* and *baryons.* The photon, π^0 meson, and η^0 meson are their own antiparticles. Only the principal mode or modes of decay are given; some particles may decay in a variety of ways.

Four categories of elementary particles

Photons. As we know, photons are quanta of electromagnetic energy which are stable in space and have zero rest mass. It is possible to think of the electromagnetic interaction as being "carried" from one charge to another by the circulation of photons between them, just as the strong nuclear interaction is carried from one nucleon to another by the circulation of mesons between them.

Since both mesons and photons actually exist, it is tempting to ask whether the gravitational and weak interactions also might have particles associated with them. Since gravitational forces are unlimited in range, the hypothetical *graviton* should have zero rest mass to avoid violating the conservation of energy. (The same argument accounts for the zero rest mass of the photon.) Also like the photon, the graviton should be stable, electrically neutral, and travel with the speed of light. Unlike the photon, the graviton should interact only feebly with matter, and so would be extremely hard to detect. There is no definite experimental evidence either for or against the existence of the graviton.

The graviton

The quantum of the weak interaction is called the *intermediate boson.* Because the weak interaction has a shorter range than the strong nuclear interaction, the rest mass of the intermediate boson should be very large, perhaps 30 or 40 proton masses. It should be charged and decay rapidly. A number of careful searches has turned up no traces of the intermediate boson, but indirect evidence for it remains strong.

The intermediate boson

Leptons. Leptons include electrons, muons, and neutrinos. There are two families of leptons, one consisting of the electron, the positron, the ν_e neutrino, and the $\bar{\nu}_e$ antineutrino, and the other consisting of the positive and negative muon, the ν_μ neutrino, and the $\bar{\nu}_\mu$ antineutrino. A conservation law applies to each lepton family: in every known process, the number of leptons of each kind remains constant, reckoning a lepton as + and an antilepton as −.

Lepton conservation

An easy way to apply these conservation laws is to assign the quantum numbers L and M to the various leptons as follows:

$L = +1$	$L = -1$	$M = +1$	$M = -1$
e^-	e^+	μ^-	μ^+
ν_e	$\bar{\nu}_e$	ν_μ	$\bar{\nu}_\mu$

All other particles have $L = 0$ and $M = 0$. Here are a few processes that involve leptons, showing that L and M have the same values before and after:

Pion decay $\quad \pi^- \rightarrow \mu^- + \bar{\nu}_\mu$
$$M = 0 \quad + 1 \quad - 1$$

Muon decay $\quad \mu^- \rightarrow e^- + \nu_\mu + \nu_e$
$$L = 0 \quad + 1 \quad \quad 0 \quad - 1$$
$$M = +1 \quad \quad 0 \quad + 1 \quad \quad 0$$

Pair production $\quad \gamma \rightarrow \quad e^- + e^+$
$$L = 0 \quad + 1 \quad - 1$$

Other conservation principles, such as those of energy, linear momentum, angular momentum, and electric charge, can be traced to certain symmetries in the natural world. What symmetries the conservation of L and M are associated with, if any, remain unknown.

Mesons. The role of mesons as carriers of the strong interaction that binds nucleons together into nuclei has been examined earlier. There seems to be no conservation principle associated specifically with the number of mesons and antimesons involved in a process, although a number of other conservation principles must be obeyed.

Baryons. Nucleons and heavier particles belong to the category of baryons. As in the case of the lepton families, a quantum number B can be assigned that is $+1$ for a baryon and -1 for an antibaryon, and it is found that in every process the value of B is conserved. An important example is the decay of the neutron, which also illustrates the conservation of L:

Baryon conservation

Neutron decay $\quad n^0 \rightarrow p^+ + e^- + \bar{\nu}_e$
$$L = 0 \quad \quad 0 \quad + 1 \quad - 1$$
$$B = +1 \quad + 1 \quad \quad 0 \quad \quad 0$$

This is the only possible way in which the neutron can decay and still conserve both energy (since the proton is the only lighter baryon) and quantum number B. The same reasoning explains why the proton is a stable particle: there is no way in which it can decay without violating either conservation of energy or conservation of B. As in the case of L and M, the ultimate significance of B is not known.

33-6 Quarks

The four leptons—the electron, the muon, and the neutrinos associated with each—seem to be truly elementary in nature, with no hint of internal structures or even of extensions in space. They are as close to being point particles as present measurements can establish. On the other hand, mesons and baryons, which are jointly called *hadrons* and are subject to the strong nuclear interaction, apparently have not only definite sizes (they are about 10^{-15} m across) but also internal structures of some kind.

What is known about the various hadrons suggests that they may be composite objects whose constituent particles are known as *quarks*. Quarks are thought to be elementary in the same sense as leptons, but unlike leptons they have never been isolated in experiments. The present status of quarks is much like that of neutrinos for 25 years after they were proposed: their existence is supported by a wealth of indirect evidence, but something in their basic character impedes their detection as independent entities. The most novel thing about quarks is that two of them are supposed to have charges of $-\frac{1}{3}e$ and the two others to have charges of $+\frac{2}{3}e$. Quarks have the baryon number $B = \frac{1}{3}$ and antiquarks the baryon number $B = -\frac{1}{3}$. Each baryon is made up of three quarks so that $B = 1$, each antibaryon of three antiquarks so that $B = -1$, and each meson of one quark and one antiquark, so that $B = 0$. Every known hadron can be accounted for on this basis, and every combination of quarks permitted by certain simple rules is associated with a known hadron. Thus the proton is believed to consist of two quarks of charge $+\frac{2}{3}e$ and one of charge $-\frac{1}{3}e$, for a total charge of $(+\frac{2}{3} + \frac{2}{3} - \frac{1}{3})e = +e$.

As originally formulated in 1963 by Murray Gell-Mann and, independently, by George Zweig, the quark hypothesis involved only three such particles, two of charge $-\frac{1}{3}e$ (the d and s quarks) and one of charge $+\frac{2}{3}e$ (the u quark). A year later another quark, also of charge $+\frac{2}{3}e$, was proposed largely by analogy with the existence of four leptons. This "charmed" quark, designated c, seemed to have no role to play until quite recently, when new hadrons were discovered that could not be accounted for on the basis of the original quarks.

A serious problem with the idea that hadrons are composed of quarks is that the presence of two quarks of the same kind in a particular particle (for instance, two u quarks in a proton) would violate the exclusion principle, to which quarks ought to be subject. To get around this problem it has been suggested that quarks have an additional property of some kind, known informally as "color," with three possible colors for each basic type of quark. This idea has turned out to be the key to explaining a number of otherwise puzzling experimental observations, such as the lifetime of the neutral pion.

So the quest for simplicity that led to the postulation of just three quarks has turned up a total of a dozen—which is still notable in view of the several hundred short-lived but different particles their combinations can account for. And it remains true that the properties of all ordinary matter can be understood on the basis of just two fundamental kinds of quark, u and d, plus two leptons, the electron and the ν_e neutrino, a truly astonishing achievement.

Important Terms

An **elementary particle** is one of the various types of particle found in nature that does not consist of a combination of other particles. The four major categories of elementary particle are the **photons**, the **leptons**, the **mesons**, and the **baryons**.

The **neutrino** is a massless, uncharged particle that is emitted in the decay of certain elementary particles. It can possess energy and both linear and angular momentum.

Nearly every elementary particle has an **antiparticle** counterpart whose electric charge and certain other properties have the opposite sign. Thus the antiparticle of the electron is the positron (charge $+e$) and that of the proton is the antiproton (charge $-e$). When a particle and an antiparticle of the same kind come together, they **annihilate** each other, with the vanished mass reappearing as photons or mesons.

Mesons and baryons, jointly known as **hadrons**, are thought to be composed of **quarks**, which are particles with fractional electric charges that have not as yet been experimentally isolated.

Multiple Choice

1. The strong interaction between nucleons is believed to arise through the exchange of
   a. neutrinos.   b. mesons.
   c. leptons.   d. baryons.

2. The particle whose properties are most nearly the same as those of the proton is the
   a. positron.   b. neutron.
   c. antiproton.   d. positive pion.

3. The neutrino does not possess
   a. charge.   b. linear momentum.
   c. angular momentum.
   d. energy.

4. The heaviest class of elementary particles consists of the
   a. photons.   b. leptons.
   c. mesons.   d. baryons.

5. The only one of the following particles that does not decay in free space is the
   a. proton.   b. neutron.
   c. muon.   d. pion.

6. The only one of the following particles that is not an elementary particle is the
   a. neutron.   b. pion.
   c. muon.   d. alpha particle.

7. The muon most closely resembles the
   a. pion.   b. electron.
   c. proton.   d. neutrino.

8. Of the following particles, the one that is its own antiparticle is the
   a. photon.   b. neutrino.
   c. neutron.   d. proton.

9. The mass of the neutrino is
   a. zero.
   b. equal to that of the electron.
   c. between that of the electron and that of the muon.
   d. greater than that of the muon.

10. A quantity not necessarily conserved in elementary-particle decays is
    a. charge.
    b. mass.
    c. lepton number.
    d. baryon number.

11. All leptons are
    a. stable.
    b. composed of quarks.
    c. subject to the weak interaction.
    d. subject to the strong interaction.

12. A particle unlikely to be composed of quarks is the
    a. muon.
    b. pion.
    c. proton.
    d. neutron.

Exercises

1. Discuss the similarities and differences between the photon and the neutrino.

2. Compare the distance traveled by a particle whose lifetime is 10^{-23} sec which is moving at nearly the speed of light with nuclear diameters.

3. A negative muon collides with a proton, and a neutron plus another particle are created. What is the other particle?

4. Can a neutron decay into a positive muon and an electron?

5. A proton can interact with a μ^--meson to form a neutron and another particle. What must this particle be?

6. Why does the Λ-hyperon not decay into a π^+- and a π^--meson?

7. Show that the laws of conservation of linear momentum and of energy make it impossible for a particle to decay into a single lighter particle.

8. According to the theory of the continuous creation of matter, the evolution of the universe can be traced to the spontaneous appearance of neutrons in free space. Which conservation laws would this process violate?

9. No particle of fractional charge has yet been observed. If none is found in the future either, does this necessarily mean that the quark hypothesis is wrong?

10. Must a neutrino be emitted in the process of electron capture by a nucleus? If so, what kind of neutrino is it?

2. Find the energy of each of the gamma-ray photons produced in the decay of a neutral pion at rest. Why must their energies be the same?

3. One way to account for the decay of the π^0 meson into two photons, which are electromagnetic quanta, despite the absence of either charge or magnetic moment is to assume that the π^0 first becomes a nucleon-antinucleon pair which can interact electromagnetically to produce two photons whose energies are consistent with the π^0 mass. (a) How long does the uncertainty principle permit such a nucleon-antinucleon pair to exist? (b) Do any conservation principles other than conservation of energy prohibit the transformation of a π^0 meson into a nucleon-antinucleon pair?

4. How much energy must a gamma-ray photon have if it is to materialize into a proton-antiproton pair with each particle having a kinetic energy of 10 MeV?

5. How much energy must a gamma-ray photon have if it is to materialize into a neutron-antineutron pair? Is this more or less than that required to form a proton-antiproton pair?

6. A 1-MeV positron collides head on with a 1-MeV electron, and the two are annihilated. What is the energy and wavelength of each of the resulting gamma-ray photons?

Problems

1. One proton strikes another, and the reaction

$$p + p \rightarrow n + p + \pi^+$$

takes place. What is the minimum energy the incident proton must have had? (Ignore momentum conservation.)

Answers to Multiple Choice

1. b	5. a	9. a
2. c	6. d	10. b
3. a	7. b	11. c
4. d	8. a	12. a

Appendix A
Useful Mathematics

A-1 Algebra

Algebra may be thought of as a generalized arithmetic in which symbols are used in place of specified numbers. The advantage of algebra is that with its help we can perform calculations without knowing in advance the numerical values of all the quantities involved. Sometimes algebra is no more than a help, perhaps by showing us how to simplify complex calculations, and sometimes it is the only way in which we can solve a problem.

The symbols of algebra are normally letters of the alphabet. If we have two quantities a and b and add them to give the sum c, we would write

$$a + b = c.$$

If we subtract b from a to give the difference d, we would write

$$a - b = d.$$

Multiplying a and b together to give e may be written $a \times b = e$, or more simply, as just

$$ab = e.$$

Whenever two algebraic quantities are written together with nothing between them, it is understood that they are to be multiplied together.

Dividing a by b to give the quotient f is usually written

$$\frac{a}{b} = f,$$

but it may sometimes be more convenient to write

$$a/b = f,$$

which has the same meaning.

Parentheses and brackets are used to show the order in which various operations are to be performed. Thus

$$\frac{(a + b)c}{d} - e = f$$

means that we are first to add a and b together, multiply their sum by c, then divide by d, and finally subtract e.

A-2 Equations

An *equation* is simply a statement that a certain quantity is equal to another one. Thus

$$7 + 2 = 9,$$

which contains only numbers, is an arithmetical equation, and

$$3x + 12 = 27,$$

which contains a symbol as well, is an algebraic equation. The symbols in an algebraic equation usually cannot have any arbitrary values if the equality is to hold. Finding the possible values of these symbols is called *solving* the equation. The *solution* of the latter equation above is

$$x = 5,$$

since only when x is 5 is it true that $3x + 12 = 27$.

In order to solve an equation a basic principle must be kept in mind.

Any operation performed on one side of an equation must be performed on the other.

An equation therefore remains valid when the same quantity, numerical or otherwise, is added to or subtracted from both sides, or when the same quantity is used to multiply or divide both sides. Other operations, for instance squaring or taking the square root, also do not alter the equality if the same thing is done to both sides. As a simple example, to solve $3x + 12 = 27$, we first subtract 12 from both sides:

$$3x + 12 - 12 = 27 - 12,$$

$$3x = 15.$$

To complete the solution we divide both sides by 3:

$$\frac{3x}{3} = \frac{15}{3},$$

$$x = 5.$$

To verify the correctness of a solution, we substitute it back in the original equation and see whether the equality is still true. Thus we can check that $x = 5$ by reducing the original algebraic equation to an arithmetical one:

$$3x + 12 = 27,$$

$$3 \cdot 5 + 12 = 27,$$

$$15 + 12 = 27,$$

$$27 = 27.$$

Problem. Solve the following equation for x:

$$\frac{3x - 42}{9} = 2(7 - x).$$

Solution. Our strategy is to assemble all the terms involving the unknown x on the left-hand side and all the purely numerical terms on the right-hand side. We begin by removing the parentheses on the right-hand side by actually multiplying $(7 - x)$ by 2. This yields

$$\frac{3x - 42}{9} = 2 \cdot 7 - 2x = 14 - 2x.$$

Next we multiply both sides by 9 to give

$$3x - 42 = 9(14 - 2x) = 126 - 18x.$$

We now add $42 + 18x$ to both sides:

$$3x - 42 + 42 + 18x = 126 - 18x + 42 + 18x,$$

$$3x + 18x = 126 + 42,$$

$$21x = 168.$$

The final step is to divide both sides by 21:

$$\frac{21x}{21} = \frac{168}{21},$$

$$x = 8.$$

To check this solution we substitute $x = 8$ in the original equation:

$$\frac{3 \cdot 8 - 42}{9} = 2(7 - 8),$$

$$\frac{24 - 42}{9} = 14 - 16,$$

$$\frac{-18}{9} = -2,$$

$$-2 = -2.$$

Two simple rules follow directly from the principle stated above. The first is,

Any term on one side of an equation may be transposed to the other side by changing its sign.

To verify this rule, we subtract b from each side of the equation

$$a + b = c$$

to obtain

$$a + b - b = c - b,$$

$$a = c - b.$$

We see that b has disappeared from the left-hand side and $-b$ is now on the right-hand side.

The second rule is,

A quantity which multiplies one side of an equation may be transposed so as to divide the other side, and vice versa.

To verify this rule, we divide both sides of the equation

$$ab = c$$

by b. The result is

$$\frac{ab}{b} = \frac{c}{b},$$

$$a = \frac{c}{b}.$$

We see that b, a multiplier on the left-hand side, is now a divisor on the right-hand side.

Problem. Solve the following equation for x:

$$4(x - 3) = 7.$$

Solution. The above rules make this problem a simple one to deal with. We proceed as follows:

$$4(x - 3) = 7,$$

$$x - 3 = \tfrac{7}{4},$$

$$x = \tfrac{7}{4} + 3 = 1.75 + 4 = 4.75.$$

A-3 Exponents

It is often necessary to multiply a quantity by itself a number of times. This process is indicated by a superscript number called the exponent, according to the following scheme:

$$A = A^1,$$

$$A \times A = A^2,$$

$$A \times A \times A = A^3,$$

$$A \times A \times A \times A = A^4,$$

$$A \times A \times A \times A \times A = A^5.$$

We read A^2 as "A squared" because it is the area of a square of length A on a side; similarly A^3 is called "A cubed" because it is the volume of a cube each of whose sides is A long. More generally we speak of A^n as "A to the nth power." Thus A^5 is read as "A to the fifth power."

When we multiply a quantity raised to some particular power (say A^n) by the same quantity raised to another power (say A^m), the result is that quantity raised to a power equal to the sum of the original exponents. That is,

$$A^n A^m = A^{(n+m)}.$$

For example,

$$A^2 A^5 = A^7,$$

which we can verify directly by writing out the terms:

$$\underbrace{(A \times A)}_{A^2} \times \underbrace{(A \times A \times A \times A \times A)}_{A^5}$$

$$= \underbrace{A \times A \times A \times A \times A \times A \times A}$$

$$= A^7.$$

From the above result we see that when a quantity raised to a particular power (say A^n) is to be multiplied by itself a total of m times, we have

$$(A^n)^m = A^{nm}.$$

For example,

$$(A^2)^3 = A^6,$$

since

$$(A^2)^3 = A^2 \times A^2 \times A^2 = A^{(2+2+2)} = A^6.$$

Reciprocal quantities are expressed in a similar way with the addition of a minus sign in the exponent, as follows:

$$\frac{1}{A} = A^{-1}, \qquad \frac{1}{A^2} = A^{-2}, \qquad \frac{1}{A^3} = A^{-3}, \qquad \frac{1}{A^4} = A^{-4}.$$

Exactly the same rules as before are used in combining quantities raised to negative powers with one another and with some quantity raised to a positive power. Thus

$$A^5 A^{-2} = A^{(5-2)} = A^3,$$

$$(A^{-1})^{-2} = A^{-1(-2)} = A^2,$$

$$(A^3)^{-4} = A^{-4 \times 3} = A^{-12},$$

$$A A^{-7} = A^{(1-7)} = A^{-6}.$$

It is important to remember that any quantity raised to the zeroth power, say A^0, is equal to 1. Hence

$$A^2 A^{-2} = A^{(2-2)} = A^0 = 1.$$

This is more easily seen if we write A^{-2} as $1/A^2$:

$$A^2 A^{-2} = A^2 \times \frac{1}{A^2} = \frac{A^2}{A^2} = 1.$$

Fractional exponents are frequently useful. The simplest case is that of the *square root* of a quantity A, commonly written $\sqrt{A}$, which when multiplied by itself equals the quantity:

$$\sqrt{A} \times \sqrt{A} = A.$$

Using exponentials we see that, because

$$(A^{1/2})^2 = A^{2 \times (1/2)} = A^1 = A,$$

we can express square roots by the exponent $\frac{1}{2}$:

$$\sqrt{A} = A^{1/2}.$$

Other roots may be expressed similarly. The *cube root* of a quantity A, written $\sqrt[3]{A}$, when multiplied by itself twice equals A. That is,

$$\sqrt[3]{A} \times \sqrt[3]{A} \times \sqrt[3]{A} = A,$$

which may be more conveniently written

$$(A^{1/3})^3 = A,$$

where $\sqrt[3]{A} = A^{1/3}$.

In general the nth root of a quantity, $\sqrt[n]{A}$, may be written $A^{1/n}$, which is a more convenient form for most purposes. Some examples may be helpful:

$$(A^4)^{1/2} = A^{(1/2) \times 4} = A^2,$$

$$(A^{1/4})^{-7} = A^{-7 \times (1/4)} + A^{-7/4},$$

$$(A^3)^{-1/3} = A^{-(1/3) \times 3} = A^{-1},$$

$$(A^{1/2})^{1/2} = A^{(1/2) \times (1/2)} = A^{1/4}.$$

A-4 Equations with Powers and Roots

An equation that involves powers and/or roots is subject to the same basic principle that governs the manipulation of simpler equations: whatever is done to one side must be done to the other. Hence the following rules:

An equation remains valid when both sides are raised to the same power, that is, when each side is multiplied by itself the same number of times as the other side.

An equation remains valid when the same root is taken of both sides.

Problem. The mass m of an object whose velocity is v is given by

$$m = \frac{m_0}{\sqrt{1 - v^2/c^2}},$$

where m_0 is the object's mass when it is at rest and c is the velocity of light. Solve this equation for v.

Solution. The first step is to transpose the $\sqrt{1 - v^2/c^2}$ to the left-hand side of the equation to give

$$m \sqrt{1 - v^2/c^2} = m_0.$$

Next the square root is removed by squaring both sides, which yields

$$m^2(1 - v^2/c^2) = m_0^2.$$

Now we proceed in the usual way to find v^2:

$$1 - \frac{v^2}{c^2} = \frac{m_0^2}{m^2},$$

$$\frac{v^2}{c^2} = 1 - \frac{m_0^2}{m^2},$$

$$v^2 = c^2(1 - m_0^2/m^2).$$

The final step is to take the square root of both sides:

$$v = c\sqrt{1 - m_0^2/m^2}.$$

Quadratic equations, which have the general form

$$ax^2 + bx + c = 0,$$

are often encountered in algebraic calculations. In such an equation, a, b, and c are constants. A quadratic equation is satisfied by the values of x given by the formulas

$$x_+ = \frac{-b + \sqrt{b^2 - 4ac}}{2a},$$

$$x_- = \frac{-b - \sqrt{b^2 - 4ac}}{2a}.$$

By looking at these formulas, we can see that the nature of the solution depends upon the value of the quantity $b^2 - 4ac$. When $b^2 = 4ac$, $\sqrt{b^2 - 4ac} = 0$ and the two solutions are equal to just $x = -b/2a$. When $b^2 > 4ac$, $\sqrt{b^2 - 4ac}$ is a real number and the solutions x_+ and x_- are different. When $b^2 < 4ac$, $\sqrt{b^2 - 4ac}$ is the square root of a negative number, and the solutions are different. The square root of a negative number is called an *imaginary number* because squaring a real number, whether positive or negative, always gives a positive number: $(+2)^2 = (-2)^2 = +4$, hence $\sqrt{-4}$ cannot be either $+2$ or -2. Imaginary numbers do not occur in the sort of physical problems treated in this book.

Problem.   Solve the quadratic equation

$$2x^2 - 3x - 9 = 0.$$

Solution.   Here $a = 2$, $b = -3$, and $c = -9$, so that

$$b^2 - 4ac = (-3)^2 - 4(2)(-9) = 9 + 72 = 81.$$

The two solutions are

$$x_+ = \frac{-b + \sqrt{b^2 - 4ac}}{2a}$$

$$= \frac{-(-3) + \sqrt{81}}{2(2)} = \frac{3 + 9}{4} = \frac{12}{4} = 3,$$

$$x_- = \frac{-b - \sqrt{b^2 - 4ac}}{2a}$$

$$= \frac{-(-3) - \sqrt{81}}{2(2)} = \frac{3 - 9}{4} = \frac{-6}{4} = -\frac{3}{2}.$$

Problem.   Solve the quadratic equation

$$x^2 - 4x + 4 = 0.$$

Solution.   Here $a = 1$, $b = -4$, and $c = 4$, so that

$$b^2 - 4ac = (-4)^2 - 4(1)(4) = 16 - 16 = 0.$$

Hence the only solution is

$$x = \frac{-b}{2a} = \frac{-(-4)}{2} = 2.$$

A-5 Trigonometry

In physics it is often necessary to know the relationship among the sides and angles of a *right triangle*, which is a triangle two of whose sides are perpendicular. As in Fig. A-1 we label θ the angle at A included between the side AB of the triangle and its hypotenuse AC. The *sine* of this angle, which is abbreviated sin θ, is the ratio between the side BC of the triangle *opposite* to θ and the hypotenuse AC. Hence

$$\sin \theta = \frac{BC}{AC} = \frac{\text{opposite side}}{\text{hypotenuse}}.$$

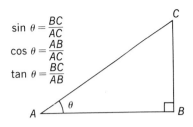

$$\sin \theta = \frac{BC}{AC}$$

$$\cos \theta = \frac{AB}{AC}$$

$$\tan \theta = \frac{BC}{AB}$$

Fig. A-1.   The trigonometric functions.

The *cosine* of the angle θ, abbreviated $\cos \theta$, is the ratio between the side AB of the triangle adjacent to θ and the hypotenuse AC. Hence

$$\cos \theta = \frac{AB}{AC} = \frac{\text{adjacent side}}{\text{hypotenuse}}.$$

The *tangent* of the angle θ, abbreviated $\tan \theta$, is the ratio between the side BC opposite to θ and the side AB adjacent to θ. Hence

$$\tan \theta = \frac{BC}{AB} = \frac{\text{opposite side}}{\text{adjacent side}}.$$

From these definitions we can obtain a useful result:

$$\frac{\sin \theta}{\cos \theta} = \frac{\text{opposite side/hypotenuse}}{\text{adjacent side/hypotenuse}}$$

$$= \frac{\text{opposite side}}{\text{adjacent side}}$$

$$= \tan \theta.$$

The tangent of an angle is equal to its sine divided by its cosine.

Numerical values of $\sin \theta$, $\cos \theta$, and $\tan \theta$ for angles from $0°$ to $90°$ are given in Appendix C. These figures may be used for angles from $90°$ to $360°$ with the help of this table:

	θ	$90° + \theta$	$180° + \theta$	$270° + \theta$
sin	$\sin \theta$	$\cos \theta$	$-\sin \theta$	$-\cos \theta$
cos	$\cos \theta$	$-\sin \theta$	$-\cos \theta$	$\sin \theta$
tan	$\tan \theta$	$-\dfrac{1}{\tan \theta}$	$\tan \theta$	$-\dfrac{1}{\tan \theta}$

For example, if we require the value of $\sin 120°$, we first note that $120° = 90° + 30°$. Then, since

$$\sin (90° + \theta) = \cos \theta,$$

we have

$$\sin 120° = \sin (90° + 30°) = \cos 30° = 0.866.$$

Similarly, to find $\tan 342°$, we begin by noting that $342° = 270° + 72°$. Then, since

$$\tan (270° + \theta) = -\frac{1}{\tan \theta},$$

we have

$$\tan 342° = \tan (270° + 72°) = -\frac{1}{\tan 72°} = -\frac{1}{3.078}$$

$$= -0.325.$$

Trigonometric functions can be treated algebraically, just as any other scalar quantity. Suppose we know the length of the side b and the angle θ in the right triangle of Fig. A-2 and wish to find the lengths of the sides a and c. From the definitions of sine and tangent we see that

$$\tan \theta = \frac{a}{b}, \qquad \sin \theta = \frac{a}{c},$$

and so

$$a = b \tan \theta, \qquad c = \frac{a}{\sin \theta}.$$

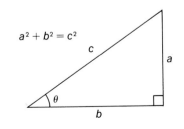

$$a^2 + b^2 = c^2$$

Fig. A-2.   The Pythagorean theorem.

Another useful relationship in a right triangle is the Pythagorean theorem, which states that the sum of the squares of the sides of such a triangle adjacent to the right angle is equal to the square of its hypotenuse. For the triangle of Fig. A-2,

$$a^2 + b^2 = c^2.$$

Hence we can always express the length of any of the sides of a right triangle in terms of the other sides:

$$a = \sqrt{c^2 - b^2},$$
$$b = \sqrt{c^2 - a^2},$$
$$c = \sqrt{a^2 + b^2}.$$

Appendix B
Powers of Ten and Logarithms

B-1 Powers of Ten

Very small and very large numbers are common in physics. For example, the mass of an electron is 0.000,000,000,000,000,000,000,000,000,000,910,9 kilogram, and the mass of the earth is 5,983,000,000,000,000,000,000,000 kilograms. Such numbers in ordinary decimal form are clumsy to write and to make calculations with, and it is hard to appreciate their precise magnitudes because of the sea of zeros.

A better method for expressing numbers makes use of powers-of-ten notation. This method is based upon the fact that all numbers may be represented by a number between 1 and 10 multiplied by a power of 10. In powers-of-ten notation the mass of an electron is written simply as 9.109×10^{-31} kilogram and the mass of the earth is written as 5.983×10^{24} kilogram.

Table B-1 contains powers of 10 from 10^{-10} to 10^{10}. Evidently positive powers of 10 (which cover numbers greater than 1) follow this pattern:

Table B-1.

$10^{-10} = 0.000,000,000,1$	$10^{0}\ = 1$
$10^{-9}\ = 0.000,000,001$	$10^{1} = 10$
$10^{-8}\ = 0.000,000,01$	$10^{2} = 100$
$10^{-7}\ = 0.000,000,1$	$10^{3} = 1000$
$10^{-6}\ = 0.000,001$	$10^{4} = 10,000$
$10^{-5}\ = 0.000,01$	$10^{5} = 100,000$
$10^{-4}\ = 0.000,1$	$10^{6} = 1,000,000$
$10^{-3}\ = 0.001$	$10^{7} = 10,000,000$
$10^{-2}\ = 0.01$	$10^{8} = 100,000,000$
$10^{-1}\ = 0.1$	$10^{9} = 1,000,000,000$
$10^{0}\ = 1$	$10^{10} = 10,000,000,000$

$$10^{0} = 1. \qquad = 1 \qquad = 1 \text{ with decimal point moved 0 places,}$$
$$10^{1} = 1.0 \qquad = 10 \qquad = 1 \text{ with decimal point moved 1 place to the right,}$$
$$10^{2} = 1.00 \qquad = 100 \qquad = 1 \text{ with decimal point moved 2 places to the right,}$$
$$10^{3} = 1.000 \qquad = 1,000 \qquad = 1 \text{ with decimal point moved 3 places to the right,}$$
$$10^{4} = 1.000\ 0 \qquad = 10,000 \qquad = 1 \text{ with decimal point moved 4 places to the right,}$$
$$10^{5} = 1.000\ 00 \qquad = 100,000 \qquad = 1 \text{ with decimal point moved 5 places to the right,}$$
$$10^{6} = 1.000\ 000 = 1,000,000 = 1 \text{ with decimal point moved 6 places to the right,}$$

and so on. The exponent of the 10 indicates how many places the decimal point is moved *to the right* from 1.000 · · ·

A similar pattern is followed by negative powers of 10, whose values always lie between 0 and 1:

$$10^0 = 1. \qquad\qquad = 1 \qquad\quad = 1 \text{ with decimal point moved 0 places,}$$

$$10^{-1} = 0\underset{\frown}{1}. \qquad\quad = 0.1 \qquad\ = 1 \text{ with decimal point moved 1 placed to the left,}$$

$$10^{-2} = 0\underset{\frown}{01}. \qquad\quad = 0.01 \qquad = 1 \text{ with decimal point moved 2 places to the left,}$$

$$10^{-3} = 0\underset{\frown}{00\ 1}. \qquad = 0.001 \qquad = 1 \text{ with decimal point moved 3 places to the left,}$$

$$10^{-4} = 0\underset{\frown}{00\ 01}. \qquad = 0.000,1 \qquad = 1 \text{ with decimal point moved 4 places to the left,}$$

$$10^{-5} = 0\underset{\frown}{00\ 001}. \qquad = 0.000,01 \quad = 1 \text{ with decimal point moved 5 places to the left,}$$

$$10^{-6} = 0\underset{\frown}{00\ 000\ 1}. = 0.000,001 = 1 \text{ with decimal point moved 6 places to the left,}$$

and so on. The exponent of the 10 now indicates how many places the decimal point is moved *to the left* from 1.

Here are a few examples of powers-of-ten notation:

$$600 = 6 \times 100 = 6 \times 10^2,$$

$$7940 = 7.94 \times 1000 = 7.94 \times 10^3,$$

$$93{,}000{,}000 = 9.3 \times 10{,}000{,}000 = 9.3 \times 10^7,$$

$$0.023 = 2.3 \times 0.01 = 2.3 \times 10^{-2},$$

$$0.000{,}035 = 3.5 \times 0.000{,}01 = 3.5 \times 10^{-5}.$$

B-2 Using Powers of Ten

Let us see how to make calculations using numbers written in powers-of-ten notation. To add or subtract numbers written in powers-of-ten notation, they must be expressed in terms of the *same* power of ten.

$$7 \times 10^4 + 2 \times 10^5 = 0.7 \times 10^5 + 2 \times 10^5$$
$$= 2.7 \times 10^5,$$

$$5 \times 10^{-2} + 3 \times 10^{-4} = 5 \times 10^{-2} + 0.03 \times 10^{-2}$$
$$= 5.03 \times 10^{-2},$$

$$8 \times 10^{-3} - 7 \times 10^{-4} = 8 \times 10^{-3} - 0.7 \times 10^{-3}$$
$$= 7.3 \times 10^{-3},$$

$$4 \times 10^5 - 1 \times 10^6 = 4 \times 10^5 - 10 \times 10^5$$
$$= -6 \times 10^5.$$

To multiply powers of ten together, add their exponents:

$$10^n \times 10^m = 10^{n+m}.$$

Be sure to take the sign of each exponent into account.

$$10^2 \times 10^3 = 10^{2+3} = 10^5$$

$$10^7 \times 10^{-3} = 10^{7-3} = 10^4$$

$$10^{-2} \times 10^{-4} = 10^{-2-4} = 10^{-6}$$

To multiply numbers written in powers-of-ten notation, multiply the decimal parts of the numbers together and add the exponents to find the power of ten of the product:

$$A \times 10^n \times B \times 10^m = AB \times 10^{n+m}.$$

If necessary, rewrite the result so the decimal part is a number between 1 and 10.

$$3 \times 10^2 \times 2 \times 10^5 = 3 \times 2 \times 10^{2+5}$$
$$= 6 \times 10^7$$

$$8 \times 10^{-5} \times 3 \times 10^7 = 8 \times 3 \times 10^{-5+7}$$
$$= 24 \times 10^2 = 2.4 \times 10^3$$

$$1.3 \times 10^{-3} \times 4 \times 10^{-5} = 1.3 \times 4 \times 10^{-3-5}$$
$$= 5.2 \times 10^{-8}$$

$$-9 \times 10^{17} \times 6 \times 10^{-18} = -9 \times 6 \times 10^{17-18}$$
$$= -54 \times 10^{-1} = -5.4$$

To divide one power of ten by another, subtract the exponent of the denominator from the exponent of the numerator:

$$\frac{10^n}{10^m} = 10^{n-m}.$$

Be sure to take the sign of each exponent into account.

$$\frac{10^5}{10^3} = 10^{5-3} = 10^2$$

$$\frac{10^{-2}}{10^4} = 10^{-2-4} = 10^{-6}$$

$$\frac{10^{-3}}{10^{-7}} = 10^{-3-(-7)} = 10^{-3+7} = 10^4$$

To divide a number written in powers-of-ten notation by another number written that way, divide the decimal parts of the numbers in the usual way and use the above rule to find the exponent of the power of ten of the quotient:

$$\frac{A \times 10^n}{B \times 10^m} = \frac{A}{B} \times 10^{n-m}.$$

If necessary, rewrite the result so the decimal part is a number between 1 and 10.

$$\frac{6 \times 10^5}{3 \times 10^2} = \frac{6}{3} \times 10^{5-2} = 2 \times 10^3$$

$$\frac{2 \times 10^{-7}}{8 \times 10^4} = \frac{2}{8} \times 10^{-7-4} = \frac{1}{4} \times 10^{-11}$$

$$= 0.25 \times 10^{-11} = 2.5 \times 10^{-12}$$

$$\frac{-7 \times 10^5}{10^{-2}} = -7 \times 10^{5-(-2)} = -7 \times 10^{5+2}$$

$$= -7 \times 10^7$$

$$\frac{5 \times 10^{-2}}{-2 \times 10^{-9}} = -\frac{5}{2} \times 10^{-2-(-9)} = -2.5 \times 10^{-2+9}$$

$$= -2.5 \times 10^7$$

To find the reciprocal of a power of ten, change the sign of the exponent:

$$\frac{1}{10^n} = 10^{-n},$$

$$\frac{1}{10^{-m}} = 10^m.$$

$$\frac{1}{10^5} = 10^{-5}$$

$$\frac{1}{10^{-3}} = 10^3$$

Hence the prescription for finding the reciprocal of a number written in powers-of-ten notation is

$$\frac{1}{A \times 10^n} = \frac{1}{A} \times 10^{-n}.$$

$$\frac{1}{2 \times 10^3} = \frac{1}{2} \times 10^{-3} = 0.5 \times 10^{-3} = 5 \times 10^{-4}$$

$$\frac{1}{4 \times 10^{-8}} = \frac{1}{4} \times 10^8 = 0.25 \times 10^8 = 2.5 \times 10^7$$

The powers-of-ten method of writing large and small numbers makes arithmetic involving such numbers relatively easy to carry out. Here is a calculation that would be very tedious if each number were kept in decimal form.

$$\frac{3800 \times 0.0054 \times 0.000,001}{430,000,000 \times 73}$$

$$= \frac{3.8 \times 10^3 \times 5.4 \times 10^{-3} \times 10^{-6}}{4.3 \times 10^8 \times 7.3 \times 10^1}$$

$$= \frac{3.8 \times 5.4}{4.3 \times 7.3} \times \frac{10^3 \times 10^{-3} \times 10^{-6}}{10^8 \times 10^1}$$

$$= 0.65 \times 10^{(3-3-6-8-1)}$$

$$= 0.65 \times 10^{-15}$$

$$= 6.5 \times 10^{-16}$$

B-3 Powers and Roots

To square a power of ten, multiply the exponent by 2; to cube a power of ten, multiply the exponent by 3:

$$(10^n)^3 = 10^{2n},$$

$$(10^n)^3 = 10^{3n}.$$

In general, to raise a power of ten to the mth power, multiply the exponent by m:

$$(10^n)^m = 10^{m \times n}.$$

Be sure to take the sign of each exponent into account.

$$(10^3)^2 = 10^{2 \times 3} = 10^6$$

$$(10^{-2})^5 = 10^{5 \times -2} = 10^{-10}$$

$$(10^{-4})^{-2} = 10^{-2 \times -4} = 10^8$$

To raise a number written in powers-of-ten notation to the mth power, multiply the decimal part of the number by itself m times and multiply the exponent of the power of ten by m:

$$(A \times 10^n)^m = A^m \times 10^{m \times n}.$$

If necessary, rewrite the result so the decimal part is a number between 1 and 10. A table of squares and cubes of numbers from 1 to 100 is given in Appendix C.

$$(2 \times 10^5)^2 = 2^2 \times 10^{2 \times 5} = 4 \times 10^{10}$$

$$(3 \times 10^{-3})^3 = 3^3 \times 10^{3 \times -3} = 27 \times 10^{-9}$$

$$= 2.7 \times 10^{-8}$$

$$(5 \times 10^{-2})^{-4} = 5^{-4} \times 10^{-4 \times -2} = \frac{1}{5^4} \times 10^8$$

$$= \frac{1}{625} \times 10^8 = 0.0016 \times 10^8$$

$$= 1.6 \times 10^5$$

A fractional exponent signifies a root. The *square root* of a quantity A, written $\sqrt{A}$, when multiplied by itself equals the quantity:

$$\sqrt{A} \times \sqrt{A} = A.$$

Because

$$(A^{1/2})^2 = A^{2 \times 1/2} = A^1 = A,$$

we see that we can express a square root by using the exponent $\frac{1}{2}$:

$$\sqrt{A} = A^{1/2}.$$

The *cube root* of A, written $\sqrt[3]{A}$, multiplied by itself twice equals A:

$$\sqrt[3]{A} \times \sqrt[3]{A} \times \sqrt[3]{A} = A.$$

Because

$$(A^{1/3})^3 = A^{3 \times 1/3} = A^1 = A,$$

we have

$$\sqrt[3]{A} = A^{1/3}.$$

In general, the mth root of a quantity may be written

$$\sqrt[m]{A} = A^{1/m}.$$

$$\sqrt{A^4} = (A^4)^{1/2} = A^{1/2 \times 4} = A^2$$

$$(\sqrt{A})^4 = (A^{1/2})^4 = A^{4 \times 1/2} = A^2$$

$$\sqrt{\sqrt{A}} = (A^{1/2})^{1/2} = A^{1/2 \times 1/2} = A^{1/4}$$

To take the mth root of a power of ten, divide the exponent by m:

$$\sqrt[m]{10^n} = (10^n)^{1/m} = 10^{n/m}.$$

$$\sqrt{10^4} = (10^4)^{1/2} = 10^{4/2} = 10^2$$

$$\sqrt[3]{10^9} = (10^9)^{1/3} = 10^{9/3} = 10^3$$

$$\sqrt[3]{10^{-9}} = (10^{-9})^{1/3} = 10^{-9/3} = 10^{-3}$$

In powers-of-ten notation, the exponent of the 10 must be an integer. Hence in taking the mth root of a power of ten, the exponent should be an integral multiple of m. Instead of, for example,

$$\sqrt{10^5} = (10^5)^{1/2} = 10^{2\,1/2}$$

which, while correct, is hardly useful, we would write

$$\sqrt{10^5} = \sqrt{10^1 \times 10^4}$$
$$= \sqrt{10} \times \sqrt{10^4}$$
$$= 3.16 \times 10^2.$$
$$\sqrt{10^{-3}} = \sqrt{10^1 \times 10^{-4}} = \sqrt{10} \times \sqrt{10^{-4}}$$
$$= 3.16 \times 10^{-2}$$
$$\sqrt[3]{10^8} = \sqrt[3]{10^2 \times 10^6} = \sqrt[3]{100} \times \sqrt[3]{10^6}$$
$$= 4.64 \times 10^2$$

To take the mth root of a number expressed in powers-of-ten notation, first write the number so the exponent of the 10 is an integral multiple of m. Then take the mth root of the decimal part of the number and divide the exponent by m to find the power of ten of the result:

$$(A \times 10^n)^{1/m} = \sqrt[m]{A} \times 10^{n/m}.$$
$$\sqrt{9 \times 10^4} = \sqrt{9} \times 10^{4/2} = 3 \times 10^2$$
$$\sqrt{9 \times 10^{-6}} = \sqrt{9} \times 10^{-6/2} = 3 \times 10^{-3}$$
$$\sqrt{4 \times 10^7} = \sqrt{40 \times 10^6} = \sqrt{40} \times 10^{6/2}$$
$$= 6.32 \times 10^3$$
$$\sqrt[3]{3 \times 10^{-5}} = \sqrt[3]{30 \times 10^{-6}} = \sqrt[3]{30} \times 10^{-6/3}$$
$$= 3.11 \times 10^{-2}$$

A table of square and cube roots of numbers from 1 to 1,000 is given in the Appendix.

B-4 Significant Figures

An advantage of powers-of-ten notation is that it gives no false impression of the degree of accuracy with which a number is stated. For instance, the average radius of the earth is 3,959 miles, but it is often taken as 4,000 miles for convenience in making rough calculations. To indicate the approximate character of the latter figure, all we have to do is write

$$r = 4.0 \times 10^3 \text{ miles.}$$

Now how large the number is and how accurate it is are both clear. The accurately known digits, plus one uncertain digit, are called *significant figures*; in the above case r has two significant figures, 4 and 0. If we require more accuracy, we would write

$$r = 3.96 \times 10^3 \text{ miles}$$

which contains three significant figures, or

$$r = 3.959 \times 10^3 \text{ miles}$$

which contains four significant figures (see Fig. B-1).

Sometimes one or more zeros in a number are significant figures, and it is proper to retain them when expressing the number in exponential notation. Thus

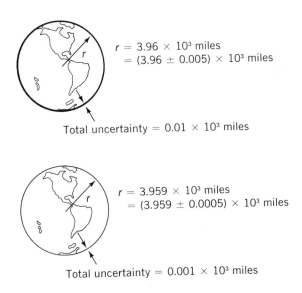

$r = 4.0 \times 10^3$ miles
$= (4.0 \pm 0.05) \times 10^3$ miles

Total uncertainty $= 0.1 \times 10^3$ miles

$r = 3.96 \times 10^3$ miles
$= (3.96 \pm 0.005) \times 10^3$ miles

Total uncertainty $= 0.01 \times 10^3$ miles

$r = 3.959 \times 10^3$ miles
$= (3.959 \pm 0.0005) \times 10^3$ miles

Total uncertainty $= 0.001 \times 10^3$ miles

Fig. B-1.

the number 1 represents any number between 0.5 and 1.5 expressed in one significant figure, whereas 1.0 narrows the range to any number between 0.95 and 1.05 and 1.00 further narrows it to any number between 0.995 and 1.005 (Fig. B-2):

$$1 = 1 \pm 0.5,$$

$$1.0 = 1.0 \pm 0.05,$$

$$1.00 = 1.00 \pm 0.005.$$

When quantities are combined arithmetically, the result is no more accurate than the quantity with the largest uncertainty. Suppose a man weighing 162 lb picks up an 0.34-lb apple. The total weight of man + apple is still 162 lb because all we know of the man's weight is that it is somewhere between 161.5 and 162.5 lb, which means a total uncertainty greater than the apple's weight. If the man's weight is instead 162.0 lb, he and the apple together weigh 162.3 lb (Fig. B-3); if his weight is 162.00 lb, he and the apple together weigh 162.34 lb.

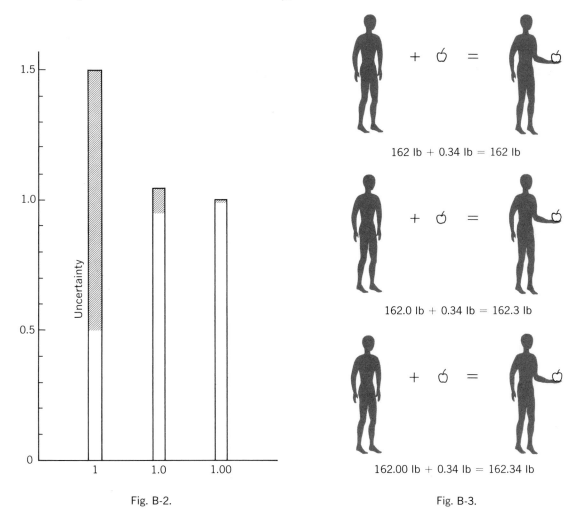

162 lb + 0.34 lb = 162 lb

162.0 lb + 0.34 lb = 162.3 lb

162.00 lb + 0.34 lb = 162.34 lb

Fig. B-2.

Fig. B-3.

In multiplication and division, the result is entitled to only as many significant figures as there are in the least accurately-known quantity. For example, if we divide 1.4×10^5 by 6.70×10^3, we are not justified in writing

$$\frac{1.4 \times 10^5}{6.70 \times 10^3} = 20.89552\cdots$$

We may properly retain only two significant figures, corresponding to the two significant figures in the numerator, and so the correct answer is just 21.

In a calculation with several steps, however, it is a good idea to keep an extra digit in the intermediate steps and to round off the result only at the end.

$$\frac{5.7 \times 10^4}{3.3 \times 10^{-2}} + \sqrt{1.8 \times 10^{12}} = 1.73 \times 10^6 + 1.34 \times 10^6$$

$$= 3.07 \times 10^6$$

$$= 3.1 \times 10^6$$

3. Evaluate the following.
$(4 \times 10^2)^2 =$
$(2 \times 10^{-6})^3 =$
$(2 \times 10^2)^{-2} =$

4. Evaluate the following.
$\sqrt{4 \times 10^{-8}} =$
$\sqrt{4 \times 10^{12}} =$
$\sqrt{1.6 \times 10^5} =$

5. Evaluate the following.
$\dfrac{6 \times 14}{7 \times 10^{-6}} =$

$\dfrac{3 \times 10^4 \times 5 \times 10^{-12}}{10^3} =$

$\dfrac{9 \times 10^{12}}{9 \times 10^{-12}} =$

$\dfrac{8 \times 10^{10} \times 3}{6 \times 10^{-4}} =$

$\dfrac{10^{-3}}{5 \times 10^4 \times 2 \times 10^2} =$

$\dfrac{5 \times 10^5 \times 2 \times 10^{-18}}{4 \times 10^4} =$

Exercises

1. Express the following numbers in decimal notation.
$2 \times 10^5 =$
$8 \times 10^{-2} =$
$7.819 \times 10^2 =$
$4.51 \times 10^8 =$
$1.003 \times 10^{-6} =$
$10^{-10} =$
$9.56 \times 10^{-5} =$

2. Express the following numbers in powers-of-ten notation.
$70 = \underline{\hspace{1cm}} \times 10$
$0.14 = \underline{\hspace{1cm}} \times 10$
$3.81 = \underline{\hspace{1cm}} \times 10$
$8400 = \underline{\hspace{1cm}} \times 10$
$1,000,000 = \underline{\hspace{1cm}} \times 10$
$0.007,890 = \underline{\hspace{1cm}} \times 10$
$351,600 = \underline{\hspace{1cm}} \times 10$

Answers

1. 200,000
0.08
781.9
451,000,000
0.000,001,003
0.000,000,000,1
0.000,0956

2. 7×10^1
1.4×10^{-1}
3.81
8.4×10^3
1×10^6
7.890×10^{-3}
3.516×10^5

3. 1.6×10^5
8×10^{-18}
2.5×10^{-5}

4. 2×10^{-4}
2×10^6
4×10^2

5. 1.2×10^7
1.5×10^{-10}
10^{24}
4×10^{14}
10^{-10}
2.5×10^{-17}

B-5 Logarithms

The logarithm of a number N is the exponent n to which a given base number a must be raised in order that $a^n = N$. That is, if

$$N = a^n, \quad \text{then} \quad n = \log_a N.$$

Here are some examples using the base $a = 10$:

$$1000 = 10^3, \quad \text{therefore} \quad \log_{10} 1000 = 3;$$
$$5 = 10^{0.699}, \quad \text{therefore} \quad \log_{10} 5 = 0.699;$$
$$0.001 = 10^{-3}, \quad \text{therefore} \quad \log_{10} 0.001 = -3.$$

Since the decimal system of numbers has the base 10, this is a convenient number to use as the base of a system of logarithms. Logarithms to the base 10 are called *common logarithms* and are denoted simply as "log N." Another widely used system of logarithms uses $e = 2.718\cdots$ as the base. Such logarithms are called *natural logarithms* (because they arise in a natural way in calculus) and are denoted "ln N." To go from one system to the other these formulas are needed:

$$\log N = 0.43429 \ln N,$$
$$\ln N = 2.3026 \log N.$$

Logarithms are defined only for positive numbers because the quantity a^n is positive whether n is positive, negative, or zero. Since n is the logarithm of a^n, n can describe only a positive number.

Let us consider a number N that is the product of two numbers x and y, so that $N = xy$. If $x = 10^n$ and $y = 10^m$, then

$$N = xy = 10^n \times 10^m = 10^{n+m},$$

$$\log N = n + m.$$

Since $n = \log x$ and $m = \log y$, $n + m = \log x + \log y$, and so

$$\log N = \log x + \log y.$$

Thus we have the general rule for the logarithm of a product:

$$\log xy = \log x + \log y.$$

Similar reasoning gives the additional rules

$$\log \frac{x}{y} = \log x - \log y,$$

$$\log x^n = n \log x.$$

Before the days of electronic calculators, logarithms were widely used to simplify arithmetical work because they permit replacing multipication and division by addition and subtraction. It is much easier, and less prone to error, to find

$$A = \frac{3.18 \times 7.53 \times 8.16}{1.22 \times 4.05}$$

by using logarithms than by multiplying and dividing the numbers themselves by hand:

$$\log A = \log 3.18 + \log 7.53 + \log 8.16 - \log 1.22 - \log 4.05$$
$$= 0.5658 + 0.8768 + 0.9117 - 0.0864 - 0.6075$$
$$= 1.6605,$$
$$A = \log^{-1} 1.6605 = 45.76.$$

Today, of course, electronic calculators are used for such routine arithmetic. However, logarithms are still handy for evaluating powers and roots. For example, to find $2.13^{1.86}$ we proceed as follows:

$$\log 2.13^{1.86} = 1.86 \log 2.13$$
$$= 1.86 \times 0.3284 = 0.6108,$$
$$2.13^{1.86} = \log^{-1} 0.6108 = 4.08.$$

To obtain $\log^{-1} 0.6108$ (the *antilogarithm* of 0.6108, which is the number whose logarithm is 0.6108) with an electronic calculator, the 0.6108 is entered and the "10^x" button pressed, since 10^x is the number whose logarithm is x. The same procedure is followed for negative exponents, for instance $1/\sqrt[4]{7}$:

$$\frac{1}{\sqrt[4]{7}} = \frac{1}{7^{1/4}} = 7^{-1/4},$$

$$\log 7^{-1/4} = -\tfrac{1}{4}\log 7 = -\tfrac{1}{4} \times 0.8451 = -0.2113,$$

$$7^{-1/4} = \log^{-1}(-0.2113) = 0.615,$$

$$\frac{1}{\sqrt[4]{7}} = 0.615.$$

Apart from their use in calculations, logarithms appear in many of the formulas of science and engineering. Thus in Chapter 13 we saw them employed in constructing the decibel scale of sound intensity. Every student of chemistry encounters logarithms in the pH scale of acidity and basicity, where if $[H^+]$ represents the concentration of H^+ ions in a solution in moles/liter, the pH of the solution is defined as

$$pH = -\log[H^+].$$

While it is naturally easiest to find logarithms and antilogarithms with an electronic calculator, a table of logarithms is not hard to use. Such a table contains the logarithms of numbers from $1.00\cdots$ to $9.9\cdots$. To use a table for numbers smaller than 1 or larger than 10 we draw upon the fact that

$$\log xy = \log x + \log y.$$

Every number can be written in powers-of-ten notation as a number between 1 and 10 multiplied by a power of 10; for instance,

$$804 = 8.04 \times 10^2.$$

From a table of logarithms, $\log 8.04 = 0.9053$, and from the definition of logarithms to the base 10, $\log 10^2 = 2$. We therefore have

$$\log 804 = \log 8.04 + \log 10^2$$
$$= 0.9053 + 2 = 2.9053.$$

Thus the part of the logarithm after the decimal point, here 9053, describes the *sequence of digits* in the

number, here 804. The other part of the logarithm, here 2, describes the *magnitude* of the number, that is, whether the number is 0.804, 8.04, 80.4, or, as in this case, 804.

Similarly, to find the number whose logarithm is, say, 3.8751, we begin by separating the logarithm into two parts, one of which we use to determine the digits in the number and the other to locate the decimal point:

$$\log^{-1} 3.8751 = \log^{-1} 3 \times \log^{-1} 0.8751$$
$$= 10^3 \times 7.500 = 7,500.$$

Logarithm tables list only positive logarithms, and so negative logarithms (which represent numbers smaller than 1) have to be expressed in a special way in order to find their antilogarithms. For instance,

$$\log 0.0281 = \log 2.81 + \log 10^{-2}$$
$$= 0.4487 - 2 = -1.5513,$$

but we cannot perform the inverse operation of finding $\log^{-1}(-1.5513)$ directly since -0.5513 does not appear in any table. The procedure is therefore to do the reverse of what was done in the last line of the above calculation. We first add to the argument of the antilogarithm (here -1.5513) the integer next larger in absolute value, and then show this integer as being subtracted from the result:

$$-1.5513 = (-1.5513 + 2) - 2 =$$
$$0.4487 - 2,$$
$$\log^{-1}(-1.5513) = \log^{-1} 0.4487 \times \log^{-1}(-2)$$
$$= 2.81 \times 10^{-2} = 0.0281.$$

Exercises

6. Find the common logarithms of the following numbers.
   2.22 =
   22.2 =
   0.222 =
   $3 \times 10^{-4} =$
   $9 \times 10^{14} =$

7. Find the antilogarithms of the following logarithms.
   0.8710 =
   3.6484 =
   8.9031 =
   −3.6484 =
   −0.1931 =

8. Evaluate the following with the help of logarithms.

   $0.0181^{1.5} =$
   $62.2^{7.13} =$
   $(8.15 \times 10^{14})^6 =$
   $\sqrt[9]{156} =$
   $(6.24 \times 10^{-4})^{1/3} =$
   $(2.71 \times 10^5)^{1/8} =$

Answers

6. 0.3464
   1.3463
   −0.6536
   −3.5229
   14.9542

7. 7.43
   4,450
   8×10^8
   2.25×10^{-4}
   0.641

8. 2.44×10^{-3}
   6.16×10^{12}
   2.93×10^{89}
   3.53
   8.55×10^{-2}
   4.78

Appendix C
Tables

C-1 The Elements

Atomic number	Element	Symbol	Atomic mass*	Atomic number	Element	Symbol	Atomic mass*
1	Hydrogen	H	1.008	27	Cobalt	Co	58.93
2	Helium	He	4.003	28	Nickel	Ni	58.71
3	Lithium	Li	6.939	29	Copper	Cu	63.54
4	Beryllium	Be	9.012	30	Zinc	Zn	65.37
5	Boron	B	10.81	31	Gallium	Ga	69.72
6	Carbon	C	12.01	32	Germanium	Ge	72.59
7	Nitrogen	N	14.01	33	Arsenic	As	74.92
8	Oxygen	O	16.00	34	Selenium	Se	78.96
9	Fluorine	F	19.00	35	Bromine	Br	79.91
10	Neon	Ne	20.18	36	Krypton	Kr	83.80
11	Sodium	Na	22.99	37	Rubidium	Rb	85.47
12	Magnesium	Mg	24.31	38	Strontium	Sr	87.62
13	Aluminum	Al	26.98	39	Yttrium	Y	88.91
14	Silicon	Si	28.09	40	Zirconium	Zr	91.22
15	Phosphorus	P	30.97	41	Niobium	Nb	92.91
16	Sulfur	S	32.06	42	Molybdenum	Mo	95.94
17	Chlorine	Cl	35.45	43	Technetium	Tc	(99)
18	Argon	Ar	39.95	44	Ruthenium	Ru	101.0
19	Potassium	K	39.10	45	Rhodium	Rh	102.9
20	Calcium	Ca	40.08	46	Palladium	Pd	106.4
21	Scandium	Sc	44.96	47	Silver	Ag	107.9
22	Titanium	Ti	47.90	48	Cadium	Cd	112.4
23	Vanadium	V	50.94	49	Indium	In	114.8
24	Chromium	Cr	52.00	50	Tin	Sn	118.7
25	Manganese	Mn	54.94	51	Antimony	Sb	121.8
26	Iron	Fe	55.85	52	Tellurium	Te	127.6
				53	Iodine	I	126.9
				54	Xenon	Xe	131.3
				55	Cesium	Cs	132.9

*The unit of mass is the u. When all the isotopes of an element are radio-active, the mass number of the most stable isotope is given in parentheses.

C-1 The Elements
(Continued)

Atomic number	Element	Symbol	Atomic mass*	Atomic number	Element	Symbol	Atomic mass*
56	Barium	Ba	137.4	81	Thallium	Tl	204.4
57	Lanthanum	La	138.9	82	Lead	Pb	207.2
58	Cerium	Ce	140.1	83	Bismuth	Bi	209.0
59	Praseodymium	Pr	140.9	84	Polonium	Po	(209)
60	Neodymium	Nd	144.2	85	Astatine	At	(210)
61	Promethium	Pm	(147)	86	Radon	Rn	(222)
62	Samarium	Sm	150.4	87	Francium	Fr	(223)
63	Europium	Eu	152.0	88	Radium	Ra	(226)
64	Gadolinium	Gd	157.3	89	Actinium	Ac	(227)
65	Terbium	Tb	158.9	90	Thorium	Th	232.0
66	Dysprosium	Dy	162.5	91	Protactinium	Pa	(231)
67	Holmium	Ho	164.9	92	Uranium	U	238.0
68	Erbium	Er	167.3	93	Neptunium	Np	(237)
69	Thulium	Tm	168.9	94	Plutonium	Pu	(244)
70	Ytterbium	Yb	173.0	95	Americium	Am	(243)
71	Lutetium	Lu	175.0	96	Curium	Cm	(247)
72	Hafnium	Hf	178.5	97	Berkelium	Bk	(247)
73	Tantalum	Ta	181.0	98	Californium	Cf	(251)
74	Tungsten	W	183.9	99	Einsteinium	Es	(254)
75	Rhenium	Re	186.2	100	Fermium	Fm	(257)
76	Osmium	Os	190.2	101	Mendelevium	Md	(256)
77	Iridium	Ir	192.2	102	Nobelium	No	(254)
78	Platinum	Pt	195.1	103	Lawrencium	Lr	(257)
79	Gold	Au	197.0	104			
80	Mercury	Hg	200.6	105			

C-2 Natural Trigonometric Functions

Angle					Angle				
Degree	Radian	Sine	Cosine	Tangent	Degree	Radian	Sine	Cosine	Tangent
0°	.000	0.000	1.000	0.000	26°	.454	.438	.899	.488
1°	.017	.018	1.000	.018	27°	.471	.454	.891	.510
2°	.035	.035	0.999	.035	28°	.489	.470	.883	.532
3°	.052	.052	.999	.052	29°	.506	.485	.875	.554
4°	.070	.070	.998	.070	30°	.524	.500	.866	.577
5°	.087	.087	.996	.088					
6°	.105	.105	.995	.105	31°	.541	.515	.857	.601
7°	.122	.122	.993	.123	32°	.559	.530	.848	.625
8°	.140	.139	.990	.141	33°	.576	.545	.839	.649
9°	.157	.156	.988	.158	34°	.593	.559	.829	.675
10°	.175	.174	.985	.176	35°	.611	.574	.819	.700
11°	.192	.191	.982	.194	36°	.628	.588	.809	.727
12°	.209	.208	.978	.213	37°	.646	.602	.799	.754
13°	.227	.225	.974	.231	38°	.663	.616	.788	.781
14°	.244	.242	.970	.249	39°	.681	.629	.777	.810
15°	.262	.259	.966	.268	40°	.698	.643	.766	.839
16°	.279	.276	.961	.287	41°	.716	.658	.755	.869
17°	.297	.292	.956	.306	42°	.733	.669	.743	.900
18°	.314	.309	.951	.325	43°	.751	.682	.731	.933
19°	.332	.326	.946	.344	44°	.768	.695	.719	.966
20°	.349	.342	.940	.364	45°	.785	.707	.707	1.000
21°	.367	.358	.934	.384	46°	.803	.719	.695	1.036
22°	.384	.375	.927	.404	47°	.820	.731	.682	1.072
23°	.401	.391	.921	.425	48°	.838	.743	.669	1.111
24°	.419	.407	.914	.445	49°	.855	.755	.656	1.150
25°	.436	.423	.906	.466	50°	.873	.766	.643	1.192

C-2 Natural Trigonometric Functions (Continued)

| Angle | | | | | Angle | | | | |
Degree	Radian	Sine	Cosine	Tangent	Degree	Radian	Sine	Cosine	Tangent
51°	.890	.777	.629	1.235	71°	1.239	.946	.326	2.904
52°	.908	.788	.616	1.280	72°	1.257	.951	.309	3.078
53°	.925	.799	.602	1.327	73°	1.274	.956	.292	3.271
54°	.942	.809	.588	1.376	74°	1.292	.961	.276	3.487
55°	.960	.819	.574	1.428	75°	1.309	.966	.259	3.732
56°	.977	.829	.559	1.483	76°	1.326	.970	.242	4.011
57°	.995	.839	.545	1.540	77°	1.344	.974	.225	4.331
58°	1.012	.848	.530	1.600	78°	1.361	.978	.208	4.705
59°	1.030	.857	.515	1.664	79°	1.379	.982	.191	5.145
60°	1.047	.866	.500	1.732	80°	1.396	.985	.174	5.671
61°	1.065	.875	.485	1.804	81°	1.414	.988	.156	6.314
62°	1.082	.883	.470	1.881	82°	1.431	.990	.139	7.115
63°	1.100	.891	.454	1.963	83°	1.449	.993	.122	8.144
64°	1.117	.899	.438	2.050	84°	1.466	.995	.105	9.514
65°	1.134	.906	.423	2.145	85°	1.484	.996	.087	11.43
66°	1.152	.914	.407	2.246	86°	1.501	.998	.070	14.30
67°	1.169	.921	.391	2.356	87°	1.518	.999	.052	19.08
68°	1.187	.927	.375	2.475	88°	1.536	.999	.035	28.64
69°	1.204	.934	.358	2.605	89°	1.553	1.000	.018	57.29
70°	1.222	.940	.342	2.747	90°	1.571	1.000	.000	∞

Glossary

Absolute Temperature Scale. *Absolute zero* is the lowest temperature possible; in the elementary kinetic theory of matter, it is that temperature at which random molecular movement would cease, although in reality a small amount of movement would still persist. Absolute zero corresponds to $-273°C$ and $-460°F$. The *absolute temperature scale* expresses temperature in °C above absolute zero; its unit is the *Kelvin,* denoted K. The Rankine scale expresses temperatures in °F above absolute zero; its unit is the degree Rankine, denoted °R.

Acceleration. The *acceleration a* of a body is the rate at which its velocity changes with time; the change in velocity may be in magnitude or direction or both.

Acceleration of Gravity. The *acceleration of gravity g* is the acceleration of a freely falling body near the earth's surface. Its value is 32 ft/s² in British units and 9.8 m/s² in metric units.

Alpha Particle. An *alpha particle* is the nucleus of a helium atom. It consists of two neutrons and two protons, and it is emitted in the radioactive decay of certain isotopes.

Alternating Current. The direction of an *alternating electric current* reverses itself periodically.

Ampere. The *ampere* (A) is the unit of electric current. It is equal to a flow of charge at the rate of 1 C/s.

Amplitude. The *amplitude* of a body undergoing simple harmonic motion is its maximum displacement on either side of its equilibrium position. The amplitude of a wave is the maximum value of the wave variable (for instance, displacement, pressure, electric field) regardless of sign.

Angular Momentum. The *angular momentum L* of a rotating body is the product $I\omega$ of its moment of inertia and angular velocity. The principle of *conservation of angular momentum* states that the total angular momentum of a system of particles remains constant when no net external torque acts upon the system.

Angular Velocity. The *angular velocity* ω of a rotating body is the angle through which it turns per unit time. The *angular acceleration* α of a rotating body is the rate of change of its angular velocity with respect to time.

Antiparticle. Nearly every elementary particle has an *antiparticle* counterpart whose electric charge and certain other properties have the opposite sign. Thus the antiparticle of the electron is the *positron* (charge $+e$) and that of the proton is the *antiproton* (charge $-e$). When a particle and an antiparticle of the same kind come together, they *annihilate* each other, with the vanished mass reappearing as photons or mesons.

Archimedes' Principle. *Archimedes' principle* states that the buoyant force on a submerged object is equal to the weight of fluid it displaces.

Atom. An *atom* is the ultimate particle of an element. Every atom consists of a very small positively-charged nucleus and a number of electrons at some distance from it. The nucleus contains nearly all the mass of the atom.

Atomic Mass Unit. Atomic and nuclear masses are expressed in *atomic mass units* (u), equal to 1.66×10^{-27} kg.

Atomic Number. The *atomic number* of an element is the number of electrons in each of its atoms or, equivalently, the number of protons in each of its atomic nuclei.

Back Emf.   A *back* (or *counter*) *emf* is induced in the rotating coils of an electric motor and is opposite in direction to the external voltage applied to them.

Beta Decay.   *Beta decay* is a type of radioactive decay in which a nucleus emits an electron or a positron.

Binding Energy.   The *binding energy* of a nucleus is the energy equivalent of the difference between its mass and the sum of the masses of its individual constituent nucleons. This amount of energy must be supplied to the nucleus if it is to be completely disintegrated. The binding energy per nucleon is least for very light and very heavy nuclei; hence the *fusion* of very light nuclei to form heavier ones and the *fission* of very heavy nuclei to form lighter ones are both processes that liberate energy.

Bohr Theory of the Atom.   According to the *Bohr theory of the atom*, an electron can circle an atomic nucleus indefinitely without radiating energy if its orbit is an integral number of electron wavelengths in circumference. The number of wavelengths that fit into a particular permitted orbit is called the *quantum number* of that orbit. The electron energies corresponding to the various quantum numbers constitute the *energy levels* of the atom, of which the lowest is the *ground state* and the rest are *excited states*.

Boyle's Law.   *Boyle's law* states that, at constant temperature, the absolute pressure of a sample of a gas is inversely proportional to its volume, so that pV = constant at that temperature regardless of changes in either p or V individually.

British Thermal Unit.   The *British thermal unit* (Btu) is the unit of heat in the British system of units. It is equal to that amount of heat required to change the temperature of 1 lb of water by 1°F.

Bulk Modulus.   The *bulk modulus* B of a material is equal to the pressure on a sample of it divided by the fractional decrease in its volume.

Capacitor.   A *capacitor* is a device that stores electrical energy in the form of an electric field. The ratio between the charge on either plate of a capacitor and the potential difference between the plates is called its *capacitance* C. The unit of capacitance is the *farad* (F) which is equal to 1 C/V.

Capacitive Reactance.   The *capacitive reactance* X_C of a capacitor is a measure of its effect on an alternating current; it is equal to $1/2\pi fC$, where C is the capacitance. If V_C is the effective voltage across a capacitor, the effective current that flows into and out of it is $i = V_C/X_C$. The unit of reactance is the ohm.

Carnot Engine.   A *Carnot engine* is an idealized engine which is not subject to such practical difficulties as friction or heat losses by conduction or radiation but which obeys all physical laws. The efficiency of a Carnot engine which absorbs heat at the absolute temperature T_1 and exhausts heat at the absolute temperature T_2 is equal to $1 - T_2/T_1$; no engine operating between the same two temperatures can be more efficient than a Carnot engine operating between them.

Celsius Temperature Scale.   In the *celsius* (centigrade) *temperature scale* the freezing point of water is assigned the value 0°C and the boiling point of water the value of 100°C.

Center of Gravity.   The *center of gravity* of a body is that point from which it can be suspended in any orientation without tending to rotate. The weight of a body can be considered as a downward force acting on its center of gravity.

Centripetal Acceleration.   The velocity of a body in uniform circular motion continually changes in direction although its magnitude remains constant. The body's acceleration is called *centripetal acceleration*, and it points toward the center of the circular path. Centripetal acceleration is proportional to the square of the body's velocity and inversely proportional to the radius of its path.

Centripetal Force.   The inward force that provides a body in uniform circular motion with its centripetal acceleration is called *centripetal force*.

Charles's Law.   *Charles's law* states that, at constant pressure, the volume of a sample of a gas is directly proportional to its absolute temperature, so that V/T = constant at that pressure regardless of changes in either V or T individually.

Coherence. Two sources of waves are *coherent* if there is a fixed phase relationship between the waves they emit during the time the waves are being observed. Interference can be observed only in waves from coherent sources.

Component of a Vector. A *component of a vector* is its projection in a specified direction.

Compound. Two or more elements may combine chemically to form a *compound*, a new substance whose properties are different from those of the elements that compose it.

Compression. When equal and opposite forces that act toward each other are applied to a body, it is said to be in *compression*.

Concave Mirror. A *concave mirror* curves inward toward its center and converges parallel light to a single *real focal point*. The distance from the mirror to the focal point is the *focal length* of the mirror.

Concurrent Forces. When the lines of action of the various forces that act on a body intersect at a common point, the forces are said to be *concurrent*; when their lines of action do not intersect, the forces are *nonconcurrent*.

Conduction. In *conduction*, heat is transferred from one place to another by successive molecular collisions.

Convection. In *convection*, heat is transferred from one place to another by the motion of a volume of hot fluid from one place to another.

Convex Mirror. A *convex mirror* curves outward toward its center and diverges parallel light as though the reflected light came from a single *virtual focal point* behind the mirror. The distance from the mirror to the focal point is the *focal length* of the mirror.

Coulomb. The *coulomb* (C) is the unit of electric charge.

Coulomb's Law. *Coulomb's law* states that the force one charge exerts upon another is directly proportional to the magnitudes of the charges and inversely proportional to the square of the distance between them. The force between like charges is repulsive, and that between unlike charges is attractive.

Covalent Bond. In a *covalent bond* between adjacent atoms of a molecule or solid, the atoms share one or more electron pairs.

Critical Point. The *critical point* is the upper limit of the vaporization curve of a substance, which cannot exist in the liquid state at a temperature above that of its critical point.

Crystalline Solid. Solids whose constituent atoms or molecules are arranged in regular, repeated patterns are called *crystalline*. When only short-range order is present, the solid is *amorphous*.

De Broglie Waves. A moving body behaves as though it has a wave nature. The waves representing such a particle are called *de Broglie waves*.

Density. The *density d* of a substance is its mass per unit volume.

Dielectric Constant. The *dielectric constant K* of a particular material is a measure of how effective it is in reducing an electric field set up across a sample of it.

Diffraction. The ability of waves to bend around the edges of obstacles in their paths is called *diffraction*. A *diffraction grating* is a series of parallel slits that produces a spectrum through the interference of light that is diffracted by them.

Dispersion. *Dispersion* refers to the splitting up of a beam of light containing different frequencies by passage through a substance whose index of refraction varies with frequency.

Domain. An assembly of atoms in a ferromagnetic material whose atomic magnetic moments are aligned is called a *domain*.

Doppler Effect. The *Doppler effect* refers to the change in frequency of a wave when there is relative motion between its source and an observer.

Effective Value. The *effective value* of an alternating current is such that a direct current of this magnitude produces heat in a resistor at the same rate as the alternating current. The relationship between i_{eff} and i_{max} is $i_{eff} = 0.707i_{max}$. The similar relationship $V_{eff} = 0.707V_{max}$ holds for the effective and maximum voltages in an ac-circuit.

Elastic Collision. A *completely elastic collision* is one in which kinetic energy is conserved. A *completely inelastic collision* is one in which the bodies stick together upon impact, which results in the maximum possible kinetic energy loss.

Elastic Limit. The *elastic limit* is the maximum stress a solid can experience without being permanently altered. Hooke's law is only valid when the elastic limit is not exceeded.

Elastic Potential Energy. A body under stress possesses *elastic potential energy* which is equal to the work done in deforming it.

Electric Charge. *Electric charge q*, like rest mass, is a fundamental property of certain of the elementary particles of which all matter is composed. There are two kinds of electric charge, *positive charge* and *negative charge*; charges of like sign repel, unlike charges attract. The unit of charge is the *coulomb* (C). All charges, of either sign, occur in multiples of the fundamental *electron charge* of 1.6×10^{-19} C. The principle of *conservation of charge* states that the net electric charge in an isolated system remains constant. Electric charge is relativistically invariant.

Electric Current. A flow of electric charge from one place to another is called an *electric current*. The unit of electric current is the *ampere* (A), which is equal to a flow of 1 C/s.

Electric Field. An *electric field E* exists wherever an electric force acts on a charged particle. The magnitude of an electric field at a point is defined as the force that would act on a charge of +1 C placed there. The unit of electric field is the volt/m, which is equal to 1 N/C.

Electrolysis. *Electrolysis* is the process by which free elements are liberated from a liquid by the passage of an electric current.

Electrolyte. A substance that separates into free ions when dissolved in water is called an *electrolyte* since the resulting solution is able to conduct electric current.

Electromagnetic Induction. *Electromagnetic induction* refers to the production of an electric field wherever magnetic lines of force are in motion.

Electromagnetic Waves. *Electromagnetic waves* consist of coupled electric and magnetic oscillations. The electric field of such a wave is perpendicular to its magnetic field, and both fields are perpendicular to the direction in which the wave travels. Radio waves, light waves, x-rays, and gamma rays are all electromagnetic waves differing only in their wavelength. Electromagnetic waves are produced by accelerated electric charges, and in free space have the velocity $c = 3.00 \times 10^8$ m/s.

Electromotive Force. The *electromotive force* (emf) of a battery, generator, or other source of electrical energy is the potential difference across its terminals when no current flows. When a current is flowing, the terminal voltage is less than the emf owing to the potential drop in the *internal resistance* of the source.

Electron. The *electron* is the least massive elementary particle found in matter. The electron has a charge of $-e$, where $e = 1.60 \times 10^{-19}$ coulomb.

Electron Volt. The *electron volt* is the energy acquired by an electron that has been accelerated by a potential difference of 1 volt. It is equal to 1.6×10^{-19} J.

Elementary particle. An *elementary particle* is one of the various types of particle found in nature that does not consist of a combination of other particles. The four major categories of elementary particles are the *photons*, the *leptons*, the *mesons*, and the *baryons*. Mesons and baryons, which are subject to the strong nuclear interaction, are jointly known as *hadrons*.

Elements. *Elements* are the simplest substances encountered in bulk. They cannot be decomposed nor transformed into one another by ordinary chemical or physical means.

Energy. *Energy E* is that which may be converted into work. When something possesses energy, it is capable of performing work or, in a general sense, of accomplishing a change in some aspect of the physical world. The metric and British units of energy are respectively the *joule* (J) and the ft·lb. The three broad categories of energy are *kinetic energy*, which is the energy something possesses by virtue of its motion; *potential energy*, which is the energy something possesses by virtue of its position in a force field; and *rest energy*, which is the energy something possesses by virtue of its mass. The principle of *conservation of energy* states that the total amount of energy in a system isolated from the rest of the universe always remains constant, although energy transformations from one form to another, including rest energy, may occur within the system.

Energy Band. The *energy bands* in a crystal are ranges of energy that correspond to energy levels in an atom. An electron in a crystal can only have an amount of energy that falls within one of its energy bands.

Equilibrium. A body not acted upon by a net force is in *translational equilibrium* and may be at rest or have a constant linear velocity. A body not acted upon by a net torque is in *rotational equilibrium* and may be at rest or have a constant angular velocity.

Equivalent Resistance. The *equivalent resistance* of a set of interconnected resistors is the value of the single resistor that can be substituted for the entire set without affecting the current that flows in the rest of any circuit of which it is a part.

Escape Velocity. The minimum velocity needed by an object to permanently escape from the gravitational attraction of an astronomical body such as a planet or star is called the *escape velocity* of the body.

Exclusion Principle. According to the Pauli *exclusion principle*, no two electrons in an atom can exist in the same quantum state.

Fahrenheit Temperature Scale. In the *fahrenheit temperature scale* the freezing point of water is assigned the value 32°F and the boiling point of water the value 212°F.

Farad. The unit of capacitance is the *farad* (F), which is equal to 1 C/V.

Foot-Pound. The *foot-pound* (ft·lb) is the unit of work and energy in the British system.

Force. A *force F* is any influence that can cause a body to be accelerated. The unit of force in the metric system is the *newton* (N), in the British system the *pound* (lb).

Force Field. A *force field* is a region of space at every point in which an appropriate test object would experience a force. Thus a *gravitational field* is a region of space in which an object by virtue of its mass is acted on by a force, and an *electric field* is a region of space in which an object by virtue of its electric charge is acted upon by a force.

Frame of Reference. A *frame of reference* is something with respect to which observations are made on something else. When a person standing beside a road sees a car moving along the road, the road is the frame of reference; when a person in the car sees the roadside moving backward past him, the car is the frame of reference. All measurements, including those of time, have meaning only when the frame of reference from which they are made is specified. All frames of reference are equally valid—there is no universal frame of reference. The *special theory of relativity* relates measurements made on an object or phenomenon from frames of reference moving at constant velocity with respect to one another.

Frequency. The *frequency f* of a body undergoing harmonic motion is the number of oscillations it makes per unit time. The frequency of a train of waves is the number of waves that pass a particular point per unit time. The usual unit of frequency is the *hertz* (Hz), which is equal to 1 cycle/s.

Frequency Modulation. In *frequency modulation*, information is contained in variations in the frequency of the carrier wave.

Friction. The term *friction* refers to the resistive forces that arise to oppose the motion of a body past another with which it is in contact. *Sliding friction* is the frictional resistance a body in motion experiences, while *static friction* is the frictional resistance a stationary body must overcome in order to be set in motion. The *coefficient of friction* is the constant of proportionality for a given pair of contacting surfaces that relates the frictional force between them to the normal force holding them together; usually the coefficient of static friction is greater than that of sliding friction. *Rolling friction* refers to the resistance a circular object experiences as it rolls over a smooth, flat surface; coefficients of rolling friction are much smaller than those of sliding friction.

Fusion, Heat of.   The *heat of fusion* of a substance is the amount of heat that must be supplied to change a unit quantity of it at its melting point from the solid to the liquid state; the same amount of heat must be removed from a unit quantity of the substance in the liquid state at its melting point to change it to a solid.

Gamma Ray.   A *gamma ray* is an energetic photon emitted by certain radioactive nuclei.

Gauge Pressure.   *Gauge pressure* is the difference between true pressure and atmospheric pressure.

Generator.   A *generator* is a device that converts mechanical energy into electrical energy.

Gravitation.   Newton's *law of universal gravitation* states that every body in the universe attracts every other body with a force directly proportional to both their masses and inversely proportional to the square of the distance separating them.

Half-life.   The time required for half of a given sample of a radioactive substance to decay is called its *half-life*.

Harmonic Motion.   *Simple harmonic motion* is an oscillatory motion that occurs whenever a force acts on a body in the opposite direction to its displacement from its normal position, with the magnitude of the force proportional to the magnitude of the displacement. The period of a simple harmonic oscillator is independent of its amplitude.

Heat.   *Heat* is a quantity whose addition to a body of matter causes its internal energy to increase and whose removal from a body of matter causes its internal energy to decrease. If the matter does not change state during the process and does no work nor has work done on it, the change in internal energy results in a corresponding change in temperature. The unit of heat in the metric system is the *kilocalorie* (kcal), that amount of heat required to change the temperature of 1 kg of water by 1°C; the unit of heat in the British system is the *British thermal unit* (Btu), that amount of heat required to change the temperature of 1 lb of water by 1°F.

Henry.   The unit of inductance is the *henry* (H), which is equal to 1 V·s/A.

Hertz.   The unit of frequency is the *hertz* (Hz), which is equal to 1 cycle/s.

Hooke's Law.   *Hooke's law* states that the amount of deformation experienced by a body under stress is proportional to the magnitude of the stress. Thus the elongation of a wire is proportional to the tension applied to it. Hooke's law is only valid when the elastic limit of the body is not exceeded.

Horsepower.   The *horsepower* is a unit of power equal to 550 ft·lb/s, which is 746 W.

Ideal Gas Law.   The equation $pV/T =$ constant, a combination of Boyle's and Charles's laws, is called the *ideal gas law* and is obeyed approximately, though not exactly, by all gases.

Image.   A *real image* of an object is formed by light rays that pass through the image; the image would therefore appear on a properly placed screen. A *virtual image* can only be seen by the eye because the light rays that seem to come from the image actually do not pass through it.

Impedance.   The *impedance Z* of an ac-circuit is analogous to the resistance of a dc-circuit. When the effective alternating potential difference V is applied to a circuit of impedance Z, the effective current that flows is $i = V/Z$. The unit of impedance is the ohm.

Impulse.   The *impulse* of a force is the product of the force and the time during which it acts. Impulse is a vector quantity having the direction of the force. When a force acts on a body that is free to move, its change in momentum equals the impulse given it by the force.

Inductance.   The *inductance L* of a circuit is the ratio between the magnitude of the self-induced emf $\mathcal{E}$ due to a changing current in it and the rate of change $\Delta i/\Delta t$ of the current. The unit of inductance is the *henry* (H), which is equal to 1 V·s/A.

Inductive Reactance.   The *inductive reactance X_L* of an inductor is a measure of its effect on an alternating current; it is equal to $2\pi f L$, where f is the frequency of the current and L the inductance. If V_L is the effective voltage across an inductor, the effective current that flows through it is $i = V_L/X_L$. The unit of reactance is the ohm.

Inertia. The term *inertia* refers to the apparent resistance a body offers to changes in its state of motion.

Interactions, Fundamental. There are only four *fundamental interactions* that are responsible for all the forces in the universe. In order of decreasing strength, these are the *strong nuclear, electromagnetic, weak nuclear,* and *gravitational* interactions. The two nuclear interactions are effective only over short distances. The electromagnetic and gravitational interactions are unlimited in range, but become weaker with increasing distance.

Interference. The interaction of different waves of the same nature is called *interference: constructive interference* occurs when the resulting composite wave has an amplitude greater than that of either of the original waves, and *destructive interference* occurs when the resulting composite wave has an amplitude less than that of either of the original waves.

Ion. An *ion* is an atom or group of atoms that carries an electric charge. An atom or group of atoms becomes a negative ion when it picks up one or more electrons in addition to its normal number, and becomes a positive ion when it loses one or more of its usual number.

Ionic Bond. Electrons are transferred between the atoms of certain solids so that the resulting crystal consists of positive and negative ions rather than of neutral atoms. Such a solid is said to be held together by *ionic bonds.* The attractive forces between ions of opposite charge balance the repulsive forces between ions of like charge in an ionic solid.

Isotope. The *isotopes* of an element have the same atomic number but different mass numbers. Thus the nuclei of the isotopes of an element all contain the same number of protons but have different numbers of neutrons.

Joule. The *joule* (J) is the unit of work and energy in the metric system. It is equal to 1 kg-m^2/s^2.

Kilocalorie. The *kilocalorie* (kcal) is the unit of heat in the metric system. It is equal to that amount of heat required to change the temperature of 1 kg of water by 1°C. 1 kcal = 4185 J.

Kilogram. The *kilogram* (kg) is the unit of mass in the metric system. One kilogram weighs 2.21 lb at the earth's surface.

Kinetic Energy. *Kinetic energy* is the energy a moving body possesses by virtue of its motion. If the body has the mass m and velocity v, its kinetic energy is $\frac{1}{2}mv^2$.

Kinetic Theory. According to the *kinetic theory of gases,* a gas consists of a great many tiny individual molecules that do not interact with one another except when collisions occur. The molecules are far apart compared with their dimensions and are in constant random motion. The ideal gas law may be derived from the kinetic theory of gases.

Kirchhoff's Rules. *Kirchhoff's rules* for network analysis are: (1) The sum of the currents flowing into a junction of three or more wires is equal to the sum of the currents flowing out of the junction; (2) The sum of the emf's around a closed conducting loop is equal to the sum of the iR potential drops around the loop.

Laminar Flow. In *laminar* (or *streamline*) *flow* every particle of fluid passing a particular point follows the same path, whereas in *turbulent flow* irregular whirls and eddies occur.

Laser. A *laser* is a device for producing a narrow, monochromatic, coherent beam of light. The term stands for *l*ight *a*mplification by *s*timulated *e*mission of *r*adiation.

Lens. A *lens* is a transparent object of regular form that can produce an image of an object placed before it. A *converging lens* brings parallel light to a single real focal point, while a *diverging lens* deviates parallel light outward as though it originated at a single virtual focal point.

Lenz's Law. *Lenz's law* states that the direction of an induced current must be such that its own magnetic field opposes the changes in flux that are inducing it.

Lines of Force. *Lines of force* are means for visualizing a force field. Their direction at any point is that in which a test body would move if released there, and their concentration in the neighborhood of a point is proportional to the magnitude of the force on a test particle at that point. (In the case of a magnetic field, the direction of a line of force at a point is that in which a moving charge would experience no force.)

Longitudinal Waves. *Longitudinal waves* occur when the individual particles of a medium vibrate back and forth in the direction in which the waves travel. Sound consists of longitudinal waves.

Magnetic Field. A *magnetic field B* exists wherever a magnetic force acts on a moving charged particle. The magnitude B of a magnetic field at a point is defined as the force that would act on a charge of $+1$ C moving at a speed of 1 m/s past that point, when the direction of the motion is such as to result in the maximum force. The unit of magnetic field is the *tesla* (T), equal to 1 N/A·m.

Magnetic Force. Electric charges in motion relative to an observer appear to exert forces upon one another that are different from the forces they exert when at rest. These differences are by custom attributed to *magnetic forces*. In reality, magnetic forces represent relativistic corrections to electric forces due to the motion of the charges involved.

Mass. The property of matter that manifests itself as inertia is called *mass m*. The *rest mass* of a body is its mass when stationary with respect to an observer. The metric unit of mass is the *kilogram* (kg), the British unit is the *slug*.

Mass Number. The *mass number* of a nucleus is the number of nucleons it contains.

Matter Waves. A moving body behaves as though it has a wave character. The waves representing such a body are *matter waves*, also called *de Broglie waves*. The wave variable in a matter wave is its *wave function*, whose square is the *probability density* of the body. The value of the probability density of a particular body at a certain place and time is proportional to the probability of finding the body at that place at that time. Matter waves may thus be regarded as waves of probability.

Mechanical Advantage. The *mechanical advantage* of a machine is the ratio between the output force it exerts and the input force that is furnished to it. The *theoretical mechanical advantage* (TMA) is its value under ideal circumstances, while the *actual mechanical advantage* (AMA) is its value when friction is taken into account.

Mechanical Equivalent of Heat. The *mechanical equivalent of heat* is the constant ratio between the energy dissipated in some mechanical process and the heat that appears as a result of the process. It is equal to 4185 J/kcal in the metric system and 778 ft·lb/Btu in the British system.

Metallic Bond. The *metallic bond* which holds metal atoms together in the solid state arises from a "gas" of freely moving electrons pervading the entire metal.

Meter. The *meter* (m) is the unit of length in the metric system. One meter is equal to 3.28 ft.

Molecule. A *molecule* is a group of atoms that stick together strongly enough to act as a single particle. A molecule of a given compound always has a certain definite structure and is complete in itself with little tendency to gain or lose atoms.

Moment Arm of a Force. The *moment arm* of a force is the perpendicular distance from its line of action to a pivot point.

Moment of Inertia. The *moment of inertia I* of a body about a given axis is the rotational analog of mass in linear motion. Its value depends upon the way in which the mass of the body is distributed about the axis.

Momentum, Linear. The linear momentum of a body is the product of its mass and velocity. Linear momentum is a vector quantity whose direction is that of the body's velocity. The principle of *conservation of linear momentum* states that the total linear momentum of a system of particles isolated from the rest of the universe remains constant regardless of what events occur within the system.

Motion, Laws of. *Newton's first law of motion* states that a body at rest will remain at rest and a body in motion will continue in motion in a straight line at constant velocity in the absence of any interaction with the rest of the universe. *Newton's second law of motion* states that the net force acting on a body is equal to the rate of change of the body's linear momentum. *Newton's third law of motion* states that, when a body exerts a force on another body, the second body exerts a force on the first body of the same magnitude but in the opposite direction.

Neutrino. The neutrino is a massless, uncharged particle that is emitted in the decay of certain elementary particles. It can possess energy and both linear and angular momentum.

Neutron. The *neutron* is an electrically neutral elementary particle, slightly heavier than the proton, which is present together with protons in atomic nuclei.

Newton. The *newton* (N) is the unit of force in the metric system. It is equal to 1 kg-m/s². One newton is equal to 0.225 pound.

Nuclear Fission. In *nuclear fission*, the absorption of neutrons by certain heavy nuclei causes them to split into smaller *fission fragments* with the release of energy. Because each fission also liberates several neutrons, a rapidly multiplying sequence of fissions called a *chain reaction* can occur if a sufficient amount of the proper material is assembled. A *nuclear reactor* is a device in which a chain reaction can be initiated and controlled.

Nuclear Fusion. In *nuclear fusion*, two light nuclei combine to form a heavier one with the evolution of energy. The sun and stars obtain their energy from fusion reactions.

Nucleus. The *nucleus* of an atom is located at its center and contains all of the positive charge and most of the mass of the atom. The nucleus consists of protons and neutrons.

Nucleon. Neutrons and protons, the constituents of atomic nuclei, are jointly called *nucleons*.

Ohm. The *ohm* (Ω) is the unit of electrical resistance. It is equal to 1 V/A.

Ohm's Law. *Ohm's law* states that the current in a metallic conductor is proportional to the potential difference between its ends. Thus in such a conductor $i = V/R$.

Pascal's Principle. *Pascal's principle* states that an external pressure exerted on a fluid is transmitted uniformly throughout its volume.

Periodic Law. The *periodic law* of chemistry states that if the elements are listed in order of atomic number, elements with similar properties recur at regular intervals. The quantum theory of the atom together with the exclusion principle is able to explain the origin of the periodic law.

Period. *The period T* of a body undergoing simple harmonic motion is the time required for it to make one complete oscillation. The period of a wave is the time required for one complete wave to pass a particular point.

Permanent Magnet. A *permanent magnet* is an object composed of a ferromagnetic material whose atomic current loops have been aligned by an external magnetic field and which remain aligned after the external field is removed.

Permeability. The *permeability* of a medium is a measure of its magnetic properties. *Paramagnetic* and *ferromagnetic* substances have higher permeabilities than free space and *diamagnetic* ones have lower permeabilities.

Phase Relationships. The *phase relationships* between the instantaneous voltage and instantaneous current in ac-circuit components are as follows: the voltage across a pure resistor is in phase with the current; the voltage across a pure inductor leads the current by $\frac{1}{4}$ cycle; the voltage across a pure capacitor lags behind the current by $\frac{1}{4}$ cycle.

Photoelectric Effect. The *photoelectric effect* refers to the emission of electrons from a metal surface when light is shined on it.

Plasma. A *plasma* is a gas composed of electrically charged particles, and its behavior depends strongly upon electromagnetic forces. Most of the matter in the universe is in the plasma state.

Polar Molecule. A *polar molecule* is one whose

charge distribution is asymmetrical, so that one end is positive and the other negative even though the molecule as a whole is electrically neutral.

Polarization. A *polarized* beam of transverse waves is one whose vibrations occur in only a single direction perpendicular to the direction in which the beam travels, so that the entire wave motion is confined to a plane called the *plane of polarization*. An *unpolarized* beam of transverse waves is one whose vibrations occur equally often in all directions perpendicular to the direction in which the beam travels.

Pole, Magnetic. The ends of a permanent magnet are called its *poles*. Magnetic lines of force leave the *north pole* of a magnet and enter its *south pole*.

Positron. A *positron* is a positively charged electron.

Potential Difference. The electrical *potential difference V* between two points is the work that must be done to take a charge of 1 C from one point to the other. The unit of potential difference is the *volt* (V), which is equal to 1 J/C.

Potential Energy. *Potential energy* is the energy a body has by virtue of its position. The gravitational potential energy of a body of mass m at a height h above a particular reference point is mgh; if its weight w is specified, its potential energy is wh. Other examples of potential energy are that of a planet with respect to the sun, that of a piece of iron with respect to a magnet, and that of a body at the end of a stretched spring with respect to its equilibrium position.

Pound. The *pound* (lb) is the unit of force in the British system.

Power. The rate at which work is done is called *power*. The unit of power in the metric system is the *watt* (w), which is equal to 1 J/s, and in the British system it is the *ft·lb/s*. One *horsepower* is equal to 746 W or 550 ft·lb/s.

Power Factor. The *power factor* of an ac-circuit is the ratio between the power consumed in the circuit and the product of the effective current and voltage there; this ratio is equal to that between the resistance and the impedance of the circuit, and is less than 1, except at resonance. The unit of apparent power $V_{eff}i_{eff}$ is the *volt-ampere*, as distinct from the watt, which is the unit of consumed power.

Poynting Vector. The *Poynting vector S* describes the flow of energy in an electromagnetic wave. Its direction is that of the wave, and its magnitude is equal to the rate at which energy is being transported by the wave per unit cross-sectional area.

Pressure. The *pressure* on a surface is the perpendicular force per unit area that acts upon it. *Gauge pressure* is the difference between true pressure and atmospheric pressure.

Proton. The *proton* is an elementary particle found in all atomic nuclei; its charge is $+e$ and its mass is 1836 times that of the electron.

Quantum Theory of the Atom. In the *quantum theory of the atom* only experimentally measurable quantities are considered, and no use is made of mechanical models inconsistent with the uncertainty principle. According to this theory, four quantum numbers are required to describe each electron in an atom. These are the *principal quantum number n*, which governs the electron's energy, the *orbital quantum number l*, which governs the magnitude of its angular momentum, the *magnetic quantum number m_l*, which governs the orientation of its angular momentum, and the *spin magnetic quantum number m_s*, which governs the orientation of its spin.

Quantum Theory of Light. The *quantum theory of light* states that light travels in tiny bursts of energy called *quanta* or *photons*. If the frequency of the light is f, each burst has the energy hf, where h is known as *Planck's constant*. The quantum theory of light complements the wave theory of light.

Quarks. Elementary particles subject to the strong nuclear interaction are believed to be composed of *quarks*, which are entities with fractional electric charges that have not as yet been experimentally isolated but whose existence is supported by indirect evidence.

Radian. The *radian* is a unit of angular measure equal to 57.30°. If a circle is drawn whose center is at the vertex of an angle, the angle in radian measure is equal to the ratio between the arc of the circle cut by the angle and the radius of the circle. A full circle contains 2π radians.

Radiation. In *radiation*, energy is transferred from one place to another in the form of electromagnetic waves, which require no material medium for their passage.

Radiation Pressure. When electromagnetic waves strike a surface, they exert *radiation pressure* on it because of their momenta.

Radioactivity. *Radioactive nuclei* spontaneously transform themselves into other nuclear species with the emission of radiation. The radiation may consist of *alpha particles,* which are the nuclei of helium atoms, or *beta particles,* which are positive or negative electrons. The emission of *gamma rays,* which are energetic photons, enables an excited nucleus to lose its excess energy.

Reflection. In *diffuse reflection* an incident beam of parallel light is spread out in many directions, while in *specular reflection* the angle of reflection is equal to the angle of incidence.

Refraction. The bending of a light beam when passing from one medium to another is called *refraction.* The quantity that governs the degree to which a light beam will be deflected in entering a medium is its *index of refraction,* defined as the ratio between the speed of light in free space and its value in the medium.

Refrigerator. A *refrigerator* is a device that transfers heat from a cold reservoir to a hot one, and it must expend energy in order to do this. In essence, it is a heat engine operating in reverse.

Relative Humidity. The *relative humidity* of a volume of air is the ratio between the amount of water vapor it contains and the amount that would be present at saturation.

Relativity. The *special theory of relativity* relates measurements made on an object or phenomenon from frames of reference moving at constant velocity with respect to one another. The *Lorentz contraction* refers to the decrease in the measured length of an object when it is moving relative to an observer. The relativistic *time dilation* refers to the fact that a clock moving with respect to an observer appears to tick less rapidly than it does to an observer traveling with the clock. The *relativity of mass* refers to the increase in the measured mass of an object when it is moving relative to an observer.

Resistance. The *resistance R* of a body of matter is a measure of the extent to which it impedes the passage of electric current. It is defined as the ratio between the potential difference applied across the ends of the body and the resulting current. The unit of resistance is the *ohm* (Ω), which is equal to 1 V/A. The resistance of a conductor is proportional to its length and to the *resistivity* ρ of the material of which it is made, and inversely proportional to its cross-sectional area.

Resolution of Vectors. A vector can be *resolved* into two or more other vectors whose sum is equal to the original vector. The new vectors are called the *components* of the original vector, and are normally chosen to be perpendicular to one another.

Resolving Power. The *resolving power* of an optical system refers to its ability to produce separate images of nearby objects. Resolving power is limited by diffraction; the larger the objective lens of an optical system, the greater its resolving power.

Resonance. *Resonance* occurs when periodic impulses are given to an object at a frequency equal to one of its natural frequencies of oscillation.

Resonant Frequency. The *resonant frequency* of an ac-circuit is that frequency for which the impedance is a minimum; its value is $f_0 = 1/2\pi \sqrt{LC}$.

Right-Hand Rule for Magnetic Field. According to the *right-hand rule for magnetic field,* when a current-carrying wire is grasped with the right hand so that the thumb points in the direction of the current, the curled fingers of that hand then point in the direction of the magnetic field.

Right-Hand Rule for Magnetic Force. Open the right hand so the fingers are together and the thumb sticks out. According to the *right-hand rule for magnetic force,* when the thumb is in the direction of motion of a positive charge and the fingers are in the direction of a magnetic field, the palm faces in the direction of the force acting on the charge. When a negative charge is involved, the force is in the opposite direction.

Scalar Quantity. A *scalar quantity* is one that has magnitude only.

Shear. When equal and opposite forces that do not act along the same line are applied to a body, it is said to be in *shear.* The *shear modulus S* of a material is equal to the shear force per unit cross-sectional area applied to a sample of it divided by the relative distortion of the sample.

Shells, Atomic. The electrons in an atom that have the same total quantum number *n* are said to occupy the same *shell.* Electrons in a given shell which have the same orbital quantum number *l* are said to occupy the same *subshell.* Shells and subshells containing the maximum number of electrons permitted by the exclusion principle are *closed.* Atoms whose subshells are all closed possess unusual stability.

Slug. The *slug* is the unit of mass in the British system. The weight of 1 slug is 32 lb.

Snell's Law. *Snell's law* states that the ratio between the sine of the angle of incidence of a light ray upon a boundary between two media and the sine of the angle of refraction is equal to the ratio of the speeds of light in the two media.

Sound. *Sound* is a longitudinal wave phenomenon that consists of successive compressions and rarefactions of the medium through which it travels. Sound intensity in air is measured in *decibels.*

Specific Heat Capacity. The *specific heat capacity c* of a substance is the amount of heat required to change the temperature of a unit quantity of it by 1°. The unit of specific heat capacity in the metric system is the kcal/kg·°C; in the British system it is the Btu/lb·°F.

Spectrum. An *absorption spectrum* results when white light is passed through a cool gas; it is a *dark line spectrum* because it appears as a series of dark lines on a bright background, with the lines representing characteristic wavelengths absorbed by the gas.

An *emission spectrum* consists of the various wavelengths of light emitted by an excited substance; it may be a *continuous spectrum,* in which all wavelengths are present, or a *bright line spectrum,* in which only a few wavelengths characteristic of the individual atoms of the substance are present.

Speed. The *speed* of a moving body is the rate at which it covers distance. Speed is the magnitude of velocity and is a scalar quantity.

Spin. Every electron has a certain intrinsic amount of angular momentum called its *spin.* The spin of an electron is as fundamental a property as its mass or electric charge. Owing to its spin, every electron acts like a tiny bar magnet. Most other elementary particles (such as the proton and neutron) also have spin associated with them.

Standing Waves. *Standing waves* in a stretched string are transverse harmonic oscillations that result in a wave pattern whose amplitude varies in a sinusoidal manner along the string. Points of zero amplitude are called *nodes.*

Sublimation. *Sublimation* is the direct conversion of a substance from the solid to the vapor state, or vice versa, without it first becoming a liquid.

Superposition, Principle of. The *principle of superposition* states that when two or more waves of the same nature travel past a given point at the same time, the amplitude at the point is the sum of the amplitudes of the individual waves.

Temperature. The *temperature T* of a body of matter

is a measure of the average kinetic energy of random motion of its constituent particles. When two bodies are in contact, heat flows from the one at the higher temperature to the one at the lower temperature.

Tension. When equal and opposite forces that act away from each other are applied to a body, it is said to be in *tension*.

Terminal Velocity. Because air resistance varies with velocity, a falling body eventually reaches a *terminal velocity* after which it ceases to be accelerated downward.

Tesla. The *tesla* (T) is the unit of magnetic field. It is equal to 1 N/A·m.

Thermal Expansion. The *coefficient of linear expansion* is the ratio between the change in length of a solid rod of a particular material and its original length per 1° change in temperature. The *coefficient of volume expansion* is the ratio between the change in volume of a sample of a particular solid or liquid and its original volume per 1° change in temperature.

Thermodynamics. The *first law of thermodynamics* states that the work output of any engine is equal to its net energy input plus any decrease in its stored energy. This law is thus a restatement of the principle of conservation of energy. The *second law of thermodynamics* states that no engine can be completely efficient in converting energy to work—some of the input energy must be wasted as heat. This law is a consequence of the tendency of all physical systems to become more and more disordered as time goes on.

Thermometer. A *thermometer* is a device for measuring temperature. The two temperature scales in common use are the *celsius* (centigrade) scale, in which the freezing point of water is assigned the value 0°C and its boiling point the value 100°C, and the *fahrenheit* scale, in which these points are assigned the values 32°F and 212°F, respectively.

Thermonuclear Energy. The energy liberated by nuclear fusion is called *thermonuclear energy*.

Torque. The *torque* τ of a force about a particular axis is the product of the magnitude of the force and the perpendicular distance from the line of action of the force to the axis. The latter distance is called the *moment arm* of the force. Torque plays the same role in rotational motion that force does in linear motion.

Transformer. An alternating current flowing in the primary coil of a *transformer* induces another alternating current in the secondary coil. The ratio of the emfs is proportional to the ratio of turns in the coils.

Transverse Waves. *Transverse waves* occur when the individual particles of a medium vibrate from side to side perpendicular to the direction in which the waves travel. The vibrations of a stretched string are transverse waves.

Triple point. The *triple point* of a substance refers to the temperature and pressure at which its solid, liquid, and vapor states can exist in equilibrium with one another.

Uncertainty Principle. The *uncertainty principle* is an expression of the limits set by the wave nature of matter on finding the position and state of motion of a moving body. According to this principle, the product of the uncertainties in simultaneous measurements of the position and momentum of a body cannot be less than $h/2\pi$.

Uniform Circular Motion. A body traveling in a circle at constant speed is said to be undergoing uniform circular motion.

Van der Waals Force. *Van der Waals forces* originate in the electric attraction between asymmetrical charge distributions in atoms and molecules. Molecular solids and liquids are held together by van der Waals forces.

Vaporization, Heat of. The *heat of vaporization* of a substance is the amount of heat that must be supplied to change a unit quantity of it at its boiling point from the liquid to the gaseous (or vapor) state; the same amount of heat must be removed from a unit quantity of the substance at its boiling point to change it into a liquid.

Vector. A *vector* is an arrowed line whose length is proportional to the magnitude of some vector quantity and whose direction is that of the quantity. A *vector diagram* is a scale drawing of the various forces, velocities, or other vector quantities involved in the motion of a body. In *vector addition*, the tail of each successive vector is placed at the head of the previous one, with their lengths and original directions kept unchanged. The *resultant* is a vector drawn from the tail of the first vector to the head of the last. A vector can be *resolved* into two or more other vectors called the *components* of the original vector. Usually the components of a vector are chosen to be in mutually perpendicular directions.

Vector Quantity.   A *vector quantity* is one that has both magnitude and direction. The symbol of a vector quantity is printed in bold-face type, for instance **A**.

Velocity.   The *velocity v* of a body is a specification of both its speed and the direction in which it is moving. Velocity is a vector quantity. The *instantaneous velocity* of a body is its velocity at a specific instant of time. The *average velocity* of a body is the total displacement through which it has moved in a time interval divided by the interval.

Viscosity.   The *viscosity* of a fluid is a measure of its internal friction.

Volt.   The unit of electrical potential difference is the *volt (V)*. It is equal to 1 J/C.

Watt.   The unit of power is the *watt* (W), which is equal to 1 J/s.

Wave Motion.   *Wave motion* is characterized by the propagation of a change in a medium, rather than by the net motion of the medium itself. The passage of a wave across the surface of a body of water, for instance, involves the motion of a pattern of alternate crests and troughs, with the individual water molecules themselves ideally executing uniform circular motion.

Wavefront.   A *wavefront* is an imaginary surface that joins points where all the waves from a source are in the same phase of oscillation. According to *Huygens' principle*, every point on a wavefront can be considered as a point source of secondary wavelets which spread out in all directions with the wave velocity of the medium. The wavefront at any time is the envelope of these wavelets.

Weight.   The *weight w* of a body is the gravitational force exerted on it by the earth. The weight of a body is proportional to its mass.

Work.   Whenever a force affects the motion of a body, the body undergoes a displacement while the force acts on it. The product of the force and the component of the displacement of the body in the direction of the force is called the *work W* done by the force on the body. Work is a measure of the change (in a general sense) a force gives rise to when it acts upon something. In the metric system the unit of work is the *joule (J)*, and in the British system it is the *foot·pound* (ft·lb).

X-rays.   *X-rays* are high-frequency electromagnetic waves emitted when fast electrons impinge on matter.

Young's Modulus.   *Young's modulus Y* of a particular material is equal to the tension or compression force per unit cross-sectional area applied to a sample of that material divided by the fractional change in the length of the sample.

Solutions to Odd-numbered Exercises and Problems

Chapter 1

Exercises

1. No. The distinction between vector and scalar quantities is simply that vector quantities have directions associated with them, and both kinds of quantity are found in the physical world.

3. Yes.

5. $368 \text{ ft} \times 0.305 \text{ m/ft} = 112 \text{ m} = 0.112 \text{ km}$.

7. $1 \text{ acre} = 4840 \text{ yd}^2 \times 9 \text{ ft}^2/\text{yd}^2 = 4.356 \times 10^4$ ft^2; $1 \text{ mi}^2 = (5280 \text{ ft})^2/(4.356 \times 10^4 \text{ ft}^2/\text{acre})$ $= 640 \text{ acres}$.

9. (a) $1 \text{ bd ft} = (12 \text{ in})^2 \times 1 \text{ in} = 144 \text{ in}^3$.
    (b) Since $1 \text{ ft}^3 = (12 \text{ in})^3 = 1728 \text{ in}^3$, 1 bd ft $= 144 \text{ in}^3/(1728 \text{ in}^3/\text{ft}^3) = 0.0833 \text{ ft}^3$. (c) Since $1 \text{ in}^3 = (2.54 \text{ cm})^3 = 16.4 \text{ cm}^3$, $1 \text{ bd ft} = 144 \text{ in}^3$ $\times 16.4 \text{ cm}^3/\text{in}^3 = 2.36 \times 10^3 \text{ cm}^3$.

11. $v = 50 \text{ km/hr} \times 0.621 \text{ mi/km} = 31 \text{ mi/hr}$.

13. Horizontal plane: $s_h = \sqrt{(70 \text{ m})^2 + (60 \text{ m})^2}$ $= 92 \text{ m}$; total displacement $= s = \sqrt{s_h^2 + s_v^2}$ $= 100 \text{ m}$.

15. $F_y = 50 \text{ lb} + 25 \text{ lb} \times \sin 45° = 67.7 \text{ lb}$.

17. $F_y = F \sin 70° = 9.4 \text{ lb}$.

19. Alpha: $v_w = v_\alpha \cos 40° = 3.83 \text{ mi/hr}$. Beta: $v_w = v_\beta \cos 50° = 3.86 \text{ mi/hr}$. Beta has the higher windward component of velocity.

21. $F = \sqrt{(100 \text{ lb})^2 + (40 \text{ lb})^2} = 108 \text{ lb}$.

Problems

1. $1 \text{ gal} = [231 \text{ in}^3 \times (2.54 \text{ cm/in})^3]/10^3 \text{ cm}^3/\text{liter}$ $= 3.785 \text{ liters}$; $1 \text{ liter} = (1/3.785) \text{ gal} = 0.264 \text{ gal}$.

3. $v = v_x/\cos 37° = 87.6 \text{ mi/hr}$.

5. $F_x = 10 \text{ lb} \times \sin 40° = 6.4 \text{ lb}$; $F_y = 10 \text{ lb}$ $\times \cos 40° = 7.7 \text{ lb}$.

7. The angle between the path of ball 1 and the $+x$ direction is specified by $\tan \theta_1 = v_y/v_x = 2$, $\theta_1 = 63°$. Similarly $\tan \theta_2 = v_y/v_x = 1.5$, $\theta_2 = 56°$. The difference in angle is $7°$.

9. Calling E the $+x$ direction and N the $+y$ direction, $s_x = 200 \text{ mi} - (100 \text{ mi} \times \sin 45°)$ $= 129 \text{ mi}$ and $s_y = -200 \text{ mi} + (100 \text{ mi} \times \cos 45°)$ $= -129 \text{ mi}$. Hence $s = \sqrt{s_x^2 + s_y^2} = 182 \text{ mi}$. The direction is NW since s_x and s_y are equal in length.

11. (a) $v_y = v \sin \theta = 39.1 \text{ mi/hr}$; $h = v_y t = 39.1$ mi/hr $\times (1/60) \text{ hr} \times 5280 \text{ ft/mi} = 3441 \text{ ft}$.
    (b) $v_x = v \cos \theta = 92.1 \text{ mi/hr}$; $s = v_x t = 92.1$ mi/hr $\times (1/60) \text{ hr} = 1.54 \text{ mi}$.

13. The two velocities are perpendicular, so $v = \sqrt{(7 \text{ kn})^2 + (3 \text{ kn})^2} = 7.6 \text{ kn}$. If θ is the angle between $\mathbf{v}$ and NW, then $\tan \theta = 3 \text{ kn}/7 \text{ kn}$ $= 0.429$, $\theta = 23°$. Since NW is itself $45°$ W of N, the direction of $\mathbf{v}$ is $45° + 23° = 68°$ W of N.

15. Here $A_x = A \sin 37° = 6 \text{ cm}$; $A_y = A \cos 37°$ $= 8 \text{ cm}$; $B_x = B \cos 37° = 8 \text{ cm}$; $B_y = -B \sin 37°$ $= -6 \text{ cm}$. (a) $R_x = A_x + B_x = 14 \text{ cm}$, $R_y = A_y + B_y = 2 \text{ cm}$; $R = \sqrt{R_x^2 + R_y^2} = 14.1$ cm. $\mathbf{R}$ points θ clockwise from the $+y$ direction where $\tan \theta = R_x/R_y = 7$, $\theta = 82°$. (b) $R_x = A_x - B_x = -2 \text{ cm}$, $R_y = A_y - B_y = 14 \text{ cm}$; $R = \sqrt{R_x^2 + R_y^2} = 14.1 \text{ cm}$. $\mathbf{R}$ points θ counterclockwise from the $+y$ direction where $\tan \theta = R_x/R_y = 0.286$, $\theta = 16°$. (c) $R_x = B_x - A_x = 2 \text{ cm}$, $R_y = B_y - A_y = -14 \text{ cm}$; $R = \sqrt{R_x^2 + R_y^2}$ $= 14.1 \text{ cm}$. $\mathbf{R}$ points θ clockwise from the $+x$ direction where $\tan \theta = R_y/R_x = 7$, $\theta = 82°$.

Chapter 2

Exercises

1. Yes.
3. $t = s/v = 500$ s $= 8.3$ min.
5. $t = s/v = 14.8$ mi/hr.
7. $t = s/v = 11.2$ hr.
9. $s = vt = 1980$ m.
11. $a = (v_f - v_0)/t = 1.2$ m/s^2.
13. (a) $v = at = 2$ m/s. (b) $v = at = 20$ m/s.
15. $v_f = v_0 + at = 30$ ft/s.
17. (a) C, D, G, H, I. (b) A, B, E. (c) E. (d) H. (e) E. (f) G, I. (g) F. (h) D. (i) C. (j) B.

Problems

1. $s = v_1 t_1 + v_2 t_2 = 240$ mi; $\bar{v} = s/t = 48$ mi/hr.
3. $v = v_{air} - v_{wind} = 144$ mi/hr; $s = vt = 720$ mi.
5. (a) $v_f = v_0 + at = 12$ ft/s. (b) $v_f = v_0 + at = -60$ ft/s.
7. $a = v/t = -12.5$ (km/hr)/s $= -3.48$ m/s^2.
9. $t = v/a = 6.4$ s.
11. The airplane's velocity is 350 mi/hr from 4:00 PM to 5:46 PM, 400 mi/hr from 5:46 PM to 6:48 PM, and 300 mi/hr from 6:48 PM to 9:04 PM.
13. The car started from rest and accelerated at 18 (mi/hr)/min for 4 min to a velocity of 72 mi/hr. The car remained at this velocity for 4 min, and then slowed down at -12 (mi/hr)/min for 2 min.

Chapter 3

Exercises

1. The squirrel should stay where it is, since if it lets go, it will fall with exactly the same acceleration as the bullet and so will be struck.
3. Yes.
5. $a = -v_0^2/2s = -12.5$ ft/s^2.
7. $s = \frac{1}{2}at^2 = 0.04$ m.

9. (a) $a = v^2/2s = 2.5$ m/s^2. (b) $t = v/a = 20$ s.
11. $a = 2s/t^2 = 9.94$ ft/s^2.
13. $s = (v_f^2 - v_0^2)/2a = 180$ ft.
15. $v = \sqrt{2gh} = 26.6$ m/s.
17. $h = \frac{1}{2}gt^2 = 78.4$ m.
19. $t = \sqrt{2h/g} = 2$ s; $v = gt = 64$ ft/s.
21. $v_f = v_0 + gt = 19.8$ m/s; $v_f = v_0 + gt = 29.6$ m/s.
23. $h = v^2/2g = 16$ ft.
25. Yes; the speed is greatest at both ends of the path and is least at the highest point.
27. At maximum range, $\theta = 45°$ and $v_0 = \sqrt{Rg} = 33$ ft/s.
29. $R = (v_0^2 \sin 2\theta)/g$, $v_0 = \sqrt{Rg/\sin 2\theta}$. Here $R = 120$ ft and $\sin 2\theta = \sin 100° = \cos 10°$ since $\sin (90° + x) = \cos x$. Hence $v_0 = 62.4$ ft/s.

Problems

1. $s = v_0 t + \frac{1}{2}at^2$, hence $a = 2s/t^2 - 2v_0/t$. Since 75 s $= 0.0208$ hr, $a = -3069$ mi/hr^2 and $v_f = v_0 + at = 16$ mi/hr.
3. $t = \sqrt{2h/g} + h/v_{sound} = 2.59$ s.
5. (a) $h_1 = v_0^2/2g = 5.10$ m, $t_1 = \sqrt{2h_1/g} = 1.02$ s; $h_2 = 20$ m $+ 5.1$ m $= 25.1$ m, $t_2 = \sqrt{2h_2/g} = 2.26$ s; $t = t_1 + t_2 = 3.28$ s. (b) $v = gt_2 = 32.1$ m/s.
7. Man: $t = h/v = 25$ s; monocle: $t = \sqrt{2h/g} = 5.6$ s; $\Delta t = 19.4$ s.
9. $t = 2\sqrt{2h/g} = 0.90$ s.
11. $t = 2\sqrt{2h/g} = 4.04$ s; $v = \sqrt{2gh} = 19.8$ m/s.
13. $t_1 =$ acceleration period $= v/a = 4$ s; $s_1 =$ acceleration distance $= \frac{1}{2}at^2 = 12$ m; $t_3 =$ deceleration period $= t_1 = 4$ s; $s_3 =$ deceleration distance $= s_1 = 12$ m. Hence constant-velocity distance is $s_2 = 50$ m $- 24$ m $= 26$ m and constant-velocity period is $t_2 = s_2/v = 4.33$ s; $t = t_1 + t_2 + t_3 = 12.33$ s.
15. $v_y = \sqrt{2gh} = 62$ ft/s; $v_x = 100$ ft/s; $v = \sqrt{v_x^2 + v_y^2} = 118$ ft/s; $\tan \theta = v_y/v_x = 0.62$, $\theta = 32°$ below horizontal.
17. (a) $v_x = \sqrt{v_N^2 + v_W^2} = 12.8$ m/s, $v_y = gt = 19.6$ m/s, $v = \sqrt{v_x^2 + v_y^2} = 23.4$ m/s. (b) $\tan \theta_1 = v_x/v_y = 0.653$, $\theta_1 = 33°$. (c) $\tan \theta_2 = v_W/v_N = 1.25$, $\theta_2 = 51°$.

19. (a) $v = \sqrt{2gh} = 139$ ft/s, $R_{max} = v^2/g = 600$ ft.
    (b) 139 ft/s; 139 ft/s.
21. (a) $R = (v_0^2/g)\sin 2\theta = 7953$ m, $T = (2v_0 \sin \theta)/g = 30.6$ s. (b) $\theta_2 = 90° - \theta_1 = 60°$; $T = (2v_0/g) \sin \theta_2 = 53$ s.

Chapter 4

Exercises

1. No. Only a net force produces an acceleration, and when a force is applied to an object, other forces may come into being that cancel it out. Thus pushing down on a book lying on a table does not accelerate the book because the table pushes back with an equal and opposite force.
3. The deceleration of a person falling onto loose earth is more gradual than if he falls onto concrete, hence the force acting on him is less.
5. The net forces are, in the order given: zero, constant, and increasing so as to be proportional to t.
7. Here $F = w = mg$, hence $a = F/m = g$.
9. No. Action and reaction forces act upon different bodies, and so a single force can certainly act upon a body.
11. A propeller works by pushing backward on the air, whose reaction force in turn pushes the propeller itself forward. No air, no reaction force, so the idea is no good.
13. $F_{max} = m_0 a_0 = 2000$ N. Hence $a = F_{max}/m = 2000$ N/3000 kg $= 0.67$ m/s^2.
15. $a = F/m = 2.86$ m/s^2; $v = at = 28.6$ m/s.
17. $a = -v^2/2s = -4.8$ m/s^2; $F = ma = -9600$ N.
19. $m = w/g = 0.816$ kg; $a = F/m = 24.5$ m/s^2.
21. $m = w/g = 3$ slugs; $a = F/m = 33$ ft/s^2.
23. Here $v_0 = 0$, so from $v_f^2 = v_0^2 + 2$ as we have $a = v_f^2/2s = 130$ ft/s^2. Since $m = w/g = 5$ slugs, $F = ma = 650$ lb.
25. $F = \mu N = 72$ lb.
27. $F = \mu N = 2$ N.
29. The truck. The coefficient of rolling friction for the truck is about 9 times greater than that for

the boxcar, hence the frictional force on it and its consequent deceleration are about 9 times greater, and the distance it goes until it stops is about 9 times smaller since $s = v^2/2a$.
31. $a = v_0^2/2s = 0.4$ m/s^2; $\mu = F/mg = ma/mg = a/g = 0.041$.
33. (a) $a = v_0^2/2s = 6.75$ ft/s^2; $\mu = F/mg = ma/mg = a/g = 0.21$.
35. $F = ma = \mu mg$, $a = \mu g = 6.86$ m/s^2, $v_0 = \sqrt{2as} = 63.1$ m/s $= 227$ km/hr. Air resistance would have helped to slow the car down.

Problems

1. (a) $v_f = 200$ km/hr $= 55.6$ m/s; $a = (v_f - v_0)/t = 18.5$ m/s$^2 = 1.89$ g. (b) $F = ma = 2.22 \times 10^5$ N.
3. $F = 3w$, $F_x = F \cos 50° = 1.93w$; $m = w/g$; $a = F_x/m = 1.93g = 61.8$ ft/s$^2 = 18.9$ m/s^2.
5. Both boxes have $a = g \sin 60° = 28$ ft/s$^2 = 8.5$ m/s^2.
7. $F = w - T = ma$, $T = w - ma = mg - ma = 8800$ N.
9. (a) $F = T - w = 600$ lb; $m = w/g = 75$ slugs; $a = F/m = 8$ ft/s^2. (b) 32 ft/s^2.
11. (a) Down; (b) no; (c) yes; $F = 20$ lb, $m = w/g = 5$ slugs; $a = F/m = 4$ ft/s^2.
13. $v_0 = 176$ ft/s, $v_f = 22$ ft/s; $a = (v_f^2 - v_0^2)/2s = 127$ ft/s^2; $m = w/g = 6.25$ slugs; $F = w + ma = 994$ lb.
15. Here $s = 0.6$ m, $h_1 = 2$ m, $h_2 = 1.8$ m, $m_1 = 40$ kg, $m_2 = ?$ Kangaroo by herself: $v_1^2 = 2gh_1 = 39.2$ (m/s)^2, $a_1 = v_1^2/2s = 32.7$ m/s^2, force exerted is $F = m_1 a_1 = 1307$ N. With baby: $v_2^2 = 2gh_2 = 35.3$ (m/s)^2, $a_2 = v_2^2/2s = 29.4$ m/s^2, hence $m_1 + m_2 = F/a_2 = 44.5$ kg and $m_2 = 4.5$ kg.
17. (a) $F = m_2 g = (m_1 + m_2)a$, $a = m_2 g/(m_1 + m_2) = 4.9$ m/s^2. (b) $F = m_2 g - \mu m_1 g = (m_1 + m_2)a$, $a = (m_2 g - \mu m_1 g)/(m_1 + m_2) = 3.92$ m/s^2.
19. (a) $F = ma = \mu mg$, $a = \mu g = 16$ ft/s^2, $v_0 = 44$ ft/s; $s = v_0^2/2a = 60.5$ ft. (b) No.
21. $N = mg \cos \theta$; $F = mg \sin \theta - \mu N = mg(\sin \theta - \mu \cos \theta) = ma$;

$a = g(\sin\theta - \mu\cos\theta) = 4.11$ m/s². Since $s = \frac{1}{2}at^2$, $t = \sqrt{2s/a} = 2.09$ s.

23. $N = mg\cos\theta$; $F = mg\sin\theta + \mu N$
$= mg(\sin\theta + \mu\cos\theta) = 398$ N upward along the plane.

25. (a) From solution to Problem 21, $a = g(\sin\theta - \mu\cos\theta) = 5.55$ m/s²; $v = \sqrt{2as} = 23.6$ m/s.
(b) $F = \mu mg = ma$, $a = \mu g = 0.98$ m/s², $s = v^2/2a = 283$ m.

27. (a) $N = w + F_y = w + F\sin\theta$, $F_x = F\cos\theta$
$= \mu N = \mu(w + F\sin\theta)$, $F(\cos\theta - \mu\sin\theta)$
$= \mu w$, $F = \mu w/(\cos\theta - \mu\sin\theta)$. (b) The force becomes infinite when $\cos\theta = \mu\sin\theta$, which means at an angle such that $\tan\theta = 1/\mu$. When $\mu = 0.25$, $\theta = 76°$.

ladder yields $\frac{1}{3}w \times 6$ ft $= \frac{2}{3}w \times (6$ ft $- x)$, $x = 3$ ft.

19. $h = 3$ ft/$\tan 30° = 5.2$ ft. [At the point of tipping over the CG is directly above the line of the lower wheels.]

21. Let x be the distance of the CG from the cross member. The weight of the cross member is 10, of the horizontal member 20. Calculating torques about the CG, $10x = 20(10$ cm $- x)$, $x = 6.7$ cm.

23. (a) $\omega_{\text{out}} = \omega_{\text{in}}N_{\text{in}}/N_{\text{out}} = 3.2$ rpm.
(b) $\tau_{\text{out}} = \text{Eff} \times \tau_{\text{in}}N_{\text{out}}/N_{\text{in}} = 40$ lb-ft.

25. When a screw of pitch p is turned through 1 rotation, the applied force moves through a circle of circumference $2\pi L$ while the screw advances the distance p. Hence MA $= s_{\text{in}}/s_{\text{out}} = 2\pi L/p$.

Chapter 5

Exercises

1. (a) Perpendicular to the wall. (b) The force on the ground is always greater than the weight of the ladder since the force has a horizontal component equal to the reaction force of the wall on the ladder as well as a vertical component equal to the ladder's weight.

3. The location of the pivot point.

5. About whatever point makes the calculations easiest; the results will be the same regardless of the point chosen.

7. MA greater than 1; MA less than 1.

9. Because the airplane is moving at constant velocity, the thrust of its engines is equal to the air resistance at that velocity. Hence the entire upward force must be due to wing lift.

11. $F = \sqrt{F_x^2 + F_y^2} = 20.6$ lb; $\tan\theta = F_y/F_x = 0.555$, $\theta = 29°$ above the horizontal.

13. $T_y = T\sin\theta = w$, $T = w/\sin\theta = 156$ N.

15. $\tau = F_x h = Fh\sin\theta = 424$ lb-ft.

17. Let x be the distance of the two men from the far end of the ladder and w be the ladder's weight. Calculating torques about the CG of the

Problems

1. Calculating torques about the ankle, $F_1 L_1 = wL_2$, $F_1 = wL_2/L_1 = 3w = 540$ lb. If L_1 were greater, F_1 would decrease.

3. $F'_{\text{horiz}} = F\sin\theta = 131$ N, $F'_{\text{vert}} = F\cos\theta - w_1 - w_2 = 653$ N. The resultant force has the magnitude $\sqrt{F'^2_{\text{horiz}} + F'^2_{\text{vert}}} = 666$ N.

5. Let T_1 and T_2 be the tensions in the two parts of the rope. If θ is the angle either part of the rope makes with the horizontal, $\tan\theta = 1.6$ ft/10 ft $= 0.16$ and $\theta = 9°$. For the weight w to be in equilibrium, $\Sigma F_y = T_{1y} + T_{2y} - w = 2T\sin r - w = 0$, $T = w/2\sin\theta = 128$ lb.

7. Here $\tan\theta = 0.52$ m/10 m $= 0.052$, $\theta = 3°$. From solution to Problem 5, $T = mg/2\sin\theta = 93.6$ N.

9. Let x be the distance of the 35-lb child from the 30-lb child and calculate torques about the center of the seesaw. Then
30 lb $\times$ 6 ft $+$ 35 lb $\times$ (6 ft $- x$) $=$ 50 lb $\times$ 6 ft, $x = 2.57$ ft.

11. If w is the weight supported by the front axle, then (6000 lb $- w$) is the weight supported by the rear axle. Calculating torques about the

CG of the truck, $w \times 8$ ft $= (6000 \text{ lb} - w) \times 4$ ft, $w = 2000$ lb.

13. The angle between the boom and the rope is specified by $\tan \theta = 4 \text{ ft}/10 \text{ ft} = 0.40$, so $\theta = 22°$. Calculating torques about the hinge pin of the boom, 80 lb $\times 5$ ft $= T \times 10$ ft $\times \sin 22°$, $T = 107$ lb.

15. (a) Let **F** be the force the upper hinge exerts on the door and w be the door's weight of 50 lb acting from its CG. Since the door's weight is supported by the upper hinge, $F_y = w = 50$ lb. To find F_x, it is easiest to calculate torques about the lower hinge: $F_x \times 8$ ft $= w \times 1.5$ ft, $F_x = 9.4$ lb. Hence $F = \sqrt{F_x^2 + F_y^2} = 51$ lb. If θ is the angle between **F** and the vertical, $\tan \theta = F_x/F_y = 0.188$, $\theta = 11°$. The force the door exerts on the upper hinge is equal and opposite to **F**, hence 51 lb at 11° away from vertically downward. (b) Since the lower hinge exerts a horizontal force on the door, it must be equal in magnitude to F_x in order that the door be in equilibrium. Hence the force is 9.4 lb.

17. Since the wall is frictionless, $F_y = w = 30$ lb and F_x is equal in magnitude to the force the ladder exerts on the wall. Calculating torques about the foot of the ladder, $F_x \times 8$ ft $= w \times 3$ ft, $F_x = 11.3$ lb.

19. (a) If $\mathbf{F}_1$ is the force on the wall, $F_{1y} = 0$. To find F_{1x} it is easiest to calculate torques about the foot of the ladder, whose length is L. Since the ladder's weight of $w = 50$ lb acts from its CG and the force the wall exerts on the ladder is equal in magnitude to F_{1x}, $F_{1x} \times L \sin 60°$ $= w \times \frac{1}{2}L \cos 60°$, $F_1 = F_{1x} = 14.4$ lb. (b) If $\mathbf{F}_2$ is the force on the ground, $F_{2y} = w$ $= 50$ lb and $F_{2x} = 14.4$ lb. Hence F_2 $= \sqrt{F_{2x}^2 + F_{2y}^2} = 52$ lb. If θ is the angle between $\mathbf{F}_2$ and the horizontal, $\tan \theta = F_{2y}/F_{2x}$ $= 3.47$ and $\theta = 74°$.

21. The horizontal force needed at the foot of the ladder is equal to the horizontal force F exerted by the wall on the top of the ladder. Calculating torques about the foot of the ladder, $F \times 20$ ft $\times \sin 60° = 150$ lb $\times 16$ ft $\times \cos 60°$, $F = 69$ lb.

Chapter 6

Exercises

1. Under no circumstances.

3. Less.

5. They are equal, since the centripetal force is provided by the gravitational force.

7. (a) The centripetal force due to the earth's rotation is greatest at the equator, hence the outward reaction forces exerted by particles of the earth are also greatest there. These reaction forces distort the earth's shape. (b) If the earth rotated faster, the distortion would be greater.

9. Shorter.

11. (a) The record's circumference is $2\pi r = 3.14$ ft, so $v = (33.3 \text{ x } 3.14 \text{ ft})/60 \text{ s} = 1.74$ ft/s. (b) $a_c = v^2/r = 6.06$ ft/s².

13. $F_c = mv^2/r = 384$ N.

15. (a) $mv^2/r = T$, $v = \sqrt{rT/m}$. Here $m = w/g$ $= 0.0938$ slug, so $v = 41.3$ ft/s. (b) $f = v/2\pi r$ $= 1.64$ rev/s.

17. (a) The tension in the string at the top of the circle is equal to the centripetal force of mv^2/r minus the weight mg of the stone, so that $T = mv^2/r - mg$. If the string is to be just taut, $T = 0$ and $mv^2r =$ mg, $v = \sqrt{rg} = 2.80$ m/s. (b) Since the mass of the stone does not matter here, the velocity is again 2.80 m/s.

19. At the bottom of the circle, the tension in the string is the sum of the centripetal force mv^2/r and the stone's weight of mg. Hence $T = mv^2/r$ $+ mg = wv^2/rg + w$ and $v = \sqrt{rg(T/w - 1)} =$ 17.9 ft/s.

21. $mv^2/r = mg$, $v = \sqrt{rg} = 11.3$ ft/s; $t = 2\pi r/v$ $= 2.22$ s.

23. $m = w/g = 62.5$ slugs per sphere, hence $F = Gm^2/r^2 = 1.3 \times 10^{-6}$ lb.

25. From Sec. 6-5, the moon's acceleration toward the earth is 2.7×10^{-3} m/s². When $t = 1$ s, $s = \frac{1}{2}at^2 = 1.4 \times 10^{-3}$ m $= 1.4$ mm. The moon never reaches the earth because its "falling" causes it to move in an orbit around the earth.

27. $a = (r_e/r)^2 g = 6.3$ m/s^2.

29. $v_{escape}/v_{orbit} = 1.4$. [This ratio is equal to $\sqrt{2}$ for all astronomical bodies.]

Problems

1. $mv_{max}^2/r = \mu mg$, $v_{max} = \sqrt{\mu gr}$. Since $f = v/2\pi r$, $f_{max} = \sqrt{\mu gr}/2\pi r = 0.863$ rev/s = 52 rev/min. Hence the coin will stay where it is at $33\frac{1}{3}$ rev/min but will fly off at 78 rev/min.

3. $v = 139$ m/s; $\tan\theta = v^2/gr$, $r = v^2/(g \tan 45°) = 1971$ m = 1.97 km.

5. $\tan\theta = v^2/gr = 0.135$, $\theta = 8°$; $h = 25$ ft $\times \sin 8° = 3.5$ ft.

7. Let R = radius of moon's orbit, r = distance of object from moon, and m the object's mass. Then $Gmm_m/r^2 = Gmm_e/(R-r)^2$, $(R-r)/r = \sqrt{m_e/m_m}$, $r = R/(1 + \sqrt{m_e/m_m}) = 3.8 \times 10^7$ m.

9. (a) From Sec. 6-5, $g = GM/r^2$ at surface of body of mass M and radius r. Hence $g_m/g = (M_m/M_e)/(r_m/r_e)^2$, $g_m = 0.012g/(0.27)^2 = 5.27$ ft/s^2. (b) $w_m = w_e g_m/g = 16.5$ lb.

11. From solution to Problem 9, $g_p = g(M_p/M_e)/(r_p/r_e)^2 = 2g/4 = g/2$.

13. $T = 2\pi r/v$, $v = \sqrt{rg}$, $g = g_0(r_e/r)^2$. Hence $r = (\sqrt{g_0}\, r_e T/2\pi)^{2/3}$. Since $T = 1$ day $\doteq 8.64 \times 10^4$ s, $r = (27.6 \times 10^{10})^{2/3}$ m $= (76 \times 10^{21})^{1/3}$ m $= 4.2 \times 10^7$ m.

15. F_1 = force of sun on $m = 1$ kg at earth's center $= Gm_s m/R_e^2$; F_2 = force of sun on $m = 1$ kg at earth's surface $= Gm_s m/(R_e - r_e)^2 = Gm_s m(1 + 2r_e/R_e)/R_e^2$, where R_e = radius of earth's orbit, r_e = radius of earth. $F_2 - F_1 = 2Gm_s mr_e/R_e^3 = 5.14 \times 10^{-7}$ N. Similarly, if R_m = radius of moon's orbit and F_3 and F_4 are respectively the force of the moon on $m = 1$ kg at the earth's center and at its surface, $F_4 - F_3 = 2Gm_m mr_e/R_m^3 = 11.5 \times 10^{-7}$ N, which is more than double $F_2 - F_1$.

Chapter 7

Exercises

1. No work is done by a net force acting upon a moving body when the force is perpendicular to the direction of the body's motion.

3. At the lowest point; at the highest points.

5. Kinetic and rest energies are always positive quantities.

7. $h = W/mg = 13.6$ m.

9. $W = 0$ because the force is perpendicular to the displacement.

11. (a) $W = Fs = 1040$ J. (b) 1040 J.

13. (a) $W = mgh = 2.35 \times 10^6$ J. (b) 10 hr $= 3.6 \times 10^4$ s; $P = W/t = 65$ W $= 0.087$ hp.

15. (a) 1 yr $= 3.15 \times 10^7$ s; $P = W/t = 6.3 \times 10^{12}$ W. (b) 1.8×10^3 W $= 2.4$ hp.

17. $P = Fv = mgv = 3.92$ kW.

19. $F = P/v = 2.33 \times 10^4$ N.

21. $(60^2 - 50^2)/(20^2 - 10^2) = 3.7$ times more work needed to increase the velocity from 50 mi/hr to 60 mi/hr than from 10 mi/hr to 20 mi/hr.

23. KE $= \frac{1}{2}(w/g)v^2 = 1.73 \times 10^5$ ft-lb.

25. $v = s/t = 8.89$ m/s, KE $= \frac{1}{2}mv^2 = 2.77 \times 10^3$ J.

27. $wh = 3200$ ft-lb.

29. KE/PE $= v^2/2gh = 0.204$, hence 79.6% of the initial PE is dissipated.

31. $Fs = \frac{1}{2}mv^2$, $v = \sqrt{2Fs/m} = 7.1$ m/s.

33. (a) Here $h = 2$ m $- 0.8$ m $= 1.2$ m and, since $mgh = \frac{1}{2}mv^2$, $v = \sqrt{2gh} = 4.85$ m/s. (b) Since v is independent of m, $v = 4.85$ m/s here also.

35. $P/A = (1400$ W/m$^2) \times (0.305$ m/ft$)^2/(746$ W/hp$) = 0.175$ hp/ft^2.

Problems

1. $W = Fs \cos\theta = 376$ J.

3. (a) $W = Fs \cos \theta = 866$ lb. (b) $P = W/t = 19.2$ ft-lb/s $= 0.035$ hp.

5. In both cases $h = 3000$ m and $W = mgh = 2.65 \times 10^6$ J.

7. $P = W/t = wh/t = 1.75 \times 10^4$ ft-lb/s $= 31.8$ hp. Since twice this power is required, $2P = 63.6$ hp.

9. $P = W/t = wh/t = 131$ ft-lb/s $= 0.24$ hp.

11. (a) $W = Fs = 900$ ft-lb. (b) PE $= wh = 800$ ft-lb. (c) 100 ft-lb was dissipated as heat due to friction in the pulleys.

13. (a) $W = Fs = 2000$ ft-lb. (b) PE $= wh = 1600$ ft-lb. (c) ΔKE $= 400$ ft-lb.

15. (a) $P = W/t = mgh/t = 2.94 \times 10^6$ W. (b) 29,400 bulbs.

17. $Fs = wh$, $F = wh/s = 1000$ lb $\times 19.5$ ft/0.5 ft $= 3.9 \times 10^4$ lb.

19. $P = 50{,}000$ hp $= 3.73 \times 10^7$ W, $V_1 = 800$ m/s, and $v_2 = 100$ m/s. If m/t is the mass of steam per second, $P = \Delta$KE$/t = \frac{1}{2}(m/t)\,(v_1{}^2 - v_2{}^2)$ and $m/t = 2P/(v_1{}^2 - v_2{}^2) = 118$ kg/s.

Chapter 8

Exercises

1. Yes; yes.

3. The momentum increases by $\sqrt{2}$ since $2 \times (\frac{1}{2}mv^2) = \frac{1}{2}m(\sqrt{2}v)^2$.

5. The definition of mass and the first law of motion are both included in the principle of conservation of linear momentum. The latter principle goes considerably further as well, since it can be applied to systems that consist of any number of bodies that interact with one another in any manner whatsoever, and it holds in three dimensions, not just in one.

7. (a) Yes. (b) In the opposite direction to that

in which the man walks. (c) The car also comes to a stop.

9. $mv = (w/g)v = 400$ slug-ft/s.

11. $mv = 10$ kg-m/s.

13. (a) $v = 242$ m/s, $mv = 3.87 \times 10^7$ kg-m/s. (b) $t = mv/F = 114$ s.

15. $\Delta v = 175$ ft/s, $F\Delta t = \Delta mv = (w/g)\Delta v$, $F = (w/g)\Delta v/\Delta t = 1833$ lb.

17. The time interval in which the projectile is accelerated in the barrel is $\Delta t =$ barrel length/average velocity $= 25$ ft/700 ft/s $= 3.6 \times 10^{-2}$ s. Hence $F = \Delta mv/\Delta t = 1.7 \times 10^5$ lb.

Problems

1. From Sec. 3-6 the range of the shell is $R = v_x T$, where T is the time of flight which is unchanged by the explosion. The shell explodes at $x = R/2$. The half that continues takes all of the horizontal component of momentum, hence its velocity v_x doubles, and it continues for an additional distance of $2(R/2)$ for a total range of $R/2 + R = 3R/2$.

3. If n is the number of bullets, $n = (mv)_{\text{leopard}}/(mv)_{\text{bullet}} = 7.4$ Hence 8 bullets are required.

5. (a) $v_2 = m_1 v_1/m_2 = w_1 v_1/w_2 = 12.5$ mi/hr. (b) $\frac{1}{2}(w_1/g)v_1{}^2 + \frac{1}{2}(w_2/g)v_2{}^2 = 4.42 \times 10^5$ ft-lb.

7. (a) $m_1 v_1 = m_1 v_1' + m_2 v_2'$, $v_1' = (m_1 v_1 - m_2 v_2')/m_1 = -3.34 \times 10^4$ m/s. The neutron moves off in the opposite direction to that of its approach. (b) KE(before) $= \frac{1}{2}m_1 v_1{}^2 =$ KE(after) $= \frac{1}{2}m_1 v_1'{}^2 + \frac{1}{2}m_2 v_2'{}^2 = 8.4 \times 10^{-18}$ J, so collision is elastic.

9. If **V** is final velocity, $V_N = m_1 v_1/(m_1 + m_2) = 0.444$ m/s; $V_W = m_2 v_2/(m_1 + m_2) = 0.889$ m/s; $V = \sqrt{V_N{}^2 + V_W{}^2} = 0.99$ m/s. If θ is angle between **V** and north, $\tan \theta = V_W/V_N = 2$, $\theta = 63°$ west of north.

11. If V is initial velocity of wooden block, $mv = (M + m)V$ and $V = mv/(M + m)$. From PE $=$ KE, $(M + m)gh = \frac{1}{2}(M + m)V^2 = m^2 v^2/2(M + m)$ and $v = (1 + M/m)\sqrt{2gh}$.

Chapter 9

Exercises

1. The solid cylinder reaches the bottom first because its moment of inertia is smaller and hence less of its initial PE becomes KE of rotation.

3. The length of the day will increase; the earth's moment of inertia will be greater when the water from the icecaps becomes uniformly distributed and hence, according to conservation of angular momentum, the angular velocity must decrease.

5. See discussion in text.

7. 2π rad/9 = 0.698 rad.

9. (a) $1' = 1°/60 = 0.01745$ rad/60 = 2.91×10^{-4} rad. (b) With such a small angle, the chord and the arc are very nearly equal. Hence $s = \theta r = 7.27 \times 10^{-3}$ cm = 0.073 mm.

11. (a) $\omega = 2\pi$ rad/12 hr = 1.45×10^{-4} rad/s.
    (b) $\omega = 2\pi$ rad/60 min = 1.75×10^{-3} rad/s.
    (c) $\omega = 2\pi$ rad/60 s = 0.105 rad/s.

13. $\omega = 3000$ rev/min = 314.16 rad/s, $\theta = \omega t = 9425$ rad.

15. $v = \omega r = 200$ rad/s $\times 4 \times 10^{-3}$ m = 0.8 m/s.

17. (a) $\omega = 126$ rad/s. (b) $r = 0.0104$ ft, $v = \omega r = 1.31$ ft/s.

19. $\omega_0 = 105$ rad/s, $\theta = 785$ rad; $\omega_0^2 = -2\alpha\theta$, $\alpha = -\omega_0^2/2\theta = -7.02$ rad/s²; $t = -\omega_0/\alpha = 15$ s.

21. (a) $\theta = \frac{1}{2}\alpha t^2$, $\alpha = 2\theta/t^2 = 10$ rad/s².
    (b) $\omega = \alpha t = 100$ rad/s.

23. $I = mL^2/12 = (w/g)L^2/12 = 0.0073$ slug-ft²; KE = $\frac{1}{2}I\omega^2 = 0.365$ ft-lb.

25. $P = 550$ ft-lb/s, $\omega = 126$ rad/s; $P = \tau\omega$, $\tau = P/\omega = 4.37$ lb-ft.

27. $\alpha = (\omega_f - \omega_0)/t = 5$ rad/s, $I = \tau/\alpha = 100$ kg·m².

Problems

1. At a particular instant, the barrel can be regarded as rotating about its point of contact with the ground. Hence in (a) $r = 2$ ft, $v = \omega r = 10$ ft/s; in (b) $r = 1$ ft, $v = 5$ ft/s; and in (c) $r = 0$, $v = 0$. Alternatively we can proceed by noting that the center of the barrel is moving at +5 ft/s. Hence the top has this velocity plus the velocity of +5 ft/s corresponding to motion relative to the center for a total of 10 ft/s, while the bottom has the velocity of the center plus its own −5 ft/s velocity relative to the center for a total of $v = 0$.

3. $v = 44$ ft/s, $\omega = v/r = 22$ rad/s.

5. $\alpha = a_T/r = 0.5$ rad/s².

7. (a) $Fs = \frac{1}{2}I\omega^2$, $\omega = \sqrt{2Fs/I} = 15.8$ rad/s.
   (b) KE = $\frac{1}{2}I\omega^2 = 2.5$ J. (c) $W = Fs = 2.5$ J.

9. (a) $I = Mk^2 = \frac{2}{5}MR^2$. (b) $k = \sqrt{I/M} = \sqrt{2/5}\,R$.

11. When the rod is released, its center of gravity falls by $h = L/2$. Hence its initial potential energy is PE = $mgh = mgL/2$. The rod's KE when it has fallen is $\frac{1}{2}I\omega^2$, where $I = mL^2/3$. Since $\frac{1}{2}I\omega^2 = mgh$, $mL^2\omega^2/6 = mgL/2$, $\omega = \sqrt{3g/L}$.

13. $I = MR^2/2$, $\tau = MgR$, and $\alpha = \tau/I = 2g/R$. Hence $a_T = \alpha R = 2g$.

15. (a) $P = 1.79 \times 10^5$ ft-lb/s, $\omega = (5600$ rev/min)/5 = 117 rad/s, $\tau = P/\omega = 1530$ lb-ft, $F = \tau/r = 1020$ lb. (b) $P = Fv$, $v = P/F = 175$ ft/s.

17. The net force on the belt is 40 lb − 15 lb = 25 lb, hence the torque exerted by the motor is $\tau = Fr = 25$ lb $\times \frac{1}{3}$ ft = 8.33 lb · ft. Since $\omega = 188$ rad/s, $P = \tau\omega = 1567$ ft·lb/s = 2.85 hp.

19. (a) $I_1 = mR_1^2 = 0.2$ kg-m², $I_2 = mR_2^2 = 0.288$ kg-m², $\omega_1 = 31.4$ rad/s, $\omega_2 = I_1\omega_1/I_2 = 21.8$ rad/s = 3.5 rev/s. (b) KE$_1 = \frac{1}{2}I_1\omega_1^2 = 98.7$ J, KE$_2 = \frac{1}{2}I_2\omega_2^2 = 68.4$ J; the decrease in KE is due to the work that had to be done to stretch the string.

21. (a) $I_2 = 2I_1$, $\omega_2 = I_1\omega_1/I_2 = \omega_1/2 = 50$ rad/s. (b) KE$_1 = \frac{1}{2}I_1\omega_1^2 = 5 \times 10^3$ ft-lb, KE$_2 = \frac{1}{2}I_2\omega_2^2 = 2.5 \times 10^3$ ft-lb; the difference of 2.5×10^3 ft-lb represents energy dissipated as heat when the moving disk was pressed against the stationary one before the two stuck together.

Chapter 10

Exercises

1. $V = 2 \text{ in}^3/(12 \text{ in/ft})^3 = 1.16 \times 10^{-3} \text{ ft}^3$; $w = dgV = 1.39$ lb; 1.39 lb $\times$ 16 oz/lb $\times$ \$135/oz = \$3002.

3. (a) $d = m/V = m/(4\pi r^3/3) = 5.52 \times 10^3 \text{ kg/m}^3$.
   (b) The interior must consist of denser materials than those at the surface; no.

5. $\Delta L = L_0 F/YA$, $A = \pi d^2/4$, hence
   $\Delta L_B = (L_{0B}/L_{0A})(d_A/d_B)^2 \Delta L_A = 2\Delta L_A$.

7. The breaking strength F is proportional to the cross-sectional area $A = \pi d^2/4$ of the rope. Hence (a) $F = 5000 \text{ lb} \times (1/4)^2/(1/2)^2 = 1250$ lb, (b) $F = 5000 \text{ lb} \times (3/4)^2/(1/2)^2 = 11{,}250$ lb.

9. $s = F/k = mg/k = 0.0784$ m.

11. $\Delta L = L_0 F/YA = 5.8 \times 10^{-4} \text{ ft} = 0.0070$ in.

13. $\Delta L = L_0 F/YA = 1.66 \times 10^{-3}$ ft.

15. $A = (10^{-3} \text{ m})^2 = 10^{-6} \text{ m}^2$, $\Delta L = L_0 F/YA$
    $= L_0 mg/YA = 2.94 \times 10^{-4}$ m.

17. $\Delta V = pV_0/B = 4 \times 10^{-8} \text{ m}^3 = 40 \text{ mm}^3$.

19. (a) $Fs = \frac{1}{2}mv^2$, $F = mv^2/2s = 7.2 \times 10^4$ N. (b)
    $F/A = 7.2 \times 10^4 \text{ N}/0.2 \text{ m}^2 = 3.6 \times 10^5 \text{ N/m}^2$, hence she is likely to survive.

Problems

1. If m_0 is the mass per atom, the volume per atom is $V_0 = m_0/d$. We imagine each atom to occupy a cube h long on each edge, so that $V_0 = h^3$ and $h = \sqrt[3]{m_0/d} = 2.57 \times 10^{-10}$ m. The foil thickness is $t = V/A = (m/A)/(m/V) = (10^{-3} \text{ kg/m}^2)/d = 5.18 \times 10^{-8}$ m, and $t/h = 202$ atoms.

3. $A = \pi(r_{\text{outside}}^2 - r_{\text{inside}}^2) = 7.07 \text{ in}^2$;
   $\Delta L = L_0 F/YA = 3.26 \times 10^{-4}$ ft.

5. $F_{\text{out}} = F_{\text{in}} \times 2\pi L/p = 2011$ lb; $\Delta L = L_0 F_{\text{out}}/YA = 7.74 \times 10^{-5}$ in.

7. Each wire must support $m = 2.5$ kg. Aluminum wire: $F/A = mg/\pi r^2 = 7.80 \times 10^6 \text{ N/m}^2$, $\Delta L = L_0 F/YA = 1.67 \times 10^{-4}$ m. Steel wire: $F/A = mg/\pi r^2 = 3.12 \times 10^7 \text{ N/m}^2$, $\Delta L = L_0 F/YA = 2.34 \times 10^{-4}$ m.

9. Each cylinder has the same cross-sectional area A and volume V. The force corresponding to an ultimate strength of U is $F = UA$ and the mass of a cylinder is $m = dV$. Hence $F/m = UA/dV = U/dL$ where L is the cylinder length. In tension, $(F/m)/L = 7.5 \times 10^4$, 5.2×10^4, and 6.4×10^4 N·m/kg for bone, aluminum, and steel; in compression the figures are the same for aluminum and steel and 10.6×10^4 for bone. Bone is clearly an excellent structural material.

11. $A = \pi dh = 0.188 \text{ in}^2$, $F = (F/A)_{\text{max}} \times A = 9.4 \times 10^3 \text{ lb} = 4.7$ tons.

13. $\theta = 0.436 \text{ rad}$, $R = 5 \times 10^{-4} \text{ m}$, $\tau = Fs = 3.0 \times 10^{-3} \text{ N} - \text{m}$; $S = 2\tau L/\pi\theta R^4 = 4.91 \times 10^{10} \text{ N/m}^2$.

Chapter 11

Exercises

1. As the steam inside the can condenses, the internal pressure falls below the external pressure of the atmosphere.

3. The water level is unchanged.

5. The heavier one, because the buoyant force is the same for both.

7. The forces are the same because the height of water is the same.

9. $p = F/A = F/\pi r^2 = 6.37 \times 10^6 \text{ N/m}^2 = 63$ atm.

11. $F = pA = p_{\text{atm}}\pi r^2 = 127$ N.

13. $p = F/A = mg/A = 3.20 \times 10^3 \text{ N/m}^2 = 0.0316$ atm.

15. $p = p_{\text{atm}} + dgh = 1.10 \times 10^8 \text{ N/m}^2$.

17. $V = m/d = w/gd = 2.58 \text{ ft}^3$; $F = Vdg = 0.21$ lb.

19. $V = m/d = w/dg = 6250 \text{ ft}^3$; $A = 2400 \text{ ft}^2$; $h = V/A = 2.60$ ft.

21. $F_{\text{out}} = F_{\text{in}} \times \text{MA}_{\text{lever}} \times A_{\text{out}}/A_{\text{in}} = 1875$ lb.

23. $p = 120$ torr $= 1.60 \times 10^4$ N/m². This is a gauge pressure, so $h = p/dg = 1.55$ m.

Problems

1. $F = pA = dghA = 2.56 \times 10^4$ lb.
3. $p = dgh = 1.78 \times 10^5$ N/m²; $\Delta d/d_0 = \Delta V/V_0$
   $= p/B = 6.8 \times 10^{-6}$; density at bottom
   therefore $\Delta d = 0.096$ kg/m³ greater than
   normal density of 1.4×10^4 kg/m³.
5. $F = Vg(d_{air} - d_{hydrogen}) - mg = 892$ N.
7. (a) $V = F/g(d_{air} - d_{helium}) = 1.45 \times 10^7$ ft³.
   (b) $D = \sqrt{4V/\pi L} = 119$ ft.
9. (a) $h = \frac{1}{2}gt^2$, $t = \sqrt{2h/g} = 0.45$ s, $v = x/t$
   $= 4.43$ m/s. (b) $p = \frac{1}{2}dv^2 = 9.8 \times 10^3$ N/m².
11. (a) $p = dgh = 3.43 \times 10^3$ N/m². (b) $p = dgh =$
    2.06×10^4 N/m².
13. (a) $\Delta m/t = dV/t = dR$; $R = 0.402$ ft³/s;
    $F = \Delta(mv)/t = dRv = 50$ lb. (b) $P = 2Fv$
    $= 6400$ ft-lb/s $= 11.6$ hp.
15. The power output of the left ventricle is $P_L =$
    $Fv = (pA)(R/A) = pR = 6.2$ W. Since $P_R =$
    $0.2 P_L = 1.2$ W, the total power output is
    $P_L + P_R = 7.4$ W.
17. (a) $v_2 = v_1 A_1/A_2 = v_1 r_1^2/r_2^2 = 1.28$ m/s. (b) Here
    $h_1 - h_2 = -4$m, so $p_2 = p_1 + dg(h_1 - h_2) + \frac{1}{2}d$
    $(v_1^2 - v_2^2) = 1.92 \times 10^5$ N/m² $= 1.92$ bars.
19. $R = 16$ liters/min $= 2.67 \times 10^{-4}$m²/s, $p/L =$
    $8\eta R/\pi r^4 = 525$ (N/m²)/m.
21. (a) The sphere's weight is $4\pi r^3 dg/3$, the drag
    force on it at the terminal velocity v is $6\pi r\eta\, v$,
    and the upward buoyant force on it is $4\pi r_3 d'g/3$.
    At the terminal velocity the sphere is in
    equilibrium and its weight must be equal to the
    sum of the other two forces, hence the quoted
    formula. (b) The quantity $(d - d')\eta$ has the
    greater value in water, hence the ball dropped in
    water has the greater terminal velocity.

Chapter 12

Exercises

1. No, but only if it obeys Hooke's law will the
   oscillations be simple harmonic in character.
3. The maximum KE and maximum PE of a
   harmonic oscillator are always equal.
5. When the object has vertical sides, so the
   restoring force is proportional to the displacement of the object above or below its
   equilibrium level.
7. $\frac{1}{2}mv^2 = \frac{1}{2}ks^2$, $v = s\sqrt{k/m} = 10$ m/s.
9. Let $s =$ maximum compression. Then PE
   (gravitational) $=$ PE (elastic), $mg(h + s) = \frac{1}{2}ks^2$,
   $k = 2mg(h + s)/s^2 = 1.5 \times 10^3$ N/m.
11. $k = F/s = 8$ lb/ft. (a) $T = 2\pi\sqrt{m/k}$
    $= 2\pi\sqrt{w/kg} = 0.56$ s.
    (b) $f = 1/T = 1.8$ Hz.
13. From solution to Problem 9 with $mg = w$,
    $k = 2w(h + s)/s^2 = 2.13 \times 10^3$ lb/ft,
    $T = 2\pi\sqrt{m/k} = 2\pi\sqrt{w/kg} = 0.30$ s.
15. $v_{max} = 2\pi fA = 2\pi A/T = \pi$ m/s.
17. $T = 60$ s/24 $= 2.5$ s $= 2\pi\sqrt{L/g}$, $g = 4\pi^2 L/T^2$
    $= 9.66$ m/s².
19. (a) Since the pendulum's support is falling with
    the same acceleration as the pendulum bob, no
    oscillations occur and $f = 0$. (b) $1/T$. (c) $1/T$.

Problems

1. $T = 2\pi\sqrt{m/k} = 2\pi\sqrt{w/kg}$, $k = 4\pi^2 w/T^2 g$
   $= 24.67$ lb/ft, $s = F/k = 2.03$ ft.
3. $1/f = 2\pi\sqrt{m/k}$, $k = 4\pi^2 mf^2 = 197$ lb/ft.
   (a) $E = \frac{1}{2}kA^2 = 24.7$ ft·lb. (b) $a_{max} = 4\pi^2 f^2 A$
   $= 493$ ft/s².
5. $f = 1/T = 25$ Hz. (a) $a_{max} = 4\pi^2 f^2 A = 247$ m/s².
   (b) $F = ma = 1.23$ N. (c) $a = 4\pi^2 f^2 s = 123$ m/s².
   (d) $F = ma = 0.617$ N.

7. (a) $T = 2\pi \sqrt{L/g} = 1.42$ s. (b) The motion of the pendulum bob is unaffected by motion of its support at constant velocity, hence $T = 1.42$ s. (c) The downward force on the bob here is $m(g - a)$, hence $T = 2\pi \sqrt{L/(g - a)} = 1.59$ s. (d) As in part (b), $T = 1.42$ s. (e) The downward force on the bob here is $m(g + a)$, hence $T = 2\pi \sqrt{L/(g + a)} = 1.29$ s.

9. From Fig. 9-9, $I = \frac{1}{3}mL^2 = \frac{1}{3}wL^2/g = 0.0833$ slug·ft^2. (a) $h = 2$ ft, $T = 2\pi \sqrt{I/mgh} = 1.81$ s. (b) $L = I/mh = Ig/wh = 2.67$ ft.

11. (a) From Fig. 9-9, $I = mL^2/3$; $h = L/2$; $T = 2\pi \sqrt{I/mgh} = 2\pi \sqrt{2L/3g} = 1.47$ s. (b) $L' = I/mh = 2L/3 = 53.3$ cm.

13. The requirement for simple harmonic motion to occur is that the acceleration of the object be proportional to its displacement from the equilibrium position and in the opposite direction. Here $a = (g/R)r$ and a is opposite to r, which fulfills this condition. From Eq. (12-3) the period is $T = 2\pi \sqrt{r/a} = 2\pi \sqrt{R/g} = 5.08 \times 10^3$ s $= 1$ hr 25 min.

Chapter 13

Exercises

1. The energy becomes dissipated as heat, and the string becomes warmer as a result.
3. It increases fourfold, since the rate of flow of energy is proportional at A^2.
5. The energy absent from locations of destructive interference is added to the wave energy at locations of constructive interference to give a greater amplitude there than if no interference has occurred. The total energy remains the same as it would be without interference, but its distribution is different.

7. (a) $\lambda = v/f = 0.136$ ft. (b) 0.628 ft. (c) 2.13 ft.
9. $\lambda = 10^{-4}$ m, $f = v/\lambda = 3000$ Hz.
11. (a) $f = v/\lambda = 0.9$ Hz. (b) $T = 1/f = 1.11$ s.
13. Air: $t_1 = s/v_{air} = 1.5106$ s; rail: $t_2 = s/v_{steel} = 0.0962$ s; $t_1 - t_2 = 1.41$ s.
15. (a) $f = 440$ Hz, $\lambda = v_{water}/f = 11.2$ ft. (b) $f = 440$ Hz, $\lambda = v_{air}/f = 2.47$ ft.
17. 1 atm $= 1.013 \times 10^5$ N/m^2, hence the amplitude here is 1.97×10^{-10} atm which is 1.97×10^{-8} % of sea-level pressure.
19. Here $v_s = 0$, so $f_L = f_S(v + v_L)/v = f_S + f_S v_L/v$ and $v_L = v(f_L - f_S)/f_S = 4.24$ ft/s.

Problems

1. Each 10-db interval represents a factor of 10 in intensity, hence 10; 100; 10,000 times.
3. $I = I_0 \log^{-1} [I \text{ (dB)}/10] = 10^{-12}$ W/m^2 $[\log^{-1}(120/10)] = 1$ W/m^2.
5. (a) Wavelength of fundamental frequency $= \lambda = 2L = 2m$, hence $v = \lambda f = 600$ m/s. (b) $f_2 = 2f_1 = 600$ Hz; $f_3 = 3f_1 = 900$ Hz; $f_4 = 4f_1 = 1200$ Hz.
7. Since $f_1 = \sqrt{T/(m/L)}/2L$, increasing the tension in one of the wires to 1.02 of its former value increases f_1 to $\sqrt{1.02}f_1 = 1.01 f_1 = 404$ Hz. Hence 4 beats/s will occur.
9. From the formula given, $v = c(f^2 - f_S^2)/(f^2 + f_S^2)$. Here $f = c/\lambda = 5.56 \times 10^{14}$ Hz, $f_S = c/\lambda_S = 4.84 \times 10^{14}$ Hz, hence $v = 0.138 c = 4.14 \times 10^7$ m/s $= 9.27 \times 10^7$ mi/hr. The fine is therefore $92.7 million less $50.

Chapter 14

Exercises

1. The work goes into the internal energy of the water, as manifested by an increase in its temperature.
3. The ice, because of its heat of fusion.

5. The highest temperature that water can have while remaining liquid is its boiling point. Increasing the rate of heat supply thus increases the rate at which steam is produced but does not change the temperature of the boiling water.

7. $T_F = \frac{9}{5}T_C + 32°$, hence $330°C = 626°F$ and $1170°C = 2138°F$.

9. If $T = T_F = T_C$, $\frac{9}{5}T + 32° = \frac{5}{9}(T - 32°)$, $T = -40°$.

11. $T_C = \frac{5}{9}(T_F - 32°) = -80°C$.

13. $Q = mc\Delta T = 17$ kcal.

15. $Q = mc\Delta T = 1.32 \times 10^4$ Btu.

17. $Q = mc\Delta T = 3.92 \times 10^{-3}$ kcal.

19. $\Delta T = Q/mc = -4°F$, final temperature is $30°F - 4°F = 26°F$.

21. $\Delta T = Q/mc = 667°C$, final temperature is $667°C + 10°C = 677°C$.

23. Heat gained = heat lost, $m_{cop}c_{cop}(T - 20°C) + m_{water}c_{water}(T - 20°C) = m_{iron}c_{iron}(120°C - T)$, $T = 20.7°C$.

25. $Q = Pt/J = 51.6$ kcal.

27. $Q/t = 120$ J/s $\times (3600$ s/hr$)/4185$ J/kcal $= 103$ kcal/hr. The mass of sweat per hour is $(103$ kcal/hr$)/(580$ kcal/kg$) = 0.178$ kg $= 178$ g.

29. From Fig. 14-14, at $p = 70$ atm and $T = 20°C$, CO_2 is in the liquid state.

Problems

1. $mgh = JQ$, $h = JQ/mg = 1.28 \times 10^4$ m.

3. $Q = mc\Delta T = 1.2$ kcal, which is 0.34% of the 350 kcal available from the digestion of the piece of pie.

5. $JmL_f = \frac{1}{2}mv^2$, $v = \sqrt{2JL_f} = 818$ m/s.

7. $Jmc\Delta T + JmL_f = \frac{1}{2}mv^2$, $v = \sqrt{2J(c\Delta T + L_f)}$ $= 327$ m/s, where ΔT is the difference between the melting point of lead and the initial temperature of $100°C$.

9. If M is the mass of the block of ice and m is the mass that melts and if $M \gg m$ (as it is here), then $JmL_f = Fs = \mu Mgs$, $\mu = JmL_f/Mgs = 0.854$.

11. $Q_1 = mL_f + mc\Delta T + mL_v = 720$ kcal, $Q_2 = mc\Delta T = 100$ kcal, so $Q_1 - Q_2 = 620$ kcal.

13. Heat gained by ice = heat lost by alcohol-water mixture, $m_{ice}c_{ice}[0°C - (-10°C)] + m_{ice}L_f + m_{ice}c_{water}(5°C - 0°C) = m_{water}c_{water}(20°C - 5°C) + m_{alcohol}c_{alcohol}(20°C - 5°C)$, $m_{ice} = 0.0263$ kg $= 26.3$ g.

15. The boiling point of nitrogen is $-196°C$, hence $\Delta T = 196°C$; heat lost by water = heat gained by nitrogen, $m_wL_{fw} + m_wc_{ice}\Delta T = m_nL_{fn}$, $m_n = 0.742$ kg.

17. To establish the physical state of the mixture, we note that if all the steam is cooled to $212°F$, the heat lost is $Q_1 = m_sc_s \times 8°F = 38.4$ Btu, and if all the water warms to $212°F$, the heat gained is $Q_2 = m_wc_w \times 12°F = 120$ Btu. We conclude that the final temperature is $212°F$ and that at least some of the steam condenses. If m is the weight of steam that condenses, heat lost by steam = heat gained by water, $Q_1 + mL_v = Q_2$, $m = (Q_2 - Q_1)/L_v = 0.084$ lb.

Chapter 15

Exercises

1. Homogeneous: salt, diamond, iron, solid carbon dioxide, gaseous carbon dioxide, helium, rust. Heterogeneous: leather, stone, blood.

3. Glass expands less than copper with a rise in temperature, so the bulb would crack as it heats up. Lead-in wires must have the same coefficient of thermal expansion as glass if they are to be used successfully.

5. The force is $F = YA\Delta L/L_0$. Here $\Delta L/L_0 = a\Delta T$, so $F = YaA\Delta T$.

7. The internal energy of a solid resides in oscillations of atoms and molecules about their equilibrium positions.

9. Intermolecular attractions tend to reduce gas pressures since some of the energy of the molecules is in the form of PE instead of being entirely KE.

11. The air in the room was originally cold and therefore had a low moisture content, even though its relative humidity may have been high. When this air is heated, its moisture content remains the same, hence its relative humidity decreases.

13. $\Delta L = aL_0\Delta T = 0.0563$ ft. The true width is 119.944 ft; to the nearest inch, 119 ft 11 in.

15. $a = \Delta L/L_0\Delta T = 8.06 \times 10^6/°C$.

17. Glycerin: $\Delta V = bV_0\Delta T = 1.275$ cm³; beaker: $\Delta V = bV_0\Delta T = 0.023$ cm³; thus $(1.275-0.023)$ cm³ = 1.252 cm³ overflows.

19. Alcohol, $V_0 = 0.5000$ qt: $\Delta V_1 = bV_0\Delta T = 0.0153$ qt; Water. $V_0 = 0.5000$ qt: $\Delta V_2 = bV_0\Delta T = 0.0030$ qt. Hence increase in volume is $\Delta V_1 + \Delta V_2 = 0.0183$ qt and profit is $5.00 \times \Delta V = \$0.0915$.

21. (a) $p_2 = p_1V_1/V_2 = 4 \times 10^5$ N/m². (b) $V_2 = p_1V_1/p_2 = 2.67$ m³. (c) $V_2 = V_1T_2/T_1 = 2.67$ m³.

23. $V_2 = V_1T_2/T_1 = 10$ ft³.

25. $T_0 = 27°C = 300$ K. (a) 2 $(KE_0) = \frac{3}{2}k(2T_0)$, so $T = 2T_0 = 600$ K $= 327°C$. (b) $\frac{1}{2}m(2v_0)^2 = \frac{3}{2}k(4T_0)$, so $T = 4T_0 = 1200$ K $= 927°C$.

27. At the same temperature, $\frac{1}{2}m_Hv_H^2 = \frac{1}{2}m_Ov_O^2$, $v_O = v_H\sqrt{m_H/m_O} = 0.25$ m/s.

Problems

1. $T_1 = 273$ K and $T_2 = 373$ K. Since $m = d_1V_1 = d_2V_2$, $d_2 = d_1T_1p_2/T_2p_1 = 1.893$ kg/m³.

3. (a) $\Delta L = aL_0\Delta T = 0.0312$ ft. (b) $A = 7.2$ in², $F = YA\Delta L/L_0 = 3.74 \times 10^4$ lb.

5. $p_1 = p_{atm} + dgh = 1.993 \times 10^5$ N/m², $p_2 = p_{atm} = 1.013 \times 10^5$ N/m², $T_1 = 278$ K, $T_2 = 293$ K; $V_2/V_1 = p_1T_2/p_2T_1 = 2.07 = D_2^3/D_1^3$, hence $D_2 = \sqrt[3]{2.07}\ D_1 = 1.27$ cm.

7. (a) $T = 273$ K, $m = (12.01 + 2 \times 16.00) \times 1.66 \times 10^{-27}$ kg $= 7.306 \times 10^{-26}$ kg; $v_{av} = \sqrt{3kT/m} = 393$ m/s. (b) $\frac{1}{2}m(2v_0)^2 = \frac{3}{2}k(4T_0)$, so $T = 4T_0 = 1092$ K $= 819°C$.

9. The average time interval between collisions is $t = s/v = 2.5 \times 10^{-10}$ s. Hence each molecule

makes an average of $1/t = 4 \times 10^9$ collisions per second.

Chapter 16

Exercises

1. Its ends must be at different temperatures.

3. The vacuum prevents heat transfer by conduction or convection, and the silvered surfaces are poor radiators.

5. Under all circumstances; the higher the temperature, the shorter the predominant wavelength.

7. The kitchen will warm up because the refrigerator exhausts more heat than it absorbs; leaving its door open means that it will run continuously and hence add even more heat to the kitchen than otherwise.

9. The wound watch evolves more heat as it dissolves since its energy content is greater.

11. $e = 1$, $T = 973$ K, $P = e\sigma AT^4 = 51$ W.

13. Eff $= 1 - T_2/T_1 = 44\%$.

15. Maximum efficiency $= 1 - T_2/T_1 = 65\%$, hence actual efficiency is 62% of maximum.

17. $T_2/T_1 = Q_2/Q_1$, $T_2 = T_1Q_2/Q_1 = 375$ K.

19. The heat of fusion of water is 144 Btu/lb, so 1 refrigeration ton = (144 Btu/lb $\times$ 2000 lb/day)/ (24 hr/day) = 12,000 Btu/hr.

Problems

1. Since $R_1/R_2 = T_1^4/T_2^4$, $T_2 = T_1\sqrt[4]{R_2/R_1} = 800$ K $= 527°C$.

3. The net power radiated is the difference between the power radiated and that absorbed, so $P = RA = Ae\sigma(T_2^4 - T_1^4)$ and $e = P/A\sigma(T_2^4 - T_1^4)$. Here $T_1 = 20°C = 293$ K and $T_2 = 34°C = 307$ K, so that $e = 0.62$. This is a reasonable

emissivity figure, so it is likely that much of the woman's metabolic heat is radiated away under the given circumstances.

5. $T_1 = 773$ K, Eff $= 1 - T_2/T_1$, $T_2 = T_1(1 - \text{Eff}) = 502$ K. If T_1' is the intake temperature for an efficiency of 50%, then $T_1' = T_2/(1 - \text{Eff}) = 1005$ K $= 732°$C.

7. $mh = Jmc\,\Delta T$, $\Delta T = h/Jc = 2.0°$F and $T_1 = 52°$F $= 512°$R, $T_2 = 50°$F $= 510°$R, Eff $= 1 - T_2/T_1 = 0.0039 = 0.39\%$.

9. For simplicity we consider a cylinder of cross-sectional area A which has a piston that the gas pushes out a distance s. The force on the piston is $F = pA$, hence $W = Fs = pAs = p\Delta V$, where $\Delta V = As$ is the increase in volume of the gas due to the piston's motion.

11. If Q_1 is the heat absorbed and Q_2 is the heat ejected, $W = Q_2 - Q_1$. Since $Q_2/Q_1 = T_2/T_1$ for a Carnot engine, $W/Q_1 = (Q_2 - Q_1)/Q_1 = Q_2/Q_1 - 1 = T_2/T_1 - 1$.

13. $T_1 = 253$ K, $T_2 = 313$ K, $W = Q[(T_2/T_1) - 1] = 992$ J/kcal for a Carnot refrigerator, which is the most efficient possible. Hence only design C is capable of operating.

7. The positive charge is located in the nucleus, the negative charge in the electrons that surround the nucleus.

9. Most of an atom consists of empty space.

11. An electric current in a metal consists of moving electrons, in an ionized gas it consists of positive and negative ions that move in opposite directions.

13. The need for extremely low temperatures.

15. A molecular ion is a molecule with a net charge. A polar molecule is electrically neutral but has a nonsymmetrical distribution of charge.

17. Water is the most notable polar liquid and dissolves ionic compounds such as NaCl and some non-ionic compounds such as HCl, sugar, and alcohol. Gasoline is a nonpolar liquid and dissolves such substances as fats and oils.

19. $F_1/F_2 = (r_2/r_1)^2$, $r_2 = r_1\sqrt{F_1/F_2} = 2$ cm.

21. (a) $F_2 = F_1(r_1/r_2)^2 = 0.025$ N. (b) 0.001 N. (c) 2.5 N.

23. $F = kq_A q_B/r^2 = 1.08 \times 10^{-4}$ N. The force is repulsive.

25. $F = kq_A q_B/r^2 = 1.125$ N. The force is repulsive.

Chapter 17

Exercises

1. The notion of electricity as a fluid can explain phenomena that involve charges whose magnitudes are large compared with the electron charge; it cannot explain the behavior of elementary particles or the structure of matter on an atomic level.

3. There are many reasons, but a convincing one is that all gravitational forces are observed to be attractive, whereas if they were electric in origin some would have to be repulsive.

5. An equal amount of charge of the opposite sign appears on the fur.

Problems

1. $F = mg = ke^2/r^2$, $r = e\sqrt{k/mg} = 5.08$ m.

3. $Gm_e m_m/r^2 = kq^2/r^2$, $q = \sqrt{Gm_e m_m/k} = 5.7 \times 10^{13}$ C.

5. Each outer charge exerts on the middle charge a force of magnitude $F = kq_A q_B/r^2 = 0.135$ N. Since both forces are directed toward the negative charge, the total force is 0.27 N.

7. (a) If r_1 is the distance from the test charge q to $q_1 = -1 \times 10^{-6}$ C, then $r_2 = (0.4$ m $- r_1)$ is the distance to $q_2 = -3 \times 10^{-6}$ C. The forces on q will be equal when $kqq_1/r_1^2 = kqq_2/r_2^2$, $r_2/r_1 = \sqrt{q_2/q_1} = (0.4$ m $- r_1)/r_1$, $r_1 = 0.4$ m$/(\sqrt{q_2/q_1} + 1) = 0.146$ m $= 14.6$ cm. (b) Since r_1 is independent of the magnitude and sign of q, $r_1 = 14.6$ cm here also.

9. The force each of the other charges exerts has the magnitude $F = kq^2/a^2$. The lines of action of the forces are 120° apart (since each of the interior angles of the triangle is 60°), so the vector sum of the forces is $F' = 2F \cos 60°$ $= 2F \times 0.5 = kq^2/a^2$. The force is parallel to the side of the triangle opposite to the $+q$ charge and points toward the $-q$ charge.

Chapter 18

Exercises

1. Compare the forces each field exerts on a charged and on an uncharged body
5. No, because at a given point an appropriate test object travels along the line of force passing through that point and it can travel in only one direction at any point.
7. (a) It will not move. (b) It will rotate until aligned with the field.
9. $E = 0$ at the center of the square, since a charge placed there will experience no force.
11. $E = F/e = V/s$, $V = Fs/e = 2.5 \times 10^3$ V.
13. (a) $W = qV$ $P = qV/t = 120$ W. (b) $W = Pt$ $= 4.32 \times 10^5$ J.
15. $W = qV = 3.2 \times 10^8$ J.
17. KE $= 8 \times 10^{-18}$ J, $\frac{1}{2}mv^2 =$ KE, $v = \sqrt{2KE/m}$ $= 4.2 \times 10^6$ m/s.
19. KE $= \frac{1}{2}mv^2 = 3.25 \times 10^{-14}$ J/1.6×10^{-19} J/eV $= 2.03 \times 10^5$ eV

Problems

1. $E = 0$ at a point between the charges and the distance s from the charge $q_1 = 1.5 \times 10^{-6}$ C. At this point $kq_1/s^2 = kq_2/(0.2 \text{ m} - s)^2$, $(0.2 \text{ m} - s)/s = \sqrt{q_2/q_1}$, $s = 0.2 \text{ m}/(\sqrt{q_2/q_1} + 1)$ $= 0.0828$ m.
3. $F = qE = mg$, $E = mg/q = 9.8$ V/m.
5. $E = V/s = 500$ V/m. The acceleration of the electron is $a = F/m = eE/m = 8.78 \times 10^{13} \text{m/s}^2$.

Since $s = 0.01 \text{ m} = \frac{1}{2}at^2$, the electron strikes the plate $t = \sqrt{2s/a} = 1.51 \times 10^{-8}$ s later. In this time it travels $d = vt = 0.151 \text{ m} = 15.1$ cm.
7. KE $= \frac{1}{2}mv^2 =$ PE$_2$ $-$ PE$_1 = ke^2/r_2 - ke^2/r_1$, $r_2 = 1/[(mv^2/2ke^2) + (1/r_1)] = 1.43$ m.
9. $W = mc^2$, $m = W/c^2 = 1.78 \times 10^{-36}$ kg.

Chapter 19

Exercises

1. No, because electrons have little mass and hence little inertia and also because they undergo frequent changes in direction during collisions inside the wire in any case.
3. With a single wire, charge would be permanently transferred from one end to the other, and soon so large a charge separation would occur that the electric field that produced the current would be cancelled out. With two wires, charge can be sent on a round trip, so to speak, and a net flow of energy from one place to another can occur without a net flow of charge.
5. Some of the power produced by the cell of high emf will be dissipated in the cell of low emf.
7. $q = it = 180$ C.
9. $i = q/t = 60 \ e/s = 9.6 \times 10^{-18}$ A.
11. (a) $q = 1.5 \text{ C/s} \times 3600 \text{ s} = 5400$ C, $W = qV = 7290$ J. For $P = 0.1 \text{ mW} = 10^{-4}$ W, $t = W/P = 7.29 \times 10^7 \text{ s} = 844 \text{ days} = 2.3$ years. (b) $i = P/V = 7.41 \times 10^{-5} \text{ A} = 74.1 \ \mu$A.
13. $P = iV = 1200$ W.
15. $R = V_1^2/P = 240 \ \Omega$; $i = V_2/R = 0.33$ A.
17. $R = \rho L/\pi r^2$, $r = \sqrt{\rho L/\pi R} = 1.427 \times 10^{-4}$ m, $d = 2r = 2.85 \times 10^{-4}$ m.
19. $d = 10^{-4}$ m, $A = \pi d^2/4 = 7.85 \times 10^{-9}$ m², $L = RA/\rho = 2.85$ m.
21. $\Delta R = \alpha R_0 \Delta T = -20 \ \Omega$, so $R = 480 \ \Omega$ at 100°C.

23. The increase in resistivity cannot be due to the expansion of the metal because its cross-sectional area expands faster than its length and so the resistivity should decrease on this basis. Also, the coefficient of linear expansion is too much smaller than the temperature coefficient of resistivity for the expansion of the metal to have any major effect on resistivity.

25. (a) The resistance of each bulb is $R = V^2/W$ = 960 Ω, $i = V/R = 0.25$ A. (b) The power dissipated by each bulb is 60 W, hence both dissipate 120 W when operated at their rated potential differences.

27. (a) $R = 12 \times 5 \ \Omega = 60 \ \Omega$. (b) $P = V^2/R = 240$ W.

29. $i_1 = \mathcal{E}/(R + r) = 5.0$ A, $i_2 = 3.3$ A. Since $P = i^2R$, $P_2/P_1 = i_2^2/i_1^2 = 0.44$.

Problems

1. $q/s = 10^{20} \ e/cm = 16$ C/cm $= 1.6 \times 10^3$ C/m, $t = s/v$, $i = q/t = v(q/s)$, $v = i/(q/s) = 6.25 \times 10^{-4}$ m/s.

3. $R = V^2/P = 14.4 \ \Omega$, $L = R\pi d^2/4\rho = 5.54$ m.

5. (a) $P = iV = 5 \times 10^3$ W. (b) 5×10^6 eV
   (c) 5×10^6 eV $\times 1.6 \times 10^{-19}$ J/eV $= 8 \times 10^{-13}$ J.

7. $W = i^2Rt = 4608$ J; $Q = W/J = mc\Delta T$, where J = mechanical equivalent of heat, so $\Delta T = W/Jmc = 0.28°C$.

9. (a) $R' = (5 \ \Omega \times 8 \ \Omega)/(5 \ \Omega + 8 \ \Omega) = 3.08 \ \Omega$; $R'' = 3.08 \ \Omega + 5 \ \Omega = 8.08 \ \Omega$; $R = (8.08 \ \Omega \times 3 \ \Omega)/(8.08 \ \Omega + 3 \ \Omega) = 2.19 \ \Omega$. (b) The current through the upper branch is $i'' = V''/R'' = 1.49$ A, hence the potential difference across the 8-Ω resistor is $V = i''r'$ = 4.59 V and the current through it is $i = V/R = 0.57$ A.

11. If i is the current through each battery $I = 3i$ is the current in the external resistor. In the loop that contains the external resistor and one of the batteries, $\mathcal{E} = ir + IR = ir + 3iR$, $i = \mathcal{E}/(3R + r) = 0.909$ A; hence $I = 3i = 2.73$ A.

13. (a) $R_1 = (5 \ \Omega \times 10 \ \Omega)/(5 \ \Omega + 10 \ \Omega) = 3.33 \ \Omega$, $R_2 = (8 \ \Omega \times 8 \ \Omega)/(8 \ \Omega + 8 \ \Omega) = 4 \ \Omega$, $R = R_1 + R_2 = 7.33 \ \Omega$. (b) The resistance of the entire circuit is $R' = 7.33 \ \Omega + 1.5 \ \Omega$ = 8.83 Ω, so the current is $i = V/R' = 2.72$ A. The potential difference across the first pair of resistors is $V_1 = iR_1 = 9.06$ V, hence in the 5-Ω resistor $i = V_1/5 \ \Omega = 1.81$ A and in the 10-Ω resistor $i = V_1/10 \ \Omega = 0.906$ A. The potential difference across the second pair of resistors is $V_2 = iR_2 = 10.88$ V. hence in each 8-Ω resistor $i = V_2/8 \ \Omega = 1.36$ A.

15. Assume currents downward in each resistor. At the top junction, $i_1 + i_2 + i_3 = 0$. Proceeding counterclockwise in the left-hand loop, 10 V + 6 V $= i_1 \times 10 \ \Omega - i_2 \times 5 \ \Omega$. Proceeding clockwise in the right-hand loop, 10 V + 10 V $= i_3 \times 20 \ \Omega - i_2 \times 5 \ \Omega$. Solving yields $i_1 = 0.857$ A, $i_2 = -1.486$ A, $i_3 = 0.629$ A.

17. (a) To find the potential difference between a and b we can disregard the 8-V battery and the 6-Ω resistor. Applying Kirchhoff's second rule to the outside loop, going clockwise and assuming a clockwise current of i, yields 10 V − 5 V $= i (1 \ \Omega + 1 \ \Omega + 3 \ \Omega + 9 \ \Omega)$, $i = 0.357$ A. In the upper part of the loop a potential drop occurs in both the battery and the 3-Ω resistor, hence $V = 5$ V $+ i (1 \ \Omega + 3 \ \Omega)$ = 6.43 V. The lower part of the loop gives the same result. Here there is a potential increase of 10 V in the battery and a potential drop in the battery's internal resistance and in the 9-Ω resistor, so $V = 10$ V $- i (1 \ \Omega + 9 \ \Omega)$ = 6.43 V. (b) Point b is at a potential of −6.43 V relative to a. No current passes through the 6-Ω resistor or the 8-V battery, so there is no iR drop in either. The 8-V battery's positive terminal is nearest to c, hence the potential of c relative to a is 8 V − 6.43 V = 1.57 V.

19. When the bridge is balanced, no current passes through the galvanometer. Hence the same

current i_1 passes from left to right through R and C, and the same current i_2 passes from left to right through A and B. In order for the junction of R, C, and G to be at the same potential as that of A, B, and G, it must be true that $i_1R = i_2A$ and that $i_1C = i_2B$. Dividing the first of these equations by the second, $R/C = A/B$, $R = AC/B$.

Chapter 20

Exercises

1. See discussion in Sec. 20-1.
3. $0°$; $90°$.
5. The electric force between the streams is attractive and the magnetic force is repulsive. Which force predominates depends upon the velocities of the particles, since the magnetic force increases with velocity whereas the electric force does not change.
7. The magnetic field causes the protons to move in curved paths and hence increases their transit times relative to what they would be if the paths were direct.
9. The magnetic field itself must be regarded as able to possess linear momentum.
11. $B = \mu_0 i/2\pi s = 10^{-4}$ T.
13. $s = \mu_0 i/2\pi B = 0.08$ m $= 8$ cm.
15. $i = 2Br/\mu_0 = 7.94 \times 10^5$ A.
17. $i = BL/\mu_0 N = 5.49 \times 10^{-3}$ A.
19. (a) $\theta = 45°$, $F = i\Delta lB \sin \theta = 0.283$ N; the force is downward according to the right-hand rule. (b) $F = 0.283$ N; the force is upward.
21. $F/l = mg/l = iB$, $i = mgB/l = 0.49$ A.
23. $R = mv/eB = 2.84 \times 10^{-3}$ m.
25. $F_{grav} = 8.9 \times 10^{-30}$ N; If v is perpendicular to $\mathbf{B}$ and $B = 3 \times 10^{-5}$ T (a typical value), $F_{mag} = evB = 1.4 \times 10^{-16}$, which is over 10^{13} times greater than F_{grav}.

27. $P = iV$, $i = P/V = 1.67$ A $= i_1 = i_2$; $F/l = \mu_0 i_1 i_2/2\pi s = 2.78 \times 10^{-4}$ N/m; the force is repulsive, since the currents are in opposite directions.

Problems

1. $\mathbf{B} = 0$ midway between the wires because the fields of the currents are opposite in direction and of the same magnitude there.
3. $B_1 = \mu_0 i/2\pi s = 2 \times 10^{-5}$ T, $B_2 = 0.67 \times 10^{-5}$ T. Since the fields are in opposite directions, $B = B_1 - B_2 = 1.33 \times 10^{-5}$ T.
5. The fields of the two layers are in opposite directions, hence the result is the same as if the solenoid had 200 turns. Thus $B = \mu_0 Ni/L = 1.26 \times 10^{-4}$ T.
7. $B_{loop} = B_{solenoid}$, $\mu_0 i_1/2r = \mu_0 i_2(N/L)$, $i_1 = 2ri_2(N/L) = 2$ A since $(N/L) = 2 \times 10^3$ turns/m.
9. If R is the orbit radius, $T = 2\pi R/v = 2\pi mv/vqB = 2\pi m/qB$, which is independent of R and v.
11. (a) The magnetic field 0.1 m from the wire is $B = \mu_0 i/2\pi s = 2 \times 10^{-4}$ T, so the force on the electron is $F = evB = 3.2 \times 10^{-16}$ N since $\mathbf{v} \perp \mathbf{B}$. The force is toward the wire since the moving electron is equivalent to a current in the opposite direction. (b) Here also $\mathbf{v} \perp \mathbf{B}$, so $F = 3.2 \times 10^{-16}$ N. The force is in the same direction as the current.
13. (a) $B = \mu_0 i/2\pi s = 1.6 \times 10^{-4}$ T, $F = evB = 1.54 \times 10^{-15}$ N. The force is directed away from the wire. (b) The force has the same magnitude but is directed toward the wire.
15. (a) The force has the same magnitude as in Problem 13(a), but is in the direction of the current. (b) The force has the same magnitude but is opposite to the direction of the current.
17. (a) From Sec. 20-6, $B = E/v$. The direction of $\mathbf{B}$ should be perpendicular to both $\mathbf{v}$ and $\mathbf{E}$ with its sense given by the right-hand force rule. (b) Magnetic fields can exert no forces

on charged particles parallel to their directions of motion, hence there is no magnetic field that can cancel the effects of **E** on the particle.

19. $v = E/B$, $R = mv/qB = mE/eB^2 = 2.27$ m.

Chapter 21

Exercises

1. Counterclockwise; zero current; clockwise.

3. The left-hand wheels have positive charges, the right-hand wheels have negative charges. See sketch of geomagnetic field in Fig. 20-3.

5. The particle streams are deflected by the earth's field so that the magnetic fields they produce are opposite in direction to the earth's field within the ring currents formed.

7. (a) An emf is induced in the wire loop and a current therefore comes into being. This current interacts with the magnetic field to produce a torque that, by Lenz's law, opposes the rotation. (b) The current in a loop of aluminum wire would be smaller because of the greater electrical resistance of aluminum, and the torgue opposing the rotation would be correspondingly smaller.

9. Alternating current is necessary in order that there be a changing magnetic field to induce a current in the secondary winding.

11. (a) The emf increases because the greater the current, the greater the iR drop in the armature and the less the potential difference across the generator. Hence the current in the magnet windings drops and, since the emf is proportional to B, it too decreases. (b) A similar argument explains the increase in the emf when the current drops.

13. $\mathcal{E} = Blv = 1.8 \times 10^{-3}$ v.

15. (a) $A = 5 \times 10^{-3}$ m², $\Phi = BA = 5 \times 10^{-6}$ Wb. (b) $\mathcal{E} = \Delta\Phi/\Delta t = 1.67 \times 10^{-6}$ V.

17. $\mathcal{E} = N\Delta\Phi/\Delta t = 6$ V

19. The induced emf is zero when the plane of the coil is perpendicular to the magnetic field, which happens twice per rotation. Hence $T = 1$ s/(2 × 100) = 0.005 s.

21. (a) $N_2 = N_1 i_1/i_2 = 2$ turns. (b) $\mathcal{E}_2 = \mathcal{E}_1 N_2/N_1 = 1.1$ V

23. $i_2 = P/\mathcal{E}_2 = 21.7$ A, $i_1 = i_2\mathcal{E}_2/\mathcal{E}_1 = 0.455$ A.

Problems

1. $V = \mathcal{E} - iR$, $i = (\mathcal{E} - V)/R = 25$ A, $P = iV = 2875$ W.

3. Total current is $i = P/V = 25$ A, $i_f = V/R_f = 1$ A, $i_a = i - i_f = 24$ A. $P_a = i_a^2 R_a = 57.6$ W, $P_f = i_f^2 R_f = 120$ W, total power dissipated is $P' = P_a + P_f = 178$ W. Efficiency = $P/(P + P') = 94\%$.

5. (a) Total current is $i = P/V = 3.5$ A, field current is $i_f = V/R_f = 0.6$ A, armature current is $i_a = 3.5$ A − 0.6 A = 2.9 A. Back emf is $\mathcal{E}_b = V - i_a R_a = 117$ V, power output is $\mathcal{E}_b i_a = 340$ W, efficiency is 340 W/420 W = 81%. (b) $R = R_a R_f/(R_a + R_f) = 1.0$ Ω, i (start) $= V/R = 120$ A. (c) Total resistance needed is 120 V/25 A = 4.8 Ω, so series resistor should be 4.8 Ω − 1.0 Ω = 3.8 Ω.

Chapter 22

Exercises

1. No.

3. Since for a parallel-plate capacitor $C = KA/4\pi kd$, $C' = C$; $W_1 = \frac{1}{2}Q^2/C$, $W_2 = \frac{1}{2}Q^2/C' = Q^2/C$, hence the work needed is $\Delta W = W_2 - W_1 = \frac{1}{2}Q^2/C$.

5. Since $V = Q/C$, when a certain charge Q is given to a capacitor, the greater C is, the smaller V is and hence the weaker the electric field E in the capacitor is. The inductance L, on the other hand, is a measure of the magnitude of and volume occupied by the

magnetic field produced by a current i in an inductor, hence the greater L is, the greater the energy of the inductor.

7. (a) From Lenz's law, the selfinduced emf will be opposite to the direction of the current.
(b) For the same reason, the emf here will be in the same direction as that of the current.

9. (a) $Q = CV = 0.025$ C. (b) $W = \frac{1}{2}CV^2 = 12.5$ J.

11. (a) Here $K = 1$, $A = 5 \times 10^{-3}$ m^2, $d = 10^{-3}$ m, so $C = A/4\pi kd = 4.42 \times 10^{-11}$ F $= 44.2$ pF.
(b) $Q = CV = 1.99 \times 10^{-9}$ C. (c) $W = \frac{1}{2}CV^2 = 4.48 \times 10^{-8}$ J.

13. $Q_1 = C_0V$, $Q_2 = KC_0V$, hence if $Q_2 = 2Q_1$, $K = 2$.

15. $C = C_1C_2/(C_1 + C_2) = 14.3 \ \mu$F.

17. $N = \sqrt{LI/\mu A} = 503$ turns.

19. $\mathcal{E} = L\Delta i/\Delta t = 200$ V.

21. $f = \frac{1}{2}\pi\sqrt{LC} = 650$ Hz.

23. $C = \frac{1}{4}\pi^2 Lf^2 = 63.3 \ \mu$F.

$= 1.2 \times 10^{-3}$ C. The charge distributes itself on the capacitors so there is the same new potential difference across each. Hence $V_1 = V_2$, $Q_1/C_1 = Q_2/C_2$, and $Q_2 = Q - Q_1$; $Q_1 = Q/(1 + C_2/C_1) = Q/3 = 4 \times 10^{-4}$ C, $Q_2 = Q - Q_1 = 8 \times 10^{-4}$ C.

9. (a) If the relative permeability is K, then $\mu = K\mu_0$ and $L = K\mu_0N^2A/l = 4.11$ mH.
(b) $i = \mathcal{E}/(R + r) = 0.286$ A. (c) $W = \frac{1}{2}Li^2 = 1.68 \times 10^{-4}$ J.

11. When $t = 1/8f$, $2\pi ft = \pi/4 = 45°$, and $q = 0.707 \ Q$, $i = 0.707 \ I$. Since $W_e = q^2/2C$, here $W_e = \frac{1}{2}Q^2/2C$ and since $W_m = Li^2/2$, here $W_m = \frac{1}{2}LI^2/2$. Because $\frac{1}{2}Q^2/2C = \frac{1}{2}LI^2/2$, $W_e = W_m$ at $t = 1/8f$.

Chapter 23

Problems

1. There are several ways to proceed, one of which is this. (a) $C = A/4\pi kd = 2.21 \times 10^{-11}$ F, $W = \frac{1}{2}CV^2 = 1.11 \times 10^{-7}$ J. The stored energy W is equal to the work Fd that would be done if the plates were released and flew together so $F = W/d = 1.11 \times 10^{-4}$ N. (b) The force is attractive since the plates have opoosite charges. (c) $w = W/Ad = 0.044$ J/m$^3 = E^2/8\pi k$.

3. $1/C = 1/C_1 + 1/C_2 + 1/C_3$, $C = 3.125 \ \mu$F, $Q = Q_1 = Q_2 = Q_3 = CV = 3.75 \times 10^{-5}$ C; $V_1 = Q_1/C_1 = 7.5$ V, $V_2 = Q_2/C_2 = 3.75$ V. $V_3 = Q_3/C_3 = 0.75$ V

5. $V_1 = V_2 = V_3 = 12$ V; $Q_1 = C_1V_1 = 6 \times 10^{-5}$ C, $Q_2 = C_2V_2 = 1.2 \times 10^{-4}$ C, $Q_3 = C_3V_3 = 6 \times 10^{-4}$ C.

7. At first, $Q_{01} = C_1V_0 = 1.2 \times 10^{-3}$ C, $Q_{02} = C_2V_0 = 2.4 \times 10^{-3}$ C. Since terminals of opposite sign are joined, the total charge that remains on the combination is $Q = Q_{02} - Q_{01}$

Exercises

1. The resistance of an ac-circuit determines how much of the electrical energy that passes through it is dissipated as heat. The capacitive reactance and inductive reactance respectively determine the extent to which the capacitors and inductors in the circuit oppose the flow of current without dissipating energy X_L and X_C vary with frequency whereas R does not, and their combined effects depend upon $X_L - X_C$, so their analogy with resistance is very limited. The impedance determines the current in the circuit when a potential difference V of a certain frequency is applied according to $i = V/Z$; thus Z has the same role as R does in a dc circuit with respect to i, though the rate of energy dissipation is equal to i^2R in both cases.

3. Under these circumstances $X_L > X_C$, so the voltage leads the current.

5. The power factor of a circuit is the ratio

between its resistance and impedance and governs the rate at which power is absorbed by the circuit for a given i and V; it varies with frequency The power factor is 0 only when $R = 0$; it is 100% at resonance, when $Z = R$.

7. Twice per cycle, which is $2f$ per second.

9. (a) $i_{max} = \sqrt{2}\, i_{eff} = 14.1$ A. (b) $V_{max} = \sqrt{2}\, V_{eff}$ $= 84.9$ V. (c) No; they are coincident only when the frequency is the resonant frequency

11. $C = \frac{1}{2}\pi f X_C = 1.59 \times 10^{-4}$ F.

13. $L = X_L / 2\pi f = 0.796$ H.

15. $R = V/i = 120\ \Omega$. To halve the current, we require $Z = 240\ \Omega$. Since $Z = \sqrt{R^2 + X_L^2}$, $X_L = 2\pi f L = \sqrt{Z^2 - R^2}$ and $L = \sqrt{Z^2 - R^2}/2\pi f$ $= 0.662$ H.

17. $X_L = V/i = 400\ \Omega$; $f = X_L/2\pi L = 424$ Hz.

19. (a) $X_C = \frac{1}{2}\pi f C = 53\ \Omega$. (b) $i = V/X_C = 0.226$ A.

21. (a) $X_C = \frac{1}{2}\pi f C = 637\ \Omega$; $Z = \sqrt{R^2 + X_C^2}$ $= 704\ \Omega$; $i = V/Z = 0.170$ A. (b) $P = i^2 R$ $= 8.7$ W.

23. Since $V_C = V_L$, the circuit is in resonance, and $V_R = 60$ V

25. (a) Since $X_L = X_C$, $X_L - X_C = 0$ and $Z = R$ $= 20\ \Omega$. (b) $\tan \phi = 0$, $\phi = 0$. (c) $V = iZ = 400$ V

27. (a) The power line must have a rating of at least 10 kW/(0.70 kW/kVA) = 14.3 kVA. (b) The minimum rating is now 10 kVA.

5. At $f = 200$ Hz, $L = X_L/2\pi f = 0.0955$ H. At $f = 60$ Hz, $X_L = 2\pi f L = 36\ \Omega$, $i = V/X_L$ $= 6.67$ A.

7. (a) The voltage leads the currents when X_C is less than X_L. (b) $i_{max} = \sqrt{2}\, i_{eff} = 7.07$ A; when $V = 0$, $i = i_{max} \sin(360° - 30°) = -i_{max}\sin 30°$ $= -3.54$ A if V is going from $+$ to $-$, or $+3.54$ A if V is going from $-$ to $+$.

9. (a) $R = V_1/i_1 = 5\ \Omega$, $Z = V_2/i_2 = 8\ \Omega$; $Z = \sqrt{R^2 + X_L^2}$, $L = \sqrt{Z^2 - R^2}/2\pi f = 16.6$ mH. (b) $P_1 = i_1^2 R = 320$ W; $P_2 = i_2^2 R = 125$ W.

11. (a) $X_L = 2\pi f L = 113\ \Omega$, $X_C = \frac{1}{2}\pi f C = 44\ \Omega$, $Z = \sqrt{R^2 + (X_L - X_C)^2} = 85\ \Omega$. (b) $i = V/Z$ $= 1.41$ A. (c) $\tan \phi = (X_L - X_C)/R = 1.38$, $\phi = 54°$, power factor = $\cos \phi = 0.588$. (d) $P = i^2 R = 99.4$ W. (e) $iV = P/\cos \phi$ $= 169$ V·A.

13. (a) $f_0 = \frac{1}{2}\pi \sqrt{LC} = 37.5$ Hz. (b) $i = V/R = 2.4$ A.

15. (a) $X_L = 2\pi f L = 38\ \Omega$, $X_C = \frac{1}{2}\pi f C = 265\ \Omega$, $Z = \sqrt{R^2 + (X_L - X_C)^2} = 235\ \Omega$, $i = V/Z$ $= 0.51$ A. (b) $P = i^2 R = 15.6$ W. (c) $V_L = iX_L$ $= 19.4$ V; $V_C = 135$ V; $V_R = iR = 30.6$ V

17. (a) $\cos \phi = P/iV = 0.625$. (b) Since $\cos \phi$ $= 0.625$, $\phi = 51°$, $\tan \phi = 1.235 = X_L/R$; $R = P/i^2 = 75\ \Omega$, $X_L = 1.235R = 92.6\ \Omega$; if $X_C = X_L$, $C = \frac{1}{2}\pi f X_L = 28.6\ \mu$F. (c) $i = V/R$ $= 1.6$ A. (d) $P = i^2 R = 192$ W. (e) 192 V·A since V and i are in phase at the resonant frequency

Problems

1. (a) $2\pi ft = 2\pi/8 = 45°$, $V = V_{max}\sin 2\pi ft$ $= V_{max}\sin 45° = 70.7$ V
(b) $2\pi ft = 2\pi/4 = 90°$, $V = V_{max}\sin 90° = 100$ V
(c) $2\pi ft = 6\pi/8 = 135°$, $V = V_{max}\sin 135°$ $= V_{max}\sin 45° = 70.7$ V (d) $2\pi ft = 2\pi/2 = 180°$, $V = V_{max}\sin 180° = 0$.

3. (a) Let Z_1 be the impedance at $f_1 = 100$ Hz and Z_2 be the impedance at $f_2 = 500$ Hz. Then $Z_1^2 = R^2 + X_{L1}^2 = R^2 + 4\pi^2 f_1^2 L^2$, $Z_2^2 = R^2 + 4\pi^2 f_2^2 L^2$, and $L =$ $\sqrt{(Z_2^2 - Z_1^2)/4\pi^2(f_2^2 - f_1^2)} = 28.1$ mH. (b) $R = \sqrt{Z_1^2 - 4\pi^2 f_1^2 L^2} = 46.8\ \Omega$.

Chapter 24

Exercises

1. The electric fields produced by electromagnetic induction are readily detected through the currents they produce in metal wires, but no such simple means is available for detecting the weak magnetic fields produced by changing electric fields.

3. Light waves consist of coupled electric and magnetic field fluctuations and need no material medium in which to occur. Sound

waves consist of pressure fluctuations in a material medium and hence cannot occur in a vacuum.

5. The variations are always in phase.

7. An advantage of Huygens' principle is that no mathematics is needed to use it to investigate the propagation of light in simple situations.

9. A light ray is always perpendicular to the wavefronts whose motion it describes.

11. Electromagnetic waves travel more slowly in material media than they do in free space.

13. The advantage is that more of the incident light is reflected; disadvantages include cost and size.

15. Since both are transparent with the same index of refraction, light passes through the bottle unaffected by the presence of the rod.

17. The flashes of color are the result of dispersion. In red light the flashes would be red only.

19. (a) $f = 1/T = 10^9$ Hz. (b) $\lambda = c/f = 0.3$ m.

21. The image is always twice as far from the man as the mirror; hence it must be moving twice as fast as the mirror, namely at 40 mi/hr.

23. (a) $v = c/n = 2.999 \times 10^8$ m/s.
(b) 1.24×10^8 m/s. (c) 1.97×10^8 m/s.
(d) 2.26×10^8 m/s.

25. $\sin r = (n_1/n_2)\sin i = 0.789$, $r = 52°$.

27. $n_2 = n_1 \sin i/\sin r = 1.51$.

29. (a) $\sin i_c = n_2/n_1 = 0.413$, $i_c = 24°$.
(b) $\sin i_c = 0.550$, $i_c = 33°$.

Problems

1. $\lambda_2/\lambda_1 = n_1/n_2$.

3. Here angle of reflection $+ 90° +$ angle of refraction $= 180°$, hence, since the angle of reflection equals the angle of incidence, the angle of refraction is $r = 30°$ and n_2 $= n_1 \sin i/\sin r = 1.73$.

5. If h is the depth of the pool, r the radius of the circle of light, and i_c the critical angle, then $d = 2r = 2h \tan i_c$. Here $\sin i_c = n_2/n_1$ $= 0.75$, $i_c = 49°$, $\tan i_c = 1.15$, and $d = 4.6$ m.

7. $h = h'n_1/n_2 = 2.7$ in.

9. $h' = hn_2/n_1 = 0.59$ m.

Chapter 25

Exercises

1. No; yes, if the object is placed between the lens and the focal point.

3. The glass tube magnifies the mercury column, which is therefore narrower than it appears.

5. To increase the image distance, the object distance must be reduced, hence the lens should be moved closer to the slide.

7. (a) Under no circumstances. (b) When the object is between the lens and the focal point.

9. (a) Inverted and larger than the object.
(b) Inverted and smaller than the object.
(c) Erect and smaller than the object.
(d) Erect and larger than the object.

11. $R_1 = -30$ cm, $R_2 = +20$ cm, $1/f$ $= (n - 1)(1/R_1 + 1/R_2)$, $f = 12$ cm.

13. $1/R_2 = 0$, $R_1 = (n - 1)f = 4.95$ in.

15. (a) Converging. (b) $p = \infty$, $q = f = 8$ cm.

17. No image is formed when the object is at the focal point.

19. $q = pf/(p - f) = 30$ cm, $m = -q/p = -1$. The image is real, inverted, and the same size as the object.

21. $f = pq/(p + q) = 12$ cm; the lens is converging since the image is real.

23. (a) $q = pf/(p - f) = 0.533$ ft $= 6.4$ in.
(b) $m = -q/p = -1/15$, hence the area covered is 60 in $\times$ 75 in.

25. $q = pf/(p - f) = -9.86$ in, $m = -q/p = 4.3$, so the egg appears 0.086 in long.

Problems

1. (a) Converging. (b) The index of refraction of
   water is 1.33, so the relative index of
   refraction is $n' = 1.60/1.33 = 1.20$. From the
   lensmaker's equation, $f/f' = (n' - 1)/(n - 1)$
   $= 1/3$, and $f' = 3f = 19.5$ in.
3. $m = 1/10 = -q/p$, $q = -p/10 = -2$ ft;
   $f = pq/(p + q) = -2.22$ ft.
5. (a) $p + q = 24$ in, $p = 24$ in $- q$; $m = -4$
   $= -q/p$, $p = q/4 = 24$ in $- q$, $q = 19.2$ in,
   $p = 4.8$ in; $f = pq/(p + q) = +3.84$ in.
   (b) $m = -2 = -q/p$, $p = q/2$; $1/f = 1/p + 1/q$
   $= 3/q$, $q = 3f = 11.52$ in, $p = q/2 = 5.76$ in,
   $p + q = 17.28$ in.
7. $q = -60$ cm and $f = 1$ m/3.33 $= 0.30$ m $= 30$
   cm, hence $p = qf/(q - f) = 20$ cm.
9. (a) Converging lens. (b) $p = 25$ cm, $q = -100$
   cm, $f = pq/(p + q) = +33$ cm.
11. (a) $q = 160$ mm, $p = qf/(q - f) = 4.1$ mm,
    $m = -q/p = -39$; the total magnification is
    $10 \times 39 = 390\times$. (b) 4.1 mm.
13. (a) Objective: $q = pf/(p - f) = 1.02$ m,
    $m_1 = -q/p = -0.020$. Eyepiece: $q = -25$ cm,
    $p = qf/(q - f) = 4.17$ m, $m_2 = -q/p = 6$. Hence
    $m = m_1 m_2 = -0.12$, $h' = mh = -4.8$ cm,
    where the minus sign signifies an inverted
    image.
15. The object distance of the first lens is $p_1 = \infty$
    and its image distance is $q_1 = f_1$. The object
    distance of the second lens is therefore
    $p_2 = -q_1 = -f_1$, and, since $1/p_2 + 1/q^2 = 1/f_2$,
    we obtain $1/q_2 = 1/f_1 + 1/f_2$. But q_2 is the
    distance from the lenses at which the original
    beam is brought to a focus. Hence $q_2 = f$ and
    we have the desired result.

Chapter 26

Exercises

1. Light from incoherent sources can interfere, but
   the resulting interference pattern shifts
   continually because there is no definite phase
   relationship between the beams from the
   sources. When the sources are coherent, there
   is such a relationship, and the pattern is
   stable and therefore discernable.
3. The wavelengths in visible light are very
   small relative to the size of a building,
   whereas those in radio waves are more nearly
   comparable.
5. Diffraction and interference are both consequences of the wave nature of light. Interference refers in general to the reinforcement
   or cancellation of waves of the same kind
   from different sources that meet at some point;
   diffraction refers in general to the "bending"
   of waves around the edge of an obstacle.
   Interference patterns occur in diffraction, as
   discussed in Sec. 26-6.
7. This is a diffraction effect.
9. The first-order spectrum contains bright lines
   for which $\sin \theta = \lambda/d$, where d is the spacing
   of the slits, and the second-order spectrum
   contains bright lines for which $\sin \theta = 2\lambda/d$.
   The second-order spectrum is deviated by more
   than the first-order spectrum, and is wider as
   well (twice as wide, in fact). A prism
   produces only a single spectrum.
11. The "missing" energy is found in the bright
    lines, which are brighter than would be the case
    without interference. See also Exercise 5 of
    Chapter 13.
13. Higher resolving power; greater light-
    gathering ability, hence ability to form images
    of faint objects.
15. The individual waves in an ordinary light
    beam are polarized, but their planes of
    polarization are random so the beam itself is
    unpolarized.
17. Rotate the Polaroid sheet and observe that
    the sky appears darker at certain orientations
    of the sheet.
19. $\Delta y = L\lambda/d$, $\lambda = d\Delta y/L = 4.33 \times 10^{-7}$ m.
21. (a) $y = 2L\lambda/d = 8.74 \times 10^{-3}$ m. (b) $y = 3L\lambda/d$
    $= 1.31 \times 10^{-2}$ m. (c) $y = 5L\lambda/2d$
    $= 1.09 \times 10^{-2}$ m.

23. (a) $d = 2.5 \times 10^{-6}$ m, $\sin\theta = \lambda/d = 0.30$, $\theta = 17°$. (b) $\sin\theta = 3\lambda/d = 0.90$, $\theta = 64°$.

25. $\lambda = c/f = 3.156 \times 10^{-2}$ m $= 0.1036$ ft; $d_0 = 1.22\,\lambda L/D$, $D = 1.22\,\lambda L/d_0 = 3.71$ ft $= 44$ in.

27. $d_0 = 1.22\,\lambda\,L/D = 47.1$ m. In practice, irregularities in the earth's atmosphere prevent this resolution from being attained.

Problems

1. If the known wavelength is λ_1 and the unknown wavelength is λ_2, $7L\lambda_1/2d = 5L\lambda_2/d$, $\lambda_2 = 0.7\lambda_1 = 4.06 \times 10^{-7}$ m.

3. Here $d = 2.5 \times 10^{-6}$ m. Since $\sin\theta = n\lambda/d$ and the largest value $\sin\theta$ can have is 1, the maximum number of images possible is $n = d/\lambda = 4$.

5. Here $d = 3.33 \times 10^{-6}$ m; $\sin\theta_1 = \lambda_1/d = 0.12$, $\theta_1 = 7°$; $\sin\theta_2 = \lambda_2/d = 0.21$, $\theta_2 = 12°$; $h_1 = L\tan\theta_1 = 0.246$ m, $h_2 = L\tan\theta_2 = 0.426$ m, $h_2 - h_1 = 0.18$ m.

7. (a) $\theta_0 = 1.22\,\lambda/D = 3.37 \times 10^{-3}$ rad $= 0.193°$; the angular diameter is $2\theta_0 = 0.386°$. (b) The apparent angular diameter of the moon is 2160 mi/240,000 mi $= 0.0090$ rad $= 0.516°$.

9. For a distant object, the image distance L equals the focal length f of the lens. The ratio f/D for the required resolution, with $d_0 = 10^{-5}$ m, is therefore $d_0/1.22\,\lambda = 15$. This is an aperture of $f/15$.

Chapter 27

Exercises

1. No; yes; yes.

3. More conspicuous.

5. The rod appears longest to the stationary observer.

7. Because the hydrogen molecule is electrically

neutral despite the very different velocities of its electrons and protons, electric charge must be an invariant quantity.

9. The density is greater because the object's mass is greater and its length and hence volume is smaller when it is in motion.

11. $L = L_0\sqrt{1 - v^2/c^2}$ in general, here $v \ll c$ so $L = L_0(1 - \tfrac{1}{2}v^2/c^2)$, $\Delta L = \tfrac{1}{2}L_0 v^2/c^2 = 1.8 \times 10^{-11}$ ft.

13. $t = 2t_0$, $\sqrt{1 - v^2/c^2} = \tfrac{1}{2}$, $v^2/c^2 = \tfrac{3}{4}$, $v = 2.6 \times 10^8$ m/s.

15. $m = 3m_0$, $\sqrt{1 - v^2/c^2} = \tfrac{1}{3}$, $v^2/c^2 = \tfrac{8}{9}$, $v = 2.8 \times 10^8$ m/s.

17. $m = 1.01\,m_0$, $\sqrt{1 - v^2/c^2} = 1/1.01$, $v^2/c^2 = 0.0197$, $v = 4.2 \times 10^7$ m/s.

Problems

1. $t = t_0/\sqrt{1 - v^2/c^2}$ in general, here $v \ll c$ so $t = t_0 + \tfrac{1}{2}t_0 v^2/c^2$, $t - t_0 = \tfrac{1}{2}t_0 v^2/c^2 = 5 \times 10^{-13}\,t_0 = 1$ s, $t_0 = 2 \times 10^{12}$ s $= 6.35 \times 10^4$ years.

Chapter 28

Exercises

1. Such wave phenomena as diffraction and interference are easier to demonstrate than such quantum phenomena as the photoelectric effect.

3. Such a transfer would not conserve both energy and momentum.

5. No.

7. Apply an electric or magnetic field and look for a deflection.

9. $f = c/\lambda = 5 \times 10^{14}$ Hz; $n = E/hf = 3$ photons.

11. $P/hf = 1.71 \times 10^{31}$ photons/s.

13. $f = eV/h = 2.41 \times 10^{18}$ Hz; X-rays.

15. $\lambda = h/mv = 3.3 \times 10^{-29}$ m.

17. (a) $\Delta v = h/2\pi m\Delta x = 1.16 \times 10^5$ m/s. (b) $\Delta v = h/2\pi m\Delta x = 63$ m/s.

Problems

1. $KE = \frac{1}{2}mv^2 = 2.23 \times 10^{-19}$ J; $f_0 = f - KE/h = 4.64 \times 10^{14}$ Hz.
3. (a) $P/hf = 4.22 \times 10^{21}$ photons/m²·s. (b) The photon velocity is c, hence there are $(4.22 \times 10^{21}$ photons/m²·s$)/c = 1.41 \times 10^{13}$ photons/m³.
5. (a) $E = hf = hc/\lambda = 9.95 \times 10^{-15}$ J.
   (b) $p = h/\lambda = 3.32 \times 10^{-23}$ kg·m/s.
7. $KE = 1$ MeV $= 1.6 \times 10^{-13}$ J $= \frac{1}{2}mv^2$, $mv = \sqrt{2mKE}$, $\lambda = h/mv = h/\sqrt{2mKE} = 2.87 \times 10^{-14}$ m.
9. $KE = 10^3$ eV $= 1.6 \times 10^{-16}$ J $= \frac{1}{2}mv^2$, $mv = \sqrt{2mKE} = 1.71 \times 10^{-23}$ kg·m/s; $\Delta mv = h/2\pi\Delta x = 1.06 \times 10^{-24}$ kg·m/s; $\Delta mv/mv = 0.062 = 6.2\%$.
11. $E_1 = h^2/8mL^2 = 3.29 \times 10^{-13}$ J $= 2.06$ MeV.
13. If the radius of the circle is r, $L = I\omega = mvr$ and $\Delta L = r\,\Delta mv$. The uncertainty in the particle's position along the circle is $\Delta x = r\,\Delta\theta$. Hence $\Delta x\,\Delta mv = \Delta L\,\Delta\theta$ and $\Delta L\,\Delta\theta \geq h/2$.

Chapter 29

Exercises

1. The Rutherford model indicates the division of the hydrogen atom into a nucleus and a relatively distant electron, and the Bohr model goes on from there to specify the motion and energy of the electron.
3. A negative total energy signifies that the electron is bound to the nucleus; the kinetic energy of the electron is of course a positive quantity.
5. A hydrogen sample contains a great many atoms, each of which can undergo a variety of possible transitions.
7. (a) Continuous emission spectrum (see Sec. 16-2). (b) Emission line spectrum. (c) Absorption line spectrum.

9. With no incident photons, there is nothing to absorb.
11. From Fig. 29-6, the energy needed to raise an electron from the ground state to the $n = 3$ state in hydrogen is 13.6 eV $-$ 1.5 eV $= 12.1$ eV.
13. The shorter the wavelength, the higher the frequency, which in turn means the greater the energy change. From Fig. 29-10 the maximum energy change in the Brackett series corresponds to a transition from $n = \infty$ to $n = 4$, so that $hf = E_i - E_f = -E_1/4^2 = 1.36 \times 10^{-19}$ J, $f = 2.06 \times 10^{14}$ Hz, and $\lambda = c/f = 1.46 \times 10^{-6}$ m.
15. $E_n = -E_1/n^2 = 13.6$ $eV/5^2 = 0.54$ eV.

Problems

1. (a) $\lambda = h/mv = 3.68 \times 10^{-63}$ m. (b) $n\lambda = 2\pi r$, $n = 2\pi r/\lambda = 2.56 \times 10^{74}$.
3. $\frac{3}{2}kT = E_1$, $T = 2E_1/3k = 1.05 \times 10^5$ °K
5. $hf = E_i - E_f = E_1(1/n_i^2 - 1/n_f^2)$, $1/\lambda = f/c = (E_1/ch)\,(1/n_i^2 - 1/n_f^2)$.
9. $KE = 40$ keV $= 6.4 \times 10^{-15}$ J; as in Problem 7, $2\pi^2 k^2 me^4/h^2 = 2.82 \times 10^3$ eV.

Chapter 30

Exercises

1. The uncertainty principle.
3. No, because in the Bohr theory the electron must revolve around the nucleus.
5. Such an electron has no vector property associated with it on the average, hence its probability cloud must vary in the same way in all directions.
7. The densest part of a probability cloud, which corresponds to the maximum probability of finding the electron, occurs at a radius proportional to n^2.

9. (a) Spin is an intrinsic property of electrons, which they always exhibit. (b) All electrons have the same spin.

11. They are similar.

13. Two electrons.

15. Both Li and Na atoms have one electron outside a closed inner electron shell.

17. $n = 1$, $\ell = 0$, $m_\ell = 0$, $m_s = \pm 1$; $n = 2$, $\ell = 0$, $m_\ell = 0$, $m_s = \pm 1$; $n = 2$, $\ell = 1$, $m_\ell = 0$, ± 1, $m_s = \pm 1$; $n = 3$, $\ell = 0$, $m_\ell = 0$, $m_s = +1$ (or -1).

19. 182 elements.

Chapter 31

Exercises

1. When the spins of the electrons are opposite.

3. Their ability to bond with each other as well as with other atoms.

5. In a polar covalent molecule the shared electrons spend more time on the average near one of the atoms.

7. (a) The number of dislocations in the structure of the iron increased. (b) The strong heating increased the energies of the individual iron atoms, which were then able to shift their positions to produce a more regular crystal structure.

9. Li atoms have a single electron outside a closed shell, and this electron is held relatively weakly since the inner electrons shield most of the full nuclear charge. In fluorine, the outer shells lack an electron of completion, and the electrons in these shells are held very tightly to the nucleus since little of the nuclear charge is shielded from them.

11. Na$^+$ ions have closed shells, whereas Na atoms each have a single, relatively easily-detached electron.

13. Only the outermost electrons of the atoms of a metal are members of its electron "gas."

15. The electromagnetic interaction is responsible for all the bonding mechanisms in solids.

17. In a NaCl molecule, the Na$^+$ and Cl$^-$ ions interact only with each other. In an NaCl crystal, each ion of a given kind has six nearest neighbors of the other kind of ion with which it interacts strongly, plus twelve neighbors of the same kind farther away, and so on. There is no reason why the equilibrium separation in each case should be the same.

19. The exclusion principle.

21. A good conductor of heat is also a good conductor of electricity, because the mechanism of transport in each case involves the electron gas in the metal.

23. Both are nonconductors at very low temperatures; germanium is the better conductor at room temperature because its forbidden band is narrower.

Problems

1. $E = 4.5$ eV $= 7.2 \times 10^{-19}$ J, $\frac{3}{2}kT = E$, $T = 2E/3k = 3.48 \times 10^4$ °K.

3. (a) 4.3 eV $- 3.8$ eV $= 0.5$ eV. (b) PE $= ke^2/r$, $r = ke^2$ /PE $= 2.88 \times 10^{-9}$ m.

Chapter 32

Exercises

1. The isotopes have the same atomic number and hence the same electron structures, therefore they have the same chemical behavior. They have different numbers of neutrons, hence different atomic masses.

3. $5n$, $5p$; $12n$, $10p$; $20n$, $16p$; $50n$, $38p$; $108n$, $72p$.

5. In both cases the liberated energy comes from the difference in mass between the initial nucleus or nuclei and the final ones.

7. The limited range of the strong nuclear force.

9. The strong and weak nuclear interactions are too limited in range to affect planetary motion. The reasons why electromagnetic forces are not significant in astronomy are outlined in Sec. 17-4.

11. In fission a large nucleus splits into smaller ones; in fusion small nuclei join to form a large one. In both, the product nucleus or nuclei have less mass than the initial nucleus or nuclei, with the missing mass being evolved as energy.

13. Helium and radon cannot be combined chemically to form radium, nor can radium be broken down into helium and radon by chemical means.

15. (a) When it contains too many neutrons to be stable. (b) When it contains too many protons to be stable.

17. $^{11}_{5}\text{B}$.

19. $^{206}_{82}\text{Pb}$.

21. One-sixteenth is left, and since $\frac{1}{2} \times \frac{1}{2} \times \frac{1}{2} \times \frac{1}{2} = \frac{1}{16}$, this means four half-lives or 6400 years.

23. $^{2}_{1}\text{H}$; $^{1}_{0}\text{n}$; $^{1}_{1}\text{H}$.

25. 6, 12, carbon.

Problems

1. $2m_{\text{H}} + 2m_{\text{n}} = 4.032980$ u, $\Delta m = 0.030377$ u $= 28.3$ MeV. There are four nucleons in $^{4}_{2}\text{He}$, hence the binding energy per nucleon is 7.1 MeV.

3. $17m_{\text{H}} + 18m_{\text{n}} = 35.288995$ u, $\Delta m = 0.32015$ u $= 298.1$ MeV. There are 35 nucleons in $^{35}_{17}\text{Cl}$, hence the binding energy per nucleon is 8.5 MeV.

5. $m_{\text{p}} + m_{\text{e}} = 1.00728$ u $+ 0.00055$ u $= 1.00783$ u. The difference between this and m_{n} is 0.00084 u $= 0.78$ MeV, which is less than the observed binding energies per nucleon in stable nuclei; hence neutrons do not decay inside nonradioactive nuclei.

7. (a) PE $= ke^2/r = 1.355 \times 10^{-13}$ J $= 0.847$ MeV. (b) $m_{\text{H}} + 2m_{\text{n}} = 3.025155$ u, $\Delta m = 0.009105$ u, hence the binding energy of $^{3}_{1}\text{H}$ is

8.477 MeV; $2m_{\text{H}} + m_{\text{n}} = 3.024315$ u, $\Delta m = 0.008285$ u, hence the binding energy of $^{3}_{2}\text{He}$ is 7.713 MeV. The difference is 0.764 MeV, almost the same as the electric potential energy, hence nuclear forces must be very nearly independent of charge.

9. $E = 22.4$ MeV $= 0.02406$ u; $m_{\text{Li}} = 2m_{\text{He}} + E - m_{\text{H}} = 6.01516$ u.

11. $E = 22.4$ MeV $= 0.02406$ amu; $m_{\text{Li}} = 2m_{\text{He}} + E - m_{\text{H}} = 6.01516$ amu.

13. $n = 10^3 (\text{J/s})/(200 \text{ MeV} \times 1.6 \times 10^{-13} \text{ J/MeV}) = 3.125 \times 10^3$ fissions/s.
(a) $m = 10^{-3}$ kg, $E = mc^2 = 9 \times 10^{13}$ J $= 2.15 \times 10^{10}$ kcal. (b) 2.15×10^4 tons of TNT.

15. $\Delta m = 3m_a - m_c = 0.007809$ u $= 7.27$ MeV.

Chapter 33

Exercises

1. Both have zero rest mass. The photon is associated with the electromagnetic interaction, whereas the neutrino is associated with the weak nuclear interaction. The neutrino has an antiparticle, whereas the photon is its own antiparticle. The photon can materialize (if it has enough energy) into an electron-positron pair, whereas the neutrino cannot materialize.

3. A neutrino.

5. A μ-neutrino.

9. No; it is possible that there is a reason why quarks cannot exist except in combination with each other as hadrons.

Problems

1. $E_{\text{min}} = (m_{\text{n}} + m_{\pi} - m_{\text{p}})c^2 = 142$ MeV.

3. $\Delta E = (2m_{\text{p}} - m_{\pi})c^2 = 2.79 \times 10^{-10}$ J, $\Delta t = h/2\pi\Delta E = 3.78 \times 10^{-25}$ s; no.

5. $2m_{\text{n}}c^2 = 1879$ MeV; more, because the neutron mass exceeds the proton mass.

Index

Physical Constants and Quantities

Quantity	Symbol	Value
Absolute zero	0 K	$-273°C = -460°F$
Acceleration of gravity at earth's surface	g	9.81 m/s^2 = 32.2 ft/s^2
Boltzmann's constant	k	1.38×10^{-23} J/K
Coulomb constant	k	8.99×10^9 N·m^2/C^2
Earth: mass	m_{earth}	5.98×10^{24} kg
radius	r_{earth}	6.38×10^6 m
Electron: charge	e	1.60×10^{-19} C
rest mass	m_e	9.11×10^{-31} kg = 0.000549 u
Gravitational constant	G	6.67×10^{-11} N·m^2/kg^2 = 3.44×10^{-8} lb·ft^2/slug^2
Neutron rest mass	m_n	1.675×10^{-27} kg = 1.008665 u
Permeability of free space	μ_0	$4\pi \times 10^{-7}$ T·m/A = 1.26×10^{-6} T·m/A
Planck's constant	h	6.63×10^{-34} J·s
Proton rest mass	m_p	1.673×10^{-27} kg = 1.007277 u
Stefan-Boltzmann constant	σ	5.67×10^{-8} w/m^2 · k^4
Velocity of light in free space	c	3.00×10^8 m/s = 1.86×10^5 mi/s

Multipliers for SI Units

a	atto-	10^{-18}	da	deka-	10^1
f	femto-	10^{-15}	h	hecto-	10^2
p	pico-	10^{-12}	k	kilo-	10^3
n	nano-	10^{-9}	M	mega-	10^6
μ	micro-	10^{-6}	G	giga-	10^9
m	milli-	10^{-3}	T	tera-	10^{12}
c	centi-	10^{-2}	P	peta-	10^{15}
d	deci-	10^{-1}	E	exa-	10^{18}